BRAIN PLASTICITY, LEARNING, AND MEMORY

ADVANCES IN BEHAVIORAL BIOLOGY

Recent Volumes in this Series

Volume 17 ENVIRONMENTS AS THERAPY FOR BRAIN DYSFUNCTION
Edited by Roger N. Walsh and William T. Greenough

Volume 18 NEURAL CONTROL OF LOCOMOTION
Edited by Richard M. Herman, Sten Grillner, Paul S. G. Stein, and Douglas G. Stuart

Volume 19 THE BIOLOGY OF THE SCHIZOPHRENIC PROCESS
Edited by Stewart Wolf and Beatrice Bishop Berle

Volume 20 THE SEPTAL NUCLEI
Edited by Jon F. DeFrance

Volume 21 COCAINE AND OTHER STIMULANTS
Edited by Everett H. Ellinwood, Jr. and M. Marlyne Kilbey

Volume 22 DISCRIMINATIVE STIMULUS PROPERTIES OF DRUGS
Edited by Harbans Lal

Volume 23 THE AGING BRAIN AND SENILE DEMENTIA
Edited by Kalidas Nandy and Ira Sherwin

Volume 24 CHOLINERGIC MECHANISMS AND PSYCHOPHARMACOLOGY
Edited by Donald J. Jenden

Volume 25 CHOLINERGIC MECHANISMS: Phylogenic Aspects, Central and Peripheral Synapses, and Clinical Significance
Edited by Giancarlo Pepeu and Herbert Ladinsky

Volume 26 CONDITIONING: Representation of Involved Neural Functions
Edited by Charles D. Woody

Volume 27 THE BASAL GANGLIA: Structure and Function
Edited by John S. McKenzie, Robert E. Kemm, and Lynette N. Wilcock

Volume 28 BRAIN PLASTICITY, LEARNING, AND MEMORY
Edited by B. E. Will, P. Schmitt, and J. C. Dalrymple-Alford

BRAIN PLASTICITY, LEARNING, AND MEMORY

Edited by

B. E. Will

Louis Pasteur University
Strasbourg, France

P. Schmitt

CNRS
Center for Neurochemistry
Strasbourg, France

and

J. C. Dalrymple-Alford

Louis Pasteur University
Strasbourg, France

PLENUM PRESS • NEW YORK AND LONDON

Library of Congress Cataloging in Publication Data

EBBS-SFECA Workshop on Brain Plasticity, Learning, and Memory (1984: Strasbourg, France)
Brain plasticity, learning, and memory.

(Advances in behavioral biology; v. 28)

Proceedings of an EBBS-SFECA Workshop on Brain Plasticity, Learning, and Memory, held Aug. 30–Sept. 1, 1984, in Strasbourg, France.
Bibliography: p.
Includes indexes.
1. Learning—Physiological aspects—Congresses. 2. Memory—Congresses. 3. Neural circuitry—Adaptation—Congresses. 3. Brain—Growth—Congresses. I. Will, Bruno. II. Schmitt, Pierre. III. Dalrymple-Alford, J. C. (John C.) IV. European Brain and Behaviour Society. V. Société française pour l'étude du comportement animal. VI. Title. VII. Series.
QP408.E33 1984 153.1 85-12139
ISBN 0-306-42063-5

Proceedings of an EBBS-SFECA workshop on Brain Plasticity, Learning, and Memory, held August 30–September 1, 1984, in Strasbourg, France

A Division of Plenum Publishing Corporation
233 Spring Street, New York, N.Y. 10013

Printed in the United States of America

PREFACE

This book is the result of the contributions presented at a conference held from August 30 to September 1, 1984 at the Université Louis Pasteur, Strasbourg, France. This meeting was organized under the joint auspices of the European Brain and Behaviour Society (EBBS) and the Société Française pour l'Etude du Comportement Animal (SFECA).

The objective of this meeting was to bring together an international group of participants to evaluate and to report on recent research in three broad and overlapping fields within the general theme of the relationships between brain plasticity and learning and memory. These three fields are "developmental plasticity " "adaptive plasticity" and "restorative plasticity." Although the boundaries between these fields are a matter of debate (see Introduction), they have been retained as the major sections of this volume, the arrangement of which roughly parallels that of the meeting. We believe and very much hope that the contents of this volume convey an internal consistency despite the diversity of the material presented.

Developmental plasticity constitutes the first part of the book in which are considered the repercussions of early external or intracerebral stimulation and of early learning on brain development. These chapters are based on studies on imprinting phenomena and visuocortical plasticity, followed by work on the modulatory rôle of noradrenalin in plasticity in immature animals. The second part of this book, on adaptive plasticity in adult animals, first presents chapters that describe some of the biochemical, morphological and electrophysiological (long-term and short-term potentiation) alterations that accompany or underlie learning and memory processes. These chapters are complemented by work on the selective activation of cerebral stuctures, together with sections on noradrenergic, neuroendocrine and peptidergic factors that modulate learning and memory in adult animals. The section on restorative plasticity presents chapters looking at responses to different kinds of brain dysfunction in young, adult and aged organisms, mostly in animals. This part of the book is completed by several chapters that deal with the "recovery of function" in animals with experimental brain lesions. These final chapters describe factors that play a rôle in the expression of lesion-induced deficits and ways to reduce such deficits. In addition, most of the poster abstracts, which summarise a considerable amount of further information on brain plasticity, learning and memory, are presented in an appendix.

We gratefully acknowledge the following organizations for financial support: Dausse Labs., Merrell Dow Research Inst., Preclinical Research Division of Sandoz Ltd., and the Conseil Général du Bas-Rhin, Strasbourg. Facilities and assistance were kindly provided by the Université Louis Pasteur and by the Mairie de Strasbourg. We are also indebted to our colleagues F. Deluzarche, C. Kelche, C. Pacteau, V. Pallage and G. Toniolo of the Laboratoire de Neurobiologie des Comportements, and M-J. Angst, A.

Boehrer, P. Carrive, B. Fritsch, B. Jardon, J-L. Moreau and G. Rudolf of the Laboratoire de Neurophysiologie for their tremendous work which contributed to the success of this meeting. We thank Professor P. Karli, acting president of the EBBS, for his assistance and for his welcoming address. Special thanks are also due to Professor J. McGaugh and Dr. S. Sara for their help in organizing the sessions on neuroendocrine and catecholaminergic modulation of learning and memory.

Bruno E. Will
Pierre Schmitt
John C. Dalrymple-Alford

March 1985

CONTENTS

INTRODUCTION

DEVELOPMENTAL PLASTICITY: THE NEUROBIOLOGY OF LEARNING AND MEMORY DURING ONTOGENY

Alterations induced by early learning and early stimulation

Noradrenergic modulation of learning and brain plasticity

ADAPTIVE PLASTICITY: THE NEUROBIOLOGY OF LEARNING AND MEMORY IN ADULT ORGANISMS

Biochemical and morphological alterations

Short-term and long-term potentiation

Selective activation of cerebral structures

Catecholaminergic modulation of learning and memory

Neuroendocrine and peptidergic modulation of learning and memory

RESTORATIVE PLASTICITY: BRAIN DAMAGE AND FUNCTIONAL RECOVERY

Early brain damage

Brain damage in adult organisms

Aging

Modulation of functional recovery

APPENDIX: POSTER ABSTRACTS

BRAIN PLASTICITY, LEARNING AND MEMORY:

HISTORICAL BACKGROUND AND CONCEPTUAL PERSPECTIVES

Bruno Will*, Pierre Schmitt**, and John Dalrymple-Alford*

* Laboratoire de Neurobiologie des Comportements
Université Louis Pasteur, 67000 Strasbourg, France
** Laboratoire de Neurophysiologie
Centre de Neurochimie du CNRS, 67000 Strasbourg, France

The relationship between brain plasticity, learning and memory has in recent years become a topic of great interest in the neurobiological sciences. Given the number of different approaches to this field, there is a clear need to bring together a body of data that derive from such approaches and to examine some of the concepts that are currently employed. In this introductory chapter we shall discuss some theoretical considerations of the concept of brain plasticity in its relationship to learning and memory. However, as the interest in the existence of a relationship between brain function and various facets of learning and memory is not new, it may be of value to take a brisk, selective look at the development of this general field from a historical perspective, and to compare past and present trends.

HISTORICAL BACKGROUND

Interspecific differences in brain and intelligence

As alluded to above, the relationship between brain anatomy and "intelligence", often regarded as a global though unspecified measure of learning and memory capacities, is an old subject of interest. Indeed, such relations were considered during classical antiquity. For example, Erasistratos who lived in Alexandria around 300 years B.C. suggested that the more convolutions the cerebrum showed, the more intelligent was the species.

One of the major concerns over the centuries has been the estimation of what might be called the phylogenetic plasticity of brain and "intelligence", that is, a concern with the evolutionary relationship between brain plasticity and "intelligence" as measured by one or more crude indices. Thus, many scientists looked for a suitable anatomical index which would create a scale such that the most intelligent species, preferably *Homo sapiens*, was ranked at the top. For instance, in the 18th century, Samuel Sömmerring (1755-1830) stated that although Man does not have the largest brain in absolute values, Man has the largest brain when measured in relation to the size of the cranial nerves (Sömmerring, 1785, 1791).

At the beginning of the nineteenth century, the french paleontologist Georges Cuvier (1769-1832) ranked *Homo sapiens* and even, to what must have been comfortingly acceptable at that time, European *Homo sapiens* at the top of a scale that represented a simplistic neuroanatomical index based on a ratio of cranial size to the size of the face. Cuvier's index was derived by measuring the angle between a so-called "facial line" and the "line of the cranial basis". In his "Leçons d'anatomie comparée", Cuvier (1800) commented that "man is the animal whose brain case is the largest and whose face is the smallest... the more animals depart from these proportions, the more stupid or the more ferocious they are "(p. 4). According to his measurements, Cuvier was able to satisfy the obvious bias of his time by placing the "adult Negro" below the "decrepit European" and just above the young Orang-Outang. However, Cuvier also extended his thinking to the study of fossilized animals and initiated paleoneurology by describing a fossilized brain case in the skull of an Anoplotherium, an extinct pachyderm whose bones were found in the region of Paris. This brain case was rather small, flattened horizontally and deprived of convolutions. From his description of the brain-cast, Cuvier seemed hardly to hesitate in estimating the Anoplotherium's intelligence : "All the laws of analogy permit us to conclude that the animal was greatly lacking in intelligence (1822, p. 44). Of course, current paleoneurological studies on the evolution of hominid species still have to rely to a great extent on an estimation of brain size and capacity from the often fragmented remains of fossil crania.

Later during the nineteenth century, several other biologists and physicians (e.g. Brandt, 1867; Snell, 1892) proposed more generalisable measures to specify the relationship between neuroanatomy and intelligence by taking into account the allometric relation existing between brain weight and body surface area, estimated from the body weight. Based on these measures and on the work of a number of people who even during this century have studied the relationship between intelligence and head or brain measurements, a causal relationship between brain size, brain complexity and intelligence is intuitively assumed when different species are compared. However, while birds and mammals have evolved enlarged brains in excess of the body size factor, this development cannot be explained in terms of the classical theory of encephalisation of functions and the evidence does not support any simple relationship between the phylogenetic hierarchy of species and a hierarchy of intellectual skill (Steele Russell, 1979).

Intraspecific differences in brain and intelligence

Gross anatomy . In investigations of the possible relationship between brain and intelligence within a given species, a common trend once again has been to look first for relationships between measures of gross anatomy and simple measures of intelligence. However, a study by Michele Vincenzo Malacarne (1744-1816) deserves special mention. Malacarne studied the intraspecific neuroanatomical differences induced by differential rearing. His experimental design was in effect prototypical of that used from the 1960's onwards by Rosenzweig's group at Berkeley. Malacarne chose as subjects two littermate dogs, and also pairs of parrots, chaffinches and blackbirds, each pair of birds being taken from the same chutch of eggs. In each pair, one individual received extensive training and the other received none. The animals were sacrificed after a few years of such a differential treatment and their brains were compared. Malacarne is reported (Bonnet, 1781) to have found that the trained animals showed more folds in the cerebellum than the untrained control animals. Although many workers might wish to question his specific findings, Malacarne's basic idea of a possible relationship between brain

Clutch

anatomy and experience has now been accepted by most neurobiologists, but the possibility of a relationship between changes in the cerebellum and some forms of learning has received special attention only recently (e.g., McCormick and Thompson, 1984; Yeo et al., 1984).

Malacarne's approach was not typical of the dominant trends of his time. Although there was a move towards a localization of brain function, especially after Paul Broca's (1825-1880) case descriptions, people continued to look for a global relationship between brain size and "intelligence". For Instance, François Magendie (1783-1855) wrote that "the volume of the brain is generally in direct proportion to the capacity of the mind. We ought not to suppose, however, that every man having a large head is necessarily a person of superior intelligence, for there are many causes of an augmentation of the volume of the head beside the size of the brain ; but it is rarely found that a man distinguished by his mental faculties has not a large head" (1831, p. 103-104).

In the 1920's and 1930's, Karl Lashley's studies on lesions to the cerebral cortex in rats represented a shift in emphasis towards the possibility of cerebral localization of cognitive functions. However, his failure to localize the memory trace in the cerebral cortex and his subsequent development of the "mass action" principle (see Eames and Oakley, this volume) reflect a combination of using the newer approach of localization of function with the old behavioral approach of looking for a global measure of intelligence, in his case based on problem solving in mazes.

A global relationship between the size of the whole brain and some general factor of intelligence is too simplistic and too crude, but relationships between some aspects of cerebral structure or function and some kinds of learning do seem more and more granted. For example, Fernando Nottebohm has recently reported that at least one brain function, namely song-learning in canary birds, is regulated by what he terms a "library principle". This principle is that a large library can house few or many books, whereas a small library can only house a small number of books. Nottebohm has reported that "birds with a large hyperstriatum ventrale pars caudale (HVc) have large or small song repertoires, but birds with a large song repertoire tend to have a large HVc, and birds with a small HVc tend to have a small repertoire, regardless of absolute brain size" (1981, p. 105). Nottebohm's statement on song learning and HVc size in canaries appears at first to be much the same as Magendie's statement, 150 years earlier, on intelligence and brain size in humans. The difference is that present-day workers now recognise the need to describe relationships between more restricted structures and functions, while accepting that in many instances functional characteristics are influenced by widespread and interacting brain structures. In addition, Nottebohm considers the size of a given brain area not only as an expression of plasticity, but also as a condition for it. Nottebohm has observed that in autumn the HVc is 50 % smaller than it is in spring. He suggests that this seasonal shrinkage of HVc and of some other nuclei that are involved in the control of song behavior in canaries is a prerequisite for the learning of a new repertoire. Furthermore, Nottebohm was concerned with describing changes that occur within the individual whereas Magendie was interested only in seemingly fixed differences that arise between individuals.

<u>Cellular modifications</u> . Changes in the size of nervous structures may of course be the result of various other changes at more elementary levels. Eugenio Tanzi (1856-?) and Santiago Ramõn Y Cajal (1852-1934) in the late 19th century were perhaps the first to propose that learning

is due to a change in neuronal connectivity. Based on phylogenetic as well as developmental considerations, Cajal believed it probable that mental exercise led to greater growth of neuronal collaterals in the stimulated regions of the brain. Rosenzweig's (1979) translation of Cajal's picturesque analogy is that he compared "the cerebral cortex to a garden planted with innumerable trees, the pyramidal cells, which thanks to intelligent cultivation, can multiply their branches, sink their roots deeper and produce flowers and fruits of ever greater variety and quality" (Cajal, 1894, p. 467-468).

However, the cerebral modifications induced by mental exercise are even larger than Cajal foresaw 90 years ago. Over the past few years it has been shown that changes in neuronal circuitry range from "gain" modifications in existing synapses elicited by repeated stimulation to the replacement or addition of connections in response to environmental or surgical manipulations (e.g., Cotman and Nieto-Sampedro, 1982).

Although Cajal was a pioneer in proposing that cerebral modifications can result from "mental exercise", it is surprising that Cajal was also a major proponent of the idea of a fixed CNS in the adult animals. Indeed, in 1928, Cajal stated that "Once development is ended, the founts of growth and regeneration of the axons and dendrites dry up irrevocably. In the adult brain, the nerve paths are something fixed, and immutable; everything may die, nothing may be regenerated" (p. 750). Because of Cajal's influence, such a conceptual fiction of non-plasticity in the fully developed brain may have contributed to retard by several decades neurological and biological research in some directions such as, for instance, approaches to the recovery of function after damage to the CNS.

As with many other conceptual fictions in the history of behavioral neurosciences, Cajal's erroneous conception of a fixed adult CNS was tenacious, although less persistent than fictions of older times in which there was much slower scientific progress. The cell doctrine of brain function, for instance, held out over centuries and the concept of an interface between mind and body located at the level of the so called "rete mirabile" (or "marvellous net") lasted from classical antiquity to the 17th century (see Clarke and Dewhurst, 1972).

There are also, however, some recent examples of inaccurate or fictional conceptions in brain plasticity, learning and memory. The generality of what has become popularly known as Kennard's principle (see Finger and Stein, 1982), i.e. there is a greater capacity for functional sparing and/or restorative plasticity in the young than in the adult organism with CNS lesions, has been challenged on the basis of numerous experimental findings (Isaacson, 1975; Schneider, 1979; Passingham et al., 1983; Will and Eclancher, 1984). A second recent example, that of rerouting of still growing nerve fibers after early CNS damage, illustrates how a hypothesis which was proved to be an erroneous explanation for the case in which it was specifically proposed (the number of uncrossed ocular fibers found after unilateral enucleation in the rat; Lund et al., 1973; Land and Lund, 1979) has nevertheless found merit in other examples of brain plasticity (e.g., Schneider, 1973; Lund, 1978; Zimmer, this volume). A third example is that of cell death and regression of axonal terminals in the developing organism. Whereas it has been implicitly assumed that cell death and regressive phenomena were a consequence of aging, it now appears that brain damage in infants rather than in adults is followed by the degeneration of more neurons (Bleier, 1969; Prendergast and Stelzner, 1976; Perry and Cowey, 1982) and that even programmed cell death and axonal regression is a developmental fate commonly observed during invertebrate and vertebrate neurogenesis (e.g., Hamburger, 1975; Greenwald and Martinez-Arias, 1984).

With such considerations in mind, we turn now to some current conceptual frameworks (or perhaps fictions).

CURRENT CONCEPTUAL FRAMEWORKS

The processes underlying learning and memory have been investigated at different levels (from the molecular level to that of social behavior) and various experimental techniques have been used at each level. Thus the task of synthesising data from different levels into a consistent framework is made difficult by the fact that data are often obtained from experiments using different methodologies and that interpretations are formulated within different conceptual frames. Despite (and because of) these difficulties, the present progress of the neurobiological sciences calls for an interdisciplinary dialogue to view the possibility of linking molecular or cellular phenomena with behavioral phenomena. One basic condition for establishing such a dialogue requires the use of terms such as plasticity, learning and memory each with a commonly agreed basis across the many levels of investigation and interpretation.

Definitions of plasticity

We do not wish to recount here the problems that remain associated with definitions of learning and memory (e.g., Bloch, 1970; Sahakian, 1970; John, 1972; Lewis, 1979; Spaer, 1980). However, we would like to discuss briefly the concept of plasticity. Although recognized for many years (e.g., Bethe, 1930), the concept of plasticity as applied to the nervous system has often been defined too vaguely or in a too limited fashion. The notion of plasticity as it is formulated by Bethe (1930) presents both drawbacks. Bethe loosely considered plasticity as the ability to adapt to changes and to meet the dangers of life. Bethe also considered plasticity as the capacity of the CNS to reorganize after insult and to restore adequate function. The latter formulation is too restrictive as there are forms of plasticity other than restorative plasticity after injury (e.g., "developmental" and "adaptive" plasticity in intact animals).

The definition of the term "plasticity" proposed by Jerzy Konorski avoids the drawbacks just mentioned. According to Konorski (1961), brain plasticity refers to the adaptive capacities of the CNS or, in other words, to its ability to modify its own structural organization and functioning. Plasticity is seen as an adaptive response to functional demand, as requiring feedback on whether or not environmental demands are being met and as permiting enduring functional changes. Konorski's definition seems accurate enough, but is difficult to apply to different levels of analysis, from molecular to behavioral.

A major step towards a definition that meets the requirement of being applicable to any level of analysis was taken by Jacques Paillard (1976). He used the vocabulary of systemic analysis in which a given system is specified by its structure, activity and function. The structure of a system (or integron) is defined by the interface between this and other systems, by its elements and by the network linking the elements together; the activity of the system is defined by the nature and dynamics of the functional relations of the structure's elements; and the function is the product of the activity of the system. Paillard reserved the term plasticity to designate "those observable variations in the activity or function of a given system that correspond to long lasting changes of the material structure of the connections between the system elements due to the effect of an external force or environmental constraint" (p.47). The

A sentence like this in an essay would get a student a sour comment from the lecturer!

benefit of Paillard's systemic approach is that it might allow the application of the concept to any level of analysis and to encourage interdisciplinary dialogue.

However, the price to pay for that benefit is that crucial problems remain concerning the choice of appropriate time and space scales which differ in important respects acrross different approaches and levels of analysis. One of these problems is linked to the fact that the stability of a given living system is often conditioned by the lability of phenomena at subsystemic levels. An example of the time scale problem is whether we should consider as plastic those cerebral or behavioral modifications induced by differential rearing that last only a few days or weeks, or that are reduced or abolished by repeated behavioral testing (e.g., Brown, 1971 ; Rosenzweig and Bennett, 1977 ; Ungerer et al., 1977). Do such changes qualify as plastic or as elastic? Plasticity, according to Paillard, refers to long-lasting changes, whereas elasticity refers to events characterized by a relatively rapid return to the original state after the cessation of the experimental manipulation. Paillard's definition also generates discussion as to whether other variations such as flexibility or strategy substitution classify as plastic events. Such considerations naturally lead us to question whether CNS plasticity is multiform.

Is CNS plasticity multiform ?

Given that there is plasticity in the nervous system, an appropriate question to ask is whether plasticity is qualitatively and quantitatively the same at different ages, as a result of different experience, between different genders or species et cetera. The organization and structure of the present volume, which gathers together data obtained by researchers interested in developmental, adaptive and restorative plasticity, reflect a concern with this general question of whether CNS plasticity is multiform.

Developmental versus adaptive plasticity. The present volume draws a crude distinction between effects occuring during early development ("developmental plasticity") and those observed in adult organisms ("adaptive plasticity")* The majority of available data demonstrates, at least at the biological level, the existence of quantitative and even qualitative differences in structural and functional plasticity between young and adult organisms. Maturation and aging not only add restrictions to the extent, latency and speed of plastic processes such as reactive sprouting following injury (e.g., Gall and Lynch, 1978), but also modify qualitatively the forms of structural changes found in the young and in adults. Immature organisms display some plastic phenomena that are no longer observable in mature organisms, examples of which are the preservation of normally transient projections (e.g., Perry and Cowey, 1982) and some forms of synaptic reorganization (e.g., Lund et al., 1973; Schneider, 1973; Lynch and Gall, 1979; Gramsbergen, this volume) following neural damage, and a reduction in transient redundancy of neuronal connections during an early stage of normal development (e.g., Mariani, 1983).

The process of "selective stabilisation of synapses" has been proposed by Jean-Pierre Changeux and his colleagues as a mechanism for the specification of neuronal networks during ontogeny and learning. Basing their analysis on electrophysiological and anatomical data collected from studies on the neuromuscular junction in the rat and chicken and on the cerebral and cerebellar cortex of mammals, these workers have suggested

* Translation of Paillard's categories, 1976

that selective stabilization of synapses is the means by which epigenesis of neuronal networks takes place (Changeux, 1983). In his recent book, Changeux (1983) states that "epigenesis exercises its selection on preformed synaptic networks. Learning is the stabilization of already established (but labile**) synaptic combinations and the elimination of others" (p. 329). However, the last of these assertions seems too general for at least two reasons. First, it has been shown that the functional restriction induced by early monocular deprivation interpreted by Changeux in terms of a reduction in synaptic redundancy does not obligatorily imply structural elimination, but rather functional inhibition (e.g., Duffy et al., 1976). Secondly, during and after early development, time, experience and prolonged artificial stimulation can cause a genuine increase in the complexity of nerve circuitry (e.g., Greenough, 1976; Rutledge, 1976 ; Uylings et al., 1978; Rosenzweig and Bennett, 1980; Buell and Coleman, 1981; Kelche and Will, 1982). Thus, although some particular forms of learning could involve a process of selective stabilization of synapses, the characteristics of plasticity limited to early developmental periods may include only some particular kinds of learning rather than learning ability in general. Some kinds of "prepared" learning such as imprinting or song learning in birds are, under normal circumstances, biologically constrained to early periods of development (but see Horn, this volume), whereas "less-prepared" kinds of learning (e.g., Seligman and Hager, 1972; Hinde and Stevenson-Hinde, 1973) are clearly not restricted to these periods. However, even if one assumes that plasticity limited to early age may directly concern some learning functions, it should be underlined that what Kasamatsu and Shirokawa (this volume) call "critical period plasticity" concerns chiefly sensory functions (which, of course, may affect learning capacity). Thus, given the present state of our knowledge, it may be hazardous to base our explanations of behavioral modifications due to learning exclusively on a single neural process.

The affirmation of a potentiality for an increase in complexity in the CNS circuitry of adult and even senescent animals does not imply that the nervous system remains equally plastic throughout the lifespan. Plasticity in many cases diminishes with age, and some of the critical factors controlling this temporal shift are now being elucidated. For instance, recent data tend to demonstrate that some aspects of the critical period for visual development may be "rejuvenated" in adult cats by local perfusion with norepinephrine (Kasamatsu and Shirokawa, this volume). It has also been established that a critical role in brain and behavior plasticity may be played by hormones (see chapters on neuroendocrine modulation). De Kosky and his collaborators have shown that young rats maintained at the elevated corticosterone levels of normal senescent rats had marked suppression of lesion-induced sprouting in the hippocampal dentate gyrus (De Kosky et al., 1984) and Nottebohm has suggested that the yearly "rejuvenation" of the canary's song repertoire may require that the increase in number of neurons (e.g., Paton et al., 1984) and other morphological aspects of the nuclei which control song behavior be preceded by a shrinkage of these nuclei, possibly under hormonal control.

Such considerations suggest that the distinction we have made between "developmental plasticity" and "adaptive plasticity" is perhaps more than a convenient one. It is far less sure whether it is correct to distinguish these two kinds of plasticity found in intact animals from "restorative plasticity" found in subjects with CNS damage.

<u>Plasticity in intact and brain damaged animals</u> . Direct damage to the CNS causing partial denervation induces a cascade of events which in

** Added by the present authors

some respects constitute a recapitulation of ontogeny (e.g., Brown, 1984). However, even destruction of brain structures is unnecessary for the initiation of plastic brain events. Various neurotoxins that prevent only the propagation of nerve impulses or interrupt axonal transport are found to induce sprouting in intact neurons or preservation of normally transient projections. Such effects were found with tetrodotoxin, botulinum toxin, d-tubocurarine, alpha-bungarotoxin and colchicine (for a review, see Cotman and Nieto-Sampedro, 1982). To interpret the results obtained with colchicine, it is generally assumed that nerves carry anti-growth factors (e.g., Diamond et al., 1976 ; Goldowitz and Cotman, 1980) and that neurons and glial cells are allowed to grow when the supply of these factors is interrupted or reduced in other neurons (what might be the case during ontogeny). The triggering of sprouting by colchicine alkaloids might explain the "serial lesion effect" observed by Guy Toniolo and collaborators (this volume) who found that hippocampectomized rats showed reduced deficits when pretreated one month before hippocampectomy with a dose of colchicine that produced reversible effects at the level of hippocampal theta activity.

Given that direct denervation is not necessary to elicit the plastic potentialities of neurons (e.g., Kelche and Will,1982), it seems important to consider, as have Cotman and collaborators (Cotman et al., 1981; Cotman and Nieto-Sampedro, 1982), whether these potentialities are normally either not expressed or perhaps expressed in a more restricted manner in intact animals. Studies on seasonal variations in the number of neuromuscular junctions in hibernating animals (Wernig and Stöver, 1979) and on the noradrenergic innervation of the guinea pig uterus during and after pregnancy (e.g., Alm et al., 1979) illustrate how physiological functions can be regulated by the disappearance and reappearance of nerve input. It is possible that a relationship exists between learning and "synaptic turnover" such as is suggested by studies on learning experiences that may occur in different housing environments (e.g., Greenough, 1976; Rosenzweig and Bennett, 1978) and those on seasonal variations of song repertoire in the canary bird (Nottebohm, 1981; Paton et al., 1984). Given these findings, it seems reasonable to consider a lesion to the nervous system as a useful tool to determine some of the processes that may occur at a much reduced scale in adult intact animals.

CONCLUDING REMARKS

We wish to comment on two points in relation to the structure of this book on brain plasticity, learning and memory. First, evidence has led us - as many others - to distinguish between developmental and adaptive plasticity in the outline of this volume. However, we recognize that the boundary between these two kinds of plasticity remains blurred as it depends on the equally vague boundary of the concept of CNS maturity. Indeed, it becomes more and more difficult to find an adequate definition for the latter concept because it depends on the choice of a given criterion (end of neurogenesis, end of myelinization et cetera, which vary in different brain areas), but also because many aspects of plasticity appear to be a life-long attribute of the nervous system. The second point is to emphasise that it is perhaps artificial to qualitatively distinguish the third category, i.e. restorative plasticity, from the two other kinds of plasticity mentioned, although it might be justified to distinguish restorative plasticity found in young animals from that observed in older animals (for a more detailed discussion, see Will and Eclancher, 1984).

Additionally, it is important for future research to determine to what extend mechanisms observed at a given systemic level of analysis (molecular, cellular, etc.) are sufficient to explain mechanisms observed at other levels of analysis, particularly when different models are chosen for different levels of analysis. For example, we need to establish whether the specific biological processes presumed to underly such learning phenomena as habituation or sensitization in relatively simple nervous systems such as those found in molluscs and flatworms (Brunelli et al., this volume) can be generalized to all kinds of learning in more complex organisms.

REFERENCES

Alm, P., Björklund, A., Owmann, C., and Thorbert G., 1979, Tyrosine hydroxylase and DOPA-decarboxylase activities in the guinea-pig uterus : further evidence for functional adrenergic denervation in association with pregnancy, Neurosci. , 4:145.

Bethe, A., 1930, Plastizität und Zentrenlehre, Handb. Norm. Path. Physiol. , 15:1175.

Bleier, R., 1969, Retrograde transsynaptic cellular degeneration in mammillary and ventral tegmental nuclei following limbic decortication in rabbits of various ages, Brain Res. , 15:365.

Bloch, V., 1970, Facts and hypotheses concerning memory consolidation processes, Brain Res. , 24:561.

Bonnet, C., 1781, "Oeuvres d'Histoire Naturelle et de Philosophie", Fauche, Neuchâtel.

Brandt, A., 1867, Sur le rapport du poids du cerveau à celui du corps chez différents animaux, in : "Bulletin de la Société Impériale des Naturalistes de Moscou", 40:525.

Brown, C.P., 1971, Cholinergic activity in rats following enriched stimulation and training : direction and duration of effects, J. Comp. Physiol. Psychol. , 75:408.

Brown, M.C., 1984, Sprouting of motor nerves in adult muscles : a recapitulation of ontogeny, Trends Neurosci. , 7:10.

Buell, S.J., and Coleman, P.D., 1981, Quantitative evidence for selective dendritic growth in normal human aging but not senile dementia, Brain Res. , 214:23.

Changeux, J.P.,"L'homme neuronal", Fayard, Paris.

Clark, E., and Dewhurst, K., 1972, "An Illustrated History of Brain Function", University of California Press, Berkeley.

Cotman, C.W. and Nieto-Sampedro, M., 1982, Brain function, synapse renewal and plasticity, Ann. Rev. Psychol. , 33:371.

Cotman, C.W., Nieto-Sampedro, M., and Harris, E.W., 1981, Synapse replacement in the nervous system of adult vertebrates, Physiol. Rev. , 61:684.

Cuvier, G., 1800, "Leçons d'anatomie comparée", vol. 2, Baudoin, Paris.

Cuvier, G., 1822, "Recherches sur les ossements fossiles", vol. 3, Dufour et d'Ocagne, Paris.

DeKoski, S.T., Scheff, S.W., and Cotman, C.W., 1984, Elevated corticosterone levels: A possible cause of reduced axon sprouting in aged animals, Neuroendocrinol. , 38:33.

Diamond, J., Cooper, E., Turner, C., and MacIntyre, L., 1976, Trophic regulation of nerve sprouting, Science , 193:371.

Duffy, F.H., Snodgrass, S.R., Burchfiel, J.L., and Conway, J.L., 1976, Bicuculline reversal of deprivation amblyopia in the cat, Nature , 260:256.

Finger, S., and Stein, D.G., 1982, "Brain Damage and Recovery", Academic Press, New York.

Gall, C., and Lynch, G., 1978, Rapid axon sprouting in the neonatal rat hippocampus, Brain Res. , 153:357.

Goldwitz, D., and Cotman, C.W., 1980, Do neurotrophic interactions control synapse formation in the adult rat brain ? Brain Res. , 181:325.

Greenough W.T., 1976, Enduring brain effects of differential experience and training, in : "Neural mechanisms of learning and memory" M.R. Rosenzweig and E.L. Bennett, eds., M.I.T. Press, Cambridge Mass.

Greenwald, I., and Martinez-Arias, A., 1984, Programmed cell death in invertebrates, Trends Neurosci. , 7:179.

Hamburger, V., 1975, Cell death in the development of the lateral motor column of the chick embryo, J. Comp. Neurol. , 160:535.

Hinde, R.A., and Stevenson-Hinde, J., 1973, "Constraints on learning", Academic Press, London.

Isaacson, R.L., 1975, The myth of recovery from early brain damage, in : "Aberrant development in infancy", N.R.Ellis, ed., Erlbaum, Potomac.

John, E.R., 1972, Switchboard versus statistical theories of learning and memory, Science , 177:850.

Kelche, C., and Will, B., 1982, Effects of postoperative environments following dorsal hippocampal lesions on dendritic branching and spines in rat occipital cortex, Brain Res. , 245:107.

Konorski, J., 1961, The physiological approach to the problem of recent memory, in : "Brain Mechanisms and Learning", A. Fessard, ed., Balkwell, Oxford.

Lewis, N.E., 1979, Psychology of active and inactive memory, Psychol. Bull. , 86:1054.

Land, P.W., and Lund, R.D., 1979, Development of the rat's uncrossed retinotectal pathway and its relation to plasticity studies, Science , 205:698.

Lund, R.D., 1978, "Development and Plasticity of the Brain", Oxford University Press, New York.

Lund, R.D., Cunningham, T.J., and Lund, J.S., 1973, Modified optic projections after unilateral eye removal in young rats, Brain Behav. Evol. , 8:51.

Lynch, G., and Gall, C., 1979, Organization and reorganization in the central nervous system: Evolving concepts of brain plasticity, in : "Human Growth", vol.3, F. Falkner and J. M. Tanner, eds., Plenum Publishing Corporation, New York.

Magendie, F., 1831, "An elementary compendium of physiology", E. Milligan, trans., J. Carfrae, Edinburgh.

Mariani, J., 1983, Elimination of synapses during the development of the central nervous system, Prog. Brain Res. , 58:383.

McCormick, D.A., and Thompson, R.F., 1984, Cerebellum: Essential involvement in the classically conditioned eyelid response, Science , 223:299.

Nottebohm, F., 1981, Laterality, seasons and space govern the learning of motor skill, Trends Neurosci. , 4:104.

Paillard, J., 1976, Réflexions sur l'usage du concept de plasticité en neurobiologie, J. Psychol.(Paris) , 1:33.

Passingham, R.E., Perry, V.H., and Wilkinson, F., 1983, The long-term effects of removal of sensorimotor cortex in infant and adult rhesus monkeys, Brain , 106:675.

Paton, J.A., O'Loughlin, B., and Nottebohm, F., 1984, Neurogenesis of interneurons in a adult canary forebrain, Soc. Neurosci. Abstracts, 10:664.

Perry, V.H., and Cowey, A., 1982, A sensitive period for ganglion cell degeneration and the formation of aberrant retinofugal connections following tectal lesions in rats, Neurosci. , 7:583.

Prendergast, J., and Stelzner, D.J., 1976, Changes in the magnocellular portion of the red nucleus following thoracic hemisection in the neonatal and adult rat, J. Comp. Neurol. , 166:163.

Ramõn Y Cajal, S., 1894, La fine structure des centres nerveux, Proc. of the Royal Soc. of London , 55:444.

Ramõn Y Cajal, S., 1928, "Studies on Degeneration and Regeneration of the Nervous System", R.M.May, trans., Oxford University Press, Oxford.

Rosenzweig, M.R., 1979, Responsiveness of Brain Size to Individual Experience: Behavioral and Evolutionary Implications, in : "Development and Evolution of Brain Size", M.E.Hahn, C.Jensen, and B.C.Dudek, eds., Academic Press, New York.

Rosenzweig, M.R. and Bennett, E.L., 1977, Effects of environmental enrichment or impoverishment on learning and on brain values in rodents, in : "Genetics, Environment and Intelligence", A. Oliverio, ed., Elsevier, Amsterdam.

Rosenzweig, M.R., and Bennett, E.L., 1978, Experiential influences on brain anatomy and brain chemistry in rodents, in : "Early Influences : Studies on the Development of Behavior and the Nervous System", G. Gottlieb, ed., Academic Press, New York.

Rosenzweig, M.R., and Bennett, E.L., 1980, How plastic is the nervous system ?, in : "The comprehensive handbook of behavioral medicine", vol.1, J. Fergusson and C. Taylor, eds., Spectrum Publications, New-York.

Rutledge, L.T., 1976, Synaptogenesis : Effects of synaptic use, in : "Neural Mechanisms of Learning and Memory", M.R. Rosenzweig and E.L. Bennett, eds., M.I.T. Press, London.

Sahakian, W.S., 1970, "Learning : Systems, Models and Theories", Rand McNally College Publishing Company, Chicago.

Schneider, G.E., 1973, Early lesions of superior colliculus: Factors affecting the formation of abnormal retinal projections, Brain Behav. Evol. , 8:73.

Schneider, G.E., 1979, Is it really better to have your brain lesion early? A revision of the "Kennard principle", Neuropsychologia , 17:557.

Seligman, M.E.P., and Hager, J.L., 1972, "Biological boundaries of learning", Appleton Century Crofts, New-York.

Snell, O., 1892, Die Abhängigkeit des Hirngewichtes von dem Körpergewicht und den geistigen Fähigkeiten, Archiv für Psychiatrie und Nervenkrankheiten , 23:436.

Sömmerring, S. Th., 1785, "Über die Körperliche Verschiedenheit des Negers vom Europäer", Barrentrapp Sohn und Wenner, Frankfurt und Mainz.

Sömmerring, S. Th., 1791, "Vom Baue des Menschlichen Körpers", Barrentrapp Sohn und Wenner, Frankfurt und Mainz.

Spaer, N.E., 1980, "L'évolution des souvenirs : Oubli et mémoire", Medsi, Paris.

Steele Russell, I., 1979, Brain size and intelligence: a comparative perspective, in : Brain, Behaviour and Evolution, D.A. Oakley and H.C. Plotkin, eds., Methuen, London.

Ungerer, A., Pallaud, B., Ropartz, Ph., and Will, B., 1977, Répercussions à court et moyen termes d'un enrichissement physique de l'environnement chez le rat : Comportement explorateur et préférence pour un type d'environnement, Biol. Behav. , 2:159.

Uylings, H.B.M., Kuypers, K., and Veltman, W.A.M., 1978, Environmental influences on the neocortex in later life, in : "Maturation of the Nervous System, Progress in Brain Research", M.A. Corner et al., eds., Elsevier, North Holland.

Wernig, A., and Stöver, H., 1979, Sprouting and repression of the nerve at the frog neuromuscular junction, Pfluegers Arch. , 379:R38.

Will, B., and Eclancher, F., 1984, Early brain damage and early environment, in : "Early Brain Damage", vol.2, S. Finger and C. R. Almli, eds., Academic Press, New York.

Yeo, C.H., Hardiman, M.L., and Glickstein, M., 1984, Discrete lesions of the cerebellar cortex abolish the classically conditioned nictitating membrane response of the rabbit, Behav. Brain Res. , 13:261.

IMPRINTING AND THE NEURAL BASIS OF MEMORY

Gabriel Horn

Department of Zoology
University of Cambridge
Downing Street
Cambridge CB2 3EJ, U.K.

INTRODUCTION

The way in which information acquired through learning processes is stored in the brain has long been a matter for speculation. During the 1960's, however, some progress was made in the experimental analysis of the memory associated with habituation. This is a behavioural change, one component of which is that an animal gradually ceases to respond to a stimulus which is presented repeatedly; the absence of a response to the stimulus is an indication of a 'memory' not to respond. The scene was set for a cellular analysis of this memory when evidence appeared of changes in neuronal activity which resembled those of behavioural habituation (see for example Horn and Hill, 1964; Bruner and Tauc, 1966; Spencer et al., 1966a,b,c). On the basis of experimental evidence it was proposed (Bruner and Tauc, 1966) that the progressive reduction of a neuronal response in _Aplysia_ is an expression of synaptic depression, and (Horn and Wright, 1970) that calcium ions are implicated in this process. Further support for both proposals came from subsequent studies of habituation in _Aplysia_ (Kandel, 1979; Klein et al., 1980). Synaptic depression occurring in specific neural pathways can account for many aspects of behavioural habituation (Thompson and Spencer, 1966; Horn, 1967). However, synaptic depression may not be the only way in which transmission may be blocked when a stimulus is repeatedly delivered (Horn, 1967; Weight and Erulkar, 1976). Furthermore, it is likely that more than one kind of memory is established in some animals, when a stimulus is repeatedly presented to them (see for example Hinde, 1960, 1970; Mackintosh, 1985). Thus this experimental procedure may lead to cellular changes which are more diverse and more complex than is often assumed. Nevertheless, in rather less than two decades substantial progress has been made in the neural analysis of certain aspects of behavioural habituation.

Another form of learning which has begun to yield to experimental analysis is imprinting. Imprinting occurs in young animals which show well-coordinated motor activity soon after birth or hatching. The domestic chick (_Gallus domesticus_) is such an animal. A visually naive young chick will approach a wide range of objects. If the chick is exposed to one object for some time, the chick may learn its characteristics and avoid other objects. This change in behaviour suggests that the chick recognizes the object to which it was originally exposed and that information about the object is stored in the chick's central nervous system. What is the basis of this storage process?

If this process involves changes in the connections between neurones then changes in protein and RNA metabolism may be expected to occur in those brain regions in which storage takes place. To examine this possibility dark-reared, visually naive chicks were exposed to a rotating illuminated box which is an effective imprinting object. Dark-reared birds were used as controls. There was an increase in the incorporation of [^{3}H]lysine into protein and also of [^{3}H]uracil into RNA in a restricted brain region of the imprinted chicks (Bateson et al., 1972). The region was the dorsal part of the cerebral hemispheres which had been divided into two parts, the forebrain roof and the forebrain base. No effects of training were observed in the forebrain base or in the midbrain.

There are of course many ways in which the trained chicks differed from the dark-reared controls and the biochemical changes may have been related to some or all of these differences. A number of control experiments was therefore undertaken to attempt to distinguish biochemical effects which were specific to the learning process from those which were 'side-effects' of the training procedure (Horn et al., 1973; Bateson et al., 1973, 1975). The results of these experiments suggested that the changes in the incorporation of radioactive uracil into RNA in the forebrain roof of trained chicks were closely related to the imprinting process.

By using an autoradiographic technique with [^{14}C]uracil as the probe molecule, an increased incorporation into RNA was found in a restricted part of the medial hyperstriatum ventrale (Horn et al., 1979). The region lay approximately halfway between the anterior and posterior poles of the cerebral hemispheres. Accordingly, this intermediate and medial part of the hyperstriatum ventrale was referred to as IMHV.

EFFECTS OF LESIONS TO IMHV

If, as the evidence was beginning to suggest, IMHV is involved in the storage of information, then destruction of the region should prevent the acquisition of a preference through imprinting, and impair the retention of an acquired preference. Both of these predictions were met by the results of two lesion studies which involved the bilateral destruction of IMHV (McCabe et al., 1981, 1982). Lesions of two other brain regions, the Wulst and the lateral cerebral area, were without effect on retention.

In each of these two lesion studies two imprinting stimuli were used. Half of the chicks were trained by exposing them to a rotating flashing red box; the other half was trained by exposing the chicks to a rotating stuffed jungle fowl (Gallus gallus). Preferences were tested by placing the chicks individually in running wheels in front of which stood either the rotating box or the fowl. The wheel rotated if the chick attempted to run towards an object. The number of rotations made in a set period of time provided a measure of approach activity. Each chick's approach activity towards the training object and, separately, to a novel object (the box or fowl whichever the chick had not previously seen) were recorded. The chicks were considered to prefer one object to the other if they approached that object relatively more vigorously than the other in the preference test. The chicks were considered to have no preference if they approached the two objects with equal vigour.

Chicks with bilateral lesions of IMHV did not selectively approach the stimulus to which they had been exposed during the training period; the chicks performed at chance, showing no preference for either object. This behaviour may have arisen for a variety of reasons. However, lesioned chicks were not obviously impaired in visual abilities or in motor activity. Furthermore, they were able to perform certain associative learning tasks (McCabe et al., 1982; Johnson and Horn, 1984). There was thus a dissociation, in chicks with lesions of IMHV, between performance on a recognition task and performance on an associative learning task. This dissociation is reminiscent of a similar dissociation found in humans and in macaque monkeys with lesions of the medial temporal lobe (Sidman et al., 1968; Weiskrantz 1982; Aggleton and Mishkin 1983a,b; Zola-Morgan and Squire 1984; Mahut, see this volume). This `syndrome´ should not be taken to imply that IMHV is analogous with structures in the medial temporal lobe (see Bradley et al., 1985); the similarities may, however, imply some rather basic difference between the memory systems involved in recognition and those involved in associative learning. This dissociation may have appeared relatively early in the evolutionary history of terrestrial vertebrates, perhaps before the divergence of birds and mammals.

The lesion studies disclosed a difference in the processing of information about the red box and the stuffed jungle fowl. In each of four lesion studies (see Horn and McCabe, 1984) an analysis of variance had been performed on the preference measures (scores) contributed by the lesioned and control groups of chicks. The preferences of chicks for the object upon which they had been trained was not significantly affected by the training object. However, when data for all the lesion studies were combined, a clear effect appeared: lesions to IMHV profoundly affected the preferences of chicks exposed to the box, but had a much smaller, though significant, effect on chicks which had been trained upon the jungle fowl (Horn and McCabe, 1984).

PREDISPOSITIONS AND PREFERENCES

The results of two further studies suggest that the neural consequences of exposing visually naive chicks to the rotating red box are different in a non-trivial way from those of exposing chicks to the jungle fowl. The drug DSP4, which lowers brain noradrenaline levels, impairs the acquisition of a preference for the box, but has only a small effect on the acquisition of a preference for the fowl (Davies et al., 1985 and this volume); exogenous testosterone enhances the acquisition of a preference for the jungle fowl, but is without effect on the acquisition of a preference for the box (Bolhuis et al., 1984). In order to interpret these results it is necessary to consider again the chicks´ behaviour. Visually naive chicks will approach a wide range of objects. If a chick remains exposed to a particular object for an adequate length of time it learns the characteristics of that object. The evidence reviewed above suggests that when the stimuli are artificial, then IMHV is involved in the learning process, probably in storage. These stimuli include red and orange rotating boxes, red cylinders and flashing bars of light. What may be happening when chicks are exposed to the jungle fowl? They run towards the fowl and if exposed to it for long enough they learn the characteristics of the individual fowl (Johnson and Horn, 1985). Chicks with bilateral lesions of IMHV also run towards the jungle fowl during the training period. However, these chicks do not selectively approach that fowl when given a choice between it and another stuffed jungle fowl (Johnson and Horn 1985). The lesioned chicks show no evidence, therefore, of recognizing the individual to which they had been exposed. These results suggest that IMHV may be implicated in the recognition of individual jungle fowls, a process which involves learning. More generally, it seems likely that IMHV is involved in the storage of information acquired through exposure learning, whether the object to which a chick is exposed is an artificial imprinting stimulus or whether it is an individual conspecific.

Although bilateral lesions to IMHV impair the preference of fowl-trained chicks, these birds do not perform at chance levels: they still selectively approach the jungle fowl in the preference test (Horn and McCabe, 1984). To what neural processes may this component of the preference be ascribed? Many animals have predispositions to respond in particular ways to particular objects. For example, objects smaller than a matchbox are ignored by ducklings or pursued by them as food, and are relatively ineffective as imprinting stimuli (Fabricius and Boyd, 1952-53); the following response is more effectively elicited in chicks by blue and red stimuli than by yellow or white stimuli (Schaefer and Hess 1959). Although in none of the experimental studies described in the above paragraphs was there evidence that untrained chicks preferred the jungle fowl to the red box, more subtle tests have revealed such a preference (Johnson et al., 1985; Bolhuis et al., 1985). It is not yet known what features of the jungle fowl are critical for the predisposition to appear.

When naive chicks are exposed to the jungle fowl, two things may happen. The predisposition, an unlearned response, is strengthened, and the chicks learn the characteristics of the individual fowl. IMHV is involved in the learning process, and some other system outside IMHV appears to be involved in the prediposition. Since the preferences of fowl-trained chicks are correlated with plasma testosterone concentrations, it is possible that the neural structures subserving the predisposition may be under the control of this hormone. There are of course other possible interpretations of the relationship between plasma testosterone and the preferences of the fowl-trained birds. For example, testosterone might make the chicks more attentive (see Andrew 1983). If this were so then some aspect of the behaviour of chicks trained on the box as well as of chicks trained on the jungle fowl should be related to plasma testosterone levels. No such relationship was found for box-trained birds, either in their behaviour during training, or in their behaviour during the preference test.

The acquisition of a preference for the red box is impaired in chicks given DSP4, so that noradrenaline may be involved in the underlying memory processes. If it is, then it is reasonable to suppose that the transmitter is also involved in the learned component of the preferences of fowl-trained chicks. Lesions to IMHV significantly reduced the preferences of these chicks for their training object by approximately 11 per cent (Horn and McCabe, 1984). The mean preference of DSP4-treated chicks trained on the jungle fowl for 80 min was in the expected direction, being less than that of their controls by a mean value of 8.52 ± 4.73 (standard error) per cent. The reduction was significant on a one-tailed t-test ($t = 1.80$, $P < 0.04$). This result (Davies et al., 1985) suggests that noradrenaline is indeed involved in the development of a preference for the jungle fowl.

At a more general level, it may be that noradrenaline plays a role in the memory which underlies recognition, but not in the memory which underlies associative learning. Thus, whereas DSP4 impairs performance on a recognition task such as imprinting, the drug appears to be without effect on the acquisition by chicks of a passive avoidance task (Davies, Bradley and Galal, personal communication). The evidence, discussed above, that the two kinds of memory may be dissociated by brain lesions is consistent with the possibility that the two kinds of memory may also be differentially affected by drugs, and that different physiological constraints may operate on the two memory systems (see Horn, 1985).

CHANGES IN IMHV ASSOCIATED WITH IMPRINTING

It has often been suggested that learning leads to changes in synaptic connections. Apart from habituation, however, the hypothesis has not been supported by experimental evidence. Certainly changes in the structure of synapses occurs in many brain regions following various kinds of experimental manipulations; but to support a claim that such changes form the basis of information storage it is essential to demonstrate that the regions play a critical role in memory. This requirement is obvious almost to the point of being banal, but it is frequently ignored.

Changes in the structure of synapses occur in the IMHV of chicks which have been trained upon artificial imprinting stimuli. The changes, which are described in detail elsewhere (see Bradley et al., 1981; and this volume) occur in the left IMHV. These results suggest an asymmetry in the functions of the left and right IMHV in the imprinting process. Evidence from brain lesion experiments suggest that this is indeed the case (Cipolla-Neto et al., 1982; Horn et al., 1983. See Horn, 1985).

In addition to changes in synapse structure, changes in the spontaneous activity of neurones occur in the left IMHV of chicks which have been trained on the box. It was found that the mean firing rate of neurones in this brain region is negatively correlated with approach to the red box during the training period; the more chicks ran towards the box the lower was the mean firing rate. No such correlation was found for units in the hyperstriatum accessorium of the box-trained chicks. Furthermore, there was no correlation between mean spontaneous firing rate of neurones in IMHV and training approach counts in chicks which had been exposed to the jungle fowl. Thus the effects seen in the box-trained chicks were regionally specific and stimulus specific (Payne and Horn 1984). It is not known whether the synaptic changes and the physiological changes in IMHV are related to each other, though this possibility cannot yet be excluded (see Payne and Horn 1984; Horn, 1985).

SUMMARY

A restricted part of the hyperstriatum ventrale is involved in the storage of information acquired by young chicks when they are exposed to a visually conspicuous artificial object. Within the region, referred to as IMHV, changes in the structure of synapses occur following exposure to such an object. This learning process is severely impaired by a drug which lowers the level of brain noradrenaline. When chicks are trained by exposing them to a complex naturalistic object, a stuffed jungle fowl, they prefer it to the box when given a preference test. The strength of this preference is correlated with the level of testosterone in the plasma. For box-trained chicks, in contrast, there is no correlation between preference for the familiar and plasma testosterone concentration. There is evidence that chicks have a predisposition to approach certain objects such as the jungle fowl, the individual characteristics of which they then learn. IMHV is necessary for this learning process, whereas the neural systems supporting the predisposition lie outside IMHV and appear to be influenced by the level of testosterone in the plasma.

REFERENCES

Aggleton, J. P. and Mishkin, M., 1983a, Visual recognition impairment following medial thalamic lesions in monkeys, Neuropsychologia, 21:189.

Aggleton, J. P. and Mishkin, M., 1983b, Memory impairments following restricted medial thalamic lesions in monkeys, Exp. Brain Res., 52:199.

Andrew, R. J., 1983, Specific short-term latency effects of oestradiol and testosterone on distractability and memory formation in the young domestic chick, in: "Hormones and Behaviour in higher vertebrates, J. Balthazart, E. Prove and R. Gilles, eds., Springer, Berlin.

Bateson, P. P. G., Horn, G. and Rose, S. P. R., 1972, Effects of prior exposure to light on the imprinting process in domestic chicks, Behaviour, 42:279.

Bateson, P. P. G., Rose, S. P. R. and Horn, G., 1973, Imprinting: lasting effects on uracil incorporation into chick brain, Science, 181:576.

Bateson, P. P. G., Horn, G. and Rose, S. P. R., 1975, Imprinting: correlations between behaviour and incorporation of [^{14}C] uracil into chick brain, Brain Res., 84:207.

Bolhuis, J. J., Davies, D. C., McCabe, B. J. and Horn, G., 1984, Constraints on imprinting in the domestic chick: possible involvement of noradrenaline (NA) and testosterone, Neurosci. Lett., Suppl. 18:S131.

Bolhuis, J., Johnson, M. H. and Horn, G, 1985, Effects of early experience on the development of filial preferences in the domestic chick, Devel. Psychobiol., in press.

Bradley, P., Horn, G. and Bateson, P., 1981, Imprinting: an electron microscopic study of chick hyperstriatum ventrale, Exp. Brain Res., 41:115.

Bradley, P., Davies, D. C. and Horn, G., 1985, Connections of the hyperstriatum ventrale in the domestic chick (Gallus domesticus), J. Anat., in press.

Bruner, J. and Tauc, L., 1966, Habituation at the synaptic level in Aplysia, Nature, 210:37.

Cipolla-Neto, J., Horn, G. and McCabe, B. J., 1982, Hemispheric asymmetry and imprinting: the effect of sequential lesions to the hyperstriatal ventrale, Exp. Brain Res., 48:22.

Davies, D. C., Horn, G. and McCabe, B. J., 1985, Noradrenaline and learning: the effects of the noradrenergic neurotoxin DSP4 on imprinting in the domestic chick, Behav. Neurosci., in press.

Fabricus, E. and Boyd, H., 1952-3, Experiments on the following reactions of ducklings, Wildfowl Trust Ann. Rep., 6:84.

Hinde, R. A., 1960, Factors governing the changes in strength of a partially inborn response, as shown by the mobbing behaviour of the chaffinch (Fringilla coelebs). III. The interaction of short-term and long-term incremental and decremental effects. Proc. R. Soc. Lond. B, 153:398.

Hinde, R. A., 1970, Behavioural habituation, in: "Short-term changes in neural activity and behaviour", G. Horn and R. A. Hinde, eds., pp. 3-40, Cambridge University Press, Cambridge.

Horn, G., 1967, Neuronal mechanisms of habituation, Nature, 215:707

Horn, G., 1985, "Memory, imprinting and the brain", Clarendon Press, Oxford.

Horn, G. and Hill, R. M., 1964, Habituation of the response to sensory stimuli of neurons in the brain stem of rabbits, Nature, 202:296.

Horn, G. and McCabe, B. J., 1984, Predispositions and preferences. Effects on imprinting of lesions to the chick brain. Anim. Behav., 32:288.

Horn, G., McCabe, B. J. and Bateson, P. P. G., 1979, An autoradiographic study of the chick brain after imprinting, Brain Res., 168:361.

Horn, G., McCabe, B. J. and Cipolla-Neto, J., 1983, Imprinting in the domestic chick: the role of each side of the hyperstriatum ventrale in acquisition and retention, Exp. Brain Res., 53:91.

Horn, G., Rose, S. P. R. and Bateson, P. P. G., 1973, Monocular imprinting and regional incorporation of tritiated uracil into the brains of intact and `split-brain´ chicks, Brain Res., 56:227.

Horn, G. and Wright, M. J., 1970, Characteristics of transmission failure in the squid stellate ganglion: a study of a simple habituating system. J. exp. Biol., 52:217.

Johnson, M. H., Bolhuis, J. J. and Horn, G., 1985, Interaction between acquired preferences and developing predispositions during imprinting, Anim. Behav., in press.

Johnson, M. H. and Horn, G., 1984, Differential effects of brain lesions on imprinting and an associative learning task, Neurosci. Lett., Suppl. 18:S131.

Johnson, M. H. and Horn, G., 1985, The role of a restricted brain region of domestic chicks in the recognition of individuals, in preparation.

Kandel, E. R., 1979, Cellular insights into behavior and learning, in: "The Harvey Lectures", Series 73, pp. 19-92, Academic Press, New York.

Klein, M., Shapiro, E. and Kandel, E. R., 1980, Synaptic plasticity of the modulation of Ca^{2+} current. J. exp. Biol., 89:117.

Mackintosh, N. J., 1985, Varieties of conditioning, in: "Second conference on the neurobiology of learning and memory", G. Lynch, J. McGaugh and N. M. Weinberger, eds., Guilford Press, New York, in press.

McCabe, B. J., Cipolla-Neto, J., Horn, G. and Bateson, P. P. G., 1982, Amnesic effects of bilateral lesions placed in the hyperstriatum ventrale of the chick after imprinting. Exp. Brain Res., 48:13.

McCabe, B. J., Horn, G. and Bateson, P. P. G., 1981, Effects of restricted lesions of the chick forebrain on the acquisition of filial preferences during imprinting, Brain Res., 205:29.

Payne, J.K. and Horn, G., 1984, Long-term consequences of exposure to an imprinting stimulus on spontaneous impulse activity in the chick brain, Behav. Brain Res., 13:155.

Schaefer, H. H. and Hess, E. R., 1959, Color preferences in imprinting objects, Z. Tierpsychol., 16:161.

Spencer, W. A., Thompson, R. F. and Neilson Jr., D. R., 1966a, Response decrement of the flexion reflex in the acute spinal cat and transient restoration by strong stimuli, J. Neurophysiol., 29:221.

Spencer, W. A., Thompson, R. F. and Neilson Jr., D. R., 1966b, Alterations in responsiveness of ascending and reflex pathways activated by iterated cutaneous afferent volleys, J. Neurophysiol., 29:240.

Spencer, W. A., Thompson, R. F. and Neilson Jr., D. R., 1966c, Decrement of ventral root electrotonus and intracellularly recorded PSPs produced by iterated cutaneous afferent volleys, J. Neurophysiol., 29:240.

Sidman, M., Stoddard, L. T. and Mohr, J. P., 1968, Some additional quantitative observations of immediate memory in a patient with bilateral hippocampal lesions, Neuropsychologia, 6:245.

Thompson, R. F. and Spencer, W. A., 1966, Habituation: a model phenomenon for the study of neuronal substrates of behavior, Psychol. Rev., 173:16.

Weiskrantz, L., 1982, Comparative aspects of studies of amnesia, Phil. Trans. R. Soc. Lond. B, 298:97.

Weight, F. F. and Erulkar, S. D., 1976, Modulation of synaptic transmitter release by repetitive postsynaptic action potentials, Science, 193:1023.

Zola-Morgan, S. and Squire, L. R., 1984, Preserved learning in monkeys with medial temporal skills: sparing of motor and cognitive skills, J. Neurosci., 4:1072.

MORPHOLOGICAL CORRELATES OF IMPRINTING

P.M. Bradley[+], G. Horn[*] and B.J. McCabe[*]

[+]Department of Anatomy
Medical School,
University of Newcastle-upon-Tyne
NE1 7RU

[*]Department of Zoology
University of Cambridge
Downing Street
Cambridge CB2 3EJ

INTRODUCTION

A particular region of the chick brain is critically involved in the storage of information acquired through the learning process of imprinting (Horn, 1981; 1985; and this volume). The region is the intermediate part of the medial hyperstriatum ventrale (IMHV). There is an increased incorporation of [^{14}C]uracil into RNA in this region after chicks have been trained by exposing them to an artificial imprinting stimulus (Horn et al., 1979). This result extended earlier studies which demonstrated that, in chicks trained in this way, there is an increased incorporation of [^{3}H]lysine into proteins in the forebrain roof which contains IMHV (Bateson et al., 1972). The results together suggest that exposure to an imprinting stimulus leads to a change in the metabolism of protein, possibly in protein synthesis, within IMHV. These findings are consistent with the hypothesis that information storage in the brain involves changes in synaptic connections (Tanzi, 1893; Cajal, 1911; Hebb, 1949) . However, the biochemical findings provide only indirect evidence for such changes. A direct test of the hypothesis is to determine whether such changes can be demonstrated in a brain region critical for storage. We have therefore used quantitative electron microscopical techniques to investigate the effects of imprinting on the structure of synapses in the IMHV of the domestic chick. The results presented here show that there is an increase in the size of the postsynaptic density of a specific set of synapses in IMHV as a result of exposing chicks to an imprinting stimulus. Since the structure of the avian IMHV may have some relevance to the interpretation of these synaptic changes, we briefly review recent studies of the structure, connections and development of this region of the brain (Bradley et al., 1985).

STRUCTURE AND CONNECTIONS OF IMHV

The avian telencephalon appears, histologically, to be a laminated structure. The dorsal laminae are collectively known as the Hyperstriatum and comprise the Hyperstriatum accessorium, intercalatus, dorsale and ventrale. Of these, the first three are most dorsal and comprise the visual Wulst. The Wulst together with the Ectostriatum are major primary telencephalic visual projection areas. Of the more ventral structures of the telencephalon, the paleostriatal complex is considered to be homologous with components of the mammalian basal ganglia (Reiner et al., 1984) while the Archistriatum contributes to descending motor pathways as well as to the avian 'limbic' system (Zeier and Karten, 1971). Parts of the Neostriatum receive inputs from auditory (Karten, 1968), somatic sensory (Delius and Bennetto, 1972) and visual (Parker and Delius, 1972) pathways.

The connections of IMHV in the chick brain have been investigated and are summarised in Figure 1. IMHV has reciprocal connections with the visual Wulst, Paleostriatum augmentatum and the dorsal part of the Archistriatum. It receives additional afferents from the Hippocampus, Area Septalis, Neostriatum and the caudal part of the Hyperstriatum ventrale. IMHV sends efferents to the posterior part of the Archistriatum and to the dorsolateral part of the cerebral hemispheres. A number of these connections are bilateral. IMHV is thus ideally placed to act as an integrative centre in the telencephalon, linking the sensory and motor systems, and subject to influences from regions of the brain thought to be concerned with motivational aspects of behaviour.

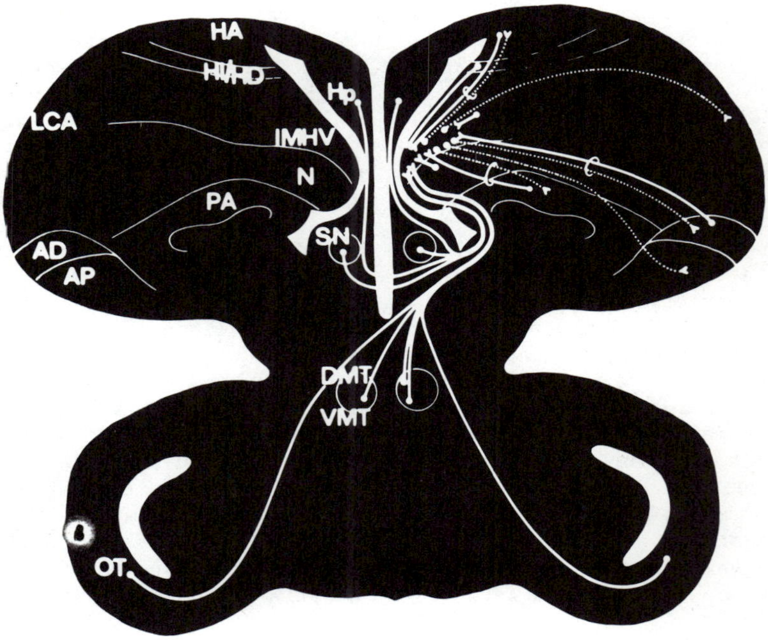

Figure 1. Summary diagram of the afferent and efferent connections of IMHV. Abbreviations: AD, dorsal archistriatum; AP, posterior archistriatum; DMT, dorsomedial thalamus; HA, hyperstriatum accessorium; HI/HD, hyperstriatum intercalatus/dorsale; Hp, hippocampus; IMHV, intermediate part of the medial hyperstriatum ventrale; LCA, lateral cerebral area; N, neostriatum; OT, optic tectum; PA, paleostriatum augmentatum; SN, septal nuclei; VMT, ventromedial thalamus (based on Bradley et al., 1985).

A Golgi study of the organisation of IMHV revealed no evidence of any clear internal lamination but showed that there are at least four distinct neuronal types present within this region (Bradley and Horn, 1982). The two major cell types were classified as either large multipolar neurons or as pyramidal-type neurones. The remainder of the neurones present were either small and bipolar or multipolar. The axons of the pyramidal-type neurones were traced beyond the boundaries of IMHV and it is likely that these neurones form a major source of the efferent projection fibres. The multipolar neurones, in contrast, possessed axons which arborised extensively within IMHV. It is not yet known how the afferents to IMHV are distributed on the various intrinsic neuronal types.

DEVELOPMENT OF IMHV

Changes in synaptic length which occur in IMHV as a result of imprinting are not the only changes which occur in this region in the early post-hatch period. During the first 30 days after hatching there are significant developmental changes within IMHV. Neuronal cell density decreases significantly with age, while between the time of hatching and 7 days of age there is a significant rise in the density, within the neuropil, of synapses on dendritic spines (Bradley, 1982). This rise is matched by an increase in the synapse:neurone ratio over the same period. The change in this ratio is partially accounted for by an increase in the length of the terminal dendritic branches of pyramidal-type neurones (Bradley, 1983). These changes are most rapid in the first few days of life and slow down thereafter. Thus, during the time in which imprinting takes place IMHV is in a labile state with new synapses being formed. IMHV is not unique in this respect since the Hyperstriatum accessorium is also undergoing developmental changes over the same period. Thus a change in synaptic morphology associated with imprinting may easily be confounded with developmental changes unless the ages of experimental and control chicks are carefully matched.

EXPERIMENTAL ANALYSIS OF SYNAPTIC CHANGES ASSOCIATED WITH IMPRINTING

In the first experiment designed to investigate morphological consequences of imprinting two groups of chicks were used (Bradley et al., 1981). Both groups were hatched and reared in darkness. When the chicks were approximately 21 h old they were exposed to an imprinting stimulus. This consisted of a horizontal red bar of light, 2 cm high and moving upwards at 47 cm/sec in a window 10 cm wide and 15 cm deep (Horn et al., 1979). The window was approximately 40 cm in front of the chick. During the training period chicks were placed in a running wheel which rotated as the chicks attempted to approach the stimulus. One group of chicks was exposed to the stimulus for 20 min (undertrained chicks). Chicks in the other group were exposed to the stimulus for a total of 140 min (overtrained birds).

After training the chicks were returned to the dark and 3 h later, approximately 30 h after hatching, all the chicks were killed by perfusion under anaesthesia with 0.1M phosphate buffer pH 7.4, followed by 2% glutaraldehyde/2% paraformaldehyde fixative. Blocks were removed from the right and left IMHV and processed for electron microscopy. Standard stereological sampling procedures were used and for each block a total of 25 micrographs printed at a final magnification of 22,200 were examined. A number of features of synapses was measured. These features were: number of synapses per unit volume, the mean percentages of micrograph area occupied per synaptic bouton and dendritic spine respectively, and the length of the postsynaptic density. In this, and in the other experiments described below, all measurements were made without knowledge of the treatment each chick had received or the side of the brain from which a block had been cut.

Of the various measures of synapse structure, only the length of the postsynaptic density was significantly affected by further training. The effect was found in synapses of the left IMHV, not the right (Figure 2). The length of the postsynaptic density was shorter in the left IMHV than the right in undertrained chicks. Further training eliminated this asymmetry, primarily through an effect on the left IMHV.

These studies were extended by including a dark-reared group of birds, and by increasing the number of birds in the undertrained group (Horn et al., in preparation). Furthermore, synapses were subdivided into two groups, those on the shafts and those on the spines of dendrites.

The only significant effect of training was in the left IMHV. The mean lengths of the postsynaptic density of spine synapses ($\overline{Sp_L}$) in this region of dark-reared chicks was 286.00 $\pm$ 12.24 nm. This mean was significantly less, by 49.2 $\pm$ 17.8 than that of overtrained birds ($P < 0.01$), and not significantly different from that of undertrained birds. $\overline{Sp_L}$ was significantly greater in overtrained than in undertrained chicks ($P < 0.01$). No other measure of synapse morphology in the left IMHV was significantly affected by training.

A second experiment (Horn et al., in preparation) was conducted to allow comparisons to be made with a study which had been designed to investigate some electrophysiological consequences of imprinting (Payne and Horn, 1984). Two groups of chicks were used. One group of 15 chicks was dark-reared throughout. The other group of 15 chicks was reared in the same way until the chicks were 21 h old. They were then exposed to a rotating flashing red box (McCabe et al., 1982) for 1 h. The chicks were trained for further 1 h periods at 24 h and 42 h post-hatch. Between training periods and before tissue removal the chicks were replaced in the dart holding incubators. Three hours after the end of training the chicks were killed and blocks of tissue were removed from the right and left hyperstriatum accessorium, as well as from the right and left IMHV. The blocks were processed for electron microscopy as described above.

There were no significant effects of training on any measure of synapse morphology in the hyperstriatum accessorium or in the right IMHV. There was, however, an effect of training on spine synapses in the left IMHV. $\overline{Sp_L}$ in the left IMHV of trained chicks was significantly greater than that in the dark-reared birds ($P < 0.05$).

The relationship between approach activity during training and the length of the postsynaptic density of spine synapses was examined. For the left IMHV the correlation between $\overline{Sp_L}$ and approach counts recorded during the first hour of training was significant ($r = 0.52$, d.f. 13, $p < 0.05$). The correlation was not significant when approach counts during the second and third hour of training were used as variables. Nor was there a significant correlation between $\overline{Sp_L}$ of the right IMHV and approach activity, or between $\overline{Sp_L}$ in the hyperstriatum accessorium and approach activity.

DISCUSSION

The most striking feature of the results presented here is the consistent involvement in imprinting of one set of structures in the left IMHV. Thus in two separate studies using different training protocols,

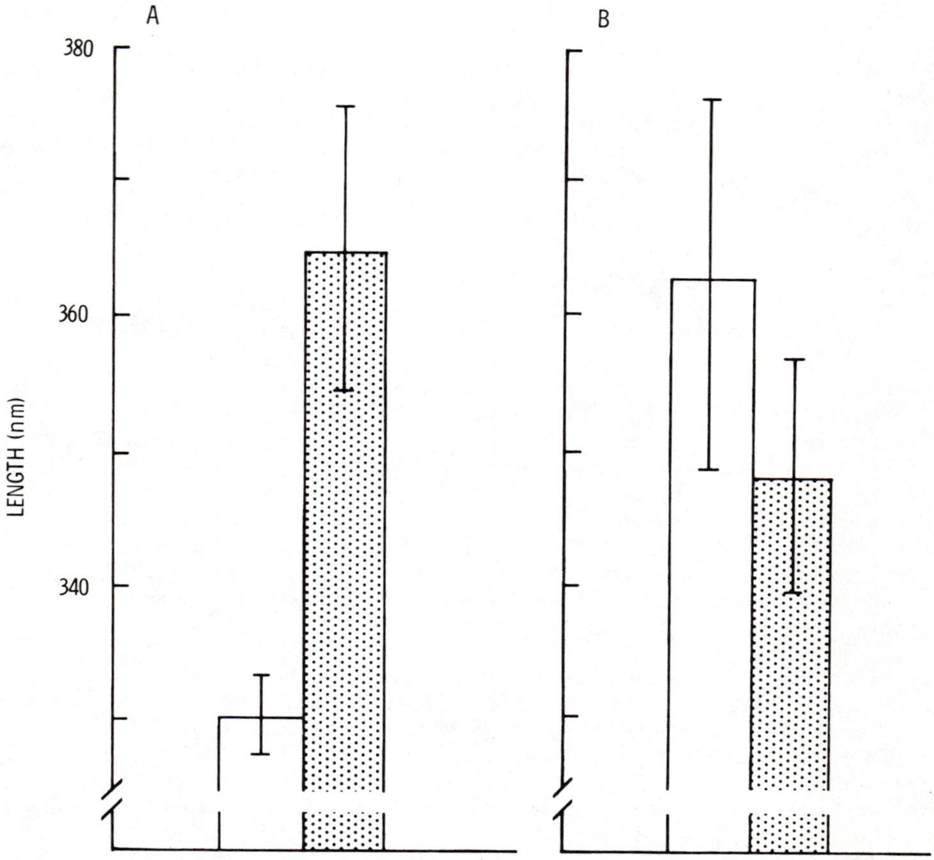

Figure 2. Length of the postsynaptic density of synaptic apposition zones in IMHV. No distinction was made between synapses on the shafts and synapses on the spines of dendrites. Means ± SEM plotted according to treatment and hemisphere (open bars, left hemisphere; dotted bars, right hemisphere). (A) Undertrained birds. The mean length of the postsynaptic density in the right hemisphere exceeded that in the left by 10.6% ($P < 0.025$). (B) Overtrained birds. Although the mean value for the right hemisphere was smaller by 15 ± 16.4 nm than the left, a relation opposite in direction to that found in the undertrained birds, the difference was not statistically significant. (After Bradley et al., 1981).

different stimuli, and chicks of different ages the only morphological measure to be significantly altered by training was the length of the postsynaptic density of synapses on dendritic spines in the left IMHV. This measure was also significantly correlated with a measure of the chick's behaviour, namely their approach activity during the first hour of training. Previous studies have shown that this approach activity is correlated with the strength of imprinting (Bateson and Jaeckel, 1974).

The evidence of a hemispheric asymmetry in the imprinting process has been followed up by lesion studies (Cipolla-Neto et al., 1982; Horn et al., 1983). The results of these studies have shown that the right and left IMHV do indeed play different roles in this process. Briefly, the left IMHV appears to be critical for information storage, but the right IMHV may serve as a 'buffer store' passing information on to some other brain region, or regions, collectively referred to as S' (Cipolla-Neto et al., 1982; Horn, 1985).

Using an entirely different learning task Rose and his collaborators (Stewart et al., 1984; and Rose, this volume) have found changes in synaptic morphology in the medial part of Hyperstriatum ventrale, a region which overlaps with IMHV, in the chick telencephalon. It would seem, therefore, that the medial part of the hyperstriatum ventrale can act as a modifiable, integrative structure in the brain of the chick and perhaps of other birds as well (Maier and Scheich, 1983). This is not to say that it is the only such region in the avian brain, nor that it is the only region involved in learning. Indeed, as stated above there is evidence that in the process of imprinting other brain regions are involved in the storage of information (Cipolla-Neto et al., 1982; Horn, 1985).

It is a widely held belief that long-term memory is dependent on the structural modification of particular sets of synapses, and the evidence presented here supports this view. Such a modification could be achieved for example by increasing receptor number (see Horn, 1962) or receptor affinity in the postsynaptic density. These changes may be associated with changes in the levels of protein phosphorylation. The present study provides evidence that the overall size of the postsynaptic density is increased, but it is not yet known in what ways if any, this change is associated with a change in the number of receptors in this membrane. Nor is it known whether the synapses affected by training are excitatory or inhibitory. However indirect evidence suggests that spine synapses are excitatory (Gray, 1959; Eccles, 1964).

Because of the methods used for quantifying the synaptic changes, it is not possible to state whether they represent small changes in synaptic length distributed amongst a large population of dendritic spines in the left IMHV, or whether they represent large changes in a small sub-set of these spine synapses. It is, however, clear that a short period of training leads to changes in a specific population of synapses within a region of the chick brain which plays a critical part in the storage of information acquired through a learning process.

REFERENCES

Bateson, P. P. G. and Jaeckel, J. B., 1974, Imprinting: correlations between activities of chicks during training and testing, Anim. Behav., 22:899.

Bateson, P. P. G., Horn, G. and Rose, S. P. R., 1972, Effects of early experience on regional incorporation of precursors into RNA and protein in the chick brain, Brain Res., 39:449.

Bradley, P., 1982, Development of two regions of the chick telencephalon, Neurosci. Lett., 10:588.

Bradley, P., 1983, Development of dendritic trees in a region of the chick brain involved in a learning process, Fidia Research Series, Frontiers in Neuroscience, Abstract book no. 1:149.

Bradley, P., Davies, D. C. and Horn, G., 1985, Connections of the hyperstriatum ventrale of the domestic chick (Gallus domesticus), J. Anat., in press.

Bradley, P. and Horn, G., 1982, A Golgi analysis of the hyperstriatum ventrale in the chick, J. Anat., 134:599.

Bradley, P., Horn, G. and Bateson, P., 1981, Imprinting: an electron microscopic study of chick hyperstriatum ventrale, Exp. Brain Res., 41:115.

Cajal, S. R., 1911, in: "Histologie due systême nerveux de l´homme et des vertébrés", Vol. 2, pp. 886-890, Maloine, Paris. (Republished 1955, Instituto Ramon y Cajal, Madrid.

Cipolla-Neto, J., Horn, G. and McCabe, B.J., 1982, Hemispheric asymmetry and imprinting: the effect of sequential lesions to the hyperstriatal ventrale. Exp. Brain Res., 48:22.

Delius, J. D. and Bennetto, K., 1972, Cutaneous sensory projections to the avian forebrain, Brain Res., 37:205.

Eccles, J. C., 1964, "The physiology of synapses", Springer, Berlin.

Gray, E. G., 1959, Axo-somatic and aco-dendritic synapses of the cerebral cortex: an electron microscope study, J. Anat. (Lond.), 93:420.

Hebb, D. O., 1949, "The organization of behavior", John Wiley, New York.

Horn, G., 1962, Some neural correlations of perception, in: "Viewpoints in biology", J. D. Carthy and C. L. Duddington, eds., 1:242.

Horn, G., 1981, Neural mechanisms of learning: an analysis of imprinting in the domestic chick, Proc. R. Soc. Lond. B, 213:101.

Horn, G., 1985, "Memory, imprinting and the brain", Clarendon Press, Oxford.

Horn, G., McCabe, B. J. and Bateson, P. P. G., 1979, An autoradiographic study of the chick brain after imprinting, Brain Res., 168:361.

Horn, G., McCabe, B. J. and Cipolla-Neto, J., 1983, Imprinting in the domestic chick: the role of each side of the hyperstriatum ventrale in acquisition and retention, Exp. Brain Res., 53:91.

Horn, G., Bradley, P. and McCabe, B. J. (1985), Changes in the structure of synapses associated with learning, in preparation.

Karten, H. J., 1968, The ascending auditory pathway in the pigeon (Columba livia). II. Telencephalic projections of the nucleus ovoidalis thalami, Brain Res., 11:134.

Maier, V. and Scheich, H. (1983). Acoustic imprinting leads to differential 2-deoxy-D-glucose uptake in the chick forebrain, Proc. Natl. Acad. Sci. USA, 80:3860.

McCabe, B. J., Cipolla-Neto, J., Horn, G. and Bateson, P. P. G., 1982, Amnesic effects of bilateral lesions placed in the hyperstriatum ventrale of the chick after imprinting, Exp. Brain Res., 48:13.

Parker, D. M. and Delius, J. D., 1972, Visual evoked potentials in the forebrain region of the pigeon, Exp. Brain Res., 14:198.

Payne, J. K. and Horn, G., 1984, Long-term consequences of exposure to an imprinting stimulus on spontaneous impulse activity in the chick brain, Behav. Brain Res., 13:155.

Reiner, A., Brauth, S. E and Karten, H. J., 1984, Evolution of the basal ganglia, Trends in Neuroscience, 7:320.

Stewart, M. G., Rose, S. P. R., King, T. S., Gabbott, P. L. A. and Bourne, R., 1984, Hemispheric asymmetry of synapses in chick medial hyperstriatum ventrale following passive avoidance training: a stereological investigation, Develop. Brain Res., 12:261.

Tanzi, E., 1893, I fatti e le induzioni nell' odierna istologia del sistema nervoso, Riv. sper. Freniat. Med. leg. Alien. ment., 19:419.

Zeier, H. and Karten, H. J., 1971, The archistriatum of the pigeon: organization of afferent and efferent connections, Brain Res., 31:313.

NORADRENALINE AND ITS POSSIBLE ROLE IN IMPRINTING

D.C. Davies, G. Horn[+], and B.J. McCabe[+]

Department of Anatomy and Cell Biology,
St. Mary's Hospital Medical School
Norfolk Place, Paddington
London W2 1PG, U.K.

[+]Department of Zoology
University of Cambridge
Downing Street,
Cambridge CB2 3EJ, U.K.

INTRODUCTION

It has often been postulated that changes in the connections of neurones underlie memory (see De Feudis and De Feudis, 1977) and that the occurrence of these changes is dependent on the presence of some substance or substances which mediate reinforcement (see Young, 1963; Griffith, 1966). It has been suggested that neurones containing catecholamines, noradrenaline in particular, have such a reinforcing function (Kety, 1967, 1970; Crow, 1968; Crow and Arbuthnott, 1972). In order to test the hypothesis that noradrenaline is implicated in learning, Anlezark et al., (1973) placed lesions in the locus coeruleus, which provides the noradrenergic innervation of the rat cerebral cortex (for review see Amaral and Sinnamon, 1977). These lesions resulted in a reduction in cortical noradrenaline levels to one third of control values and impaired the ability of the lesioned rats to learn a simple approach task, which was reinforced by the reward of food. However, in spite of the promise of this early work, a substantial number of subsequent studies failed to demonstrate an impairment of learning ability after depletion of cerebral noradrenaline levels by either electrocoagulation or chemical lesions of the locus coeruleus, or the dorsal noradrenergic bundle (for review see Mason, 1984).

The diverse effects on behaviour of depleting the brain of noradrenaline are difficult to interpret. An effect on learning may be indirect, resulting from changes in arousal, motivation, locomotor ability and so on. In addition, lesions may exert their own actions on the nervous system which are independent of their effects on noradrenergic neurones. Differences in the training task, the method of reducing noradrenaline levels, the extent to which these levels are reduced, the age at which treatment is given and the interval between treatment and testing may also contribute to the apparent conflict of results.

More convincing evidence has been provided for a role of noradrenaline in the modification of neuronal function in the kitten visual cortex (see Kasamatsu, 1983). Although it is clear that noradrenaline is not unique in this respect (Shaw and Cynader, 1984) the evidence is strong that it is involved in this form of neuronal plasticity. This evidence taken together with the suggestion by Kety (1970) and Crow and Arbuthnott (1972) that noradrenaline may facilitate learning through a primary effect on synapses, prompted the investigation of a possible role for noradrenaline in imprinting in the domestic chick. There are a number of reasons why imprinting is a good model in which to investigate this possibility. Firstly, imprinting involves learning processes (Sluckin, 1972; Horn, 1985): as a result of being exposed to a visually conspicuous object young chicks selectively approach it and avoid other conspicuous objects. Secondly, following imprinting structural (Bradley et al., 1981; Horn et al., in preparation; Bradley et al., this volume), electrophysiological (Payne and Horn, 1982; 1984) and biochemical (Horn et al., 1979; Kohsaka et al., 1979) changes occur in a restricted region of the chick telencephalon. This region, the intermediate part of the medial hyperstriatum ventrale is referred to as IMHV. Finally, catecholamine-containing varicose fibres are present in IMHV in the day-old chick (unpublished observations).

CATECHOLAMINE CONCENTRATIONS IN THE FOREBRAIN: CHANGES WITH AGE AND IMPRINTING

The first step in the analysis of the role of catecholamines in imprinting was to measure the concentrations of dopamine and noradrenaline in discrete regions of the forebrain of young chicks at and shortly after hatching. The three forebrain samples investigated were (1) a sample from the Wulst, (2) a medial forebrain sample comprising mainly IMHV and (3) a basal forebrain sample comprising mainly paleostriatum augmentatum. These regions were chosen because (i) the Wulst is a visual projection area (see Webster, 1974), (ii) the paleostriatum augmentatum is thought to be involved in the control of motor activity (Karten and Dubbledam, 1973; Rieke, 1980) and (iii) IMHV is involved in the acquisition and retention of preferences through imprinting (Horn, 1985).

Chicks were hatched in darkness. Immediately after hatching, chicks of either sex were randomly assigned to 'dark-reared' or 'imprinted' groups. Forebrain samples were taken from both hemispheres of dark-reared chicks (n = 48) at 0, 12, 22, 26, 29 and 50 h post-hatch. Eight chicks were killed by decapitation at each of these ages. The imprinted chicks (n = 32) were reared in darkness until 21 h post-hatch when they were exposed individually for 0.5 h to diffuse overhead illumination from a 25 W bulb. This short period of 'priming' enhances the efficacy of subsequent training (Bateson and Wainwright, 1972; Cherfas, 1978). The chicks were then replaced in the lightproof holding incubator. Forebrain samples were taken bilaterally from 8 chicks, 0.5 h after the end of priming. Training of the remaining 24 chicks commenced at 22 h post-hatch. Each chick was placed in a running wheel 50 cm away from and facing the training stimulus, a rectangular box illuminated red on its two larger (18 x 18 cm) surfaces with its remaining surfaces painted black. The stimulus rotated for 50s in each minute and was turned off for the remaining 10s. A stimulus and 3 running wheels were enclosed in a light-proof cabinet. More complete details of the training apparatus are given elsewhere (McCabe et al., 1982). The chicks were trained for 2 x 1 h periods, separated by 1 h in the dark. They were then replaced in the holding incubator and forebrain samples were taken from 8 chicks at 1 h (26 h post-hatch), 3 h (29 h post-hatch) and 24 h (50 h post-hatch) after training. The ages at forebrain sample removal and training schedule are summarised in Fig. 1. The chicks were not given a preference test (McCabe et al., 1981) after training in case the testing procedure affected forebrain catecholamine levels. However, the training regime used is known to produce strong preferences for the training stimulus (McCabe et al., 1982).

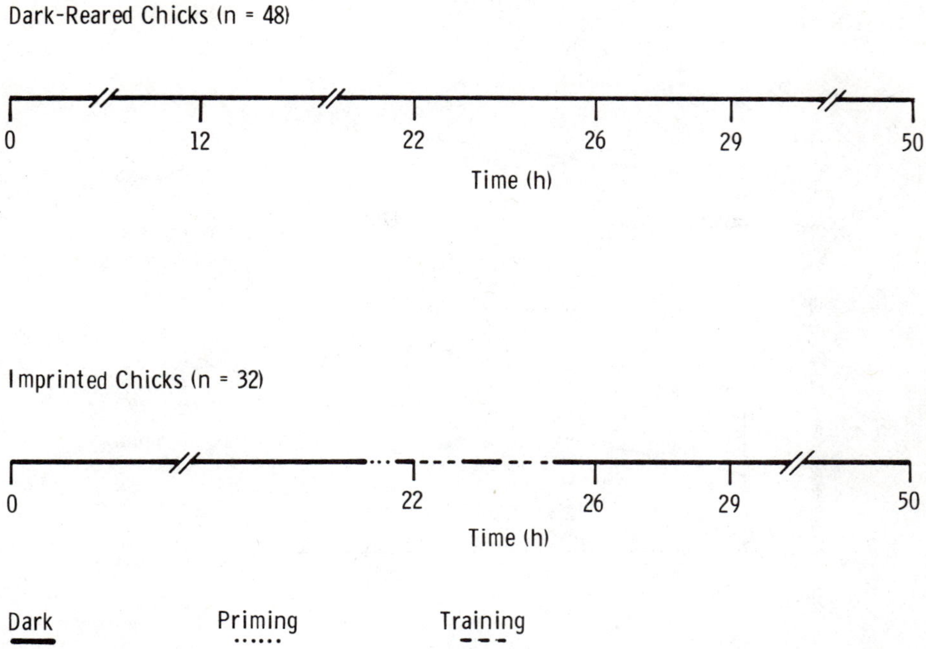

Figure 1. The ages at which forebrain samples were removed in dark-reared and imprinted chicks.

The dopamine and noradrenaline concentrations of the samples were determined using a modification of the catchol-O-methyl transferase radiochemical assays of Coyle and Henry (1973) and Cuello et al., (1973). Assays for adrenaline were not included in the above design since in preliminary measurements on day-old, light-reared, socially experienced chicks, this amine was undetectable in the Wulst and medial forebrain samples and its concentration was so low in basal forebrain samples that accurate and reproducible quantification was impossible. With the exception of newly-hatched (0 h) chicks, which were still wet, forebrain samples were removed without knowledge of the chicks' age or experience. The assays for dopamine and noradrenaline were also performed 'blind'.

Analysis of variance revealed no hemispheric asymmetry or sex differences in the concentrations of dopamine and noradrenaline in any of the three forebrain regions investigated. Data from left and right hemispheres and from both sexes were therefore combined.

Dopamine concentrations differed significantly ($P < 0.001$) between the three forebrain regions investigated. The mean dopamine concentrations at hatching are shown in Table 1. These levels did not change with age in dark-reared chicks and were not affected by training.

TABLE 1

Mean ($\pm$ SEM) concentrations of dopamine in samples of the chick forebrain at hatching.

Wulst	9.13 $\pm$ 1.13 ng/g wet wt.
Medial forebrain	16.66 $\pm$ 2.56 ng/g wet wt.
Basal forebrain	121.19 $\pm$ 33.08 ng/g wet wt.

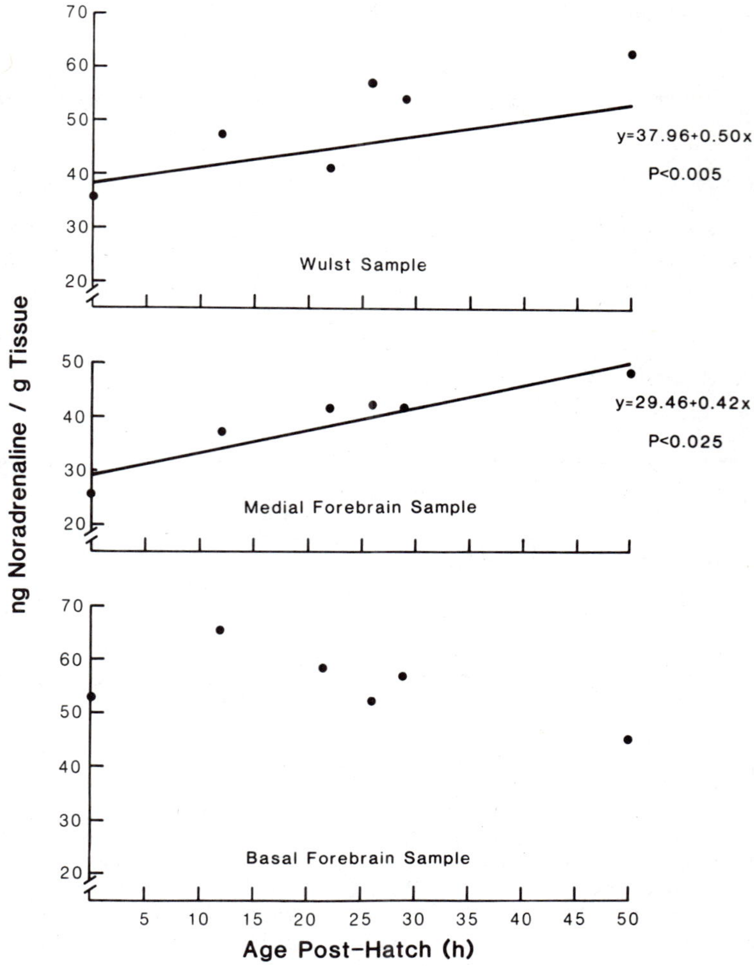

Figure 2. Post-hatch concentrations of noradrenaline in the forebrain of dark-reared chicks. (After Davies et al., 1983).

The noradrenaline concentrations in the three regions investigated differed significantly from each other. In dark-reared chicks there was a significant ($P < 0.001$) increase in noradrenaline concentration with age both in the Wulst and medial forebrain samples (Fig. 2) but not in basal forebrain samples. An analysis of variance also disclosed a significantly higher level of noradrenaline in imprinted compared with dark-reared chicks ($P < 0.05$). The effect was of similar magnitude in all three regions of the telencephalon and the combined data are presented in Fig. 3.

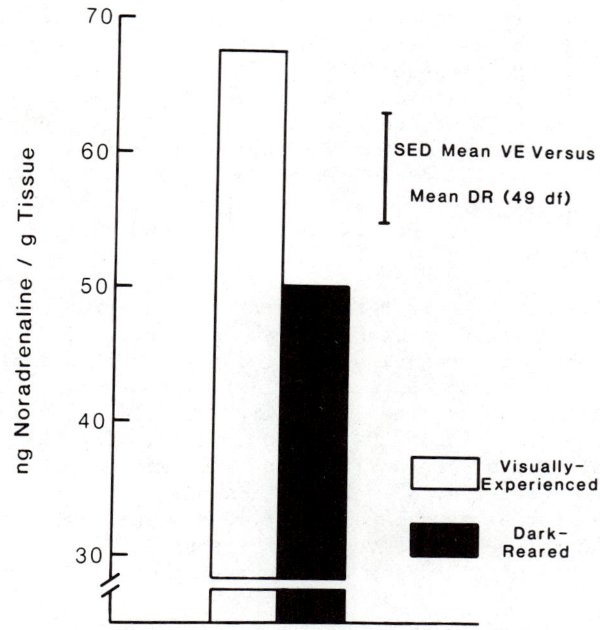

Figure 3. Noradrenaline concentrations in trained and dark-reared chicks. The diagram shows combined data from the Wulst, medial and basal forebrain samples. (After Davies et al., 1983).

Thus in contrast to the data for dopamine, there was an increase in noradrenaline concentration in the Wulst and medial forebrain samples over the first 50 h post-hatch. There was also a further increase in all regions of the forebrain investigated as a consequence of training. Since the blood-brain barrier is poorly developed in young chicks (Waelsch, 1955), the possibility existed that changes in the concentrations of peripheral noradrenaline, resulting from the stress of handling and the training procedures, could have accounted for the training-related increase in forebrain noradrenaline. This possibility is unlikely since the mean plasma noradrenaline level in 50 h old dark-reared chicks (37.76 $\pm$ 4.05 SEM, ng/ml, n = 10) was not siginficantly different from that of imprinted chicks of the same age (36.33 $\pm$ 6.58 SEM, ng/ml, n = 8).

THE EFFECTS OF DSP4 ON IMPRINTING

The results described above are consistent with, but do not provide clear evidence for, a role of noradrenaline in imprinting. We therefore sought to deplete the forebrain of this amine to determine whether this procedure would impair imprinting (Davies et al., 1985). 6-OHDA has been widely used to deplete the brain of noradrenaline (see Mason, 1984). However, gross behavioural side-effects follow the administration of this drug in kittens (Kasamatsu and Pettigrew, 1979), and also in chicks (unpublished observations). Another neurotoxin N-(2-chlorethyl)-N-ethyl-2-bromobenzylamine hydrochloride (DSP4) which did not have such side-effects in chicks was therefore used. Jonsson et al. (1982) have also shown that in comparison to 6-OHDA, DSP4 is a more selective neurotoxin for noradrenergic as opposed to dopaminergic neurons. Thus the effect of DSP4 on imprinting was examined in the chick.

Fertile eggs were incubated, hatched and selected as in the previous experiment (Davies et al., 1983). Chicks of each sex were assigned randomly to either ´DSP4-treated´ (n = 36) or ´control´ (n = 36) groups. At hatching, DSP4-treated chicks were injected i.p. with 50 ng/kg DSP4 in 0.1 ml distilled water; controls received an injection of 0.1 ml of distilled water. The chicks were then placed in individual compartments in a dark holding incubator. At 30 h post-hatch the chicks were given a second dose of either DSP4 or distilled water. The chicks were then returned to the holding incubator. All subsequent experimental procedures were performed ´blind´.

Approximately 60 h after hatching the chicks were ´primed´ and placed in the training apparatus as described above. However, in this experiment half of the chicks were trained on the red box and the remainder were trained on an illuminated, stuffed jungle fowl (McCabe et al., 1981). As in the case of the red box, the jungle fowl rotated at 28 rpm and was turned on for 50s and off for 10s in each minute. The chicks were trained for 2 periods of 40 min with an intervening 50 min period in the dark holding incubator (see Fig. 4). During training, the pre-recorded maternal call of a domestic hen was relayed via a loudspeaker into the training cabinet, to enhance the effectiveness of training (Horn et al., 1979). The chicks were then returned to the holding incubator.

Two hours after training the preference of each chick for its training stimulus was determined. Each chick was exposed in balanced order, to the training and the novel stimulus (i.e. either the box or the fowl, whichever it had not previously seen) and the revolutions of the running wheel towards each stimulus was recorded. A preference score (McCabe et al., 1981), which provides an index of the attractiveness of the training stimulus compared to the novel stimulus was then calculated as follows:

$$\text{Preference Score} = \frac{\text{Approach counts to training stimulus} \times 100}{\text{Approach counts to training stimulus} + \text{Approach counts to novel stimulus}}$$

Chicks which approached only the training stimulus during the test scored 100, chicks which approached only the novel stimulus scored 0 and chicks which approached both stimuli equally (i.e. showed no preference) scored 50.

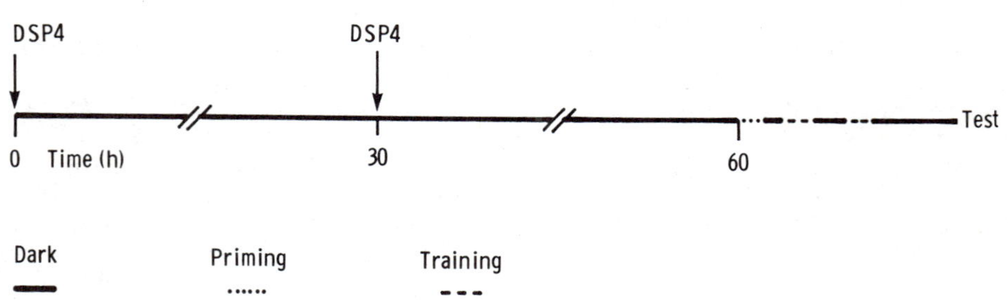

Figure 4. Training schedule for DSP4-treated chicks.

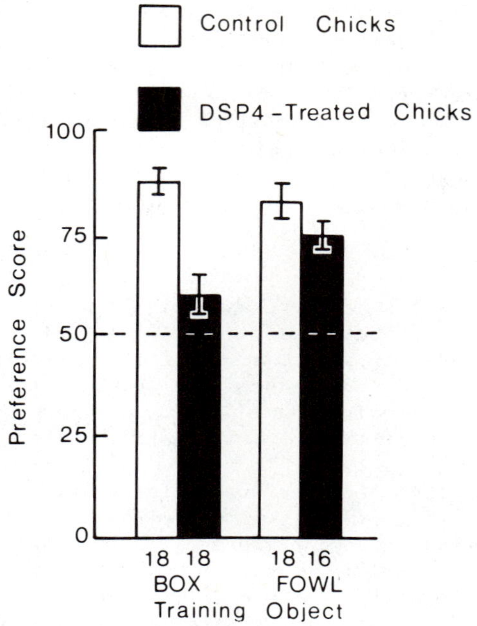

Figure 5. The effect of DSP4 treatment on preference score. (After Davies et al., 1985).

The mean preference score (Fig. 5) of control chicks trained on the red box was not significantly different from that of controls trained on the jungle fowl. Both mean scores were significantly greater than 50 ($P < 0.001$). The mean preference score of DSP4-treated chicks trained on the jungle fowl was also significantly greater than 50 ($P < 0.001$), though it was reduced slightly, by 8.52 ± 4.73 ($P < 0.04$ 1-tailed) below that of the fowl-trained controls. In contrast, the mean preference score of DSP4-treated chicks trained on the red box was markedly reduced. Their score was (i) not significantly different from a chance score of 50, (ii) significantly less than that of their box-trained controls ($P < 0.001$) and (iii) significantly less than that of the DSP4-treated chicks trained on the jungle fowl ($P < 0.05$). Thus DSP4-treated chicks trained on the red box did not acquire a preference for their training stimulus. In a parallel experiment (Davies et al., 1985) using box-trained chicks there was a significant correlation between preference score and the concentration of noradrenaline in the medial forebrain sample ($r = 0.61$; d.f. = 9; $P < 0.05$) and the basal forebrain sample ($r = 0.82$, d.f. = 9; $P < 0.01$). The IMHV of the medial forebrain sample is reciprocally connected to the paleostriatum augmentatum of the basal forebrain sample (Bradley et al., 1985). There was no correlation between preference score and noradrenaline concentration in the Wulst of the box-trained birds and no correlation between preference score and noradrenaline concentration in any forebrain sample in fowl-trained birds.

When a drug impairs acquisition it is possible that the effect is a consequence of a general debilitating action of the drug. It is reasonable to discount such a non-specific effect in this instance since (a) approach activity in the preference test was not affected by DSP4 treatment, (b) DSP4 had no effect on the latency of the chicks to peck accurately at a rocking bead, (c) DSP4 impaired the preferences of chicks trained on the box, but had only a slight effect on chicks trained on the jungle fowl.

The drug treatment regime described above has previously been shown to result in a significant ($P < 0.001$ for each region) reduction in noradrenaline concentration in the three forebrain regions examined (Davies et al., 1985). The percentage reduction, approximately 65%, was similar in the Wulst, medial and basal forebrain samples. There was no effect of training or of stimulus on noradrenaline concentration. Thus DSP4 reduced noradrenaline concentrations to a similar extent in all three regions of the forebrain investigated in chicks trained on the red box and in chicks trained on the jungle fowl.

DISCUSSION

The evidence that has been presented in this paper is suggestive of a role for noradrenaline in the imprinting process: (i) noradrenaline concentration, but not the concentration of dopamine, increased with age and with imprinting, (ii) DSP4 reduced the level of noradrenaline in three forebrain samples to approximately 35% of that found in control chicks, (iii) DSP4 profoundly impaired the acquisition of a preference for an artificial imprinting object, and had a small but statistically significant effect on the acquisition of a preference for a naturalistic object, the stuffed jungle fowl, (iv) for chicks trained on the box there was a significant correlation between preference for box and the concentration of noradrenaline in the medial forebrain sample. This sample, as described above, consists mainly of IMHV, the region which is critical for imprinting on the box, and plays a small part in the acquisition of a preference for the jungle fowl (see Horn, this volume).

If, as the evidence suggests, noradrenaline is implicated in the learning process of imprinting, its mechanism of action is not known. However, changes in the structure of synapses are known to be associated with imprinting (see Bradley et al., this volume) and there is evidence that neuronal development (Felton et al., 1982) and the formation of synapses (Parnavelas and Blue, 1982) in the cerebral cortex of young rats are influenced by depleting the forebrain of noradrenaline.

REFERENCES

Amaral, D. G. and Sinnamon, H. M., 1977, The locus coeruleus: neurobiology of a central noradrenergic nucleus, Progress in Neurobiology, 9:147.

Anlezark, G. M., Crow, T. J. and Greenway, A. P., 1973, Impaired learning and decreased cortical norepinephrine after bilateral locus coeruleus lesions, Science, 181:682.

Bateson, P. P. G. and Wainwright, A. A. P., 1972, The effects of prior exposure to light on the imprinting process in domestic chicks, Behaviour, 42:279.

Bradley, P., Davies, D. C. and Horn, G., 1985, Connections of the hyperstriatum ventrale of the domestic chick (Gallus domesticus). J. Anat., in press.

Bradley, P., Horn, G. and Bateson, P., 1981, Imprinting: an electron microscopic study of chick hyperstriatum ventrale. Exp. Brain Res., 41:115.

Cherfas, J. J., 1978, Simultaneous colour discrimination in chicks is improved by brief exposure to light, Anim. Behav., 26:1-5.

Coyle, J. T. and Henry, D., 1973, Catecholamines in fetal and newborn rat, J. Neurochem., 21:61.

Crow, T. J., 1968, Cortical synapses and reinforcement: an hypothesis, Nature, 219:736.

Crow, T. J. and Arbuthnott, G. W., 1972, Function of catecholamine-containing neurones in mammalian central nervous system, Nature New Biol., 238:245.

Cuello, A. C., Hiley, R. and Iversen, L. L., 1973, Use of catechol-O-methyl transferase for the enzyme radiochemical assay of dopamine, J. Neurochemistry, 21:1337.

Davies, D. C., Horn, G. and McCabe, B. J., 1983, Changes in telencephalic catecholamine levels in the domestic chick. Effects of age and visual experience, Develop. Brain Res., 10:251.

Davies, D.C., Horn, G. and McCabe, B. J., 1985, Noradrenaline and learning: the effects of the noradrenergic neurotoxin DSP4 on imprinting in the domestic chick, Behav. Neurosci., in press.

DeFeudis, F. V. and DeFeudis, P. A. F., 1977, "Elements of the behavioural code", Academic Press, London.

Felton, D. L., Hallman, H. and Jonsson, G., 1982, Evidence for a neurotrophic role of noradrenaline in the postnatal development of rat cerebral cortex, J. Neurocytology, 11:119.

Griffith, J. S., 1966, A theory of the nature of memory, Nature, 211:1160.

Horn, G., 1985, "Memory, imprinting and the brain", Clarendon Press, Oxford.

Horn, G., McCabe, B. J. and Bateson, P. P. G., 1979, An autoradiographic study of the chick brain after imprinting, Brain Res., 168:361.

Horn, G., Bradley, P. and McCabe, B. J., Changes in the structure of synapses associated with learning, in preparation.

Jonsson, G., Hallman, H. and Sundström, E., 1982, Effects of the noradrenaline neurotoxin DSP4 on the postnatal development of central noradrenaline neurons in the rat, Neuroscience, 7:2895.

Karten, H. J. and Dubbeldam, J. L., 1973, The organisation and projections of the paleostriatal complex in the pigeon (Columba livia), J. comp. Neurol., 148:61.

Kasamatsu, T., 1983, Neuronal plasticity maintained by the central norepinephrine system in the cat visual cortex, Prog. Psychobiol. physiol. Psychol., 10:1.

Kasamatsu, T. and Pettigrew, J. D., 1979, Preservation of binocularity after monocular deprivation in the striate cortex of kittens treated with 6-hydroxydopamine, J. comp. Neurol., 185:139.

Kety, S. S., 1967, The central physiological and pharmacological effects of the biogenic amines and their correlations with behaviour, in: "The neurosciences: a study program", G. C. Quarton, T. Melnechuk and F. O. Smith, eds., pp. 444-451, Rockefeller University Press, New York.

Kety, S. S., 1970, The biogenic amines in the central nervous system: their possible roles in arousal, emotion and learning, in: "The neurosciences: second study program", F. O. Schmitt, ed., pp. 324-336, Rockefeller University Press, New York.

Kohsaka, S., Takamatsu, K., Aoki, E. and Tsukada, Y., 1979, Metabolic mapping of chick brain after imprinting using [^{14}C]2-deoxyglucose technique, Brain Res., 172:539.

Mason, S. T., 1984, "Catecholamines and behaviour", Cambridge University Press, Cambridge.

McCabe, B. J., Cipolla-Neto, J., Horn, G. and Bateson, P. P. G., 1982, Amnesic effects of bilateral lesions placed in the hyperstriatum ventrale of the chick after imprinting, Exp. Brain Res., 48:13.

McCabe, B. J., Horn, G. and Bateson, P. P. G., 1981, Effects of restricted lesions of the chick forebrain on the acquisition of filial preferences during imprinting, Brain Res., 205:29.

Parnavelas, J. G. and Blue, M. E., 1982, The role of the noradrenergic system on the formations of synapses in the visual cortex of the rat, Developmental Brain Res., 3:140.

Payne, J. K. and Horn, G., 1982, Differential effects of exposure to an imprinting stimulus on `spontaneous´ neuronal activity in two regions of the chick brain, Brain Res., 232:191.

Payne, J. K. and Horn, G., 1984, Long-term consequences of exposure to an imprinting stimulus on `spontaneous´ neuronal activity in two regions of the chick brain, Brain Res., 232:191.

Rieke, G. K., 1980, Kainic acid lesions of pigeon paleostriatum: a model for study of movement disorders, Physiol. Behav., 24:683.

Shaw, C. and Cynader, M., 1984, Disruption of cortical activity prevents ocular dominance changes in monocularly deprived kittens, Nature, 308:731.

Sluckin, W., 1972, "Imprinting and early learning", Methuen, London.

Waelsch, H., 1955, Blood brain barrier and gas exchange, in: "Biochemistry of the developing nervous system", H. Waelsch, ed., pp. 187-201, Academic Press, New York.

Webster, K. E., 1974, Changing concepts of the organization of the central visual pathways in birds, in: "Essays on the nervous system", R. Bellairs and E. G. Gray, eds., pp. 258-298, A Festschrift for Professor J. Z. Young, Clarendon Press, Oxford.

Young, J. Z., 1963, Some essentials of neural memory systems. Paired centres that regulate and address the signals of the results of action, Nature, 198:626.

THE CELL BIOLOGICAL CONSEQUENCES OF PASSIVE AVOIDANCE TRAINING IN THE CHICK

Steven P.R. Rose

Brain Research Group
Open University
Milton Keynes MK7 6AA
U.K.

INTRODUCTION

In the late 1960s, Pat Bateson, Gabriel Horn and I began to work together on the cell biology of learning and memory, using imprinting in the young chick as a model. Other papers in this volume describe the elegant way in which the imprinting work has been continued by the Cambridge group. Here, I will review the results of the use of an alternative form of early learning in the young chick, based on a one trial passive avoidance task. The point is that, for the biochemist there are certain disadvantages to studying the sequelae of imprinting as a model for memory. The process of training by exposure to the stimulus is a relatively long one, and this makes it harder to distinguish the cellular events associated with putative memory storage processes from those which are the concomitants of the experience of training. Birds also show a degree of variability in their behaviour during training and the strength of their later preference which can be exploited experimentally but which also adds to the complexity of the analysis.

By contrast, the one trial passive avoidance learning (PAL) task developed by Cherkin and Lee-Teng (Cherkin 1969) and by Gibbs (Gibbs and Ng, 1977), exploits the tendency of young chicks to explore their environment by pecking at small bright objects in their field of view. Such pecking is clearly part of the ontogenetic learning about sources of food and drink, and unpleasant tasting objects once explored by pecking are thereafter avoided. In the experimental protocol, pairs of day old birds are placed in small pens, and after a period of equilibration and three 'pretraining' trials with a small (2.5 mm dia) white bead are offered a bright chromed bead (4 mm dia, attached to a thin rod as handle). More than 80% of birds will peck at least once at this bead within a 10 second presentation period. If the bead has been coated with a bitter tasting substance - we use methylanthranilate (MeA) - they show a characteristic 'disgust' response, backing away, shaking their heads, swallowing vigorously and wiping their beaks on the floor of the pen. More than 70% of the birds which have pecked at the MeA coated bead will actively avoid a similar, but dry bead when presented in a further 10 sec trial at times (in our experiments) from 15 mins to 48 hr subsequently. By contrast, of birds which in the training trial have pecked at a water coated bead, upwards of 90% will subsequently peck at a dry bead. We are thus able to directly compare cellular events occuring in two groups of birds, which we call MeA-trained (M) and water-control (W), for a number of biochemical and morphological measures, and at various times after training.

This training protocol, described in full most recently by Lossner and Rose (1983), enables us to compare directly two groups of birds which are distinguished by specific different experiences and show a clear distinction in subsequent behaviour; that is M-birds have learned to avoid the offered bead, W-birds continue to show their ontogenetically 'natural' tendency to peck at it.

In what follows, I review our evidence on the sites of change following training, the biochemical nature of the changes, and the relationship of these changes to memory formation.

The sites of change

Our initial studies followed a similar strategy to those in imprinting; we looked at the distribution of a number of biochemical markers in grossly dissected 'roof' and 'base' regions of the brain following training; as in imprinting, changes were found in 'roof', but some effects were also apparent in the 'base'. To investigate the locus of these changes more precisely, we have used the 2-deoxyglucose (2DG) autoradiographic technique described by Sokoloff et al (1977). 2DG is a general metabolic marker for glucose utilization. Enhanced glucose utilization is likely to be a consequence of increased neuronal activity whether because of increased firing rate or metabolic mobilization for biochemical and structural reshaping.

Chicks were pretrained, and just prior to training injected pericardially with 10μCi ^{14}C 2DG. They were trained on either W or M beads, tested 30 min later, the brains immediately frozen and coronal and parasaggital autoradiographic sections prepared (Kossut & Rose 1984). Optical density measurements on 13 identified anatomical structures in brains from 7M and 7W birds were made using a computerised digital image analysis system. (In this, as in all our experiments, the morphological, autoradiographic or biochemical measurements are made blind with respect to the treatment of the birds.) Optical densities in each autoradiogram section were normalized by comparison with reference structures chosen in each section, and the modal values for serial autoradiograms through each structure in each bird were compared (Fig 1). The overall pattern of labelling was (not unexpectedly) similar in W and M birds but quantitative analysis showed that 3 of the 13 regions examined, the palaeostriatum augmentatum (PA), hyperstriatum ventrale, posterior region (HVp) and lobus parolfactorius (LPO) showed increased metabolic activity, of 13% ($p<0.001$), 10% ($p<0.05$) and 11% ($p<0.05$) respectively, in the M-birds.

The hyperstriatum ventrale is located in the forebrain roof. It includes the region (IMHV) found by Horn et al (1979) to show increased uptake of uracil into RNA during exposure to the flashing light which forms the imprinting stimulus. Lesions in the HV are known to impair the performance of PAL and related tasks (Benowitz & Lee-Teng, 1973; Benowitz 1974) and lesions in the IMHV affect either capacity to imprint or recall for the imprinting stimulus, depending on where they are placed (Horn; this volume). The PA & LPO are located in the forebrain base. Both receive projections from the HV (Bradley & Horn 1979; Benowitz 1980) though the functional relationships of the LPO seem obscure.

This experiment locates regions of enhanced metabolic activity during training and its immediate sequelae in the M- birds. It should be immediately apparent, though, that the changes could be consequent on aspects of the training experience - for instance the taste of the MeA -

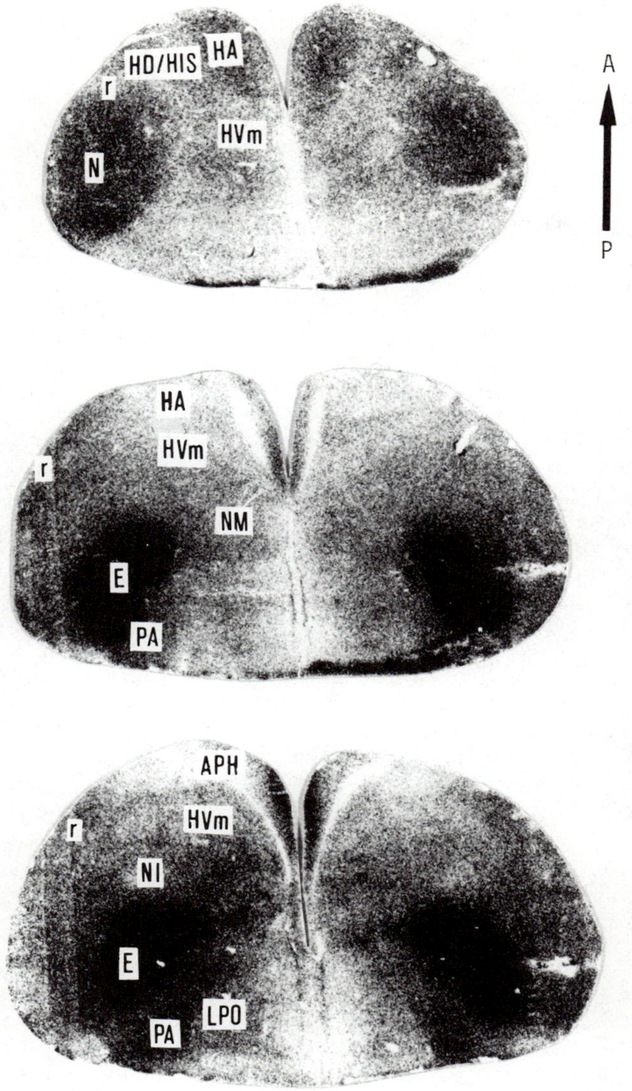

Figure 1. Series of coronal sections from the brain of a single W bird. 2-DG autoradiograms were prepared, structures identified and o.d. measurements made as described in the text. Abbreviations for the several brain regions are as follows:

APH	Area parahippocampalis
E	Ectostriatum
HA	Hyperstriatum accessorium
HD/HIS	Hyperstriatum dorsale/ intercalatus
HVm	Hyperstriatum ventrale (mediale)
HVp	Hyperstriatum ventrale (posteriore)
LPO	Lobus Parolfactorius
N	Neostriatum laterale
NI	Neostriatum intermediale
NM	Neostriatum mediale
PA	Palaeostriatum augmentatum
r	Reference zone for o.d. measurement

as well as mobilization of these neuronal regions in the interests of memory storage. Also, the 2DG serves as a marker for short term changes in metabolic activity. Can these be precursors to longer term phenomena? In particular, most theories of memory storage processes are based on the assumption that they involve forms of synaptic remodelling. Based on the 2DG results, we have therefore explored the HV, PA and LPO for evidence of longer term changes in synaptic morphology as a consequence of training. In addition, in these experiments we were also able to take account of increasing evidence for the lateralization of memory trace formation in the chick (Rogers & Anson, 1979; Horn, 1981; Horn et al 1983).

Chicks were trained on water or MeA as in the 2DG experiments but were held for 24 h after training, to allow time for any possible morphological restructuring to take place. At the end of this period, the birds were perfused, brain regions located stereotaxically, and blocks of the HV, LPO and PA from left and right hemispheres taken separately for electron microscopy. Sections from 8M and 8W birds were randomly sampled, and photographed or analysed directly through a projection system. A variety of synaptic measurements were made using a Kontron Videoplan system, enabling us to measure stereologically the following synaptic parameters: length of postsynaptic thickening; number of synapses per unit volume of neuropil, volume density of the presynaptic bouton; number of synaptic vesicles per unit volume of neuropil and presynaptic bouton. The details are described in Stewart et al (1984b).

The salient morphological observations were as follows. First, in both groups of birds, W and M, there were hemispheric asymmetries in a number of synaptic measurements. For instance, in the MHV of the W-birds, the length of the postsynaptic thickening was some 12% greater than in the same region in the left hemisphere ($p<0.02$). Such differences were not wholly unexpected, and are comparable to those reported by Bradley and Horn (1981). (See also Rogers 1982.)

However, this asymmetry in the W birds was absent in the M birds: the postsynaptic thickenings in the right MHV were reduced in length to that in the left MHV. An analogous finding was made in the LPO (Stewart et al 1984a). There were also striking changes in the left hemisphere consequent on training. The number of synapses per unit volume of MHV neuropil increased by 23% compared with the right hemisphere, and most strikingly the number of vesicles per synapse in the left hemisphere MHV increased in the M birds by comparison with the right hemisphere by 61%; comparing left hemisphere vesicle number in M birds with those in the comparable region in the control bird, the increase was 72% (Fig 2). All these differences were highly significant ($p<0.02$ on ANOVA).

These changes refer to synaptic profiles of the MHV. By contrast, despite the increased metabolic activity shown by the use of the 2DG marker, we were unable to find lasting structural changes in the PA (though this may be due to the fact that we sampled for electron microscopy only a small portion of a large and not necessarily homogenous structure). In the LPO changes so far observed were confined to the loss of asymmetry of postsynaptic thickening described above.

The implications of these observations are striking. Following training, birds which have pecked at a MeA coated bead show enhanced metabolic activity in three brain structures, HV, PA and LPO, and subsequently in MHV and LPO there are long lasting changes in the length of synaptic apposition zone, and in the MHV, changes the relative volume of the synapses and above all in the number of vesicles per synapse. It is obviously very tempting to see such changes as morphological evidence for a change in - presumably a strenthening of - synaptic connectivity

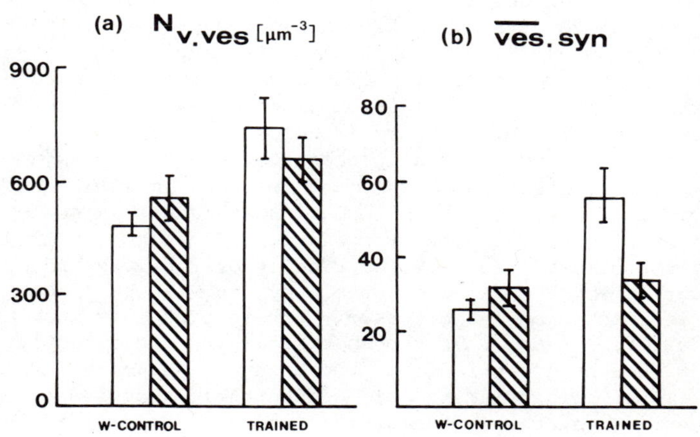

Figure 2. (a) Synaptic vesicle numerical density ($N_{V.ves}$) in the right and left MHV of W-control and MeA-trained chicks. Open bars left hemisphere; cross-hatched bars right hemisphere. Each bar is a mean value of measurements from 8 birds and the vertical lines represent S.E.M.s. The effect of training on hemisphere is statistically significant.

(b) The data in Figure 2a has been transformed in order to permit the mean synaptic vesicle number per synapse ($\overline{ves}$.syn) to be expressed. The size of the difference between $\overline{ves}$. syn in left and right hemispheres of trained chicks (61 percent) is such that an ANOVA now shows that differences between hemispheres are statistically significant, in addition to the effect of training on hemispheres.

in these regions consequent to training. Before considering these implications further, however, it is necessary to say something more about the biochemistry of the processes involved.

The biochemistry of passive avoidance training

The changes in synaptic structures discussed in the preceding section may or may not require a net synthesis and lasting change in the concentrations of biochemical components; they could presumably result merely from the reorganization of existing membrane components and subunits. However, even if this were the case, a biochemical mobilisation of enzyme systems associated with such reorganization would be required. The enhanced glucose metabolism indicated by the 2DG data also implies an activation of at least some biochemical processes. Our earlier imprinting studies had shown enhanced incorporation of precursors into RNA and protein, as well as changes in the enzymes of the cholinergic system (Rose 1977). One might expect to find analogous changes in passive avoidance but the one-trial nature of the task enables more pertinent questions about the time course of biochemical changes also to be asked. It is well known that the consolidation process involves more than one phase of short term or labile memory (Gibbs & Ng 1977). I have elsewhere pointed out that for a biochemical process to be regarded as a correspondent of memory formation, its time course must be compatible with that of the phenomenologically observed memory formation processes (Rose 1982). In the chick, following training, there is a period of up to a minute or so during which the memory trace is electrically disruptible and of up to an hour or so during which it is disruptible by drugs such as ouabain. Long term memory is blocked by inhibitors of protein synthesis such as cycloheximide (Gibbs & Ng 1977) though this may not be directly a consequence of the drug's effect on protein synthesis but of damage resulting from the changed amino acid concentrations caused by the blocking of protein synthesis (Hambley & Rogers 1979).

At any event, it is appropriate to look for biochemical markers both of short and longer term sequelae of exposure to training on the PAL task.

Short term changes

A variety of considerations led us to suspect that amongst the markers for short term change would be alteration in aspects of neurotransmitter/ receptor systems. Enhancement or inhibition of neurotransmitter function could in principle be achieved by alteration in the level of transmitter, resulting from modifying synthetic or degradative enzyme activity, or transmitter availability, perhaps achieved by altering the proportion bound in synaptic vesicles at the synapse, or transmitter interaction with its receptor, achieved by altering either the amount or the binding capacity of receptor. The The results are summarised in Table 1. Thirty minutes after training, there PA and LPO are not, at present, known. However, HV (especially MHV) and PA are rich in muscarinic cholinergic receptors, as indicated by autoradiographic mapping of the distribution of the binding ligands quinuclidinyl benzilate (QNB) and propyl benzylcholine mustard (PrBCM); (Coulter, 1982). Binding of the presumed nicotinic ligand α-bungarotoxin is one to two orders of magnitude lower. There is also a wide distribution of serotonin binding sites. The enzymes of acetylcholine synthesis and degradation are also especially enriched in the forebrain roof of the chick.

On this admittedly rather circumstantial basis we examined the effect of training on the specific binding of the muscarinic ligand QNB, of the nicotinic ligand α-bungarotoxin and of serotonin to chick brain regions at various times after training (Aleksidze et al 1981; Rose et al 1980). The results are summarised in Table 1. Thirty minutes after training, there

Table I Changes in ligand binding in chick forebrain roof following passive avoidance training.

Time after training (min)	QNB	M/W x 100 αBT	5HT
10	110	-	-
30	122***	83*	114
180	101	98	115*

*$p < 0.05$

***$p < 0.001$

Maximal specific binding of ligand to brain homogenates was measured for QNB (quinuclidinyl benzilate, for muscarinic cholinergic receptors), αBT (α-bungarotoxin, for nicotinic cholinergic receptors) and 5-HT (serotonin, for serotonergic receptors). Significant changes in ligand binding at various times after the training are indicated (From Rose et al 1980, Aleksidze et al 1981).

is a 22% increase in QNB binding activity in the forebrain roof of the M compared with the W birds. This increase is transient; it is not apparent 10 min after the onset of training, and has disappeared by 3h after training. By contrast, the binding of α-bungarotoxin shows a comparable 17% decrease at 30 minutes in the M birds, but is at control level again by 3h. Serotonin binding levels show an increase at 30 min which persists for the full 3 hr of the experiment. The reciprocal increase in QNB binding and decrease in α-bungarotoxin binding are thus compatible with their association with the short term phase of memory storage, the serotonin binding increase with some more prolonged process.

Longer term processes

We have, however, paid most attention to potential candidates for biochemical changes subserving long-term processes. Modification of synaptic membrane structure is likely, for instance, to involve changed production of both proteins and glycoproteins - generally regarded as serving as key recognition molecules at the synaptic membrane.

If a half-hour pulse of ^{14}C-leucine is given by intraperitoneal injection to chicks at varying times after training, an increased incorporation occurs into TCA-precipitable proteins of the anterior forebrain roof of M compared with W birds. This increased incorporation persists for at least 24 hr after the training experience, but has dropped back to control (W) levels by 48 hr. The increase of incorporation is into both soluble and insoluble proteins but, with the half hour pulse, the major effect is on soluble proteins. We have made some progress in identifying the proteins involved. For instance, one of the major constituents to show an increase in incorporation is the microtubular protein tubulin, whose concentration, determined by a colchicine binding assay, also shows a similar increase persisting for 24 hr, but is at control levels by 48 hr. (Mileusnic et al 1981). Calmodulin, the calcium-binding protein, also shows increased turnover (P. Tillson & Rose, unpublished results) but actin does not (Tillson & Rose 1981).

In order to investigate this increase in protein production further we have recently developed an in vitro protein synthesizing system derived from chick brain slices which can incorporate ^{14}C-leucine into tissue protein at the rate of 2-3 nmol/mg protein per hour over a period of at least 90 min (Schliebs & Rose, 1984). If slices are prepared from the chick brain an hour after training and the in vitro rate of protein synthesis measured, there is a 23% increase ($p<0.01$) in synthetic rate in the forebrain roof of M compared with W birds (3.44 against 2.80 nmol leu/mg prot/hr). This observation of enhanced in vitro incorporation (also predominantly into soluble protein) which matches the in vivo observation, is important because it rules out the possibility that increased incorporation observed in vivo is merely an artefactual consequence of changed bloodflow, amino acid uptake or whatever (see Rose & Haywood 1976 for discussion of this). But also, one of the problems of in vivo studies has been that analysis of differentially radioactively labelled protein has been limited by the amount of radioactivity which can be got into the cells across the blood brain barrier, even in the chick. With the availability of the in vitro system this limitation has been transcended and we are now in a much stronger position to analyse the protein products of the training experience. The persistence of the increase into an in vitro assay system must also imply a retained mobilization of protein synthetic machinery, presumably involving increased transcriptional rates and the activation of ribosomal systems, as a consequence of training. (In imprinting one of the earliest steps to occur after onset of stimulation is an enhanced nuclear RNA polymerase activity; Haywood et al 1970).

Our study of the involvement of glycoproteins has followed a parallel path to that of proteins. The chosen precursor is ^{3}H- or ^{14}C-labelled fucose, which is selectively incorporated into glycoproteins, but the analysis of its fate is harder in some respects than that of protein because incorporation is slow and fucose uptake across the blood brain barrier certainly one limiting factor. With an experimental design which involved injecting the radiolabelled fucose at the time of training and analysing incorporation at various periods thereafter we were able to show an enhanced incorporation into particulate glycoproteins of the anterior forebrain roof; this enhanced incorporation was concentrated to the synaptic membranes (Sukumar et al 1981; Burgoyne & Rose 1981). If the experimental design was changed so as to inject the ^{14}C-fucose some time after training, effects were also found in the forebrain base (Rose & Harding 1984). Proteins are fucosylated in the Golgi apparatus by pathways involving the enzymes fucosyl transferase and fucokinase. We were able to show that training increased the fucokinase activity in the forebrain base of M birds compared with W (Lossner & Rose 1983). As with protein, an in vitro glycoprotein synthesizing system can be produced, and enhanced incorporation of ^{14}C-fucose into the glycoprotein of the right base observed in slices prepared from birds an hour after training (McCabe & Rose 1984). Further analysis of the synaptic membrane glycoproteins is underway. We need to know whether the increased production is confined to the glycoprotein of the junctional complex and if so, at which side of the cleft the glycoproteins lie. Our prospects will be enhanced if we can find an in vitro incorporation system which can generate labelled glycoprotein in amounts adequate for analysis and if we can go on from there to raise antibodies to specific membrane constituents.

Relationships between cell biological changes and memory processes.

The morphological and biochemical changes discussed above are all sequelae of exposure to the experience of pecking at a methylanthranilate - as opposed to a water-coated-bead on a training trial. There are many aspects to this experience - including of course the taste of the MeA, differential arousal, possibly fear, and so forth which distinguish W from M birds, and I have pointed out above and elsewhere (e.g. Rose, 1981, 1982) that to move from observation of sequelae of training to claims about relationships between these sequelae and memory storage processes requires rigorous controls.

One approach would be to attempt to decouple the experience of the taste of MeA from the association with pecking at the bead. For instance if the bird's eyes are covered and MeA introduced into the beak, then the increase observed in QNB binding 30 min after training does not occur (Rose et al 1980). But there are of course obvious other differences about the two types of experience. In the same series of experiments, we also observed that intracerebral injection of ouabain or cycloheximide, drugs which interfer with short or long term phases of memory formation, also prevent the training-induced increase in QNB binding. Such data too is suggestive, but scarcely conclusive.

Perhaps the most powerful attempt we have made to separate out the sequelae of the training experience from those of memory formation, however, takes advantage of the fact that brief subconvulsive transcranial electroshock administered within a minute after training renders the birds apparently amnestic when they are tested for recall (that is, avoidance of the bead) any time from half an hour to at least 24 hrs. subsequently. If the administration of the electroshock is delayed to 10 min after training, there is no such amnesia; birds avoid the bead on testing subsequently. If the electroshock is administered to birds trained on the water bead, there is no observable change in their pecking at the bead on test. We could thus carry out an experiment utilising six groups of birds:

(a) Controls, unshocked which have pecked a water coated bead;
(b) Controls, immediate shocked;
(c) Controls, delayed shocked;
(d) Trained, unshocked which have pecked a MeA coated bead;
(e) Trained, immediate shocked;
(f) Trained, delayed shocked.

Groups (d) and (f) avoid the bead on test, Group (e) pecks at the bead, as do control groups (a), (b) and (c). Groups (e) and (f) have had identical experiences; they have pecked MeA-coated beads and been shocked; the only difference is in the time of shock relative to test, and hence (e) behaves as if it is amnestic and (f) shows recall.

A biochemical change that is consequent upon some general aspect of training should appear in Groups (d), (e) and (f). If it is part of the memory storage process, it should be present in (d) and (f) compared with (a) and (c), but be absent in (e) which should be identical to (b).

We chose to use this experimental design with fucose incorporation in glycoproteins as our biochemical marker, since the elevated fucose incorporation persists beyond the immediate training period. To avoid the possibility that any immediate consequences of training and electroshock - such as blood flow changes - were responsible for any biochemical effect, the precursor was not administered until after all behavioural measures, including testing, had been carried out. Birds were trained, shocked or not, tested 30 min after training, and ^{14}C-fucose injected intraperitoneally after a further 15 min. Birds were killed after a 3 hr incorporation period and fucose incorporation studied in several brain regions. The measure thus refers to cellular events taking place, not during training itself, but during a subsequent consolidation phase of memory formation.

The results of these experiments are summarized in Fig 3. There were no differences of incorporation in W birds as a result of the shock. Unshocked M birds showed the expected increase in fucose incorporation in base and anterior roof. M Birds rendered amnestic by immediate shock showed no increase in incorporation over W birds; delayed shocked and therefore not amnestic M birds showed a comparable increase in incorporation to the unshocked M birds.

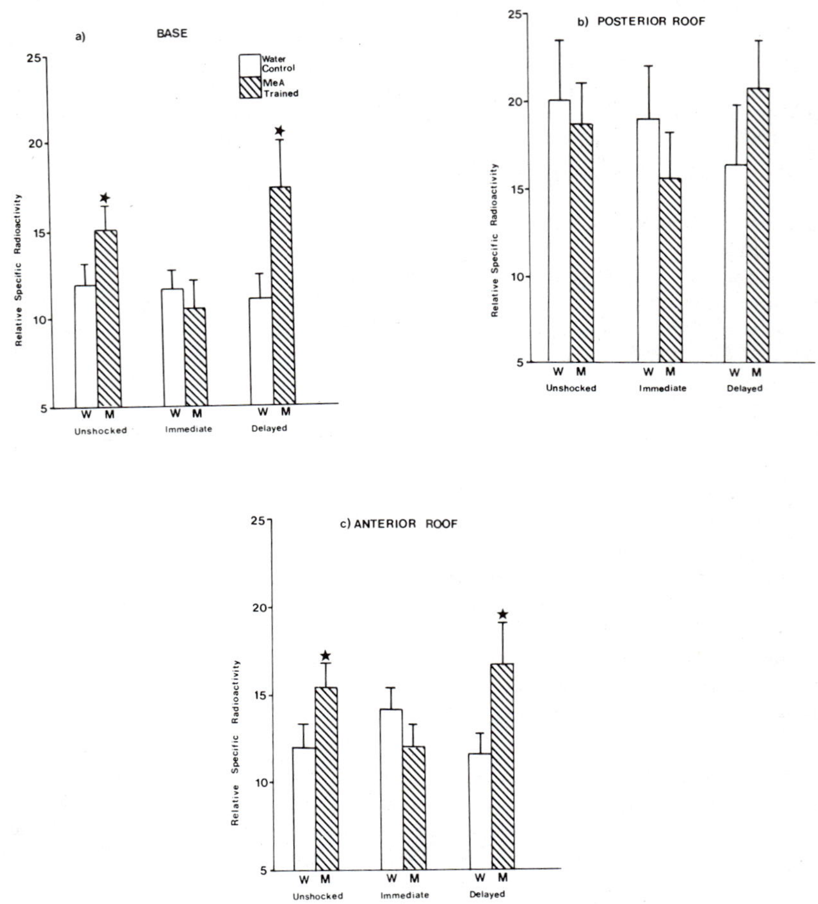

Figure 3. ^{3}H-fucose incorporation into 3 brain regions following passive avoidance learning and immediate or delayed electroshock. Histograms show relative specific radioactivity for W-U and M-U, W-I and M-I, and W-D and M-D conditions; n = 9 for each condition; error bars show s.e.m. (a) Forebrain base; (b) Posterior roof; (c) Anterior roof. Significant differences from corresponding water control ($p<0.05$) shown by asterisks. Abreviations: W-U Water control, unshocked; W-I Water control, immediate shocked; W-D Water control, 10' delayed shocked; M-U Methylanthranilate trained, immediate shocked; M-D Methylanthranilate trained, 10' delayed shocked.

The most parsimonious interpretation of this result must be that the experience of training, without the memory consolidation process occuring, is not sufficient to result in increased fucose incorporation in the base and anterior forebrain roof. Hence this increase is likely to be part of the biochemical process involved in that consolidation; that is, it is a necessary and specific part of the consolidation process.

The fact that we have shown this reasonably convincingly for one biochemical process, that of fucose incorporation, does not allow us to conclude that all other biochemical or morphological sequelae of training we observe are equally specific. This is a problem we have met before in the context of the markers of imprinting (e.g. Rose 1977). Bar infinite time to repeat all possible controls with all possible markers, we can only offer the relative security produced by assembling a series of control experiments and hope that parsimony rules, O.K. If we are wrong, we will doubtless learn the hard way.

Acknowledgements

The experiments described here have been conducted with my colleagues in the Brain Research Group and valued visitors to our laboratories. I thank all those whose work is cited, both for experimental collaboration and many fruitful discussions. I also thank Pat Bateson for continued interactive discourse over the years. This work was carried out under grants from the MRC (1500BB/AMV/FR-15M) and NIH (1 R01 NS19030-1) which are gratefully acknowledged.

REFERENCES

Aleksidze, N., Potempska, A., Murphy, S. and Rose, S.P.R. 1981, Passive avoidance in the young chick affects forebrain α-bungarotoxin and serotonin binding. Abstr. 8th ISN p.382.

Benowitz, L. 1974, Conditions for the bilateral transfer of monocular learning in the chick, Brain Res. 65:203.

Benowitz, L. 1980, Functional organization of the avian telencephalon, in: "Comparative Neurology of the Telencephalon" S. Ebbeson, ed., Plenum Press, New York.

Benowitz, L. and Lee-Teng, E. 1973, Contrasting effects of three forebrain ablations on discrimination learning and reversal in chickens J. Comp. Physiol. Psychol. 84: 391.

Bradley, P., and Horn, G. 1979, Efferent connections of hyperstriatum ventrale in the chick J. Anat. 128: 414.

Burgoyne, R.D. and Rose, S.P.R. 1982, Subcellular localization of increased incorporation of ^{3}H-fucose following passive avoidance learning in chicks Neurosci. Lett. 19: 343.

Cherkin, A. 1969, Kinetics of memory consolidation: role of amnestic treatment parameters Proc. Nat. Acad. Sci (U.S.) 63: 1094.

Coulter, J.C. 1982, Chick brain cholinergic receptors, Thesis, Open Univ.

Gibbs, M.E. and Ng., K.T. 1977, Psychobiology of memory: towards a model of memory formation Biobehav. Rev. 1: 113.

Hambley, J.W., and Rogers, L.J. 1979, Retarded learning induced by amino acids in the neonatal chick, Neurosci., 4: 677.

Haywood, J., Rose, S.P.R. and Bateson, P.P.G. 1970, Effects of an imprinting procedure on RNA polymerase activity in the chick brain, Nature 228: 373.

Horn, G. 1981, Neural mechanisms of learning: an analysis of imprinting in the domestic chick, Proc. Roy. Soc. B. 213: 101.

Horn, G., McCabe, B.J. and Cipolla-Neto, J. 1983, Imprinting in the domestic chick: the role of each side of the hyperstriatum ventrale in acquisition and retention. Exp. Brain Res. 53: 91.

Kossut, M., and Rose, S.P.R. 1984, Differential 2-deoxyglucose uptake into chick brain structures during passive avoidance training, Neurosci. 12: 971.

Lossner, B. and Rose, S.P.R. 1983, Passive avoidance training increases fucokinase activity in right forebrain base of day-old chicks, J. Neurochem. 41:1357.

McCabe, N.R. and Rose, S.P.R. (1984) Passive avoidance training increases fucose incorporation into glycoproteins in chick forebrain slices in vitro. Neurochem Res. In press.

Mileusnic, R., Rose, S.P.R., and Tillson, P. 1980, Passive avoidance training results in changes in concentration of and incorporation into colchicine binding proteins in the chick forebrain J. Neurochem. 34: 1007.

Rogers, L.J. 1982, Light experience and asymmetry of brain function in chickens Nature 297: 223.

Rogers, L.J. and Anson, J.N. 1979, Lateralization of function in the chick forebrain Pharmacol. Biochem. Behav. 10:679.

Rose, S.P.R. 1977, Early visual experience, learning and neurochemical plasticity in the rat and the chick. Phil. Trans. Roy. Soc. London. (B) 278:307.

Rose, S.P.R. 1981, What should a biochemistry of learning and memory be about? Neurosci. 6: 811.

Rose, S.P.R. 1982, From causations to translations: a dialectical solution to a reductionist enigma in "Towards a Liberatory Biology", S.P.R. Rose, ed., Allison & Busby, London.

Rose, S.P.R. and Harding, S. 1984, Training increases ^{3}H-fucose incorporation in chick brain only if followed by memory storage. Neurosci. 12:663.

Rose, S.P.R. and Haywood, J. 1976, Experience, learning and brain metabolism in "Biochemical Correlates of Brain Structure and Function", A.N. Davison, ed., Academic Press, London.

Rose, S.P.R., Gibbs, M.E. and Hambley, J.W. 1980, Transient increase in forebrain muscarinic cholinergic receptors following passive avoidance learning in the young chick. Neurosci. 5:169.

Schliebs, R., Rose, S.P.R. and Stewart, M.G. 1984, Effect of passive avoidance training on in vitro proteins synthesis in forebrain slices of day old chicks, J. Neurochem. In press.

Sokoloff, L., Reivich, M., Kennedy, C., Des Rosiers, M.H., Patlak, C.S., Pettigrew, K.D., Sakurada, O., and Shinohara, M. 1977, The ^{14}C-deoxyglucose method for the measurement of local cerebral glucose untilisation: theory, procedure and normal values in the conscious and anaesthetized albino rat, J. Neurochem. 28:897.

Stewart, M.G., Rose, S.P.R. and King, T.S. 1984a, A stereological investigation of synapses in chick palaeostriatum augmentatum and lobus parolfactorius following training on a passive avoidance task, Acta Stereologica. 2: Supp 1 227.

Stewart, M.G., Rose, S.P.R., King, T.S., Gabbott, P.L.A. and Bourne, R., 1984b, Hemispheric asymmetry of synapses in chick medial hyperstriatum ventrale following passive avoidance training: a stereological investigation, Devel. Brain Res.12:261.

Sukumar, R., Rose, S.P.R., and Burgoyne, R.D. 1980, Increased incorporation of ^{2}H-fucose into chick brain glycoproteins following training on a passive avoidance task. J. Neurochem. 34: 1000.

INFLUENCE OF DEVELOPMENTAL FACTORS ON IMPRINTING

Hans-Joachim Bischof

Lehrstuhl Verhaltensphysiologie
Fakultät Biologie, Universität Bielefeld
Postf. 8640, 4800 Bielefeld 1, F.R.G.

INTRODUCTION

Since Konrad Lorenz (1935) published his famous "Kumpan" paper, where he defined imprinting as a special learning process, a wealth of literature on the topic has accumulated (for reviews see Bateson 1966, Klinghammer 1967, Immelmann 1972, Hess 1973, Hoffman and Ratner 1973, Immelmann and Suomi (1981). As a consequence, the definition of "imprinting" has been changed, as not all of the criteria originally proposed by Lorenz could be applied to all those "imprinting - like" phenomena that needed to be included in a useful definition of this learning paradigm.

Thus Immelmann and Suomi (1981) defined imprinting as "an early learning process with very stable results". The two criteria, namely that learning occurs only during a sensitive period early in development of an animal and that the information acquired in the course of this sensitive period is very stably stored, are special features of imprinting and cannot be found to the same degree in other learning paradigms (Immelmann and Suomi 1981).

Whereas the existence of a sensitive period for the acquisition of external influences is generally accepted by students of imprinting and the discussion at present concentrates more on the question of determining the limits of this sensitive period and the influence of experimental conditions on its onset and duration, there is no such general agreement concerning the stability of storage. From filial imprinting, some authors argued that at the end of the sensitive period a bird has become familiar with the object to which it has been exposed. This familiarity with one object leads to an "unwillingness" to learn new features subsequently (Bateson 1979). If this unwillingness can be erased by appropriate experimental conditions, the bird can learn new features and become attached to a new object. However, in filial as well in sexual imprinting, it was recently demonstrated that this new preference is erased if the bird is exposed after some time to a double choice test between the first and the second object: in this case it prefers the first object again (Cherfas and Scott 1981, Immelmann 1979). This is not an effect of primacy and recency of stimuli, as Bateson (1983) points out, but demonstrates that there must be some sort of internal protection of the information previously stored. The latter is what is meant by the notion of permanent or irreversible storage.

In the development of the visual cortex, which shares many features with imprinting (Bischof 1983), such permanent storage may be explained by the fact that the storage of information takes place during a period of development of cortical tissue. As LeVay and Stryker (1979) demonstrated, the development of the physiological characteristics of the adult visual cortex coincides with morphological segregation of geniculocortical projections into columns. They speculated that the last step following this segregation is the myelination of the acquired wiring. This subsequently protects the visual cortex connectivity from alterations by outside stimulation. The same view was expressed by Wolff (1981).

In this paper, I want to discuss the question whether such developmental processes may be involved in imprinting also. I will present some recent behavioural data on the participation of the development of the zebra finch visual pathway in the onset of the sensitive period for sexual imprinting. Furthermore, I present a speculative model which shows a reciprocal dependency between the occurrence of a sensitive period and the stability of storage. This model is illustrated by a simple computer simulation. Finally, I discuss the relevance of the data concerning the development of the visual system of the zebra finch for the speculations mentioned above.

SEXUAL IMPRINTING IN ZEBRA FINCHES

Sexual imprinting is an early learning process by which the young zebra finch learns the features of the object to which he courts as an adult. Careful studies of sexual imprinting in zebra finches were provided by Immelmann (1972, 1975, 1979, Immelmann and Suomi 1981). He demonstrated that the sensitive period for the acquisition of the features of the imprinting object was longer and occurred later than that obtained by Hess (1973) for filial imprinting (Fig. 1). The peak of sensitivity for sexual imprinting is at about 13 to 15 days posthatching. This sensitivity is reduced considerably before the zebra finch reaches sexual maturity, and thus long before the bird has any courting experience. However, in one series of experiments Immelmann demonstrated that sensitivity does not reach zero, but descends asymptotically. Sometimes, imprinting can take place as late as 70 days of age (Immelmann and Suomi 1981).

Storage of information acquired in the course of the sensitive period is indeed very stable in the zebra finch. Immelmann caged adult zebra finches, which had been imprinted to a foster species, together with conspecifics for a long time. These birds eventually courted to conspecific females and even reared clutches. When, after this experience with conspecifics, the finches were tested in a double choice test with a conspecific and a female of the foster species, the birds preferred the foster species again. Thus, information stored during the sensitive period had not been forgotten in the

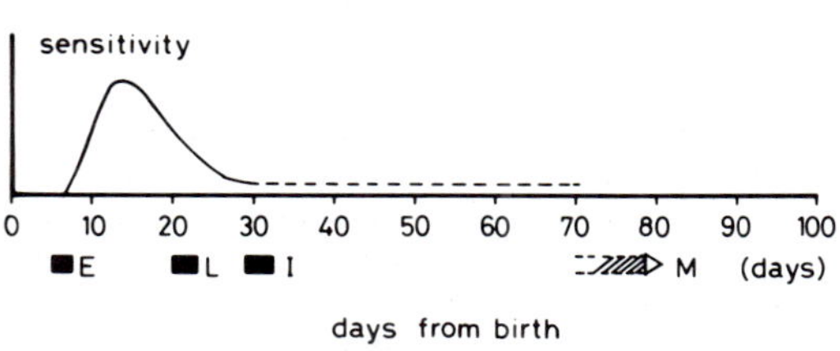

Fig.1: Idealized sensitivity diagram for sexual imprinting in zebra finches. Data from Immelmann and Suomi (1981). E = Eye opening, L = Leaving of the nest, I = Independence. Further explanations see text.

meantime. Obviously, it was masked for the time the birds lived with conspecifics and came to the surface again in a situation of choice.

THE ONSET OF THE SENSITIVE PERIOD FOR SEXUAL IMPRINTING

Immelmann (1972) investigated the phase of sensitivity to the imprinting stimulus by separating young birds at different times from their parents and transferring them to a foster species, the Bengalese finch. Using this procedure, it was found that 13 days of age was the earliest age of transfer, in which a stable preference for the first object could be demonstrated when the birds were adult. As the birds need some time of visual experience before that time to learn the features of the imprinting object, susceptibility to external stimulation should start earlier than this age. Immelmann proposed that the sensitive period may start between day 5 from hatching, when the eyes begin to open, and day 10, when the eyes are fully open. Bischof and Lassek (in press) looked at the development of the gaping reactions of young birds to determine the age at which they can use visual stimuli as triggers for gaping. Whereas the birds react to acoustic and tactile stimuli from the first day of life, gaping to visual stimuli does not occur before 10 days of age (fig.2). Thus, imprinting to visual stimuli, which seem to be the most important for the elicitation of courtship (Bischof, unpubl. results), cannot start earlier than 10 days. At day 15, the gaping reaction towards stuffed dummies is replaced by a reaction of fear; however, gaping towards the parents remains until the young birds reach independence at about 30 days of age. Thus, at 15 days they are able to discriminate between a dummy and their parents. Some of the birds, as revealed by Immelmann's studies (1972), are already stably imprinted at this time. However, most of the birds are imprinted at later age. Thus, storage of new information is still possible beyond 15 days of age, but under normal conditions the development of fear prevents the bird from attending to and learning such new information (Bischof and Lassek in press).

We have found (Herrmann and Bischof in prep.) that there are some correlations between the development of the visual system of the zebra finch and the behavioural data mentioned above. As can be seen from fig.3, the myelination of the thalamic nucleus of the tectofugal pathway, the nucleus rotundus, does not start before 5 days of age. Accordingly, the myelination of the telencephalic station of this pathway, the Ectostriatum, does not start before day 10. We suggest that the stabilization of the wiring of the nuclei, which may be indicated by the myelination of the axons (Wolff 1981), starts in nucleus rotundus when the eyes open, and in the Ectostriatum at the time when the birds can react to visual stimuli. Whether this is only an arbitrary coincidence in time or there are functional relationships,

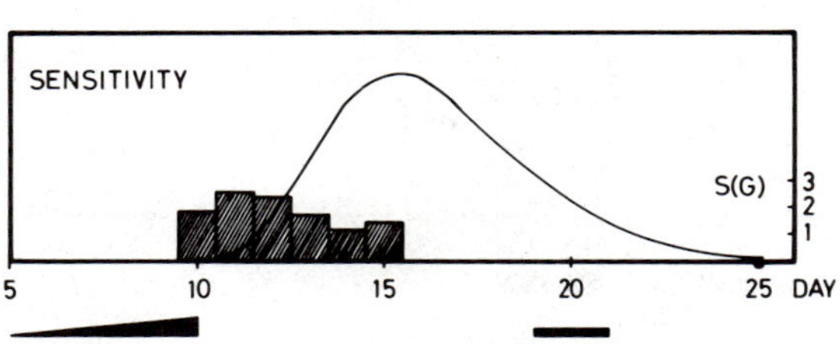

Fig.2: Gaping responses of young zebra finches and sensitivity curve for sexual imprinting. S (G) = Gaping score, hatched bars: Gaping responses.

cannot be decided from these experiments. To investigate this, we just perform anatomical and behavioural experiments with visually deprived animals. However, one conclusion of the results presented above is that the beginning of the sensitive period is most likely determined by the developmental state of the visual nuclei, and that the learning of the features of the parents coincides in time with the development of the visual pathway.

THE TIME COURSE OF THE SENSITIVE PERIOD FOR SEXUAL IMPRINTING: EFFECTS OF IRREVERSIBLE STORAGE

As shown in the preceding section, developmental aspects surely play a role for the onset of the sensitive period. It demonstrates that under normal conditions the development of fear prevents the animal from acquiring new information from the environment.

However, this fear reaction has to be overcome by the birds in the case of the cross fostering experiments made by Immelmann, as they depend on feeding and thus have to gape to the new parents after transfer even if they are older than 15 days of age. As the sensitivity to environmental stimulation goes down despite of the elimination or at least diminution of the fear components (Fig.1), there must be another internal mechanism which protects the acquired data. This internal mechanism, however, cannot be fixed to an appointed time of development, as the course of decline can be essentially altered by experimental manipulations (Bateson 1978).

One possibility of how such flexibility may be achieved is that the storage process itself has some influence on the termination of sensitivity. For example, the increasing amount of stored information may lead to a decrease of the proportion of information which can be stored subsequently. This is easily demonstrated by a simple computer model which simulates the filling of limited stores by a random process (Fig.4). In the model, information from the outside flows in a chain through three concatenated stores, but with certain constraints. These stores would represent several levels of processing in the brain from the periphery to more central stations. The random procedure chooses one of the elements of the first store and puts the information in, if the element is still empty. Otherwise this information is lost. Thus, the probability of storage depends on the filling of the store: the more information is already stored, the lower is the probability for new storage.

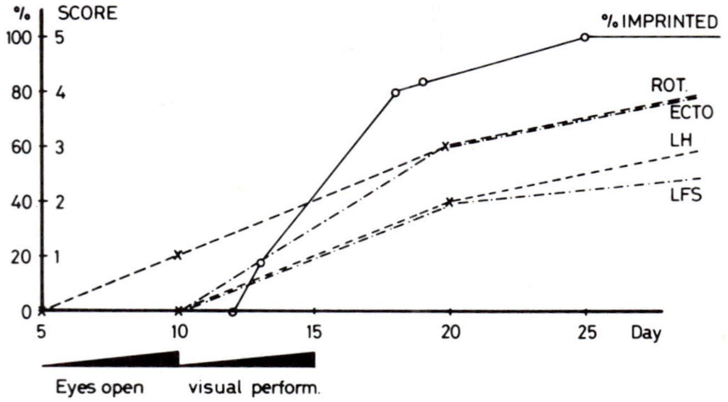

Fig.3: Myelination of different brain areas. Left ordinate scale: % of birds imprinted at that time (Data from Immelmann and Suomi 1981). Right ordinate scale: Myelination score. Rot = n. rotundus, E = Ectostriatum, LH = Lamina hyperstriatica, LFS = Lamina frontalis superior.

The probability of transfer of information from one store to the next is constrained by the filling of the preceding store. The more that one is filled, the higher is the probability of transfer to the next one. The procedures are the same for store two and three, respectively (Fig.5).

In the computer simulation, the above process is replicated 275 times. Then the computer plots the percentage of stored bits for each store; this is done for 230 time bins (Fig.5). As the percentage of storage within each time bin can interpreted as an indicator of the "sensitivity" of the storage medium for information arriving from the outside, the diagrams can be directly compared with the sensitivity curves obtained in imprinting experiments (Fig.1). The shape of the resulting curves for the second and the third store are very similar to those obtained in experiments of sexual imprinting. In each case "sensitivity" to new information rises quickly. Then, after reaching a peak, the curve declines to zero as a consequence of the amount of the information already stored. From more peripheral to more central stores, the peak of the curve shifts to the right and the shape is more and more flattened.

From the similarity of the sensitivity diagram obtained in the imprinting experiment and the curves obtained from the computer simulation, one can speculate that a storage process may be involved in imprinting which has the two characteristics mentioned above of limited space and permanent storage. The length of the sensitive period can be easily altered in the model by varying the speed of information input or can be prolonged by assuming that only part of the incoming information can get access to the storage medium.

The question now is whether physiological correlates of such a storage process can be found. If one takes the development of the visual cortex of the cat as a model for what may happen in the zebra finch brain in the course of imprinting (Bischof 1983), one can argue that a given brain area may be capable for permanent plastic changes as long as this area can be morphologically altered.

In the zebra finch, Teuchert et al. (1982) observed degeneration of axon terminals in different brain nuclei of the visual pathway. Degeneration occurred in a sequence from peripheral to central stations. As these authors argue, this degeneration may be a prerequisite for subsequent rewiring of connections under the influence of external stimultion as it occurs in imprinting. This process of rewiring is finished when a certain balance of inhibitory and excitatory inputs to a given neuron is reached. Afterwards, by myelination of the involved areas, the newly installed connections are stabilized and consequently cannot be altered by external influences later (Wolff 1981).

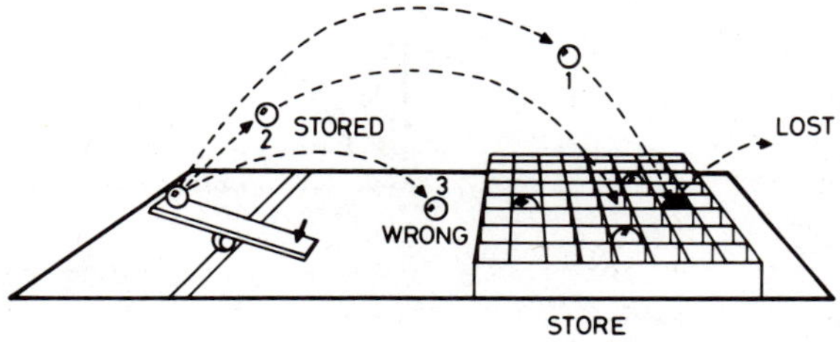

Fig.4: Randomly filling of a store. Explanations see text.

If myelination follows the rules we applied to the computer model, the amount of myelination per time should follow a similar time course to that we obtained in the simulation. To examine this, we differentiated the myelination curves obtained in the experiments already mentioned above (Herrmann and Bischof in prep.) to estimate the speed of myelination for different brain areas at different times. Fig.6 shows that these diagrams are similar to those of the computer simulation and those obtained in sexual imprinting.

However, the other feature we obtained from the simulation, the delay of development from peripheral to more central stations of a given pathway, is not easily detectable in the course of myelination of the different nuclei. This does not affect the parallels mentioned above, as the time shift in the model can be made smaller by altering the coupling conditions between the stores in a sense that more throughput of information is allowed from the very beginning of the simulation.

It is not possible from our data to speculate which of the brain areas is involved in imprinting. All telencephalic areas we examine have their

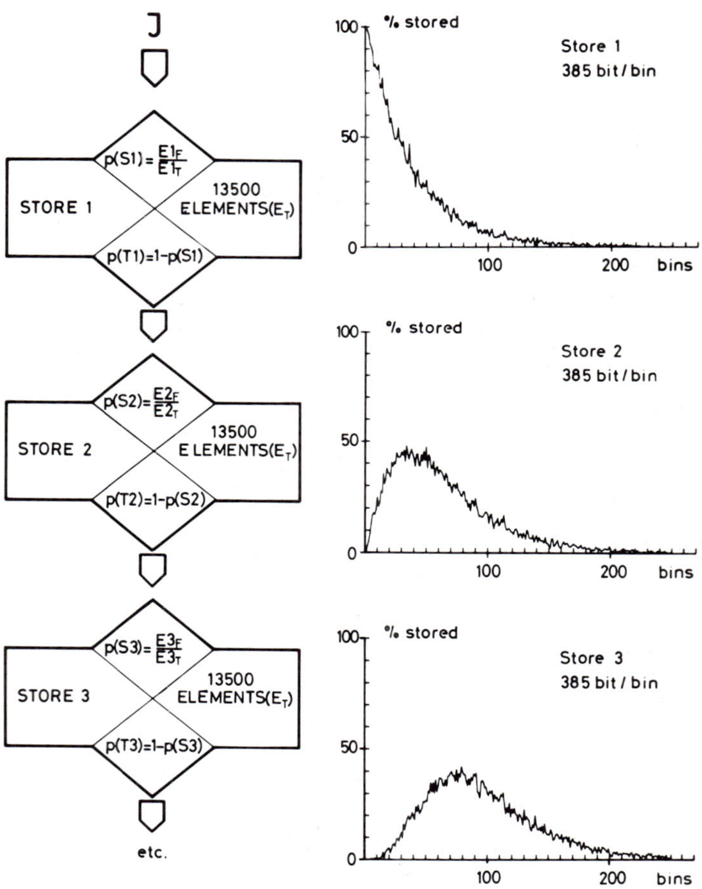

Fig.5: Left: Mathematical formulation of the computer model. Right: Sensitivity diagrams of the different stores. E1F = Elements full, E1T = Elements total, p (S1) = probab. of storage, p (T1) = probab. of transfer.

fasted rate of development at the time when imprinting occurs. There are two exceptions to this rule, the nucleus robustus archistriatalis (RA, not shown in the Fig.6), and the hyperstriatum ventrale pars caudale (HVc), two nuclei involved in control of song in songbirds (Nottebohm 1980). Interestingly, the development of HVc coincides in time with the development of song in the zebra finch (Böhner, pers. comm.). As song learning is in many aspects similar to imprinting, this may be an additional indication that stable learning as demonstrated by imprinting and song learning may be coupled to developmental processes.

Again, one has to stress that it is not possible to prove a direct link between the sensitive period for imprinting and the development of the different brain nuclei from our data; for direct evidence one has to perform deprivation experiments in which the sensitive period for sexual imprinting and, if the above mentioned considerations are correct, the development of different brain areas are shifted forwards.

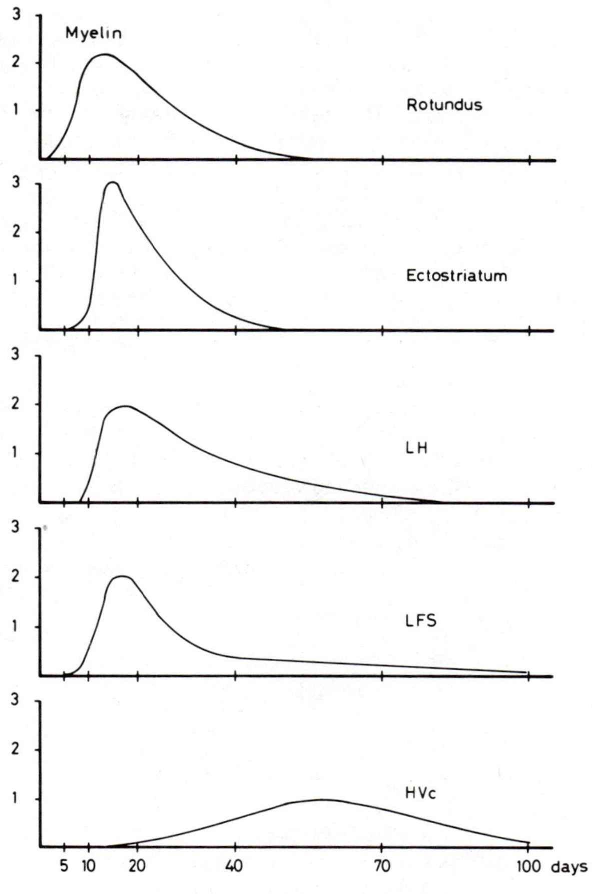

Fig.6: Rate of myelination in different brain areas. Ordinate scale is arbitrary. HVc = Hyperstriatum ventrale pars caudale. For other abbreviations see Fig.3.

CONCLUSIONS

I would like to stress here that the arguments presented above are an attempt to show certain time-dependent parallels which represent another way of thinking about the physiological processes involved in imprinting. The main argument is that imprinting is not different from other learning paradigms in terms of the features which are to be learned and the process of acquisition of the data (Immelmann 1975), but that the stability of storage, which is an important feature of imprinting, is due to the developmental state of the brain areas involved. As long as the neuronal tissue is still developing, it seems to be more vulnerable to external stimulation. This may lead to morphological changes of the neuronal connectivity in the young animal (for example, elimination or construction of new synapses), which may be more stable than the submicroscopic or biochemical changes observed in adult animals. Myelination might be a process which stabilizes the newly acquired connections and prevents them from further alterations; however, this stability could also be reached by a balance of excitatory and inhibitory inputs to a given neuron, as was proposed by Wolff (1981).

If a comparable stability of storage is required in older animals, as for example in song learning, the related neuronal area is held for a longer time in a state of early development, as demonstrated by the delayed myelination of RA and HVc. Nottebohm and Goldman (1983) have found that even new neurons appear in HVc, evidence which is consistent with the present.

If we accept the assumptions presented above, the occurrence of a sensive period may be explained simply by the fact that the remaining "plastic" tissue is reduced as a consequence of a stabilization process, which in turn depends on external stimulation. As shown in the model, it may be that a certain small amount of neuronal tissue remains alterable for a long time and may also be accessible for further imprinting processes, but not as easily as in the young animals. Bateson (1983) argues that imprinting-like processes can readily occur in adult organisms under certain conditions such as stress; this view would fit in the speculations above if one presumes that under stress conditions the input of external information is so massive that part of it gets access to the remaining brain tissue that can be altered.

Admittedly, the view presented above is simplistic and obviously cannot be a complete description of what occurs in imprinting phenomena. The assumptions may represent some basic process which is complicated by many other conditions. However, the physiological basis of features such as, for example the influence of motivation and arousal on the learning process, or the way the bird selects between the different stimuli occurring in the environment, is still poorly understood. Some speculations concerning their influence on imprinting are discussed by Bischof (1983).

REFERENCES

Bateson, P.P.G., 1966, The characteristics and context of imprinting. Biol. Rev., 41: 177

Bateson, P.P.G., 1978, Early experience and sexual preference, *in*: "Biological Determinants of Sexual Behaviour", J.B. Hutchinson, ed., John Wiley, London.

Bateson, P.P.G., 1981, The control of sensitivity to the environment during development, *in*: "Behavioural Development. The Bielefeld Interdisciplinary Project", K. Immelmann, G.W. Barlow, L. Petrinovich, M. Main, eds., Cambridge Univ. Press, Cambridge, Mass.

Bateson, P.P.G., 1983, The interpretation of sensitive periods, in: "The Behavior of Human Infants", A. Oliverio, M. Zapella, eds., Plenum, New York.

Bischof, H.-J., 1983, Imprinting and cortical plasticity: a comparative review, Neurosci. Biobehav. Rev., 7: 213

Bischof, H.-J., and Lassek, R., 1985, The gaping reaction and the development of fear in young zebra finches (Taeniopygia guttata castanotis), Z. Tierpsychologie: in press

Hess, E., 1973, "Imprinting", Van Nostrand Reinhold, New York.

Immelmann, K., 1972, The influence of early experience upon the development of social behaviour in estrildide finches, Proc. XV. Int. Orn. Congr. 1972: 317

Immelmann, K., 1975, Ecological significance of imprinting and early learning, Ann. Rev. Ecol. System., 6: 15

Immelmann, K., 1979, Genetical constraints on early learning: a perspective from sexual imprinting in birds, in: "Theoretical Advances in Behaviour Genetics", J.R. Royce, L.P. Mos, eds., Sijthoff and Noordhoff, Germantown, Maryld.

Immelmann, K., and Suomi, S., 1981, Sensitive phases in development, in: "Behavioural Development. The Bielefeld Interdisciplinary Project", K. Immelmann, G.W. Barlow, L. Petrinovich, M. Main, eds., Cambridge Univ. Press, Cambridge, Mass.

Klinghammer, E., 1967, Factors influencing choice of mate in altricial birds, in: "Early Behavior. Comparative and Developmental Approaches", H.E. Stevenson, E.H. Hess, H.J. Rheingold, eds., John Wiley, New York.

Le Vay, S., and Stryker, M.P., 1979, The development of ocular dominance columns in the cat, in: "Society for Neuroscience Symposia IV. Aspects of Developmental Neurobiology", J.A. Ferrendelli, eds., Soc. Neuroscience, Bethesda.

Lorenz, K., 1935, Der Kumpan in der Umwelt des Vogels, J. Orn., 83: 132 and 289

Nottebohm, F., 1980, Brain pathways for vocal learning in birds: a review of the first ten years, Progress in Psychobiol. and Physiol. Psychol., 9: 85

Nottebohm, F., and Goldman, S.A., 1983, Connectivity and kinetics of neurons born in adulthood, Soc. of Neuroscience Conf. Abstr., 40.2.

Teuchert, G., Wolff, J.R., and Immelmann, K., 1982, Physiologische Degeneration in der Ontogenese des ZNS von Vögeln: eine Einflußnahme auf die sensible Phase für Prägung?, Verh. Dtsch. Zool. Ges. 1982, 259.

Wolff, J.R., 1981, Some morphogenetic aspects of the development of the central nervous system, in: "Behavioural Development: The Bielefeld Interdisciplinary Project", K. Immelmann, G.W. Barlow, L. Petrinovich, M. Main, eds., Cambridge Univ. Press, Cambridge, Mass.

EXTRARETINAL FACTORS CONTROLLING THE DEVELOPMENT OF NEURONAL SELECTIVITY

Michel Imbert

Laboratoire des Neurosciences de la Vision
Université Pierre et Marie Curie
4, place Jussieu - 75230 Paris Cedex 05 France

A fundamental problem for developmental psychobiology resides in knowing the degree to which individual experience participates in the formation of the brain, and to what extent it affects perception in the adult. Dealing with such questions leads us to identify what in the development of the nervous system is brought about by the expression of the genetic program, and what is acquired by interaction with the environment.

Since its first formulation by W. Roux (1905) it has generally been accepted that the brain is formed during two successive periods in the development of the organism. During the first period, events pertinent to embryology take place, such as cellular multiplication and migration, neuronal differentiation, cell death and the establishment of the first interneuronal connections. These events seem to follow rigid rules primarily dictated by genetic factors (Sidman, 1972). During the second period, mainly post-natal, a stabilization of the synaptic contacts takes place. This phase includes the selective elimination of certain neuronal contacts. Epigenetic factors, such as the functioning of the brain itself and processes which are dependant on the experience the organism has of its environment, are considered to be essential for the second phase.

The mammalian visual system is a good model for the study of interactive processes operating throughout ontogeny. A spectacular example of selective stabilization of synaptic contacts is found in the development of the trigger features of neurons in the visual cortex of the kitten or in their modification due to restricted conditions during rearing. If the visual experience of an immature animal is biased by more or less severe monocular or binocular deprivation one can observe important functional modifications ; the critical period of susceptibility of ocular dominance and orientation selectivity to visual "manipulation" starts from the 3rd week of post-natal life and ends around the 12th week (Hubel and Wiesel, 1970) . The role played by visual experience (use and disuse of vision, and binocular competition) in producing cortical functional changes has been extensively studied (Blakemore, 1978 ; Hirsch and Leventhal, 1978) . However until very recently little attention has been given to extraretinal factors involved in experiencing the visual world.

The presence in the primary visual cortex of the adult cat of highly specific neurones, encoding precise features of the environment suggest that early functioning might produce matching between neuronal transmission in cortical microcircuits and particular attributes of the peripheral stimuli. The hypothesis whereby nervous activity allows the architecture of the brain to be shaped in accordance with its sensory environment should be of value because it would permit the development of specialized neurones without the need for complex genetic programming. However, the evidence for instructional mechanisms, representing an elementary form of learning, is still a subject of controversy (see review: Frégnac and Imbert, 1984).

In this chapter I present some of our recent results showing that the development of the two main visual cortical receptive field properties namely orientation selectivity and binocular integration, depends both on vision and factors related to ocular motility.

The specific orientational properties of visual cortical neurones, as originally described by D. Hubel and T. Wiesel, have been observed in very young kittens. Most studies agree that 25 to 35 % of visually responsive cells present adult standards of orientation selectivity at the end of the second week. In kittens deprived of visual experience, either by dark rearing or by binocular lid suture, one can observe a rapid decrease of the proportion of oriented cells.

Figure 1 represents the time course of the development of orientation selectivity. The percentage of orientation selective neurones among visual cortical cells is plotted against postnatal age given in days. The graph represents normalized data taken from several authors. The first orientation selective neurones appear before the eyes open, i.e. around 8 days of age, and an intrinsic process of maturation develops independently of visual experience. Figure 2 is a summary of our own results on the development of orientation selectivity in area 17 of kittens either reared normally or in complete darkness since birth. The proportion of oriented neurones, either broadly tuned (I) or with adult specificity (S) is plotted with postnatal age in each rearing condition : normally reared (NR) and dark reared (DR).

It should be emphazized that as soon as we are able to record cortical neurones that are visually activated, some of them are already truly oriented. That is to say, we have not observed any initial period during which there were only non-oriented neurones present. It seems that specific selective neurones develop immediately at the time of eye opening. In kittens reared in complete darkness during the first 6 weeks of life, the cortical representation of orientation selectivity almost completely disappears. However, the ocular dominance distribution appears to be normal when compared with normally reared kittens of the same age, i.e. most cells are activated through both eyes.

In dark reared kittens a few hours of visual experience are sufficient to elicit a rapid increase in the proportion of orientation selective cells. This is represented by the vertical arrow at the right side of Figure 2. A period of 6 hours of visual experience is enough to restore orientation selectivity to a level comparable to that normally found in normally- reared kittens of the same age. This brief delayed visual experience leads to the expression of a rapid process of maturation which has been masked up to this point by the absence of visual experience.

A series of experiments have been undertaken by Buisseret et al. (1978) in order to determine the factors involved in this "reconstructive effect", according to an expression coined by Barlow (1975) to describe the role of visual experience for cortical specificity. The period of

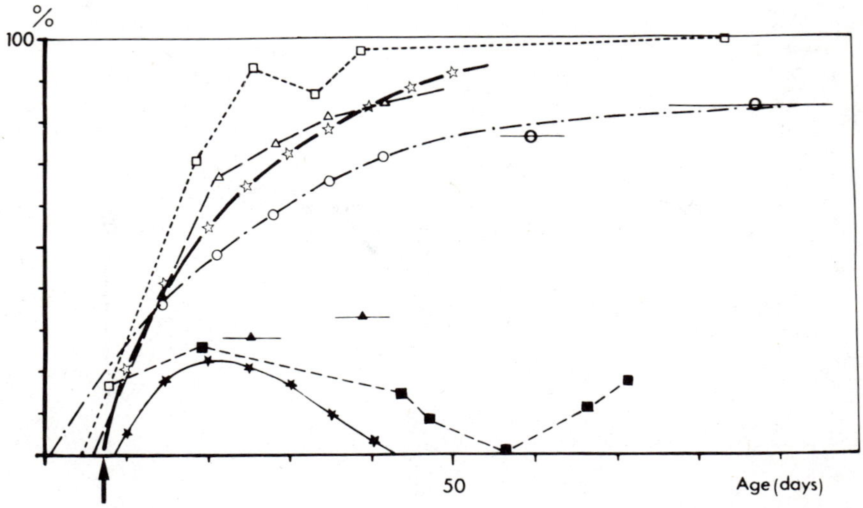

Fig. 1. Kinetics of development of orientation selectivity in normally-reared and visually-deprived kitten area 17. Percentage of orientation-selective neurones among visual cortical cells is plotted with postnatal age given in days. Each mark on abcissa indicates tens of days. Open symbols represent normally reared kittens □, data from Blakemore and Van Sluyters (1975); ★, data from Frégnac (1979); △, data from Bonds (1979); ○, data from Tsumoto and Suda (1982), calculated from Table 2 and Fig. 6 ▽, data from Albus and Wolf (1984). Filled symbols represent kittens visually deprived, either by dark rearing or by binocular lid suture ■, data from Blakemore and Van Sluyters (1975); ★, data from Frégnac (1979); ▼, data from Bonds (1979). Black arrow under abcissa indicates age around which eye opening occurs. Reprinted with permission from Frégnac and Imbert, 1984.

"consolidation", which is the time spent between the end of the session of active visual experience and the time of the second electrophysiological exploration, is not necessary. When the second acute experiment is performed immediately after the session of active visual experience, a large proportion of oriented neurones still appears. If visual experience is limited to a purely passive visual stimulation obtained for example by immobilizing the kitten with a muscular relaxant, there is no recovery of orientation selectivity. In contrast, when motor activity is permitted, even limited to ocular motor activation, orientation selectivity recovery is almost complete. These observations demonstrate that ocular motility plays an essential role in the recovery of orientation selectivity of visual cortical neurones (Buisseret et al., 1978) . It thus seems likely that the extraocular muscle afferents could be active in the process of specification, an hypothesis that has been verified by taking advantage of an anatomical particularity. In cats, most of

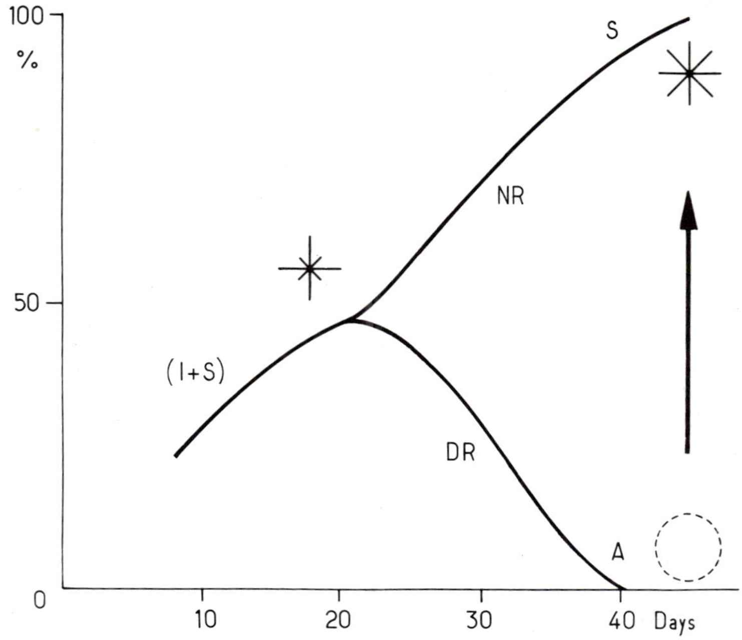

Fig. 2. Effects of visual experience on the development of orientation selectivity in kitten area 17. The proportion of oriented neurones among visual cells, either broadly tuned (I) or with almost adult specificity (S) , is plotted against postnatal age for each rearing condition. The beginning of the period of susceptibility to binocular deprivation is around 18 days. Six hours of visual experience delayed after 6 weeks of dark rearing produce a dramatic increase in specificity (I + S) as indicated by arrow. All orientation preferences are represented, in contrast to the over representation of horizontal and vertical orientation found in kittens less than 3 weeks of age. NR : normal rearing ; DR : dark rearing ; A : aspecific. Adapted from Imbert and Frégnac (1983).

the proprioceptive fibers from extraocular muscles run extraorbitally in the ophthalmic branch of the trigeminal nerve ; only a minor fraction runs in the motor nerves. Thus, it is possible to separate the afferent proprioceptive inflow from the motor inflow by severing this branch before it enters the Gasser ganglion.

The ophthalmic branch of dark-reared kittens was cut bilaterally, and the animal returned to the dark-room for a few days. These animals were then exposed to a normal visual environment for 6 hours and returned to the dark-room for 12 hours of consolidation, before electrophysiological studies (Trotter et al., 1981). Figure 3 shows the relative percentages of orientation-selective, immature, and non-specific visual cells in primary visual cortex of kittens submitted to various rearing conditions. Comparison of the relative proportions of orientation selective cells in each situation suggests that a short visual exposure presented after 6 weeks of dark rearing restores orientation selectivity to a level which depends on the extraocular inflow available during visuomotor experience.

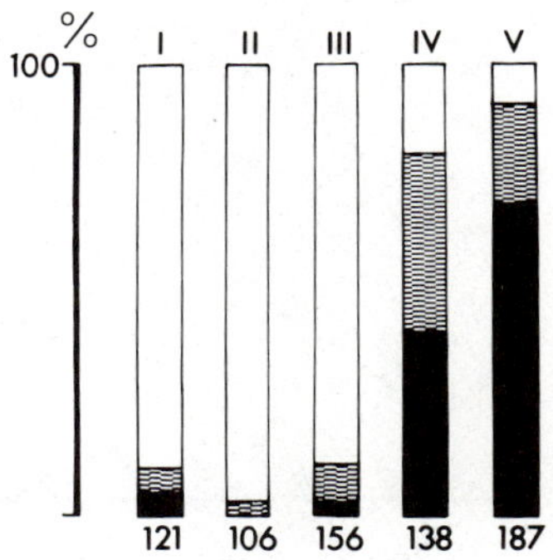

Fig. 3. Extraocular proprioception and cortical effects of delayed visual experience. Columns show relative percentages of orientation-selective (black area) , immature (shaded area) , and non selective (empty area) visual cells in primary visual cortex of kittens submitted to various rearing conditions. Column I, functional state observed in 4- to 6-week-old dark-reared kittens (data from Frégnac and Imbert, 1978) . Column II-IV, cortex effects of 6 hr of binocular vision in 6-week-old dark-reared kitten 1) when experienced under paralysis maintained by intravenous perfusion of Flaxedil or d-tubocurarine (column II data from Buisseret et al., 1978) ; 2) when experienced without extraocular proprioceptive feedback ; in this situation bilateral section of ophthalmic branch of trigeminal nerve was performed a few days before visual exposure (column II data from Buisseret and Gary-Bobo, 1979) ; 3) when experienced with unrestrained body and eye movements (column IV data from Trotter et al., 1981) . Column V, functional state observed in 4- to 6-week-old normally reared kittens (data from Frégnac and Imbert, 1978) . Number of visual cells recorded in each rearing condition is indicated below each column. From Frégnac and Imbert (1984) , with permission.

The ophthalmic branch of the trigeminal nerve innervates various other receptors besides the stretch receptors located in the extraocular muscles, namely the cutaneous receptors and the numerous orbital nociceptors. However, it is unlikely that these afferents play a very important

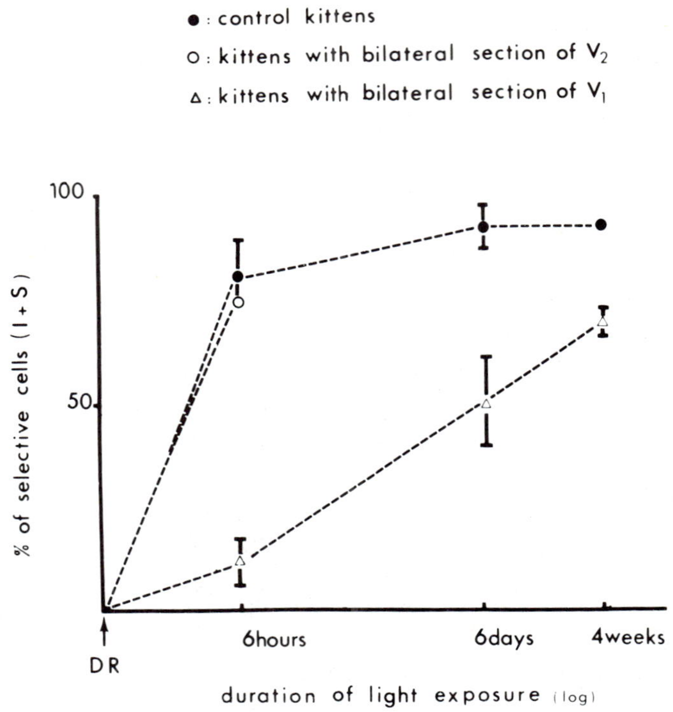

Fig. 4. Rate of recovery of orientation selectivity in kittens dark-reared (DR) from birth to the age of 6 weeks and allowed thereafter binocular visual experience in 3 experimental situations. The percentage of orientation selective units (I + S); immature + specific cells among the total population of recorded visual cells in each situation is plotted as a function of the duration of the exposure (log scale) : kittens with bilateral section of ophthalmic branches (△); kittens with bilateral section of maxillary branches (○) ; intact kittens (●) ; V1 and V2, respectively ophthalmic and maxillary branches of the trigeminal nerve. The zero exposure data (DR) are taken from Buisseret and Gary-Bobo (1979), Imbert and Buisseret (1975). The control point at 4 weeks is derived from extrapolation of kinetic curves of immature and specific populations. Vertical bars are the extreme values in each group of kittens. From Trotter et al. (1981) , with permission.

role in the restoration of orientation selectivity. Sectioning the maxillary branches of the trigeminal nerves, which eliminates a contingent of cutaneous and nociceptive fibres equivalent to those contained in the ophthalmic branches, does not hinder the extraocular proprioceptive fibers which only run in the ophthalmic branch. Under these conditions, 6 hours of active visual experience promotes a degree of orientation selectivity of about the same order as that obtained in control kittens. Figure 4, taken from Trotter et al. (1981) , illustrates the rate of increase in orientation selectivity in kittens dark reared from birth to the age of 6 weeks and thereafter allowed thereafter binocular experience in these three experimental conditions. The proportion of oriented cells is plotted as a function of the duration of the exposure in a logarithmic scale 6 hours, 6 days and 4 weeks ; kittens with bilateral section of the ophthalmic branch of the Vth nerve (empty triangles); kittens with bilateral section of the maxillary branch (empty circles) ; intact kittens (filled circles). One can note an impairment of orientation selectivity after bilateral section of the ophthalmic afferents and no effect after interruption of maxillary afferents. Orientation selectivity in VI operated kittens is established slowly and incompletely even when visual exposure is extended to 4 weeks.

In order to study this aspect further, Trotter et al. (1981) have studied the effects of partial muscular deafferentation. They performed unilateral sections of the ophthalmic branch at different post-natal ages in kittens which were then either normally reared or dark reared. When the unilateral section was performed between the fourth and the eight week, a significant decrease in the number of binocularly activated neurones was found. This effect was observed only when a post-surgical delay of about a month was permitted. On the other hand, this effect does not seem to come from a kind of imbalance of the visual experience of the two eyes, which could have possibly been produced by uncoordinated ocular movements resulting from the muscular deafferentation, since the same result was obtained when the kitten was reared in complete darkness. Therefore, this striking disorganization of binocular integration of visual cortical cells, observed in the above mentioned conditions, seems to occur totally independently of vision since it appears in kittens deprived of visual experience. Two points deserve mention. First, a delay of about a month is necessary between the proprioceptive deafferentation and the electrophysiological exploration in order to observe an effect on binocularly activated neurons in the striate cortex, suggesting that muscular afferents could influence cortical activity through a rather long and complicated pathway "en cascade". Secondly, proprioceptive manipulation affects visual neurones only if it is practiced within a limited period of time which coincides with the peak of the critical period for monocular deprivation.

Yves Trotter has followed up the same kind of questions by combining the proprioceptive deafferentation of one eye with active monocular experience for 6 hours in 6-week old dark-reared kittens. In intact 6 week old dark reared kittens monocularly exposed to 6 hours of visual experience, one can observe two phenomena. First there is a modification in the distribution of ocular dominance : most neurons are driven by the experienced eye. Secondly, the orientation selectivity recovers to a proportion not far removed from that obtained after six hours of binocular experience. When monocular active visual experience is given to kittens previously given a unilateral section of the ophthalmic branch of the trigeminal nerve, these two aspects are affected in that fewer neurones are "captured" by the experienced eye and very few neurones recover their orientation selectivity. These effects are more or less independent of the relative laterality of the proprioceptive and retinal information (Trotter et al., 1983) .

Two major conclusions result from the studies briefly reviewed above. One conclusion is that the remarkable performance of orientation detection precedes the complete formation of the cortical neuropile. It should be recalled that as soon as we were able to record cortical neurones that are visually activated, about half of them were already truly oriented. Specific selective neurones develop from the time of eye opening until the end of the third week, even without visual experience. The second conclusion concerns the role played by the proprioceptive afferents, ensuing from ocular movements, within the processes of maintenance and development of the visual properties of the cortical neurones. Numerous experiments using visual deprivation imply that there is not only a selective bias of the visual input, but also a significant alteration of the normal motor output of the organism during perception. The fact that there exists an interaction between visual and proprioceptive information, which we have demonstrated to occur at a relatively low level of visual information processing, gives a physiological substrate to some theories (Held and Hein, 1963) concerning perception, suggesting that early experience may unite the visual system and the control systems of visually initiated motor responses.

REFERENCES

Albus, K. and Wolf, R., 1984, Early postnatal development of neuronal function in the kitten's visual cortex : a laminar analysis, J. Physiol., London, 348: 153.

Barlow, H.B., 1975, Visual experience and cortical development. Nature, London, 258: 199.

Blakemore, C. and Van Sluyters, R.C., 1975, Innate and environmental factors in the development of the kitten's visual cortex. J. Physiol. London, 248: 663.

Blakemore, C., 1978, Maturation and modification in the developing visual system, in: "Handbook of Sensory Physiology", Springer Verlag, Berlin, Heidelberg, New York, Vol. 8, pp 337.

Bonds, A.B., 1979, Development of orientation tuning in the visual cortex of kittens, in: "Developmental Neurobiology of Vision", R.D. Freeman, ed., Plenum, New York, vol. 27, pp 31.

Buisseret, P., Gary-Bobo, E. and Imbert, M., 1978, Evidence that ocular motility is involved in the recovery of the orientational properties of visual cortical neurones. Nature,London, 272: 816.

Frégnac, Y., 1979, Development of orientation selectivity in the primary visual cortex of normally and dark reared kittens. I- Kinetics. Biol. Cybern., 34: 187.

Frégnac, Y. and Imbert, M., 1984, Development of neuronal selectivity in primary visual cortex of cat. Physiol. Rev., 64: 325.

Held, R. and Hein, A., 1963, Movement-produced stimulation in the development of visually guided behavior. J. Comp. Physiol. Psychol., 56:872.

Hubel, D.H. and Wiesel, T.N., 1970, The period of susceptibility to the physiological effects of unilateral eye closure in kittens. J. Physiol., London, 206: 419.

Imbert, M. and Frégnac, Y., 1983, Specification of cortical neurones by visuomotor experience, in: "Progress in Brain Research", J.P. Changeux et al. (eds. , Vol. 58, pp 427.

Imbert, M. and Buisseret, P., 1975, Receptive field characteristics and plastic properties of visual corteical cells in kittens reared with or without visual experience. Exp. Brain Res., 22: 25.

Roux, W., 1905, Die Entwickelungsmechanick, ein neur Zweig der biologischen Wissenschaft. Leipzig, Englemann.

Sidman, R., 1972, Cell interaction in developing mammalian central nervous system, in: "Cell interaction Proc.", 3rd Lepetit Colloq. Amsterdam, North-Holland, pp 1.

Trotter, Y., Frégnac, Y. and Buisseret, P., 1981, Recovery of orientation selectivity in kitten primary visual cortex is slowed down by bilateral section of ophthalmic trigeminal afferents. Dev. Brain Res,, 1: 450.

Trotter, Y., Frégnac, Y. and Buisseret, P., 1983, Synergie de la vision et de la proprioception extraoculaire dans les mécanismes de plasticité fonctionnelle du cortex visuel primaire du chaton. C.R. Acad. Sci., Série C, 296: 665.

Trotter, Y., Frégnac, Y. and Buisseret, P., 1981, Période de sensibilité du cortex visuel primaire du chat à la suppression unilatérale des afférences proprioceptives extraoculaires. C.R. Acad. Sci., Série D, 293: 245.

Tsumoto, T. and K. Suda, 1982, Laminar differences in development of afferent innervation to striate cortex neurones in kittens. Exp. Brain Res., 45: 433.

CHANGES IN THE AVIAN VISUAL WULST

FOLLOWING EARLY MONOCULAR DEPRIVATION

Paola Bagnoli, Giovanni Casini and Roberto Alesci

Istituto di Fisiologia dell' Università e
Istituto di Neurofisiologia del CNR
via S. Zeno 31-51, 56100 Pisa, Italy

Since the work of Wiesel and Hubel (1963), the mammalian visual system has been extensively used as a model for studying developmental plasticity in the brain. Recently, the avian visual system has also been used for research examining the neuronal correlates of experience-dependent changes. In particular the visual Wulst was shown to be involved in plastic changes occurring as a result of early monocular deprivation (Burkhalter et al, 1982).

Similarities in function between the mammalian striate cortex and the avian Wulst have been proposed (Cohen and Karten, 1974; Hodos, 1976; Pettigrew, 1978; Karten, 1979). In birds, which have a totally crossed retinal projection, binocular convergence takes place in the visual Wulst by means of recrossing fibers at the level of the supraoptic decussation (DSO). Electrophysiological and morphological abnormalities have been observed in the avian Wulst following early monocular deprivation. Indeed, a dramatic loss of binocularly driven neurons was shown by single unit recordings in the Wulst of owls raised with one eye sutured (Pettigrew and Konishi, 1976). In addition, a [^{14}C]-2 Deoxyglucose (DG) study, performed to localize visual structures with altered functional activity, demonstrated that, in monocularly deprived pigeons exposed with both eyes free after DG injection, the visual Wulst was the only structure asymmetrically labeled (Cuenod et al, 1981; Burkhalter et al, 1982). The Wulst appeared more intensely labeled on the side contralateral to the deprived eye as compared to the other side. This asymmetry was shown to depend upon monocular deprivation since it was present irrespective of any acute exposure of the deprived eye to the light during the DG experiment. The altered functional activity in the visual Wulst contralateral to the deprived eye seems to be related to changes in the commissural input from the undeprived eye. Indeed, DSO section 10 days before the DG injection was shown to reduce the DG asymmetry of the visual Wulst (Burkhalter et al, 1982).

At the behavioural level, significant deficits in pattern discrimination learning and interocular transfer have been shown in pigeons monocularly deprived immediately after hatching (Burkhalter and Cuenod, 1978). These behavioural changes were postulated to be the result of synaptic competition between the projections of the two eyes in the visual Wulst because: i) binocular deprivation did not affect pigeon discrimination learning; ii) early DSO section prevented the learning deficits induced by monocular deprivation. Taken together, these results support the hypothesis of functional analogy, including plasticity traits, between the mammalian striate cortex and the avian Wulst.

Recently, we investigated the possible neurochemical changes in the pigeon Wulst associated with monocular deprivation. In particular, we asked the question whether biochemical systems involved in synaptic transmission might be modulated by altered environmental conditions during development.

The presence in the Wulst of cholinergic and GABA-ergic systems has been suggested (Vischer et al, 1982). A high density of quinuclidinyl benzilate (QNB) binding sites and a medium level of choline acetyltransferase (ChAT) activity have been found in the hyperstriatal laminae. In addition, a low concentration of γ-aminobutyric acid (GABA) and a low glutamate decarboxylase (GAD) activity were measured in the hyperstriatum. In contrast, a high density of muscimol binding sites was found. This low ratio of GABA and GAD activity versus binding sites favors the presence of Wulst afferents or interneurons versus efferents using GABA as transmitter.

Evidence that thalamic afferents might provide cholinergic and GABA-ergic input comes from a lesion study of the dorsolateral visual thalamus (Henke, 1981). Indeed, ChAT activity, GABA pool and GAD activity were reduced at specific locations one month after kainate injection or electrolytic lesion in the dorsolateral thalamic nuclei.

To further investigate the origin of cholinergic and GABA-ergic projections to the visual Wulst, transmitted related retrograde tracing was used (Cuenod et al, 1982). We investigated the possibility that cholinergic and GABA-ergic pathways might be selectively labeled after injection of their radiolabeled putative transmitters in the terminal field. A selective retrograde transport of radioactivity to the nucleus dorsolateralis anterior thalami, pars magnocellularis (DLAmc) has been observed after injection of [^{3}H] choline into the Wulst (Bagnoli et al, 1981). On the contrary, [^{3}H] GABA injection into the same region led to perikaryal labeling of neuronal populations mainly confined to the ipsilateral thalamic visual relay, nucleus dorsolateralis anterior thalami, pars lateralis ventralis (DLLv) (Bagnoli et al, 1983a). None of the other regions which were retrogradely labeled after Wulst HRP injections (Bagnoli and Burkhalter, 1983) showed perikaryal labeling. The fact that only restricted neuronal populations in the dorsolateral thalamus can be labeled after either [^{3}H] GABA or [^{3}H] choline injections supports the argument of a specific terminal uptake of labeled material.

The biochemical and neuroanatomical evidence described above suggests the presence of cholinergic and GABA-ergic projections from the dorsolateral thalamus to the visual Wulst in the pigeon. The cholinergic and GABA-ergic systems indicated by these results were shown to be affected by manipulating visual input. Indeed, following early monocular deprivation, ChAT activity increased in restricted areas of the visual Wulst contralateral to the deprived eye, whereas GAD activity decreased bilaterally (Bagnoli et al, 1982). Whether this higher ChAT activity might be related to functional modifications in the Wulst as described by the DG method, and/or to the behavioural deficits of monocularly deprived pigeons is unknown. It is also unknown whether these effects on ChAT activity might be attributed to a cholinergic projection from the dorsolateral thalamus to the visual Wulst or to other hyperstriatal cholinergic systems.

In addition to acetylcholine (ACh) and GABA, a third transmitter substance has been localized in the Wulst, whose level was also shown to be affected by manipulations during early development. Dense catecholaminergic (CA) innervation in the Wulst has been recently demonstrated by means of glyoxylic-acid histofluorescence (Bagnoli and Casini, 1984). In contrast to the uniform pattern of CA innervation of the somatosensory anterior Wulst, the visual portion shows regional variations in pattern and density of CA terminals. This organization fits well with the distribution of monoaminergic (MA) afferents in the mammalian striate cortex where they exibit a laminar termination pattern (Foote et al, 1983). The presence of dense CA afferents in those hyperstriatal laminae which receive visual input suggests a possible convergence of CA terminals and visual fibers on common target cells in the Wulst.

Although glyoxylic-acid tecnique does not permit the distinction of dopamine from norepinephrine (NE), additional evidence suggests that NE is the most likely transmitter. Moderate levels of NE (Bagnoli et al, 1983b) and dihydroalprenol binding sites have been found in the Wulst (Vischer et al, 1982). On the contrary, among the visual areas, the Wulst was found to contain the lowest level of spiroperidol binding (Vischer et al, 1982).

The noradrenergic innervation to the Wulst presumably arises from NE containing neurons of the Locus coeruleus (LoC) since NE content in the Wulst was shown to be drastically reduced by electrolytic lesion of the pontine tegmentum (Bagnoli et al, 1983b).

We studied the development of noradrenergic systems in the Wulst by measuring the endogenous level of NE at different posthatching times (Bagnoli et al, 1983b). White and black columns in fig. 1 refer to endogenous NE levels in the left and right Wulst of pigeons at different developmental stages. A high level of NE was measured immediately after hatching. This content was shown to significantly decrease by 45% in the first 6 days posthatching to reach a level similar to that measured in the adult.

A similar maturation curve was reported in the rat where an early peak of endogenous NE observed in the neocortex a few days after birth was followed by a drop at around 1 week of age (Levitt and Moore, 1979). This drop was interpreted to be the result of a decrease in the density of noradrenergic terminals induced by the formation of more numerous synapses of some other type. In particular, the decrease in the NE content in the pigeon Wulst which was shown to occur at 6 days posthatching, might reflect a decrease in the density of noradrenergic terminals induced by the activation of visual projections to the Wulst. Indeed,photosensitive lamellae in the retinal outer segments can first be observed at around 4-6 days posthatching (Bagnoli et al, in press). At the same time the first retinal responses to light can be recorded (Bagnoli et al, 1983c). This temporal correlation between the reduction in the NE content and the activation of visual afferents to the Wulst suggests a causal relationship between the two systems.

In adult animals, the monoamine level of brain regions has been shown to reflect the density of MA nerve terminals (Sachs and Jonsson, 1973). How the monoamine content of a developing system might be related to fiber density, however, remains unclear. On the one hand, axonal populations in neonatal cortex have been shown to store large quantities of exogenous catecholamines (Levitt and Moore, 1979), thus causing an overestimation of real terminal density. On the other hand, developing MA terminals have been shown to contain a lower transmitter content than adult MA axons due to poorer storage capacity of the transmitter (Jonsson and Sachs, 1972; Levitt and Moore, 1979). In any event, dense MA terminals have been identified in the neocortex during the late stage of embryogenesis either by biochemical assays or fluorescence histochemistry (Coyle and Molliver, 1976; Lidov et al, 1978).

The early development of MA systems has led to speculation that it may serve a regulatory function in brain development. In particular, MA afferents are established in the rat cortex before the geniculo-cortical pathway and callosal projections (Schlumpf et al, 1980). In addition, removal of noradrenergic input to the neocortex was shown to result in a greater number of synapses during early life and thus in the earlier development of cortical

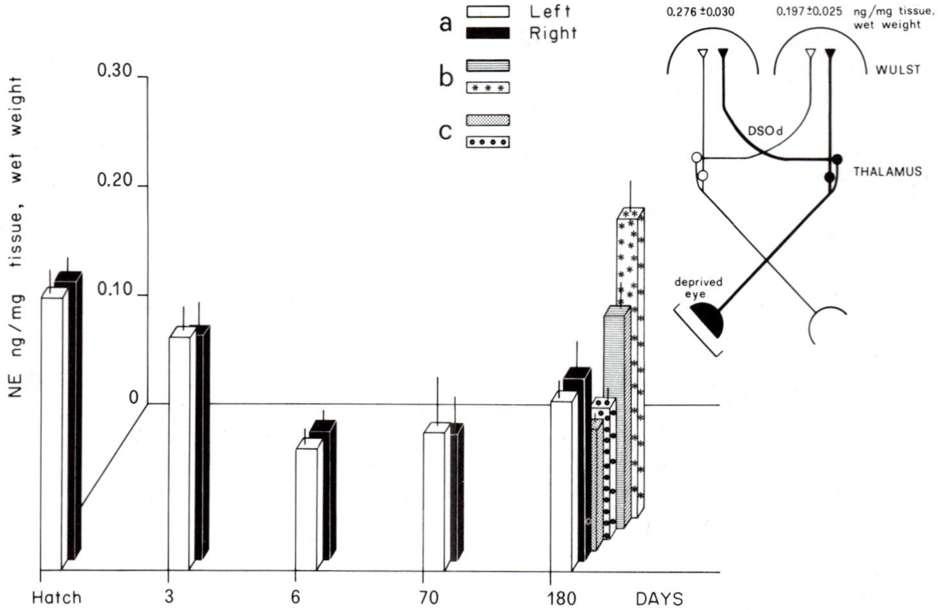

Fig. 1. Endogenous NE levels in the pigeon Wulst as determined by the Renzini's two step chromatographic trihydoxindole fluorimetric assay (Renzini et al, 1970; Ghione et al, 1979; Pellegrini, 1982). Wulst samples included hyperstriatum accessorium, nucleus intercalatus hyperstriati accessorii and hyperstriatum intercalatus superior according to the nomenclature of Karten and Hodos (1967). The white and black columns refer to the endogenous NE levels in the left and right Wulst of undeprived pigeons at different developmental stages. The striped and asterisked columns refer to the NE contents in the Wulst contralateral and ipsilateral to the deprived eye in early monocularly deprived pigeons (180 days old, right eye covered) respectively. The stippled and circled columns refer to the NE levels in the Wulst contralateral and ipsilateral to the deprived eye in adult monocularly deprived pigeons (180 days old, right eye covered), respectively. The results are expressed as mean values (ng/mg tissue, wet weight) ± S.E.M. of NE contents. In the insert: schematic diagram of the retino-thalamo-hyperstriatal pathway. Numbers above the Wulst refer to the NE content in early monocularly deprived pigeons.

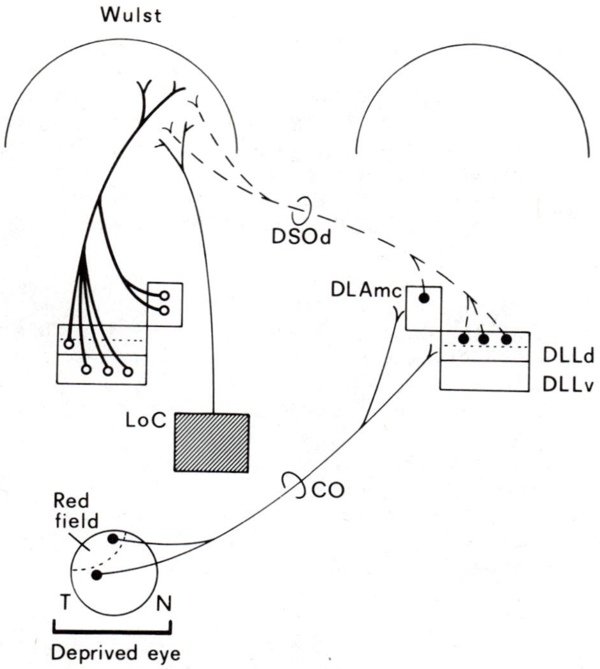

Fig. 2. Schematic representation of the retino-thalamo-hyperstriatal pathway and noradrenergic afferents from the LoC. Visual afferents from the temporal retina (T) of the deprived eye cross the midline in the CO and terminate in the dorsal thalamic nuclei, predominantly in DLAmc and dorsal division of DLL (right side: filled dots). Commissural fibers from visual thalamic nuclei cross the midline at the level of DSOd and terminate in the Wulst ipsilateral to the deprived eye. Open dots on the left side represent the thalamic origin of the ipsilateral visual pathway to the Wulst. Abbreviations: CO=Chiasma opticum; DLLv=Nucleus dorsolateralis anterior thalami, pars lateralis ventralis; DLLd=Nucleus dorsolateralis anterior thalami, pars lateralis dorsalis; DLAmc=Nucleus dorsolateralis anterior thalami, pars magnocellularis; DSOd=Decussatio supraoptica dorsalis; LoC=Locus coeruleus; N=Nasal retina.

connections. It has been suggested that the early and transient presence of noradrenergic boutons might trigger a form of postsynaptic specialization which would later receive specific thalamic terminations (Blue and Parnavelas, 1982).

The presence of dense noradrenergic terminals in the pigeon visual Wulst which largely overlap visual afferents either during the developmental or adult stage prompted us to investigate the effect of monocular deprivation on the Wulst noradrenergic system. Our results demonstrated that noradrenergic systems in the visual Wulst are affected by monocular deprivation during development (Bagnoli et al, 1983b). Early, monocular deprivation resulted in an asymmetry of the NE content in the Wulst, the side ipsilateral to the deprived eye (asterisked column in fig. 1) displaying higher levels than the contralateral one (striped column in fig. 1). This effect seems to be induced by monocular deprivation only during development because monocularly deprived adults (circled column in fig. 1) displayed no significant alteration in the NE content of the Wulst as compared to normal animals (stippled column in fig. 1).

Visual experience of an imprinting stimulus was shown to increase NE levels in 3 regions of the chick telencephalon: i) the Wulst; ii) the intermediate part of the medial hyperstriatum ventrale; and iii) the paleostriatum augmentatum (Davies et al, 1983). In addition, early visual deprivation has been reported to influence either the activities of enzymes involved in amine metabolism (Bigl et al, 1974; Biesold et al, 1977) or ß-adrenergic receptors (Schliebs et al, 1982) in some visual areas of the rat. Moreover, plastic changes of the LoC projection were shown to occur in rat visual centers following early eye removal (Nakamura et al, 1984). Indeed, an increase in the density of the projection from the LoC to the lateral geniculate nucleus and visual cortex was found in rat with an eye removed at 30 days.

In our model, the increase in the NE content of the pigeon Wulst ipsilateral to the deprived eye might be explained by assuming that noradrenergic fibers and visual afferents compete during development for the same terminal field. The commissural system which crosses in the DSO and connects the deprived eye to the ipsilateral Wulst seems to be more sensitive to early visual deprivation and appears similar to the uncrossed optic fibers in mammals (Sherman and Spear, 1982). Indeed, as shown in fig. 2, these recrossed fibers convey to the Wulst information from the frontal and binocular visual field and early DSO section was shown to prevent both discrimination deficits of the deprived eye and the DG asymmetry induced in the Wulst by monocular deprivation (Burkhalter et al, 1982; Burkhalter and Cuenod, 1978). A reduced functional activity of the commissural system following early visual deprivation might prevent the normal maturation of noradrenergic fibers. The regression of noradrenergic terminals might not take place and the NE level in the Wulst of adult pigeons monocularly deprived at hatching should be similar to that measured in the Wulst of newly hatched animals.

Taken together, the results presented here point to the avian visual Wulst as a structure affected by early monocular deprivation at the cellular and metabolic level.

REFERENCES

Bagnoli, P., and Burkhalter, A., 1983, Organization of the afferent projection to pigeon Wulst, J. Comp. Neurol., 214:103.

Bagnoli, P., and Casini, G., 1984, Regional distribution of catecholaminergic terminals in the pigeon visual system, Brain Res., in press.

Bagnoli, P., Beaudet, A., Stella, M., and Cuénod, M., 1981, Selective retrograde labeling of cholinergic neurons with [^{3}H]-choline, J. Neurosci. 1:691.

Bagnoli, P., Burkhalter, A., Vischer, A., Henke, H., and Cuénod, M., 1982, Effects of early monocular deprivation on choline acetyltransferase and glutamic acid decarboxylase in pigeon visual Wulst, Brain Res., 247:289.

Bagnoli, P., Burkhalter, A., Streit, P., and Cuénod, M., 1983a, [^{3}H]-GABA selective retrograde labeling of neurons in the pigeon thalamo-Wulst

pathway, Arch. Ital. Biol., 121:47.
Bagnoli, P., Barsellotti, R., Pellegrini, M., and Alesci, R., 1983b, Norepinephrine level in developing pigeon brain: effect of monocular deprivation on the Wulst noradrenergic system, Develop. Brain Res., 10:243.
Bagnoli, P., Porciatti, V., and Barsellotti, R., 1983c, The development of spatial resolution in the pigeon retina, Perception, 12:A20.
Bagnoli, P., Porciatti, V., Lanfranchi, A., and Bedini, C., 1984, Developing pigeon retina: light evoked responses and ultrastucture of outer segments and synapses, J. Comp. Neurol., in press.
Biesold, D., Bigl, V., and Uzbekov, M., 1977, Effects of sensory deprivation upon the development of transmitter systems in the visual system, Act. Nerv. Super. (Praha), 19:154.
Bigl, V., Biesold, D., and Weisz, K., 1974, The influence of functional alteration on monoamine oxidase and catecol-O-methyltransferase in the visual pathway of rats, J. Neurochem;, 22:509.
Blue, M.E., and Pernavelas, J.G., 1982, The effect of neonatal 6-Hydroxydopamine treatment on synaptogenesis in the visual cortex of the rat, J. Comp. Neurol., 205:199.
Burkhalter, A., and Cuenod, M., 1978, Changes in pattern discrimination learning induced by visual deprivation in normal and commissurotomized pigeons, Exp. Brain Res., 31:369.
Burkhalter, A., Streit, P., Bagnoli, P., Vischer, A., Henke, H., and Cuenod, M., 1982, Deprivation induced functional modifications in the pigeon visual system, in "Neuronal Plasticity and Memory Formation", IBRO Monograph Series Vol. 9, C. Aimone Marsan and H. Matties, eds., Raven Press, N.Y.
Cohen, D.H., and Karten, H.J., 1974, The structural organization of avian brain: an overview, in "Birds Brain and Behaviour", I.J. Goodman and M.W. Schein, eds., Academic Press, N.Y.
Coyle, J.T., and Molliver, M.E., 1976, Major innervation of newborn rat cortex by monoaminergic neurons, Science, 196:444.
Cuénod, M., Bagnoli, P., Beaudet, A., Burkhalter, A., Henke, H., Knüsel, B., and Vischer, A., 1981, Behavioural and biochemical changes induced by early monocular deprivation in the pigeon, Doc. Ophthalmol. Proc. Ser., 30:230.
Cuénod, M., Bagnoli, P., Beaudet, A., Rustioni, A., Wiklund, L., and Streit, P., 1982, Transmitter-specific retrograde labeling of neurons, in: "Cytochemical Methods in Neuroanatomy", S.L. Palay and V. Chan-Palay eds., A.R. Liss, Inc., N.Y.
Davies, D.C., Horn, G., and McCabe, B.J., 1983, Changes in telencephalic catecholamine levels in the domestic chick. Effect of age and visual experience. Develop. Brain Res., 10:251.
Foote, S.L., Bloom, F.E., and Aston-Jones, G., 1983, Nucleus Locus coeruleus: new evidence of anatomical and physiological specificity, Phys. Rev., 63:844.
Ghione, S., Pellegrini, M., and Clerico, A., 1979, A sensitive and accurate plasma noradrenaline assay using a modified fluorimetric method, Clin. Chim. Acta, 97:89.
Henke, H., 1981, "The central part of the avian visual system", Ph. D. Diss. Univ. of Zürich, Switzerland.
Hodos, W., 1976, Vision and the visual system: a bird's eye view, in: "Progress in Psychobiology and Physiological Psychology", Vol.6, J.M. Sprague and A.N. Epstein, eds., Academic Press, N.Y.
Jonsson, G., and Sachs, C., 1972, Neurochemical properties of adrenergic nerves regenerated after 6-hydroxydopamine, J. Neurochem., 19:2577.
Karten, H.J., 1979, Visual lemniscal pathways in birds, in: "Neural Mechanisms of Behaviour in the Pigeon", A.M. Granda and J.H. Maxwell, eds., Plenum Press, N.Y.
Karten, H.J., and Hodos, W.A., 1967, " A Stereotaxic Atlas of the Brain of the Pigeon (Columba Livia)", John Hopkins Press, Baltimore.

Levitt, P., and Moore, R.Y., 1979, Development of the noradrenergic innervation of neocortex, Brain Res., 162:243.
Lidov, H.G.W., Molliver, M.E., and Zecevic, N.R., 1978, Characterization of the monoaminergic innervation of immature rat neocortex: a histofluorescence analysis, J. Comp. Neurol., 181:663.
Nakamura, S., Shirokawa, T., and Sakaguchi, T., 1984, Increased projection from the Locus coeruleus to the lateral geniculate nucleus and visual cortex in young and adult rats following unilateral enucleation, Neurosci. Lett., 49:77.
Pellegrini, M., 1982, Description of a rapid fluorimetric method for the determination of catecholamines in plasma (90 min.), Elettromedicali e Strumentazione Scientifica, 11/12:55.
Pettigrew, J.D., 1978, Comparison of the retinotopic prganization of the visual Wulst in nocturnal and diurnal raptors, with a note on the evolution of frontal vision, in: "Frontiers in Visual Science", S.F. Cool and E.I. Smith, eds., Springer Verlag, N.Y.
Pettigrew, J.D., and Konishi, M., 1976, Effect of monocular deprivation on binocular neurons in the owl's visual Wulst, Nature, 264:753.
Renzini, V., Brunori, C.A., and Valori, C., 1970, A sensitive and specific fluorimetric method for the determination of noradrenaline and adrenaline in human plasma, Clin. Chim; Acta, 30:587.
Sachs, C., and Jonsson, G., 1973, Quantitative microfluorimetric and neurochemical studies on adrenergic nerves after axotomy, J. Histochem. Cytochem;, 21:902.
Schliebs, R., Burgoyne, R.D., and Bigl, V., 1982, The effects of visual deprivation on ß-adrenergic receptors in the visual centres of the rat brain, J. Neurochem;, 38:1038.
Schlumpf, M., Lichtensteiger, W., Shoemaker, W., and Bloom, F.E., 1980, Fetal monoamine systems: early stages and cortical projections, in: "Biogenic Amines in Development", H. Parvez and S. Parvez, eds., Elsevier/North-Holland Biomed. Press, Amsterdam.
Vischer, A., Henke, H., and Cuenod, M., 1982, Neurotransmitter receptor ligand binding and enzyme regional distribution in the pigeon, J. Neurochem., 38:1372.
Wiesel, T.N., and Hubel, D.H., 1963, Single-cell responses in striate cortex of kittens deprived of vision in one eye, J. Neurophysiol., 26:1003.

ARE BETA ADRENORECEPTORS INVOLVED IN VISUOCORTICAL PLASTICITY?

T. Kasamatsu and T. Shirokawa*

Smith-Kettlewell Institute of Visual Sciences
2200 Webster Street
San Francisco, CA 94115
and
School of Optometry, University of California*
Berkeley, California 94720

In everyday life we adults learn many things, usually with great effort. On the other hand, children sometimes learn a few things very quickly without much effort, performing far better than adults. A good example is learning a spoken language which is different from one's mother tongue. What can account for this age-dependent difference in learning? What makes performance of the immature brain more efficient than that of the mature one? There is no reason at the moment to believe the presence of totally independent neuronal machinery operating in the two brains. It seems to be practical, however, to treat them separately and devise paradigms which may test likely mechanisms or factors at the cellular level unique to each experimental condition. Thus, changes that occur in neuronal connections in the young brain during early postnatal or posthatching life, namely, critical period plasticity (e.g. Erzurumlu and Killackey, 1982; Kasamatsu, 1983), may be understood as an example of "involuntary learning" by an immature neuronal network during a process of its self-organization. This type of learning is possible for only young individuals and totally depends, within the ground framework defined by genetic constraint, on the sum of "experience" received by them. One cannot help drawing an analogy between critical period plasticity and state-dependent learning in adults. The particular strength of our paradigm is, as will be shown below, that neuronal changes induced in early days may be later studied by a straightforward application of single-unit recordings, since the changes in neuronal connections last long beyond the duration of altered experience, showing little decay under usual conditions.

For the last several years, we have been studying the roles of the noradrenaline (NA) system in the brain for the regulation of critical period plasticity, using binocular cells in the cat visual cortex as a model system (e.g., Kasamatsu, 1982 a, b; 1983; Kasamatsu et al., 1984 for review). Other contributors to this volume review the involvement of the central NA system in brain plasticity, learning and memory in much wider scope than here. Several reviews are available on the topic of neuronal plasticity in the visual cortex (e.g. Hubel et al., 1977; Pettigrew, 1978; Blakemore, 1978; Movshon and Van Sluyters, 1981; Sherman and Spear, 1982; Frégnac and Imbert, 1984).

NA HYPOTHESIS

We studied changes in ocular dominance which usually follow brief monocular eyelid suture in young kittens (Wiesel and Hubel, 1963; Hubel and Wiesel, 1970) using the following two indices: (1) shift in the distribution pattern of cells having different ocular dominance toward the open, used eye and (2) decrease in the proportion of binocularly driven cells. Our early results showed that a normally functioning catecholamine system, especially NA terminals within the visual cortex, is needed for the effects of monocular deprivation to be expressed in the immature visual cortex (Kasamatsu and Pettigrew, 1976; 1979; Kasamatsu et al., 1979; Pettigrew and Kasamatsu, 1978). Accordingly, we have proposed that an intact NA system is necessary for maintaining the high level of visuocortical plasticity which is typically seen in the immature neocortex of young kittens.

This NA hypothesis, however, has been challenged recently by a few groups of investigators who claimed that the shift in ocular dominance was observed in the visual cortex whose NA content was later proved to be significantly lower than normal (Adrien et al., 1982; Bear and Daniels, 1983; Bear et al., 1983; Daw et al., 1983; 1984). They used various ways other than our original methods of direct perfusion of the visual cortex with an NA-related neurotoxin, 6-hydroxydopamine (6-OHDA) (e.g. Kostrzewa and Jacobowitz, 1974; Jonsson, 1980), to reduce the endogenous content of NA in the kitten brain. They injected 6-OHDA into either the lateral ventricle or the locus coeruleus (LC) area of several week-old kittens (Adrien et al., 1982), the peritoneal space of newborn kittens (Bear and Daniels, 1983; Bear et al., 1983), or the lateral hypothalamus (Daw et al., 1983; 1984). The last authors also placed massive electrolytic lesions at one side of the lateral hypothalamus including the ascending bundle of the NA projection. It should be noted that the consequences of the treatments in their studies were not necessarily confined to the visual cortex where the primary effects of monocular deprivation are present. In their studies physiological assays of plasticity were carried out a few weeks to months after NA depletion which renders it, in our opinion, particularly problematic for interpreting accurately their findings.

RESEARCH STRATEGIES

There are basically two ways to clarify this conceptual confusion which has been caused by a flood of "negative" results obtained by others: (1) to examine directly whether the logic behind the main theme of our NA hypothesis for visuocortical plasticity itself is ungrounded and (2) to devise studies that may decide, one way or the other, which interpretation of the "conflicting" results is incorrect or at least insufficient. Knowing that the two approaches are both needed and not mutually exclusive of each other, we decided at this time to address the first of the above alternatives, which seems to be more urgent, constructive and rewarding. Accordingly, we would like to review below two sets of recent results, which we believe not only have restored our faith in the original NA hypothesis but also advanced it further by suggesting the involvement of beta adrenoreceptors as a next step in the cascade of cellular events underlying visuocortical plasticity.

1) Evidence for Involvement of Beta Adrenoreceptors

Provided that the original NA hypothesis is basically correct, we may now ask about types of NA-related receptors involved in visuocortical plasticity. We perfused, directly and continuously, the visual cortex of 5- to 7-week-old kittens with beta-adrenergic receptor antagonists such as propranolol and sotalol in conjunction with monocular lid suture which lasted for a week. We found many binocular cells in the beta-blocker-perfused cortex of monocularly deprived kittens (Kasamatsu, 1979; Shirokawa and Kasamatsu, 1984). The results were interpreted as suggesting that the cortex perfused with a beta blocker failed to register the usual changes in ocularity due to a lack of plasticity. We further noted that in monocularly deprived kittens the predominance of binocular cells was maintained in proportion to the logarithmic concentration of the two beta blockers perfused into the visual cortex (Shirokawa and Kasamatsu, in preparation). The present results were not only essentially consistent with our original findings with 6-OHDA but also indicated that the activation of beta-adrenergic receptors is crucial in the visuocortical plasticity.

In a straightforward biochemical study, we perfused the cortex with ^{3}H-propranolol to examine the spatial distribution of the recovered radioactivity in the cortex at the end of the perfusion which lasted for a week. We obtained a nearly exponential decline in radioactivity with increasing distance from the site of perfusion. This curve allowed us to calculate the "average dilution factor" of propranolol at a given site within the perfused cortex. The dilution factor was about 1/170 in an area ~ 2 mm from the perfusion site and about 1/1,000 at ~ 3 mm, where the radioactivity levelled off. Applying this dilution factor of 1/170, we estimated the approximate concentration of propranolol needed at the recording site for the half-maximum effect of blocking the expected shift in ocular dominance as 5.8×10^{-7} M or less (Shirokawa and Kasamatsu, in preparation). Essentially the same results were obtained with sotalol, except for the fact that maximum binocularity attained with sotalol (~40%) was lower than that with propranolol (67%) at the highest concentration used in the present study (10^{-2} M in a minipump). Needless to say, the possibility that a "local anesthetic effect" of propranolol might have caused the state of reduced binocularity as typically seen in the cortex of binocularly deprived cats was ruled out in the present study.

2) Effects of Electrical Stimulation of the Locus Coeruleus

Despite the arguments against the NA hypothesis based on "negative" results mentioned above, the fact remains that exogenous NA directly perfused into the aplastic cortex of either 6-OHDA-pretreated kittens or adult cats restored susceptibility of the visual cortex to monocular vision (Pettigrew and Kasamatsu, 1978; Kasamatsu et al., 1979). This fact is hardly explained, without referring to the NA system, by merely assuming nonspecific effects of 6-OHDA in the visual cortex, unless one also assumes equally nonspecific action of NA which somehow counteracts unspecified and nonspecific effects of 6-OHDA. Especially, the restoration of the plasticity by NA to the aplastic, adult cortex is harder to be refuted than anything mentioned above, since in this paradigm no pretreatments with toxins were involved.

We further extended this line of research to study the role of endogenous NA released from NA terminals within the visual cortex, leaving the NA terminals and related receptors in the cortex intact. By electrically stimulating the NA-containing somata of the NA cells in the cat LC (Watabe et al., 1982), we attempted to restore in the mature visual cortex the susceptibility to brief monocular deprivation which is normally found in only the young visual cortex. During brief monocular vision the LC was stimulated with a short train of square pulses, which mimicked low spontaneous activity of LC cells. Animals were kept in the dark at all other times (light/dark = 2/22). The daily monocular exposure was repeated for 6 consecutive days. This rearing condition by itself, without accompanying LC stimulation, did not cause any change in ocular dominance. When the LC stimulation was combined with monocular exposure, however, we found on the seventh day a significant decrease in the proportion of binocular cells in the mature cortex (Kasamatsu et al., 1982; 1985). The extent of this reduction of binocularity was comparable with that previously obtained by exogenous NA in the adult cortex over a 1-3 week period of monocular lid suture (Kasamatsu et al., 1979). We also showed that the decrease in binocularity was not attainable under comparable conditions if the cortex had been perfused with 6-OHDA to destroy the NA terminals prior to the start of the LC stimulation (Kasamatsu et al., 1985).

The present results suggest that the plasticity was partially restored by endogenous NA released from the NA terminals within the visual cortex in response to electrical activation of the somata of NA cells in the LC. Although we are short of direct evidence for release of endogenous NA upon electrical stimulation of the LC in cats, such evidence has been available in the rat neocortex under comparable conditions (Korf et al., 1973; Tanaka et al., 1976).

CONCLUSIONS

The two studies described above have not only provided further support for our NA hypothesis but also clearly extended it by showing the involvement of beta-adrenergic receptors, as a next step in the cascade of physiological actions within the visual cortex in the expression of visuocortical plasticity.

Assuming that the NA hypothesis is valid, the current confusion caused by several "negative" results may be resolved as follows: the significant decrease in endogenous NA in the visual cortex accompanied by that in other regions of the brain, was by itself not sufficient to indicate a dysfunction of the NA system in the visual cortex. This explanation is, in fact, consistent with our recent finding that beta adrenoreceptor binding showed a high plateau during the postnatal critical period, while endogenous NA itself continually increased toward adulthood (Jonsson and Kasamatsu, 1983).

The "negative" results thus contributed to a shift in our quest for the biochemical processes that underly visuocortical plasticity to examine the degree of <u>activated</u> beta adrenoreceptors instead of examining the absolute amount of endogenous NA. It remains to be established, however, whether the NA-beta adrenoreceptor system is only one of many factors that are needed to modulate the plasticity rather than the indispensable factor.

ACKNOWLEDGEMENTS

We thank Ms. S. Ovington, Ms. M. King & Ms. N. Post for their excellent secretarial service. Supported by USPHS grant EY-05549. TK's trip to Strasbourg to participate in the Conference was made possible through a fund from the Smith-Kettlewell Eye Research Foundation.

REFERENCES

Adrien, J., Buisseret, P., Frégnac, Y., Gary-Bobo, E., Imbert, M., Tassin, J.-P., and Trotter, Y., 1982, Noradrénaline et plasticité du cortex visuel du chaton: un réexamen, <u>C.R. Acad. Sci. Paris</u>, 295:745.

Bear, M. F., and Daniels, J. D., 1983, The plastic reponse to monocular deprivation persists in kitten visual cortex after chronic depletion of norepinephrine, J. Neurosci., 3:407.

Bear, M. F., Paradiso, M. A., Schwartz, M., Nelson, S. B., Carnes, K. M., and Daniels, J. D., 1983, Two methods of catecholamine depletion in kitten visual cortex yield different effects on plasticity, Nature, 302:245.

Blakemore, C., 1978, Maturation and modification in the developing visual system, in: "Handbook of Sensory Physiology VIII: Perception", H. W. Leibowitz and H.-L. Teuber, eds., Springer-Verlag, Berlin.

Daw, N. W., Rader, R. K., Robertson, T. W., and Videen, T. O., 1983, Do short term and long term depletion of noradrenaline have different effects on visual deprivation in the kitten visual cortex?, Soc. Neurosci. Abstr., 9:1217.

Daw, N. W., Robertson, T. W., Rader, R. K., Videen, T. O., and Coscia, C. J., 1984, Substantial reduction of cortical noradrenaline by lesions of adrenergic pathways does not prevent effects of monocular deprivation, J. Neurosci., 4:1354

Erzurumlu, R. S., and Killackey, H. P., 1982, Critical and sensitive periods in neurobiology, in: "Current Topics in Developmental Biology", A. A. Moscona, A. Monroy, and R. K. Hunt, eds., Academic Press, New York.

Frégnac, Y., and Imbert, M., 1984, Development of neuronal selectivity in primary visual cortex of cat, Physiol. Rev., 64:325.

Hubel, D. H., and Wiesel, T. N., 1970, The period of susceptibility to the physiological effects of unilateral eye closure in kittens, J. Physiol. Lond., 206:419.

Hubel, D. H., Wiesel, T. N., and Le Vay, S., 1977, Plasticity of ocular dominance columns in monkey striate cortex, Phil. Trans. R. Soc. Lond. B., 278:377.

Jonsson, G., 1980, Chemical neurotoxins as denervation tools in neurobiology, Ann. Rev. Neurosci., 3:169.

Jonsson, G., and Kasamatsu, T., 1983, Maturation of monoamine neurotransmitters and receptors in cat occipital cortex during postnatal critical period, Exp. Brain Res., 50:449.

Kasamatsu, T., 1979, Involvement of the β-adrenergic receptor in cortical plasticity, ARVO Abstr. Suppl. Invest. Ophth. Vis. Sci., 18:135.

Kasamatsu, T., 1982a, A role of the central norepinephrine system in regulation of neuronal plasticity in cat visual cortex, Biomed. Res. Suppl. 3, in: "Neurotransmitters in the Retina and the Visual Centers", A. Kaneko, N. Tsukahara, and K. Uchizono, eds., Biomedical Research Foundation, Tokyo.

Kasamatsu, T., 1982b, Enhancement of neuronal plasticity by activating the norepinephrine system in the brain: A remedy for amblyopia, Human Neurobiol., 1:49.

Kasamatsu, T., 1983, Neuronal plasticity maintained by the central norepinephrine system in the cat visual cortex, in: "Progress in Psychobiology and Physiological Psychology", J. M. Sprague, and A. N. Epstein, eds., Academic Press, New York.

Kasamatsu, T., Itakura, T., Jonsson, G., Heggelund, P., Pettigrew, J. D., Nakai, K., Watabe, K., Kuppermann, B. D., and Ary, M., 1984, Neuronal plasticity in cat visual cortex: A proposed role for the central noradrenaline system, in: "Monoamine Innervation of Cerebral Cortex", L. Descarries, T. Reader, and H. H. Jasper, eds., Alan R. Liss, Inc., New York.

Kasamatsu, T., and Pettigrew, J. D., 1976, Depletion of brain catecholamines: failure of ocular dominance shift after monocular occlusion in kittens, Science, 194:206.

Kasamatsu, T., and Pettigrew, J. D., 1979, Preservation of binocularity after monocular deprivation in the striate cortex of kittens treated with 6-hydroxydopamine, J. Comp. Neurol., 185:139.

Kasamatsu, T., Pettigrew, J. D., and Ary, M., 1979, Restoration of visual cortical plasticity by local microperfusion of norepinephrine, J. Comp. Neurol., 185:163.

Kasamatsu, T., Watabe, K., Heggelund, P., and Schöller, E., 1985, Plasticity in the cat visual cortex restored by electrical stimulation of the locus coeruleus, Neuroscience Res., (in press).

Kasamatsu, T., Watabe, K., Schöller, E., and Heggelund, P., 1982, Activation of the central noradrenaline system: A remedy for experimental amblyopia, First World Cong. IBRO Abstr., Neurosci. Suppl., 7:S113.

Kostrzewa, R. M., and Jacobowitz, D. M., 1974, Pharmacological actions of 6-hydroxydopamine, Pharmacol. Rev., 62:199.

Korf, J., Roth, R. H., and Aghajanian, G. K., 1973, Alterations in turnover and endogenous levels of norepinephrine in cerebral cortex following electrical stimulation and acute axotomy of cerebral noradrenergic pathways, Eur. J. Pharmacol., 23:276.

Movshon, J. A., and Van Sluyters, R. C., 1981, Visual Neural Development, Ann. Rev. Psychol., 32: 477-522.

Pettigrew, J. D., 1978, The paradox of the critical period of striate cortex, in: "Neural Plasticity", C. W. Cotman, ed., Raven Press, New York.

Pettigrew, J. D., and Kasamatsu, T., 1978, Local perfusion of noradrenaline maintains visual cortical plasticity, Nature, 271:761.

Sherman, S. M., and Spear, P. D., 1982, Organization of visual pathways in normal and visually deprived cats, Physiol. Rev., 62:738.

Shirokawa, T., and Kasamatsu, T., 1984, β-Adreneragic receptor mediates neuronal plasticity in visual cortex, ARVO Abstr. Suppl. Invest. Ophth. Vis. Sci., 25:214.

Tanaka, C., Inagaki, C., and Fujiwara, H., 1976, Labelled noradrenaline release from rat cerebral cortex following electrical stimulation of locus coeruleus, Brain Res., 106:384.

Watabe, K., Nakai, K., and Kasamatsu, T., 1982, Visual afferents to norepinephrine-containing neurons in cat locus coeruleus, Exp. Brain Res., 48:66.

Wiesel, T. N., and Hubel, D. H., 1963, Single-cell responses in striate cortex of kittens deprived of vision in one eye, J. Neurophysiol., 26:1003.

ROLE OF THE LOCUS COERULEUS SYSTEM IN BEHAVIORAL PLASTICITY

Lydia Velley (1), Eliane Kempf (2), Jeanne Velly (3) and Bernard Cardo (1)

(1) Laboratoire de Psychophysiologie, Univ. Bordeaux I, Avenue Facultés, 33405 Talence France, (2) Centre Neurochimie, 5 Rue Blaise Pascal, 67084 Strasbourg France (3) Institut Pharmacologie, 11 Rue Human 67000 Strasbourg

INTRODUCTION

In previous studies we showed that electrical stimulation of the lateral hypothalamus of 15 day old rats significantly enhanced the learning of various tasks examined 4 weeks later. We first demonstrated that this stimulation improved the acquisition and extinction of a food-reinforced task (Velley and Cardo, 1977). In subsequent experiments, we observed the same effect on the acquisition and reversal of a light-dark discrimination whether appetitively or aversively reinforced as well as on the acquisition of a one way avoidance task (Velley et al., 1981b).

The next step in these investigations was to determine whether the same improvement could be obtained after stimulation of other brain regions, or if the phenomenon was limited to the lateral hypothalamus. We found that such an increase in learning ability was dependent on the site of stimulation. Stimulation of the parietal cortex,the nucleus accumbens or the substantia nigra had no effect on learning, but stimulation of the nucleus locus coeruleus (LC) produced the same improvement as stimulation of the lateral hypothalamus (Fig.1)(Velley and Cardo, 1979). The aim of the experiments summarized below was to examine further the role of the LC in these behavioral effects.

STIMULATION STUDIES

Detailed descriptions of methods have been published elsewhere (Velley and Cardo,1977,1979; Velley et al.,1981,1983). In all experiments male Sprague-Dawley rats were used. Stimulation electrodes were bilaterally implanted in the LC under stereotaxic control when the rats were 13 days of age.Each electrode consisted of two twisted platinum-irridium wires, 0.09 mm in diameter. Stimulation was applied on days 15 and 16, in 15 min sessions separated by an interval of 30 min. The stimulation parameters were as follows: 2580 stimulations per hour, duration 200 ms, sinusoidal current 100 Hz,60 μA (peak to peak). Each rat was stimulated for a total of 4 hours, 2 hrs on day 15 and 2 hrs on day 16. The electrodes were removed 1 day later under light anaesthesia. Three groups of rats were used : stimulated rats (S), implanted but not stimulated rats (I) and non-implanted rats (C).

Fig.1 summarizes the results of the LC stimulation on the acquisition and extinction of an food reinforced skinnerian task performed 4 weeks later (Velley, 1983). For the acquisition session (A), each rat was placed in the test chamber for 1 hr on a continuous reinforcement schedule (CRF) without previous training or shaping. Two measures of the animal's performance were taken,the number of lever presses per 5 min interval (A) and the latency (in seconds), to obtain the tenth pellet (B). The rats underwent one daily 15 min session until their performance reached at least 25 responses/5 min. These sessions were followed 24 hr later by a 30 min extinction session immediately after a 5 min session with reinforcement (C). All behavioral tests were conducted blind. In all experiments, the t test and analysis of variance with repeated measures were used to analyze the data.

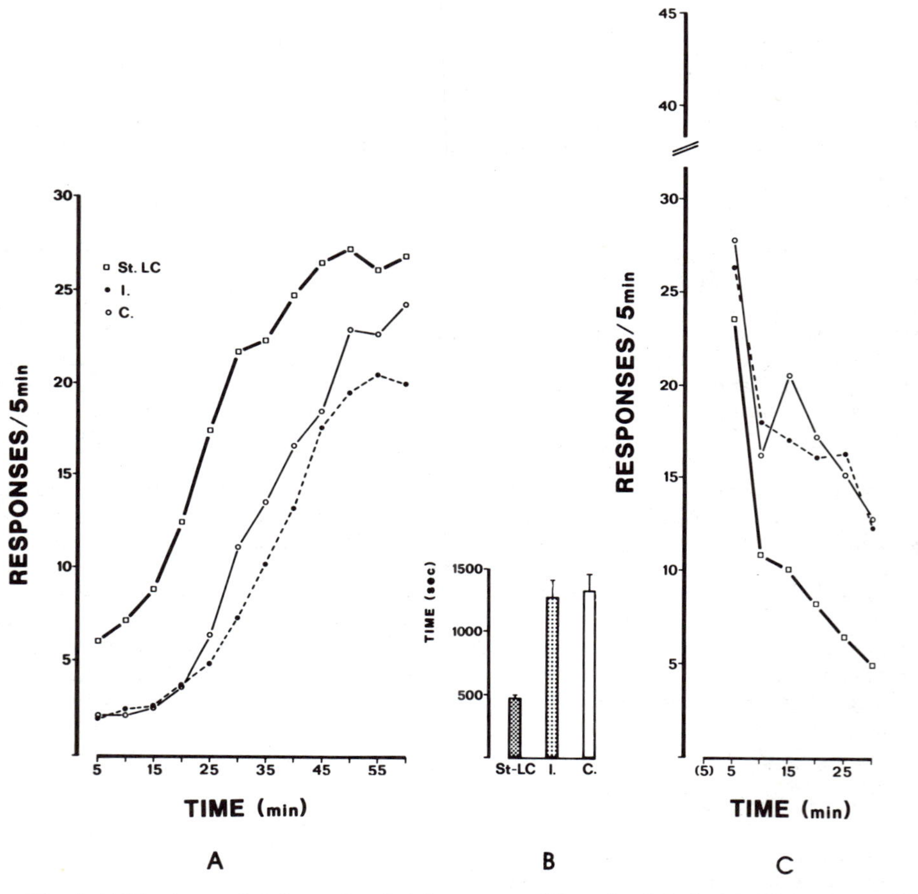

Fig.1.Effects of LC stimulation at 15 and 16 days of age on the acquisition and extinction of a CRF task performed 4 weeks later. A : acquisition curves of S.LC (N=20), I (N=18), and C (N=20) rats during the 1 hr session (I vs C,n.s.; S vs I,p < 0.0001).B : latency of the 10th response in the acquisition (I vs C, n.s.; S vs I, $p < 0.0001$).C : Extinction curves of S.LC (N=8),I (N=10),C (N=10) rats. (5) indicates a 5 min session with reinforcement before the extinction session (S vs I, vs C, n.s.). For the extinction curves : I vs C, n.s.; S.LC vs I, $p < 0.0001$). S.LC : rats stimulated in the LC; I : rats implanted but not stimulated; C : non-implanted rats.

These results showed that the stimulation of the LC four weeks previously produces a significant improvement of performance in the acquisition of an CRF task appetitively reinforced.

This enhancement appeared at the beginning of acquisition as shown by the difference in the number of responses during the first 5 min of the test (Fig.1A), and by the difference in the mean latencies for the 10th pellet (Fig.1B). Extinction in S rats was also more rapid than that of the two other groups, in spite of the fact that during the first five min with continuous reinforcement the performance of the 3 groups was not significantly different.

LESION EXPERIMENTS

The findings in our stimulation studies strongly suggested that the LC system plays an important part in the behavioral effects observed. However, given the widespread projections of the LC (Lindvall and Björklund, 1974, Moore and Bloom, 1979), the effects could be due either to a long term pre- or post-synaptic modification in all the regions innervated by the nucleus, or to a modification located in a particular region of the brain. The largest ascending projection of the LC is the dorsal noradrenergic bundle (DNB). Consequently, in a first experiment the DNB of 25 days old rats was bilaterally destroyed at the level of mesencephalon by local injections of 6-hydroxydopamine (4 μg in 2 μl on each side). During the same operation 2 electrodes were bilaterally implanted in the LC. The stimulation was applied 8 days later. As in previous experiments, the acquisition of the operant task took place 4 weeks later. Four groups of rats were included : unlesioned and stimulated (NL-S), lesioned and stimulated (L-S), unlesioned and non-stimulated (NL-NS) and lesioned and non-stimulated (L-NS).

As shown in fig.2,acquisition was fastest in the NL-S group. The performance of the L-NS group was the same as that of the NL-NS group. The overall performance of the L-S rats was significantly lower than that of the NL-S rats (A) but better than that of L-NS and NL-NS groups. However, during the first 20 min of acquisition the response rates of NL-S and L-S groups were the same and the mean latency of the tenth response was not different in the 2 groups (B).

The residual content of noradrenaline (NA) in the hippocampus and cortex of the rats of the two lesioned groups varied between 1.1 and 3.1%. Taken together these data showed that the lesion of the DNB do not have a clear and consistent effect since the main effect of the stimulation, i.e. the enhancement of performance at the beginning of the acquisition, was not disturbed by the DNB lesion.

Consequently, in another experiment we lesioned the LC proper by local injection of 6-hydroxydopamine (4μg in 1μl on each side). Two weeks later, the stimulation electrodes were implanted in the region of the LC. All the other aspects of the experiment (stimulation and behavioral testing) were the same as in the first experiment. Likewise, 4 groups of rats were used : L-S, NL-S, L-NS and NL-NS.

The results of the acquisition session are summarized in fig 3. Stimulation alone again produced a significant improvement of performance. However, in contrast with the small effect of the DNB lesion, the destruction of the LC proper, without stimulation (L-NS group) produced very poor performance during the acquisition session. The overall performance of the L-S group was significantly poorest than that of the NL-S group, but significantly better than the performance of L-NS group. The residual content of NA after neurochemical lesion of the LC was measured in the hippocampus, the cortex and the brainstem.

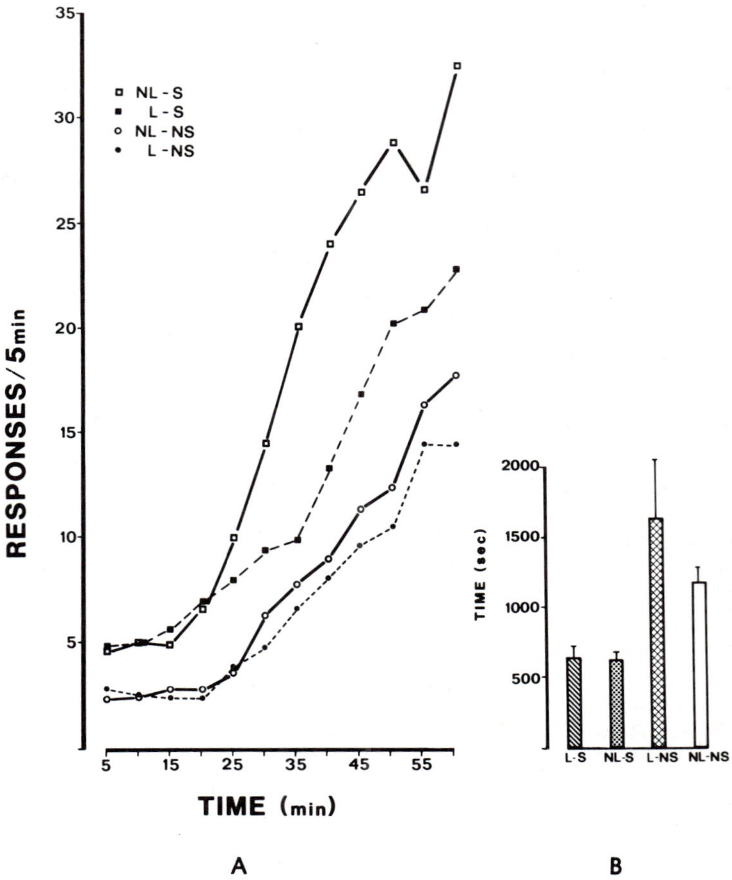

Fig.2 :Effects of DNB lesion on the learning enhancement produced by LC stimulation. Number of rats :NL-S,8;L-NS,8;L-S,7;NL-NS, 8. A : acquisition curves for the 4 groups of rats (statistical significance : NL-S vs NL-NS, $p < 0.002$; L-NS vs NL-NS, n.s.; L-S vs NL-S, $p < 0.05$). B : mean latencies of the 10th response (L-S vs NL-S, n.s.; L-S vs NL-NS, $p < 0.001$).

Although the loss of NA in the 3 structures was relatively small compared to that found after DNB lesions, the differences between the NA content of the control group (NL-NS) and of the two lesioned groups (L-NS and L-S) were significant (% of control; cortex, 37.2%; hippocampus, 60%; brainstem, 72.3%).

In conclusion these results suggest that the LC system is involved in the long term effect of the stimulation, but that the rostral projections passing through the dorsal bundle, in front of the lesion, are not critically involved in the observed effect (Velley et al.,1983).

BIOCHEMICAL STUDIES

Given the biochemical homogeneity of the coerulean system, we attempted to find some long term neurochemical modifications,which might eventually be found to be related to the behavioral effects. To this end we stimulated the LC in exactly the same conditions as for the behavioral experiments and examined NA levels and turnover in the cortex, hippocampus and LC four days and four weeks after the stimulation.

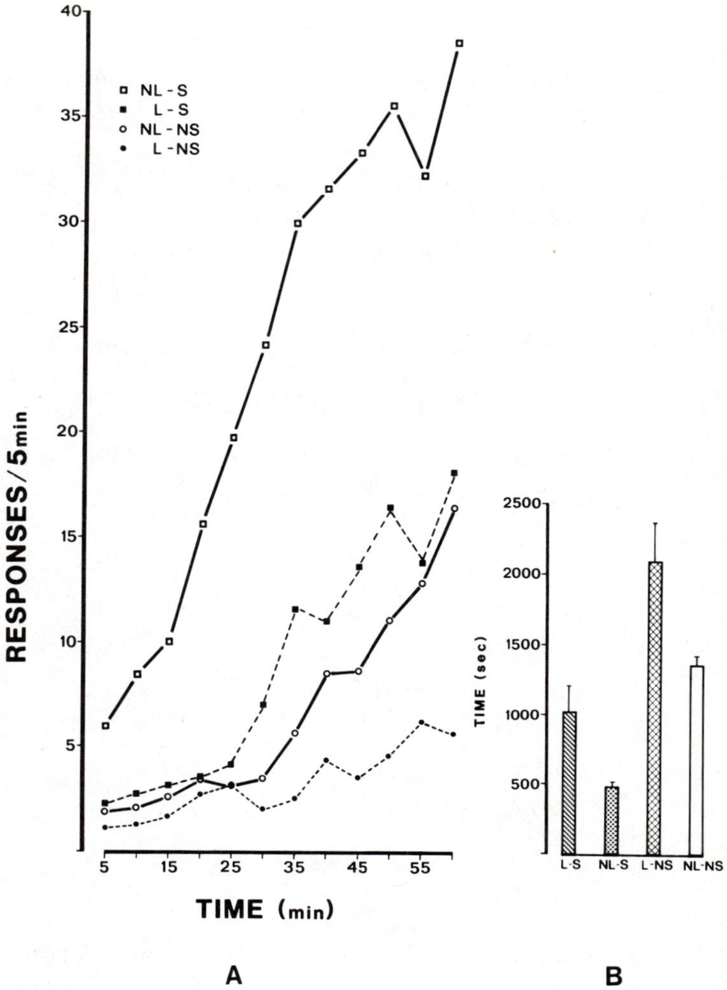

Fig.3 :Effect of a LC lesion on the learning enhancement produced by the stimulation of the LC region. Number of rats : NL-S, 7; L-S, 9; NL-NS, 8; L-NS, 10. A : acquisition curves for the 4 groups of rats; (statistical significance : NL-S vs NL-NS, $p < 0.001$; L-NS vs NL-NS, n.s.; L-S vs NL-S, $p < 0.001$; L-S vs L-NS, $p < 0.02$). B : mean latencies of the 10th response (NL-S vs NL-NS, $p < 0.01$; L-NS vs NL-NS, $p = 0.05$; L-S vs NL-S, $p < 0.001$; L-S vs L-NS : $p < 0.01$).

Levels of NA in the 3 structures studied were not modified either 4 days or 4 weeks after the stimulation. However,four days after the stimulation , the turnover of NA was significantly increased in the LC but not in the hippocampus or in the cortex (Table 1). This observation was confirmed by another experiment showing that, at this time, tyrosine hydroxylase activity was increased by 48% in the LC.

Table 1. Regional brain noradrenaline levels and turnover in non stimulated and stimulated rats 4 days after the stimulation.

		NA steady state level (ng/g WW ± S.D.)	NA decline after aMT (Slope ± S.D.)	Turnover rate (ng/g/h)
Hippocampus	NS	212 ± 27 (8)	0.092 ± 0.029 (16)	45 ± 15
	S	235 ± 61 (8)	0.115 ± 0.035 (16)	62 ± 25
Cortex	NS	156 ± 19 (8)	0.164 ± 0.029 (16)	59 ± 12
	S	193 ± 41 (8)	0.160 ± 0.027 (16)	71 ± 19
		ng/LOCUS		ng/LOCUS/h
Locus	NS	4.51 ± 1.06 (8)	0.076 ± 0.030 (16)	0.78 ± 0.35
Coeruleus	S	5.09 ± 1.43 (8)	0.142 ± 0.036*** (16)	1.61 ± 0.61

NS : non stimulated ; S : stimulated

*** $P < 0.001$ ( ) Number of rats are given in parentheses.

Table 2. Regional brain noradrenaline level and turnover in non stimulated and stimulated rats 28 days after the stimulation.

		NA steady state level (ng/g/WW ± S.D.)	NA decline after MT (Slope ± S.D.)	Turnover rate (ng/g/h)
Hippocampus	NS	422 ± 59 (7)	0.113 ± 0.024 (15)	109 ± 27
	S	358 ± 21 (8)	0.034 ± 0.018*** (16)	28 ± 15
Cortex	NS	246 ± 31 (7)	0.131 ± 0.029 (15)	74 ± 18
	S	234 ± 50 (8)	0.147 ± 0.031 (16)	79 ± 20
		ng/LOCUS		ng/LOCUS/h
Locus	NS	9.83 ± 1.45 (7)	0.109 ± 0.027 (15)	2.46 ± 0.70
Coeruleus	S	8.83 ± 1.76 (8)	0.039 ± 0.029*** (16)	0.79 ± 0.17

NS : non stimulated ; S : stimulated.

*** $P < 0.001$ ( ) Number of rats are given in parentheses.

In contrast, 4 weeks after the stimulation, a significant decrease of the NA turnover was observed in the LC and in the hippocampus (Table 2) (Velley, 1983).

Modulation of rat brain α-adrenoceptor populations 4 weeks after stimulation of the LC

The aim of these experiments was to find out whether adrenergic receptors were involved in some way in the long-lasting behavioral modification. In a first experiment, we investigated the characteristics of the α_1, α_2 and β-adrenergic receptors of the cerebral cortex 4 weeks after the stimulation (Velley et al., 1981a). No significant alteration in the number of β-receptors ((^{3}H)-dihydroalprenolol binding sites) was observed, but the stimulation produced a slight (14%) but significant increase in the number of α_1-receptors ((^{3}H) WB 4101 binding sites) as well as a rise (36%) in the number of α_2-receptors ((^{3}H) clonidine binding sites).

In order to verify these biochemical data by a pharmacological approach, we compared in S and control rats in the open-field test, the effect of clonidine injection on the locomotor activity. To stimulate only the α_2-adrenoceptors, very small doses of the drug were used (1, 2.5, 5 and 10 µg/kg). In stimulated and control animals the well known sedation produced by clonidine was the same, but only S rats exhibited a delayed hyperactivity which began 24 h after the injection. For the 5 µg/kg dose, this rebound of activity was still detectable 8 days after the injection. More recently, using the same procedure, we demonstrated that this late hyperactivity was due to the activation of α_1-adrenoceptors, as it was suppressed by prazosin, an α_1-adrenoceptor antagonist (Velley et al., 1982). Although the binding data cannot be directly compared to the pharmacological results, taken together, these observations strongly suggest that the stimulation of the LC produces modifications of the adrenoceptor populations and more importantly, that these modifications last at least 4 weeks.

Thus, the possibility that the long term behavioral effects were due, at least in part, to a modulation of α-receptor populations cannot be excluded. The binding data prompted us to study the long term effect of the LC stimulation on the characteristics of the α_1, α_2 and β-adrenoceptors, not only in the cortex but also in the hippocampus, the hypothalamus and the brainstem. The ligands used were : (^{125}I)-iodocyanopindolol (ICYP) for β-adrenoceptors (Engel et al., 1981), (^{3}H)-prazosin for α_1-adrenoceptors and (^{3}H)-yohimbine for the α_2-adrenoceptors. The preliminary results were as follows. In agreement with our previous data (Velley et al., 1981a), no significant alteration in the number of β-receptors was shown in any of the 4 structures analyzed. In the cortex, the number of α_1 and α_2-adrenoceptors was increased (+62% and +86% respectively). In the hippocampus and in the hypothalamus, only the number of α_2-adrenoceptors was increased. In the hippocampus, the increase was small but significant (+33%). In the hypothalamus, the increase was very large (+127%) (Fig. 4). Lastly, no modification of β, α_1 and α_2-receptors was observed in the brainstem.

Although these results have to be confirmed, they strongly suggest that 4 weeks after the LC stimulation only some adrenoceptor populations are modified and that these modifications are observed only in some brain regions.

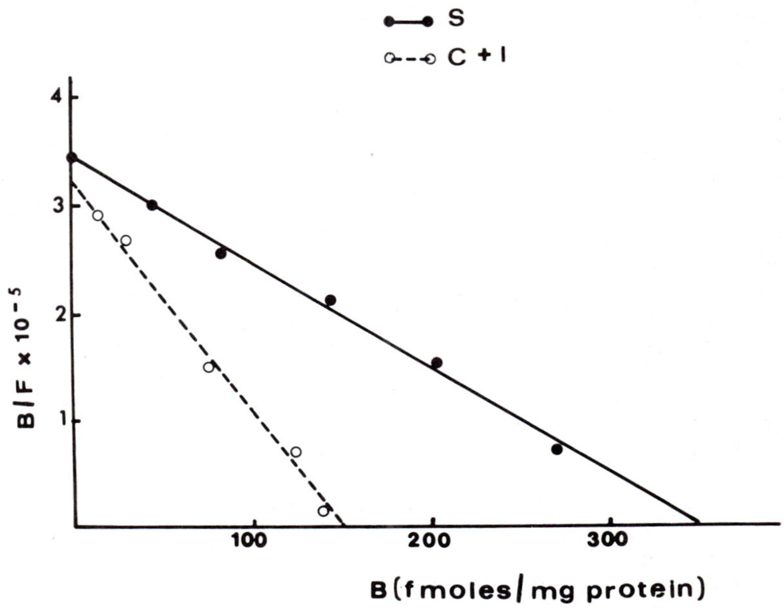

Fig.4 :Scatchard saturation plots of (^{3}H)-yohimbine binding sites in the hypothalamus :(C+I) control + implanted rats : K_D, 4.7 nM B_{max} , 153 fmoles/mg protein. S : stimulated animals : K_D, 10 nM, B_{max}, 348 fmoles/mg protein.

DISCUSSION

Three main points emerge from the preceding data : 1) The stimulation of the LC produces, 4 weeks later, a significant improvement of performance in the acquisition and extinction of an appetitively reinforced operant task. 2) Whereas the neurochemical lesion of the LC proper suppresses this effect it does not seem that the dorsal bundle projections of the LC are a critical factor in these behavioral effects. 3) The same stimulation applied under the same conditions induces biochemical modifications of the noradrenergic system, i.e. a turnover decrease and an long-term increase of $\alpha 2$-receptors in the hypothalamus, the hippocampus and the cortex, and of $\alpha 1$-receptors in the cortex.

Furthermore, the enhancement of performance of the pre-stimulated rats is probably not explained by a modification of the activity levels since these rats were neither hyperactive nor hypoactive in an open-field test (Velley et al.,1982). Moreover, stimulation of the LC did not produce any modification of food-consumption (Velley, 1983).

Some years ago, several theories postulated that the LC system was necessary or even the sole brain system responsible for behavioral functions such as ingestive behavior, arousal, sleep, reinforcement, learning and attention (rev. in Amaral and Sinnamon, 1977, Clark, 1979, McNaughton and Mason,1980). Two of these hypotheses,i.e. the learning and the attention hypotheses, might explain our data. Crow (1968) suggested that the LC system may play a role in learning but, despite earlier reports (Anlezark et al.,1973), more recent work has failed to confirm a noradrenergic component in the acquisition of many behavioral tasks (Amaral and Foss, 1975 ;Heybach et al.,1978; Koob et al.,1978; Roberts et al., 1976 ; Sessions et al.,1976). The second theory that could explain our data is the attentional hypothesis proposed by Mason (rev.in Mason and Iversen, 1979). This author studied in various learning tasks the behavior of rats in which the dorsal noradrenergic bundle was destroyed by 6-hydroxydopamine. His findings allowed Mason to suggest that this fiber system controls the capacity to ignore irrelevant stimuli in a normal rat. However, this attentional hypothesis was not supported by the findings of the effects of dorsal bundle lesions (Pisa and Fibiger, 1980; Owen et al., 1982) or by studies on the effects of LC nucleus lesions (Crow et al.,1978). These negative data were in agreement with various observations showing that the LC system was not essential for complex behaviors. On the contrary, the anatomical organization and the physiological properties of this system predict a "modulatory" role. This general modulatory function can be analyzed in two slightly different ways.

Firstly, numerous electrophysiological data suggest a role for the LC nucleus in the control of sensory inputs (Cedarbaum and Aghajanian, 1978; Foote et al., 1980; Segal and Bloom, 1976; Woodward et al.,1979). Specifically, Segal and Bloom (1976) have proposed that the main function of the LC projection is to increase the signal to noise ratio. More recently, Aston-Jones and Bloom (1981) found in awake rats that LC neurons exhibit strong responses to environmental stimuli and concluded that the LC system "may function to facilitate transition between behavioral states". It is noteworthy that in our experiments, the enhanced performance of the stimulated rats was observed only during behavioral transitions: the enhancement was first seen at the beginning of the first learning session (Fig.1A). Likewise, after the removal of the reinforcement during the extinction session, the most significant decrease of responses of treated rats took place during the first 10 min of the session (Fig. 1C). However, when the task was well learned, no difference between stimulated and control rats was observed (Fig.1C, first point of the abscissae). Thus, it seems that the behavioral enhancement appeared only when a parameter of the experimental situation was modified. If we suppose that the functioning of the LC is improved in stimulated rats, perhaps by the modification of the β-adrenoceptor populations, we might say that at each transitional point in a learning situation the processing of novel stimuli will be more efficient in these rats than in unstimulated control animals.

Another hypothesis that is consistent with our experimental data was clearly formulated by Amaral and Sinnamon (1977). These authors took into account the anatomical and physiological similarities between the LC and the sympathetic ganglia and proposed that a major function of the LC is to dampen the organism's response to stressors. This hypothesis is supported by the well known data on the reactivity of the LC system to stressfull situations (Stone, 1975; Anisman, 1978; Tsuda et al., 1982; Weiss et al.,1981). For example, various observations show that the novel stimuli and more generally, all the transitional states are stressfull and produce an activation of the pituitary-adrenal axis.

For example the introduction of an animal in a new environment induces an increase of the corticosterone (Bassett and Cairncross, 1973; Pfister, 1979; File and Peet, 1980) as well as an increase of ACTH (Hennessy et al.,1979). The same effect is observed after the omission of the reinforcement during extinction (Coover et al.,1971 ; Davis et al., 1976; Levine et al.,1972). Thus, given the fact that the enhanced performance in the stimulated rats appeared only during the transitions between behavioral states, it is possible that these effects were due only to a reduced stress reaction, with a consequently better adaptation to a new experimental situation by the stimulated rats.

Furthermore, in order to test this possibility more directly, we recently compared the reactivity of the pituitary-adrenal axis of stimulated and implanted rats, introduced for the first time in an open-field during 10 min. The first results showed that although no behavioral difference was observed between the two groups, the increase of ACTH response of the stimulated rats to the stressfull situation was significantly smaller than the increase found in implanted rats (basal levels of blood ACTH in pg/ml : S: 13 ± 3, I:17 ± 2 ; blood levels of ACTH after 10 min exposure in the open field in pg/ml : S: 133 ± 13, I: 415 ± 67). Thus, stress-reduction hypothesis cannot be excluded and requires confirmation.

In conclusion, the long-term behavioral effects of LC stimulation may be explained either by an improved ability to detect novel and significant stimuli or by a reduced reaction to the stress produced by the same stimuli. The data presently available do not allow us to choose between these two hypotheses. Furthermore, these two possibilities are perhaps not exclusive : the LC system appears to have diffuse and widespread projections with little regional specificity. The functions of this system may therefore be rather general. Consequently, it may be that the ability to detect novel stimuli, and the reaction to the stress are simultaneously modified by the stimulation of the LC. This integrated hypothesis requires further investigation .

Acknowledgements : This work was supported by the I.N.S.E.R.M. (grant n° 82.6.017), the D.G.R.S.T. (grant n° 82.J.1428) and partially by the C.N.R.S. (U.A.n°339). We thank S. Sara for helpful comments on this manuscript.

REFERENCES

Amaral, D.G., and Foss, J.A., 1975, Locus coeruleus lesions and learning, Science, 188:377.

Amaral, D.G., and Sinnamon, H.M., 1977, The locus coeruleus: Neurobiology of a central noradrenergic nucleus, Progr. Neurobiol., 9:147.

Anisman, H., 1978, Neurochemical changes elicited by stress. Behavioral correlates, in : "Psychopharmacology of aversively motivated behavior", A. Anisman and G. Bignami, eds., Plenum Press, N.Y.

Anlezark, G.M., Crow, T.J., and Greenway, A.P., 1973, Impaired learning and decreased cortical norepinephrine after bilateral locus coeruleus lesions, Science, 181:682.

Aston-Jones, G., and Bloom, F.E., 1981, Norepinephrine-containing locus coeruleus neurons in behaving rats exhibit pronounced responses to non-noxious environmental stimuli, J. Neurosci., 1:887.

Bassett, J.R., and Cairncross, K.D., 1973, Parameters of novelty, shock predictability and response contingency in corticosterone release in the rat, Physiol. Behav., 10:901.

Cedarbaum, J.M., and Aghajanian, C.K., 1978, Afferent projections to the rat locus coeruleus as determined by a retrograde tracing technique, J. Comp. Neurol., 178:1.

Clark, T.K., 1979, The locus coeruleus in behavioral regulation : evidence for behavior-specific versus general involvement, Behav. Neur. Biol., 25:271.

Coover, G.D., Goldman, L., and Levine, S., 1971, Plasma corticosterone increases produced by extinction of operant behavior in rats, Physiol. Behav., 6:261.

Crow, T.J., 1968, Cortical synapses and reinforcement : A hypothesis, Nature, 219:736.

Crow, T.J., Deakin, J.F.W., File, S.E., Longden, A., and Wendlandt, S., 1978, The locus coeruleus noradrenergic system. Evidence against a role in attention, habituation, anxiety and motor activity, Brain Res., 155:249.

Davis, H., Memmott, J., McFadden, L., and Levine, S., 1976, Pituitary-adrenal activity under different appetitive extinction procedures. Physiol. Behav., 17:687.

Engel, G., Hoyer, D., Berthold, R., and Wagner, H., 1981, (I)^{125}iodocyanopindolol, a new ligand for beta-adrenoceptors : identification and quantification of subclasses of beta-adrenoceptors in guinea-pig, Naunyn-Schmiedeberg Arch. Pharmacol., 317:277.

File, S.A., and Peet, L.A., 1980, The sensitivity of the rat corticosterone response to environmental manipulations and to chronic chlordiazepoxyde treatment, Physiol. Behav., 25:753.

Foote, S.L., Aston-Jones, G., and Bloom, F.E., 1980, Impulse activity of locus coeruleus neurons in awake rats and monkeys is a function of sensory stimulation and arousal, Proc.Nat. Acad. Sci. U.S., 77:3033.

Hennessy, M.B., Heybach, J.P., Vernicos, J., and Levine, S., 1979, Plasma corticosterone concentrations sensitively reflect levels of stimulus intensity in the rat, Physiol. Behav., 22:821.

Heybach, J.P., Coover, G.D., and Lints, C.E., 1978, Behavioral effects of neurotoxic lesions of the ascending monoamine pathways in the rat brain, J. comp. physiol. Psychol., 92:58.

Koob, G., Kelley, A.E., and Mason, S.T., 1978, Locus coeruleus lesions : learning and extinction, Physiol. Behav., 20:709.

Levine, S., Goldman, L., and Cooper, G.D., 1972, Expectancy and the pituitary-adrenal system, in : "Ciba Foundation Symposium",8: Elsevier, Amsterdam.

Lindvall, O., and Björklund, A., 1974, The organization of the ascending catecholamine neuron systems in the rat brain, Acta Physiol. scand., suppl. 412:1.

Mason, S.T., and Iversen, S.D., 1979, Theories of the dorsal bundle extinction effect, Brain Res. Rev., 1:107.

McNaughton, N., and Mason, S.T., 1980, The neurophysiology and neuropharmacology of the dorsal ascending noradrenergic bundle : a review, Progr. Neurobiol., 14:157.

Moore, R.Y., and Bloom, F.E., 1979, Central catecholamine neuron systems: Anatomy and physiology of the norepinephrine and epinephrine systems, Ann. Rev. Neurosci., 2:113.

Owen, S.R., Boarder, M.R., Gray, J.A., and Fillenz, M., 1982, Acquisition and extinction of continuously and partially reinforced running in rats with lesions of the dorsal noradrenergic bundle, Behav. Brain Res., 5:11.

Pfister, H.P., 1979, The glucocorticosterone response to novelty as a psychological stressor, Physiol. Behav., 23:649.

Pisa, M., and Fibiger, H.C., 1980, Noradrenaline and discrimination learning : Failure to support the attentional hypothesis, Soc. Neurosci. Abstr., 6:724.

Roberts, D.C.S., Price, M.T.C., and Fibiger, H.C., 1976, The dorsal tegmental noradrenergic projection : an analysis of its role in maze learning, J. comp. physiol. Psychol., 90:363.

Segal, M., and Bloom, F.E., 1976, The action of norepinephrine in the rat hippocampus. IV. The effect of locus coeruleus stimulation on evoked hippocampal unit activity, Brain Res., 107:513.

Sessions, C.R., Kant, G.J., and Koob, G.F., 1976, Locus coeruleus lesions and learning in the rat, Physiol. Behav., 17:853.

Stone, E.A., 1975, Stress and catecholamine, in : "Catecholamine and Behavior", A.J. Friedhoff, ed., Vol.2, Plenum Press, N.Y.

Tsuda, A., Tanaka, M., Kohno, Y., Nishikawa, T., Iimori, K., Nakagawa,R., Hoaki, Y.,Ida, Y., and Nagasaki, N., 1982,Marked enhancement of noradrenaline turnover in extensive brain regions after activity stress in rats, Physiol. Behav., 29:337.

Velley, L., and Cardo, B., 1977, A long-term effect of an early stimulation of the lateral hypothalamus on the acquisition and extinction of a food reinforced operant conditioning in the rat, Neurosci. Lett., 5:221.

Velley, L., and Cardo, B., 1979, Long-term improvement of learning after early electrical stimulation of some central nervous structures : is the effect structure and age-dependent ? Brain Res. Bull., 4:459.

Velley, L., Cardo, B., and Bockaert, J., 1981a, Modulation of rat brain adrenoceptor populations four weeks after stimulation of the nucleus locus coeruleus, Psychopharmacology, 74:226.

Velley, L., Chassaing, J.M., and Cardo, B., 1981b, Learning improvement of appetitively or aversively reinforced light-dark discrimination and reversal four weeks after electrical stimulation of the lateral hypothalamus of the rat, Brain Res. Bull., 6:377.

Velley, L., Kempf, E., and Cardo, B., 1982, Locomotor activity of rats after stimulation of the nucleus locus coeruleus region or after lesion of the dorsal noradrenergic bundle, Psychopharmacology,78:239

Velley, L.,1983, Rôle des systèmes aminergiques centraux dans la maîtrise des conduites acquises. Thèse d'Etat, Université de Bordeaux I.

Velley, L., Nassif, S., Kempf, E., and Cardo, B., 1983, Enhancement of learning four weeks after stimulation of the nucleus locus coeruleus in the rat : Differential effects of dorsal noradrenergic bundle lesion and lesion of the locus coeruleus proper, Brain Res., 265:273

Weiss, J.M., Goodman, P.A., Losito, B.G., Corrigan, S., Charry, J.M., and Bailey, W.H., 1981, Behavioral depression produced by an uncontrolable stressor : Relationship to norepinephrine, dopamine and serotonin levels in various regions of the rat brain, Brain Res. Rev., 3:167.

Woodward, D.J., Moises, H.C., Waterhouse, B.D., Hoffer, B.J., and Freedman R., 1979, Modulatory actions of norepinephrine in the central nervous system, Fed. Proc., 38:2109.

ROLE OF SEROTONIN AND cAMP ON SHORT-TERM PLASTIC CHANGES UNDERLYING SIMPLE FORMS OF LEARNING PROCESS

M. Brunelli, G. Demontis and G. Traina

Istituto di Fisiologia, Università di Pisa and Istituto di Neurofisiologia CNR, Pisa

INTRODUCTION

In recent years, invertebrate nervous systems have proved to be experimentally advantageous preparations for studying the cellular and molecular mechanisms that underlie learning, because the neuronal networks of invertebrate ganglia are made up by cells large enough to be analysed by different methodologies. Moreover, it is possible to test the effects on neuronal functions of a variety of molecules by injecting them into the neuronal somata.

Short-term sensitization, an elementary form of learning in which one strong nociceptive stimulus can vigorously enhance responses to weak test stimuli, has been extensively studied in the gill and syphon withdrawal defensive reflex in Aplysia. A light touch on the mantle surface produces a gill and syphon withdrawal into the mantle cavity and repetitive tactile stimulation depresses the amplitude of this withdrawal response (habituation). When a strong shock is applied to the head or the tail the withdrawal of both non-habituated and habituated response (dishabituation) clearly increases for 15-60 min. Repeated shock stimulation applied everyday can enhance the response for as long as two weeks or more (long-term sensitization). Analysis at cellular level has demonstrated that such behavioral sensitization of the syphon and gill withdrawal reflex results from an increase in the transmitter release at the sensory motoneuron synapse (heterosynaptic facilitation) (Kandel et al.,1976).

ROLE OF cAMP ON SENSITIZATION

An explanation of the chain of molecular events underlying the augmented release of transmitter during heterosynaptic facilitation was provided in a key study by Brunelli et al.(1976). Serotonin (5-HT) applied in the perfusion bath mimicked the synaptic potentiation, whereas the 5-HT blocking agent cinanserin prevented the effect.

Intracellular iontophoretic injection of cyclic adenosine monophosphate (cAMP) into sensory neurons strongly potentiates the transmitter output as revealed by the clearcut increase of the monosynaptic excitatory post-synaptic potential (EPSP) recorded from the follower neurons. Similar results have been obtained with the cAMP analog dibutyryl cAMP. Cyclic guanosine monophosphate (cGMP), 5'AMP and adenosine were ineffective in potentiating synaptic transmission.

These data suggest that noxious shocks delivered to the head or tail trigger a sensitizing pathway made up by neurons with a serotonin-like transmitter that make contact with the terminals of the sensory neuron and activate an adenylate-cyclase that enhances intracellular levels of cAMP which, in turn, induce a potentiation of transmitter release. This view has recently been supported by the observation that 5 min. exposure to 5-HT produces a clear increase of intracellular levels of cAMP in individual sensory neurons (Bernier et al., 1982). Furthermore, it has been possible to identify the neurons mediating the sensitization. These cells are L28, a single neuron, and L29 a cluster of electrically coupled cells. Their electrical stimulation has been found to provoke a presynaptic facilitation of sensory-motoneuron synapses (Hawkins et al., 1981).

As there was no corresponding evidence of a change in the amount of transmitter available in the terminals a series of experiments has been performed to elucidate the mechanism by which transmitter release is potentiated.

There is good evidence that an increase in cAMP leads to the activation of a cAMP-dependent protein kinase (PK). The catalytic subunit of PK would induce phosphorylation of specific proteins which in turn affect the gating of ionic channels. The intrasomatic injection of the catalytic subunit of the cAMP-dependent PK into the cell body of individual tetraethylammonium (TEA) pretreated sensory neurons induces a broadening of the action potential (AP) and an enhancement of the EPSP amplitude of motor cells, indicating a potentiation of transmitter output (Castellucci et al., 1980).

Klein and Kandel (1978, 1981) have analysed the changes of membrane ionic currents of sensory neurons in isolated abdominal ganglia treated with TEA to inhibit the delayed K current. TEA-treated neurons were found to exhibit a spike enlargement. In this cellular preparation, slight changes of Ca and K current, factors responsible for the falling phase of AP, might result in a further modification of spike duration. In such TEA-pretreated preparations a further spike broadening has been observed: 1) with 5-HT perfusion; 2) with electrical stimuli applied to the sensitized pathway; 3) with intracellular iontophoretic injection of cAMP into sensory neuron and with a phosphodiesterase inhibitor isobuthylmethylxanthyne (IBMX); 4) with intracellular stimulation of L29 neurons. The increase in duration of sensory neuron AP following 5-HT application seems therefore due to a modulation of outward current.

Siegelbaum et al.(1982), using the "patch clamp technique" have found that 5-HT seems to decrease the possibility of opening of novel K channels without any change of conductance or selectivity. Intracellular cAMP injection caused a lasting closure of K channels which were active while the cell was at resting potential. The blockage of K channels would induce a spike broadening of 10-20% allowing an increase of Ca current which might account for a potentiation of transmitter release. All these data taken together indicate that modulation of ionic current depends upon mediation of cell membrane phosphorylated proteins. To test whether 5-HT might induce a true intracellular increase of cAMP levels we decreased the content of endogenous cAMP by using an inhibitor of adenylate cyclase activity (RMI 12330A). Such a substance is capable of blocking the cAMP increase following 5-HT treatment. Fifteen min. preincubation with 0.1 mM RMI 12330A inhibited the facilitation of EPSP observed in motor neuron L7 after syphon nerve stimulation, an effect that was reversed by washing out the cyclase inhibitor. Similarly, the injection of cyclase inhibitor into the whole animal clearly depressed behavioral sensitization. These findings provided direct evidence that heterosynaptic facilitation and behavioral sensitization are mediated by an intracellular increase of cAMP level in the sensory neuron terminal (Belardetti et al., 1983).

The injection of a specific cAMP-dependent PK inhibitor purified from rabbit muscle reduced and prevented the spike broadening that accompanies the facilitation in TEA pretreated sensory neurons. In addition, the injection of PK inhibitor narrowed the spike broadening that resulted from incubation with 5-HT. The latter result suggests that the persistence of facilitation does not reside in a lasting phosphorylated state but rather in a sustained elevation of kinase activity (Castellucci et al., 1982). The concentration of cAMP might remain enhanced for both a prolonged potentiation of cyclase and for a decreased degradation of the nucleotide due to a reduction of phosphodiesterase activity. Experiments using nitrocellulose filters to separate free cAMP from cAMP linked with binding proteins argue against the hypothesis that cAMP might persist in a form protected from phosphodiesterase activity. Recently, Castellucci et al.(1983) observed that the intracellular injection of GDPβS (an analog of GDP which inhibits the cyclase) (Eckstein et al., 1979) into TEA pretreated sensory neurons narrows the spike enlargement observed after 5-HT application. The effect was reversed completely by intracellular injection of cAMP and partially by application of Forskolin, a cyclase activator (GDPβS would act by preventing the sustained activation of the G-cyclase complex).

The cyclase complex is made up of three components: a receptor, the regulatory G protein and a catalytic subunit. The G protein binds guanosine triphosphate (GTP), and the G-GTP complex associates with the catalytic subunit and activates it. GDPβS substitute GTP on the G unit, and prevents the sustained activation of the catalytic subunit. It seems possible that short-term sensitization might be determined by elevation of cAMP level resulting from a persistent cyclase activity, and in this scheme, G protein appeares to be important, timing how long the enzyme remains active. Therefore short-term sensitization might decay as soon as the cyclase complex is inactivated. It is possible therefore to speculate about the lasting potentiation observed during heterosynaptic facilitation. In the presynaptic membrane of sensory neurons the sequence of 5-HT - cAMP - phosphorylated protein is followed by the closure of a K channel, thereby increasing the Ca influx which produces a major Ca/calmodulin complex. Since cyclase activity is stimulated by Ca/calmodulin a potentiation of phosphorylated protein is set up. This in turn can facilitate in a sustained way the release of vesicles through the reduction of their link with their binding matrix. Thus, a prolonged potentiation of synaptic efficacy results. This mechanism is supported by recent experiments in which using the phosphorylated protein antibody, Synaptin I, it has been possible to localize the labelled protein around the vesicles inside the presynaptic membrane in various vertebrate nervous systems (De Camilli and Navone, 1984). It is still undetermined whether phosphodiesterase activity is affected by the facilitation process.

Hawkins et al.(1983) have demonstrated that when an AP in one sensory neuron is paired with tail stimulation in Aplysia, a greater spike broadening occurs when compared with unpaired sensory neuron. Kandel (1983) proposed the following scheme for associative learning: the Ca ion which accompanies each AP might bind, by means of the Ca/calmodulin complex, to the catalytic subunit of the cyclase. As a result of this binding, the cyclase may change so that it could synthesize more cAMP in response to 5-HT. Or alternatively, one can say that the Ca/calmodulin system could prepare the vesicles for release by reducing the connections between the vesicles and their matrix. The cAMP - phosphorylated protein can amplify this effect by subsequent potentiation of the Ca influx which leads to the closure of the K channel. If this hypothesis proves to be true, we can say that, at least at the molecular level, classical conditioning and sensitization are not different processes because they utilize similar molecular keys.

SHORT-TERM CHANGES IN BEHAVIORAL ACTIVITY OF THE LEECH HIRUDO m.

At this point a more general question arises. Is it possible to generalize the mechanism of short-term synaptic plasticity observed in Aplysia nervous system to that of other organisms? Are there different systems for modulating cellular dialogue?

In order to answer such questions, we have tested the cellular and molecular mechanisms of plastic changes underlying modifications of elementary forms of behavior in the leech Hirudo m. for the leech nervous system offers relevant advantages. This nervous system consists of a chain of segmental ganglia connected by fiber bundles (connectives). In each ganglion, it is possible to recognize a precise number of identified neurons accessible to electrophysiological and neurochemical analysis. Among the most common type of cells studied we mention: the mechanosensory neurons T,P,N which respond to light touch, pressure and nociceptive stimuli respectively; two giant cells Retzius cells (Rz cells), and the motor neurons (L for longitudinal muscles, AE for annulus erector, HN for circulation).

First, we focused our research on exploring the possibility of finding a useful model that would represent a simple form of behavioral activity which could exhibit short-term changes of the non associative type of learning: habituation and sensitization (or dishabituation). Then we investigated, at the cellular level, the mechanism of such learning processes. We studied the non-associative learning of the following behavioral activities: the generalized body shortening reflex and swimming.

Shortening reflex: photic and tactile stimuli brought on the body surface are transmitted through segmental roots to impinge upon a fast conducting system (FCS) which runs in the Faivre nerve (medial connective). The FCS is distributed along the entire nerve cord and is made up by the axon (Rhode fiber) of a single cell (S neuron) located in each ganglion and electrotonically coupled to each other (Bagnoli et al., 1975). This system is connected with longitudinal body muscles and seems to mediate the shortening of the whole animal. Thus the activity of the FCS represents a good index of the shortening reflex.

In this system we have obtained evidence that 5-HT and cyclic nucleotides play a fundamental role in short-term changes of dishabituation.

Repetitive tactile stimulation of skin patches provoked a decrease in the activity of the S cell (habituation). Nociceptive stimulation brought onto a different area potentiated the depressed response. Both dopamine and 5-HT applied in the perfusion bath induced a large and lasting enhancement of cAMP levels in each ganglion. The effect of 5-HT treatment is prevented by the serotonin blocking agent methysergide. The dopamine effect, on the contrary, is mediated by 5-HT neurons because application of high Mg solutions inhibits the increase of cAMP levels (Biondi et al., 1982). As in the Aplysia model it has been possible to observe, with electrophysiological experiments, that 5-HT clearly potentiates the FCS activity, normally habituated by repetitive tactile stimulation (dishabituation); Methysergide and imidazole (a phosphodiesterase activator) prevented such a facilitation. In addition, dibutyryl cAMP potentiated the activity of such cells. All these effects lasted 30-60 min. Control experiments excluded the possibility that this potentiation was due to a postsynaptic activation of some interneurons or to a post-tetanic potentiation (Belardetti et al., 1982). These data therefore suggested that, in the leech, short-term changes underlying the non associative type of learning are mediated by serotonergic neurons via the intracellular activation of a second messenger of the nucleotide type.
Swimming behavior: the swim activity was found to depend upon the level of 5-HT in the blood stream. There are two different kind of serotonergic cells controlling swimming behavior: the Retzius cells (segmentally paired neurons) and cells 21 and 61 which seem to act synaptically onto swim related neurons (Kristan et al., 1983). Therefore we studied the modulatory role of 5-HT on non-associative learning processes taking leech swim activity as the behavioral model.

In brainless animals, light tactile or electrical stimuli are capable of triggering swim behavior. Repetitive stimulation induced an increase of latency between the onset of stimulus and the beginning of swim behavior (habituation). Strong nociceptive stimulation shortened the delayed latency. Similar effects were obtained with 5-HT injections (dishabituation). Dopamine also reduced the latency for swimming behavior, but in a less predictable way. Experiments are in progress to test whether 5-HT may act through the mediation of cAMP.

PLASTIC CHANGES OF THE LEECH SEGMENTAL SENSORY NEURONS

In an attempt to study the sites and the mechanisms where the short-term changes take place, experiments have been performed on single neurons of the neural networks. We began by analysing intracellularly the activity of single mechanosensory T cells in preparations of isolated ganglia linked to skin patches by segmental roots. Repetitive tactile stimulation does not change the discharge of that T cell directly connected to the receptive field stimulated, while the T neurons innervating the segment adjacent to the stimulated skin area show a decrement in their activity. Surprisingly, the application of 5-HT reduces the post-hyperpolarization which follows the repetitive discharge of such mechanosensory neurons. It is well known that the firing of sensory neurons generates a hyperpolarization (HP), set up by a Na/K electrogenic pump and by an increase in a Ca-dependent K conductance (g_k). This change of membrane potential makes possible a conduction block at branch points of the sensory axons. In T neurons,

HP is sustained by a prevalent Na/K pump; in P neurons, both Na/K and g_k are responsible for HP and in N cells g_k is the agent responsible for the displacement of membrane potential. Our findings have shown that 5-HT application or serotonergic cells stimulation are capable of reducing for a prolonged time the HP which follows repetitive discharge of T neurons.

By applying 5-HT for 10 min., a dose-dependent, reversible reduction of HP following T cell stimulation occured. Pretreatment with methysergide prevented such an effect. No changes of T cell membrane resting potential, input resistance or action potential shape were observed in the presence of serotonin. 5-HT does not change the excitability of T cells. Dopamine is also capable of depressing the HP following repetitive activity of T cells, but the total number of sensory cells spike tends to drop down indicating a change in the excitability of the T cell. 5-HT does not affect the HP which follows N neurons activation, while only a weak effect is observed on P cells. (Fig. 1).

The effects described following 5-HT application might be due to: A) Inhibition of K current (as in Aplysia model for the closure of K channel); or B) Block of Na/K electrogenic pump for direct action on ATPase. Because the major effect is observed on T neurons, the second hypothesis seems to be more likely, but we still need additional information to clarify the true mechanism. In order to obtain a depression of HP in a more physiological way, we tried to stimulate serotonergic neurons. In each segmental ganglion, 7 serotonin containing neurons, two of which are Rz cells, have been identified. They are electrically coupled and function as a whole. We observed that electrical stimulation of Rz cells produced a reduction of T cell hyperpolarization. Preincubation with serotonin antagonist methysergide prevented such an effect. The reduction of HP in T neurons was partially inhibited by pretreatment with the cyclase inhibitor 12330A.

Furthermore, Forskolin (a cyclase activator) induced a strong depression of HP in T neurons. The effect was potentiated in presence of theophylline and was build up in about 30 min. Also, iontophoretic injection of cAMP into sensory neurons was effective in depressing HP. These findings suggest that the cascade of events, 5-HT - cAMP - phosphorylated proteins -change of ionic currents, might represent the basic mechanism responsible for plastic changes in different systems, that is they may be the elementary language which the nerve cell has to modulate neuronal dialogue.

As a consequence of the depression of the T cell hyperpolarization after 5-HT application or Rz cell stimulation, we observed a relief of conduction block. Such a relief allows the previously blocked impulses to run along the axonal branches up to the terminal inducing a potentiation of transmitter output to the target cells. Accordingly, this mechanism seems to exert a modulation of sensory information. Thus, it is possible to control the synaptic efficacy of many terminals and thereby short-term plastic changes of the extensive cellular territory below the branching points. Fig.2 shows a schematic drawing of the molecular events which might occur at the level of the axonal branching points of sensory neurons. The presynaptic bouton impinging upon the sensory neuron would release 5-HT. This neurotransmitter would activate the catalytic subunit of adenylate cyclase increasing the cAMP level. This in turn would activate a protein kinase, which causes increased protein phosphorylation. At this point we lack sufficient information to know which mechanism is set up at the membrane level. Two main hypotheses are suggested: A) Protein phosphorylation

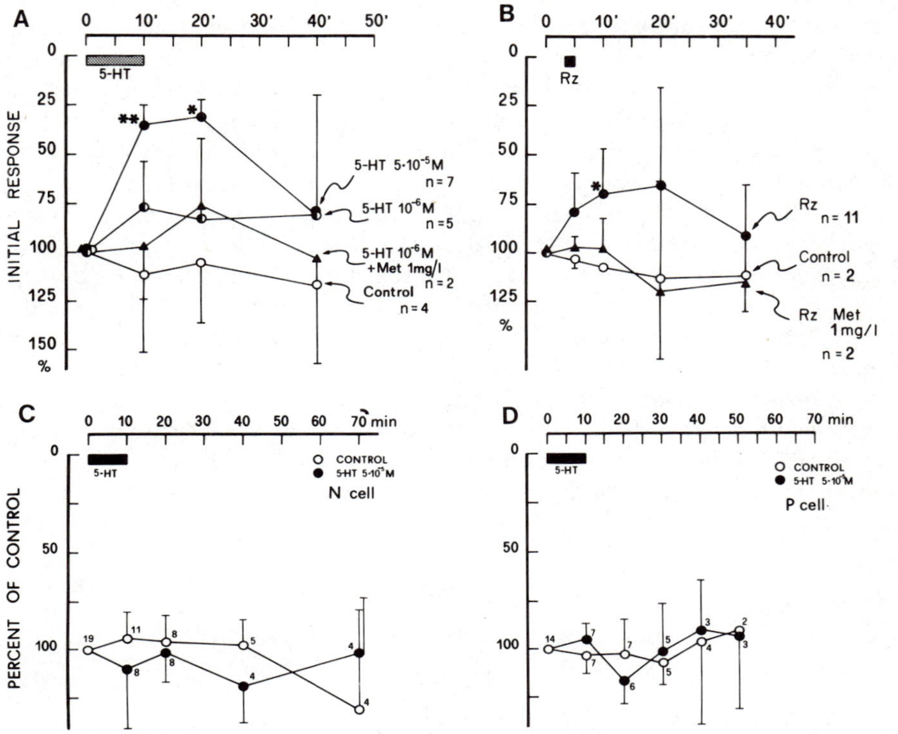

Fig. 1 Graphs summarizing the effect of 5-HT on the hyperpolarization which follows the discharge of single sensory neuron. In A, the effects of 5-HT on T neuron are pooled. T cell was fired with train of depolarizing pulse. The value of the amplitude of the hyperpolarization of the first trial of each series was taken as 100%. All the following values were calibrated in relation to the first one. The control values (without 5-HT application - open circles) were stable over time. 5-HT application clearly reduced the amplitude of HP. With 50 μM (filled circles) the depression of HP was highly significant after 10-20 min. With 1 μM (half-filled circles) that effect was weaker. Pretreatment with methysergide (1mg/L) blocked the depression of HP (triangles). In B, the effects of Retzius cell stimulation on the HP of T cells is summarized. The stimulation of Retzius cell was capable of reducing the HP of T neurons for up to 20-30 min. (filled circles), while HP of T neurons of control (open circles) and methysergide pretreated (triangles) preparations did not show any significant change. In C, the 5-HT application (filled circles) did not affect the HP following firing of N sensory neurons. Similarly in D, 5-HT application (filled circles) did not change the HP after firing discharge in P neurons.

might induce, in analogy with the Aplysia model, a closure of a K channel. The K outward current is thus reduced, less amount of K is outside and a minor activation of the pump would occur. B) Protein phosphorylation might act directly on the electrogenic pump, affecting membrane ATPase type enzymes. A series of experiments is in progress to test this hypothesis.

Furthermore, we have drawn some interesting considerations about the connections existing between Rz cells and T cells. Intracellular stimulation of Rz cells do not evoke any monosynaptic changes in the membrane conductance of the T sensory cells except for a few unpredictable polysynaptic IPSPs. But when T cells show a specific pattern of activity, such as a hyperpolarization after repetitive discharge, Rz cells are effective in controlling such behavior. Rz cell stimulation activates a chain of events which lead to a mechanism that modulates specific functions of the follower neurons. This specific action after Rz cell stimulation might be due either to the fact that 5-HT exerts a neurohormonal action or that there exists an unconventional synapse which can broaden the concept of cellular dialogue. This mechanism could explain how it is possible for a neuron that produces two or more neuropeptides and/or neurotransmitters, to release only one of them when receiving the trigger by a specific pathway related to specific function.

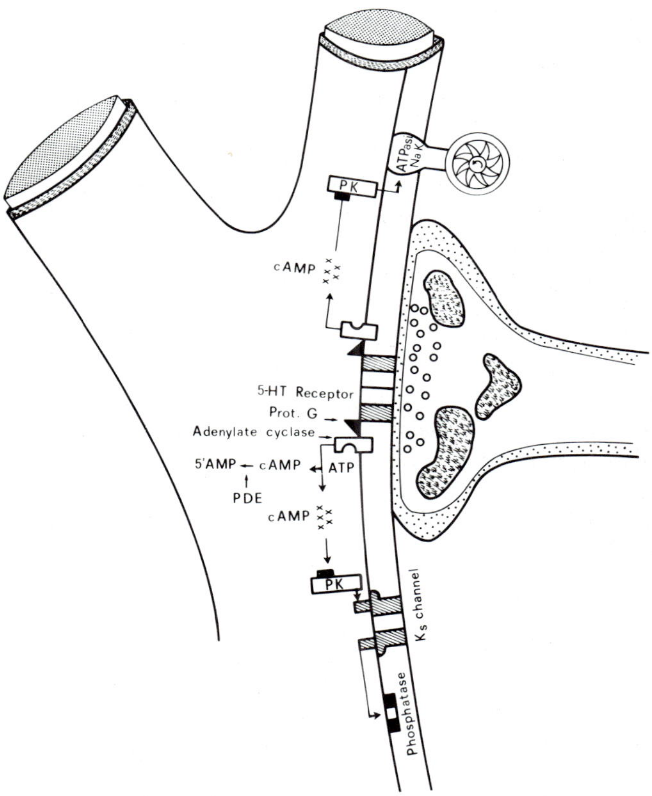

Fig. 2 Schematic drawing of the cascade of events occurring at the level of the T neuron axonal branching when serotonergic cells are activated.

In an attempt to sort out the cellular interconnections between the two cells we combined injections of horse radish peroxidase (HRP) in a Rz cell with Lucifer Yellow in an ipsilateral T cell. Many bundles of intermingled fibers have been observed in narrow areas indicating overlapping spots and fine interconnections. We are trying to analyse at the electron microscopic level the type of connections that exist between the two marked cells.

In conclusion it seems that similar short-term plastic changes in behavior share the following sequence characterics: 5-HT - second messenger (cAMP) - membrane phosphorylated protein - changes of ionic current. This sequence might represent the elementary language of different forms of learning processes (classical conditioning and sensitization) and therefore of plastic changes. As far as we know this basic mechanism might be widespread and it is still uncertain whether more complicated forms of learning processes and memory might be generated by a complex assembly of the neural networks described in this simple molecular mechanism or by different, still unknown plastic mechanisms. Most importantly, however, we have demonstrated that the studies on simple nervous systems represent a useful approach to unravel some of the cellular and molecular mysteries of fundamental changes occurring as a result of learning.

REFERENCES

Bagnoli P., Brunelli M., Magni F. and Pellegrino M., 1975, The neuron of the fast conducting system in Hirudo m.: identification and synaptic connection with primary afferent neurons, Arch. Ital.Biol., 113:21

Belardetti F., Biondi C., Colombaioni L., Brunelli M., Trevisani A., 1982, Role of serotonin and cAMP on facilitation of the fast conducting system activity in the leech Hirudo m., Brain Res., 246:89

Belardetti F., Biondi C., Brunelli M., Fabri M. and Trevisani A., 1983, Heterosynaptic facilitation and behavioral sensitization are inhibited by lowering endogenous cAMP in Aplysia, Brain Res., 288:95

Belardetti F., Brunelli M., Demontis G. and Sonetti D., 1984, Serotonin and Retzius cell depress the hyperpolarization following impulses of leech touch cell, Brain Res., 300:91

Bernier L., Castellucci V.F., Kandel E.R., Schwartz J.H., 1982, Facilitatory transmitter cause a selective and prolonged increase in Adenosine 3'-5' monophosphate in sensory neuron mediating gill and syphon withdrawal reflex in Aplysia, J. Neuroscience, 2:1682

Biondi C., Belardetti F., Brunelli M., Portolan A. and Trevisani A., 1982, Increased synthesis of cAMP and short-term plastic changes in the segsegmental ganglia of the leech, Hirudo m., Cell.Mol.Neurobiol., 2:81

Brunelli M., Castellucci V.F. and Kandel E.R., 1976, Synaptic facilitation and behavioral sensitization in Aplysia: possible role of serotonin and cAMP, Science, 194:1178

Castellucci V.F., Kandel E.R., Schwartz J.H., Wilson F.D., Nairn A.C., and Greengard P., 1980, Intracellular injection of the catalytic subunit of cAMP dependent protein kinase simulates facilitation of transmitter release underlying behavioral sensitization in Aplysia, Proc.Natl. Acad.Sci.USA, 77:7492

Castellucci V.F., Nairn A., Greengard P., Schwartz J.H. and Kandel E.R., 1982, Inhibitor of adenosine 3',5' monophosphate-dependent protein kinase blocks presynaptic facilitation in Aplysia, J.Neurosci., 2:1673

Castellucci V.F., Bernier L., Schwartz J.H. and Kandel E.R., 1983, Persistent activation of adenylate cyclase underlies the time course of short-term sensitization in Aplysia, Soc.Neurosci.Abstr., 9:169.

De Camilli P. and Navone F., 1984, Immunocytochemistry as a tool in the study of neurotransmitter actions. Trends in Pharmacol.Sci., 7:300

Eckstein F., Cassel D., Levkovitz H., Lowe M., Selinger Z., 1979, Guanosine 5'-O-(2-Thiodiphosphate) an inhibitor of adenylate cyclase stimulation by guanine nucletides and fluoride ions, J. Biol. Chem., 254:982

Hawkins R.D., Castellucci V.F., Kandel E.R., 1981, Interneurons involved in mediation and modulation of gill-withdrawal reflex in Aplysia.II Identified neurons produce heterosynaptic facilitation contributing to behavioral sensitization, J. Neurophysiol., 45:315

Hawkins R.D., Abrams T.W., Carew T.J., Kandel E.R., 1983, A cellular mechanism of classical conditioning in Aplysia: activity-dependent amplification of presynaptic facilitation, Science, 219:397

Kandel E.R.,Brunelli M., Byrne J. and Castellucci V.F., 1976, A common presynaptic locus for the synaptic changes underlying short-term habituation and sensitization of the gill withdrawal reflex in Aplysia, in: "Cold Spring Harbor Laboratory Symposia on Quant.Biol." vol.XL Editor, Cold Spring Harbor Laboratory.

Kandel E.R., Abrams T., Bernier L., Carew T.J., Hawkins R.D. and Schwartz J.H. 1983, Classical conditioning and sensitization share common aspects of the same molecular cascade in Aplysia, in: "Cold Spring Harbor Symposia on Quantitative Biology" vol. XLVIII, Editor Cold Spring Harbor

Klein M. and Kandel E.R., 1978, Modulation of voltage Ca current:mechanism for behavioral sensitization in Aplysia californica, Proc.Natl.Acad. Sci.USA, 75:3512

Klein M. and Kandel E.R., 1981, Mechanism of Ca current modulation underlying presynaptic facilitation and behavioral sensitization in Aplysia, Proc.Natl.Acad.Sci.Usa, 77:6912

Kristan W.B., 1983, The neurobiology of swimming in the leech, Trends in Neuroscience, 6:84

Siegelbaum S.A., Camardo J.S., Kandel E.R., 1982, Serotonin and cAMP close single K channel in Aplysia sensory neurons, Nature, 299:413

PROTEIN KINASE C AND SUBSTRATE PROTEIN F1 (47 kD, 4.5 pI): RELATION TO SYNAPTIC PLASTICITY AND GROWTH

Aryeh Routtenberg

Northwestern University
Cresap Neuroscience Laboratory
Evanston, Illinois 60201

I. INTRODUCTION

The reactivity of synaptic relationships is regulated by post-translational modification of brain proteins (Routtenberg, 1982a). Information storage would occur when such molecular alterations are persistent. This hypothesis was first advanced following the discovery that training-related activation of the brain of the behaving organism selectively increased the *in vitro* phosphorylation of particular brain proteins (Routtenberg et al., 1975).

There is some utility in studying the problem of memory directly by analyzing brain molecules after learning has occurred, since the pattern of brain activation by the internal and external environment is truly physiological. There is the formidable task, however, of specifying the particular environmental events that directly relate to the specific change at the molecular level. Such difficulties stand in the way of relating molecular events to information storage mechanisms at the synaptic level.

It is clearly desirable, therefore, to seek a model of learning and memory that would directly link synaptic plasticity to molecular events. Our specific goal was to determine whether a direct relation could be forged between the phosphorylation state of identified brain phosphoproteins and the altered state of synaptic strength measured electrophysiologically.

To achieve this we have studied the enhancement of synaptic reactivity in the intact hippocampus. High frequency stimulation of the perforant path produces long-term enhancement (LTE) or potentiation (LTP) of the evoked synaptic potential in the dentate gyrus. [I shall refer to short-term alterations in reactivity lasting minutes as LTP (Bliss and Lomo, 1973) and long-term alterations lasting hours or days as LTE (see McNaughton, 1982).] The molecular mechanisms required to maintain LTP may be quite different from the molecular mechanisms required to maintain LTE. A similar enhanced synaptic response can nonetheless be observed immediately after the first potentiating train and after several hours, persisting for several days or weeks in chronic preparations (Bliss and Gardner-Medwin, 1973; Barnes, 1979). The relation of this model to information storage has been discussed (Berger, 1984).

In this chapter the evidence that points to a role of a specific phosphoprotein, Protein F1, in the synaptic plasticity of LTP and LTE is first reviewed. The identification of the endogenous F1 kinase as protein kinase C is next considered. Finally, I review the morphological evidence that indicates that a specific class of synapse, the synaptic spinule, is altered during LTP. It is then proposed that Protein F1 and kinase C may participate in growth of the synaptic spinule by a coordinated increase in the size of the synaptic terminal and the post-synaptic dendritic spine. Regulation of this growth by Protein F1 kinase C is thus considered to be an essential ingredient in the cell biology of neuronal plasticity.

II. THE ROLE OF PROTEIN F1 IN PLASTICITY

The results to be briefly reviewed in this section derive from our study of brain protein phosphorylation following LTP (Routtenberg et al., 1983, 1985). In these studies the intact rat hippocampal formation was used. This selection was made, in part, because prior reports using the in vitro hippocampal slice (Browning et al., 1979; Bar et al., 1980; Tielen et al., 1983; Hoch, et al, 1984) have yielded inconsistent results. [These inconsistencies may be due to certain inherent problems in using the slice preparation to study neurochemistry (Routtenberg, 1984a, for further discussion of this issue). For example, 35-50% of the tissue taken from the slice preparation is proteolyzed and in a variable state of deterioration. In contrast, in the intact preparation, the afferent and efferent connections of the hippocampal formation are intact, as is the blood supply to this region. This integrity of neuronal and vascular elements is, in our view, a critical requirement for properly evaluating, from the biochemical point of view, the subsequent protein-specific alterations following potentiation].

To produce LTP, stimulating electrodes were placed in the perforant path to evoke synaptic responses recorded with a micropipette placed in the dentate gyrus of urethane-anesthetized rats. The region of granule cell excitation was defined both by the presence of a negative-going population spike superimposed on a positive-going population EPSP and the reversal of these field potentials during laminar profile analysis (Lomo, 1971, for example). We have therefore defined the dorsal hippocampal tissue specimen to be dissected and subsequently used for in vitro phosphorylation analysis on the basis of electrophysiological criteria (Routtenberg, 1984a, for a mapping of this input).

After a stable baseline population EPSP and spike response was established, eight trains of eight 0.4 msec cathodal pulses were delivered (1 train every 30 sec) to produce LTP or enhancement of the synaptic response (Wilson, Levy and Steward, 1981). Then the change from baseline was monitored for selected durations, at which time, the brain was rapidly frozen (< 5 sec to elimination of brain activity) in situ. Procedures for preparation of the tissue, in vitro phosphorylation and densitometric analysis of phosphate content of electrophoretically separated phosphoproteins have been described in detail elsewhere (Routtenberg et al., 1981; Morgan and Routtenberg, 1981).

Protein F1 Phosphorylation Selectively Increased Following Synaptic Enhancement

Five min after LTP a significant and selective increase in the phosphate content of the 47kD protein (F1) was observed (Routtenberg et al., 1983, 1985). No alteration was detected in the other phosphoproteins studied. The specific change in F1 phosphorylation was further restricted to the LTP condition: no alteration was observed when non-potentiated responses in dentate gyrus were evoked by stimulation of the perforant pathway at low frequency. Thus, the changes in F1 phosphorylation can be related to the **plasticity** of synaptic transmission rather than to synaptic transmission *per se*.

The molecular mechanism for this plasticity may, on the one hand, involve those processes directly related to the recorded synaptic potential (e.g., ion channel proteins, receptor proteins). Such proposals have been made (e.g., Kandel and Schwartz, 1982; Lynch and Baudry, 1984). On the other hand, the molecular change may involve a primary process that leaves the molecules specifically mediating synaptic transmission unchanged or changed only secondarily to some other process. As will be suggested in what follows this primary process may involve growth of the synaptic complex.

Long-term Alterations in Protein F1 Phosphorylation Following Enhancement

In the studies discussed earlier an increase in Protein F1 phosphorylation 5 min after the initiation of LTP was observed. In recent studies using the same urethane - anesthetized preparation, the increase in Protein F1 phosphorylation was present 1 hr after the initiation of LTP (Colley and Routtenberg, in preparation). In our view, this is a sufficiently long period of time to designate the synaptic change as LTE.

Given this persistent change in Protein F1 phosphorylation 1 hr after high frequency stimulation, we wished to determine whether a change in F1 phosphorylation lasting several days could be observed. It had previously been shown, in fact, that electrophysiological changes persist for days or possibly months (Bliss and Gardner-Medwin, 1973; Douglas and Goddard, 1975; Barnes, 1979) after the initiation of potentiating procedures. We wished to determine whether parallel long term alterations in Protein F1, or possibly some other phosphoprotein, would be detected.

To investigate this issue, in a collaborative study with C. Barnes and B. McNaughton, preparations were fitted with a chronically implanted stimulating electrode in the perforant path and a chronic recording semi-microelectrode in the dentate gyrus hilus. Three days after the initiation of LTE procedures, which were nearly identical to the procedures used to induce LTP in our earlier acute studies, the brain was rapidly frozen and subject to *in vitro* phosphorylation analysis.

Three days after LTE a selective increase in F1 phosphorylation was observed (Lovinger et al., 1985). No other phosphoprotein demonstrated a significant change. Thus, the change in Protein F1 phosphorylation is selective and persists at least for three days. A significant relation existed between Protein F1 phosphorylation and the growth of the potentiated response ($r=+0.764$, $p<.01$). Thus, the greater the persistence of the LTE three days after its initiation, the greater the F1 phosphorylation.

This latter analysis provides a new perspective on the significance of the change in Protein F1 phosphorylation: that phosphorylation of this molecule, rather than enhancing plasticity, may function to prevent decay of the synaptic growth caused by potentiation. A particular mechanism for preventing this decay will be discussed in a later section.

Identification of Protein F1 and Protein F1 Kinase

1. Protein F1 may be brain-specific and, synaptically-localized. We have sought to identify the Protein F1 substrate using two-dimensional gel electrophoresis. This "protein fingerprinting" has proven a powerful tool for identification of the Protein F1 substrate. In this method, proteins are first separated on the basis of their charge or pH. In the second dimension they are separated on the basis of their mass or molecular weight. We use both equilibrium and nonequilibirium pH gel electrophoresis (EPHGE and NEPHGE) in the first dimension to scan the entire pH range of brain phosphoproteins. We have found that the F1 protein whose phosphorylation is increased by LTP has a molecular weight of 47 kD and an isoelectric point of 4.5 (Nelson et al., 1984). Because it is also sensitive to ACTH and is phosphorylated by protein kinase C dependent processes (see what follows) it appears to have identical properties to the Protein B-50 described by Zwiers et al. (1982). The identification of Protein F1 as B-50 would mean that Protein F1 is then a brain specific protein, enriched in the synaptic membrane fraction (Oestereicher et al., 1981; 1983; Kristjansson et al., 1982). In a recent survey of monkey cortical brain regions done in our laboratory (Nelson et al., 1983), Protein F1 was concentrated in those locations shown to be involved in the representation and storage of visual information (Mishkin, 1982). Thus, Protein F1 is enriched in those brain regions which may be associated with information storage processes.

2. Protein F1 protein kinase is protein kinase C. There is recent evidence from our laboratory that Protein F1 kinase may be the same as protein kinase C, the novel Ca2+/phospholipid dependent kinase first described by Takai et al. (1977). We have demonstrated that Protein F1 phosphorylation is stimulated by calcium, by phospholipid, by phorbol esters and by exogenous kinase C (Akers and Routtenberg, 1985; Nelson et al., 1984). As recently suggested (Routtenberg, 1984a, 1984b) this finding places the regulation of synaptic reactivity within the framework of a newly discovered but critical multifunctional protein kinase C enzyme (see Nishizuka, 1984a, 1984b, 1984c for recent reviews). Because this enzyme is related to cellular growth mechanism (Akers and Routtenberg, 1984; Boynton et al., 1983), it is attractive to think that protein kinase C may engage this growth mechanism during the expression of synaptic plasticity.

Synaptic Enhancement Predicted by Protein F1 Phosphorylation State

If the change in synaptic strength following LTP or LTE depends on the state of phosphorylation of Protein F1, then one would predict that the extent of synaptic enhancement would be directly related to Protein F1 phosphate content. In support of this prediction, we found (Nelson et al, 1984) that a direct relation exists between Protein F1 phosphorylation and the change in spike amplitude at 5 min ($r=+0.904$, $p<.01$) and at 1 hr ($r=+0.704$, $p<.01$).

Since the increase in the population spike amplitude corresponds to an increase in the number of cells firing, this means that LTP causes cells to fire that did not do so prior to potentiation. As yet, the evidence using intracellular microelectrode techniques has not uncovered the mechanism of change. Conflicting claims exist as to the cellular and synaptic location of the altered synaptic communication (Barrionuevo, Kelso and Brown, 1983; Lynch et al., 1983; Taube and Schwartzkroin, 1983).

III. DENDRITIC SPINE MORPHOLOGY AFTER LTP: RELATION TO PROTEIN KINASE C AND GROWTH

The site of the cellular mechanisms for LTP may be closely associated with the cellular location of Protein F1 phosphorylation. Since Protein F1 is enriched in synaptic membranes one may suspect that LTP involves a synaptic event. It has been sugggested (Sorensen, Kleine and Mahler, 1981) that this protein (as B-50) is presynaptic, hence alterations in the presynaptic terminal following LTP would be expected. In addition, Ellis, Katz and Pfenninger (1983) have demonstrated that a phosphoprotein similar in molecular weight to Protein F1 is present in a highly purified growth cone preparation. This fraction only contains presynaptic growing axon tips, and appears free of postsynaptic material. Thus, it may be reasonable to suppose a presynaptic location for Protein F1. Whether it would be exclusively presynaptic remains to be determined. The possibility that Protein F1 is present within the post-synaptic dendritic spine has not been excluded.

Alterations in post-synaptic membrane dimensions, in fact, do occur following LTP. van Harreveld and Fifkova (1973) and Desmond and Levy (1983) have both demonstrated alterations in the size and number of dendritic spines following LTP. Desmond and Levy (1983) have also shown that the increase in the number of dendritic spines is in the precise region where activation occurs as defined by current source density analysis. This strengthens the view that the significant morphological alterations observed occur within those dendritic spines that have been activated during LTP.

It is of interest that this change in post-synaptic morphology is restricted to a class of dendritic spines that is either U-shaped or possess a synaptic spinule (Desmond and Levy, 1983). A dendritic spinule was first observed within the molecular layer of the dentate gyrus (Westrum and Blackstad, 1962). It was re-defined by Tarrant and Routtenberg (1977) to include the entire synaptic complex and was termed the "synaptic spinule". The synaptic spinule was described as a parallel apposition of pre- and post-synaptic membrane, with the post-synaptic membrane invaginating into the presynaptic terminal. The invagination occurs such that the post-synaptic density is perforated, as was first described by Peters and Kaiserman-Abramof (1971).

It is now proposed that the altered physiology of LTP is directly related to an increase in the dimensions of the spinule as the amount of apposed pre-synaptic and post-synaptic membrane is increased. Several related proposals that invoke synaptic spinule growth have appeared (Nieto-Sampedro, et al., 1982; Carlin and Siekevitz, 1983; Dyson and Jones, 1984). A schematic diagram of this proposal is shown in Figure 1. [It is apparent from serial sections (Tarrant & Routtenberg, 1979; Routtenberg, unpublished observations) that U-shaped spines are typically part of the synaptic spinule. Because the Desmond and Levy study used single sections, an increase in the number of spines counted may actually reflect an enlargement of existing spines].

It was suggested that the source of presynaptic membrane was from presynaptic vesicles and the source of postsynaptic membrane was from spine apparatus (Tarrant and Routtenberg, 1977, 1979; see Figure 1). The present proposed hypothesis suggests that the change in synaptic dimensions following LTP is mediated by an increased fusion of vesicle membrane and spine apparatus membrane.

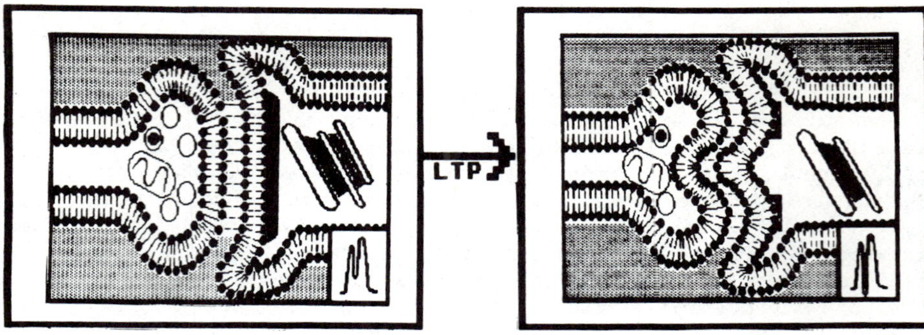

Fig. 1 Highly schematic model of dendritic spine morphological change after LTP that accompanies population EPSP and population spike increase (inset in lower right hand corner). As shown here after LTP the amount of membrane apposition increases giving rise to an entity designated the synaptic spinule (see text for further detail). It is also possible that existing synaptic spinules may increase their dimensions following LTP. Note also that the source of this membrane may derive from presynaptic vesicles (round circles in the axon terminal) and from spine apparatus (membrane saccules in the dendritic spine).

How might this increased spine dimension persist? One attractive possibility is that the recycling of membrane following fusion is prevented, thereby increasing synaptic dimensions. It is interesting that Protein F1 kinase - the C kinase - can be associated with the membrane fusion process (Moskowitz, Schook and Puszkin, 1983; Rink, Sanchez and Hallam, 1983; Nishizuka et al. 1984) raising the possibility that Protein F1 could regulate growth of pre- or post-synaptic membrane by fusion of membrane from intracellular origin.

Since LTP increases both spine dimension and Protein F1 phosphorylation, the following model is proposed: F1 phosphorylation is linked to preventing the normal recycling of membrane back into vesicles or into spine apparatus, thereby causing an increase in synaptic dimensions of synaptic spinules. The increased membrane apposition could increase the efficacy of the synaptic relationship, such that a synaptic input or set of inputs that were not capable of firing a cell previously would now be able to activate the cell with the same synaptic input. This then would account for the increase in population EPSP and population spike amplitude observed following LTP.

Based on our observation that the extent of Protein F1 phosphorylation directly relates to the extent of plasticity seen electrophysiologically, the present hypothesis predicts that the level of F1 phosphorylation would be directly related to the extent of increase in spine dimension observed by Desmond and Levy (1983).

IV. THE ROLE OF GROWTH IN THE EXPRESSION OF PLASTICITY

In a paper entitled "Plasticity recapitulates ontogeny" presented in January 1972, I suggested that certain brain proteins involved in synaptogenesis might play an important role in synaptic plasticity in the adult. In the present context, it would be interesting to know, therefore, whether Protein F1 is present in developing synapses. Studies from our laboratory (Cain and Routtenberg, unpublished observations) indicated that Protein F1 is present at birth. Moreover, as noted earlier, the report of Ellis et al. (1983) suggests the possibility that growth cones from 17-day old rat embryos may contain Protein F1.

A link between the synaptic plasticity of LTP and growing presynaptic terminals such as are found in developing organisms would be most attractive. It is unlikely, however, that such growth in the adult would be of the same magnitude or extent as occurs during development. Llinas (1979) and Llinas and Sugimori (1982) have suggested, for example, that synaptic plasticity in adults may be a controlled or isometric form of the growth process that unfolds during the development of the nervous system. Interestingly, during the outward growth of filopodia the extension of protoplasm appears to require calcium, as does LTP (Lynch et al., 1983; see also Morris et al., 1983). It has been suggested by Llinas (1979) that this growth occurs by the addition of membrane to the growing axonal tip, as was suggested here for hippocampal synapses following LTP. Thus, the mechanisms underlying the plasticity of LTP may, in a modified and controlled form, recapitulate the growth mechanisms necessary for synaptogenesis.

It is attractive to think that the mechanism that regulates the isometric growth of plasticity during LTP is closely linked to protein kinase C phosphorylation of Protein F1. Protein kinase C is associated with growth (e.g., Boynton et al., 1983), platelet activation and secretion (Kawahara et al., 1980) and morphological change (Jaken and Black, 1981). It has been proposed that following LTP intracellular calcium and diacylglycerol concentrations are elevated and lead to protein kinase C activation (Routtenberg, 1984b). Given that plasticity is under tight regulatory control, it may be that when a threshold level of both intracellular calcium and of diacylglycerol is reached, only then is protein kinase C activated. The tight control of synaptic growth in the adult is temporarily relaxed when plasticity is expressed. As emphasized by Nishizuka (1984a) it is the synergistic stimulation of protein kinase C by calcium and by diacylglycerol that is important for the full activation of this enzyme. It is suggested, then, that when this synergism occurs the process of synaptic plasticity is set into motion.

Given that there is a threshold to the process of plasticity, one would expect that alterations in Protein F1 phosphorylation would not be linear. That is, only when a threshold of both calcium and diacylglycerol is reached will kinase C activation occur. In several different experiments we have observed that the level of Protein F1 phosphorylation markedly increases at a certain calcium concentration (5×10^{-5}M), at a certain level of naloxone binding (regions high in naloxone binding are associated with information storage processes), and at a certain level of change in synaptic efficacy following LTP (Akers et al., 1983). We proposed, therefore, that Protein F1 phosphorylation may trigger mechanisms leading to a change in synaptic efficacy. The exponential nature of the relationship between functional change ([Ca^{2+}], naloxone binding, LTP) and Protein F1 phosphorylation may be taken to support this proposal.

It is apparent from such considerations that an important mechanism yet to be identified in the expression of synaptic plasticity is one which regulates the activation of Protein F1. Since Protein F1 is regulated by protein kinase C, it is reasonable to suppose that it is the control of kinase C that may regulate plasticity. That control of this molecule is a key element in cell viability is clearly (and poignantly) illustrated by its capacity for activation by the tumour promoting phorbol esters (Castagna et al., 1982; Ashendel et. al., 1983). We have shown parri passu that Protein F1 is stimulated in its phosphorylation by phorbol ester. Endogenous compounds with similar structure to these tumour promotors may be important in promoting a regulated isometric plasticity in particular brain pathways.

ACKNOWLEDGEMENT

Gratitude is expressed to Florence Sales for preparation of this manuscript. The studies were carried out by my students, Ray Akers, Patti Colley, David Lovinger and Bob Nelson who provided much healthy discussion (i.e., they did not always agree with the author's point of view) of the issues raised by these results. Support for the research contained within this review was provided by MH25281 and AFOSR 83-0335.

REFERENCES

Acosta-Uriquidi, J., Alkon, D.L., and Neary, J.T., 1984, Ca++-dependent protein kinase injection in a photoreceptor mimics biophysical effects of associative learning, Science, 224: 1254.

Akers, R.F., Cain, S.T., Gonzales-Mariscal, G., Lovinger, D.M., Nelson, R.B., and Routtenberg, A., 1983, Hypothesis: A 47kD phosphoprotein (F1) serves as molecular trigger for synaptic plasticity, Soc. Neurosci., 9: 1030.

Akers, R., and Routtenberg, A., 1984, Brain protein phosphorylation: Selective action of insulin, Life Sci., 35, 809.

Akers, R., and Routtenberg, A., 1985, Protein kinase C phosphorylates a 47kD protein directly related to synaptic plasticity, Brain Res., in press.

Ashendel, C.L., Staller, J.M., and Boutwell, R.K., 1983, Protein kinase activity associated with a phorbol ester receptor purified from mouse brain, Cancer Res., 43: 4333.

Bar, P.R., Schotman, P., Gispen, W.H., Lopes da Silva, F.H., and Tielen, A.M., 1980, Changes in synaptic membrane phosphorylation after tetanic stimulation in the dentate area of the rat hippocampal slice, Brain Res., 8: 478.

Barnes, C.A., 1979, Memory deficits associated with senescence: A behavioral and electrophysiological study, J. Comp. Physiol. Psychol., 93: 74.

Barrionuevo, G., Kelso, S., and Brown, T.H., 1983, Voltage-clamp analysis of long-term synaptic potentiation, Neurosci. Soc., 9: 103.

Berger, T.W., 1984, Long-term potentiation of hippocampal synaptic transmission affects rate of behavioral learning, Science, 224: 627.

Bliss, T.V.P., and Gardner-Medwin, A.R., 1973, Long-lasting potentiation of synaptic transmission in the dentate area of unanaesthetized rabbit following stimulation of the perforant path, J. Physiol, 232: 357.

Bliss, T.V.P., and Lomo, T., 1973, Long-lasting potentiation of synaptic transmission in the dentate area of the anesthetized rabbit following stimulation of the perforant path, J. Physiol., 232: 334.

Boynton, A.L., Whitfield, J.F., and Kleine, L.P., 1983, Ca++/phospholipid dependent protein kinase correlates to the ability of transformed liver cells to proliferate in Ca++ deficient medium, Biochem. Biophys. Res. Comm., 115: 383.

Browning, M., Dunwiddie, M., Bennett, W., Gispen, W.H., and Lynch, G., 1979, Synaptic phosphoproteins: Specific changes after repetitive stimulation of the hippocampal slice, Science, 205: 60.

Carlin, R.K. and Siekevitz, P., 1983, Plasticity in the central nervous system: Do synapses divide? Proc. Natl. Acad. Sci., 80: 3517.

Castagna, M., Takai, Y., Kaibuchi, K., Sano, K., Kikkawa, U., Nishizuka, Y., 1982, Direct activation of calcium-activated, phospholipid-dependent protein kinase by tumor-promoting phorbol esters, J. Biol. Chem., 257: 7847.

Desmond, N.L., and Levy, W.B., 1983, Synaptic correlates of associative potentiation/depression: An ultrastructural study in the hippocampus, Brain Res., 265(1): 21.

Douglas, R.M., and Goddard, G.V., 1975, Long-term potentiation of the perforant path-granule cell synapse in the rat hippocampus, Brain Res., 86: 205.

Dyson, S.E. and Jones, D.G., 1984, Synaptic remodelling during development and maturation: Junction differentiation and splitting as a mechanism for modifying connectivity, Develop. Brain Res., 13: 125.

Ellis, L., Katz, F., and Pfenninger, K.H., 1983, Characterization of phosphoproteins of isolated nerve growth cone particles, J. Cell Biol., 97: 242.

Hoch, D.B., Dingledine, R.J., and Wilson, J.E., 1984, Long-term potentiation in the hippocampal slice: Possible involvement of pyruvate dehydrogenase, Brain Res., 28: 1.

Jaken, S., Black, P.H., 1981, Correlation between a specific molecular weight form of plasminogen activator and metabolic activity of 3T3 cells, J. Cell Biol., 90: 721.

Kandel, E.R., and Schwartz, J.H., 1982, Molecular biology of learning: Modulation of transmitter release, Science, 218: 433.

Kawahara, Y., Takai, Y., Minakuchi, R., Sano, K., and Nishizuka, Y., 1980, Possible involvement of Ca++-activated, phospholipid-dependent protein kinase in platelet activation. Biochem. J., 88: 913.

Kristjansson, G.I., Zwiers, H., Oestreicher, A.B., and Gispen, W.H., 1982, Evidence that the synaptic phosphoprotein B-50 is localized exclusively in nerve tissue, J. Neurochem., 39: 371.

Levy, W.B., and Desmond, N.L., 1983, Synaptic correlates of associative potentiation/depression: An ultrastructural study in the hippocampus. Brain Res., 265: 21.

Llinas, R., 1979, The role of calcium in neuronal function. in: "The Neurosciences: Fourth Study Program." F.O. Schmitt and F. G. Worden (Eds.), M.I.T. Press, Cambridge, Mass.

Llinas, R., and Sugimori, M., 1982, Calcium conductances in purkinje cell dendrites: Their role in development and integration. in: Prog. in Brain Research., Cuenod, M., Kreutzberg, G.W. and Bloom, F. E. (Eds.). , Vol. 51, Elsevier/North Holland Biomedical Press, B.V., Amsterdam, The Netherlands.

Lomo, T., 1971, Patterns of activation in a monosynaptic cortical pathway: The perforant path input to the dentate area of the hippocampal formation, Exp. Brain Research, 12: 18.

Lovinger, D., Akers, R., Nelson, R., Barnes, C., McNaughton, B., and Routtenberg, A. Protein F1 (47kD,4.5pI) in vitro phosphorylation increased by and directly related to three day growth of long term synaptic enhancement, Soc. Neurosci., 1984, 10, 77.

Lynch, G., and Baudry, M., 1984, The biochemistry of memory: A new and specific hypothesis, Science, 224: 1057.

Lynch, G., Larson, J., Kelso, S., Barrionuevo, G., and Schottler, P., 1983, Intracellular injections of EGTA block induction of hippocampal long-term potentiation, Nature, 305: 719.

Mishkin, M., 1982, A memory system in monkeys. Phil. Trans. Roy. Soc., Lond., B298: 85.

McNaughton, B.L., 1982, Long-term synaptic enhancement and short-term potentiation in rat fascia dentata act through different mechanisms. J. Physiol., 324: 249.

Morgan, D.G., and Routtenberg, A., 1981, Brain pyruvate dehydrogenase: Phosphorylation and enzyme activity altered by a training experience, Science, 214: 470.

Morris, M.E., Krnjevic, K., and Ropert, N., 1983, Changes in free Ca++ recorded inside hippocampal neurons in response to fimbrial stimulation, Soc. Neurosci., 9: 395.
Moskowitz, N., Schook, W., and Puszkin, S., 1982, Interaction of brain synaptic vesicles induced by endogenous Ca++-dependent phospholipase A2, Science, 216: 305.
Nelson, R., Friedman, D., O'Niell, J.B., Lewis, M., and Routtenberg, A., 1983, Protein phosphorylation and opioid receptor gradients in monkey cerebral cortex: phosphorylation state of a 47kD phosphoprotein, Soc. Neurosci., 9: 585.
Nelson, R., Akers, R.F., and Routtenberg, A., 1984, Does protein kinase C activity regulate neural plasticity and its time-dependent processes? Soc. Neurosci., 10: 1180.
Nieto-Sampedro, M., Hoff, S.F., and Cotman, C.W., 1982, Perforated postsynaptic densities: Probable intermediates in synapse turnover, Proc. Natl. Acad. Sci., 79: 5718.
Nishizuka, Y., 1984, The role of protein kinase C in cell surface signal transduction and tumour promotion Nature, 308: 693. (a)
Nishizuka, Y., 1984, Protein kinases in signal transduction, TIBS, 9: 163. (b)
Nishizuka, Y., 1984, Turnover of inositol phospholipids and signal transduction, Science, 225: 1365. (c)
Nishizuka, Y., Takai, Y., Kishimoto, A., Kikkawa, U., and Kaibuchi, K., 1984, Phospholipid turnover in hormone action, Recent Prog. in Hormone Research, 40: 301.
Oestreicher, A.B., Zwiers, H., Schotman, P., and Gispen, W.H., 1981, Immunohistochemical localization of phosphoprotein (B-50) isolated from rat brain synaptosomal plasma membranes, Brain Res. Bull., 6: 145.
Oestreicher, A.B., Van Dongen, C.J., Zwiers, H., and Gispen, W.H., 1983, Affinity-purified anti-B-50 antibody: Interference with the function of the phosphoprotein B-50 in synaptic plasma membranes, J. Neurochem., 41: 331.
Peters, A., and Kaiserman-Abramof, I.R., 1969, The small pyramidal neuron of the rat cerebral cortex, Z. Zellforsch. mikrosk. Anat., 10: 487.
Rink, T.J., Sanchez, A., and Hallam, T.J., 1983, Diacylglycerol and phorbol ester stimulate secretion without raising cytoplasmic free calcium in human platelets, Nature, 305: 317.
Routtenberg, A., 1972, Plasticity recapitulates ontogeny. Talk given at The Fifth Annual Winter Conference on Brain Research at Vail, Colorado.
Routtenberg, A., 1982, Memory formation as a post-translational modification of brain proteins, In: Mechanisms and Models of Neural Plasticity. C.A. Marsden and H. Matthies (Eds.). Proc. VIth Intl. Neurobiol. IBRO Symposium on Learning and Memory, Vol. 9, Raven Press, New York, pp. 17-24.
Routtenberg, A., 1984, Brain Phosphoproteins Kinase C and Protein F1: Protagonists of plasticity in particular pathways. In: Memory Neurobiology. Lynch, G., McGaugh, J., and N. Weinberger (Eds.). The Guilford Press, New York. (a)
Routtenberg, A., 1985, Phosphoprotein regulation of memory formation: Enhancement and control of synaptic plasticity by protein kinase C and Protein F1, New York Acad. Sci., in press. (b)
Routtenberg, A., Ehrlich, Y.H., and Rabjohns, R., 1975, Effect of a training experience on phosphorylation of a specific protein in neocortical and subcortical membrane preparations, Fed. Proc., 34: 293.

Routtenberg, A., Morgan, D.G., Conway, R., Schmidt, M.J., and Ghetti, B., 1981, Human brain protein phosphorylation in vitro: Cyclic AMP stimulation of electrophoretically-separated substrates, Brain Res., 222: 323.

Routtenberg, A., Lovinger, D., Cain, S., Akers, R., and Steward, O., 1983, Effects of long-term potentiation of perforant path synapses in the intact hippocampus on in vitro phosphorylation of a 47kD protein (F1), Fed. Proc., 42: 755.

Routtenberg, A., Lovinger, D., and Steward, O., 1985, Selective increase in the phosphorylation of a 47kD protein (F1) directly related to long-term potentiation, Behav. Neural Biol., 43: 3.

Sorenson, R.G., Kleine, L.P., and Mahler, H.R., 1981, Pre-synaptic localization of phosphoprotein B-50, Brain Res. Bull., 7: 56.

Takai, Y., Kishimoto, A., Iwasa, Y., Kawahara, Y., Mori, T., Nishizuka, Y., 1977, Calcium-dependent activation of a multifunctional protein kinase by membrane phospholipids, J. Biol. Chem., 254: 3692.

Tarrant, S.B., and Routtenberg, A., 1977, The synaptic spinule in the dendritic spine: Electron microscopic study of the hippocampal dentate gyrus. Tissue & Cell, 9: 461.

Tarrant, S.B., and Routtenberg, A., 1979, Postsynaptic membrane and spine apparatus: Proximity in dendritic spines. Neuroscience Letters, 11: 289.

Taube, J.S., and Schwartzkrain, P., 1983, Intracellular tests of possible LTP mechanisms. Soc. Neurosci., 9: 287.

Tielen, A.M., De Graan, P.N.E., Mollevanger, W.J., Lopes da Silva, F. H., and Gispen, W. H., 1983, Quantitative relationship between post-tetanic biochemical and electrophysiological changes in rat hippocampal slices, Brain Res., 277: 189.

Van Harreveld, A., and Fifkova, E., 1975, Swelling of dendritic spines in the fascia dentata after stimulation of the perforant path fibers as a mechanism of post-tetanic potentiation, Exp. Neurol., 49: 736.

Westrum, L.E., and Blackstad, T.W., 1962, An electron microscopic study of the stratum radiatum of the rat hippocampus (regio superior CA 1) with particular emphasis on synaptology, J. comp. Neurol., 119: 281.

Wilson, R.C., Levy, W.B., and Steward, O., 1981, Changes in translation of synaptic excitation to dentate granule cell discharge accompanying long-term potentiation. II. An evaluation of mechanisms utilizing dentate gyrus dually innervated by surviving and sprouted crossed temporodentate inputs. J. Neurophys., 46: 324.

Zwiers, H., Jolles, J., Aloyo, V.J., Oestreicher, A.B., and Gispen, W.H., 1982, ACTH and synaptic membrane phosphorylation in rat brain. In: Brain phosphoproteins: Characterization and function. W.H. Gispen and A. Routtenberg (Eds.). Prog. in Br. Res., Vol. 56, Elsevier/North Holland Biomedical Press, Amsterdam, The Netherlands, pp. 405-417.

lian cortical neurons which were previously considered hyperdiploid (Bregnard et al., 1977), but are now believed to remain diploid throughout development (Hobi et al., 1984). In attempting to resolve the issue by incorporation studies, we realized that the cephalopod brain was too "sluggish" in DNA synthesis to warrant a continued effort. We were then led to concentrate our attention to rat brain.

In the initial experiments (Giuditta et al., 1978 a; Perrone Capano et al., 1980; 1982) we observed that approximately half of the radioactive DNA present in brain five hours after injection of ^{3}H-thymidine was lost from the tissue in the ensuing eight hours. A comparable turnover was present in a purified fraction of neuronal perikarya, but not in the liver. The differential behaviour of liver and brain did not seem to support the concept of a repair process, since there was no reason to believe that brain DNA was more sensitive to radiation damage than liver DNA. A comparable differential loss of newly-sythesized brain DNA had previously been reported in the comparison between neonatal and adult mice (Kimberlin et al., 1974). In the latter group of animals radioactive brain DNA decayed at a faster rate despite the fact that the concentration of labelled DNA was more than 30-fold smaller. The observation was certainly at odds with the idea of a radiation damage. We were also led to doubt the general validity of such an explanation by the consideration that while radiation damage may possibly be invoked to explain the decay of radioactive DNA, the initial process of ^{3}H-thymidine incorporation has to be explained otherwise, since no radiation damage may occur before incorporation of the radioactive precursor. The possibility that brain DNA is indeed subjected to a true turnover was also supported by the observation that different subcellular fractions prepared from rat brain displayed different decay rates for radioactive DNA. The fractions included large and small nuclei, mitochondria and synaptosomes (Giuditta, unpublished data).

BRAIN DNA SYNTHESIS AND LEARNING

To clarify the nature of the apparent brain turnover, we decided to follow a physiological approach and to examine whether the synthesis of brain DNA was related to brain activity. We thought that a positive outcome would definitely dispose of the repair hypothesis, bringing us closer to an understanding of the role of the metabolic fraction of brain DNA. In the first experiment we analyzed the possible effects induced by electroconvulsive treatment (ECS). We knew that ECS elicits a massive stimulation of the brain, that it induces retrograde amnesia and, on the biochemical side, that it

TABLE 1

Inhibition of ^{3}H-thymidine incorporation into brain cortex DNA by ECS.

Precursor Administration	Incorporation (%) Control	ECS	Inhibition (%)
Intracerebral	2.51±0.27(9)	1.51±0.09(9)*	40
Intraperitoneal	3.35±0.27(4)	2.31±0.26(5)*	31

*, P<0.005, with respect to control value, (Student's t test). In parenthesis, number of rats. Incorporation time, 30 minutes.

triggers a marked but reversible inhibition of brain protein synthesis (Vesco and Giuditta, 1968). We observed that in rats subjected to a single ECS the synthesis of brain DNA was significantly decreased (Table 1). The inhibition was most marked soon after ECS and declined to control values after 10-20 minutes (Giuditta et al., 1978 b). The results were encouraging, since they could not be reconciled with the idea that brain DNA synthesis reflects solely the occurrence of repair processes. In addition, in view of the amnesic effect of ECS, they suggested the possible involvement of brain DNA synthesis in learning. We were quite intrigued at that time to discover that brain DNA synthesis had indeed been related to learning in mice (Reinis, 1972; Reinis and Lamble, 1972; for a review, see Giuditta, 1983). The data of these experiments, however, appeared to have remained largely neglected or, worse, to have been disposed of as "obviously" resulting from artifacts. Of course, we had a different outlook and decided that it was worthwhile to examine the problem further using a different species and different learning paradigms.

Our initial results with rats trained for avoidance learning in a shuttle-box (Ambrosini et al., 1980; Scaroni et al., 1981; 1983) showed that after 4.5 hours of incorporation, DNA specific radioactivity remained the same in liver in all animal groups, irrespective of experimental condition, while stress and learning induced significant increments in the cerebral cortex. An involvment of stress was suggested by the observation that both in rats having learned and in rats having not learned the concentation of radioactive DNA was inversely related to the number of avoidances, i.e. directly related to the amount of stress (number of footshocks received). In addition, the regression line obtained in the learning group differed significantly from the regression line obtained in the non-learning group (Figure 1). In particular, at each level of stress, learning induced an increment in the incorporation of ^{3}H-thymidine into cortical DNA which was superimposed to the level of incorporation observed in non-learning rats displaying the same number of avoidances.

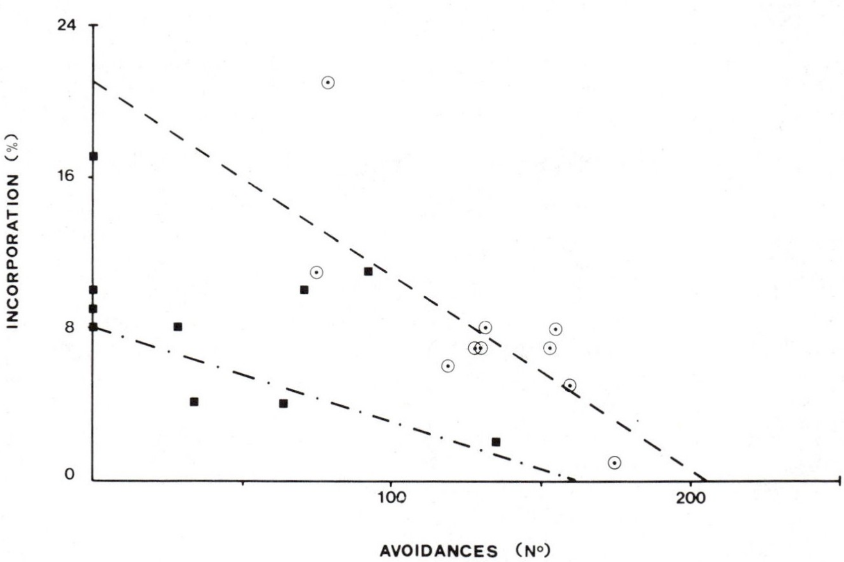

Fig. 1. Regression lines obtained by plotting values of percent incorporation of ^{3}H-thymidine into cerebral cortex DNA versus total number of avoidances scored during the incorporation period. ⊙, learning rats; ■, non-learning rats. From Scaroni et al., (1983).

Opposite results were obtained in an appetitive learning situation (Perrone Capano et al., 1981; D'Onofrio et al., 1982; Giuditta et al., submitted). In this experiment we utilized the reverse handedness paradigm in which rats are forced to use their unfamiliar paw to retrieve food pellets from a glass cylinder (Hyden and Egyhazi, 1964). In this case the period of incorporation lasted one hour and training took place in the second half of the incorporation period. In comparison to passive control rats (remaining in their home cage), learning animals displayed significantly lower values of percent incorporation in most brain regions, with the exception of the sensory-motor cortex (Table 2). Decrements of considerably lower magnitude were observed in active control rats, i.e. in animals left free to retrieve food with their familiar paw. Separation of a purified fraction of neuronal perikarya from the hippocampus and from the remaining brain revealed that the condition of reverse handedness induced even more drastic decrements in the specific radioactivity of neuronal DNA (Table 2). Using the same learning paradigm we also analyzed the possible effect of training on the degree of repetitiveness of labelled brain DNA. To this aim, purified brain DNA was heat denatured and left to renature at three different Cot values (DNA concentration x renaturation time). As shown in Table 3, radioactive DNA was

TABLE 2

Inhibition of ^{3}H- thymidine incorporation into brain DNA in rats learning the reverse handedness paradigm.

Brain Region		Incorporation (%)	Inhibition † (%)	Neuronal DNA (dpm/ g)	Inhibition † (%)
Hippocampus	PC	3.6±0.27(10)		978±79	
	AC	3.3±0.18(20)	8	584±28*	60
	L	2.8±0.11(20)*	22	366±49*	63
Entorhinal cortex	PC	4.4±0.17(23)			
	AC	3.5±0.16(29)*	21		
	L	3.0±0.15(32)*	32		
Visual cortex	PC	3.8±0.30(20)			
	AC	3.4±0.16(26)	11		
	L	2.6±0.16(32)*o	32		
Sensory-motor cortex	PC	2.4±0.11(21)			
	AC	2.3±0.16(26)	4		
	L	2.0±0.09(32)	17		
Remaining brain	PC	6.1±0.29(11)		514±67	
	AC	5.3±0.24(18)	13	421±61	18
	L	4.4±0.19(20)*o	28	160±20*o	69

PC, passive control rats; AC, active control rats; L, learning rats.
*, different from PC ($P<0.05$); o, different from AC ($P<0.05$) using Sheffé's test. In parenthesis, number of rats. †, calculated with respect to PC.

TABLE 3

Relative specific radioactivity (RSR) and percent recovery of renatured DNA purified from the remaining brain.

Cot value	Recovery (%)	PC (RSR)	AC (RSR)	L (RSR)
2	18.1±2.5	1.19±0.06	0.90 ± 0.04^{o}	0.92 ± 0.03^{oo}
100	43.5±0.9	1.09±0.01	1.29 ± 0.03^{oo}	$1.10\pm0.03^{*}$
50.000	86.3±0.5	1.03±0.01	0.95 ± 0.01^{oo}	0.93 ± 0.01^{oo}

RSR is the ratio of the specific radioactivity of renatured DNA over the specific radioactivity of total imput DNA. Other details as in Table 2. Differences from PC are indicated by ^{o}($P<0.01$) and ^{oo}($P<0.005$); differences from AC by * ($P<0.005$).

found to be widely dispersed among highly repetitive, moderately repetitive and single copy DNA fractions. Training in the reverse handedness paradigm induced significant shifts in the distribution of radioactivity among the fractions. Significant variations were also observed in the group of active control rats. It may be concluded that newly-syntesized brain DNA is not a copy of a single or of a few segments of the genome, as it could be expected of amplification processes, but is rather of widely heterogeneous composition.

Perhaps the main conclusion to be drawn from the above results concerns the association existing between brain DNA synthesis and the animal's experience and behaviour. On the basis of these data and of comparable work (Reinis and Lamble, 1972; Ashapkin et al., 1983) it seems difficult to maintain that brain DNA turnover is concerned solely with repair processes. What is then the role played by newly-synthesized DNA in learning? Our data indicate that neuronal DNA is involved and that significant behaviour-dependent changes occur in the nuclear and mitochondrial fractions prepared from brain (Giuditta et al., submitted). We know furthermore that two-way avoidance training enhances brain DNA synthesis (Ambrosini et al., 1980; Scaroni et al., 1983), while reverse handedness decreases it (Perrone Capano et al., 1981). The latter observation imposes serious constraints to possible hypothesis made at the molecular level. We have already mentioned that amplification events appear to be of rather limited relevance in explaining the available data. Other authors have emphasized the role of reverse transcription in learning (Salganik et al., 1983) or have suggested that the increased DNA synthesis observed during acquisition of an active avoidance task may be due to a transcription-dependent DNA repair (Ashapkin et al., 1983). While the contribution of these two possible mechanisms remains to be elucidated, attention should be paid to the decrease of radioactive brain DNA observed in the reverse handedness paradigm.

TRANSPOSITION HYPOTHESIS

On purely speculative grounds, we have been attracted by the possibility that brain DNA synthesis might be reflecting the occurrence of on-going transposition events (Lewin, 1983). Reshuffling of the neural genome might conceivably be related to the modified experiential inputs and behavioural outputs associated to learning. The striking potential versatility of a combinatorial mechanism such as this should not be underestimated as a possible molecular counterpart of the multitude of associations a brain can make among information bits. If this is the case, it may not be unreasonable to expect that the steady-state level of transposition events occurring in the neural genome be stimulated by a complex learning problem involving a considerable amount of stress (shuttle-box training) and might conversely be inhibited by an appetitive learning task involving little or no stress. It is worth mentioning that in recent experiments carrried out with rats subjected to habituation, we have found that DNA synthesis is likewise decreased in selected brain regions (Lamberti et al., 1983). Habituation being a relatively simple and unstressful form of learning, the latter result may also be in line with the transposition hypothesis.

BRAIN DNA TURNOVER, LEARNING AND SLEEP

There is, however, an additional observation that needs to be considered when making hypotheses on the nature of newly-synthesized brain DNA. The observation regards the fate of radioactive brain DNA in the period of sleep which follows training (Ambrosini et al., 1983; Giuditta et al., in press). Our experimental design was based on the accepted role of sleep as a period of information processing for the brain (McGrath and Cohen, 1978) and on the assumption that the molecular correlates of newly-acquired information might undergo a comparable degree of processing in the immediate posttraining sleep period. Two-way active avoidance in rats appeared to be an appropriate learning task to test our hypothesis, since this learning paradigm has been shown to induce a significant increase in paradoxical sleep (PS) in the immediate posttraining period (Leconte et al., 1973). In our experiment, rats prepared a week in advance for EEG recording were given an intraventricular injection of ^{3}H-thymidine and, half an hour later, were subjected to shuttle-box active avoidance training for 4 hours. In the following three hours they were left free to rest while their EEG activity was being recorded. Control rats were treated likewise but were not exposed to the training procedure. When the concentration of radioactive DNA was compared to the degree of learning as well as to various sleep parameters, it became clear that in the group of trained rats which had not learned the avoidance task, a striking inverse correlation was present between the DNA specific radioactivity of four different brain regions (cortex, hippocampus, cerebellum and brain stem) and the amount of PS determined in the posttraining period. No such relationship was present in learning animals nor in the control rats. Since DNA synthesis was likely to have largely occurred early in the incorporation period, the results were interpreted to indicate a loss of radioactive DNA synthesized during training. The loss appeared to require a threshold amount of PS to become effective. On the whole, the data were consistent with the interpetration that brain DNA is synthesized in concomitance with the acquisition step and is retained in the tissue only if an appropriate adaptive response is forthcoming. In the absence of such response (non-learning animals) newly-synthesized brain DNA is disposed of during the following sleep period. The data may also be viewed as providing experimental support to the hypothesis that PS is involved in the shedding away of irrelevant brain information (Crick and Mitchinson, 1983).

Can these results be accomodated in the transposition hypothesis? On a speculative vein, it may be thought that the loss of radioactive DNA observed in non-learning rats reflects a process of deletion of DNA segments initially transposed during the acquisition step. This reverse mechanism might be considered an appropriate molecular counterpart of the process of information disposal, in the same way as transposition might be the molecular correlate of acquisition. On a more cautious vein, we believe it is too early to decide what mechanism modulates brain DNA synthesis, particularly in view of the lack of satisfactory information on the nature of the DNA segments involved in learning and in their subsequent modification in the posttraining sleep period.

REFERENCES

Ambrosini, M.V., Federici, F., Giovannini, E., Giuditta, A., Principato, G., and Scaroni, R., 1980, Biosintesi di DNA cerebrale durante l'apprendimento, 32° Congr. Soc. Ital. Fisiol., Abstracts, 75.

Ambrosini, M. V., Scaroni, R., Chiurulla, C., and Giuditta, A., 1983, Influence of sleep on cerebral DNA synthesized during learning. Neurosc. Lett. Suppl., 14:S8.

Ashapkin, V.V., Romanov, G.A., Tushmalova, N.A. and Vanyushin, B.F., 1983, Selective DNA synthesis in the rat brain induced by learning. Biokhimija, 48:355.

Bregnard, A., Kuenzle, C.C. and Rush, F., 1977, Cytophotometric and autoradiographic evidence for post-natal synthesis in neurons of the rat cerebral cortex, Exp. Cell Res., 107:151.

Briggs, M.H. and Kitto, G.B., 1962, The molecular basis of memory and learning, Psychol. Rev., 69:537.

Brown, I.R. and Church, R.B., 1971, RNA transcription from non-repetitive DNA in the mouse, Biochem. Biophys. Res. Commun., 42:850.

Chaudari, N. and Hahn, W.E., 1983, Genetic expression in the developing brain, Science, 220:924.

Crick, F. and Mitchinson, G., 1983, The function of dream sleep, Nature, 304:111.

De Marianis, B., Olmo, E. and Giuditta A., 1979, Excess DNA in the nuclei of the suboesophageal region of octopus brain, J. Compar. Neurol., 186: 293.

D'Onofrio, G., Perrone Capano, C., Toniatti, C., Hyden, H. and Giuditta, A., 1982, Effetto di un apprendimento di tipo appetitivo sulla sintesi del DNA cerebrale del ratto, 28° Congr. Soc. Ital. Biochim., Abstracts, 469.

Gaito, J., 1961, A biochemical approach to learning and memory, Physiol. Rev., 68:288.

Giuditta, A., 1983, Role of DNA in brain activity, in "Handbook of Neurochemistry", 2nd Edition, A. Lajtha, ed., Plenum Press, New York, 5:251.

Giuditta, A., Abrescia, P. and Rutigliano, B., 1978 a, Metabolic DNA in brain, Neurosci. Lett. Suppl., 1:S82.

Giuditta, A., Abrescia, P. and Rutigliano, B., 1978 b, Effect of electroshock on thymidine incorporation into rat brain DNA, J. Neurochem., 31:983.

Giuditta, A., Ambrosini, M.V., Scaroni, R., Chiurulla, C. and Sadile A., Effect of sleep on cerebral DNA synthesized during shuttle-box avoidance training, Physiol. Behav., in press.

Giuditta, A., Libonati, M., Packard, A. and Prozzo, N., 1971, Nuclear counts in the brain lobes of Octopus vulgaris as a function of body size, Brain Res., 25:55.

Grouse, L., Chilton, M.D. and McCarthy, B.J., 1972, Hybridization of RNA with unique sequences of mouse DNA, Biochem., 11:798.

Hahn, W.E. and Laird, C.D., 1971, Transcription of non repeated DNA in mouse brain, Science, 173:158.

Hobi, R., Studer, M., Ruch, F. and Kuenzle, C.C., 1984, The DNA content of cerebral cortex neurons. Determinations by cytophotometry and high performance liquid chromatography, Brain Res., 305:209.

Hyden, H., 1959, Biochemical changes in glial cells and nerve cells at varying activity, in "Proceed. Fourth Intern. Congr. Biochem.: Biochemistry of the Central Nervous System", Pergamon Press, Oxford, 3:64.

Hyden, H. and Egyhazy, E., Changes in RNA content and base composition in cortical neurons of rats in a learning experiment involving transfer of handedness, Proceed. Nat. Acad Sci. US, 52:1030.

Kaplan, B.B., 1983, RNA-DNA Hybridization: Analysis of gene expression, in "Handbook of Neurochemistry", 2nd Edition, A. Lajtha, ed., Plenum Press, New York, 2:1.

Kimberlin, R. H., Shirt, D.B. and Collins, S.C., 1974, The turnover of

isotopically labelled DNA in vivo in developing, adult and scrapie-affected mouse brain, J. Neurochem., 23:241.
Lamberti, C., Sadile, A., Cerbone, A., Menna, T., Buono, C. and Giuditta, A., 1983, Brain DNA synthesis during non-associative learning, 3rd Capo Boi Conference on Neuroscience, Abstracts, 98.
Leconte, P., Hennevin, E. et Block, V., 1973, Analyse des effet d'un apprentissage et de son niveau d'acquisition sur le sommeil paradoxal consecutif, Brain Res., 49:367.
Lewin, B., 1983, "Genes", Wiley, New York.
McGrath, M.J. and Cohen, D.B., 1978, REM sleep facilitation of adaptive waking behaviour: a review of the literature, Physiol. Bull., 85:24.
Merits, I. and Cain, J., 1969, Rapid loss of labeled DNA from rat brain due to radiation damage, Biochim. Biophys. Acta, 174:315.
Merits, I. and Cain, J., 1970, Loss of labeled DNA from rat brain following injection of precursors with high specific radioactivity, Biochim. Biophys. Acta, 209:327.
Perrone Capano, C., D'Onofrio, G. and Giuditta, A., 1980, DNA turnover in rat cerebral cortex, 3rd Meeting Europ. Soc. Neurochem., Abstracts, 628.
Perrone Capano, C., D'Onofrio, G. and Giuditta, A., 1982, DNA turnover in rat cerebral cortex, J. Neurochem., 38:52.
Perrone Capano, C., Giuditta, A., Vitale, A. and Hyden, H., 1981, Behaviour-dependent change in cerebral DNA turnover, 5th Europ. Neurosc. Congr., Neurosc. Lett. Suppl., 7:S244.
Reinis, S. and Lamble, R.W., 1972, Labeling of brain DNA by ^{3}H-thymidine during learning, Physiol. Chem. Phys., 4:335.
Reinis, S., 1972, Autoradiographic study of ^{3}H-thymidine incorporation into brain DNA during learning, Physiol. Chem. Phys., 4:391.
Salganik, R.I., Parvez, H., Tomson, V.P. and Shumskaya, I.A., 1983, Probable role of reverse transcription in learning: correlation between hippocampal RNA-dependent DNA synthesis and learning ability in rats, Neurosc. Lett., 36:317.
Scaroni, R., Ambrosini, M.V., Principato, G.B., Federici, F., Giovannini, E. and Giuditta, A., 1981, Cerebral DNA synthesis during acquisition of an active avoidance task, 8th Meeting Intern. Soc. Neurochem., Abstracts, 406.
Scaroni, R., Ambrosini, M.V., Principato, G.B., Federici, F., Ambrosi, G. and Giuditta, A., 1983, Synthesis of brain DNA during acquisition of an active avoidance task, Physiol. Behav., 30:577.
Vesco, C. and Giuditta, A., 1967, Pattern of RNA synthesis in rabbit brain, Biochem. Biophys. Acta, 142:385.
Vesco, C. and Giuditta, A., 1968, Disaggregation of brain polysomes induced by electroconvulsive treatment, J. Neurochem., 15:81.

HIPPOCAMPAL MOSSY FIBER DISTRIBUTION AND TWO-WAY AVOIDANCE LEARNING IN RATS AND MICE

H. Schwegler, B. Heimrich, W.E. Crusio, and H.P. Lipp*

Institut für Anthropologie und Humangenetik, D-6900 Heidelberg

*Anatomisches Institut der Universität CH-8057 Zürich

INTRODUCTION

The genetic basis of animal learning has been studied extensively in the last two decades. Extensive overviews have been provided by Fuller and Thompson (1978) and Wimer and Wimer (1984). Our group in Heidelberg has previously studied these problems by investigating different behavioral parameters such as two-way avoidance performance, water-maze learning, and operant conditioning in mice (Buselmaier et al., 1982; Schwegler and Buselmaier, 1982). Using a classical mendelian and a half-diallel cross we were able to establish a clear genetic basis for shuttle-box and water-T-maze learning in mice. The next step was to determine the neurobiological correlates of these differences in learning.

The notion that heritable variations in specific brain parameters cause variations in learning abilities in animals has had an intuitive appeal but little experimental support. Genetically dependent variations of brain mechanisms have been demonstrated by the use of neurochemical (Will, 1977) and neurophysiological techniques (Izquierdo, 1972; Lipp, 1979). Thus far, morphological studies have been concentrated on volumetric measures of the mouse brain. For example, Wimer and Wimer (1971) and Roderick et al. (1976) found covariations between neocortical/ hippocampal volume ratios, exploratory behavior and passive shock avoidance. Furthermore, it has been tried to correlate brain weight and different learning parameters (Jensen, 1977). In addition to the volumetric measures one might expect more discrete variation in neural circuitry patterns underlying behavioral differences. The hippocampal formation seemed a very promising structure to study, as it is known to show strain differences in mossy fiber distribution patterns (Barber et al., 1974), in the density of dentate granule cells (West et al., 1978; Wimer et al., 1983), and in its lamination (Vaughn et al., 1977; Nowakowski, 1983). As lesions of the hippocampus and adjacent brain areas lead to an improvement of shuttle-box learning (e.g. O'Keefe and Nadel, 1978) we felt it fruitful to look for correlations between two-way avoidance learning and hippocampal parameters in different strains of rodent species.

First, we investigated the hippocampi of rats selectively bred for good or poor avoidance acquisition. Second, we looked for associations between behavior and morphology in mouse inbred strains known to differ in their learning performance and/or their hippocampal circuitry patterns (Schwegler and Lipp, 1983). Our third genetic approach examined behavioral and neural variables in a random-bred mouse population to find out whether the association discovered in rats and mice was also present here (Schwegler et al., 1981). These studies demonstrated that the number of intra- and infrapyramidal mossy fiber synapses (iip-MF) was strongly negatively correlated with shuttle-box acquisition and performance. Moreover, early postnatal treatment with thyroxine to produce hyperthyroidism led to an impairment of shuttle-box learning and to an increase of the number of these synapses as shown in rats and mice (Lipp et al., 1984). Finally, we also studied the genetic architecture of the mossy fiber distribution.

METHODS

Animals

In order to check the correlation between behavior and morphology we used Roman High-Avoidance (RHA) and Roman Low-Avoidance (RLA) rats which have been selectively bred for avoidance acquisition for more than 40 generations, 7 mouse strains: C3H/HeJ, DBA/2, C57BL/6, BALB/c, SM/J (inbred), NMRI, ICR (partially inbred) from our own stocks, and a random-bred mouse population (Albino Füllinsdorf). Thyroxine experiments were performed using RHA rats and DBA and BALB/c mice with different doses and control groups (for details see Lipp and Schwegler, 1982). For the classical genetic analysis of synaptic distribution patterns we used DBA (P_1) and C3H (P_2), F_1 (C3H x DBA), F_2, B_1 (F_1 x DBA), and B_2 (C3H x F_1). The same design was employed to investigate the genetics of the hippocampal lamination. Here we used BALB/c (showing exclusively supra- and intrapyramidal mossy fiber synapses) and SM/J (showing supra- and infrapyramidal pattern) (Barber et al., 1974).

Apparatus and Procedure

The shuttle-box and testing procedure have been described elsewhere (Buselmaier et al., 1982). In brief, we used a light as the conditioned stimulus (5 sec for rats, 10 sec for mice) and scrambled shock as unconditioned stimulus. The intertrial period was 30 sec for rats and 10 sec for mice. Rats were tested for acquisition (4 consecutive correct trials within the first 30 trials), mice were trained for performance (5 days, 80 trials per day).

Morphometry and Statistics

To visualize the hippocampal synaptic fields we used Timm's silver sulfide stain of group IIb metals, especially zinc, which are associated with synaptic enzymes (Danscher and Zimmer, 1978). Since the hippocampal terminal fields are arranged in laminae or clusters they can be seen as colored bands or patches (see Schwegler and Lipp, 1983), the most prominent ones among these being the clusters of the darkly staining mossy fiber terminals. These fibers are the axons of the dentate granule cells which project predominantly upon the most proximal parts of the apical dendrites of the pyramidal neurons (suprapyramidal mossy fibers). A smaller part takes its course within or below the pyramidal cell layer (intra- and infrapyramidal mossy fibers) (Fig. 4). This projection is variable and depending on genotype. Most rat and mouse strains show both intra- and infrapyramidal fibers, except BALB/c and SM/J. The various hippocampal fields of 10 (or 5 in thyroxine experiments and genetic analyses) consecutive frozen horizontal sections from the midsepto-temporal level were drawn by means of a projection microscope. Areas were measured on a graphics tablet connected to a desk computer. The areas covered by the iip-MF were determined by overlying the regions in question with a point grid and counting the numbers of the points over the silver deposit. For each animal, areas of the strata, the mossy fiber terminal fields, and the whole CA3 were summed for all sections and the relative sizes of the various areas were expressed as percentages of the CA3.

The results of the classical cross between DBA and C3H were analyzed with Collins' nonparametric method (Collins, 1967, 1968; Tully and Hirsch, 1982) and with parametric quantitative-genetic methods, i.e. Kerbusch's searching procedure for transformations and models (Kerbusch et al., 1981; Crusio, 1984) to detect possible major-gene effects. The F_2 between BALB/c and SM/J and the backcross B_1 (F_1 x BALB/c) were analyzed with a chi-square test for a mendelian two-gene model.

RESULTS

Selectively-bred Rats

In the first study we compared the hippocampi of RHA- and RLA-rats in order to check for a correlated response in morphology to the behavioral selection. In 6 RHA and 7 RLA rats we determined the areas of CA3 terminal fields. No structural anomalies could be detected in either line, but they revealed one apparent difference: the iip-MF projection in RLA rats, although relatively small, was almost twice as large as in RHA (3.6% vs. 2.1%, $p < 0.001$, Mann-Whitney U-test) (Fig. 1). Among the other terminal fields we found significant differences in the relative size of the stratum oriens (RLA < RHA) and in the stratum lacunosum-moleculare (RLA < RHA).

Unfortunately, no behavioral measures of these rats were taken. However, since the selection procedure was still being continued at the time of our experiment and since inbreeding has been avoided (cf. van der Staay et al., 1983) a strong behavioral difference between these lines can be assumed. Individual correlations between trials to criterion and all hippocampal variables were calculated for saline treated RHA rats. The highest correlation was found with the iip-MF score (r=0.91; df=9; $p < 0.01$). The multiple correlation coefficient between avoidance and all hippocampal variables was 0.985 ($p < 0.01$). Thus, almost all behavioral variability can be accounted for by the variation in the hippocampal parameters. We estimated the contribution of each hippocampal variable to the behavioral variability using multiple linear regression techniques. Here, the three most important factors showing significant partial regression coefficients were iip-MF ($p < 0.001$), stratum pyramidale ($p < 0.01$), and stratum radiatum ($p < 0.05$).

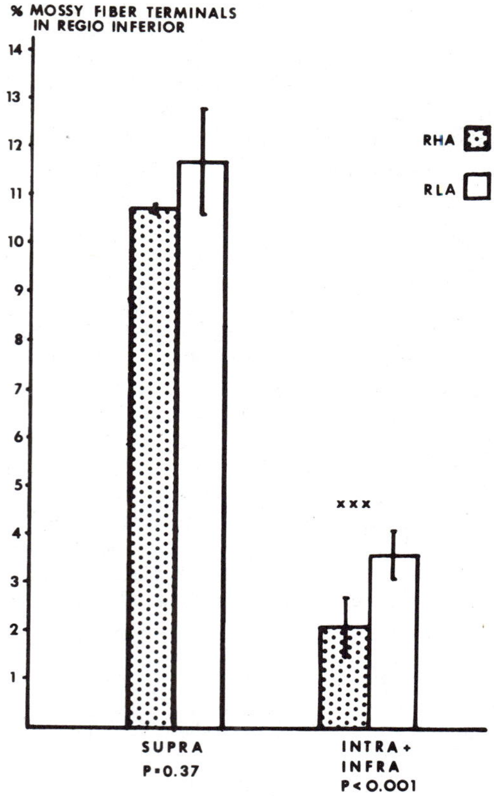

Fig. 1. Percentage of supra- and intra/infrapyramidal mossy fiber synapses in RHA and RLA rats (vertical bars: S.E.M.)

Mouse strains

We were now interested in looking for a similar relation between two-way avoidance and hippocampal parameters in mice. Visual inspection of the different mouse strains revealed differences in the mossy fiber distribution patterns. It was even possible to recognize a strain by its own specific pattern (Fig. 2). Morphometric analyses revealed a highly significant negative correlation between the shuttle-box performance on the 5th day of training and the iip-MF score ($r=-0.97$, $df=5$, $p < 0.01$) although it should be noted that this was not a correlation between individual values but between strain means (Fig. 3). In later experiments, where we used 4 animals of each inbred strain (except ICR), we could confirm these results with individual values. For these 24 mice we found a correlation of $r=-0.81$ ($p < 0.001$).

Thus, these studies on mouse strains confirmed the relationship between the iip-MF score and two-way avoidance acquisition found with selectively-bred rats.

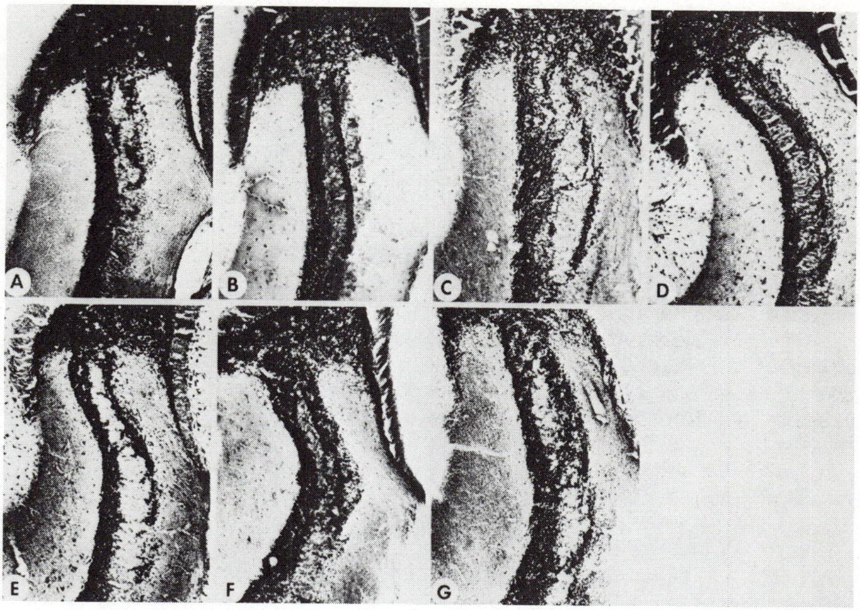

Fig. 2. Genetic variation in the distribution of intra- and infrapyramidal mossy fiber projections among seven mouse strains. A = DBA, B = BALB/c, C = NMRI, D = C57BL/6, E = SM/J, F = C3H, G = ICR. Note the different distribution pattern of BALB/c and SM/J. BALB/c have exclusively supra- and intrapyramidal mossy fiber synapses, SM/J supra- and infrapyramidal synapses (from Lipp and Schwegler, 1982).

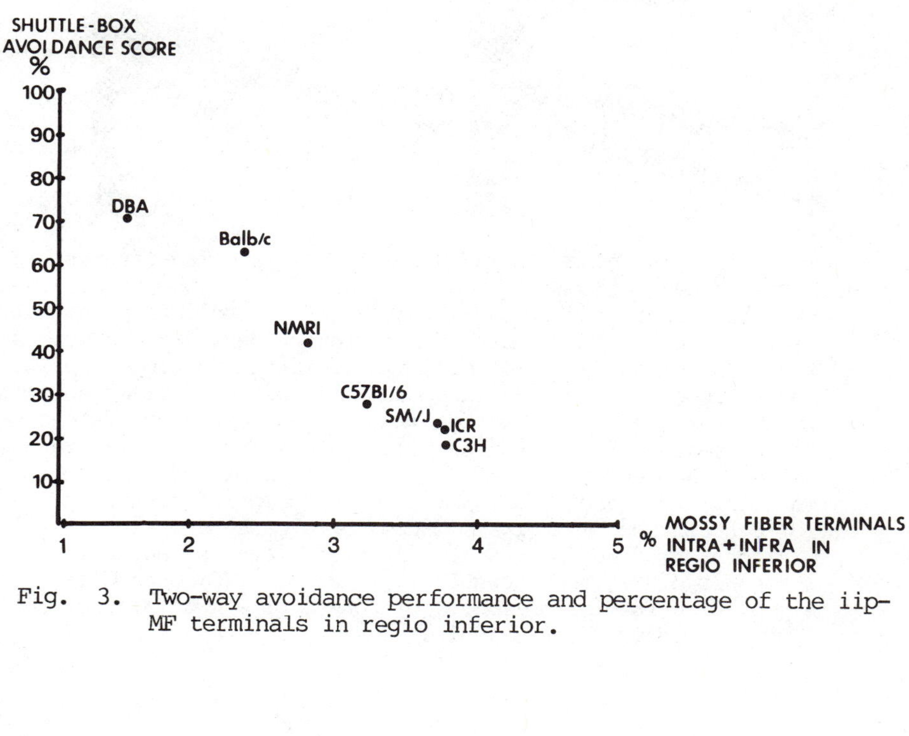

Fig. 3. Two-way avoidance performance and percentage of the iip-MF terminals in regio inferior.

Outbred Mouse Strain

The purpose of this experiment was to check whether the described correlation still holds independently from the genetic background of the individual. Thus, we examined 20 mice from the randomly bred strain (albino Füllinsdorf) first for 4 shuttle box parameters (avoidance, escape, intertrial activity, and "freezing") and for morphometric measures with the corresponding conditions as those used for the inbred mice.

After the analysis a correlation matrix was calculated including all behavioral and morphological data. These data were further analyzed with linear multiple regression techniques. This procedure provided an estimation of the independent contributions of the morphological variables to the determination of a given behavioral score (for details see Schwegler and Lipp, 1983).

The shuttle-box results showed a considerable variability among the 20 animals. The highest avoidance score was 46.7%, the lowest

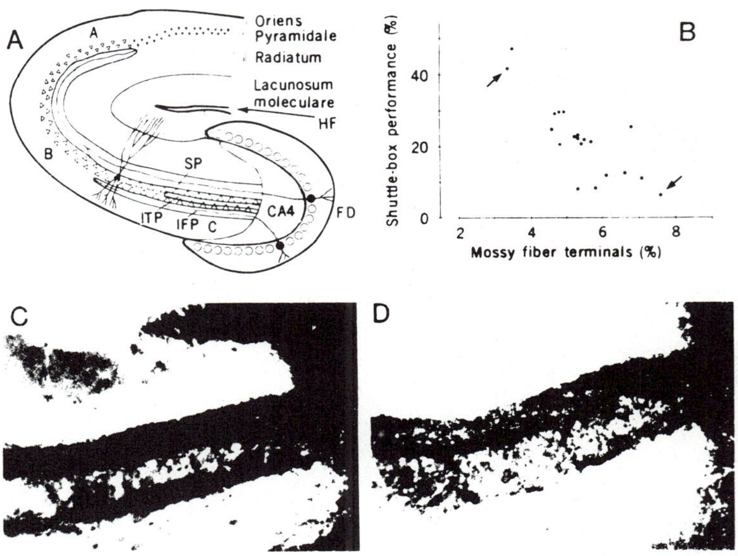

Fig. 4. Morphology and behavior (A) Schematic diagram made from a Timm-stained horizontal cross section of the hippocampus. Stippled areas: supra- (SP), intra- (ITP), and infrapyramidal (IFP) and hilar (CA4) mossy fiber projections. Strata oriens and radiatum contain the terminals of intrinsic hippocampal connections, and stratum lacunosum-moleculare receives entorhinal input from the cortex. Dotted lines: planimetric boundaries of the regio inferior to the regio superior and the fascia dentata. Abbreviations: A, B, and C, zones of the regio inferior; FD, fascia dentata: HF, hippocampal fissure. (B) Correlation plot of avoidance performance and the iip-MF score (r=-0.80, p 0.01). Arrows: scores of animals whose hippocampi are shown in (C) and (D). (C) Zone C of a poorly performing animal. (D) Zone C of a good avoider (From Schwegler et al., 1981).

5% (Fig. 4). Compared to the inbred mice the avoidance scores were rather low. Highly significant correlations were found between avoidance and freezing ($r=-0.91$, $p < 0.01$) and avoidance and intertrial activity ($r=+0.94$, $p < 0.001$).

The hippocampi did not reveal any morphological differences except for the mossy fiber distribution where we found a considerable variability (Fig. 4). Two anatomical parameters correlated significantly with avoidance performance namely iip-MF ($r=-0.80$, $p < 0.001$) and stratum pyramidale ($r=-0.65$, $p < 0.01$). Because of the interrelations between the behavioral data there was a similar negative correlation of iip-MF with the intertrial activity ($r=-0.76$, $p < 0.01$) and a positive correlation with freezing ($r=0.82$, $p < 0.01$).

The multiple regression analysis showed that after correction for the covariation among the anatomical variables only the relative size of the iip-MF projection had significant partial regressions with behavior. Thus, there is only one relation allowing the prediction of a behavioral score independent from the proportions of the other hippocampal fields: the intra-and infrapyramidal mossy fiber synapses.

Though suggestive for a causal relation, the data were still correlative. It cannot be excluded that the iip-MF projection is only a very tight association with no functional meaning. Causality could only be inferred if this neural parameter could be manipulated so as to lead to predicted alterations in behavior.

Thyroxine-treated Rats and Mice

The number of iip-MF synapses can be enlarged experimentally following early postnatal hyperthyroidism. We thus injected 51 RHA rat pups and 12 BALB/c and 45 DBA mouse pups with different doses of thyroxine during their first days of life. 24 RHA rats, 6 BALB/c and 15 DBA mice received saline injections. 15 DBA remained untreated in order to check for a possible effect of the saline treatment. These strains normally have few iip-MF synapses and show good avoidance acquisition (rats) and performance (mice). Thyroxine treatment induced a largely dose-dependent hyperplasia of the iip-MF distribution which was highly correlated with the measured shuttle-box scores of the animals (53 rats survived treatment) (rats: $r=+0.74$, $df=51$, $p < 0.001$; BALB/c: $r=-0.79$, $df=16$, $p < 0.01$; DBA: $r=-0.73$, $df=58$, $p < 0.001$). In all thyroxine treated rats and mice shuttle-box learning was impaired. In rats, the amount of failures and the number of trials to criterion were increased (Fig. 5). In mice, we found a reduction of the shuttle-box avoidance score of about 25% (Fig. 6). The behavioral changes were neither caused by freezing responses to shock nor by reduced locomotor activity. Our data imply that a reduction of shuttle box learning ability is produced by an increased number of intra- and infrapyramidal mossy fiber synapses.

Genetic Analysis of Hippocampal Variation

As seen above, genetic variability exists for two different parameters concerning the mossy fiber distribution: firstly, the number of iip-MF synapses, secondly, the distribution pattern caused by differences in hippocampal lamination.

In the first cross-breeding experiment we used DBA and C3H mice that showed considerable differences in the number of iip-MF synapses: DBA (P_1) has about one third of C3H (P_2). We examined the hippocampi of these two parental strains (14 animals each), their F_1 (14) and F_2 (53) hybrids, and their backcrosses B_1 (14) and B_2 (14) (Fig. 7). A quantitative-genetic analysis was performed using Kerbusch's and Collins' methods.

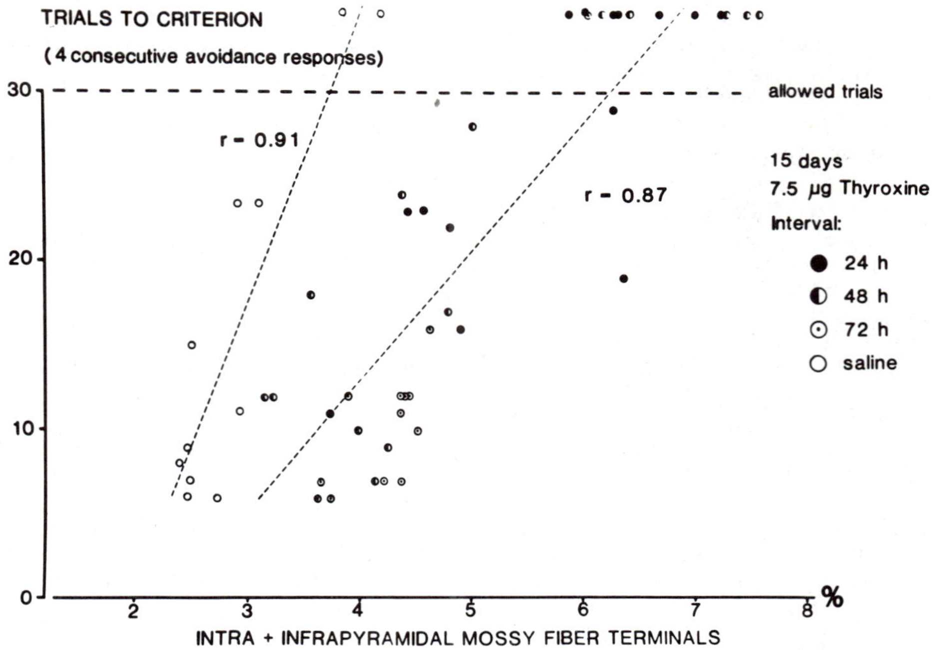

Fig. 5. Plot of all thyroxine- and saline-treated RHA-rats. Open circles: saline controls, variably filled circles: treatment groups. Because of the non-linear relation, regression was computed separately for controls and thyroxine rats.

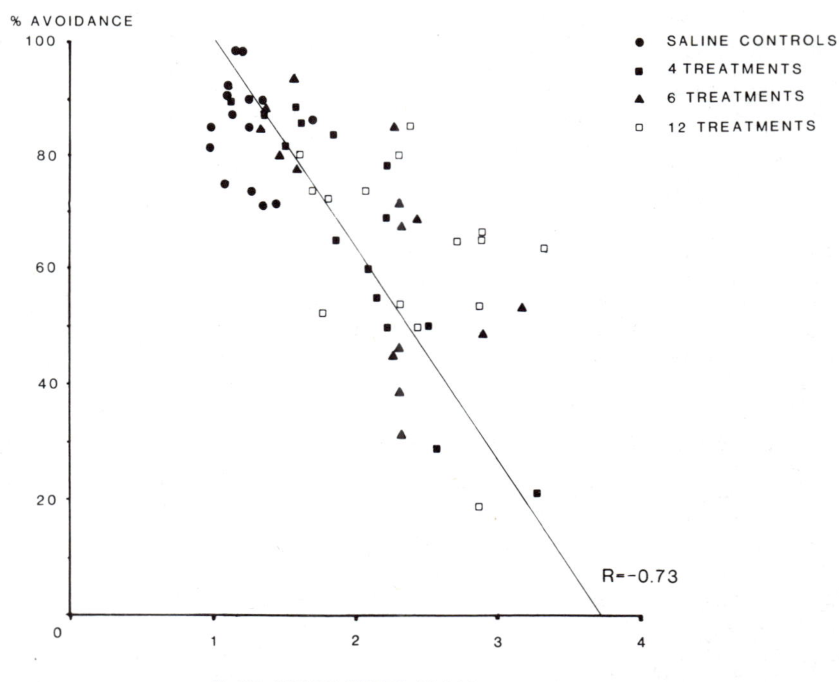

Fig. 6. Plot of all thyroxine-treated and saline-treated DBA/2 mice.

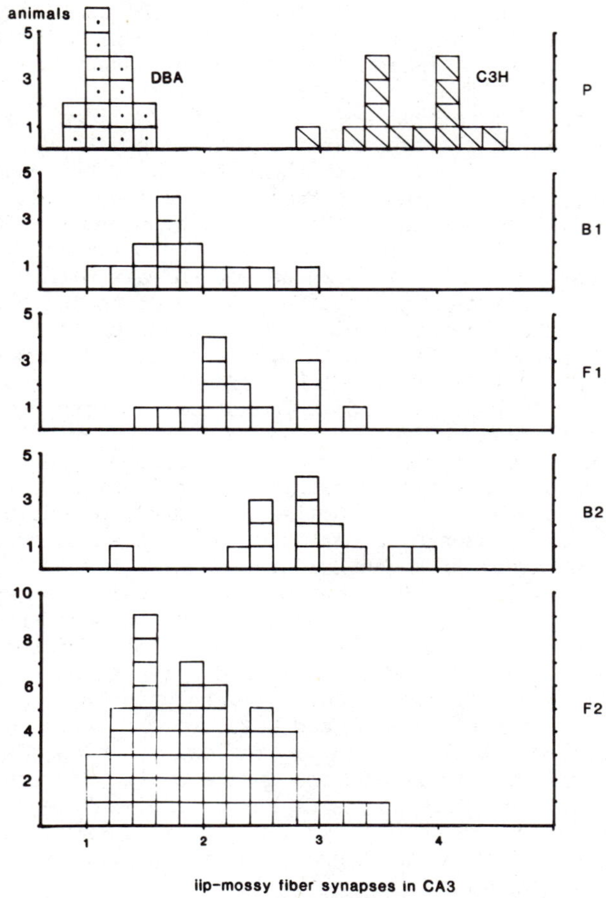

Fig. 7. Distribution of the numbers of mossy fiber synapses in C3H, DBA, F_1, F_2, B_1, B_2.

With 6 generations available, 6 genetic parameters may be fitted to the generation means: m = general mean, d = additive genetic effect, h = dominance deviation, and three parameters accounting for epistatic, that is interlocular or non-allelic interactions i.e. between homozygous combinations at different loci (i), between homozygous and heterozygous combinations (j), and between heterozygous combinations (l). Usually there is no need for a complex model containing all these parameters to account for the observed generation means. Kerbusch's method proceeds as follows. After a possible transformation to meet the usual assumptions of normality and homoscedasticity, weighted least square estimates for the genetic parameters are obtained from the observed generation means for all possible models. The number of possible models is reduced by assuming the presence of m in all models and by taking into account the interaction parameters only if the parameters of the corresponding main effects are also part of the model (thus, for instance, parameter i is considered only if d is also included in the model). The fit of the model is evaluated by means of a chi-square test applied to the expected and observed generation means. For each possible number of degrees of freedom the model with the lowest chi-square is selected as the best-fitting one of that subset of all possible models. For the resulting subset, F ratios of the two chi-squares (after division by the corresponding df) denoting the fit of two models may be formed for all combinations, allowing a comparison of the models with each other. The most parsimonious model, that is, the model explaining the highest possible amount of the variation present in the data with as few parameters as possible, is selected as the best-fitting model.

Collins' nonparametric method is based on the idea that segregating generations (B_1, B_2, F_2) are a mixture of genotypes that, under a monogenous hypothesis, are identical to those found in the nonsegregating generations (P_1, P_2, F_1). Under certain assumptions (Tully and Hirsch, 1982) this idea may be extended to a digenic hypothesis. For a particular phenotype the range of variation is divided in an arbitrary number of classes. For each class the frequencies of animals with scores in that class are obtained for each generation. Based on the observed P_1, P_2, and F_1 frequencies, expected frequencies are calculated for the segregating generations. For the F_2, for instance, the expectation under a monogenic hypothesis is $1/2\ F_1 + 1/4\ P_1 + 1/4\ P_2$. Then, the expected frequencies may be compared with the observed (by means of a G-test) to test the single-gene hypothesis.

On the square-root scale, Kerbusch's procedure rendered, according our preliminary data, a model containing the following four parameters: m=3.01, d=-0.76, h=0.41, i=0.40 (χ^2=3.368, df=2, $p > 0.10$). Due to the presence of epistasis, a partitioning of the variance was not allowed and no estimate of k (number of effective factors) could be obtained. We conclude from this parametric analysis that epistasis is present and that, contrary to what one should expect from the parental and F_1 values, dominance is in the direction of high scores. Both a one-locus (G=61.135, df=2, $p < 0.001$) and a two-locus model (G=11.701, df=2, $p < 0.01$) were rejected by Collins' analysis.

The second parameter we analyzed has been the iip-MF distribution pattern. BALB/c have supra- and intrapyramidal synapses, SM/J show a supra- and infrapyramidal MF pattern as consequence of the differentiation of the CA3 pyramidal neurons: BALB/c reveals an "outside-in" sequence, SM/J an "inside-out" sequence (Vaughn et al., 1977). It is suggested that altered temporal patterns of pyramidal neuron generation plays a role in determining the hippocampal lamination.

The F_1 mice exhibited the SM/J pattern, i.e. supra- and infrapyramidal synapses. The distribution of the F_2 and B_1 suggested a two-gene model with the SM/J pattern being dominant over the BALB/c one. Out of 67 F_2 mice 35 had the SM/J pattern, 6 had the BALB/c distribution, and 26 showed both patterns in various hippocampal sections (χ^2=0.997, df=2, $p > 0.50$, based on a hypothetical 9 : 6 : 1 distribution of the F_2). The backcross (n=16) F_1 x BALB/c revealed the following distribution: 4 mice showed the BALB/c pattern, 4 the SM/J pattern, and 8 showed both patterns in various sections (χ^2=0.00, df=2, $p > 0.99$), based on a hypothetical 1 : 1 : 2 distribution of the B_1).

CONCLUSIONS

There is substantial evidence that the proportion of mossy fiber terminals synapsing upon the basal dendrites of hippocampal pyramidal neurons correlates strongly with acquisition and performance of two-way avoidance in rats and mice. Considering our experiments, one can conclude that about 70% of the behavioral variability can be explained by the hippocampal parameters. This covariation probably depends on hereditary and environmental factors which control the development of neuronal circuitry within the hippocampal formation.

A polygenic system underlies the differences in the number of iip-MF synapses, whereas a two-gene model is implied for the lamination of the mossy fiber distribution pattern. Postnatal treatment with thyroid hormone causes an increase of the number of mossy fiber synapses. This increase is accompanied by an impairment in active-avoidance acquisition and performance. The iip-MF factor seems to be powerful enough to express itself against a genetic background exclusively established in favor of good avoidance. Thus, a causal relationship of avoidance learning with the number of iip-MF synapses is rendered even more likely. However, the exact mechanism underlying this relationship is yet to be determined.

ACKNOWLEDGMENTS

We wish to thank F. Vogel for critically reading the manuscript. The work was supported by the Deutsche Forschungsgemeinschaft (Schw 252/3) and the Swiss National Foundation for Scientific Research (Grant 3.516). WEC is a recipient of a NATO Science Fellowship awarded by the Netherlands Organization for the Advancement of Pure Research (Z.W.O.).

REFERENCES

Barber, R.P., Vaughn, J.E., Wimer, R.E., and Wimer, C.C., 1974, Genetically-associated variations in the distribution of dentate granule cell synapses upon the pyramidal dendrites in mouse hippocampus, J. Comp. Neurol. 156:417.

Buselmaier, W., Vierling, T., Balzereit, W., and Schwegler, H., 1982, Genetic analysis of avoidance learning by different psychological testing systems with inbred mice as model organisms, Psychol. Res. 40:317.

Collins, R.L., 1967, A general nonparametric theory of genetic analysis. I. Application to the classical cross, Genetics 56:551.

Collins, R.L., 1968, A general nonparametric theory of genetic analysis. II. Digenic models with linkage for the classical cross, Genetics 60:169.

Crusio, W.E., 1984, "Olfaction and behavioral responses to novelty in mice: A quantitative-genetic analysis," Thesis, Nijmegen.

Danscher, G., and Zimmer, J., 1978, An improved sulfide silver method for light and electron microscopic localization of heavy metals in biological tissues, Histochemistry 55:27.

Fuller, J.L., and Thompson, W.R., 1978, "Foundations of Behavior Genetics", CV Mosby, St. Louis.

Izquierdo, J., 1972, Hippocampal physiology: experiments on regulation of its electrical activity and the mechanism of seizure and on a hypothesis of learning, Behav. Biol. 7:669.

Jensen, C., 1977, Generality of learning differences in brain weight-selected mice, J. Comp. Physiol. Psychol. 91:629.

Kerbusch, J.M.L., van der Staay, F.J., and Hendricks, N., 1981, A searching procedure for transformations and models in a classical mendelian cross breeding study, Behav. Genet. 11:239.

Lipp, H.P., 1979, Differential hypothalamic self-stimulation behavior in RHA and RLA rats, Brain Res. Bull. 4:553.

Lipp, H.P., and Schwegler, H., 1982, Hippocampal mossy fibers and avoidance learning: 325, in: "Genetics of the brain," I. Lieblich, ed., Elsevier Biomedical Press, Amsterdam.

Lipp, H.P., Schwegler, H., and Driscoll, P., 1984, Postnatal modification of hippocampal circuitry alters avoidance learning in adult rats, Science 225:80.

Nowakowski, R.S., 1983, Single gene inheritance of an abnormality in area CA3 of the hippocampus of BALB/c mice, Soc. Neurosci. Abstr. 9:833.

O'Keefe, J., and Nadel, L., 1978, "The hippocampus as a cognitive map", Clarendon Pres, Oxford UK.

Roderick, T.H., Wimer, R.E., and Wimer, C.C., 1976, Genetic manipulation of neuroanatomical traits, in: "Knowing, Thinking, and Believing," L. Petrinovich and J.L. McGaugh, eds., Plenum Publ. Corp. NY.

Schwegler, H., and Buselmaier, W., 1982, Behavior genetic analysis of water-T-maze learning in inbred strains of mice, their hybrids and selected second generation crosses, Psychol. Res. 43:335.

Schwegler, H., Lipp, H.P., van der Loos, H., and Buselmaier, W., 1981, Individual hippocampal mossy fiber distribution in mice correlates with two-way avoidance performance, Science 214:817.

Schwegler, H., and Lipp, H.P., 1983, Heriditary covariations of neural circuitry and behavior: correlations between the proportions of hippocampal synaptic fields in the regio inferior and two-way avoidance in mice and rats, Behav. Brain Res. 7:1.

Tully, T., and Hirsch, J., 1982, Behavior genetic analysis of Phormia regina. II. Detection of a single major-gene effect from behavioral variation for central excitatory state (CES) using hybrid crosses, Anim. Behav. 30:1193.

Van der Staay, F.J., Raaijmakers, W., and Kerbusch, J.M.L., 1983, Rapid acquisition of two-way avoidance in inbred Roman low-avoidance rats, Behav. Genet. 13:435.

Vaughn, J.E., Mathews, D.A., Barber, R.P., Wimer, C.C., and Wimer, R.E., 1977, Genetically-associated variations in development of hippocampal pyramidal neurons may produce differences in mossy fiber connectivity, J. Comp. Neurol. 173:41.

West, J.M., Danscher, G., and Gydesen, H., 1978, A determination of the volumes of the layers of the rat hippocampal region, Cell Tiss. Res. 188:345.

Will, B.E., 1977, Neurochemical correlates of individual differences in animal learning capacity, Behav. Biol. 19:143.

Wimer, C.C., and Wimer, R.E., 1971, Some behavioral differences associated with relative size of hippocampus in the mouse, J. Comp. Physiol. Psychol. 76:57.

Wimer, R.E., and Wimer, C.C., 1984, Animal behavior genetics: A search for the biological foundations of behavior, Ann. Rev. Psychol., in press.

Wimer, R.E., Wimer, C.C., and Wimer, J.S., 1983, An association between granule cell density in the dentate gyrus and two-way avoidance conditioning in the house mouse, Behav. Neurosci. 97:844.

WHAT CAN THE LONG-TERM POTENTIATION PROCEDURE TELL US ABOUT THE NEURAL MECHANISMS OF LEARNING AND MEMORY?

Serge Laroche

Département de Psychophysiologie
LPN 2, C.N.R.S.
91190 Gif-sur-Yvette, France

INTRODUCTION

The theoretical background of most physiological models of learning and memory rests on the idea that persistent synaptic modifications in neural networks are the basis of information storage. The neurophysiological approaches to the study of learning and memory have thus long been concerned with such questions as how and under what circumstances do synaptic changes occur in the brain, as well as what could be the nature of the change itself.

In the network concept of memory traces, encoding is assumed to depend on the establishment of specific spatio-temporal patterns of activation in neuronal networks. According to the well known ideas promoted by Hebb (1949), a widely postulated mechanism involves a lasting increase in the permissive properties of the synapses. That is, relatively permanent fixation of acquired information is the result of a long-lasting change in some synaptic parameters.

In reviewing the several hypotheses that can be advanced in light of our present knowledge concerning the possible underlying physical mechanism(s), Bloch and Laroche (1984) have presented a number of arguments to support the hypothesis that the consolidation of the memory trace results from an increase in the transmission factor at the synapses involved in the initial activation and the structuration of relevant networks by experience. In this context, during memory registration and processing (active state of memory, see Lewis, 1979), neurodynamic engrams are conceived of as a pattern of persistent activation in a connective network, reactivated during retrieval as well. In absence of excitation from a subset of relevant input stimuli (inactive memory), the topological network, defined by the activity pattern along the connective structure, is potentially maintained by permanent marking of the connection points within the network (Bloch and Laroche, 1984).

The purpose of this chapter is to discuss the possibility that the phenomenon of long-term potentiation (LTP) be viewed as a candidate for a mechanism involved in the strengthening of synaptic connections within memory networks.

On physiological grounds, considerable experimental effort has focused on simplifying preparations in order to demonstrate forms of neural plasticity at a cellular or at a more integrative level of network connectivity. The detailed description of their properties and characteristics will set the limits to the validity of the analogy to mnemonic processes. A well known example of this model system approach is the study of LTP in the mammalian hippocampus.

LTP is considered here as an experimental model of an enduring increase in synaptic efficiency following the repeated activation of presynaptic fibers. This form of post-excitatory facilitation was discovered in synaptic elements of granular cells dendrites in the dentate gyrus which receive inputs from the entorhinal cortex via perforant path axons (Bliss and Lømo, 1973). The phenomenology as well as thorough examination of available data on the proposed mechanism(s) of LTP, can be found in the extensive reviews by Swanson et al. (1982), Voronin (1983), and Eccles (1983).

There are at least three basic properties that distinguish LTP from other forms of post-excitatory facilitation and which may be relevant to the present discussion.

The first is its extremely long duration. In fact, it has been shown to persist for days, or even weeks, after brief episodes of tetanizing stimuli (Bliss and Gardner-Medwin, 1973; Douglas and Goddard, 1975). It is important, however, to emphasize that LTP is not permanent. The change in synaptic efficacy following an LTP procedure fades out over time (Barnes, 1979), with no physiologically detectable consequences for further LTP production (Laroche and Bloch, 1982; de Jonge and Racine, 1983).

A second characteristic is that lasting enhancement of synaptic efficiency can be produced by short (tenth of ms) high-frequency stimulus trains (Douglas, 1977) which mimic the bursts of activity occurring spontaneously in hippocampal neurons (Ranck, 1973). However rough this parallel, it is often evoked when discussing the circumstances under which LTP in natural conditions would occur.

A third property is that, though LTP appears specific to the tetanized pathway (Andersen et al., 1977; McNaughton and Barnes, 1977; but see Yamamoto and Chujo, 1978; Misgeld et al., 1979; and discussion in Voronin, 1983), it requires a number of co-active convergent synapses to be set into an active state (McNaughton et al., 1978; Levy and Steward, 1979; Lee, 1983). This cooperative property has been extensively discussed in the context of the principle put forth by Hebb (1949). Indeed it appears to be essentially similar to the requirement of a correlated pre- and post-synaptic activation for an association between input events to be formed (see discussions in Barnes and McNaughton, 1980; Goddard, 1981; Swanson et al., 1982).

Because LTP exhibits these three properties, and because it appears to be associated with biochemical (Browning et al., 1979; Bär et al., 1980; Duffy et al., 1981; Routtenberg, this volume) as well as morphological (Van Harreveld and Fifkova, 1975; Lee et al., 1980) modifications of pre- and post-synaptic elements (see Bliss and Dolphin, 1982), it has been increasingly proposed as a potential substrate of memory.

These numerous findings leave LTP as an evidently good experimental preparation to evoke biochemical and structural substrates of network plasticity. Network plasticity is not, however, in itself, a model of memory, but a function upon which memorizing activity could, at least in part, be based. Indeed, LTP-like phenomena could be conceived of as a candidate mechanism through which neurodynamic engrams could be established and transformed into morphological ones.

The point is to ask whether an elementary operation in a network such as a change in activation leads to a change in LTP (corresponding to an change in a synaptic transmission factor), is actually used in the brain to store information. Long before LTP was discovered, this kind of functioning was conjectured to fulfill persistent activation in chains of autoactive neurons (Fessard and Szabo, 1961). To approach that question, it is at least necessary to know when the change occurs in the learning process, what are the neural processes that generate the change, and how long does it or its consequences persist?

To date, the properties of LTP, as well as its suspected underlying synaptic support, seem to fulfill, at least theoretically, to the requirements for a hypothetical model of a physiological mechanism subserving mnemonic processes.

DOES LTP DEVELOP AS A CONSEQUENCE OF LEARNING ?

A first logical requirement in support of this hypothesis would be that learning specifically induces changes in the amplitude of field potentials, since this measure is the usual test of LTP. For example, consider a typical electrophysiological picture of an LTP experiment as

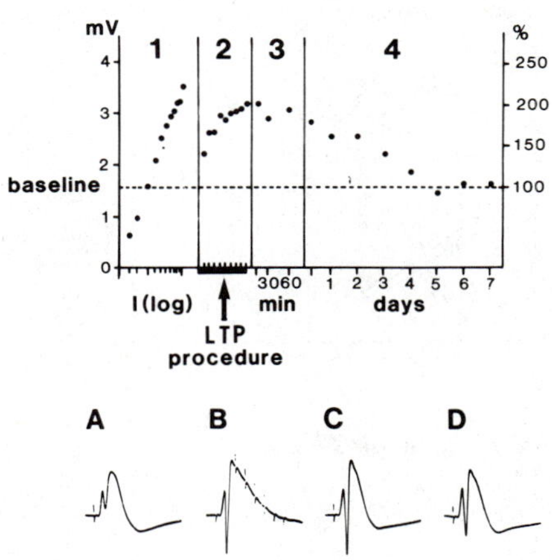

Fig. 1. Long-term potentiation of dentate field potentials. (1) Mean input/output curve showing the amplitude of the population spike to perforant path single test pulses of increasing intensity delivered during a 5-day control period. The dashed line indicates response amplitude at the intensity used in the following phases (baseline, 100 %). (2) Arrow shows the tetanization episode: 10 high-frequency (hf) stimulus trains to the perforant path (400 Hz, 20 msec, 8 pulses per train) with 5-min intervals. Plotted is mean amplitude of the responses obtained after each hf train. (3) Mean amount of LTP over the hour following tetanization. (4) Decay time-course of LTP. Inserts show examples of extracellular field potentials, to a test pulse before tetanization (A), to the tenth hf stimulus train (B), and to single test pulses delivered 2 min (C) and 2 days (D) after tetanization.

illustrated in Fig. 1. The test consists in measuring the magnitude of dentate field potentials elicited by single test pulses to perforant path fibers, before and at various delays after tetanization. At a given intensity, the 10 short high-frequency stimulus trains to the perforant path result in a considerable and long-lasting (3 days) growth in the amplitude of the population spike component of the potential elicited by test pulses identical to those delivered during the pre-tetanization period. The amount of LTP increases cumulatively with successive high-frequency stimulus trains until a plateau is reached, and then decays gradually towards the baseline level determined before tetanization.

Table 1 summarizes a number of studies in which the efficiency of synaptic transmission was investigated with the field potential technique during or after a learning experience. Though positive results have been noted, examination of table 1 shows that the overall picture is rather conflicting, not only concerning the change itself, but also in regard to the detailed time course of the change when it seems to occur.

In our own studies, we have recorded dentate field potentials elicited by perforant path single pulses, during both instrumental learning (lever-pressing for food reward) and classical conditioning (tone-shock association) in rats (Laroche and Bergis, unpublished data).

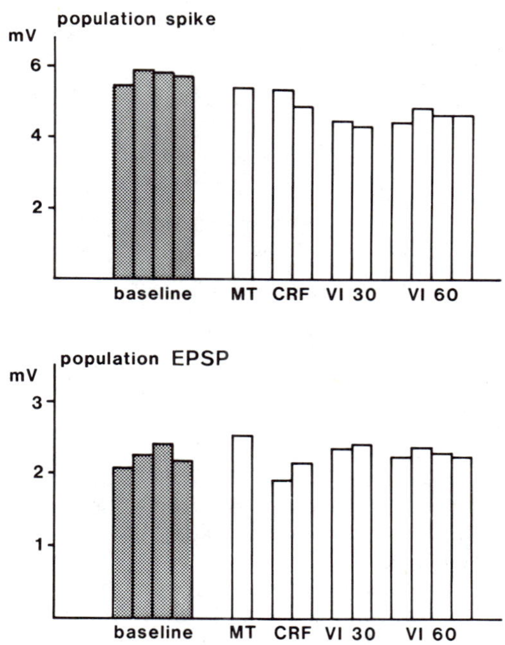

Fig. 2. Averaged amplitude of dentate field responses during instrumental learning (lever-pressing for food reward). Baseline (dotted bars) corresponds to the averaged magnitude of dentate responses recorded during four days before training sessions. During the first session (MT), animals had free access to food in the magazine. They then experienced 2 sessions of continuous reinforcement (CRF), and finally, 2 sessions of 30 sec (VI30) followed by 4 sessions of 60 sec (VI60) variable interval schedules of reinforcement.

Table 1. Characteristics of hippocampal field potentials in learning experiments

Authors	species	Task	Locus of analysis	Stimulation procedure	Direction of the change: Population EPSP	Direction of the change: Population spike	correlation with behaviour	time of analysis
Segal M. (1977)	Rat	Classical conditioning	Comm → CA1	Double pulse	1st response: no change 2nd response: ↗ (sometimes ↘)		during presentation of the CS	CS period
Jaffard R. Jeantet Y. (1981)	Mouse	Food-reinforced lever pressing (CRF)	Comm → CA1	Double pulse	1st response: ? 2nd response: ↗ 1 h post-training session	?	Low retention performance (Kamin effect)	5 min, 1-3-6-24 h posttraining session
Buzsaki G. et al. (1981)	Rat	Water-reinforced lever pressing and running task	Comm → CA1	Single pulse	no change no change	no change ↗	Drinking - grooming lever pressing - running	during lever pressing running for reward drinking grooming
			PP → DG	Single pulse	no change	↘	lever pressing running for reward	
Ruthrich H. et al.(1982)	Rat	Brightness discrimination	PP → DG	Single pulse	no change	↗ 4 h and 24 h posttraining	High retention performance	1 min, 1-4-24 h posttraining session
Thompson R.F. (1982)	Rabbit	Classical conditioning of the nictitating membrane	PP → DG	Single pulse	?	↘ Early training ↗ overtraining	In parallel with behavioural response	CS period/ 30 sec posttrial
		Pseudo-conditioning			?	↘ Early training and overtraining		
Reymann K.G. Ott T. (1983)	Rat	Escape behaviour	PP → DG	Single pulse	↗	no change in amplitude ↘ latency	Escape learning? footshock application?	posttraining session

We measured the magnitude of both the population spike, formed by the nearly synchronous discharges of granular cells (Andersen et al., 1971), and population excitatory post-synaptic potential (EPSP), which reflects the synaptic current at the activated synapses (Lømo, 1971). No significant change in the amplitude of the population EPSP occurred during the different phases of instrumental learning (Fig. 2), while a tendency for a decrease in population spike amplitude was noted at different steps of the experiment, possibly in correlation with attention shifts, though this hypothesis will have to be tested in more appropriate paradigms.

The results were similar when dentate evoked potentials were examined (either 4 minutes post-trial or 24 hours post-session) after the presentation of sensory stimuli or after paired CS-US trials in the classical conditioning experiment.

There is thus much discrepancy in the literature, and the lack of a coherent picture does not seem to depend on particular variables. Among the variables which differ largely in available studies are the locus of analysis, stimulation procedure (single versus double pulses), time of analysis (with reference to behavioral events or related stimuli) and behavioral task. These data have not provided, thus far, unequivocal support for the hypothesis linking LTP phenomena to a general learning process. Clearly more studies are needed to evaluate the necessary and sufficient conditions for a change in field potential to occur at precise loci as a result of a particular learning experience. Moreover, the analysis of evoked potentials is perhaps not the most appropriate parameter to use in testing this hypothesis. Indeed, the implicit hypothesis is that a learning experience should mimic the tetanization of fibers in its effects on monosynaptically driven field potentials. The tetanization of a pathway of fibers used in LTP experiments is obviously an artificial procedure. Thus, the fact that the usual electrophysiological description of LTP does not entirely parallel that of electrophysiological data from learning experiments, does not necessarily mean that no functional change in some synaptic parameters has occurred as a consequence of learning.

DOES LTP MODEL A SYNAPTIC ALTERATION THAT OCCURS AS A CONSEQUENCE OF LEARNING ?

Learning and synaptic remodelling

The suggestion that synaptic reorganization is induced by environmental manipulations stems from diverse conceptual frameworks and is based upon experimental evidence from diverse paradigms. For example, alterations in various morphological features of synaptic elements (including dendritic field size, frequency of dendritic spines, synaptic density length, the curvature of synaptic contacts etc.) are caused by visual deprivation during development in neonates or exposure to enriched environment in early life as well as in adulthood (for reviews see Rosenzweig and Bennett, 1976; Thompson et al., 1983; Greenough, 1984).

Evidence is emerging which indicates that synapse remodelling also occurs as a result of specific learning experiences in adult organisms (Wenzel et al., 1980; Vrensen and Cardozo, 1981; Chang and Greenough, 1982; Bottjer and Arnold, 1984) and in early learning of the characteristics of visual cues in imprinting (see Horn, 1981; Bradley et al., this volume) or one-trial avoidance learning (see Rose, this volume).

It is noteworthy that most of the changes that have been detected at the structural or ultrastructural levels of analysis concern the output or post-synaptic side of the system to which terminal and/or collateral sprouting may represent models of a potential mechanism(s) for a presynaptic correspondent (for reviews see Tsukahara, 1981; Cotman and Neito-Sampedro, 1982).

In this paper, no attempt is made to provide an extensive coverage of the literature which may be found in the reviews cited above; rather, emphasis will be placed on the circuits the synapses of which have the potential to exhibit LTP. Indirect evidence in favour of a parallel between LTP mechanisms and memory mechanisms involves the following electrophysiological, biochemical amd morphological data.

Electrophysiological data

One of the first indirect arguments came from experiments by Barnes (1979) and Barnes and McNaughton (1980) who compared the performances in a spatial discrimination task and the susceptibility of the perforant path to dentate granule cell synapses to display LTP after repeated tetanization in young and senescent rats. A positive correlation was found between the asymptotic level of performance and the persistence of LTP, both between and within groups of age.

In our own attempt to address this issue, we have demonstrated that an experimental treatment, namely, a stimulation of the mesencephalic reticular formation (MRF), has similar facilitatory influences on behavioral learning, on the development of associative changes in hippocampal neuronal activity during classical conditioning and on LTP at the perforant path to dentate granular cell synapses.

1. Behavioral studies provided strong evidence that a low intensity post-trial MRF stimulation facilitates learning in a variety of experimental situations, including active avoidance (Bloch et al., 1966), positively reinforced one-trial discrimination (Bloch et al., 1970), extinction of bar-pressing or maze learning (see Deweer, 1976).

   Recently, we have extended these findings to the learning of a classically conditioned tone-shock association in rats (Laroche and Bergis, unpublished data). The animals were initially trained to press a lever for food in an operant chamber (60 sec variable interval schedule of reinforcement: VI 60) until they achieved a stable rate of responding. In the next phase of the experiment, the rats first experienced a series of successive 10 sec tone and light presentations with the VI 60 schedule for food in effect. Then, the tone (CS+) was paired with a brief footshock (US) while the light (CS-) was not. This procedure resulted in a reduction of food-reinforced responding in the presence of the tone (conditioned suppression). The animals in one group received an MRF stimulation 10 sec after each CS-US pairings. It was found that post-trial MRF stimulation facilitated learning of the tone-shock association, as illustrated in Fig. 3.

2. It is now well documented that marked changes in hippocampal unit or multiunit activity occur progressively in response to a sensory stimulus paired with US during classical conditioning (e.g. Olds et al., 1972; Berger et al., 1976). Moreover, we have recently shown that these associative changes in neuronal activity in the hippocampus may be viewed as an index of the predictive value acquired by the CS rather than as a correlate of a specific behavioral response that it elicits as a result of its pairing with US (Neuenschwander-El Massioui et al., this volume).

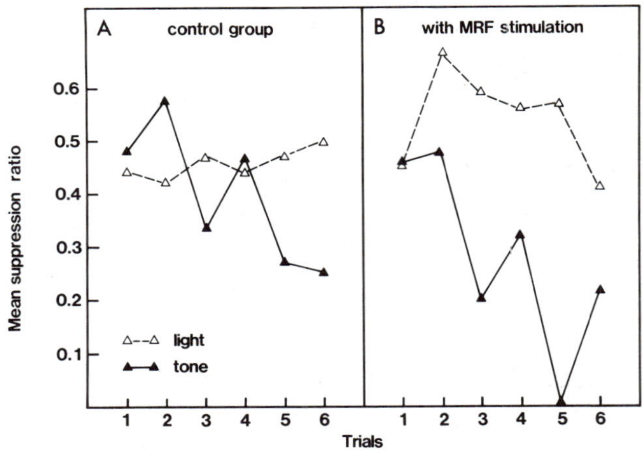

Fig. 3. Effect of post-trial MRF stimulation in classical conditioning in rats. Data are from only the first training session with 6 light (CS-) and 6 tone (CS+)-shock paired trials presentations. The first trial to which a significant conditioned suppression appeared to the tone was the sixth in control animals (A) and the third in the group of animals receiving an m.r.f stimulation after each paired trial (B).

We examined the effect of the same post-trial MRF stimulation on the development of classically conditioned increases in multiunit activity at various levels of hippocampal circuitry (Bloch and Laroche, 1981; Laroche et al., 1983). We found that this stimulation, delivered 10 sec after CS-US paired trials, facilitates the development of associative changes in hippocampal (CA3) and dentate multiunit activity and results in a long-term enhancement of the responsiveness of hippocampal cells to the CS. Furthermore, this post-trial stimulation has no effect in the entorhinal cortex, a structure in which no associative changes in multiunit activity had been detected as conditioning proceeded, showing that the facilitatory effects of post-trial MRF stimulation are specific to regions displaying learning-induced changes in multiunit activity, at least within the neuronal circuit under study.

3. Similarly, a low intensity MRF stimulation enhances the magnitude of LTP at the synapses of the perforant path on dentate granular cells, and prolongs its duration by several days, while having no effect in itself on dentate field potentials (Laroche and Bloch, 1982; Bloch and Laroche, 1985). In this series of LTP experiments, the MRF stimulation was delivered, as in our behavioral studies, 10 sec after each high-frequency stimulus train to perforant path fibers. More generally, it is interesting to note that other treatments known to disrupt or facilitate memory (e.g. electroconvulsive seizures, neuropeptide vasopressin, amphetamine) have also been shown to produce deleterious or enhancing influences on LTP, respectively (Hesse and Teyler, 1976; Chepkova and Skrebitsky, 1982; Gold et al., 1984).

An important point in examining this parallel with behavioral data is that in our studies the effectiveness of post-event MRF stimulation on LTP is time-dependent and that the temporal gradient of efficacy is similar to those found after acquisition, or reactivation of memory.

Biochemical and morphological data

In this field of research, an indirect but striking indication of a parallel between memory and LTP-associated synaptic alterations is the demonstration that a learning experience induces changes in synaptic features that are prevailingly associated with the tetanization of perforant path fibers in LTP experiments.

First, Fifkova and Van Harreveld (1978) found an enlargement of dendritic spines confined to the region of perforant path terminals after classical conditioning in the mouse. Enlargement of dendritic spines in the very same region has also been shown to occur following an LTP procedure (Van Harreveld and Fifkova, 1975; Fifkova et al., 1982). Second, Thompson et al. (1983) found an increase in glutamate receptor binding in hippocampal synaptic membranes after classical conditioning of the rabbit eyelid response. A similar process has been hypothesized to accompany the production of LTP since tetanization causes an increased uptake of labelled glutamate (Baudry et al., 1980).

Interestingly, these convergent data provide two examples of post-synaptically located alterations. It will be of great interest in learning experiments to look for other possible changes that are like the known biochemical or structural modifications that accompany LTP production. Of special importance might be their presynaptic (e.g. Browning et al., 1979; Dolphin et al., 1982) and/or post-synaptic (e.g. Lee et al., 1980; Desmond and Levy, 1983; Greenough, 1984) location. Further, in terms of hypotheses of changing network connectivity in the context of learning and memory, such experiments might allow an evaluation of the respective validity of the synaptogenesis versus stabilization of existing synapses.

The use of an LTP procedure after conditioning

If any synaptic modification involved in the elaboration or maintenance of associative cellular changes during classical conditioning resembles one or more of those associated with LTP, then it might be detected as an actual change in LTP itself, provoked by learning and modulated by a memory facilitating agent. Under this rationale, an LTP procedure was conducted by means of 10 high-frequency (400 Hz) short duration (20 msec) stimulus trains (5-min intertrain intervals) delivered to the perforant path in a single series, 2 days after the last of 6 tone (CS)-footshock (US) conditioning sessions (Laroche et al., 1983). Animals in one group received a low intensity MRF stimulation after each conditioning trial. As previously shown, the development of associative changes in dentate multiunit activity was facilitated by post-trial MRF stimulation.

Experiencing the tone-shock association resulted in an increased LTP of the perforant path to dentate granule cell synapses (Fig. 4). This LTP strengthening effect was further enhanced when conditioning had been facilitated by post-trial MRF stimulation (Fig. 4).

Thus, the development of associative changes in the dentate gyrus is accompanied by a lasting change in a synaptic transmission factor clearly detected by the LTP procedure. Moreover, this transmission factor is still more enhanced when memory registration has been facilitated by a modulatory agent. Since artificially-elicited synaptic LTP benefits from conditioning, it might be inferred that both processes share, at least partly, common underlying synaptic mechanism(s).

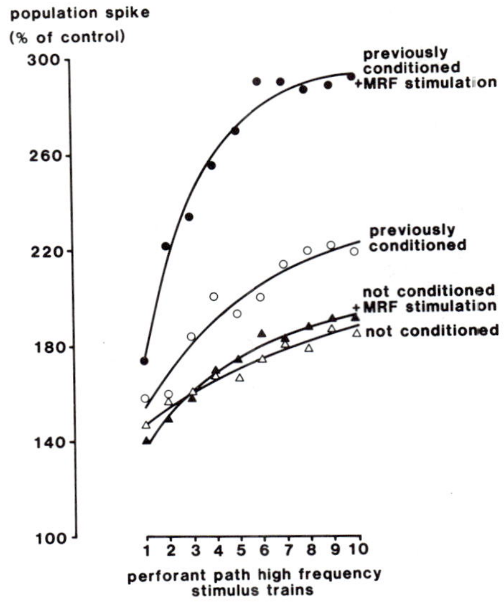

Fig. 4. Effect of classical conditioning (tone-shock association) and its facilitation by post-trial MRF stimulation on subsequent LTP of the population spike in the dentate gyrus.

Another argument in favor of this interpretation is the finding that differential conditioning can be facilitated by prior induction of LTP at the perforant path to dentate granule cell synapses (Berger, 1984). In this situation, artificially-elicited LTP may prepare the circuitry so that conditioning subsequently benefits from facilitated synaptic transmission.

CONCLUSION

On the basis of the above discussion, it would appear, thus far, tenuous to conclude that the storage of information relies on an LTP phenomenon. Alternatively, the above electrophysiological, biochemical and morphological considerations, though not individually conclusive, provide compelling evidence suggesting that LTP, as a laboratory preparation to evoke synaptic plasticity, partly models some of the mechanisms responsible for the strengthening of synaptic connections within memory networks. Furthermore, it seems difficult to assume that the firing activity triggered by learning is equivalent to the powerful synchronous activity provoked by the tetanization used in LTP experiments. This view implies that LTP is a "candidate mnemonic device" as discussed by Teyler and Discenna (1984), although in a more restricted sense.

If we accept the analogy between the mechanisms underlying tetanization-induced LTP and those responsible for learning-induced lasting enhancement in synaptic strengths, then we may speculate that, during the formation of memory traces, an ensemble of input events may create a certain specific configuration of activation in the networks which will lead to an enhancement of the permissive properties of a few constitutive synapses.

If the topology of this activation represents a coding in active memory, the LTP procedure could then be useful to reveal a process upon which depends a more permanent marking in network connectivity through biochemical and morphological changes (i.e. synaptic remodelling).

Interestingly, on the basis of the assembly concept promoted by Hebb (1949) and further elaborated by John (1972), mathematical formalization of the firing behavior of neurons grouped in assemblies within topological networks has led to the prediction that a cooperative activity among neural constitutive elements supports persisting activity in small assemblies among which correlative interactions may ultimately result in persistent changes in individual synaptic strengths (Shaw, 1978; Shaw et al., 1982). In the latter paper, dendritic bundling, largely prevalent in cortical areas (see Roney et al., 1979), was conjectured to represent an anatomical candidate support for small electronically coupled assemblies that exhibit such cooperative behavior. A comparable machinery appears to exist in the laminated hippocampal regions (Serfling and Schuster, 1983; Kosaka, 1983; see also Dudek et al., 1983) in which en passage synapses may cause synchronous activation of many such modules. By analogy, it is possible to consider the induction of hippocampal LTP as an artificial exaggeration of correlative interactions in widespread rising nets, leading to an enhancement of a large number of individual synaptic strengths.

Perhaps the bursting patterns of pyramidal cells occurring during hippocampal sharp-waves may reflect a type of synchronous activation in natural conditions (see Buzsáki, this volume). Moreover, it is relevant that artificial tetanization increases the incidence and amplitude of hippocampal sharp-waves (Buzsáki, 1984). Ephaptic interactions such as those described by Richardson et al. (1984) in the in vitro hippocampal slice preparation may perhaps be involved as well.

Relevant to the analogy discussed above are the data of Ott and associates (1982), which show that stimulation of the perforant path can be used as a conditioned stimulus in avoidance learning provided that the parameters of stimulation allow LTP to develop. In this case, when the CS is directly delivered to the pathway involved in learning, the amount of LTP is highly correlated with the level of performance.

In this context, LTP may be viewed as a model of an intermediate mechanism which may prepare lasting changes in synaptic function involved in the specification of neural networks by experience. The change is assumed to occur during the phase of information processing (referred to as the period of consolidation, see Bloch, 1970) and to be generated by the perseverative activation triggered in neural networks by incoming information.

We have discussed elsewhere (Bloch and Laroche, 1984) the role perseverative processes, whatever the neural mechanism(s) involved, may play in storage and retrieval of information, and have argued that MRF activation, like other modulators, may exert its effects on the points of connection in the network within which excitation is maintained. Yet, during this perseverative process, MRF stimulation is assumed to set the network in a particular state so that its processing capacity is improved. The "now print" signal is inherent to this state. Hypermnesic agents such as MRF stimulation may thus exert their facilitatory effect on a perseverative process engaged in the stabilization of relevant networks through lasting changes in synaptic function.

A number of requirements can be identified if the LTP model is to be compatible with behavioral data. It is now widely accepted that memory is multidimensional, i.e., memories are an ensemble of attributes that define an event (Underwood, 1969; see also Spear, 1978). This implies that the trace of an event cannot be localized in a particular brain area but is rather distributed throughout the brain, synaptic reorganization occurring in multiple relays in the relevant networks.

On the basis of the analogy discussed here, we might thus expect various brain areas to display LTP. The question of the distribution of LTP in the central nervous system has incited several commentaries in recent reviews (see Swanson et al., 1982; Voronin, 1983; Teyler and Discenna, 1984), particularly in attempting to assess the parallel between the regions displaying LTP and those whose disruption leads to learning or memory deficits.

In a recent study, LTP has been reported in almost all regions of the limbic forebrain (Racine et al., 1983). However, such a cartography in other brain regions is still to be developed. For instance, in the neocortex, LTP of cortical evoked potentials occurs after stimulation of the underlying white matter in vitro (Lee, 1982). LTP of transcallosal responses has also been found in the anterior neocortex in rats (Wilson and Racine, 1981). Of added interest is the finding (Komatsu et al., 1981), in an in vitro slice preparation of striate cortex from young kittens, that LTP in the region of geniculo-cortical projections to the visual cortex, appears prevalently during the same critical period as that during which visual cortical neurons display plastic modifications in response to visual experience. Also significant is the demonstration that the brachium of the inferior colliculus to the magnocellular division of the medial geniculate nucleus supports LTP (Gerren and Weinberger, 1983), and that the same region of the auditory system develops persistent cellular changes during conditioning (see Weinberger, 1980). LTP has also been demonstrated in the superior cervical ganglion after stimulation of preganglionic nerves (Brown and McAfee, 1982), and in the optic tectum of the goldfish after stimulation of the optic nerves (Lewis et al., 1981).

Nevertheless, a more precise cartography of the regions and pathways exhibiting LTP appears crucial. Specific experiments would further allow the distinction between the intrinsic plasticity of a connective system on the one hand, and its ability to be modified in the context of learning and memory on the other. However, it is conceivable that differences in macro- and/or micro-organization in brain structures set limits for the generalization of the hippocampal LTP phenomenon. If so, any general feature of changing network connectivity as a result of experience could possibly involve different underlying mechanisms and thus be revealed experimentally by different means according to the functional properties and organization of constituting cells and synapses.

The issue of the persistence of the change is also often commented upon when assessing the compatibility of the LTP model with memory. However, it should be noted that, first, there is a lack of data concerning the correlation between LTP duration and the level and quality of performance in retention. Second, and perhaps more importantly, the multidimensional view of memory makes it unlikely that the decay time-course of the change at a particular connective point examined, is directly parallel to the time-course of change in memory inferred through retention performance.

In summary, we have stressed the possibility that the persistence of spatio-temporal patterns of activation in neural networks (active memory) triggered by the original learning situation may result in a more permanent marking of network connectivity so that the topological structure can later be reinstalled when pertinent signals trigger the readout of the trace. LTP is thought to model at least some of the mechanisms involved in the specification of relevant networks by experience in the process of active/inactive memory shifting. Perhaps one of the merits of the LTP procedure is that it may enlighten the future direction of studies of the nature and kinetics of this process.

REFERENCES

Andersen, P., Bliss, T.V.P., and Skrede, K.K., 1971, Unit analysis of hippocampal population spikes, Exp. Brain Res. , 13:208.

Andersen, P., Sundberg, S.H., Sveen, O., and Wigstrom, H., 1977, Specific long-lasting potentiation of synaptic transmission in hippocampal slices, Nature , 266:736.

Bär, P.R., Schotman, P., Gispen, W.H., Tielen, A.M., and Lopes da Silva, F.H., 1980, Changes in synaptic membrane phosphorylation after tetanic stimulation in the dentate area of the rat hippocampal slice, Brain Res. , 198:478.

Barnes, C.A., 1979, Memory deficits associated with senescence: a neurophysiological and behavioral study in the rat, J. Comp. Physiol. Psychol. , 93:74.

Barnes, C.A., and McNaughton, B.L., 1980, Spatial memory and hippocampal synaptic plasticity in senescent and middle-aged rats, in :"The Psychobiology of Aging: Problems and Perspectives", D. Stein, ed., Elsevier, Amsterdam.

Baudry, M., Oliver, M., Creager, R., Wieraszko, A., and Lynch, G., 1980, Increase in glutamate receptors following repetitive electrical stimulation in hippocampal slices, Life Sci. , 27:325.

Berger, T.W., Alger, B., and Thompson, R.F., 1976, Neuronal substrate of classical conditioning in the hippocampus, Science , 192:483.

Berger, T.W., 1984, Long-term potentiation of hippocampal synaptic transmission affects rate of behavioral learning, Science , 224:627.

Bliss, T.V.P., and Dolphin, A.C., 1982, What is the mechanism of long-term potentiation in the hippocampus ?, Trends Neurosci. , 5:289.

Bliss, T.V.P., and Gardner-Medwin, A.R., 1973, Long-lasting potentiation of synaptic transmission in the dentate area of the unanaesthetized rabbit following stimulation of the perforant path, J. Physiol. (Lond.) , 232:357.

Bliss, T.V.P., and Lømo, T., 1973, Long-lasting potentiation of synaptic transmission in the dentate area of the anaesthetized rabbit following stimulation of the perforant path, J. Physiol. (Lond.) , 232:331.

Bloch, V., 1970, Facts and hypotheses concerning memory consolidation processes, Brain Res. , 24:561.

Bloch, V., Denti, A., and Schmaltz, G., 1966, Effets de la stimulation réticulaire sur la phase de consolidation de la trace mnésique, J. Physiol. (Paris) , 58:469.

Bloch, V., Deweer, B., and Hennevin, E., 1970, Suppression de l'amnésie rétrograde et consolidation d'un apprentissage à essai unique par stimulation réticulaire, Physiol. Behav. , 5:1235.

Bloch, V., and Laroche, S., 1981, Conditioning of hippocampal cells: its acceleration and long-term facilitation by post-trial reticular stimulation, Behav. Brain Res. , 3:23.

Bloch, V., and Laroche, S., 1984, Facts and hypotheses related to the search of the engram, in : "Neurobiology of Learning and Memory", G. Lynch, J.L. McGaugh, and N.M. Weinberger, eds., Guilford, New York.

Bloch, V., and Laroche, S., 1985, Enhancement of long-term potentiation in the rat dentate gyrus by post-trial stimulation of the reticular formation, J. Physiol. (Lond.) , in press.

Bottjer, S.W., and Arnold, A.P., 1984, Hormones and structural plasticity in the adult brain, Trends Neurosci. , 7:168.

Brown, T.H., and McAfee, D.A., 1982, Long-term synaptic potentiation in the superior cervical ganglion, Science , 215:1411.

Browning, M., Dunwiddie, T., Bennett, W.F., Gispen, W.H., and Lynch, G., 1979, Synaptic phosphoproteins: specific changes after repetitive stimulation of the hippocampal slice, Science , 203:60.
Buzsáki, G., 1984, Long-term changes of hippocampal sharp-waves following high frequency afferent activation, Brain Res. , 300:179.
Buzsáki, G., Grastyán, E., Czopf, J., Kellényi, L., and Prohaska, O., 1981, Changes in neuronal transmission in the rat hippocampus during behavior, Brain Res. , 225:235.
Chang, F.F., and Greenough, W.T., 1982, Lateralized effects of monocular training on dendritic branching in adult split-brain rats, Brain Res., 232:283.
Chepkova, A.N., and Skrebitsky, V.G., 1982, Effects of some adrenergic drugs and neuropeptides on long-term potentiation in hippocampal slices, in : "Neuronal Plasticity and Memory Formation", IBRO Monograph Series, Vol. 9, C. Ajmone Marsan, and H. Matthies, eds., Raven, New York.
Cotman, C.W., and Neito-Sampedro, M., 1982, Brain function, synapse renewal, and plasticity, Ann. Rev. Psychol. , 33:371.
de Jonge, M.C., and Racine, R.J., 1983, Effects of repetitive induction of long-term potentiation in the dentate gyrus of the rat, Soc. Neurosci. Abstr. , 9:102.
Desmond, N.L., and Levy, W.B., 1983, Synaptic correlates of associative potentiation/depression: an ultrastructural study in the hippocampus, Brain Res. , 265:21.
Deweer, B., 1976, Selective facilitative effect of post-trial reticular stimulation in discriminative learning in the rat, Behav. Proc. , 1:243.
Dolphin, A.C., Errington, M.L., and Bliss, T.V.P., 1982, Long-term potentiation of the perforant path in vivo is associated with increased glutamate release, Nature , 297:496.
Douglas, R.M., 1977, Long-lasting synaptic potentiation in the rat dentate gyrus following brief high-frequency stimulation, Brain Res. , 126:361.
Douglas, R.M., and Goddard, G.V., 1975, Long-term potentiation of the perforant path-granule cell synapse in the rat hippocampus, Brain Res. , 86:205.
Dudek, F.E., Andrew, R.D., MacVicar, B.A., Snow, R.W., and Taylor, C.P., 1983, Recent evidence for and possible significance of gap junctions and electronic synapses in the mammalian brain, in : "Basic Mechanisms of Neuronal Hyperexcitability", Alan R. Liss, New York.
Duffy, C., Teyler, T.J., and Shashoua, V.E., 1981, Long-term potentiation in the hippocampal slice: evidence for stimulated secretion of newly synthetized proteins, Science , 212:1148.
Eccles, J.C., 1983, Calcium in long-term potentiation as a model for memory, Neurosci. , 10:1071.
Fessard, A., and Szabo, T.H., 1961, La facilitation de post-activation comme facteur de plasticité dans l'établissement des liaisons temporaires, in : "Brain Mechanisms and Learning", A. Fessard, R.W. Gerard, J. Konorski, and J.F. Delafresnaye, eds., Blackwell, Oxford.
Fifkova, E., Anderson, C.L., Young, S.J., and Van Harreveld, A., 1982, Effect of anisomycin on stimulation-induced changes in dendritic spines of the dentate granule cells, J. Neurocytol. , 11:183.
Fifkova, E., and Van Harreveld, A., 1978, Changes in dendritic spines of the dentate molecular layer during conditioning, Soc. Neurosci. Abstr. , 4:257.
Gerren, R.A., and Weinberger, N.M., 1983, Long term potentiation in the magnocellular medial geniculate nucleus of the anesthetized cat, Brain Res. , 265:138.
Goddard, G.V., 1981, Progress on the physiological bases of memory, in : "Brain and Behaviour", Advances in Physiological Sciences, Vol. 17, G. Adam, I. Meszaros, and E.I. Banyai, eds., Pergamon, Akademiai Kiado, Budapest.

Gold, P.E., Delanoy, R.L., and Merrin, J., 1984, Modulation of long-term potentiation by peripherally administered amphetamine and epinephrine, Brain Res. , 305:103.

Greenough, W.T., 1984, Structural correlates of information storage in the mammalian brain: a review and hypothesis, Trends Neurosci. , 7:229.

Hebb, D.O., 1949, "The Organization of Behavior", Wiley, New York.

Hesse, G.W., and Teyler, T.J., 1976, Reversible loss of hippocampal long-term potentiation following electroconvulsive seizures, Nature , 264:562.

Horn, G., 1981, Neural mechanisms of learning: an analysis of imprinting in the domestic chick, Proc. R. Soc. Lond. , 213:101.

Jaffard, R., and Jeantet, Y., 1981, Posttraining changes in excitability of the commissural path-CA1 pyramidal cell synapses in the hippocampus of mice, Brain Res. , 220:167.

John, E.R., 1972, Switchboard versus statistical theories of learning and memory, Science , 177:850.

Komatsu, Y., Toyama, K., Maeda, J., and Sakaguchi, H., 1981, Long-term potentiation investigated in a slice preparation of striate cortex of young kittens, Neurosci. Lett. , 26:269.

Kosaka, T., 1983, Axon initial segments of the granule cell in the rat dentate gyrus: synaptic contacts on bundles of axon initial segments, Brain Res. , 274:129.

Laroche, S., Bergis, O.-E., and Bloch, V., 1983, Posttrial reticular facilitation of dentate multiunit conditioning is followed by an increased long-term potentiation, Soc. Neurosci. Abstr. , 9:645.

Laroche, S., and Bloch, V., 1982, Conditioning of hippocampal cells and long-term potentiation: an approach to mechanisms of post-trial memory facilitation, in : "Neuronal Plasticity and Memory Formation", IBRO Monograph Series, Vol. 9, C. Ajmone Marsan, and H. Matthies, eds., Raven, New York.

Laroche, S., Falcou, R., and Bloch, V., 1983, Post-trial reticular facilitation of associative changes in multiunit activity: comparison between dentate gyrus and entorhinal cortex, Behav. Brain Res. , 9:381.

Lee, K.S., 1982, Sustained enhancement of evoked potentials following brief, high-frequency stimulation of the cerebral cortex in vitro, Brain Res. , 239:617.

Lee, K.S., 1983, Cooperativity among afferents for the induction of long-term potentiation in the CA1 region of the hippocampus, J. Neurosci. , 3:1369.

Lee, K.S., Schottler, F., Oliver, M., and Lynch, G., 1980, Brief bursts of high-frequency stimulation produce two types of structural change in rat hippocampus, J. Neurophysiol. , 44:247.

Levy, W.B., and Steward, O., 1979, Synapses as associative memory elements in the hippocampal formation, Brain Res. , 175:233.

Lewis, D., Teyler, T., and Shashoua, V., 1981, Development of long-term potentiation in the in vitro goldfish optic tectum, Soc. Neurosci. Abstr. , 7:66.

Lewis, D.J., 1979, Psychobiology of active and inactive memory, Psychol. Bull. , 86:1054.

Lømo, T., 1971, Patterns of activation in a monosynaptic cortical pathway: the perforant path input to the dentate area of the hippocampal formation, Exp. Brain Res. , 12:18.

McNaughton, B.L., and Barnes, C.A., 1977, Physiological identification and analysis of dentate granule cell responses to stimulation of the medial and lateral perforant pathway in the rat, J. Comp. Neurol. , 175:439.

McNaughton, B.L., Douglas, R.M., and Goddard, G.V., 1978, Synaptic enhancement in fascia dentata: cooperativity among coactive afferents, Brain Res. , 157:277.

Misgeld, U., Sarvey, J.M., and Klee, M.R., 1979, Heterosynaptic postactivation potentiation in hippocampal CA3 neurons: long-term changes in the postsynaptic potentials, Exp. Brain Res. , 37:217.
Olds, J., Disterhoft, J.F., Segal, M., Kornblith, C.L., and Hirsh, R., 1972, Learning centers of rat brain mapped by measuring latencies of conditioned unit responses, J. Neurophysiol. , 35:202.
Ott, T., Ruthrich, H., Reymann, K., Lindenau, L., and Matthies, H., 1982, Direct evidence for the participation of changes in synaptic efficacy in the development of behavioral plasticity, in : "Neuronal Plasticity and Memory Formation", IBRO Monograph Series, Vol. 9, C. Ajmone Marsan, and H. Matthies, eds., Raven, New York.
Racine, R.J., Milgram, N.W., and Hafner, S., 1983, Long-term potentiation phenomena in the rat limbic forebrain, Brain Res. , 260:217.
Ranck, J.B., 1973, Studies in single neurons in dorsal hippocampal formation and septum in unrestrained rats, Exp. Neurol. , 41:461.
Reymann, K.G., and Ott, T., 1983, Footshock-induced modification of a monosynaptic evoked potential in the hippocampus, Physiol. Behav. , 31:615.
Richardson, T.L., Turner, R.W., and Miller, J.J., 1984, Extracellular fields influence transmembrane potentials and synchronization of hippocampal neuronal activity, Brain Res. , 294:255.
Roney, K.J., Scheibel, A.B., and Shaw, G.L., 1979, Dendritic bundles: survey of anatomical experiments and physiological theories, Brain Res. Rev. , 1:225.
Rosenzweig, M.R., and Bennett, E.L., 1976, Enriched environments: facts, factors and fantasies, in : "Knowing, Thinking and Believing", J.L. McGaugh, and L. Petrinovich, eds., Plenum, New York.
Rüthrich, H., Matthies, H., and Ott, T., 1982, Long-term changes in synaptic excitability of hippocampal cell populations as a result of training, in : "Neuronal Plasticity and Memory Formation", IBRO Monograph Series, Vol. 9, C. Ajmone Marsan, and H. Matthies, eds., Raven, New York.
Segal, M., 1977, Excitability changes in rat hippocampus during conditioning, Exp. Neurol. , 55:67.
Serfling, R., and Schuster, T., 1983, Light-microscopic investigations on the problem of a dendritic bundling in the rat hippocampus, J. Hirnforsch. , 24:1.
Shaw, G.L., 1978, Space-time correlations of neuronal firing related to memory storage capacity, Brain Res. Bull. , 3:107.
Shaw, G.L., Harth, E., and Scheibel, A.B., 1982, Cooperativity in brain function: assemblies of approximately 30 neurons, Exp. Neurol. , 77:324.
Spear, N.E., 1978, "The Processing of Memories: Forgetting and Retention", Erlbaum, Hillsdale.
Swanson, L.W., Teyler, T.J., and Thompson, R.F., 1982, Hippocampal long-term potentiation: mechanisms and implications for memory, Neurosci. Res. Prog. Bull. , 20:613.
Teyler, T.J., and Discenna, P., 1984, Long-term potentiation as a candidate mnemonic device, Brain Res. Rev. , 7:15.
Thompson, R.F., 1982, Increases in the monosynaptic population spike response of dentate gyrus during classical conditioning, Neurosci. Res. Prog. Bull. , 20:693.
Thompson, R.F., Berger, T.W., and Madden IV, J., 1983, Cellular processes of learning and memory in the mammalian CNS, Ann. Rev. Neurosci. , 6:447.
Thompson, R.F., Manounas, L.A., Lynch, G., and Baudry, M., 1983, Increased glutamate receptor binding in hippocampus following classical conditioning of the rabbit eyelid response, Soc. Neurosci. Abstr. , 9:830.
Tsukahara, N., 1981, Synaptic plasticity in the mammalian central nervous system, Ann. Rev. Neurosci. , 4:351.

Underwood, B.J., 1969, "Attributes of memory", Psychol. Rev. , 76:559-573.
Van Harreveld, A., and Fifkova, E., 1975, Swelling of dendritic spines in the fascia dentata after stimulation of the perforant fibers as a mechanism of post-tetanic potentiation, Exp. Neurol. , 49:736.
Voronin, L.L., 1983, Long-term potentiation in the hippocampus, Neurosci., 10:1051.
Vrensen, G., and Cardozo, J.N., 1981, Changes in size and shape of synaptic connections after visual training: an ultrastructural approach of synaptic plasticity, Brain Res. , 218:79.
Weinberger, N.M., 1980, Neurophysiological studies of learning in association with the pupillary dilatation conditioned reflex, in : "Neural Mechanisms of Goal-Directed Behavior and Learning", R.F. Thompson, L.H. Hicks, and V.B. Shvyrkov, eds., Academic, New York.
Wenzel, S., Kammerer, E., Kirsche, W., Matthies, H., and Wenzel, M., 1980, Electron microscopic and morphometric studies on synaptic plasticity in the hippocampus of the rat following conditioning, J. Hirnforsch., 21:647.
Wilson, D.A., and Racine, R.J., 1981, Transient and long-term potentiation in the neocortex of the rat, Soc. Neurosci. Abstr. , 7:69.
Yamamoto, C., and Chujo, T., 1978, Long-term potentiation in thin hippocampal section studied by intracellular and extracellular recordings, Exp. Neurol. , 58:242.

WHAT DOES THE "LTP MODEL OF MEMORY" MODEL?

György Buzsáki

Department of Physiology
University Medical School
7643 Pécs, Hungary

INTRODUCTION

In several areas of the brain, and particularly in the hippocampal formation, short bursts of high frequency impulses result in potentiation of the responses to subsequent single pulse stimuli (Bliss and Lømo, 1973). The potentiation may last for days, or even weeks, hence the designation long-term potentiation (LTP). Some consider this use-dependent enhancement of synaptic efficacy to be a model system for the investigation of the biological substrate of memory (Bliss and Dolphin, 1982; Goddard 1980; Voronin, 1983).

However, the finding that LTP can be produced in nervous tissue does not imply that the nervous system utilizes this potential during naturally occurring activity. In order to accept that LTP serves as a useful paradigm to search for the engram we must ask the question: What physiological state(s) does the LTP model simulate? What behaviors and associated electrophysiological and neurochemical changes in the brain favor effective facilitation of synaptic efficacy?

In order to produce LTP, presynaptic stimulation must be strong, at a high frequency and preferably many convergent afferents must be synchronously activated (Douglas, 1977; Levy and Steward, 1979; McNaughton et al., 1978). It is the thesis of this chapter that under physiological conditions these requirements may be met during the occurrence of hippocampal sharp waves (SPW) associated with reinforcing behaviors such as consummatory activity and slow wave sleep.

HIPPOCAMPAL SHARP WAVES

The most characteristic EEG pattern of the hippocampus is the theta rhythm which in the rat occurs during exploratory activity. Another, albeit less studied, spontaneous EEG pattern is the irregularly occuring sharp wave of 40-120 ms duration. SPWs never occur during behaviors accompanied by theta waves but are consistently recorded from the CA 1 region when the animal is immobile, engaged in consummatory behaviors or is in slow wave sleep (Buzsáki et al, 1983; Suzuki and Smith, 1982). SPWs occur synchronously in both hippocampi, either in isolation or in groups of 2 to 7 spikes in either an ascending or descending pattern. Depth profile experiments indicated that SPWs were positive in stratum oriens and stratum pyramidale of CA 1 and negative in stratum radiatum with a sudden polarity reversal just below the pyramidal cell layer while maximum negativity occured in the middle third of stratum radiatum (up to 3.5 mV). The profiles of the SPWs and the field potentials evoked by stimulation of the Schaffer collaterals displayed essentially the same pattern.

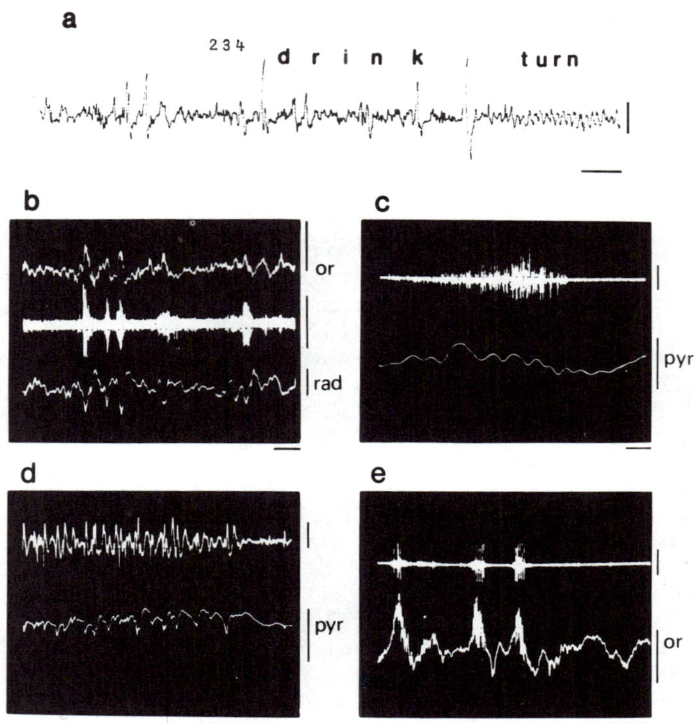

Fig. 1. Relationship between sharp waves (SPW) and cellular activity. a: EEG recorded from str. radiatum of CA 1. Note presence of SPWs only during drinking. b: multiple unit activity in CA 1 pyramidal layer and EEG recorded from str. oriens (or) and str. radiatum (rad) during drinking. c and d: multiple unit bursts during SPW at faster sweeps. e: interneuron in str. oriens. Calibrations: 20 µV (unit), 1 mV (EEG), 1 s (a), 0.25 s (b, e), 10 ms (c, d). EEG filtered (1-100 Hz) in a, b and c, wideband in d and e. Positivity down in a and up in b to e.

SPWs are invariably correlated with the synchronous discharge of a great number of pyramidal cells in both CA 1 and CA 3 regions (Fig. 1). When a microelectrode is positioned in the pyramidal layer of CA 1 a series of small negative population spikes ride on the positive SPW (Fig. 1d). In between bursts of SPWs, activity of the pyramidal cells is suppressed.

Interneurons in strata oriens, pyramidale and radiatum of CA 1 also fire synchronously with SPWs (Fig. 1e). Although interneurons may also discharge between the SPWs at a low frequency, some interneurons produce bursts of 10-25 action potentials during a SPW, with firing rates up to 800 Hz (Buzsáki et al., 1983).

Based on the above findings we proposed that the SPWs represent a synchronous depolarization of the apical dendrites of numerous CA 1 pyramidal cells via the Schaffer collaterals of the CA 3 pyramidal neurons (Buzsáki et al., 1983). Preliminary data from my laboratory indicate that large amplitude SPWs are also generated in the inner third of the molecular layer of the dentate gyrus and in the subiculum. These SPWs occur in synchrony with SPWs of the CA 1 and are associated with population discharges of the granule cells and subicular pyramidal cells. The CA 3 region might act as a trigger zone via mutual excitation of a population of pyramidal cells.

Behavioral correlates of hippocampal SPWs

Hippocampal SPWs are strictly related to sensorimotor behavior. They occur, in order of frequency, during slow wave sleep, sitting still, drinking, eating, face washing and grooming. The incidence of SPWs during awake immobility ranged from 0.1/sec to 6/sec. Sensory stimulation (e.g. sound stimuli or moving objects) dramatically suppressed the SPWs, indicating that observing motor correlates alone would not reveal the conditions which efficiently produce SPWs.

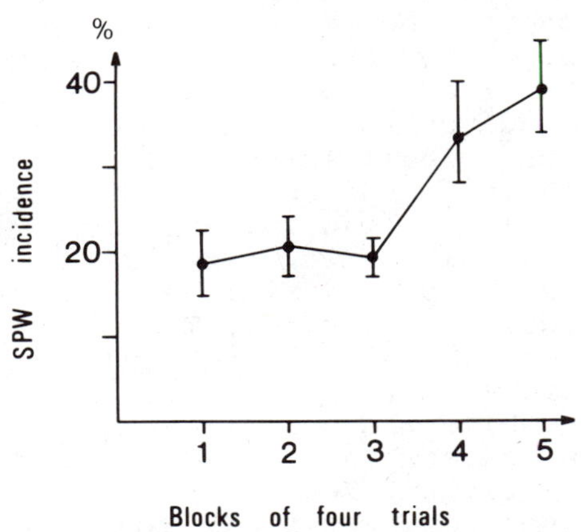

Fig. 2. Correlation between satiation and the probability of occurrence of hippocampal SPWs. Data based on 6 sessions (20 trials each) in 26 rats. Vertical bars indicate standard errors of the mean.

In an attempt to obtain further correlates of hippocampal SPWs in a reproducable behavioral situation we trained rats to lap water from a drinking tube on a variable interval (40 sec) schedule (Buzsåki 1983a). Twenty trials were given in a session. SPWs derived from stratum radiatum of CA 1 were discriminated and counted during the initial 10 sec of drinking. The occurrence of SPWs showed a considerable intersubject variability, but in a given rat the number of SPWs in successive sessions was stable. The frequency of SPWs in a given trial varied within the sessions. About twice as many SPWs occurred during the late drinking trials than during the early ones (Fig. 2). When the rat was water deprived for 96 h, rather than 24 h, many trials occurred without SPWs. Thus, this experiment demonstrated a direct relationship between the satiation of the animal and the incidence of SPWs.

Excitability changes during the SPWs

Changes in responsiveness of hippocampal cells can be tested by stimulating various afferent pathways. For example, CA 3 pyramidal cells can be activated by stimulating the perforant path to the dentate gyrus. Synchronous activation of the CA 3 pyramidal cells, in turn, may produce an evoked response in CA 1 via the Schaffer collaterals. The amplitude of this trisynaptically evoked potential shows a striking covariance with ongoing behavior (Buzsåki et al. 1983; Winson and Abzug 1978). This "Schaffer" potential is quite large during slow wave sleep, drinking and immobility and is small or absent completely during theta correlated behaviors. The variability of the potential is quite large, and the first response may be followed by a second potential with a fixed latency of about 15-25 ms after the first CA 1 response. The presence of the first Schaffer potential is a prerequisite for the late response, but the amplitude of the latter may be greater than that of the former. The late potential appears to be virtually all-or-none (Fig. 3).

Closer examination of the relationship between the evoked potentials and the ongoing EEG revealed that stimulus volleys delivered during bursts of SPWs resulted in large amplitude responses. In contrast, small amplitude or no responses were evoked in their absence. We therefore decided to explicitly examine the SPW-associated excitability changes in the hippocampus. The data reported here are taken from 14 rats with chronically implanted recording electrodes in the CA 1 pyramidal layer and granule cell layer. Stimulating electrodes were placed in both ipsi- and contralateral angular bundles and the contralateral CA 3 region. The rats were trained to run in an activity wheel for water reward (Buzsåki et al, 1983). The EEG from the pyramidal layer electrode was passed through an amplitude discriminator. The output pulses derived from the discriminated SPWs could be used to trigger the stimulator or to prevent stimulation for 0.5 sec. Stimuli were triggered manually while the animal was running in the wheel or during drinking in between the SPWs, and by the SPWs themselves during SPW activity. The interstimulus interval varied between 15 and 60 seconds. A typical experiment is shown in Fig. 4. As reported earlier (Buzsåki et al., 1981), population discharges of pyramidal neurons and granule cells showed predictable changes during running-drinking transitions. The present experiment revealed that a great deal of this

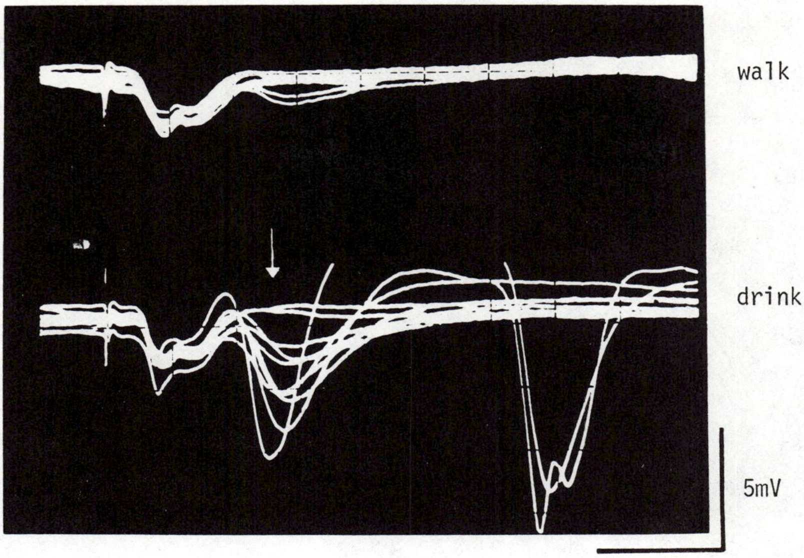

Fig. 3. Multisynaptically evoked potentials in stratum radiatum of CA 1 in response to single pulse stimulation of the angular bundle. The early response is a volume conducted potential from the dentate gyrus. Ten superimposed sweeps each. Note the absence of the "Schaffer" response (arrow) during walking.

difference could be traced back to the sudden increase of cellular excitability during the SPWs. Population spike responses to all afferent pathways tested so far (commissural and perforant path inputs) were significantly greater during the SPWs as compared to responses obtained during drinking between the SPWs or during running. The long latency evoked responses (e.g. the trisynaptic "Schaffer" potential) were especially increased. The increase of monosynaptically evoked population spikes varied between 150 and 380% and the long-latency responses increased between 190 and 910% as compared to the responses in the absence of SPWS. In some instances a SPW-triggered volley induced the appearance of a new component. For example, stimulation of the commissural input during a SPW discharged CA 1 pyramidal cells followed by discharges of the granule cells. Occasionally, as a result of the synchronous firing of the granule cells, the CA 1 neurons were reactivated via the dentate-CA3 - CA 1 path (Fig. 5.). Such late responses were never observed in the absence of SPWs. The late CA 1 population spike was often larger in amplitude than the monosynaptically evoked response.

The above experiments revealed that in the hippocampal formation the effectiveness of afferent activation suddenly increases from 1.5 to 9 times during hippocampal SPWs. Such an enormous change in excitability is certainly not without physiological implications.

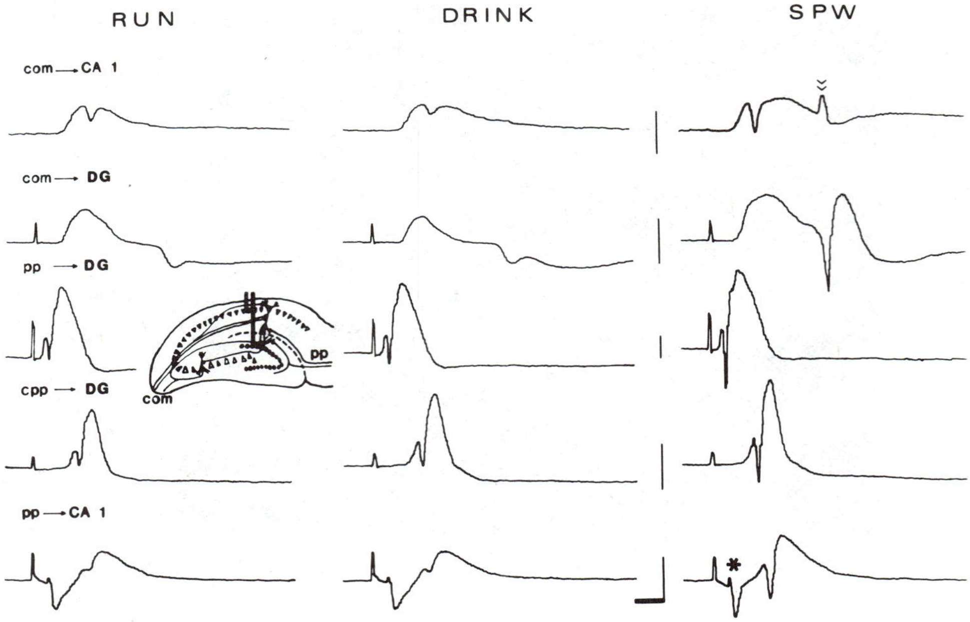

Fig. 4. Excitability changes of the intrahippocampal circuitry during behavior. Inset shows the positions of the recording microelectrodes in the pyramidal layer and granule cell layer. All traces (averages of 8 potentials) were recorded in a single rat. Site of stimulation is indicated above the artifact. com: commissural input, pp: perforant path input; cpp: contralateral perforant path. Run: running in the wheel; drink: stimulating during drinking in the absence of sharp waves (SPW); SPW: sharp wave-triggered potentials during drinking. Note high amplitude population spikes during SPWs. Asterisk: volume conducted population spike from dentate gyrus (DG). Double arrowhead: multisynaptically evoked population spike generated in dentate gyrus. Calibrations: 5 ms, 5 mV.

LONG-TERM POTENTIATION OF HIPPOCAMPAL SPWs

Considerable evidence supports the view that hippocampal SPWs are population postsynaptic potentials of CA 1 pyramidal cells and dentate granule cells resulting from synchronized discharges of a number of CA 3 cells. Since high frequency activation of the Schaffer collateral-CA 1 pyramidal cell synapse results in increased evoked responses to subsequent single pulses, we asked whether such activation would significantly affect the drinking-associated spontaneous SPWs.

The Schaffer collateral/commissural system was stimulated by 20 msec trains at 400/sec at intervals of 5 min. The intensity of pulses was adjusted to a level below or just above the population spike threshold. The discriminated SPWs were counted

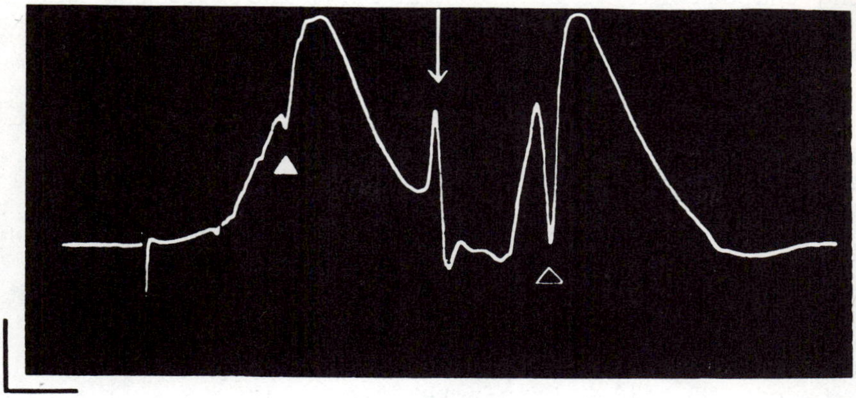

Fig. 5. Mono- and multisynaptically evoked responses in the pyramidal layer of CA 1. Single pulse stimulation of the comissural input during a SPW. Note that the multisynaptically evoked population spike (black triangle) is larger than the monosynaptically evoked (white triangle). Arrow indicates volume conducted population spike from the dentate gyrus. Calibrations: 5 ms, 2 mV.

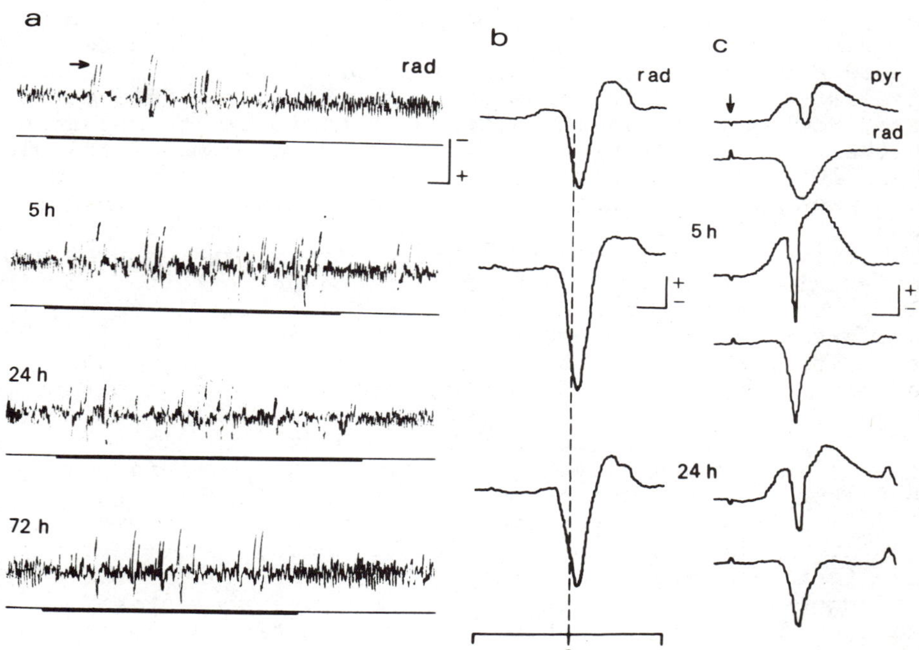

Fig. 6. Long-term potentiation of hippocampal SPWs. a: representative polygraph records before and at various intervals after high frequency stimulation (100 Hz, HFS). Marker channel indicates drinking. b: averaged SPWs before and after HFS. Each trace is an average of 64 SPWs. c: long-term potentiation of evoked responses in strata pyramidale (pyr) and radiatum (rad). Calibrations: 1 mV (a), 0.5 mV (b), 2 mV (c); 1 s (a), 50 ms (b), 5 ms (c). (Reprinted with permission from Buzsáki, 1984a).

e initial 10 sec of drinking. Two sessions were given The control level of the number of SPWs was measured ısecutive sessions.

ɔllowing the high frequency trains both the extracellu- naptic potential and the population spike evoked by sti- ıon of the Schaffer collateral/commissural inputs showed ɩcement from 24 to 96 h. Parallel with the potentiation of the evoked responses both the incidence (0-300%) and the amplitude (0-80%) of the spontaneous SPWs increased significantly in the first (5 h) session. Potentiation of the SPWs lasted from 5 to 47 h (Fig. 6) (Buzsaki, 1984a).

This experiment demonstrated that high frequency activation causes long-lasting changes in the spontaneous activity of hippocampal cells. Long-term changes in cellular excitability might have occurred either in the CA 3 region resulting in more synchronous firing of the pyramidal cells or in the Schaffer collateral-CA1 pyramidal cell synapse, or both. Further experiments, involving recordings of multiple unit activity in CA 3, are needed to determine the major site of plastic changes.

HIPPOCAMPAL SPWs AND THE LTP MODEL OF MEMORY

Long-term potentiation is a promising candidate in the construction of models of memory. Its long-lasting nature and its demonstrated physiological, biochemical and anatomical consequences provide opportunities for developing an hypothesis of plastic changes that represent memory consolidation. To date, however, no direct evidence is available to determine whether LTP is a laboratory artifact or, as a model, simulates physiological processes underlying memory trace formation. Such presumed physiological states should share the basic conditions required for LTP, such as the necessity of strong repetitive synaptic bombardement, preferably in bursts, and cooperative activity of a number of converging afferent fibers. Also, these conditions must be present at times when an important event must be remembered and when the behavior leading to it is reinforced. I suggest that the cellular mechanisms underlying hippocampal SPWs might satisfy these requirements.

It is widely accepted that events and behavioral acts are remembered because they are reinforced. Most reinforcers must be "taken in" or experienced in some way, in order to be effective. The consummatory-response theory of reinforcement asserts that it is the behavior of eating, drinking, etc. that gives rise to the reinforcing effect (Sheffield et al., 1954). It was suggested that the source of the reinforcement effect is the species-specific responses such as eating and drinking that help the organism to adapt to its environment (Glickman and Schiff, 1967). Brain stimulation reward has been explained in a similar way, since it was discovered that the same hypothalamic stimulation that reinforces lever pressing can also elicit responses such as eating and drinking when the appropriate stimulus is available (Valenstein et al., 1969). Thus, at least some psychological views are in line with the suggestion that consolidation of information in long-term memory occurs during reinforcing consummatory behaviors which we have shown to be accompanied by hippocampal SPWs. Indeed, Grastyán et al. (1968) reported that during the rebound of

stimulation of rewarding hypothalamic loci large amplitu
spike-like activity frequently occurred in the cat hippoc

Of all physiological patterns of the hippocampus, th cellular mechanisms of the SPWs share the most with the L model of memory. During a SPW, at least several dozens of neurons fire in bursts. During theta activity, associated with preparatory behaviors, the activity of hippocampal pyramidal neurons is greatly suppressed (Buzsáki et al, 1983; Ranck, 1973). So far no estimations as to the number of pyramidal cells discharging concurrently with the SPWs are available. Such measurements would require recording from a great number of individually isolated cells. It probably involves a relatively small portion of all pyramidal and granule cells because following administration of picrotoxin, a GABA inhibitor, SPWs grew considerably, presumably due to the recruitment of more neurons (unpublished data). Also, SPW-triggered stimulation evoked the largest amplitude population spike, suggesting that not all neurons are refractory during the SPW. On the other hand, high frequency bursts of CA 1 interneurons during the SPW reflect intensive synaptic stimulation. Repetitive firing of hippocampal interneurons above 400/sec are often seen after strong afferent stimulation, but such high rates of activity have never been reported under physiological conditions, except in association with the SPWs (Buzsáki et al., 1983). Such strong synaptic drive may result from both feed-forward and recurrent excitation of numerous pyramidal cells (Buzsáki, 1984b).

Since SPWs are bilaterally synchronous, it is assumed that population discharges of CA 3 neurons in both hippocampi occur in a synchronized manner. Thus, CA 1 pyramidal cells, for example, may be excited via the Schaffer collaterals and commissural fibers in a cooperative manner.

To summarize, the excitability of hippocampal neurons increases considerably during hippocampal SPWs. Population discharges of pyramidal neurons during the SPWs produce strong and repetitive activation of target cells. These conditions meet the requirements of the LTP model. No direct evidence supports the involvement of SPWs in memory consolidation. However, if the LTP mechanism has any relevance to physiological states and to memory, then the present paper suggests that such synaptic modifications most probably occur during hippocampal sharp waves.

Acknowledgement. I thank Drs. T.V.P. Bliss, G. Czéh, F.H. Gage, E. Grastyán and W.B. Levy for their comments on an earlier version of this paper.

REFERENCES

Bliss, T.V.P. and Dolphin, A.C., 1982. What is the mechanism of long-term potentiation in the hippocampus? Trends Neurosci. 5: 289.

Bliss, T.V.P. and T. Lømo, T., 1973, Long-lasting potentiation of synaptic transmission in the dentate area of the anaesthetized rabbit following stimulation of the perforant path, J. Physiol. Lond. 232:331.

Buzsáki, G., 1984a, Long-term changes of hippocampal sharp--waves following high frequency afferent activation, Brain Res. 300:179.

Buzsáki, G. 1984b, Feed-forward inhibition in the hippocampal formation, Progr. Neurobiol. 22:131.

Buzsáki, G., Leung, L.S., and Vanderwolf, C.H., 1983, Cellular basis of hippocampal EEG in the behaving rat, Brain Res. Rev. 6:139.

Buzsáki, G., Grastyán, E., Czopf, J., Kellényi, L. and Prohaska, O., 1981. Changes in neuronal transmission in the rat hippocampus during behavior, Brain Res. 225:235.

Douglas, R.M., 1977, Long-lasting synaptic potentiation in the rat dentate gyrus following brief frequency stimulation, Brain Res. 126:361.

Eccles, J.C., 1983. Calcium in long-term potentiation as a model for memory, Neurosci. 10:1071.

Glickman, S.W. and Schiff, B.B., 1967, A biological theory of reinforcement, Psychol. Rev. 74:81.

Goddard, G.V., 1980, Component properties of memory machines: Hebb revisited, in: "The Nature of Thought: Essays in Honour of D.O. Hebb", P.W. Jusczyk and R.M. Klein, eds., Lawrence Erlbaum, Hillsdale, NJ.

Grastyán, E., Szabó, I., Molnár, P., and Kolta, P., 1968, Rebound, reinforcement and self-stimulation, Comm. Behav. Biol. 2:235.

Jouvet, M., Michel, F., and Courjon, J., 1959, L'activite electricque du rhinencephale an cours du Sommeil chez le chat, C.R. Soc. Biol. (Paris) 153:101.

Levy, W.B., and Steward, O., 1979, Synapses as associative elements in the hippocampal formation. Brain Res. 175:233.

McNaughton, B.L., Douglas, R.M., and Goddard, G.V., 1978. Synaptic enhancement in fascia dentata: cooperativity among coactive afferents, Brain Res. 157:277.

Ranck, J.B., Jr., 1973, Studies on single neurons in dorsal hippocampal formation and septum in unrestratined rats. I. Behavioral correlates and firing repertories, Exp. Neurol. 42:461.

Sheffield, F.D., Roby, T.B., and Campbell, B.A., 1954, Drive reduction versus consummatory behavior as determinants of reinforcement, J. comp. physiol. Psychol. 47:349.

Suzuki, S.S., and Smith, G.K., 1982, Spontaneous EEG spikes the hippocampus of normal behaving rat. Soc. Neurosci. Abstr. 8:1051.

Valenstein, E.S., Cox, V.C., and Kakolewski, J.W., 1969, The hypothalamus and motivated behavior, in: "Reinforcement and behavior" J.T. Tapp, ed. Academic Press, New York.

Voronin, L.L., 1983. Long-term potentiation in the hippocampus, Neurosci. 10:1051.

Winson, J. and Abzug, C., 1978, Neuronal transmission through hippocampal pathways dependent on behavior. J. Neurophysiol. 41:716.

THE CHOLINERGIC SEPTO-HIPPOCAMPAL PATHWAY, LEARNING AND MEMORY

Robert Jaffard, Daniel Galey, Jacques Micheau and Thomas Durkin

Laboratoire de Psychophysiologie, L.A. CNRS n° 339
Université de Bordeaux I, avenue des Facultés
33405 Talence Cedex France

INTRODUCTION

This paper summarizes the results of parallel behavioral, neurochemical and pharmacological experiments aimed at studying the role of septo-hippocampal cholinergic neurones in learning and memory processes in mice. A large body of evidence from pharmacological experiments supports the association of central cholinergic mechanisms with many aspects of behavior (De Feudis, 1974) and in particular with information encoding storage and/or retrieval (Deutsch, 1971; Biederman, 1974; Matthies, 1974; Squire and Davis, 1981). The hypothesis that the hippocampus plays a key role in memory processes (Isaacson, 1974) has focussed considerable attention on the possibility that the cholinergic neurones of the septo-hippocampal pathway (Lewis and Shute, 1967) may mediate control on some mechanisms involved in mnemonic function (Routtenberg, 1984). To circumvent the difficulties deriving from the frequently used approach of pharmacological studies utilising peripheral drug injections which have often generated discordant results on this problem, we have adopted the research strategy of investigating the modulation of learning and memory processes in selected inbred mouse strains combined with direct neurochemical methods to investigate hippocampal cholinergic activity.

Behavioral analysis of strain differences in learning and memory

An initial series of experiments was designed to compare learning and memory performance among the three inbred mouse strains BALB/c, C57BL/6 and C57BR. Previous studies had shown that performance on tests requiring retention of appetitive learning depended on the delay between a partial acquisition session and the retention test (Jaffard et al., 1974). Thus comparisons were made between strains of the speed of acquisition in a single session (massed trials) and of retention abilities after different training to test intervals. The task used was an appetitive operant conditioning (CRF) in which the lever and food cup were separated by a small partition (Destrade et al.,1973). Results showed that the rate of acquisition in a single session of 30 minutes was strain-dependent, being faster for mice of the two C57 strains in comparison to the BALB/c strain and that this difference could not be ascribed to any differences in general locomotor activity (Jaffard et al.,1976). In a second experiment we studied the effect of the inter-trial interval on performance during the retention session. Two groups of mice of each strain underwent a first learning session (partial acquisition) during which they were allowed 16 responses. The retention test sessions were conducted either 2.5 min or 24 hr later. The major finding of this experiment was that the retention interval effect on performance was strain-dependent (Jaffard et al., 1977). Thus a highly significant improvement of performance was observed in BALB/c between 2.5 min and 24 hr ("reminiscence" effect or "inverse forgetting") whereas the 24 hr interval led to a decrease in performance of the C57 subjects (forgetting)(see in fig.1).Thus it appeared from these results that there was an inverse relationship between the rate of acquisition and performance when the retention test was delayed for 24 hr : C57 subjects, which exhibited faster rates of acquisition when learning was either uninterrupted or only briefly interrupted showed a disruption of performance when the interval between sessions was increased to 24 hr, the inverse being true for BALB/c.

The results of these studies using an appetitive CRF operant conditioning were generalised to other learning tasks when the same behavioral protocol was applied in comparative genotypic investigations involving passive rather than active responses and negative rather than positive reinforcement (Jaffard, 1978). The synthesis of these behavioral observations led us to suggest that 1) BALB/c mice have a more efficient long term memory than C57, this difference being most evident whenever the ratio between the complexity of the task and the length of the acquisition session (i.e. the quantity of information to be processed) is high and 2) their poor performance on short term retention may be ascribed to a transitional state of impaired retrieval lasting at least 1 hour following training (Jaffard and Destrade, 1982).

Strain-dependent behavioral effects of post-training electrical stimulation of the hippocampus

Previous experiments in our laboratory have also shown that a mild electrical stimulation of the dorsal hippocampus performed within 10 min of the end of acquisition improved performance in the retention test 24 hr later (Destrade et al., 1973; Destrade and Cardo, 1974). Since the effectiveness of this treatment was shown to be time-dependent, we suggested that it affects retention by altering memory consolidation (McGaugh, 1966). The fact that such processes might be less efficient in C57 when compared to BALB/c led us to conduct a comparative study of the effect of this treatment among the three strains of mice. We were interested to know whether such post-training hippocampal stimulation might reduce, maintain or amplify the previously observed strain differences in long-term memory and thus provide some information about the mechanisms involved.

Animals of each strain were bilaterally implanted under general anaesthesia with bipolar electrodes in the CA1 field of the dorsal hippocampus. Eight days later, the threshold intensity (S) of a sinusoidal current (100 Hz) that produced after-discharges was determined for each mouse. In all the experiments reported here, the intensity of stimulation used was half of this threshold value (S/2) and was found not to be statistically different for the three strains used (Jaffard et al., 1977). As described above, mice were subjected to a short acquisition session of the appetitive lever press conditioning during which they were allowed 16 responses. Thirty seconds after the end of this training session, experimental groups (ST) of each strain were subjected to hippocampal stimulation for 80 sec at S/2 intensity; the control groups received no stimulation (NST).Twenty-four hours later, performance was measured in a second session (retention test) of 30-min duration.

The results showed that the effect of hippocampal stimulation on performance was strain-dependent (Fig.1). For BALB/c mice, stimulation

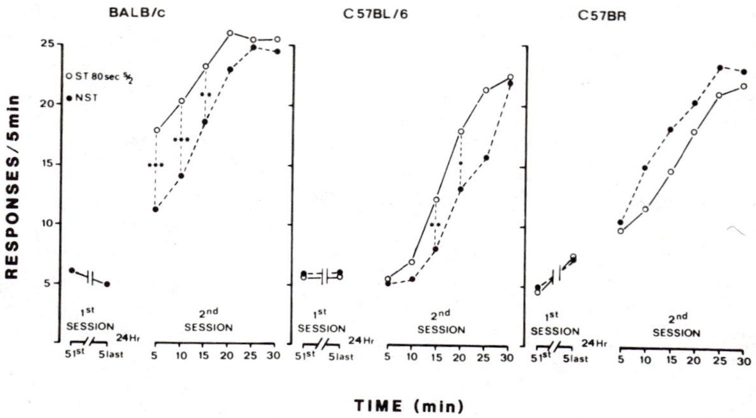

Fig.1 : Effect of hippocampal stimulation after the first learning session on performance recorded during the 24hr-delayed session in the three strains of mice. Within each strain, performance (mean ± S.E.M.) of mice stimulated (ST) 30 sec after the end of the first session was compared to performance of non-stimulated (NST) mice. Statistical comparisons : *** $p < 0.001$; ** $p < 0.02$; * $p < 0.05$.

dramatically improved performance, this facilitative effect being significant from the start (first 5 min) of testing ($t(18) = 6.63$; $p < 0.001$). For C57BL/6 mice, a slight but significant effect was observed on the whole retention curve,but not during the first 10 min of the session. For C57BR mice, no significant differences were found between the ST and NST groups.

The facilitative effect of hippocampal stimulation on long-term retention seems to be related to the capacity of each strain to improve its performance during the 24 hr period, and thus, presumably, to the pre-existing effectiveness of some processes involved in memory formation. In this sense, it may be suggested that hippocampal stimulation magnifies some of the mechanisms normally involved in memory consolidation, these mechanisms being more active for BALB/c than for C57.

Hippocampal cholinergic activity : strain differences and effects of electrical stimulation

The accumulating evidence for a role of the activity of the septo-hippocampal cholinergic neurones in memory processes led us to conduct (in collaboration with Dr.A. Ebel and her group in Strasbourg) a series of parallel comparative neurochemical studies on the activity of biochemical markers for hippocampal cholinergic activity in the three mouse strains to establish a physiological substrate for the strain-dependent behavioral differences.

An initial study compared the specific activity of the enzymes of the cholinergic system in the hippocampus of mice of each strain. Non implanted animals of the three lines were maintained under the same conditions as in the behavioral studies already described. Cholinergic activity in the dorsal hippocampus was evaluated by testing the main enzymes involved in acetylcholine synthesis (choline acetyltransferase : ChAT; EC 2.3.1.6.) and degradation (acetylcholinesterase : AChE; EC 3.1.1.7.). Whereas no significant differences were observed for the specific activity of AChE between strains, the specific activity of ChAT was found to be significantly greater in the dorsal hippocampus of BALB/c mice than that of C57BL/6 (+34%; $p < 0.001$) or C57BR mice (+47%; $p < 0.001$) thus suggesting a possible direct correlation between hippocampal ChAT activity and long term memory performance (Jaffard et al.,1977).

A second experiment tested the effect of hippocampal stimulation on cholinergic activity. ChAT and AChE activity were determined at different intervals after 80 sec of bilateral half-threshold (S/2) electrical stimulation of the CA1 field of the dorsal hippocampus.The time course of the enzymatic activity in BALB/c mice as a function of time after stimulation is represented in Figure 2. For this strain, maximum changes in ChAT activity were observed 3 hr after stimulation (+47%; $p < 0.001$); a sharp fall was then observed and the enzyme activity was minimal 6 hr and 24 hr after the treatment (-37.8% and -32%; $p < 0.001$) and returned to control levels sometime before the sixth day.In C57BL/6 mice, the increase in the ChAT activity after a 3-hr delay was much smaller (+21%; $p < 0.05$) and no significant decrease was observed 24 hr after the treatment. No changes (at 3 and 24 hr) were observed in C57BR. There was no significant modification of AChE activity in any of the strains (Jaffard et al., 1977).

These results led us to suggest that the efficiency of post-training stimulation on subsequent retention would be related to the amplitude of the induced changes in cholinergic activity. Moreover, the observed hippocampal ChAT activity changes after stimulation in BALB/c led us to examine whether, as suggested by interstrain comparisons, these experimentally induced variations would result in changes in learning and memory abilities.

Accordingly, different groups of implanted BALB/c mice received hippocampal electrical stimulation <u>before</u> acquisition and then after different delays (0-24 hrs) were submitted to partial acquisition (16 responses) of the appetitive operant conditioning task and subsequently tested for retention 24 hr later. A comparison of the results of the neurochemical study with those of the behavioral study (Fig.2) suggest a close relationship between the relative changes in ChAT activity and performance on the retention test.

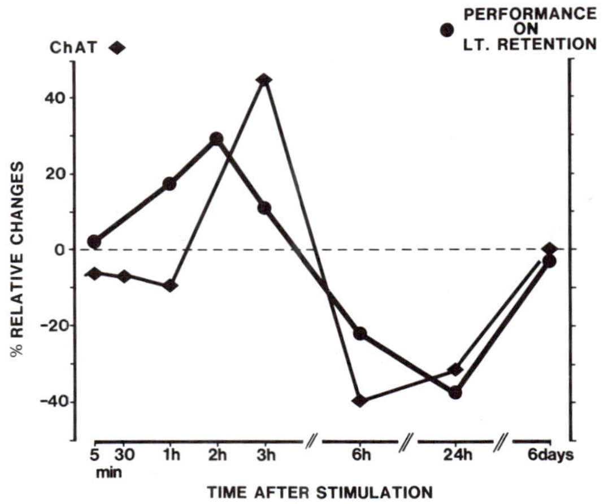

Fig.2 : A schema showing the correlation between the percentage change in ChAT activity (diamonds) and subsequent performance in a 24 hr retention test (circles). Pre-stimulation was applied to the dorsal hippocampus of BALB/c mice at different times before partial acquisition of an appetitive operant conditioning task.

The higher the enzyme activity at the time of or just after the acquisition session the better was performance in retention and an opposite picture of poorer performance in retention was observed for the longer 6 to 24 hrs delays.

The neurochemical and behavioral interstrain comparisons performed on normal animals suggest a correlation between the capacity of acetylcholine-synthesizing activity and long-term retention ability. The same correlation is found if one measures acetylcholine turnover rate (Durkin et al., 1977). The behavioral effects of pretraining stimulation aimed at inducing a decrease in ChAT activity in BALB/c makes this correlation much more evident. Accordingly, such a decrease is associated with lower performance on long-term (24 hr) retention tests so that BALB/c subjects performed as poorly as C57 mice. In other words, a genotypic (as in C57 vs BALB/c mice) or an experimentally induced decrease of ChAT activity (as in BALB/c mice) suppressed reminiscence i.e. the time-dependent improvement of performance on retention tests (Jaffard et al., 1979).

However, while it is clear from numerous experiments that ChAT may exhibit changes in its specific activity after a variety of treatments, this enzyme is generally considered to be present in brain in functional excess and, therefore, not to be a rate-limiting step in the synthesis of acetylcholine (Kuhar and Murrin, 1978). Therefore, it is not known to what degree these changes in ChAT specific activity reflect changes in acetylcholine availability at synapses. Consequently the conclusion that the observed behavioral facilitations and impairments were the consequences of these variations in enzyme activity (Fig.2) was questionable. Accordingly, we tested whether the impaired retention performance we observed when stimulation was applied 24 hrs before acquisition could be reversed by i.c.v. injections of physostigmine which should compensate the presumed reductions of acetylcholine availability.

Reversal of behavioral impairments by physostigmine

The experiment using pre-stimulation of the hippocampus of BALB/c mice 24 hrs before acquisition was repeated but in this study a group of these mice also received an i.c.v. injection of 1 µg of physostigmine 20 minutes before the acquisition session. As shown in Fig. 4A prestimulation induced an impairment of retention which was reversed by physostigmine. It may also be seen that, at least at this dose, physostigmine had no effect on control non-stimulated mice (Micheau et al., 1983), a result that is congruent with observations showing that the effects of cholinergic drugs on memory depend on the activity status of cholinergic synapses at the time of treatment (Bartus et al.,1982).

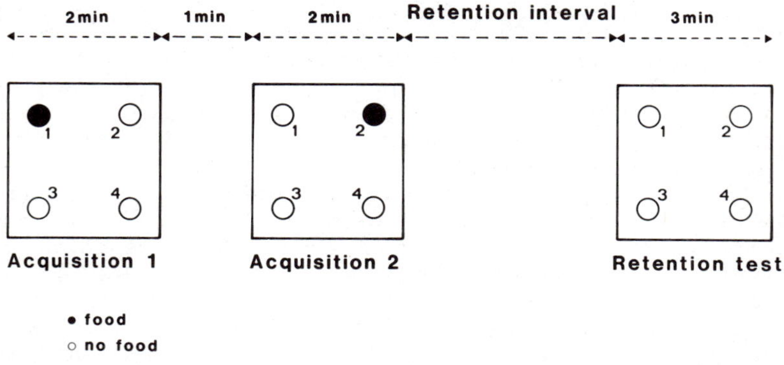

Fig.3 : Protocol for the spatial memory task in the four-hole board.

We have obtained similar results by using a very simple spatial learning task in an apparatus that consists of a board with four holes (Fig.3). For acquisition,a food reward was successively placed into two adjacent holes. On retention, 24 hrs later, food was absent and the relative time spent in exploring the previously reinforced holes was used as an index of retention. As can be seen from fig.4B pre-stimulation impaired retention relative to control mice since the scores of pre-stimulated mice were not significantly different from chance. As previously shown in the lever-press conditioning test, i.c.v. injections of physostigmine reversed the deficit produced by the hippocampal pre-stimulation.

Effects of electrical prestimulation on short and long-term memory performance

The fact that physostigmine reversed the impairment of retention in stimulated mice strongly suggested that hippocampal cholinergic mechanisms underly this behavioral phenomenon. However, the question arises of to what degree these changes in retention are related to either learning or memory processes. More precisely, since stimulation was applied before the acquisition session, the changes observed in retention could result from changes in the learning capacity during training rather than to changes in processes related to long-term memory formation.

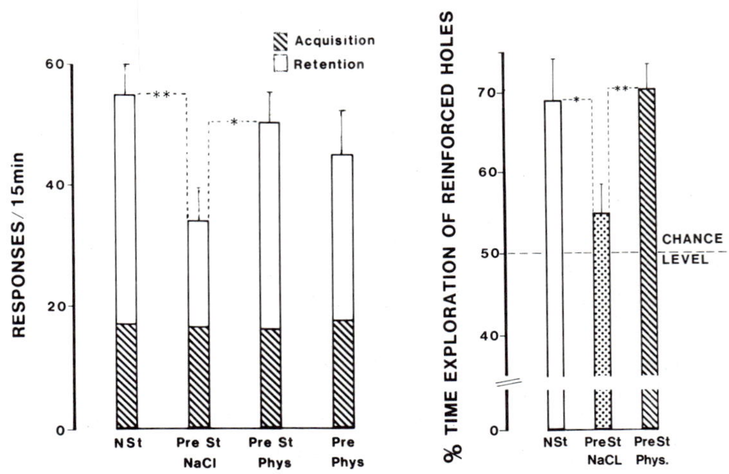

Fig.4 : A : Performance of BALB/c mice during acquisition and 24 hr delayed retention of an appetitive operant conditioning in groups of 1) control non-stimulated mice (NST); 2) mice pre-stimulated at S/2 80 sec in the dorsal hippocampus 24 hrs prior to acquisition and receiving an i.c.v. injection of either 1 µl saline (Pre St.NaCl) or 1 µg/µl physostigmine (Pre St.Phys) 20 min prior to partial acquisition; 3) control non-stimulated mice receiving the i.c.v. injection of physostigmine (Pre-Phys) 20 min prior to partial acquisition. ** $p < 0.01$; * $p < 0.05$.

B : Long-term retention performance (percentage time exploring holes baited in the acquisition session 24 hrs earlier) in 1) control non-stimulated BALB/c mice (NST); 2) mice which had received pre-stimulation of the dorsal hippocampus (S/2, 80 sec) 24 hours prior to the acquisition session (Pre.St NaCl); and 3) mice treated as for 2) but also receiving an i.c.v. injection of physostigmine (1 µg/ µl) 20 min prior to acquisition (Pre-St.Phys). ** $p < 0.01$;* $p < 0.05$.

In order to test this hypothesis, stimulation was again applied 24 hrs before acquisition but this time the retention test was performed either immediately or 24 hrs after acquisition.

The results obtained in the operant lever-press conditioning task are summarized in Fig. 5. Stimulation applied 24 hrs before acquisition which decreased ChAT activity in the previous experiment and which we have already found to impair long-term retention, improved short-term retention with the present protocol (open bars).

The pre-stimulated BALB/c mice exhibited the same behavioral profile as C57 strains which exhibit genetically lower ChAT activity levels and which also performed better on the short-term than on the long-term retention of this task (Fig. 5).

The same results were obtained using the hole board task. Pre-stimulation in BALB/c was again found to significantly improve short-term retention while impairing long-term retention. This increase in short-term retention performance coupled with an associated decrease in long-term retention suggests that response execution and memory consolidation are apparently incompatible processes and are, at least in part, related to hippocampal cholinergic activity at the time of training. Thus, ChAT activity might determine which of these two reciprocal functions predominates. This hypothesis agrees well with that of Routtenberg (1972) according to which, memory storage and retrieval are initiated by two reciprocal and inhibitory systems, hippocampal theta being the primary indicator of the system related to memory storage. . Indeed, several findings suggest the existence of a relationship between hippocampal theta activity, activation of the cholinergic septo-hippocampal pathway (Teitelbaum et al., 1975; Vanderwolf et al., 1978) and memory consolidation (Landfield, 1977; Wetzel et al., 1977; Destrade, 1981).

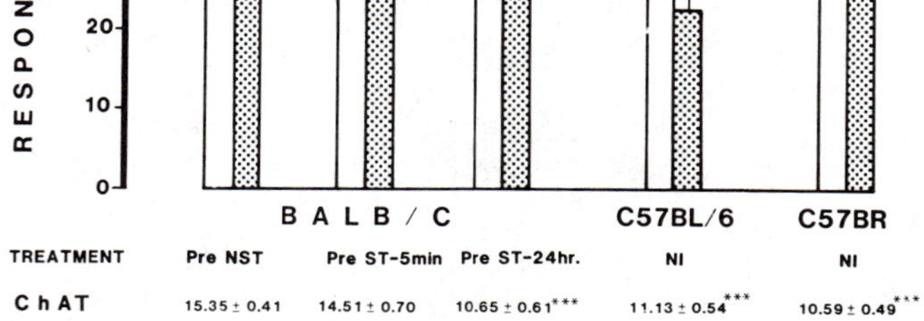

Fig.5 : Performance (mean ± SEM) recorded in retention tests of appetitive operant conditioning in non stimulated (NST) and pre-stimulated (PreST) BALB/c mice (left) and non implanted (NI) C57 mice (right) and corresponding variations (indicated by an arrow) of ChAT activity (μmol/g prot/hr). For behavioral results inter-group comparisons were made for each retention interval. For neurochemical results each group was compared to the BALB/c NST group (* $p < 0.05$; ** $p < 0.01$; *** $p < 0.001$).

We concluded from this first series of neurochemical and pharmacological studies that the hippocampal ChAT activity at the time of training may be an index of the efficiency of some processes involved in long-term memory formation. A more active acetylcholine synthesizing activity in the hippocampus may be related to a greater acetylcholine availability, thus favoring the establishment of long-term memory, perhaps by releasing greater amounts of acetylcholine during or immediately after the learning session. In these conditions, one can expect that treatments aimed at inducing an increase in acetylcholine release during the immediate post-training period would improve subsequent long-term retention. This prediction was tested by using two types of treatments administered immediately after acquisition. Firstly, electrical stimulation was applied directly to the cell bodies of the cholinergic neurons in the medial septum. Secondly, drugs known to trans-synaptically alter the activity of these cholinergic neurons were injected to the lateral septum.

Post-training electrical stimulation of the medial septum

Electrical stimulation of the medial septum at 7.5 or 30 µA, 30 sec after the end of the partial acquisition in the Skinner box, improved performance in the retention test performed 24 hrs later with a temporal gradient of effectiveness of less than 15 min (Fig. 6) (Galey et al., 1983).

In this experiment, a parallel examination of individual electrode tip locations and performance suggested that the treatment was much more efficient when the electrode was strictly located to the sites from which cholinergic neurons originate. Thus, we undertook another experiment in which subjects were first submitted to the behavioral test and after a rest period of 8 days, received the same stimulation for a second time in order to measure the rate of uptake of choline by the high affinity choline uptake mechanism in hippocampal synaptosomal fractions.

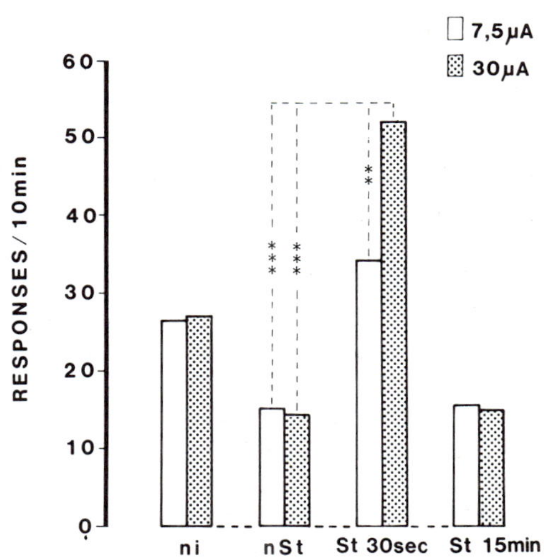

Fig. 6 : The effect of post-training medial septal stimulation on the performance of BALB/c mice in long term (24 hrs) retention of an appetitive operant conditioning in a Skinner box; in 1) control non-implanted mice (NI); 2) implanted non-stimulated mice (NST); 3) mice stimulated (100 Hz, 80 sec) in the medial septum 30 sec and 4) 15 min following the partial acquisition session. *** $p < 0.001$; ** $p < 0.01$.

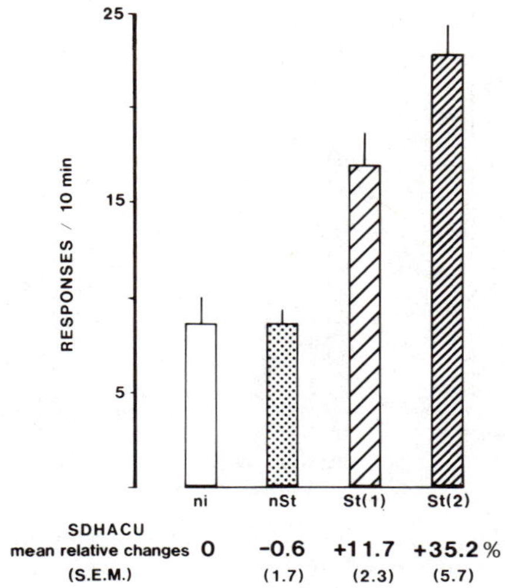

Fig. 7 : A schema showing the correlation between the amplitude of the increase in sodium dependent high affinity choline uptake (SDHACU) in the hippocampus and performance in the 24 hr retention test following post-trial (30 sec) stimulation (100 Hz, 30 μA, 80 sec) of the medial septum in C57BL/6 mice.

The observed facilitation of long term retention performance in C57BL/6 mice appears to be positively correlated with the relative amplitude of the increase in high affinity choline uptake (Fig. 7) suggesting, once again, that the greater the magnitude of hippocampal cholinergic activation after training the better the performance during the delayed retention test.

These data may be interpreted as being in agreement with previous observations that learning, per se, induces changes in the distribution of acetylcholine in sub-cellular fractions of rat hippocampus (Matthies et al., 1974) and results in an increase in choline uptake in hippocampal synaptosomal fractions (Raaijmakers, 1978). Moreover, using the same task, we found that learning the CRF task induced delayed changes in hippocampal ChAT activity (Jaffard et al.,1980).

Trans-synaptic modulation of the activity of the cholinergic neurones of the septo-hippocampal pathway

Acute intraseptal injection of the dopaminergic antagonist, haloperidol, increases the turnover rate of acetylcholine in the hippocampus of the rat (reviewed in Costa et al., 1983). This result has been interpreted as indicating that A10 dopaminergic afferents terminating in the lateral septal nucleus mediate a tonic inhibitory control on the activity of the cholinergic neurones of the medial septum and diagonal band. Similar experiments by the same group using intraseptal injections of ß-endorphin showed that this substance induced effects opposite to those of haloperidol, producing a decrease in the turnover rate of acetylcholine in the hippocampus (Wood et al., 1979). The greater specificity of these pharmacological procedures in comparison to the technique of electrical stimulation led us to examine the consequences for learning and memory of such central pharmacological modulation of the cholinergic septo-hippocampal pathway.

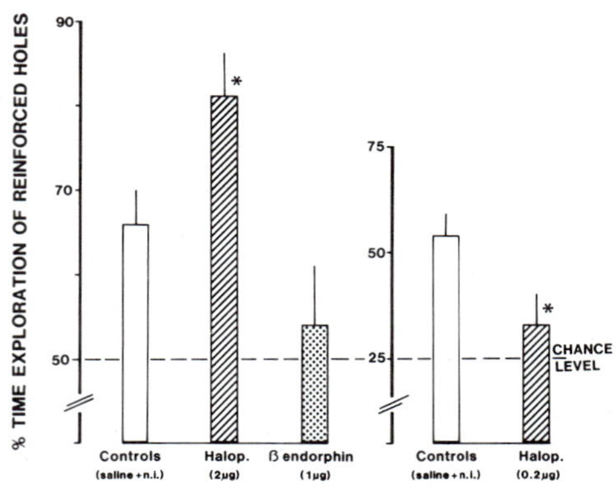

Fig. 8 : The 24 hr retention performance (percentage time exploring previously reinforced holes) of BALB/c mice in the four-hole board spatial learning task in groups of 1) control (saline injected and non-injected) mice;2) mice receiving 2 µg/0.5 µl haloperidol; 3) 1 µg/0.5 µl ß-endorphin and 4) 0.2 µg/0.5 µl haloperidol injected bilaterally into the lateral septum 30 sec following the acquisition session. * $p < 0.05$.

Post-training bilateral injections of haloperidol (2µg/0.5µl) or ß-endorphin (1µg/0.5µl) into the lateral septal nuclei were performed following an acquisition session of the hole-board task. As may be seen from the results of these preliminary studies shown in Fig.8, haloperidol (2µg) improved subsequent retention performance while ß-endorphin impaired performance.It may also be noted that the lower dose of haloperidol (0.2 µg) appeared to disrupt retention performance, possibly through a presynaptic effect.

Hippocampal cholinergic activity, spatial mapping and proactive inhibition

Besides an involvement in memory consolidation mechanisms other-possible suggested functions of the cholinergic septo-hippocampal pathway are its role in learning processes involving either behavioral inhibition (Isaacson and Kimble, 1972; Douglas, 1972) or the encoding and use of spatial information (Sutherland et al., 1982) or both. These possibilities have been the object of recent studies from our laboratory. As mentioned previously it has been shown (Robinson et al., 1979) that A10 dopaminergic afferents to the lateral septum exert a tonic inhibitory control on the cholinergic septo-hippocampal neurones; thus a specific lesion of these afferents should induce a long-term increase (disinhibition) of septo-hippocampal cholinergic activity. We found that bilateral injections of 6-hydroxydopamine (6-OHDA: 3 µg/1µl) into the lateral septum produced an increase in the rate of high affinity choline uptake by hippocampal synaptosomes measured 12-13 days after surgery (table 1). It may also be seen that the amplitude of the increase in high affinity choline uptake was strain-dependent, being greater in subjects of the C57BL/6 strain than in BALB/c (cf Durkin et al., 1983). For this reason, further behavioral experiments were conducted only on the C57BL/6 strain since the 6-OHDA lesion was found to induce behavioral facilitation of greater amplitude in C57BL/6 than in BALB/c mice. Between 6 and 12 days after surgery C57BL/6 mice were successively tested in a T-maze for spontaneous alternation, for food-rewarded spatial discrimination learning and its reversal and for a light-dark discrimination and its reversal.

Table 1. Comparison of the rates of sodium-dependent high affinity choline uptake (pmoles/4 min/mg protein, mean ± SEM, 37°C, 0.25 μM) by crude synaptosomal preparations of hippocampus taken from control, vehicle injected and 6-OHDA septal-lesioned mice of BALB/c and C57BL/6 strains at 12-13 days post-operation.

	BALB/c	C57BL/6
Control	18.51 ± 0.07	19.64 ± 0.18
Vehicle	17.95 ± 0.44	19.77 ± 0.15
6-OHDA	19.15 ± 0.44*	23.01 ± 0.43***

* $p < 0.05$ 6-OHDA vs vehicle
*** $p < 0.001$ 6-OHDA vs vehicle or control.

In mice with 6-OHDA lesions spontaneous alternation was significantly higher (77.9%) than in control mice (68.0%; $p < 0.01$) and there was a significant reduction in the mean number of trials needed to achieve a criterion of 5 consecutive correct responses on both acquisition and reversal of the spatial discrimination task (Fig. 9).

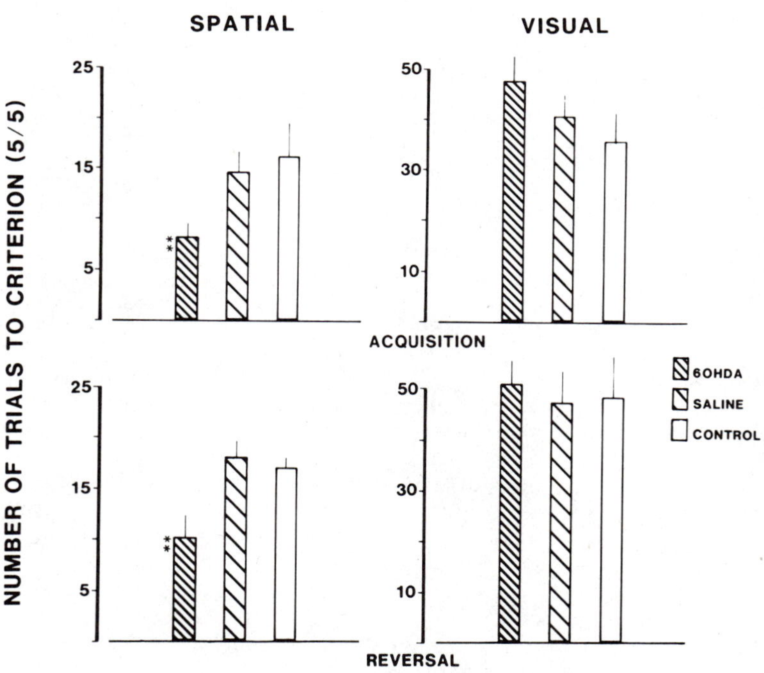

Fig. 9 : Mean (±SEM) number of trials necessary to achieve the criterion of 5 correct responses out of 5 consecutive trials in discrimination training and reversal in the spatial (left) and in the visual discrimination procedure (right) in control, vehicle injected and 6-OHDA lesioned C57BL/6 mice.
** significantly different from the two other groups $p < 0.01$.

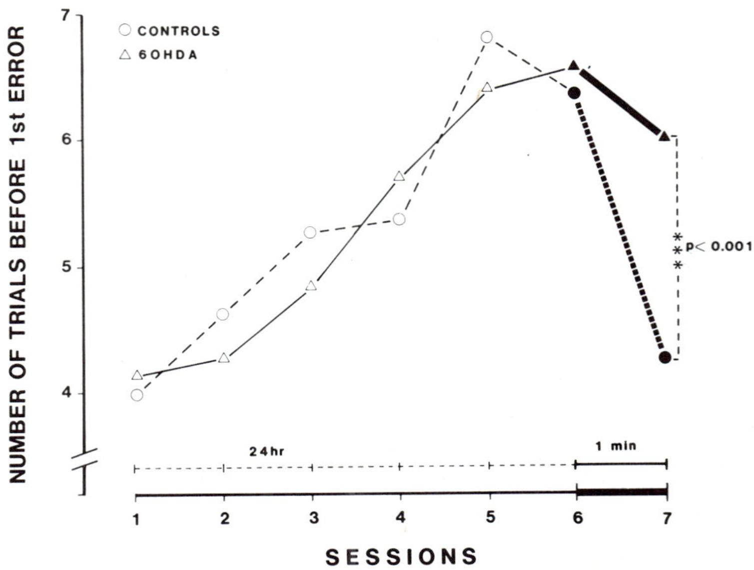

Fig. 10 : Mean number of arms visited in the 8-arm radial maze before making the first error in each daily (1-6 days) session for groups of control (non-injected and vehicle injected) and 6-OHDA septal-lesioned C57BL/6 mice. Session 7 was conducted 1 minute after the end of session 6.

However, as shown in the right hand part of figure 9, there were no between-group differences in the non-spatial version of the task (Galey et al., 1984).

As our results seemed to indicate that the improvement specifically results from the spatial requirements of the tasks, we decided to examine the effects of the lesion on spatial working memory in an elevated 8 arm radial maze.In the first part of the experiment, we were unable to reveal any obvious improvement in performance since both learning curves and the asymptotic level of performance were the same after the animals had received one daily session for 6 days (Fig. 10).

However, when animals were tested in two successive sessions at a brief (1 min) interval, we observed that only control mice were impaired. Indeed, in previous experiments, we found (see also Roberts and Dale, 1981) that in such testing conditions there was a marked proactive interference effect,(i.e., choices made on session 1 interfere with subsequent performance on session 2) but this phenomenon was not observed in our experimental subjects (Fig.10, session 7). This result suggests that the lesion-induced disinhibition of septo-hippocampal cholinergic activity prevented or reduced the vulnerability to proactive interference, an interpretation congruent with our observations of facilitated spatial reversal learning in the T-maze (see above).

CONCLUSION

Taken together the behavioral, pharmacological and neurochemical results that we have obtained are evidence that the cholinergic input to the hippocampus plays a key role in the modulation of learning and memory processes.

Firstly, genetically determined or experimentally induced variations in hippocampal ChAT activity result in differences in the ability to retain information and suggest that acetylcholine availability at the time of training determines the efficiency of processes involved in long-term memory formation. Thus, direct or indirect immediate post-training activation (within 10 min) of the septo-hippocampal cholinergic system through either electrical stimulation of the medial septal nucleus or intraseptal injections of drugs such as haloperidol result in an improvement of performance on subsequent long-term retention. This initial septo-hippocampal cholinergic activation would represent only the first step in the sequence of processes which regulates the storage of information because delayed post-training (30 min) stimulation of other structures projecting to the hippocampus (i.e. entorhinal cortex) also facilitates subsequent retention (see Destrade et al.,this volume).Secondly,increased post-training cholinergic activity could simultaneously be responsible for the transient impairment (inhibition) of immediate retrieval at least for tasks where subjects had not received previous training in other similar variants of the task (Skinner, 4 hole-board). In contrast this increased cholinergic activity improves both acquisition and short-term retention in tasks requiring the inhibition of previously learned responses (T-maze spatial discrimination and reversal) or in tasks requiring the inhibition of previously stored information (alternation tested with the sequential test procedure or during the second of two shortly spaced sessions in the radial maze). The synthesis of these two observations leads us to propose that, in both cases, the role of the septo-hippocampal cholinergic system is to increase inhibition on afferent or stored information and thus to protect memory processes from both retroactive inhibition (during consolidation) and proactive inhibition (operating at the time of new acquisition or retrieval).

This research was supported by the C.N.R.S (L.A.n°339) and by a grant from INSERM (n° 81.6.014).

REFERENCES

Bartus, R.T., Dean, R.L., Beer, B., and Lippa, A.S., 1982, The cholinergic hypothesis of geriatric memory dysfunction, Science, 217:408.

Biederman, G.B., 1974, The search for the chemistry of memory. Recent trends and the logic of investigation in the role of cholinergic and adrenergic transmitters, Progr. Neurobiol., 2:291.

Costa, E., Panula, P., Thompson, H., and Cheney, D., 1983, The trans-synaptic regulation of the septal-hippocampal cholinergic neurones, Life Sci., 32:165.

De Feudis, F.V., 1974, "Central cholinergic systems and behavior". Academic Press, London.

Destrade, C., 1981, Effects of post-trial hypothalamic stimulation driving two types of hippocampal rhythmical slow activity (RSA) on memory processes in mice, Neurosci. Lett., 7:236.

Destrade, C., and Cardo, B., 1974, Effects of post-trial hippocampal stimulation on time-dependent improvement of performance in mice, Brain Res., 78:447.

Destrade, C., Soumireu-Mourat, B., and Cardo, B. 1973, Effects of post-trial hippocampal stimulation on acquisition of operant behavior in the mouse, Behav. Biol., 8:713.

Deutsch, J.A., 1971, The cholinergic synapses and the site of memory, Science, 174:788.

Douglas, R.J., 1972, Pavlovian conditioning and the brain, in : "Inhibition and learning", R.A. Boakes and M.S. Halliday, eds., Academic Press, London.

Durkin, T., Ayad, G., Ebel, A., and Mandel, P., 1977, Regional acetylcholine turnover rates in the brains of three inbred strains of mice: correlations with some interstrain behavioral differences, Brain Res., 136:475.

Durkin, T., Hashem-Zadeh, H., Mandel, P., Kempf, J., and Ebel, A., 1983, Genotypic variation in the dopaminergic inhibitory control of striatal and hippocampal cholinergic activity in mice, Pharmacol. Biochem. Behav., 19:63.

Galey, D., Jeantet, Y., Destrade, C., and Jaffard, R., 1983, Facilitation of memory consolidation by post-training electrical stimulation of the medial septal nucleus : Is it mediated by changes in rhythmic slow activity? Behav. Neural Biol., 38:240.

Galey, D., Durkin, T., Sifakis, G., and Jaffard, R., 1984, Amélioration de conduites spatiales spontanées et acquises après lésion des afférences dopaminergiques septales chez la souris : relations possibles avec l'activité cholinergique hippocampique, C.R.Acad.Sci. Paris, Série III, in press.

Isaacson, R.L. and Kimble, D.P., 1972, Lesions of the limbic system : their effects upon hypotheses and frustration. Behav. Biol., 7:767.

Jaffard, R., 1978, "Mécanismes cholinergiques hippocampiques associés aux processus d'apprentissage et de mémorisation : Etude comportementale et neurochimique sur trois lignées de souris consanguines". Thèse de Doctorat d'Etat, Université Bordeaux I.

Jaffard, R. and Destrade, C., 1982, Learning and memory processes as related to genotypic or experimental variations of hippocampal cholinergic activity in inbred strains of mice. in "The Genetics of the Brain". I.Lieblich, Ed., Elsevier Biomedical Press, Amsterdam.

Jaffard, R. and Galey, D., 1980, Retention of appetitive operant conditioning in mice as a function of training-to-test interval: Effects of pretest electrical stimulation of the hippocampus. Physiol. Behav., 25:845.

Jaffard, R., Ebel, A., Destrade, C., Ayad, G., Mandel, P. and Cardo, B., 1976, The role of hippocampal cholinergic mechanisms in the acquisition of a bar-press response. Pharmacol. Biochem. Behav., 5:371.

Jaffard, R., Destrade, C., Soumireu-Mourat, B. and Cardo, B., 1974, Time-dependent improvement of performance on appetitive tasks in mice. Behav. Biol., 11:89.

Jaffard, R., Destrade, C., Durkin, T. and Ebel, A., 1979, Memory formation as related to genotypic or experimental variations of hippocampal cholinergic activity in mice. Physiol. Behav., 22:1093.

Jaffard, R., Destrade, C., Soumireu-Mourat, B., Durkin, T. and Ebel,A. 1980, Changes in hippocampal cholinergic activity following learning in mice. Neurosci. Lett., 19:349.

Jaffard, R., Ebel, A., Destrade, C., Durkin, T., Mandel, P. and Cardo, B., 1980, Effects of hippocampal electrical stimulation on long-term memory and on cholinergic mechanisms in three inbred strains of mice. Brain Res., 133:277.

Landfield, P.W., 1977, Different effects of post-trial driving or blocking of the theta rhythm on avoidance learning in rats. Physiol. Behav., 18:439-445.

Lewis, P.R. and Shute, C.C.D., 1967, The cholinergic limbic system : Projections to hippocampal formation, medial cortex, nuclei of the ascending cholinergic reticular system and the subfornical organ and supraoptic crest. Brain, 90:521.

Matthies, H., 1974, The biochemical basis of learning and memory. Life Sci., 15:2017.

Matthies, H., Rauca, C. and Liebmann, H., 1974, Changes in acetylcholine content of different brain regions of the rat during a learning experiment. J. Neurochem., 23:1109.

McGaugh, J.L., 1966, Time-dependent processes in memory storage. Science, 153:1351.

Micheau, J., Destrade, C., Jaffard, R. and Caudarella, M., 1983, Effects of physostigmine on memory deficits induced by pretraining electrical stimulation of the dorsal hippocampus in mice. Communication at 13th annual meeting of Society for Neuroscience, Boston, Ma, Nov. 6-11, 1983. Soc. Neurosci. Abstr., V.9, part. 2, p. 826.

Raaijmakers, W.G.M., 1978, Active avoidance learning and choline uptake in rat hippocampal synaptosomes. Neurosci.Lett., suppl.1, 86.

Roberts, W.A. and Dale, R.H., 1981, Remembrance of places lasts : proactive inhibition and patterns of choice in rat spatial memory. Learning and Motivation, 12:261.

Robinson, S., Malthe-Sorenssen, D., Wood, P. and Commissiong, J., 1979, Dopaminergic control of the septal-hippocampal cholinergic pathway. J. Pharmacol. Exp. Ther., 208:476.

Routtenberg, A., 1972, Memory as an input-output reciprocity : An integrative neurobiological theory. Ann. N.Y. Acad. Sci., 193:159.

Routtenberg, A., 1984, The CA3 pyramidal cell in the hippocampus : site of intrinsic expression and extrinsic control of memory formation. in "Neuropsychology of memory", L.R. Squire and N. Butters, Eds, The Guilford Press, New York.

Squire, L.R. and Davis, H.P., 1981, The pharmacology of memory : a neurobiological perspective. Ann. Rev. Pharmacol. Toxicol., 21:323.

Sutherland, R.J., Whishaw, I.Q. and Regehr, J.C., 1982, Cholinergic receptor blockade impairs spatial localization by use of distal cues in the rat. J. comp. physiol. Psychol., 96:563.

Teitelbaum, H., Lee, J.F. and Johannessen, J.N., 1975, Behaviorally evoked hippocampal theta waves: A cholinergic response. Science, 188:1114.

Vanderwolf, C.H., Kramis, R. and Robinson, T.E., 1978, Hippocampal electrical activity during waking behavior and sleep : Analysis using centrally acting drugs. in : Ciba Foundation Symposium 58 (new series), "Functions of the septo-hippocampal system". Elsevier Excerpta Medica, North-Holland, Amsterdam.

Wetzel, W., Ott, T. and Matthies, H., 1977, Post-training hippocampal rhythmic slow activity (theta) elicited by septal stimulation improves memory consolidation in rats. Behav. Biol., 21:32.

Wood, P., Cheney, D. and Costa, E., 1979, An investigation of whether septal γ-aminobutyrate-containing interneurones are involved in the reduction in the turnover rate of acetylcholine elicited by substance P and ß-endorphin in the hippocampus. Neuroscience, 4:1479.

SEQUENTIAL INTERVENTION OF DIFFERENT LIMBIC STRUCTURES IN MEMORY PROCESSES

Claude Destrade, Monique Gauthier and Jamila Sif

Laboratoire de Psychophysiologie, L.A. CNRS n° 339

Univ. Bordeaux I, av. des Facultés 33405 Talence, France

INTRODUCTION

The use of electrical brain stimulation (EBS) to explore brain functions was pioneered by Fritsch and Hitzig (1870). In the early studies, EBS was typically used to elicit a number of species-specific behavior patterns (see review in Doty, 1969). Much later it was discovered that EBS could serve as a conditioned stimulus for shock avoidance response (Loucks, 1933) as well as serving as a positive or negative reinforcer capable of modifying behavior (Olds, 1962).

More recently, EBS has been used as a means of studying mnemonic processes (Gengerelli and Cullen, 1954; Olds, 1956; Mahut, 1962). In fact, in 1961, Glickman put forward the hypothesis that stimulation of different brain structures can interfere with perseverative neural processes involved in memory consolidation.

In 1966, Bloch and coworkers found that retention in rats was facilitated by post-trial stimulation of the midbrain reticular formation at very low current intensities (Bloch et al.,1966). This facilitation effect, which has also been observed with stimulation of hippocampus, lateral hypothalamus and septum (see review in Kesner and Wilburn, 1974; Huston et al., 1977; Kesner, 1982), suggests an action of the electrical stimulation on neuronal mechanisms which are brought into play during the post-trial interval and which are presumed to underlie consolidation or information processing (Bloch, 1970, 1976; Jaffard et al., 1974; Destrade and Cardo, 1974).

During the last decade we have used post-training electrical stimulation of a number of neural structures in order to study, at the behavioral level, the time-course of stimulation effectiveness of different regions of the limbic system in memory formation.

At the same time, with T. Ott, in Magdeburg, we have been looking for the neurophysiological and anatomical correlates of the behavioral changes that we have observed.

BEHAVIORAL EFFECTS OF ELECTRICAL STIMULATION OF LIMBIC STRUCTURES

The behavioral experiments were carried out using male mice of the BALB/c strain, 40-50 days old at the time of surgery. Eight days prior to surgery, they were housed in individual cages with _ad libitum_ access to food and water in a climatized room maintained on a light-dark cycle

(12h-12h). The animals were operated under general anesthesia as described previously (Destrade and Jaffard, 1978; Gauthier et al.,1982). They were bilaterally implanted with bipolar electrodes (insulated platinum wire 90 μm in diameter) in one of the following brain structures: dorsal hippocampus, septum, hypothalamus or entorhinal cortex.

Two weeks after surgery, a food deprivation schedule was initiated gradually over 4 days so that, at the time of the experiment, the mice weighed 82 to 84% of their ad libitum weight and during testing were fed 1 h after each learning session. As previously described (Jaffard et al.,1974), the learning task consisted of partial acquisition and 24 h delayed retention of operant conditioning in a Skinner box (CRF schedule with 5 mg reward). The Skinner box (14 x 14 x 18 cm high) had a food dispenser which differed somewhat from the standard ones in that it was equipped with a device which permitted the removal of any pellets that were not eaten (Destrade et al.,1973). The lever and the food cup were separated by a 5-cm long partition, so the animal had to turn around to press the lever or to find food. Photocells detected the animal's presence in front of the lever or the food cup. Lever-presses which were not followed by an approach towards the food-cup within 10 sec ("complete" responses) were not counted and all the animals underwent an initial acquisition session in which they were allowed 16 such "complete" responses.

Figure 1 summarizes the behavioral paradigm. The EBS was administered in a waiting box, adjacent to the Skinner box, immediately after (30 sec) or at different time intervals ("training-stimulation" intervals) after the initial acquisition session.The experimental animals received bilateral, discontinuous (200 msec ON, 200 msec OFF) brain sine-wave (100 Hz) stimulation, delivered for an 80 sec period which consisted of 10 trains of 4 sec of stimulation alternating with 4 sec of no stimulation. The current intensities ranged from 5 to 30 μA (peak to peak). EEG was monitored for each animal to verify that no afterdischarges were present. Retention testing was always carried out 24 hr after the acquisition period.

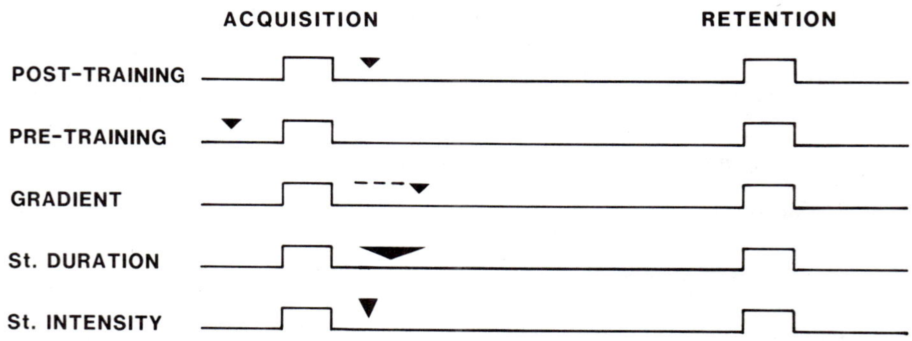

Fig. 1 : Behavioral paradigm.The brain stimulation (▼) is applied prior - or after - the initial acquisition session; the time-interval between acquisition and retention is 24 hours.

The facilitatory effect of EBS was analyzed by comparing the performance of stimulated animals with that of control, non stimulated or sham operated mice. As indicated in figure 1, the study was completed by comparing the effects of pretraining versus post-training stimulation and by looking at the influence of duration and intensity of the stimulation.

Figure 2 shows the time-dependent effect of post-training stimulation (5 to 10 μA) of the CA1 field of the dorsal hippocampus. Hippocampal CA1 stimulation improved retention (measured by "complete" responses during the 5 first min of the retention) when applied at 30 sec, 200 sec and 400 sec after the acquisition session.

The facilitatory effect on retention was observed with stimulation durations ranging from 20 sec to 160 sec (Destrade et al.,1976); on the other hand, current intensities inducing hippocampal afterdischarges impaired the consolidation processes (Soumireu-Mourat et al.,1975).

The short time gradient of effectiveness of the stimulation of CA1 field of the dorsal hippocampus correlated very well with the activation of the cholinergic septo-hippocampal pathway (see Jaffard et al., this volume) in that we found that post-training stimulation of the medial septum improved the retention performance within a 10 min training stimulation interval (Galey et al., 1983).

On the other hand, an indirect activation of the septo-hippocampal pathway by stimulation of the lateral hypothalamus shows the same short time-gradient of effectiveness (Destrade and Jaffard, 1978; Destrade, 1982).

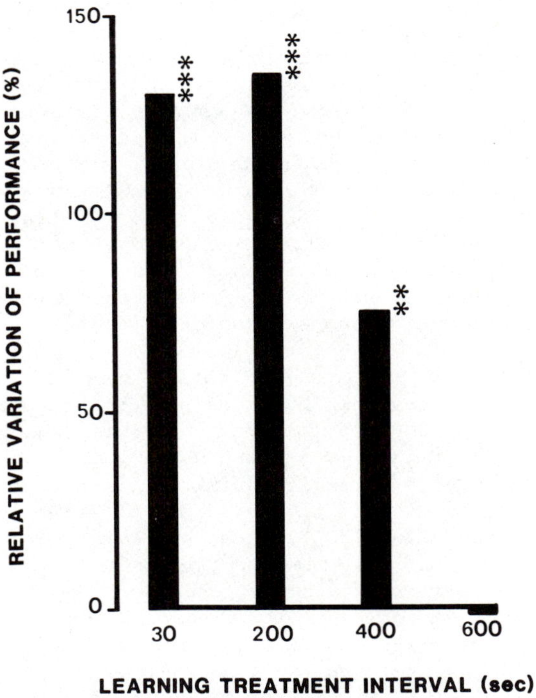

Fig. 2 : Time-dependent effects of post-training stimulation of the CA1 field of the dorsal hippocampus. The performance in retention of stimulated animals was compared with that of non stimulated control animals. *** $p < 0.001$; ** $p < 0.01$.

Furthermore we have found in BALB/c mice that the effects of electrical stimulation of the lateral hypothalamus on memory storage processes and on hippocampal choline acetyltransferase activity (Destrade et al., 1977) paralleled those observed after stimulation of the CA1 dorsal hippocampus (Jaffard et al.,1977).

ELECTROPHYSIOLOGICAL CORRELATES

The electrophysiological data obtained in collaboration with T. Ott, confirmed the behavioral results. Our working hypothesis was that the hippocampal rhythmic slow activity (RSA) is related to memory storage (Landfield et al.,1973). On the other hand, hippocampal EEG patterns, similar to the waves making up normal RSA, can be generated by appropriate electrical stimulation of specific septal, hypothalamic or brainstem sites (Landfield, 1977; Kramis and Vanderwolf, 1980; Destrade and Ott, 1981). In this manner, we demonstrated, in BALB/c mice, the possibility of driving hippocampal RSA using diencephalic electrical stimulation with the same parameters as in learning experiments. For instance electrical stimulation of the medial forebrain bundle (MFB) drove hippocampal RSA with a low frequency range (6-8 Hz). MFB stimulation improved the retention performance when applied 30 sec after acquisition, but no longer had any effect on retention when the learning-treatment interval exceeded 5 min (Destrade, 1982).

These data supported indirectly the hypothesis that the brief time-gradient for the effectiveness of post-training stimulation of the major anatomical subareas of the limbic system - hippocampus, septum, lateral hypothalamus, midbrain reticular formation -, is associated with the specific activation of the septo-hippocampal pathway during and immediately after learning. Our pharmacological studies aimed at characterizing the putative transmitter involves in the generation of the low frequency RSA supported the conclusion that the slow RSA is a sign of an activation of hippocampal muscarinic receptors (Ott et al.,1983).

However, a great deal of evidence in the literature shows that the brain remains active for a few hours following learning. For instance, if we assume that brain activity during sleep is related to learning then, as Bloch et al. (1977) have pointed out,the increase in paradoxical sleep episodes in rats and mice (Destrade et al.,1978) after an incomplete learning session suggests that memory-processing is not completed 10 min after acquisition. This delayed activity could indicate the involvement of limbic structures other than those directly connected with the septo-hippocampal complex.

The first result we obtained with stimulation of the dorsomedial hypothalamus (DMH) supported this hypothesis. Fast hippocampal RSA (9-12 Hz) could be evoked by 100 Hz stimulation of this region (Destrade and Ott, 1981). However, unlike MFB stimulation, the post-training stimulation of the DMH after a partial acquisition in the Skinner box, improved the retention with a long time-gradient of effectiveness : up to 1 hr post-training (Destrade, 1982). Moreover, taking into consideration previous data, we can assume that the DMH-driven RSA is mediated by fibers which pass through the cingulate and entorhinal cortices (Destrade and Ott, 1982) and activate glutaminergic synapses in the hippocampus (Destrade and Ott, 1980). Glutamic acid is a major candidate as a neurotransmitter of the perforant path (White et al.,1977), which connects the entorhinal cortex with the hippocampus (Hjorth-Simonsen and Jeune, 1972; Steward, 1976).

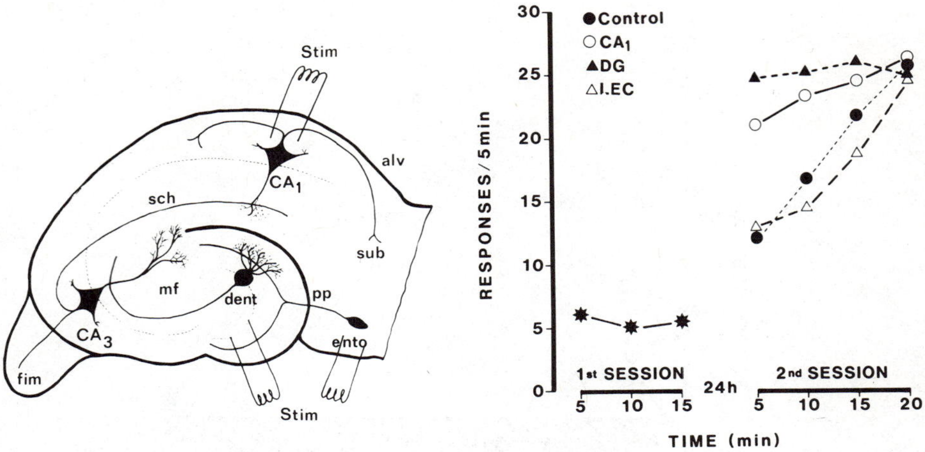

Fig. 3 : Effects of post-training stimulation of the hippocampus (CA1), dentate gyrus (DG) and lateral entorhinal cortex (LEC). The left part of the figure shows the placements of stimulating electrodes. The right part shows the effects on retention (2nd session) of a 80 sec electrical stimulation applied 30 sec after the end of the acquisition (1st session). No difference was found between the 4 groups in acquisition. In retention, CA1 and DG groups showed significant improvement in performance compared to control group ($p < .001$ in both cases); there was no difference between the control group and the LEC group stimulated 30 sec after the acquisition.

LATE POST-LEARNING PARTICIPATION OF THE ENTORHINAL CORTEX IN MEMORY PROCESSES

The effects of post-training stimulation of the entorhinal cortex were studied using the same behavioral paradigm as described. As shown in figure 3 (left part) stimulating electrodes were bilaterally implanted in the lateral part of the entorhinal cortex (LEC), in CA1 field of the hippocampus (CA1), or in the dentate gyrus (DG), in order to compare the effects of post-training stimulation of these three limbic structures.

In the first experiment, stimulation was applied 30 sec after the first learning session (16 "complete" responses) in the Skinner box. Twenty four hours later, the performance of each stimulated group was compared with that of the control group. As shown in the right part of the figure 3, only stimulation of the CA1 and DG produced a significant improvement of retention. Thus LEC stimulation had no effect when applied 30 sec following acquisition.

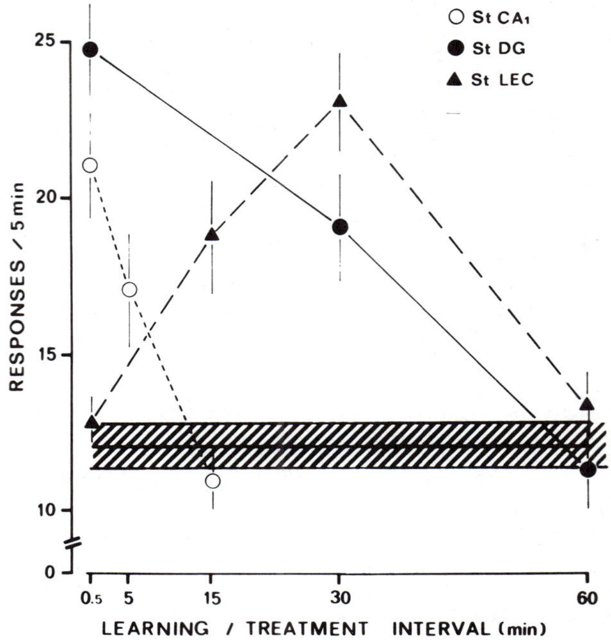

Fig.4 : Time-dependent effect of post-training stimulation of hippocampal CA1 field, dentate gyrus and lateral entorhinal cortex. The time intervals between learning and stimulation varied between 0.5 and 60 min. Performances (±S.E.M.) of stimulated animals during the first 5 min of retention session were compared with those of non-stimulated animals (hatched zone).

A second experiment compared the time-gradient of effectiveness of DG with CA1 stimulation and tested the possibility of a late post-training effect of LEC stimulation using time intervals that ranged from 30 sec to 1 hr post-training. Surprisingly, it was found that LEC stimulation improved retention when delayed between 15 min and 30 min after training. CA1 stimulation was effective only a few min after training, whereas DG stimulation was effective up until at least 30 min after initial acquisition (Fig. 4).

CONCLUSIONS

TIME COURSE OF INTERVENTION OF LIMBIC STRUCTURES IN MEMORY PROCESSES

In summary,different types of time-dependent gradients emerge from our behavioral and electrophysiological results.

On the one hand, we found that post-training stimulation of the CA1 field of dorsal hippocampus, medial septum and lateral hypothalamus have a short gradient of effectiveness : less than 10 min as a general rule.

These results confirm other reports (Huston et al.,1977; Landfield, 1977; Landfield et al.,1973; Major and White, 1978; Wetzel et al.,1977). There is now a good deal of evidence that these short gradients indicate an activation of the septo-hippocampal cholinergic pathway probably involved in immediate initial information processing (see also Jaffard et al., this volume).

On the other hand, considering the time course of effectiveness of dorsomedial hypothalamic and dentate gyrus stimulation (which induced a facilitation of long-term retention when applied during the first 30 min post-training) and the delayed effect of lateral entorhinal cortex stimulation (at 15 min and 30 min) we suggest that these results reveal the involvement of cortico-hippocampal mechanisms in secondary information processing interacting with stored cortical memories. This assertion agrees with the recent suggestion of Teyler and Discenna (1984) that the hippocampus represents "a coordinate system capable of reciprocally addressing neocortical loci in space and time".

In other words, we propose that the differential time-dependent effects of stimulation of limbic structures that we have observed might reflect the different delays of intervention (and/or interaction) of cortical and subcortical areas in memory formation.

However, as the entorhinal cortex is the major cortical structure which projects to the hippocampus and the dentate gyrus, the mechanisms involved in the behavioral effects of delayed lateral entorhinal cortex stimulation remain unclear. Entorhinal stimulation could either involve some kind of reverberatory reactivation of the hippocampus through the perforant path, or indicate a specific involvement of the LEC in late information processing. Recently, to clarify these issues, an attempt was made to dissociate entorhinal and hippocampal mechanisms in a behavioral study (Gauthier and Destrade, 1984). The facilitatory effect of subseizure electrical stimulation of the LEC was examined on retention 24 hr later in animals with perforant path lesions. The results show that the effect of LEC stimulation (delayed by 30 min after acquisition) persisted in spite of bilateral lesions of the perforant path. Compared to control animals, the perforant path lesioned LEC stimulated animals showed, as did the non-lesioned LEC stimulated mice, a clear improvement in retention; moreover, they exhibited a faster extinction of conditioning. These latter results suggest a functional dissociation between hippocampal and cortical mechanisms involved in memory facilitation produced by stimulation of the lateral entorhinal cortex; they argue in favor of a specific involvement of this cortical region in delayed information processing.However, it is difficult to see, from a functional-anatomical point of view, how electrical stimulation of the entorhinal cortex could bring into play mechanisms that are strictly limited to the stimulated area. It is more likely that electrical stimulation reveals the recruitement of cortical and subcortical regions connected with the entorhinal cortex. In order to characterize functionally the neural pathways activated by LEC stimulation,we examined the uptake of 2-deoxyglucose (2-DG) following stimulation of LEC in intact animals and after lesion of the perforant path (Destrade et al.,1984). In intact animals, significant increases in 2-DG radioactivity was observed in the hippocampus (dentate gyrus and CA3). Moreover, demonstrable labelling was found in brain areas beyond the hippocampal formation : piriform cortex, amygdala and cingulate cortex (retrosplenial area); less dense but significant increases were noted in the Diagonal Band of Broca, the medial and lateral septal nuclei and medial forebrain bundle.

After unilateral perforant path lesion (Figure 5), the metabolic activity disappeared ipsilaterally in subiculum, hippocampus and in some thalamic nuclei, but all extra-hippocampal labelling was unchanged.

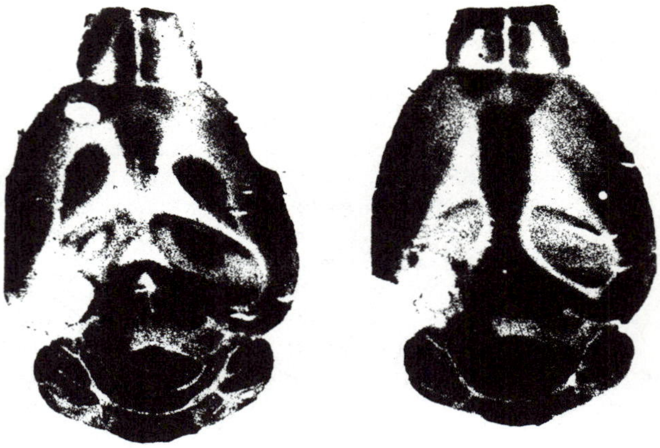

Fig. 5 : ^{14}C-2-deoxyglucose (2-DG) autoradiography of BALB/c mouse brain (horizontal sections) with unilateral lesion of the perforant path (PP). The lateral entorhinal cortex was stimulated ipsilaterally to the PP lesioned side (100 Hz, 200 msec on, 200 msec off, 25 μA) for 5 min before and 30 min following an injection of 2-DG. Labelling was absent in ipsilateral hippocampus (left section) but the 2-DG radioactivity was present in cingulate cortex (right section). Note that the colliculus posterior also showed evidence of metabolic activation ipsilaterally to the PP lesioned side, but we were unable to determine the activated pathway.

These data, considered along with our previous behavioral results, suggest that LEC stimulation may act on mnemonic processes by the recruitement of extra-hippocampal structures directly or indirectly connected with the entorhinal cortex. We speculate that the cingulate cortex, whose posterior part is apparently a special region participating in the organization of attention and memory (Baleydier and Mauguiere, 1980) could be especially involved in cortical mechanisms related to long-term information storage and processing.

This research was supported by the CNRS (L.A. n°339) and by a grant from the INSERM (n° 81.6.014).

REFERENCES

Baleydier, C., and Mauguiere, F., 1980, The duality of the cingulate gyrus in monkey, Brain, 103:525.

Bloch, V., 1970, Facts and hypotheses concerning memory consolidation processes, Brain Res., 24:561.

Bloch, V., 1976, Brain activation and memory consolidation, in : "Neural mechanisms of learning and memory", M.R. Rosenzweig and E.L. Bennett, eds., The MIT Press.

Bloch, V., Denti, A., and Schmaltz, G., 1966, Effets de la stimulation réticulaire sur la phase de consolidation de la trace mnésique, J. Physiol.(Paris) 58:469.

Bloch, V., Hennevin, E., and Leconte, P., 1977, Interaction between post-trial reticular stimulation and subsequent paradoxical sleep in memory consolidation processes. in : "Neurobiology of sleep and memory", R.R. Drucker-Colin and J.L. McGaugh, eds., Academic Press, New York.

Destrade, C., 1982, Two types of diencephalically driven RSA (theta) as a means of studying memory formation in mice. Brain Res., 234:486.

Destrade, C., and Cardo, B., 1974, Effects of post-trial hippocampal stimulation on time-dependent improvement of performance in mice. Brain Res., 78:447.

Destrade, C., Hennevin, E., Leconte, P., and Soumireu-Mourat, B., 1978, Relationship between paradoxical sleep and time-dependent improvement of performance in BALB/c mice, Neurosci. Lett., 7:239.

Destrade, C., and Jaffard, R., 1978, Post-trial hippocampal and lateral hypothalamic electrical stimulation. Facilitation on long-term memory of appetitive and avoidance learning tasks, Behav. Biol., 22:354.

Destrade, C., Jaffard, R., Cardo, B., Ebel, A., and Mandel, P., 1977, Effects of hippocampal and lateral hypothalamic electrical stimulation on hippocampal cholinergic mechanisms in BALB/c inbred mice, Neurosci. Lett., 4:181.

Destrade, C., Jaffard, R., Deminière, J.M., and Cardo, B., 1976, Effets de la stimulation de l'hippocampe sur la réminiscence chez deux lignées de souris, Physiol. Behav., 16:237.

Destrade, C., and Ott, T., 1980, Blockade of high frequency rhythmical slow activity by intrahippocampal injection of a glutamic acid antagonist, Neurosci. Lett., 18:73.

Destrade, C., and Ott, T., 1981, Quantitative analysis of hippocampal rhythmical slow activity induced by stimulation of different diencephalic structures, Neuroscience, 6:1089.

Destrade, C., and Ott, T., 1982, Is a retrosplenial (cingulate) pathway involved in the mediation of high frequency hippocampal rhythmical slow activity (theta) ? Brain Res., 252:29.

Destrade, C., Sif, J., Gauthier, M., and Calas, A., 1984, A functional anatomical study of neural pathways involved in memory mechanisms using 14-C-2-Deoxy-D-glucose : Effects of entorhinal cortex electrical stimulation, Society Neurosci. Abstr., 10:134.

Destrade, C., Soumireu-Mourat, B., and Cardo, B., 1973, Effects of post-trial hippocampal stimulation on acquisition of operant behavior in the mouse, Behav. Biol., 8:713.

Doty, R.W., 1969, Electrical stimulation of the brain in behavioral context, Annual Rev. Psychol., 20:289.

Fritsch, G., and Hitzig, E., 1870, Ueber die elektrische erregbarkeit des Grosshirms, Arch. Anat. Physiol., 37:300.

Galey, D., Jeantet, Y., Destrade, C. and Jaffard, R., 1983, Facilitation of memory consolidation by post-training electrical stimulation of the medial septal nucleus : is it mediated by changes in rhythmic slow activity ? Behav. Neural Biol., 38:240.

Gauthier, M., and Destrade, C., 1984, Late post-learning effect of entorhinal cortex electrical stimulation persists despite destruction of the perforant path, Brain Res., 310:174.

Gauthier, M., Destrade, C., and Soumireu-Mourat, B., 1982, Late post-learning participation of entorhinal cortex in Memory processes, Brain Res., 233:255.

Gengerelli, J.A., and Cullen, J.W., 1954, Studies in the neurophysiology of learning. I. Effect of brain stimulation rate during runs on maze performance in the white rat, J.comp. physiol. Psychol., 47:204.

Glickman, S.E., 1961, Perseverative neural processes and consolidation of the memory trace, Psychol. Bull., 58:218.
Hjorth-Simonsen, A., and Jeune, B., 1972, Origin and termination of the hippocampal perforant path in the rat studied by silver impregnation, J. Comp. Neurol., 144:215.
Huston, J.P., Mueller, C., and Mondadori, C., 1977, Memory facilitation by post-trial hypothalamic stimulation and other reinforcers : a central theory of reinforcement, Biobehav. Rev., 1:143.
Jaffard, R., Destrade, C., Soumireu-Mourat, B., and Cardo, B., 1974, Time-dependent improvement of performance on appetitive tasks in mice, Behav. Biol., 11:89.
Jaffard, R., Ebel, A., Destrade, C., Durkin, T., Mandel, P., and Cardo, B., 1977, Effects of hippocampal electrical stimulation in long-term memory and on cholinergic mechanisms in three inbred strains of mice, Brain Res., 133:277.
Kesner, R.P., 1982, Brain stimulation : effects on memory, Behav. Neural Biol., 36:315.
Kesner, P.R., and Wilburn, M.W., 1974, A review of electrical stimulation of the brain in context of learning and retention, Behav. Biol., 10:259.
Kramis, R., and Vanderwolf, C.H., 1980, Frequency-specific RSA-like hippocampal patterns elicited by septal, hypothalamic, and brainstem electrical stimulation, Brain Res., 192:383.
Landfield, P.W., 1977, Different effects of post-trial driving or blocking of the theta rhythm on avoidance learning in rats, Physiol. Behav., 18:439.
Landfield, P.W., Tusa, R.J., and McGaugh, J.L., 1973, Effects of post-trial hippocampal stimulation on memory storage and EEG activity, Behav. Biol., 8:485.
Loucks, R.B., 1933, Preliminary report of a technique for stimulation or destruction of tissues beneath the integument and the establishing of conditioned reactions with faradization of the cerebral cortex, J. comp. Psychol., 16:439.
Mahut, H., 1962, Effects of subcortical electrical stimulation on learning in the rat, J. comp. physiol. Psychol., 55:472.
Major, R., and White, N., 1978, Memory facilitation by self-stimulation reinforcement mediated by the nigro-neostriatal bundle, Physiol. Behav., 20:723.
Olds, J., 1956, Runway and maze behavior controlled by basomedial forebrain stimulation in the rat, J. comp. physiol. Psychol., 49:507.
Olds, J.,1962, Hypothalamic substrates of reward, Physiol. Rev., 42:554.
Ott, T., Malisch, R., and Destrade, C., 1983, Nachweis der transmitterspezifischen Modulation der langsamen rhythmischen Aktivität des Hippokampus, Biomed. Biochim. Acta, 42:967.
Soumireu-Mourat, B., Destrade, C., and Cardo, B., 1975, Effects of seizure and subseizure post-trial hippocampal stimulation on appetitive operant behavior in mice, Behav. Biol., 15:303.
Steward, O., 1976, Topographic organization of the projections from the entorhinal area to the hippocampal formation in the rat, J. Comp. Neurol., 167:285.
Teyler, T.J., and Discenna, P., 1984, The topological anatomy of the hippocampus : a clue to its function, Brain Res. Bull., 12:711.
Wetzel, W., Ott, T., and Matthies, H., 1977, Post-training hippocampal rhythmic slow activity ("theta") elicited by septal stimulation improves memory consolidation in rats, Behav. Biol., 21:32.
White, W., Nadler, J.V., Hamberger, A., Cotman, C.W., and Cummins, J.T., 1977, Glutamate as transmitter of hippocampal perforant path, Nature (Lond.), 270:356.

POST-LEARNING PARADOXICAL SLEEP :

A CRITICAL PERIOD WHEN NEW MEMORY IS REACTIVATED ?

Elizabeth Hennevin and Bernard Hars

Université de Paris-Sud and Département de Psychophysiologie
L.P.N., Centre National de la Recherche Scientifique
91190 Gif-sur-Yvette, France

Though a state of deep sleep, it appears near to wakefulness in some of its characteristics ... Of course, such peculiarities of paradoxical sleep (PS) have puzzled many people and diverse hypotheses concerning its function have been proposed (Hennevin and Leconte, 1971). But, the only one which has been really tested is the hypothesis of an involvement of PS in memory processes.

SLEEP AND MEMORY

The beneficial effect of sleep on retention of information acquired during wakefulness was demonstrated long ago in man and when PS was discovered, an active involvement of this state of sleep in memory mechanisms was assumed. Indeed, the considerable brain activity recorded during PS - with, for instance, intense activity of cortical and reticular neurons, high rate of protein synthesis, increased cerebral metabolism - suggests the existence of an active functional process. Moreover, because of the sensory isolation of the organism during PS, this process must involve information that has been previously registered during wakefulness. On an other hand, ontogenetic studies have shown that in all mammalian species the amount of PS is greater during early life, i.e. when considerable learning can be presumed to occur. Human clinical findings have indicated that severe mental retardates spend the same total time asleep but have lower PS values than normal subjects. For more details, see Hennevin and Leconte (1971). Lastly in studies on the influences of differential rearing on sleep-waking parameters there are several reports that environmental enrichment - known to enhance performances in different learning tasks - selectively increases PS (Lambert and Truong-Ngoc, 1976; Gutwein and Fishbein, 1980). All these observations lend support to the hypothesis of a relationship between PS and cognitive activities.

During the past 15 years, many studies - reviewed in Fishbein and Gutwein (1977); Hennevin and Leconte (1977); Mc Grath and Cohen (1978); Pearlman (1979) - have been performed to test this hypothesis. Two experimental procedures have been used : one consists of studying the effects of PS deprivation (PSD) on memory; the second consists of studying the effects of learning on characteristics of subsequent PS.

Postlearning PSD experiments in animals provide consistent support for the hypothesis of an involvement of PS in memory, especially when one accepts that the effects of PSD may depend upon the type of training task used. Drawing upon Seligman's distinction between prepared and unprepared learning, Greenberg and Pearlman (1974) have suggested that PS is functionally unrelated to the memory of prepared learning which involves little change in the animal's behavioral repertoire, whereas PS is necessary for the memory of unprepared learning that requires assimilation of unusual information and adaptative changes in behavior. Most experimental results are consistent with this distinction. In rats, simple tasks such as inhibitory avoidance, position habits, simple bar pressing, are unaffected by subsequent PSD; in contrast, PSD disrupts retention performance in a variety of more complex tasks as shuttle box avoidance, brightness discrimination and bar press task with a low rate reward schedule. For further details, see Pearlman (1981).

Concerning the animals studies that focused on the characteristics of sleep following training sessions, these have provided convincing evidence of a post-learning PS increase phenomenon. In our laboratory, we have accumulated data, in different species, demonstrating that a great variety of tasks produces, during the period following learning, an augmentation in the amount of PS, without alteration of slow-wave sleep. Some issues that we treat briefly here have been more thoroughly discussed in Hennevin and Leconte (1977), Bloch et al.(1979), Pacteau (1982). Our results showed that PS augmentation is truly related to acquisition, indeed:

. it did not result from non-specific effects such as fatigue, fear or excitement produced by the learning conditions.

. it was related to learning achievement, since the increase was present only if a new response was really acquired. Pseudoconditioned or unsuccess animals exhibited no PS increase.

. it was related to critical stages of acquisition. In the course of distributed trial training, PS elevations always occurred just prior to the stabilization of performance. When the learning task was mastered, PS duration returned to its normal value; however, if the animals were then confronted with a new situation requiring new learning, PS augmentation reappeared.

All these findings are in agreement with other investigators' data, cited further. There remains one debated question, concerning the temporal characteristics of PS increases. In all our experiments and in some other studies (Lucero, 1970; Leconte and Hennevin, 1973; Leconte et al., 1973; Hennevin et al., 1974; Leconte et al., 1974; Smith et al., 1974; Lecas, 1976; Destrade et al., 1978) PS augmentation appeared within the first hours following learning. By contrast, some authors observed delayed and prolonged PS increases (Fishbein et al., 1974; Smith et al., 1974, 1977, 1980). Until now, the reasons for this discrepancy remain unknown; one of the likely factors is the spontaneous sleep organization of animals, which depends upon species and even strains but also environmental conditions.

Whatever that may be, if we consider both the retention impairments induced by PSD and the existence of post-training PS increases, it stands to reason that during PS following learning some processes that have beneficial effects on memorization occur.

Of course questions arise concerning the type of the processes involved. Two alternative hypotheses may be considered.

The first one supposes that during post-learning PS neuronal materials that have been utilized during wakefulness for the acquisition activity are restored. Actually, there is a good deal of evidence that during PS some synthetic processes in the brain are increased above the waking level (Adam and Oswald, 1977). With such a viewpoint, one may assume that if this post-learning repair is prevented by PS deprivation, then neuronal metabolism would be disrupted and consequently, further training performances would be impaired. In regard to the post-learning PS increases, they would reflect this restorative process, specially important at critical periods of acquisition when more neuronal materials are consumed. Thus, the post-learning PS augmentation is considered here as a delayed consequence of the brain activity previously implicated in registration of information. In contrast, for the second hypothesis PS increases would reflect memory mechanisms that are actively processing during the phases of PS consecutive to learning for the elaboration of the memory trace. We report below a body of data, obtained in our laboratory, that lends support to this information processing hypothesis.

PARADOXICAL SLEEP AND INFORMATION PROCESSING

A first argument can be found in the recent demonstration made by Bloch and Maho (in press) that learning can occur during PS. Until now, despite many attempts, acquisition had never been shown during sleep, but it must be noted that these unsuccessful experiments were performed during slow-wave sleep. The lack of interest in PS and learning likely comes from the apparently enhanced isolation of the subject from environment, since responses to unfamiliar stimuli are decreased during this state. In order to test the possibility of learning during PS, Bloch and Maho decided to circumvent the inhibition of peripheral input and output gates. They bypassed these gates by using a conditioning paradigm in which both CS and UCS were direct brain stimulation, the response being recorded intracranially also. They replaced the usual sensory CS by electrical stimulation of medial geniculate nucleus and the usual UCS foot-shock by stimulation of nociceptive region of the dorso-lateral grey (previously assessed by switch-off behavior); the response studied was a multiunit activity increase in the hippocampus CA3. Stimuli were carefully chosen for their ability to evoke hippocampal conditioning during wakefulness but their inability to provoke arousal from drowsiness. With such stimuli, conditioning was established during PS, with a slope of the learning curve very similar to the one observed during waking, whereas no conditioning could be established during slow-wave sleep. Furthermore, the acquisition implemented during PS facilitated subsequent conditioning during wakefulness. As new associations can be built up during PS, it must be presumed that cerebral conditions required for information processing are present at this time. Thus, during PS the brain appears quite able to acquire and process new information.

Consistent with this view are the following data which lead to consider that similar dynamic processes act just after registration of information and during post-learning PS. It is well known that the registration of information is followed by a brief but critical period of time during which the memory trace initiated by the recent experience is vulnerable and can be perturbed or facilitated. The idea seems now widely accepted (Bloch, 1973) that during this period - called the consolidation phase - the newly registered information is being processed for its subsequent efficient retrieval. Some years ago we found (Bloch et al., 1977) an interdependence between the mechanisms implicated during this consolidation phase and those taking place during subsequent PS. It has often been demonstrated that a mild stimulation of the mesencephalic reticular formation (MRF) delivered immediately after learning enhances retention performances. Similar MRF stimulation appeared to largely substitute for PS by decreasing the post-training PS elevation and by abolishing most of the impairment produced by PS deprivation. Indeed, in a maze learning experiment with one trial each day, rats stimulated post-trial presented only a small PS increase after one of the trials, whereas those not-stimulated exhibited two large PS augmentations. In the same maze learning task, a 3 hour sleep deprivation after each trial impaired acquisition; but when the reticular stimulation preceded this sleep deprivation, then acquisition was no longer impaired. Thus, after MRF stimulation a PS period seems no longer necessary in post-learning sleep. That leads to assume that reticular stimulation has facilitated some neurophysiological events - known as critical for information processing - which are normally replayed during subsequent PS phases.

Another treatment that also appeared able to alleviate the deleterious effects of PSD is reminder treatment (Hars and Hennevin, 1983). In practice, reminders are events that have accompanied acquisition of a memory. Thus, a reminder procedure consists of presenting some fractional portion of the conditions of original training during the retention interval. As in the preceding experiment, rats were sleep deprived after each daily trial of a maze learning task but, on this occasion, one group was submitted to a reminder treatment; just before each trial, these animals were exposed for 90 sec. to background stimuli of the experimental environment. This pre-trial exposure to contextual cues totally counteracted learning impairments induced by post-trial PSD. A control experiment showed that this reinstatement of performance was not due to some preparatory excitation induced by pretrial manipulation and that the reminder per se did not contain sufficient information to allow some supplementary learning. This reversibility of PSD deficit by reminder suggests that similar processes occur during post-learning PS and exposure to a reminder.

To account for the reminder effect, we shall borrow from the viewpoint of Spear (1973), that a memory is a multidimensional representation of an episode, a configuration of attributes corresponding to individual events noticed during original situation. Presumably a memory attribute is aroused when a cue similar or sufficiently similar to one of these events is further noticed; arousal of a sufficient number or kind of attributes elicits retrieval of the entire memory. A good deal of evidence (Lewis, 1979; Hars, 1980) suggests that a retrieved memory is in an active state, presenting the same characteristics as a memory representing an episode immediately after episode's occurrence : both are accessible, vulnerable and susceptible of further elaboration, as opposed to memories in a passive, or inactive, state. Therefore, within this framework, reminders appear as agents of memory reactivation that facilitate retrieval by increasing accessibility to a memory.

The evidence cited above leads us to assume that during post-learning PS, memory of newly acquired information might be in a state like that during original learning, consolidation period and reactivation treatment. In other words memory might be accessible and susceptible of being processed. This assumption is strongly reinforced by the results of our latter experiments showing that a reactivating treatment introduced during post-learning PS facilitates subsequent performance tested during wakefulness.

PARADOXICAL SLEEP AND REACTIVATION

To use a reactivating treatment during PS, we had first to resolve two problems : the cue must be suitable to be introduced during sleep and it must be verified as an effective reactivator of memory. In previous experiments we had shown that spontaneous forgetting of a maze habit provided a very convenient paradigm to give evidence of effectiveness of a reactivating treatment (Deweer et al., 1980). However, in these studies, alleviation of forgetting was obtained by pretest exposure to contextual cues. Of course, this cueing procedure could not be used during sleep, so the learning situation was somewhat modified. Rats were trained in a food reinforced maze for 5 trials, with 1 trial each day; entries into blind-alleys were associated with mild, non painful electric shocks to the ear. The number of shocks in a blind-alley depended on time spent in it. 25

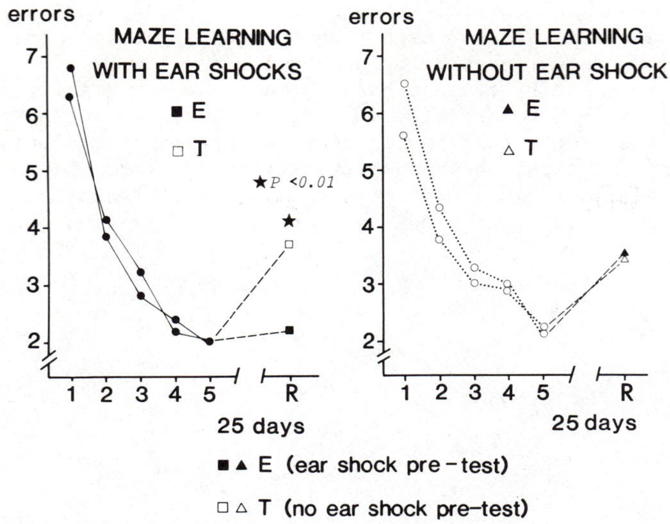

Fig. 1. Alleviation of Forgetting by Pre-Test Cueing.

For five days rats were submitted to one daily trial of a maze learning (1 to 5). A retention test (R) was performed 25 days later.
Left: ear shocks were associated with blind-alleys during learning. Note the significant forgetting in group T and the alleviation of forgetting by ear shocks delivered to awake animals just before retention test (group E).

Right: ear shocks were not associated with blind-alleys during learning. Note the absence of effect of pre-test ear shocks.

days later, just before retention, similar ear shocks were delivered to animals, awake in their home cages. This pre-test cueing alleviated forgetting resulting from 25 days without training (see fig. 1). A control experiment demonstrated that pretest ear shocks had no effect on subsequent retention performance when they had not been previously associated with learning (see fig. 1). So, they did not act in some non specific way.

In the next experiment we attempted to reactivate, during post-learning PS, a just acquired memory, using the reminder procedure of the previous experiment. As previously, rats were trained for 5 consecutive days in a maze with ear shocks signalling blind alleys. After each daily trial, all animals were returned to their individual home cages where vigilance states monitored by ECoG and EMG criteria. Animals of the experimental group (group E) received ear shocks during the first 6 phases of PS; the shocks used were below awakening threshold. Control animals (group T) received no shock. Comparison of learning performances of the 2 groups showed no significant difference for error number. Thus, the improvement of performance by ear shocks obtained in the preceding experiment was not found again. In effect, such a discrepancy has already been observed using the same maze learning task. On several occasions, a reminder treatment very efficient in decreasing number of errors when acting on a memory weakened, for instance by a long interval of retention, was found inefficient when given in the course of normal acquisition (Spear and

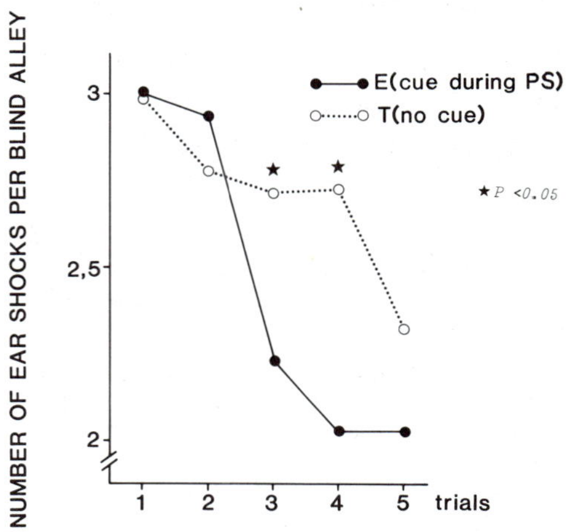

Fig. 2. Effect of Cueing During Post-Trial PS on Maze Learning.

During each daily trial (1 to 5) of a maze learning, ear shocks were delivered in blind-alleys. After each trial similar ear shocks were delivered as cueing treatment during PS in group E, whereas group T received no shock. Note that time spent in blind-alley is shorter for group E.

Gordon, 1981; Sara and Deweer, 1982). It was hypothesized that in this latter case the retrievability of memory was probably maximum. Nevertheless, in our experiment, the number of ear shocks received per blind-alley - that indicated time spent in blind-alleys - was significantly lower in group E on trials 3 and 4 (see fig. 2). Hence it appeared that shock as a signal for a blind-alley was more quickly acquired by the group which received this shock during PS. Our interpretation of this result is that ear shocks during PS have reactivated the specific memory of the shock-blind alley association.

To confirm this possibility of reactivation during PS, in the following experiment we used another learning task where, in contrast with the maze paradigm, the memory measured by performance was that specifically reminded by the cue.

Rats were submitted to an active avoidance conditioning in a shuttle-box with mild electric stimulations on the ear as the discriminative stimulus preceding a footshock. These ear shocks constituted the further cueing treatment. For 3 consecutive days, animals were submitted to 1 daily session of 20 trials. After each session they were returned to their individual home cages where vigilance states were detected as in the previous experiment. They were divided into 3 groups :

. Group E received cueing treatment during post-learning PS. Ear shocks, below arousal threshold, were delivered during the first 6 phases of PS.

. Group C received cueing treatment during wakefulness. This group was yoked with group E, receiving the same number of ear shocks at the same temporal intervals but this time during waking periods.

. Group T received no ear shock during intersession intervals. Comparison of the 3 learning curves (fig. 3a) showed that during the 2nd and 3rd conditioning sessions group E performed significantly better than the 2 other groups. At the 2nd session, performances of group C tended to be better than those of group T, but the difference failed to reach significance. In order to refine these observations, each conditioning session was divided into 2 blocks of 10 trials each (see fig. 3b). Between the end of the 1st session and the beginning of the 2nd, a significant performance decrement appeared in group T; the decrement was not significant in group C; group E performances did not decrease at all. Thus, ear shocks introduced during post-learning PS improved subsequent performances during wakefulness, by abolishing the normal warm-up decrement. There was no apparent effect on intra-session acquisition curves. Therefore, the cue seemed to act on the retrievability of the previously acquired information rather than on the ability to acquire new information.

The beneficial effect of cueing during PS leads to some preliminary considerations.

. It demonstrates that an external cue is able to "penetrate" during PS despite the input inhibition. The ability of an organism asleep to detect and recognize external information might be permitted by the use of an already significant stimulus. But if the brain in PS is able to discriminate a significant stimulus from the whole set of inputs, this implies that external stimuli can be processed during PS even though the organism is behaviorally unresponsive.

. This result is not compatible with the conventional concept of state-dependent learning, that information may be retrievable only under the same brain state as that of acquisition. There are no specific data concerning state-dependent reactivation but changes in state that occur between sleep and waking might be expected to cause state-dependent deficits in the transfer of reactivation effects. Besides, such an hypothesis could account for the well-known failure to recall dreams. However, our data clearly indicate that a reactivating event experienced during PS can transfer positively to a learning situation during wakefulness. This possibility of a transfer of information from PS to waking is consistent with the previously mentioned study of Bloch and Maho (in press) showing that conditioning during PS facilitates subsequent acquisition during waking.

On an other hand, cueing during wakefulness appears far less effective than during PS. This raises the question concerning memory processes which might be particular to PS. It was hypothesized that sensitivity of a memory to any treatment implies that this memory is in an active state (Lewis, 1979). So, the effectiveness of cueing during PS suggests that memory is

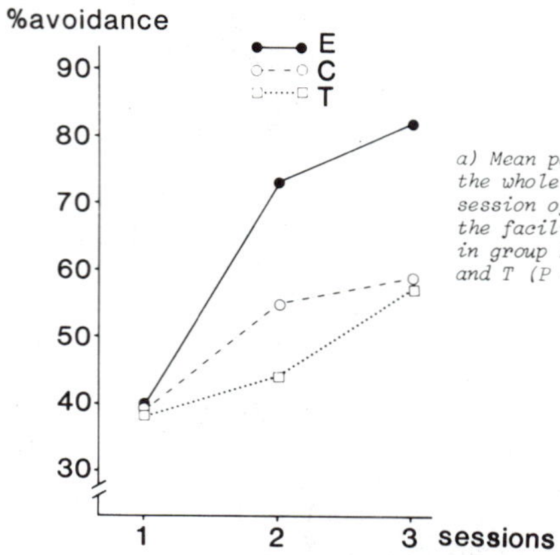

a) Mean percent of avoidance for the whole 20 trials of each session of conditioning. Note the facilitation of performance in group E compared to groups C and T (P <0.005)

b) Mean percent of avoidance for the first ten and the last ten trials of each session. Note the performance decrement exhibited by group T between the end of the first and the beginning of the second session ; this decrement is not significant in group C; in group E there is no decrement at all.

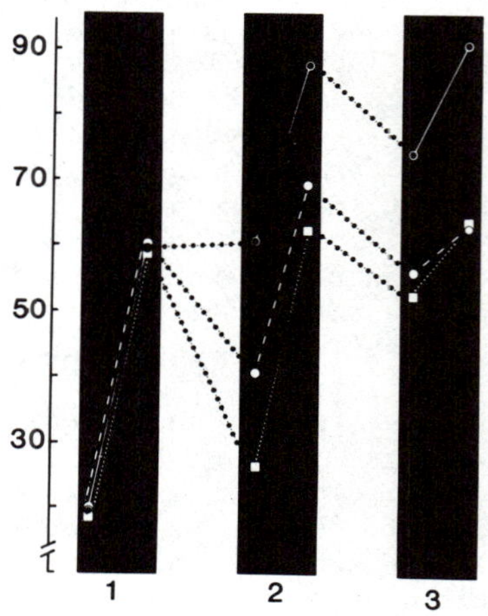

Fig. 3. Effect of Cueing During Post-Learning PS on Acquisition of a Shuttle-Box Avoidance.

During 3 conditioning sessions, with a 24 hr. inter-session interval, ear shocks were used as discriminative stimulus preceding a footshock. In the inter-session interval, similar ear shocks were delivered as cueing treatment either during PS (group E) or wakefulness (group C), whereas group T received no shock.

in active state at that time. Within this framework, two cases may be considered. Either the memory was in active state before cueing or it is the cueing that triggered memory change from passive to active state.

In the first instance, newly acquired memory is considered to be completely avaible during post-acquisition PS and the cueing is then not really a reactivator. To account for beneficial effect of cueing treatment, it may be argued that it provided the animals with some supplementary information. This idea of new learning resulting from cueing treatment per se has long been discussed in the memory literature relative to alleviation of memory deficits, arising from diverse sources, by reminder. However in most reactivation studies, the presence of appropriate control conditions points out that improvements observed in cued subjects do not result from additional learning provided by cueing (Spear, 1973). Moreover, the possibility of reestablishing the lability of old memories by such cueing treatments cannot be accounted for in terms of additional learning.

In the second instance, cueing is considered as the reactivator of the memory trace. Either the reactivation of previously learned information during PS might, in itself, be sufficient to facilitate retrieval of this information during subsequent waking. Or, the reactivation might be only a first step : the reactivated memory is in a labile state and thus able to benefit from the dynamic processes occurring during PS. These two propositions are not necessarily mutually exclusive as the positive effect of reactivation by itself could add to the beneficial effect of processing during PS.

CONCLUSION

Taken together all these data support the notion of the occurrence during post-learning PS of dynamic processes that are beneficial to memorization. These processes appear similar to those occurring at critical periods of memorization when the memory trace is known to be accessible and susceptible to further elaboration. Our recent results strongly support this view by showing that access and recycling of a newly acquired memory are possible during PS. Moreover, these operations which have been elicited by a reminder improve memory. Hence, it may be assumed that the beneficial processes which occur normally during post-learning PS involve memory reactivation. Reminder improves this spontaneous PS reactivation, by enhancing its frequency and/or its effectiveness.

To refine this idea of memory reactivation during PS a question remains relative to its triggering. As previously suggested (Spear and Gordon, 1981; Hars and Hennevin, 1983), it is possible that some attributes of new memory are maintained in active state after acquisition, become proeminent during PS (because of absence of interference by external stimuli for instance) and thus induce a reactivation of the entire memory. It is also possible that some physiological aspects or cognitive events of PS are sufficiently similar to those associated with learning to act as internal cues inducing reactivation of the memory trace. In either case, memory reactivation during post-learning PS would allow for some form of strengthening of recent memory and thus enhance its later retrieval.

REFERENCES

Adam, K., and Oswald, I., 1977, Sleep is for tissue restoration, J. Roy. Coll. Phycns., 11:376.

Bloch, V., 1973, L'activité cérébrale et la fixation mnésique, Arch. Ital. Biol., 3:577.

Bloch, V., Hennevin, E., and Leconte, P., 1977, Interaction between post-trial reticular stimulation and subsequent paradoxical sleep in memory consolidation processes, in: "Neurobiology of sleep and memory", R.R. Drucker-Colin and J.L. McGaugh, ed., Academic Press, New York.

Bloch, V., Hennevin, E., and Leconte, P., 1979, Relationship between paradoxical sleep and memory processes, in: "Brain mechanisms in memory and learning : from the single neuron to man", M.A. Brazier, ed., Raven Press, New York.

Bloch, V., and Maho, C., in press, Acquisition of classical conditioning during REM sleep in rats, in: "Sleep Research" (Proceedings of the 4th International Meeting of the Sleep Research Society, Bologna, Italy, 1983).

Destrade, C., Hennevin, E., Leconte, P. and Soumireu-Mcurat, B., 1978, Relationship between paradoxical sleep and time dependent improvement of performance in BALB-c mice, Neurosci. Lett., 7:239.

Deweer, B., Sara, S., and Hars, B., 1980, Contextual cues and memory retrieval in rats: alleviation of forgetting by a pre-test exposure to background stimuli, Anim. Learn. Behav., 8:265.

Fishbein, W., and Gutwein, B. M., 1977, Paradoxical sleep and memory storage processes, Behav. Biol., 19:425.

Fishbein, W., Kastaniotis, C., and Chattman, D., 1974, Paradoxical sleep: prolonged augmentation following learning, Brain Res., 79:61.

Greenberg, R., and Pearlman, C., 1974, Cutting the REM nerve: an approach to the adaptive role of REM sleep, Perspect. Biol. Med., 17:513.

Gutwein, B.M., and Fishbein, W., 1980, Paradoxical sleep and memory (1): selective alterations following enriched and impoverished environmental rearing, Brain Res. Bull., 5:9.

Hars, B., 1980, Consolidation, rappel, réactivation, Année Psychol., 80: 237.

Hars, B., and Hennevin, E., 1983, Reminder abolishes impairment of learning induced by paradoxical sleep retardation, Physiol. Behav., 30:831.

Hennevin, E., and Leconte, P., 1971, La fonction du sommeil paradoxal: faits et hypothèses, Année Psychol., 2:489.

Hennevin, E., and Leconte, P., 1977, Etude des relations entre le sommeil paradoxal et les processus d'acquisition, Physiol. Behav., 18:307.

Hennevin, E., Leconte, P., and Bloch, V., 1974, Augmentation du sommeil paradoxal provoquée par l'acquisition, l'extinction et la réacquisition d'un apprentissage à renforcement positif, Brain Res., 70:43.

Lambert, J. F., and Truong-Ngoc, A., 1976, Influence de l'environnement instrumental et social sur la structure d'un échantillon du cycle veille-sommeil chez le rat Wistar mâle: corrélations avec les modifications de l'excitabilité du système réticulo-cortical, Agressologie, 17:19.

Lecas, J. C., 1976, Changes in paradoxical sleep accompanying instrumental learning in the cat, Neurosci. Lett., 3:349.

Leconte, P., and Hennevin, E., 1973, Caractéristiques temporelles de l'augmentation de sommeil paradoxal consécutif à l'apprentissage chez le rat, Physiol. Behav., 11:677.

Leconte, P., Hennevin, E., and Bloch, V., 1973, Analyse des effets d'un apprentissage et de son niveau d'acquisition sur le sommeil paradoxal consécutif, Brain Res., 49:367.

Leconte, P., Hennevin, E., and Bloch, V., 1974, Duration of paradoxical sleep necessary for the acquisition of conditioned avoidance in the Rat, Physiol. Behav., 13:675.
Lewis, D. J., 1979, Psychobiology of active and inactive memory, Psychol. Bull., 86:1054.
Lucero, M., 1970, Lengthening of REM sleep duration consecutive to learning in the Rat, Brain Res., 20:319.
Mc Grath, M. J., and Cohen, D. B., 1978, REM sleep facilitation of adaptive waking behavior : a review of the literature, Psychol. Bull., 85:24.
Pacteau, C., 1982, Sommeil paradoxal et acquisition chez la souris BALB/c : relations entre les composantes cognitives de l'apprentissage et le phénomène d'augmentation du sommeil paradoxal, Thesis (Université Paris-Sud).
Pearlman, C. A., 1979, REM Sleep and information processing : evidence from animal studies, Neurosci. Biobehav. Rev., 3:57.
Pearlman, C. A., 1981, Rat models of the adaptive function of REM sleep, in: "Sleep, Dreams and Memory", W. Fishbein, ed., Spectrum Publications, New York.
Sara, S., and Deweer, B., 1982, Memory retrieval enhanced by amphetamine after a long retention interval, Behav. Neural Biol., 36:146.
Smith, C., Kitahama, K., Valatx, J. L., and Jouvet, M., 1974, Increased paradoxical sleep in mice during acquisition of a shock avoidance task, Brain Res., 77:221.
Smith, C., Lowe, D., and Smith, M.J., 1977, Increases in paradoxical and slow sleep during the acquisition of an appetitive task in rats, Physiol. Psychol., 5:364.
Smith, C., Young, J., and Young, W., 1980, Prolonged increases in paradoxical sleep during and after avoidance-task acquisition, Sleep, 3:67.
Spear, N. E., 1973, Retrieval of memory in animals, Psychol. Rev., 80:163.
Spear, N. E., and Gordon, W. C., 1981, Sleep, dreaming and the retrieval of memories, in: "Sleep, Dreams and Memory", W. Fishbein, ed., Spectrum Publications, New York.

MODULARITY OF THE PROSENCEPHALON: THE VERTICAL SYSTEMS

Ivan Divac and Jesper Mogensen

Institute of Neurophysiology
University of Copenhagen
School of Medicine

ABSTRACT

The present chapter offers in a capsular form a concept of organization of the non-limbic division of the mammalian forebrain. We first draw attention to certain structural and functional properties of the prosencephalon and conclude that interrelated parts of the thalamus, isocortex and neostriatum constitute functional systems. In the next step we attribute to these systems distinct properties which can be related to well established psychological phenomena. Finally, one of testable predictions is presented.

INTRODUCTION

The present chapter attempts to make explicit certain assumptions which we partly assimilated and partly generated during our studies of the brain. These assumptions have been fashioned into a concept about the mode of action of the non-limbic division of the prosencephalon. The concept will only be summarily presented here; a more comprehensive description will be published elsewhere.

THE VERTICAL PROSENCEPHALIC SYSTEMS

The non-limbic division of the prosencephalon consists of many systems each composed of interconnected parts of three tiers: the thalamus, the isocortex and the neostriatum. The existence of these vertical prosencephalic systems (VPSs) is evinced by data obtained with anatomical (reviews in Divac, 1972; Grofova, 1979), 2-deoxyglucose (Collins and Divac, 1984; Divac, 1983; Divac and Diemer, 1980), and various neurobehavioral and electrophysiological approaches (see reviews in Divac, 1968, 1972; Oberg and Divac,1979; Rosvold,1972), which all indicate that different VPSs mediate different functions. The divisions of each tier, represented by cortical areas, thalamic nuclei and neostriatal regions, seem to be composed of modules. The interconnected modules of the three tiers form VPSs. The isoprosencephalic VPSs may be divided into three classes. Some systems process modality-specific information. The complexity of such information varies, ranging from 'simple', such as color and edges (e.g. Zeki, 1983) to complex such as faces (Desimone et al.,

1984 ; Perrett et al., 1982). The VPSs in so-called association regions seem to represent highly abstract, typically cross-modal concepts, for example 'being to the left' (Niki, 1974), 'justice' (presumably in humans only) etc. At least some VPSs of the association kind may also exert control on the sensory VPSs (e.g. Spinelli and Pribram,1967). Finally, the VPSs of the premotor and motor cortical areas belong to the effector systems which will be ignored presently, as are the claustral and the pontocerebellar circuitry. It is not known at present to what degree VPSs are prewired genetically, i.e. to what extent they acquire properties with experience.

Hodological studies have established patterns of cortico-cortical connections: distant VPSs can be directly connected through this tier. Within the dorsal-thalamic and neostriatal tiers, however, no long connections have been found. In spite of the existence of long cortico-cortical connections, studies of spread of focal epileptic discharges suggest an unexpected mutual independence of different cortical areas, and consequently of VPSs (Collins and Divac, 1984; Divac and Diemer, 1980).

We conclude that the non-limbic division of the forebrain consists of largely, but not entirely, independent systems which are composed of parts of the thalamus, isocortex and neostriatum. These systems have different functions because their sources of inputs as well as their targets of output differ, and in spite of apparently uniform microcircuitry within each tier.

HYPOTHETICAL PROPERTIES OF VPSs

We suggest that each VPS has unique properties, i.e. a unique function. The function of a VPS becomes manifest when this VPS is activated and allowed access to the motor system and to presumed substrate of consciousness. Different mechanisms may take part in the activation of VPSs: primary sensory events contribute to activation of sensory VPSs, whereas sensory convergence and abstraction may take part in the activation of the association VPSs. Neural mechanisms involved in motivation and reinforcement, previous (one aspect of memory) as well as current, must also play a role in the selection of VPSs. Finally, neural systems which mediate attention and consciousness are assumed also to influence the activation of VPSs. Independent neural mechanisms seem to play gating roles, allowing access of the activated VPSs through different channels to the presumed substrate of consciousness and to the motor system. In this way consciousness is informed about what is going on in the isoprosencephalon (Libet 1982 ; Libet et al., 1983). The output of VPSs towards the motor system contributes information about what should be done, i.e. about the target and the goal of responding. The output of nonmotor VPSs is characterized by motor equivalence, i.e. the precise somatotopy and topography of the responses are not decided upon by sensory and association VPSs. Current anatomical knowledge offers some candidates for the hypothesized roles of recruitment and gating (see i.g. Crick, 1984; Divac, 1979).

Having described the composition of VPSs, their relations with other systems, and likely elements of their recruitment (activation and release), we procede to postulate the roles of VPSs in the regulation of behavior. There are grounds to believe that only one or a few VPSs obtain access to the motor system and consciousness at a time. The minimal time during which a VPS is recruited appears to be about 50 msec (e.g. Allan, 1978; Crick, 1984); the maximal duration is not known. A VPS which was proven adequate under particular circumstances (in the sense that its activity was coincident with optimal reinforcement) will be more easily retrieved next time under similar circumstances.

INFERENCES

These speculations suggest first that what we experience is determined proximally by the operational VPSs (c.f. Mountcastle 1975). We believe that under normal conditions the non-motor VPSs are responsible for signal detection (also on the most abstract levels), for the actual content of consciousness, and for certain aspects of responding. In other words, we suggest that perception on all levels of abstraction as well as appropriate behavior is ensured by selection of VPSs. Secondly, on the prosencephalic level learning may consist, at least in part, of selection of appropriate VPSs in appropriate sequence and constellations. These constellations may be equivalent to Crick's (1984) 'transient assemblies' and Abeles' (1982) 'synfire chains'. It should be emphasized, however, that we talk about assemblies of VPSs (interconnected modules) rather than of individual neurons. Functional properties of some VPSs may be prewired and little influenced by experience, whereas other VPSs could be modified or wired in postnatally, along the lines of current physiological theories of learning. The adequacy of recruited VPSs is tested by reinforcement contingencies. The VPSs which provide the most consistent positive reinforcement tend to be most frequently and for the longest time recruited. Thirdly, memory involving the forebrain may consist of facilitated retrieval of adequate VPSs. An entirely different mechanism may involve wiring of initially 'blank' or modifiable VPSs. Therefore the 'prewired' VPSs may not contain an 'engram', i.e. a representation of the learned association. Instead, the engram may be 'stored' in the selection mechanisms. Some of the morphological and biochemical changes in the isocortex which are commonly attributed to learning may instead reflect the changes related to retrieval and gating. Altogether different formations and processes may be involved in other kinds of learning and memory. For example a 'simple' sensory-motor conditioning such as the blink reflex, in which the questions of what should be done or of the target in the environment do not arise, may be mediated by other circuits of the nervous system (e.g. Thompson et al. 1983).

DISCUSSION

Our concept is in full agreement with Fodor's (1983) claim that the human mind is modular. Certain observations of behavior, which cannot be readily explained with current notions of brain functions, appear now more easily understandable. These are : (1) sudden transition from chance behavior to proficient performance in discrimination learning (Lashley 1929) and in 'insightful behavior' (Kohler 1925, Maier and Schneirla 1935). These phenomena can be interpreted as a sign of 'finding' the adequate VPS; (2) apparent use of hypotheses (Krechevsky 1932, 1933, Levine 1965, 1971, Tumblin and Gholson 1981), which may reflect control of behavior by a VPS recruited prior to the reinforcement event; and (3) faster learning by a group of subjects of a task which offers redundant cues, presumably because different subjects can successfully use different VPSs (Restle 1957). The occurrence of some perceptual phenomena can also be understood within the present concept. These are: ambiguous figures, eidetic memory, perceptions influenced by expectation, certain types of illusions, and dreams. Thus, the changing perception of ambiguous figures could be attributed to alternating recruitment of VPSs which cannot simultaneously be recruited (possibly due to a mechanism such as lateral inhibition). Eidetic memory may be attributed to long lasting recruitment of sensory VPSs. Peculiar interplay among different activation mechanisms may account for illusions, selective perception, and for dreams. Erroneous activation of some VPSs may be responsible for hallucinations and obsessive ideas (the latter possibly reflect persistent activity of the VPSs in 'associative' systems). Finally, if the recruited VPSs in humans fail to report to the embodiment of consciousness (or to the language system) but are adequate for guiding the responding in the given situation, the person may behave normally but be unable to report what and why he or she did during the 'absence'.

Our concept predicts that in the early stages of problem-solving the activity of the brain consists of the search for the adequate VPSs, which are switched on and off until the goal is reached. In these stages widespread activity of the brain should be predicted. In an overtrained task, however, when one or only a few VPSs remain operative, activity should be detectable only in a restricted part of the forebrain. Some electrophysiological data are in agreement with this prediction (e.g. reviews in John 1967, Galambos and Hillyard 1981). Patterns of regional blood flow and metabolism agree with the first part of the prediction (Roland 1982, Phelps et al. 1981), but no observations have been made with these techniques during overtrained behavior.

CONCLUSION

According to the present concept: (1) VPSs are input-output devices which control behavior only when activated and allowed to affect the motor systems. They are relatively uniform (in terms of microcircuitry and cellular interactions). (2) The selection mechanisms mediate certain aspects of attention, learning and memory. Impaired functions of the selection mechanisms may be responsible for pathological conditions characterised by memory losses, hallucinations, and obsessive ideas.

The present view on functioning of the non-limbic forebrain offers certain unorthodox interpretations and predictions. Only the future will decide, however, which aspects of our concept will survive, if any, and how they should dovetail with notions of functions of other parts of the brain, e.g. the limbic (e.g. Mishkin et al., in press) and motor systems.

ACKNOWLEDGEMENT

Danish Medical Research Council provided a stipend for J.M. and supported financially participation of I.D. at the EBBS-SFECA Workshop, R. Gunilla E. Oberg commented on and Margit Løvgreen typed the manuscript. We are grateful for all of this help.

REFERENCES

Abeles, M., 1982, "Local Cortical Circuits". Springer, Berlin.

Allan, L.G., 1978, The attention switching model: Implications for research in schizophrenia. J. Psychiat. Res. 14:195.

Collins, R.C., and Divac, I., 1984, Neostriatal participation in prosencephalic systems, evidence from deoxyglucose autoradiography, in: "Advances in Neurology", R.G. Hassler and J.F. Christ , eds., Raven Press, New York.

Crick, F., 1984, Function of the thalamic reticular complex : The searchlight hypothesis. Proc. Natl. Acad. Sci. U.S.A., 81:4586.

Desimone, R., Albright, T.D., Gross, C.G., and Bruce, C., 1984, Stimulus-selective properties of inferior temporal neurona in the macaque. J. Neurosci., 4:2051.

Divac, I., 1968, Functions of the caudate nucleus. Acta Biol. Exp. (Warsaw), 28:107.

Divac, I., 1978, Neostriatum and functions of prefrontal cortex. Acta Neurobiol. Exp., 32:461.

Divac, I., 1979, Patterns of subcortico-cortical projections as revealed by somatopetal horseradish peroxidase tracing. Neuroscience, 4:455.
Divac, I., 1983, Two levels of functional heterogeneity of the neostriatum. Neuroscience, 10:1151.
Divac, I., 1984, The neostriatum viewed orthogonally, in: "Functions of the Basal Ganglia", CIBA Foundation Symposium 107, Pitman, London.
Divac, I., and Diemer, N.H., 1980, The prefrontal system in the rat visualized by means of labelled deoxyglucose. Further evidence for functional heterogeneity of the neostriatum. J. Comp. Neurol., 190:1.
Fodor, J.A., 1983, "The Modularity of Mind". MIT Press, Cambridge, MA.
Galambos, R., and Hillyard, S.A., 1981, Electrophysiological approaches to human cognitive processing. Neurosci. Res. Prog. Bull., 20:141.
Grofova, I., 1979, Extrinsic connections of the neostriatum, in: "The Neostriatum", I. Divac and R.G.E. Oberg, eds., Pergamon Press, Oxford.
John, E.R., 1967, "Mechanisms of Memory". Academic Press, New York.
Kohler, W., 1925, "The Mentality of Apes". Harcourt, Brace, New York.
Krechevsky, I., 1932, "Hypotheses" in rats. Psychol. Rev., 39:516.
Krechevsky, I., 1933, Hereditary nature of "hypotheses". J. Comp. Psychol., 16:99.
Lashley, K.S., 1929, "Brain Mechanisms and Intelligence". University Chicago Press, Chicago.
Levine, M., 1965, Hypothesis behavior, in: "Behavior of Nonhuman Primates", H.F. Harlow and F. Stollnitz, eds., Academic Press, New York.
Levine, M., 1971, Hypothesis theory and nonlearning despite ideal S-R reforcement contingencies. Psychol. Rev., 78:130.
Libet, B., 1982, Brain stimulation in the study of neuronal functions for conscious sensory experiences. Human Neurobiol., 1:235.
Libet, B., Gleason, C.A., Wright, E.W., and Pearl, D.K., 1983, Time of conscious intention to act in relation to onset of cerebral activity (readiness-potential). Brain, 106:623.
Maier, N.R.F., and Schneirla, T.C., 1935, "Principles of Animal Psychology". McGraw Hill, New York.
Mishkin, M., Malamut, B.L., and Bachevalier, J., 1984, Memories and habits: two neuronal systems, in: "The Neurobiology of Learning and Memory". G. Lynch, J.L. McGaugh, and N.M. Weinberger, eds., Guilford Press, New York.
Mountcastle, V.B., 1975, The view from within: pathways to the study of perceptions. Johns Hopkins Med. J., 136:109.
Niki, H., 1974, Differential activity of prefrontal units during right and left response trials. Brain Res., 70:346.
Oberg, R.G.E., and Divac, I., 1979, "Cognitive" functions of the neostriatum, in: "The Neostriatum", I. Divac and R.G.E. Oberg, eds., Pergamon, Oxford.
Perrett, D.I., Rolls, E., and Caan, W., 1982, Visual neurones responsive to faces in the monkey temporal cortex. Exp. Brain Res., 47:329.
Phelps, M.E., Mazziotta, J.C., Kuhl, D.E., Nuwer, M., Packwood, J., Metter, J., and Engel, J., 1981, Tomographic mapping of human cerebral metabolism: visual stimulation and deprivation. Neurology, 31:517.
Restle, F., 1957, Discrimination of cues in mazes: a resolution of the "place-vs.-response" question. Psychol. Rev., 64:217.
Roland, P.E., 1982, Cortical regulation of selective attention in man. A regional cerebral blood flow study. J. Neurophysiol., 48:1059.
Rosvold, H.E., 1972, The frontal lobe system: cortical-subcortical interrelationships. Acta Neurobiol. Exp., 32:439.
Spinelli, D.N., and Pribram, K.H., 1967, Changes in visual recovery function and unit activity produced by frontal cortex stimulation. Electroencephalogr. Clin. Neurophysiol., 22:143.
Thompson, R.F., Berger, T.W., and Madden, J., 1983, Cellular processes of learning and memory in the mammalian CNS. Ann. Rev. Neurosci., 6:447.

Tumblin, A., and Gholson, B., 1981, Hypothesis theory and the development of conceptual learning. Psychol. Bull., 90:102.

Zeki, S., 1983, Colour coding in the cerebral cortex: the responses of wavelength-selective and colour-coded cells in monkey visual cortex to changes in wavelength composition. Neuroscience, 9:767.

SELECTIVE ATTENTION, MEMORY, AND THE LOCUS COERULEUS

Susan J. Sara

Departement de Psychophysiologie
Laboratoire de Physiologie Nerveuse
C. N. R. S.
Gif/Yvette, France

The role of locus coeruleus (LC) in attention, arousal and behavioral plasticity has attracted the interest of many investigators studying LC function at different levels-- from cellular to behavioral. These studies have used many different techniques including electrical stimulation and intracellular recording, neurotoxic lesions and behavioral analysis, pharmacological intervention and neurochemical assays. The common interest is in how the noradrenergic projections to the forebrain are involved in cognitive function, particularly regulation of attention and arousal, but also in developmental plasticity, learning and memory. Those who have used a behavioral approach to understanding LC function have relied heavily on neuroanatomical and electrophysiological data to provide their hypotheses (see below; also Robbins and Everitt; Velley et al., this volume). At the same time, those who have studied the effects of noradrenaline (NE) on cellular activity and reactivity, or the effects of exogenous stimulation on the activity of LC neurons, have proposed an important role for the LC system in the regulation of attention, arousal, and behavioral plasticity (Aston-Jones; Segal; Kasamatsu, this volume).

The view that the LC system is involved in selective attention is largely based on studies of the electophysiological effects of NE on target neurons in cortex, hippocampus and cerebellum. These studies, carried out independently by Segal and Bloom (1974, see Segal, this volume) and by Waterhouse et al. (1980), have shown that NE can bias subthreshold sensory input to a detectable level, or enhance neuronal responses to significant or preferred stimuli. This enhanced signal to noise ratio should then improve selective attention, according to these investigators and others.

This functional interpretation of electrophysiological results raises two important questions. What are the behavioral or environmental conditions which govern the activity of the LC and what is the significance for behavior of LC activation (or inactivation)? These two questions, along with the question of the role of NE in developmental and behavioral plasticity are discussed in this and the following chapters.

The first question has been addressed by Aston-Jones and Bloom (1981) who found that LC neurons were inhibited during grooming and consumatory behavior and that they responded robustly to environmental stimuli of all modalities. The authors take these data to support the role of the LC in attention, especially attention to changes in environmental input (see Aston-Jones, this volume). A demonstration of habituation of this LC orienting response and enhancement of LC responses when the stimuli have been paired with reinforcement would add some support to this view.

The second question is whether the NE-induced change in signal to noise ratio really represents an increase in selective attention and whether this increase in attention has a significance for behavioral adaptation. To address this question, NE activity should be manipulated in freely moving animals and behavioral analysis must validly measure attention. By far the preferred technique for manipulating NE activity has been the use of rats which have been chronically depleted of cortical and hippocampal NE through selective neurotoxic lesions (see Robbins, 1984 for review). We have used the selective neurotoxin DSP4 to deplete forebrain NE and while there was the robust avoidance deficit usually seen with NE-depleted animals, we did not observe deficits in situations which make strong attentional demands on the animals. These include acquistion of a complex maze, its retrieval after a long retention interval and reversal learning of the same task. If anything, lesioned animals were better, at least in the initial stages of the learning, and in the reversal (Sara et al., 1984). Such a lack of effect of neurotoxic lesions of the NE system on seemingly complex cognitive behavior is not uncommon. In fact, post lesion sparing of function has led some authors to reject the hypothesis that the LC forebrain projection is important in selective attention, even though the hypothesis is so attractive in the light of the electrophysiological evidence (e.g. Pisa and Fibiger, 1983). While it is well-established that NE remains depleted for a long time after neurotoxic lesions, it is possible that there is considerable functional recovery of the NE system, a question which is further considered by Robbins and Everitt and by Aston-Jones (this volume). It is also possible that after lesion, the signal-to-noise enhancing properties of NE in target cortical structures is accomplished vicariously by another mechanism, which would account for the difficulty in finding learning and memory deficits. There have been few studies of electrophysiological effects of these neurotoxic lesions, but the question of functional recovery might be studied by examining the post-lesion time course of any such effects.

Another way of addressing the question of the behavioral significance of LC modulation of neuronal reactivity would be to use acute stimulation of the LC, electrically or pharmacologically, in a behavioral situation which makes attentional demands. There is, to our knowledge, only one study which has used direct stimulation of the LC in freely moving animals (Segal and Edelson, 1979). Priming stimulation of the LC facilitated acquistion of brightness discrimination, but the effect was seen only on latency measures. We are planning further studies of this type using behavioral tasks such as complex discriminations and forgetting paradigms. We have found that priming stimulation of the mesencephalic reticular formation (MRF) alleviated forgetting of a complex maze path after a long retention interval (Sara, et al., 1980). It is possible that this stimulation activates LC, since there are fibers projecting from the area of stimulation, in the lateral tegmentum, to LC (Morgane & Jacobs, 1979). Very recent results from our laboratory have shown, however, that this MRF stimulation increases NE in the hippocampus, but not in the hypothalmus, suggesting that it is fibers of the dorsal noradrenrgic bundle, which pass close by the point of our electrode placement, that are being stimulated.

In parallel with this converging view from electrophysiological and animal behavioral studies that the noradrenergic LC forebrain projections are involved in selective attention, there is an emerging clinical interest in the role of this system in memory processes. Attentional deficits are at least partially responsible for some clinical amnesias, according to many clinicians beginning with Talland (1965). Much of our own work on recovery from amnesia and alleviation of forgetting in rats has strongly implicated attentional processes in memory, especially in retrieval of weak memories. Regulation of attention might well be considered as a cognitive function, intimately involved in stimulus selection and encoding during both acquisition and retrieval of memory (see Sara, 1984).

Clinical evidence is growing in favor of a role for the LC-noradrenergic forebrain projections in some diseases characterised by cognitive dysfunction (see van Dongen, 1981 for review). Post mortem examination of the brains of patients suffering from Alzheimer's disease reveals massive loss of LC cells with resultant decrease in forebrain NE (Iversen et al., 1983). Korsakoff patients have lower levels of noradrenaline (NE) metabolites in cerebral spinal fluid and this is correlated with poor performance in memory tasks. Memory is improved in these patients by the NE receptor agonist, clonidine (McEntee and Mair, 1980). In both of these diseases, the most telling symptom is the profound memory loss. It has recently been reported that many drugs such as vincamine, hydergine, and piracetam, used to treat memory disorders in the clinic, increase neuronal firing in LC (Olpe et al., 1983). If the damage in the LC system is, indeed, responsible for some of the memory impairment in these patients, then the resultant increase in forebrain NE activity could be at least partially responsible for the improvement which is sometimes seen with these drugs.

Piracetam, a drug which is currently in clinical use to improve cognitive functioning in geriatric patients, alleviated hypoxia and ECS-induced amnesia in rats (Sara and David-Remacle, 1975). This substance and a closely related analog also alleviated forgetting in two different appetitive tasks, an operant bar press response for water reinforcement (Sara et al., 1979) and the appetitive maze task discussed above (Sara, 1979). This early work on memory facilitation with piracetam and analogs was not at all based upon a noradrenergic hypothesis, but the recent discovery of an increase in LC neuronal firing in rats treated with piracetam, mentioned above (Olpe, et al., 1983) suggests that these behavioral studies support such an hypothesis.

CONTEXTUAL CUE REMINDER EFFECT

In several different paradigms including ECS- or hypoxia-induced amnesia of inhibitory avoidance learning (e.g. Sara and David-Remacle, 1975), spontaneous forgetting of inhibitory avoidance or of an appetitive maze task (Deweer et el., 1980; Deweer and Sara, 1984), a brief exposure (just prior to the retention test) to the context in which the training took place was sufficient to alleviate the forgetting or experimentally-induced amnesia. There is no information in this reminder procedure concerning response contingencies. It is possible that after association with the reinforcement during training, the context elicits an arousal response, physiologically similar to the orienting response to novelty. This conditioning to the context is an old Pavlovian idea, later reiterated by Konorski, who held that the establishment of a preparatory arousal response to context was necessary before the specific CR could be elaborated (see Sara, 1985 for further discussion).

If this conditioned arousal response includes an increase in firing of LC neurons, as seen in the orienting response (Aston-Jones, this volume), then this would provide a possible mechanism of action of the contextual cue reminder effect. Increase in firing of LC neurons releases NE in the forebrain. As described by Segal (this volume), this increase in NE enhances signal-to-noise ratio in cortex and hippocampus by increasing responses to conditioned or significant stimuli. Selective attention to the discriminative stimuli within the maze could be the result of this conditioned increase in LC activity. This increase in attention would be reflected in the superior maze performance of rats exposed to the contextual cue before the retention test.

CONTEXTUAL CUE REMINDER AND INCREASE IN FOREBRAIN NE ACTIVITY

In a more direct attempt to evaluate the hypothesis that the contextual cue reminder elicits an increase in LC firing, we measured forebrain amines and metabolites after rats had been exposed to a context in which they had received a mild footshock 24 h before (5 shocks of 1 mA, 1 sec duration). There was an increase in hippocampal 3-methyl 4-hydroxyl phenylglycol (MHPG), the major metabolite of NE (Sara, Guibert and Leviel, unpublished results). It is important to note that the neurochemical response to the context occurred 24 h after the footshocks. More recently, the same conditioned response to the context was obtained in rats which had received sweetened condensed milk in this context, 3 days previously (Sara, Guibert and Leviel, unpublished results). These results, although preliminary, add support to the hypothesis concerning the mechanism of action of the contextual cue reminder. But such experiments do not prove that the increase in forebrain NE activity is a sufficient or even a necessary condition for the reminder effect.

PHARMACOLOGICAL FACILITATION OF MEMORY RETRIEVAL

Pharmacological tools provide another means of approaching this question. If an increase in NE activity elicited by the context is, indeed, important as a preliminary step in focussing attention during memory retrieval, then there are two strategies which could be used to demonstrate this. First increasing the NE release pharmacologically should produce facilitory effects similar to the contextual cue reminder, in the same paradigms. Secondly, blocking the physiological effects of this increase of NE release, for example, by producing a post-synaptic receptor blockade, should prevent the contextual cue reminder from facilitating memory retrieval.

Most of our studies of pharmacological effects on memory retrieval and interactions with reminders have used the same spontaneous maze forgetting paradigm mentioned above. The procedure has been described in detail elsewhere (e.g. Sara et al., 1980; Deweer et al., 1980). Rats are trained to run for food reward in a 6-unit maze consisting of left-right choices. They are submitted to 5 pretraining trials and 5 training trials, one trial a day. Thus there are 10 context-reinforcement pairings. After a 3-4 week interval, the rats are tested for retention of the maze path. All treatments, pharmacological or reminder, are administered just prior to the retention test. Forgetting is defined as a significant increase in the number of errors as compared to the last training trial.

Amphetamine, administered systemically at low doses (.25-.5 mg/kg) alleviated forgetting in this paradigm (Sara and Deweer, 1982). When injected into the ventricles (icv), the facilitation was not as robust as with the systemic injection, but this might be a problem of appropriate dose, rather than an indication of peripheral involvement, since the systemic effect was only seen within a narrow range of doses.

Systemic injections of apomorphine, which is widely used as a dopamine (DA) receptor agonist, did not alleviate forgetting, but surprisingly, icv injections within a limited dose range (10 ug) significantly reduced the number of errors. The dose-response effects on spontaneous behavior with these icv injections were quite different from those of systemic injections and subsequent analysis revealed that the neurochemical effects were also quite different. An icv injection of apomorphine increased NE in the hippocampus and cortex, while having no effect on striatal DA at the dose which alleviated forgetting in the maze. Larger doses (40 ug), which had no memory facilitating effect, had no effect on NE levels in the brain (Sara et al., 1984). The memory retrieval facilitation of this treatment appears to depend upon the increase in forebrain NE.

Haloperidol, a dopamine receptor antagonist, also increases firing of locus coeruleus neurons, probably because it is also an antagonist of NE alpha-2 presynaptic receptors (Olpe et al., 1979; Dinan et al., 1984). We have found that pretest treatment with this neuroleptic alleviates forgetting of the maze task in normal rats (Sara, in press). Yohimbine, which is a specific NE alpha-2 antagonist (Goldberg and Robertson, 1983), facilitates memory retrieval in the same paradigm, thus supporting the interpretation of the haloperidol results in terms of NE alpha 2 action (Sara, 1984).

In an effort to block the reminder effect, propranolol, a beta noradrenergic receptor antagonist, was injected before the reminder treatment, but with no effect. An alpha antagonist might be more likely to have an effect, since the excitatory facilitation by NE is thought to be mediated by the alpha receptor. Such studies, using alpha-1 receptor antagonists in conjunction with the contextual cue reminder, are planned.

CONCLUSION

Memory retrieval is a cognitive process relying heavily on selective attention to relevant stimuli in the retrieval environment. When the animal is exposed to elements of the context before retrieva, this process is facilitated. We have borrowed from electrophysiological studies showing that LC neurons respond to significant stimuli and that the resultant increase in forebrain NE enhances signal/noise, to suggest that this contextual cue reminder effect might be mediated by conditioned activation of this system. Results showing a conditioned increase in cortical and hippocampal NE activity in response to the context in which reinforcement has been delivered, although preliminary, lend some support to this hypothesis.

Pharmacological studies show that all of the drugs which facilitate memory retrieval in the various paradigms we have studied, enhance NE activity, at the behaviorally active doses. The pharmacological effects on maze performance are very similar to the effect of the contextual cue reminder, lending support to the hypothesis that the contextual cue reminder elicits an activation of the LC system which then improves discrimination in the maze by the enhancement of selective attention.

As other contributors to this volume have emphasized, there is a need for different levels of analysis to understand the significance of NE action on target neurons, for cognitive function. Pharmacological manipulation of LC activity accompanied by astute behavioral analysis can be a very powerful tool in this attempt, especially when used in conjunction with electrophysiological and neurochemical verification of drug activity and behavior effects.

References

Aston-Jones, G. and Bloom, F., 1981, Norepinepherine-containing locus coeruleus neurons in behaving rats exhibit pronounced responses to non noxious environmental stimuli, J. Neurosc , 1:887.

Deweer, B., Sara, S. and Hars, B., 1980, Contextual cues and memory retrieval in rats : alleviation of forgetting by a pretest exposure to background stimuli, Anim. Learn. Behav. , 8: 265.

Deweer, B. and Sara, S.J., 1984, Background stimuli as a reminder after spontaneous forgetting: role of duration of cuing and cuing-test interval, Anim. Learn. Behav. , 12:238.

Dinan, T., Shaver, R. and Aston-Jones, G.,1983, Haloperidol alters activity of locus coeruleus neurons, Soc. Neurosci. Abs. , 2:1000.

Goldberg, M. and Robertson, D., 1983, Yohimbine: a pharmacological probe for study of the -adrenoreceptors, Pharmacol. Rev. , 35:143.

Iversen, L.I., Rossor, M.N., Hills, G.P., Roth, M., Mounjoy, C.Q., Foote, S.L., Morrison, J.H. and Bloom, F.E., 1983, Loss of pigmented dopamide beta hydroxylase positive cells from locus coeruleus in senile dementia of Alzheimer's type, Neurosci. Lett. , 39:95.

McEntee, W. and Mair, R, 1980, Memory enhancement in Korsakoff's psychosis by clonidine: further evidence for a noradrenergic deficit, Ann.Neurol. , 7:466.

Morgane, P. and Jacobs, M.,1979, Raphe projections to the locus coeruleus in the rat, Brain Res. Bul. , 4:519.

Olpe, H.R., Jones, R. and Steinmann, M., 1983, The locus coeruleus : actions of psychoactive drugs, Experientia , 39:242.

Pisa, M. and Fibiger, H., 1983, Evidence against a role of the rat's dorsal noradrenergic bundle in selective attention and place memory, Brain Res., 272:319.

Robbins, T.W., 1984, Cortical noradrenaline, attention and arousal, Psychol. Med. , 14:13.

Sara, S.J., 1980, Memory retrieval deficits : alleviation by etiracetam, a nootropic drug, Psychopharm. , 68:235.

Sara, S.J., 1985, Noradrenergic modulation of selective attention : its role in memory retrieval, in "Memory Dysfunctions: An Integration of Animal and Human Research from Clinical and Preclinical Perspectives," D. Olton, E. Gamzu and S. Corkin, eds., Ann. New York Acad. Sci. in press.

Sara, S.J. and David-Remacle, M., 1975, Recovery from electroconvulsive shock-induced amnesia by exposure to the training environment: pharmacological enhancement by piracetam, Psychopharm. , 36:59.

Sara, S.J. and Deweer, B., 1982, Memory retrieval enhanced by amphetamine after a long retention interval, Behav. Neural Biol. , 36:146.

Sara, S.J., Dekeyne, A., Raguin, L. and Brouillet, E., 1984, Approach and avoidance discrimination learning and retention after treatment with the noradrenergic neurotoxin DSP4, Neurosci. Lett. , Suppl. 18, 119.

Sara, S.J., David, M., Weyers, M. and Guirgea, C., 1979, Piracetam facilitates retrieval but does not impair extinction of bar pressing in rats, Psychopharm. , 61:(1), 71.

Sara, S.J., Deweer, B. and Hars, B., 1980, Reticular stimulation facilitates retrieval of a "forgotten" maze habit, Neurosci. Lett. , 18:211.

Sara, S.J., Grecksch, G. and Leviel, V., 1984, Intracerebral ventricular

apomorphine alleviates spontaneous forgetting and increases cortical noradrenaline, Behav. Brain Res., 13:42.
Segal, M. and Edelson, A., 1978, Effects of priming stimulation of catecholamine containing nuclei in rat brain on runway performance, Brain Res. Bull., 3:203.
Segal, M. and Bloom, F., 1976, Norepinephrine in the rat hippocampus III Stimulation of nucleus locus coeruleus in the awake rat, Brain Res., 107:499.
Talland, G., 1965, "Deranged memory," Academic Press.
Van Dongen, P., 1981, The human locus coeruleus in neurology and psychiatry. Progr. Neurobiol., 17:97.
Waterhouse, B.D., Moises, H. and Woodward, D., 1980, Locus coeruleus stimulation potentiates somatosensory cortical neuronal responses to afferent synaptic inputs, Soc. Neurosci. Abst., 6:448.

NORADRENALINE AND SELECTIVE ATTENTION

T.W. Robbins and B.J. Everitt

Departments of Experimental Psychology and Anatomy
University of Cambridge
Cambridge, U.K.

INTRODUCTION

There have been a number of hypotheses, derived from both neurobiological and psychological experiments, about the functions of the locus ceruleus and its projections to the cortex via the dorsal noradrenergic bundle (DNAB). The work of Kasamatsu and colleagues (eg Kasamatsu, 1983) has revealed a role for cortical NA in neuronal plasticity within the visual cortex. Electrophysiological studies, on the other hand, have indicated two major conclusions (eg Segal and Bloom, 1976; Aston-Jones and Bloom, 1981). First, that release of NA in hippocampus, neocortex and cerebellum under certain circumstances enhances the response to signals impinging on those regions via other, more direct afferents. This has been likened to an enhancement of signal-to-noise ratio in the neuronal transmission of evoked responses (eg Segal and Bloom, 1976). Secondly, that NA cells within the locus ceruleus respond rather non-specifically to all forms of environmental input and, moreover, are most active during times of waking (Aston-Jones and Bloom, 1981; Livingstone and Hubel, 1981). Both of the latter findings implicate the locus ceruleus in arousal-like processes, as had earlier been hypothesized (Jouvet, 1974), although how such processes might contribute to neural plasticity is unclear.

Behavioural approaches to locus ceruleus function have led to rather different, though related, hypotheses. Thus, Kety (1970) and Crow (1968) suggested that cortical NA was involved in associative learning by mediating respectively arousal, or reinforcement. Mason and Iversen (1979) were among the first to make selective and profound lesions of the DNAB, resulting in over 90% cortical NA depletion, by using the selective neurotoxin 6-hydroxydopamine (6-OHDA). They argued from the behavioural effects of such lesions that the locus ceruleus was involved in selective attention, that is in processing certain events and ignoring the rest. Although their hypothesis was based initially on rather indirect behavioural evidence, it evidently made contact with the electrophysiological data of Segal and Bloom (1976) which could be interpreted behaviourally as a reflection of the screening-out of irrelevant information.

Finally, Gray (1982) has argued a rather different function for the locus ceruleus, especially its projection to the hippocampus. In contrast to other accounts, he emphasizes a role of noradrenergic transmission in mediating the behavioural inhibition of response output caused by anxiogenic stimuli, such as those predicting aversive events, and novelty.

SELECTIVE ATTENTION HYPOTHESIS

Here, we focus on the evidence, particularly from our own work, which is relevant to the attentional hypothesis. It is worth stating from the outset that, taking a lesson from studies in human experimental psychology (eg Eysenck, 1982), there are probably several different types of attention revealed by different contexts and test procedures. This complexity can be captured to some extent by using different tests of attention in animals.

Latent inhibition, non-reversal shift and blocking

Several procedures derived from contemporary learning theory can be used to study attentional bias, the selective ignoring of certain inputs. Mason and Lin (1980) cleverly capitalized on this in making more specific tests of the selective attention hypothesis. They used procedures of latent inhibition and non-reversal shift to test the attentional ability of rats with 6-OHDA-induced lesions of the DNAB. Both of these tests measure the capacity of animals to ignore irrelevant stimuli, either when made irrelevant by their pre-exposure in the absence of reinforcement (latent inhibition) or as stimuli only randomly correlated with reward during the acquisition of a discrimination along a more relevant stimulus dimension (non-reversal shift) (see also Mackintosh, 1983). If an animal has not successfully ignored the "irrelevant" stimuli, this failure could be to its advantage when the stimuli suddenly became relevant. Therefore, if DNAB lesioned rats have attentional deficits they should learn more rapidly than intact rats to discriminate stimuli previously made irrelevant by prior, non-reinforced exposure (ie. attenuated latent inhibition) and also stimuli made relevant to reward, after having just been irrelevant (ie. attenuated non-reversal shift). What is particularly telling is that both of these predicted results would correspond to improvements in performance which would be sufficient to rule out the possible contribution of primary sensory, motivational or motor impairments, thus suggesting a pure attentional loss.

Mason and Lin (1980) found some evidence of attenuation of latent inhibition and non-reversal shift, but the findings were not sufficiently clear-cut to conclude strongly in favour of the attentional hypothesis. For example, both of the DNAB groups (pre-exposed and non pre-exposed) were worse than the corresponding sham-operated groups in their acquisition of the successive discrimination used for assessing the degree of latent inhibition that had occurred. In similar vein, the DNAB-lesioned rats were impaired in acquiring even the initial, simultaneous visual discrimination prior to being required to shift to a spatial one.

We had run into similar interpretative problems while designing our own version of a non-reversal shift task (Robbins et al, 1982). A discrimination of temporal frequency was used so that each rat had to discriminate between fast and slow stimuli in either the auditory or visual modality. Both modalities could be presented simultaneously, with only one being fully predictive of reward, prior to the non-reversal shift to the other modality. However, in several experiments we demonstrated impaired acquisition by DNAB-6-OHDA-lesioned rats of the fast-slow discrimination, even with only one modality presented (Everitt et al, 1983). This difference in acquisition makes difficult any interpretation of faster shifting to the other modality, although in fact we subsequently found that DNAB-lesioned rats were no different from controls in shifting from one modality to the other (Robbins, T.W., Everitt, B.J., Gaskin, M. and Fray, P.J., unpublished results).

Furthermore, when we designed an appetitive tone/light discrimination task which DNAB-lesioned rats were able to solve as rapidly as controls, we found no evidence of attenuated latent inhibition by DNAB-lesioned rats: both DNAB-lesioned and control groups showed equal degrees of latent inhibition under these circumstances (Kilbourn, A., Robbins, T.W. and Everitt, B.J., unpublished results). This general failure to repeat the finding of Mason and Lin of attenuated latent inhibition following central NA depletion has now been confirmed by others using different means to deplete NA (eg. DSP-4, Mohammed et al, 1983) and aversive procedures (Tsaltas et al, in press; Lorden et al, 1983) except under very special circumstances (Lorden et al, 1983) which overstretch the normal theoretical interpretation of latent inhibition.

Lorden et al (1980) have employed another test borrowed from learning theorists, namely blocking, to measure the extent to which DNAB-lesioned rats attend to redundant (as contrasted with irrelevant) information. In a blocking experiment, prior conditioning of a given conditioned stimulus (CS) to an unconditioned stimulus such as electric shock apparently prevents conditioning accruing to an added stimulus compounded with the original CS (see Mackintosh, 1983). DNAB-lesioned rats were reported to exhibit attenuated blocking, by showing some evidence of conditioning to the added element of a tone-light compound stimulus, in a conditioned suppression procedure. In view of the general lack of effect of DNAB lesions upon latent inhibition, this is a potentially very important result because it suggests that latent inhibition and blocking do indeed measure different forms of attention, which can be dissociated by DNAB lesions. However, the attenuation of blocking by DNAB lesions requires replication because we have been unable to replicate this result using an appetitive instrumental task (Robbins, T.W. and Everitt, B.J., unpublished results). Moreover, our attempts to use an aversive analogue may be confounded by a familiar problem; - DNAB-lesioned rats apparently show impaired acquisition of conditioned suppression to an auditory CS, whether conditioning occurs "on-the-baseline" or "off-the-baseline" (Cole, B.J. and Robbins, T.W., unpublished results).

Distraction and Surprise

Although the evidence reviewed above is hardly favourable to the selective attention hypothesis, it is nevertheless difficult to discount some kind of attentional impairment in explaining some of the effects of DNAB lesions. For example, the deficit observed in acquisition of the temporal frequency discrimination only became significant at stringent criteria of acquisition. In addition, it was not seen if the rats were trained to criterion prior to surgery (Everitt et al, 1983). Therefore, simple effects upon sensory, spatial, or motivational processes probably cannot account for the results and one is left again to distinguish between an associative deficit and an attentional one. Although performance of the discrimination was not altered by the DNAB lesion, it was possible to separate the groups by using shorter periods between the discrimination trials, the DNAB group being worse (Everitt et al, 1983). Whether this manipulation interfered with performance by presenting the discriminative stimuli at unpredictable times or by proactive interference, or some other type of effect is for the moment unclear, but it is important to note that the general result was replicated in a later experiment, using a completely different apparatus and discrimination task (Carli et al, 1983). In the latter experiment, we found that the DNAB-lesioned rats were spending longer in the vicinity of the food tray and so this conceivably could have interfered with effective orientation towards the sensory cues. But it remains difficult to define such an impairment in attentional terms - the DNAB-lesioned rats could be said to be <u>less</u> "distracted" by the discriminative stimuli from the acts of eating and other post-reinforcement behaviour. Alternatively, they could be said to be <u>more</u> distracted from the demands of the task. Thus, the attentional hypothesis can handle any result - a far from happy state of affairs and one which will tend to persist so long as "selective attention" avoids precise definitions in a task- or context-dependent way.

A similar theoretical agony is evident in the published literature. Thus, using novel sandpaper and flashing lights to interfere with maze-running performance, Roberts et al (1978) found a quite dramatic lengthening in latency to traverse the maze by the DNAB-lesioned group. This result is probably paralleled by analogous disruptions of performance in the runway (Anlezark et al, 1973; Owen et al, 1977) and in the open field in the context of a "food neophobia" test (Britton et al, 1981) following either electrolytic lesions of the locus ceruleus or 6-OHDA lesions of the DNAB. However, in contrast to this evidence of over-distractibility following cortical NA depletion, Crow et al (1978) reported reduced distraction by an auditory tone of rats licking a water spout. This confused picture is exemplified in particular by Mason and Fibiger's (1979) report of either enhanced, or reduced, or equivalent distraction produced by stimuli varying in modality or physical location, upon lever-pressing performance in an operant chamber.

We, too, have had some puzzling, but potentially important results when studying the distractibility of DNAB-lesioned rats. For example, we have seen no differential effect of either light or tone when introduced as distractors for the first time into the discrimination of temporal frequency of the alternate modality described above (Everitt et al, 1983). Indeed, we found rats to be remarkably resistant to distraction under those conditions and occasionally (in the case of controls only) performance actually improved when the distractor was introduced (Everitt et al, 1983). We were able, however, to discern impaired performance in DNAB-lesioned rats under certain circumstances, using another discrimination task (Carli et al, 1983). This was based on the 5-choice serial reaction time task much used in studies of human attention. Rats were trained to discriminate the location of a briefly-presented visual stimulus at one of 5 positions, randomly determined. DNAB lesions had no effect on discrimination *per se*, even when the stimuli were dimmed. However, they were impaired when shorter inter-trial intervals were introduced (see above) and also when a distractor (loud white noise) was presented just prior to the onset of the visual target stimuli. Both the sham-operated and DNAB-lesioned rats reacted equivalently to the distracting noise in certain respects. Thus, both groups exhibited increases in the number of inappropriate, premature responses ("false alarms") induced by the noise. However, although the accuracy of discrimination of the visual stimuli was unaffected in control rats, this accuracy was significantly reduced in rats with DNAB lesions. Thus, it seemed that rats with DNAB lesions were more distracted by white noise, but only under special circumstances. It was in fact hypothesized that DNAB lesioned rats were only made more distractible under conditions of elevated arousal or activation. Although the white noise introduced equivalent amounts of activation in the control and DNAB groups (in terms of premature responses), this activation increment uncovered a deficit in stimulus or response selection in the rats with cortical NA depletion. In speculative terms, the locus ceruleus was suggested to contribute to an "upper arousal" or "supervisory" mechanism which is engaged at times of sub- or supra-arousal of a "lower arousal mechanism" and hence protects the animal from the deleterious effects of the hyper-arousal produced by the white noise (Carli et al, 1983; Robbins, 1984).

The latter hypothesis makes several predictions (Robbins, 1984) which will not concern us here. What is becoming evident though is that the hypotheses concerning functions of the locus coruleus listed in the Introduction are not as distinct as they at first appear, and are inter-related. This explains why it has been so difficult to rule out any particular hypothesis, because they are generally vague and are insufficiently well-specified. Indeed, with evidence available from solely our own laboratory, we could have mounted a case for each of the main theories! This can be further emphasized by the results of some of our most recent work.

RECENT FINDINGS

It appears from our work that many impairments following DNAB lesions are revealed under conditions of elevated arousal. We have, in our further analyses, recently concentrated on aversive paradigms in order to explore this more fully. Our preliminary results underscore the point made above. For example, in a choice test of food neophobia in a novel open-field, the DNAB-lesioned rats have a stronger tendency than sham-operated rats to choose the palatable, but novel food option (Sahakian et al, 1983) (although they may take no longer than usual to make this choice). At first glance, this is difficult to explain according to the attentional hypothesis and more compatible with the anxiety account or the hypothesis of impaired response choice under high arousal. In particular, diminished anxiety when confronted by novelty seems a likely explanation of the DNAB-lesion induced change in behaviour. However, in contradiction of the anxiety hypothesis, an anxiolytic drug, chlordiazepoxide, fails to "release" responding to the novel of the two foods, although it does reduce the time spent in choosing between them (Cole, B.J. and Robbins, T.W., unpublished results).

In addition, although DNAB lesions apparently reduce conditioned suppression, they only do so if the lesions are made prior to training rather than afterwards, when there is no deficit (Cole, B.J. and Robbins, T.W., unpublished results). The latter result is difficult for the anxiety hypothesis to explain as anxiety is presumably still maintaining the conditioned suppression. An alternative account of the reduced conditioned suppression to a specific CS might be that the DNAB-lesioned rat has attended more to the contextual stimuli in the test chamber than to the CS because of a putative broader range of attention. However, this prediction of the attentional hypothesis has been tested and found not to hold: DNAB lesioned rats apparently show less evidence of contextual conditioning as well (Cole, B.J. and Robbins, T.W., unpublished results), perhaps more in keeping with the original hypothesis of impaired learning.

CONCLUSIONS

The approach of lesioning the DNAB with 6-OHDA to investigate the behavioural functions of the locus ceruleus can be criticised on several grounds because of the potential capacity of this cortically projecting system to undergo post-lesion recovery. Hence, negative results, for example concerning latent inhibition or blocking, are perhaps more difficult to interpret and should be given less weight than the positive ones. However, even positive results deserve more critical scrutiny if obtained several weeks after surgery at a time when the many mechanisms of neuronal compensation may have produced the effects secondarily, rather than as direct consequences of cortical NA depletion. While it is obviously useful and important to employ novel procedures for studying the function of the ceruleo-cortical NA projection such as acute pharmacological interventions at the level of the locus ceruleus, a case can be made that the use of 6-OHDA to lesion the DNAB is perhaps the most powerful, as well as one of the most practicable methods available for studying the DNAB. If the site of infusion is just rostral to the locus ceruleus, almost total depletion of cortical NA can be achieved (with some additional hypothalamic depletion). The lateral tegmental noradrenergic projections via the ventral NA bundle

(VNAB) is largely spared although it is difficult completely to avoid destruction of some A1 axons. Induction of tyrosine hydroxylase (Acheson et al, 1980) and proliferation of cortical adrenoceptors (Farik et al, 1981) would seem unlikely by themselves to enable complete recovery from such profound and widespread cortical NA depletion. Furthermore, local sprouting of NA fibres at the site of the lesions in the mesencephalon, of course, could not be expected to restore cortical NA functions.

Even if neural recovery (if not intrinsic to the ceruleo-cortical NA system, then by compensatory changes in other systems) were able to account for some of the failures we have had in testing the attentional hypothesis, it could not explain some of the longer-lasting deficits seen in a number of our own experiments and those of others. For example, the reduced aversion to novel food in the neophobia test occurs as little as 2 weeks, or as long as 2 months after surgery (often in the same animals performing in the attentional tasks, Robbins, T.W. et al, unpublished results; Sahakian et al, 1983). The occurrence of such stable alterations in behaviour suggests that the dynamic, compensatory neural changes in response to the lesions are not always sufficient to produce behavioural recovery.

The ability to show substantial and reliable effects of 6-OHDA lesions of the DNAB using the neophobia test also suggests that the search for purely attentional deficits may have been misguided. The lack of large effects in attentional tests may reflect the inappropriateness of the test or the true nature of the deficit, rather than an incomplete or ineffective lesion of the DNAB. In fact, the results emphasize that we must develop our existing behavioural hypotheses to a much greater extent, or invent new ones, if we are to understand the true significance of the electrophysiological results which have given impetus to this investigation of functions of the locus ceruleus.

REFERENCES

Acheson, A., Zigmond, M. J., and Stricker, E. M., 1980, Compensatory increase in tyrosine hydroxylase activity in rat brain after intraventricular injections of 6-hydroxydopamine, Science, N.Y., 207:537.

Anlezark, G. M., Crow, T. J., and Greenway, A. P., 1973, Impaired learning and decreased cortical norepinephrine after bilateral locus coeruleus lesions, Science, N.Y., 181:682.

Aston-Jones, G., and Bloom, F. E., 1981, Norepinephrine-containing locus coeruleus neurones in behaving rats exhibit pronounced responses to non-noxious environmental stimuli, J. Neurosci., 1:887.

Britton, D. R., Ksir, C., and Koob, G. F., 1981, Effects of dorsal noradrenergic bundle lesions on measures of conflict, learning and locomotion, Abst. Am. Soc. Neurosci., 7:113.10.

Carli, M., Robbins, T. W., Evenden, J., and Everitt, B. J., 1983, Effects of lesions to ascending noradrenergic neurons on performance of a 5-choice serial reaction task in rats: implications for theories of dorsal noradrenergic bundle function based on selective attention and arousal, Behav. Brain Res., 9:361.

Crow, T.J., 1968, Cortical synapses and reinforcement, Nature, 219:736.

Crow, T. J., Deakin, J. F. W., File, S. E., Longden, A., and Wendlandt, S., 1978, The locus coeruleus noradrenergic system: evidence against a role in attention, habituation, anxiety and motor activity, Brain. Res., 155:249.

Everitt, B. J., Robbins, T. W., Gaskin, M., and Fray, P. J., 1983, The effects of lesions to ascending noradrenergic neurones on discrimination learning and performance in the rat, Neurosci., 10:397.

Eysenck, M. W., 1982,"Attention and arousal",Springer, Berlin.
Gray, J. A., 1982,"The neuropsychology of anxiety",Oxford University Press, Oxford.
Harik, S. I., Duckrow, R. B., La Manna, J. C., Rosenthal, M., Sharma, V. K., and Bannerjee, S. P., 1981, Cerebral compensation for chronic noradrenergic denervation induced by locus coeruleus lesion: recovery of receptor binding, isoproterenol-induced adenylate cyclase activity and oxidative metabolism, J. Neurosci., 1:641.
Jouvet, M., 1974, Monoaminergic regulation of the sleep-waking cycle in the cat, in: "The Neurosciences, Third Study Program," F. O. Schmidt and F. G. Worden, eds., MIT Press Cambridge, Mass.
Kasamatsu, T., 1983, Neuronal plasticity maintained by the central norepinephrine system in the cat visual cortex, in: "Progress in Psychobiology and Physiological Psychology," Vol. 10, J. M. Sprague and A. N. Epstein, eds., Academic Press, New York.
Kety, S., 1970, The biogenic amines in the central neurons system: their possible roles in arousal, emotion and learning, in: "The Neurosciences, Second Study Program," F. O. Schmidt, ed., Rockefeller University Press, N.Y.
Livingstone, M. S., and Hubel, D. H., 1981, Effects of sleep and arousal on the processing of visual information in the cat, Nature (Lond.), 291:554.
Lorden, J. F., Rickert, E. J., Dawson, R. Jr., and Pelleymounter, M. A., 1980, Forebrain norepinephrine and the selective processing of information, Brain Res., 190:569.
Lorden, J. F., Rickert, E. J., and Berry, D. W., 1983, Forebrain monoamines and associative learning: I.latent inhibition and conditioned inhibition, Behav. Brain Res., 9:181.
Mackintosh, N. J., Conditioning and Associative Learning, The Clarendon Press, Oxford, 1983.
Mason, S. T., and Fibiger, H. C., 1978, Evidence for a role of brain noradrenaline in attention and stimulus sampling, Brain Res., 159:421.
Mason, S. T., and Iversen, S. D., 1979, Theories of the dorsal bundle extinction effect, Brain Res. Rev., 1:107.
Mason, S. T., and Lin, D., 1980, Dorsal noradrenergic bundle and selective attention, J. comp. physiol. Psychol., 94:819.
Mohammed, A. K., Everitt, B. J., Dunn, L. T., Robbins, T. W., Kilbourn, A., and Archer, T., 1983, Noradrenaline and the latent inhibition effect in rats, Prog. Neuropsychopharm. and Biol. Psychiat. Supplement, Abstract 338:226.
Owen, S., Boarder, M. R., and Gray, J. A., 1977, The effects of depletion of forebrain noradrenaline on the runway behaviour of rats, Expl. Brain Res., 28:R32-3.
Robbins, T. W., 1984, Cortical noradrenaline, attention and arousal, Psychol. Med.., 14:13.
Robbins, T. W., Everitt, B. J., Fray, P. J., Gaskin, M., Carli, M., and de la Riva, C., 1982, The roles of the central catecholamines in attention and learning, in: "Behavioural Models and the Analysis of Drug Action," M. Y. Spiegelstein and A. Levy, eds., Elsevier, North Holland.
Roberts, D. C. S., Price, M. T. C., and Fibiger, H. C., 1978, The dorsal tegmental noradrenergic projection: an analysis of its role in maze-learning, J. comp. physiol. Psychol., 90:363.
Sahakian, B. J., Winn, P., Robbins, T. W., Deeley, R. J., Everitt, B. J., Dunn, L. T., Wallace, M., and James, W. P. T., 1983, Changes in body weight and food-related behaviour induced by destruction of the ventral or dorsal noradrenergic bundle, Neurosci., 10:1405.

Segal, M., and Bloom, F.E., 1976, The action of norepinephrine in the rat hippocampus IV. The effect of locus coeruleus stimulation on evoked hippocampal activity, Brain Res., 107:513.
Tsaltas, E., Preston, G. C., Rawlins, J. N. P., Winocur, G., and Gray, J. A., 1984, Dorsal bundle lesions do not affect latent inhibition of conditioned suppression, Psychopharm., in press.

MODULATION OF SPONTANEOUS AND SENSORY-EVOKED DISCHARGE OF LOCUS COERULEUS NEURONS BY BEHAVIORAL STATE: FUNCTIONAL IMPLICATIONS

Gary Aston-Jones

Department of Biology
New York University
New York, NY 10003

INTRODUCTION—STEPS TO ACHIEVING A CELLULAR PHYSIOLOGIC UNDERSTANDING OF LOCUS COERULEUS

Although many hypotheses have been generated concerning the role of the norepinephrine (NE)-containing locus coeruleus (LC) system in behavior, there is as yet no single theory to unify the vast array of observations relevant to this nucleus. In order to more fully elucidate the overall role of the LC in brain and behavioral processes, my collaborators and I have sought a more complete understanding of the cellular physiologic properties of this system (see Aston-Jones et al., 1984). There are at least 3 essential questions in achieving such a cellular physiologic understanding: (1) where do LC neurons project, (2) what effect does NE released from LC terminals have on target neurons, and (3) when are LC neurons active (and presumably releasing NE) during ongoing behavior. For the first of these questions, I assume that the reader is familiar with the vast efferent trajectories of the LC system (for review, see Foote et al., 1983). In sum, for the present purpose one may consider the LC projection system to be extremely global, innervating nearly all parts of the neuraxis with few exceptions. Regarding the second question, the effects of NE on LC target neuronal activity are described for one sample target system by Segal in the present volume. These and other studies have found that while NE is inhibitory overall, it may also enhance the selectivity of target cell discharge, so that in the presence of this neurotransmitter neurons in many brain regions respond with increased preference to their most strongly determined inputs. Thus, for auditory cortex neurons in waking monkeys, for example, NE inhibits spontaneous discharge to a greater extent than discharge driven by preferred auditory stimuli (Foote et al., 1975). This effect has been termed an enhancement of "signal-to-noise" ratios in target cell activity, and has been replicated for many brain regions (for a recent review, see Foote et al., 1983).

While a great deal of information regarding the efferent anatomy and postsynaptic physiology of the LC system has been available for some time, knowledge pertinent to the third question above, i.e., the sensory/behavioral conditions that determine impulse activity in these neurons (and presumably release of NE from their terminals) has, until recently, remained obscure. The present chapter addresses this question by briefly reviewing our studies of LC neuronal discharge in behaving animals (performed in collaboration with Drs. Stephen Foote and Floyd Bloom). In these experiments, we used species (rats and monkeys) which allow analysis specifically of NE-containing LC neurons.

Following this review, functional implications of these findings in the context of overall cellular physiologic properties of the LC system are considered, and a new working hypothesis of LC function is presented. Finally, implications of this hypothesis for behavioral studies of the LC system are discussed.

SPONTANEOUS LC DISCHARGE AND BEHAVIORAL STATE

The spontaneous discharge rate of LC neurons varies with certain naturally occurring behaviors, including sleep, grooming, consumption and orientation. In the case of sleep, there is close correspondence between stages of the sleep-waking cycle and LC discharge: These cells tonically fire fastest during waking, are slower during slow-wave sleep, and are virtually silent during paradoxical sleep (PS) (Aston-Jones and Bloom, 1981a; Hobson et al., 1975). The marked quiescence of LC neurons during PS deserves special note, since nearly every other brain structure examined to date (with the exception of certain raphe nuclei; see Jacobs et. al., 1984) exhibits activity during this stage closely resembling that in waking--hence the term "paradoxical" sleep (Steriade and Hobson, 1976). Similar results have been observed previously for a subpopulation of cat LC neurons, perhaps corresponding to NE-containing cells in this species as well (Hobson et al., 1975). In the rat, the nearly total lack of activity in this nucleus during PS is evident not only from the consistent quiescence of single neurons, but also when several neurons in the densely packed noradrenergic cell group are recorded simultaneously: The entire population typically becomes silent, with a prominent decrease in "background noise" as well.

Further analysis revealed that, in addition to these changes in average discharge rates for sleep-waking cycle stages, LC neurons also alter their impulse activity within stages of the sleep-waking cycle, in anticipation of the following stage. Thus, during waking LC neurons gradually decrease in activity before the onset of slow-wave sleep, and likewise for slow-wave sleep before the onset of PS (Aston-Jones and Bloom, 1981a; Hobson et al., 1975). However, if waking rather than PS follows slow-wave sleep, LC neurons typically emit suddenly increased activity a few hundred msec prior to waking. The one exception to such anticipatory discharge in LC neurons occurs for the PS-to-waking transition. LC neurons return to activity characteristic of waking either coincident with or slightly after the cessation of PS (and the onset of waking) as measured by the electroencephalogram (EEG). Thus, while LC activity during other stage transitions is consistent with a possible role of this nucleus in generating the subsequent stage, it seems that this nucleus is not responsible for the termination of PS, but rather that its activity during this stage is controlled by other brain areas more primarily involved in generating the transition to waking (Aston-Jones and Bloom, 1981a).

In addition to recordings of unit activity from LC during natural sleep and waking, we were able to simultaneously record spontaneously occurring field potentials (FPs) in LC (Aston-Jones and Bloom, 1981a). Such FPs presumably reflect the concerted response of a group of neurons near the recording electrode to incoming excitatory afferent activity (Steriade and Hobson, 1976). This is supported by our observation that FPs in LC occur synchronously in time with bursts of unit activity recorded simultaneously from the same electrode. Most interestingly, however, while this close association between FPs and heightened discharge is apparent during waking and slow-wave sleep, it does not hold during PS. In fact, these two measures exhibit a marked dissociation during this stage of sleep, with FPs occurring at their highest spontaneous rate while impulse activity was virtually nil. This result indicates that LC neurons may be actively inhibited during PS, while receiving barrages of excitatory input reflected in rapid FP rates (Aston-Jones and Bloom, 1981a). This idea is supported by more recent observations of others (Aghajanian et al., 1983; Williams et al., 1982) that LC neurons in vitro exhibit regular spontaneous discharge, not requiring exogenous excitatory influences. Thus, periods of quiescence (such as PS) are likely to reflect active inhibition of these cells. This in turn implies that the LC plays a critical but passive role in the maintenance of PS. That is, in order for PS to occur or continue, it seems that the LC is prevented from discharging in response to intense excitatory inputs. Thus, it may be that heightened LC discharge would disrupt certain behavioral states such as sleep. In this way, LC neurons resemble motoneurons, whose discharge is also actively inhibited during PS (despite receiving strong barrages of excitatory influence), presumably to prevent disruption of this state (Chase, 1980).

In addition to sleep, LC discharge is decreased during two other spontaneously occurring behaviors: During both grooming (but not similarly intense motor activity such as struggling during restraint) and consuming a sweet water solution, LC discharge is diminished compared to adjacent times of similar EEG arousal (Aston-Jones and Bloom, 1981a). These results indicate that LC discharge is reduced not only for epochs of low arousal (such as during sleep), but also during certain specific behaviors when animals are quite highly aroused. One property common to sleep, grooming and consumption is the low level of vigilance (defined here as the degree of surveillance of the external environment) accompanying these behaviors, e.g., such activities are relatively resistant to disruption by external stimuli. Thus, we conclude that spontaneous LC discharge varies with the degree of vigilance typifying different behavioral states, rather than the overall level of arousal only (Aston-Jones and Bloom, 1981a).

LC discharge also varies strongly with orienting behaviors. In fact, the highest discharge rates observed for LC neurons in our studies are consistently associated with apparently spontaneous (or evoked) orienting responses, especially when tonic ongoing behavioral activity (such as sleep, grooming or consumption) is disrupted and the animal suddenly orients towards the external environment. Such phasic transitions in behavioral state are closely accompanied by a short but robust period of LC discharge (e.g., a burst of 5-7 spikes within 200 msec).

LC SENSORY RESPONSIVENESS AND BEHAVIORAL STATE

In addition to the above spontaneous fluctuations in LC discharge for various behaviors, we found that these neurons in unanesthetized behaving animals are very responsive to low-level, non-noxious environmental stimuli (Aston-Jones and Bloom, 1981b; Foote et al., 1980). This is in marked contrast to the generally tonic and unvarying discharge found for these cells in anesthetized animals, in which responses are elicited only to strongly noxious stimuli (see Foote et al., 1983 for review). In awake animals, LC activity is quite phasic, yielding short-latency (15-50 msec) responses to stimuli in every modality tested (auditory, visual and somatosensory). These responses are similar for the various sensory modalities, and consist of a brief excitation followed by diminished activity lasting a few hundred msec.

As found for spontaneous bursts of discharge in LC, sensory-evoked activity is accompanied by FPs, temporally synchronized with evoked unit activity (Aston-Jones et al., 1981b). Thus, both spontaneously occurring and sensory-evoked barrages of excitatory input to LC may be recorded as FP events. As also found for spontaneous activity, sensory-evoked unit and FP responses are time-locked during waking and slow-wave sleep, but become markedly dissociated during PS, when unit responses cease to occur but FPs continue to be elicited. This result is consistent with our general conclusion that LC neurons are actively inhibited during PS.

During the course of monitoring LC responses to trains of sensory stimuli, fluctuations in the response magnitudes were apparent (Aston-Jones and Bloom, 1981b). On first analysis, these fluctuations resemble habituation and dishabituation, with initial responses being large, followed by reduced response magnitudes for a few stimulus trials, which in turn are followed by markedly larger responses, and so on. However, further examination revealed that response magnitudes vary closely with spontaneous fluctuations in the animals' behavioral state. Largest responses are elicited by stimuli presented during sleep which cause an abrupt transition to waking. Also, responses elicited during continuous waking are larger than those evoked during uninterrupted slow-wave sleep, while no responses occur to stimuli presented during uninterrupted PS. Similar to results for sleep, response magnitudes during uninterrupted grooming or consumption of sweet water are reduced compared to other epochs of waking, whereas stimuli that disrupt such activity and generate orienting behavior elicit strong responses. Thus, as found for spontaneous discharge rates, sensory-response magnitudes in LC vary with levels of vigilance associated with different behavioral states. Habituation independent of behavioral state was not found to occur: Response magnitudes for stimuli following even 100 or more presentations were similar to those for initial stimuli when presented during similar behavioral states.

Thus, it appears that both spontaneous and sensory-evoked LC discharge fluctuate with behavioral state, and that the most significant behavioral state-variable associated with altered LC excitability is vigilance, or the degree of attention to the external environment. Times of high vigilance, as occurs for sudden waking after sleep, interruption of ongoing behavior such as grooming or consumption, or presentation of an unexpected stimulus that evokes an orienting behavioral response by the animal, are associated with the most intense spontaneous or sensory-evoked discharge in LC, and represent times of highest excitability for LC neurons. Conversely, periods of low vigilance, such as during sleep, grooming or consumptive behaviors, are associated with diminished LC excitability. The time of least excitability for LC is during PS, when the neurons appear to be under a strong tonic inhibitory influence.

A NEW HYPOTHESIS OF LC FUNCTION BASED ON CELLULAR PHYSIOLOGIC ATTRIBUTES

The results described here for LC impulse activity in behaving animals have a number of functional implications when considered in light of other findings relevant to a cellular physiologic understanding of the LC system. As outlined at the beginning of this chapter, such an understanding requires knowledge of the efferent projections of LC neurons, the effects of NE released from LC terminals on target neuron activity, and the conditions under which LC neurons are active and releasing their transmitter. The broad efferent trajectory of the LC system implies that its function is a very general one. This notion is supported by our physiologic studies, revealing that LC neurons are markedly homogeneous in their discharge characteristics, e.g., LC neurons throughout the nucleus exhibit very similar rates and patterns of spontaneous or sensory-evoked impulse activity (Aston-Jones and Bloom, 1981a,b). Postsynaptically, NE influences target cells so as to enhance strong responses to other of their afferents, while simultaneously reducing spontaneous or low-level activity. Such an enhancement of postsynaptic "signal-to-noise" ratios can lead to relatively increased selectivity of target cell discharge to favor specific aspects of their response profiles, as discussed in this volume for the work of Segal and collaborators.

In the context of these previous findings, the specific conditions of LC activation in unanesthetized behaving animals lead us to a global hypothesis for the function of the LC system in brain and behavioral activities, consistent with a role of this system in vigilance or attentional processes (Aston-Jones and Bloom, 1981a,b; Aston-Jones et al., 1984). First, we propose that the LC is strongly influenced by two classes of afferents: excitatory inputs mediating sensory-evoked (or spontaneous orienting) activity in LC neurons, and a more tonically active set of inhibitory afferents serving to modulate overall LC excitability in accordance with behavioral state or vigilance. The level of LC activity at any time may be a consequence of the relative influence of each of these two classes of inputs. Strong tonic inhibition (such as found during PS) may serve to prevent LC neurons from responding to environmental stimuli. This may be necessary since NE released from LC terminals in response to a sensory event could globally enhance strongly determined activity in target neurons (e.g., activity associated with salient sensory stimuili) and thereby promote transmission of corresponding relevant information through the nervous system for expression in behavior. That is, increased LC activity may interrupt low priority, internally generated or vegetative behaviors (such as sleep, grooming or consumption) that are inconsistent with phasic behavioral responding to the wide range of current environmental events, and instead engage a mode of brain and behavioral activities characterized by high vigilance or attention to environmental stimuli. This theoretic framework is consistent with our observation that LC activity is most intense just before interruption of low-vigilance behaviors such as sleep, grooming or consumption, giving rise to alert orienting behaviors. Intense LC activation may occur when either tonic inhibition of LC neurons (reflecting on-going vegetative processes) has suddenly decreased, or when excitation impinges on these cells (in response to a strong, unexpected sensory event) that is sufficiently intense to overcome concurrent tonic inhibitory influences. In this way, the LC may serve as a gate to determine the relative influences of two mutually exclusive sets of behavioral programs. In more general terms, the LC may function to influence the overall orientation of behavior or mode of sensory-motor activities, to favor either tonic vegetative behavioral programs, or phasic adaptive responding to current environmental stimuli.

Note that in this theoretic framework, the information as to what external event receives the most attention (i.e., the information guiding "selective attention") is not present within LC itself. Rather, activation of LC by a strong external stimulus releases NE in a global fashion, and this NE "sets the stage" in LC terminal areas so that the most relevant or intense stimuli will preferentially influence overall brain activity, and thereby control attention and behavior as well. In this way, the LC system may serve as a simple but important gating mechanism for determining the global level of vigilance or attention to environmental stimuli, while the interaction of NE in LC terminal areas with particular sensory afferent information further shapes or specifies the selectivity of such attention.

IMPLICATIONS FOR BEHAVIORAL STUDIES OF LC FUNCTION

The above hypothesis implies certain predictions for behavior in the absence of LC (as with lesions), or in the presence of artificially high LC activity (stimulation). One such prediction is that animals lacking LC should be less distractible by environmental stimuli during non-vigilant behaviors such as sleep, grooming or consumption. That is, we predict that such low-vigilance behaviors would be less disruptable by environmental events without the influence of LC. Note that it would be important to test distractability <u>during</u> such non-vigilant behaviors, as the LC may be less important for distraction by stimuli when vigilance is already high.

The above findings also indicate that the LC may be important in learning the significance of certain sensory events. The effect of NE to enhance the selectivity of target cell responsiveness to strong or preferred inputs implies that without LC, sensory selectivity in brain (and behavioral) activities should be reduced. Note, however, that such an effect of LC lesions may only be apparent when close differential discrimination is required, since the enhancement of selectivity by NE may not be necessary for discrimination between widely differing sensory cues. Recent results of Robbins and colleagues for learning deficits following LC lesions (reviewed in the present volume) are consistent with this notion. Such enhancement of response selectivity in LC target cells may also be necessary for aquisition of a subtle sensory cue after training has been established using a strong cue. Note again that it may be important that the new cue be a low-level or subtle one, since NE's augmentation of postsynaptic "signal-to-noise" ratios may not be necessary for attaining the significance of an another strong stimulus.

There are various possible reasons why some effects of LC lesions have not been forthcoming in behavioral studies. One is that CNS systems such as the LC are notoriously plastic and exhibit substantial recovery following lesion. This may apply also to non-LC systems, taking over the functions of LC (perhaps even partially), and thereby obscuring LC lesion effects. It may be, therefore, that behavioral tests of functional hypotheses for the LC system would be best carried out with pharmacologically specific but acute inactivation of LC, as for example by local injection of clonidine. Secondly, the proper questions or behavioral tests may not have been posed, e.g., distractibility during non-vigilant behaviors has not been tested. Results described herein for the cellular physiologic attributes of LC deserve careful consideration in behavioral studies employing LC lesions or stimulation. Knowledge of the behavioral conditions during which LC neurons are normally most active should give useful direction to the design of behavioral tasks for elucidating effects of such LC manipulations. Future studies along these lines should prove to be very valuable in testing behavioral hypotheses of LC function.

REFERENCES

Aghajanian, G. K., C. P.Vandermaelen and R. Andrade, 1983, Intracellular studies on the role of calcium in regulating the activity and reactivity of locus coeruleus neurons in vivo, Brain Res., 273: 237.

Aston-Jones, G. and F. E. Bloom, 1981a, Activity of norepinephrine-containing locus coeruleus neurons in behaving rats anticipates fluctuations in the sleep-waking cycle, J. Neurosci., 1: 876.

Aston-Jones, G. and Bloom, F. E., 1981b, Norepinephrine-containing locus coeruleus neurons in behaving rats exhibit pronounced responses to non-noxious environmental stimuli, J. Neurosci., 1: 887.

Aston-Jones, G., S. L. Foote and F. E. Bloom, 1984, Anatomy and physiology of locus coeruleus neurons: functional implications, in: "Norepinephrine" ("Frontiers in Clinical Neuroscience, Vol.2"), M. Ziegler and C. Lake, eds., Williams and Wilkins, Baltimore

Chase, M., 1980, The motor functions of the reticular system are multifaceted and state-determined, in: "The Reticular System Revisited", J. Hobson and M. Brazier, eds., Raven Press, New York.

Foote, S. L., F. E. Bloom and G. Aston-Jones, 1983, The nucleus locus coeruleus: new evidence of anatomical and physiological specificity, Physiological Rev., 63: 844.

Foote, S. L., G. Aston-Jones and F,. E. Bloom, 1980, Impulse activity of locus coeruleus neurons in awake rats and squirrel monkeys is a function of sensory stimulation and arousal, Proc. Natl. Acad. Sci. (USA), 77: 3033.

Hobson, J., R. McCarley and P. Wyzinski, 1975, Sleep cycle oscillation: reciprocal discharge by two brainstem groups, Science, 189: 55.

Jacobs, B.L., J. Heym and G. F. Steinfels, 1984, Physiological and behavioral analysis of raphe unit activity, in: "Handbook of Psychopharmacology, Vol. 18", L. L. Iversen, S. D. Iversen and S. H. Snyder, eds., Plenum, New York.

Steriade, M. and J. Hobson, 1976, Neuronal activity during the sleep-waking cycle, Prog. Neurobiol., 6: 155.

Williams, J, T. Egan and R. North, 1982, Enkephalin opens potassium channels on mammalian central neurones, Nature, 299: 74.

MECHANISMS OF ACTION OF NORADRENALINE IN THE BRAIN

M. Segal

Center for Neurosciences and Behavioral Research
The Weizmann Institute of Science
Rehovot, 76100 Israel

The noradrenaline (NA) innervation of the brain has been studied extensively over the past decade. Early views suggested that the nucleus locus coeruleus (LC) has a widespread, diffuse, slow conducting projection system with little anatomical or functional specificity. These views were supported by the observations that (a) the LC projects into nearly every brain structure in a rather diffuse manner, (b) LC fibers are small, unmyelinated and slowly conducting and therefore unlikely to convey specific information (c) action potential discharge rates of LC neurons vary considerably with sleep-wakefulness cycles being nearly silent in REM sleep and (d) iontophoretic application of NA produces in nearly every brain area a long lasting beta receptor-mediated suppression of spontaneous firing rates. It is now clear, however, that these conclusions represent an oversimplification of the actual evidence and that LC functions might be more specific and complex. LC innervation of the forebrain is actually quite heterogeneous with some structures being heavily invested with NA fibers while others are relatively devoid of them. Furthermore, within a given structure, e.g. the neocortex, there is a distinct laminar distribution of NA fibers and receptors. The physiological properties of rat LC neurons, characterized in a recent series of studies, also indicate that LC might be involved in more complex functions; First, LC neurons respond vigorously to sensory stimuli with a short latency excitation. Secondly, their responses depend on the vigilance state of the rat (Aston-Jones and Bloom, 1981). Third, their axonal conduction velocity is not as slow as originally thought, especially in large mammals (i.e. monkeys). Finally, there are strong indications that NA of LC origin plays an important role in developmental plasticity (Kasamatsu, this volume). These observations mitigate against the view of the LC as only being involved in a global function associated with sleep-wakefulness cycles.

The mechanisms of action of NA in the brain have been analyzed at the behavioral level, where effects of stimulation or lesion of the LC on behavior are assessed; at the physiological level where the effects of stimulation of LC or iontophoretic application of NA on firing properties of single cells or cell assemblies are assessed; and at the biophysical level where effects of NA on single species of membrane currents are analyzed. A meaningful theory on the action of NA should integrate the three levels of analysis. The following review will concentrate on the physiological and biophysical analysis of LC action. A number of other authors address the behavioral significance of LC activation elsewhere (Mason, 1981; Sara 1985 and this volume).

EXTRACELLULAR RECORDINGS: PHYSIOLOGICAL MECHANISMS

Most of the early studies employing iontophoresis of NA onto neurons throughout the brain reported that it causes a beta adrenergic suppression of spontaneous action potential discharges. (see Foote et al., 1983). Excitatory, alpha adrenergic actions are found in cortex, geniculate (Rogawski and Aghajanian, 1980) and spinal cord (see Marshall, 1983). The effects of electrical stimulation of the LC on spontaneous activity of postsynaptic neurons were tested in the cerebellum (Hoffer et al., 1973), hippocampus (Segal and Bloom, 1976), neocortex (Phillis and Kostepoulus, 1977), septum (Segal, 1976), spinal cord (Sasa et al., 1976) and other deep nuclei (see Foote et al., 1983). The stimulation produces, in general, a characteristic slow suppression of spontaneous action potential discharge rate. This response is different from the typical IPSP seen in GABA-mediated inhibitory connections. The noradrenergic response can be potentiated by NA uptake inhibitors and blocked by beta antagonists or by drugs which interfere with noradrenergic transmission (Segal and Bloom, 1976).

While suppression of spontaneous activity is the most common response to LC stimulation, it is by no means the only action and perhaps not the main one. In the awake animal, LC stimulation suppresses spontaneous activity in the hippocampus but can facilitate an excitatory response in that structure to a significant, conditioned stimulus (Segal and Bloom, 1976). In the cerebellum (Woodward et al., 1979), low dose of NA or a LC stimulation can potentiate reactivity of purkinje cells to iontophoretic administration of GABA and other amino acids. A similar effect is seen in the somatosensory region of the neocortex (Waterhouse et al., 1981). The apparent dual action of NA was studied extensively in the rat visual cortex (Madar, 1983; Madar and Segal, 1980). This system offers the advantage of studying functionally identifiable neurons whose response patterns to visual stimuli can be quantified and any changes in these can be easily detected. Stimulation of LC or iontophoretic application of NA suppressed spontaneous activity of complex cells via activation of a beta-1 adrenoreceptor. Stimulation of LC and application of NA enhanced specific activity of both simple and complex cells. This specific activity is evoked by a light bar, moving in a certain direction and orientation. The response pattern to the visual stimulus was modified during LC simulation; the duration of response was shortened but its magnitude was, in many cases, enhanced. This enhancement was more evident when the bar of light was moving in the preferred direction and orientation. Some of these effects were found in response to iontophoretic application of NA.

It appears that NA of LC origin can modulate visual reactivity at two levels; in the geniculate, NA facilitates visual responses nonselectively by acting on alpha adrenoreceptors (Rogawski and Aghajanian, 1980). In the visual cortex, the effects of NA involve both a beta adrenergic suppression of spontaneous, background activity and an alpha adrenergic selective enhancement of reactivity to some aspects of visual stimuli. The concomitant action of LC in the geniculate and visual cortex brings about the observed increase in signal (visual responses) to noise (background activity) ratio. This will result in sharper and more defined receptive fields, and a narrower tuning curve for orientation and direction selectivity. These improved visual responses will result in a more refined visual acuity and perhaps behavioral visual functions. Such conditions are observed when the organism attends to its environment. Indeed, behavioral studies have indicated that visual cortex neurons of an aroused cat have sharper receptive fields (Livingstone and Hubel, 1981).

It is conceivable that LC acts simultaneously at different levels of information processing also in other neural systems. In the hippocampal circuit, the medial septum and dentate gyrus are heavily invested with NA fibers. While the relative contribution of each relay station to the observed effects of NA of LC origin in the hippocampus proper is not totally understood, it is clear that evoked responses in the hippocampus are markedly affected by behavioral states and by stimuli known to activate the LC (Segal, 1977). These effects cannot be explained by an action of NA at a single site in the hippocampal circuit.

INTRACELLULAR RECORDINGS: BIOPHYSICAL MECHANISMS

Early intracellular studies done in the intact brain have indicated that NA can hyperpolarize recorded neurons by 5-10mV (Hoffer et al., 1973). This hyperpolarization can account for the suppression of spontaneous spike discharge rate. It was suggested that NA acts by increasing intracellular CAMP which acts as a second messenger to phosphorilate some membrane proteins and cause the observed conductance and potential changes. There is indeed a large body of evidence to suggest that NA causes an increase in CAMP. The relations of this to the observed physiological changes remain unclear.

Studies in spinal cord neurons indicate that NA may reduce both NA^{+} and K^{+} conductance as its reversal potential was about -15mV and did not involve a change in Cl^{-} -conductance (Marshall 1983). The observed increase in input resistance would accordingly enhance reactivity to afferent stimulation; an injected synaptic current would produce a larger potential change when input resistance is higher.

Another possible mechanism activated by NA involves the Na-K electrogenic pump. Evidence in favor of such a mechanism has been accumulated in both in-vivo and in-vitro preparations (Phillis and Wu, 1981). The hippocampal slice preparation was used for the analysis of the action of NA (Segal, 1981). Topical application of NA produces a transient, 3-5mV hyperpolarizing response associated with little conductance changes. This response could not be reversed either by somatic hyperpolarization or depolarization but was markedly reduced in low temperature. Also, the response was blocked by ouabain. These experiments support the contention that NA activates a Na-K pump that hyperpolarizes the recorded neurone and causes suppression of spike discharge. This mechanism does not explain the several fold increase in population spike response to stimulation of the schaffer collateral system seen after beta adrenergic stimulation of the in-vitro hippocampus (Mueller et al., 1981). Activation of an alpha adrenoreceptor suppressed this population spike response. It was suggested that the increase in spike discharge underlying the population spike, is not preceded by an increase in the EPSP. Using intracellular recording we found that NA caused a marked, beta mediated, increase in the depolarizing response to iontophoretic application of glutamate and a subsequent increase in spike discharge during the depolarization (Segal, 1982). A possible mechanism underlying this effect was proposed recently by two independent groups (Haas and Konnerth, 1983; Madison and Nicoll, 1982). They suggest that NA, acting on a beta receptor, blocks the after-hyperpolarization (AHP) that follows a burst discharge. This AHP reflects activation of a Ca_2dependent K current which is triggered by an increase in intracellular Ca^{2+} in a voltage dependent manner. The effects of NA on AHP is different from a depressive effect of NA on I(Ca) observed in the sympathetic ganglion (Horn and McAfee, 1979) since in the hippocampus the Ca^{2+} spike is only minimally affected by NA compared to its effects on the AHP. The functional significance of this blocked AHP is not entirely clear but since glutamate may activate AHP, the blockade of this mechanism by NA may result in a larger depolarizing response as observed in our experiments. This blockade of AHP is probably not responsible for the large increase in population spike response to orthodromic stimulation because activation of AHP is a slower process than the initial action potential discharge underlying the population spike.

Two other mechanisms of action of NA have been suggested recently. These involve a blockade of rectification and a reduction in GABA inhibition (Langmoen et al., 1981; Jahr and Nicoll, 1982). It is conceivable that NA has at least three and perhaps more modes of action. The relevance of any of these to the physiological actions of NA seen in the intact brain is still not clear. There are several differences between the responses to NA in the intact brain and the slice: The alpha facilitatory and beta depressive actions seen in vivo are not maintained in vitro, where the facilitatory action is mediated by a beta adenoreceptor. Also, stimulation of LC in the awake rat can produce a pronounced and long lasting suppression of spontaneous activity. Such an effect is not seen in the slice preparation in response to NA. Needless to say, there are a number of physiological and biochemical differences between the slice and the brain; the slice contains about half as much ATP as the hippocampus in situ and there is a good reason to believe that the slice is somewhat hypoxic. These conditions may not favor activation of an energy dependent process by NA and instead may amplify detection of other mechanisms acted upon by NA. Altogether, NA acts to suppress spontaneous activity and enhance evoked responses using one or more of the mechanisms suggested here. The action of NA on the local circuit and its interaction with linked brain regions may bring about the observed behavioral consequences of activation of LC.

REFERENCES

Aston-Jones, G. and Bloom, F.E., 1981, Norepinephrine-containing locus coeruleus neurons in behaving rats exhibit pronounced responses to non noxious environmental stimuli. J. Neurosci. 1:887.

Foote, S.L., Bloom, F.E. and Aston Jones, G., 1983, Nucleus Locus coeruleus: new evidence of anatomical and physiological specificity. Physiol. Rev., 63:844.

Haas, H.L. and Konnerth, A., 1983, Histamine and noradrenaline decrease calcium-activated potassium conductance in hippocampal pyramidal cells. Nature, 302:432.

Hoffer, B.J., Siggins, G.R., Oliver, A.P. and Bloom, F.E., 1973, Activation of the pathway from locus coeruleus to rat cerebellar purkinje neurons: pharmacological evidence of noradrenergic central inhibition. J. Pharmacol. Exp. Ther., 184:553.

Horn, J.P. and McAfee, D.A., 1979, Norepinephrine inhibits calcium dependent potentials in rat sympathetic neurons. Science, 204:1233.

Jahr, C.E. and Nicoll, R.A., 1982, Noradrenergic modulation of dendrodendritic inhibition in the olfactory bulb. Nature, 297:227.

Langmoen, I.A., Segal, M. and Andersen, P., 1981, Mechanisms of norepinephrine actions on hippocampal pyramidal cells in vitro. Brain Res., 208:349.

Livingstone, M.S. and Hubel, D.H., 1981, Effects of sleep and arousal on the processing of visual information in the cat. Nature (Lond), 291:554.

Madar, Y., 1983, The noradrenergic action in the visual cortex. Unpublished Ph.D. dissertation. The Weizmann Institute of Science, Rehovot.

Madar, Y. and Segal, M., 1980, The functional role of the noradrenergic system in the visual cortex. Activation of the noradrenergic pathway. Exp. Brain Res., 41:A14.

Madison, D.V. and Nicoll, R.A., 1982, Noradrenaline blocks accomodation of pyramidal cell discharge in the hippocampus. Nature (Lond), 299:636.

Marshall, K.C., 1983, Catecholamines and their actions in the spinal cord, in: "Spinal Cord Pharmacology", R.A. Davidoff, ed., Dekker, N.Y.

Mason, S.T., 1981, Noradrenaline in the brain: Progress in theories of behavioral function. Prog. Neurobiol., 16:263.

Meuller, A.L., Hoffer, B.J. and Dunwiddie, T.V., 1981, Noradrenergic responses in rat hippocampus: Evidence for mediation by alpha and beta receptors in the in vitro slice. Brain Res., 214:113.

Phillis, J.W. and Kostepoulus, G.K., 1977, Activation of a noradrenergic pathway from the brain stem to rat cerebral cortex. Gen. Pharmacol., 8:207.

Phillis, J.W. and Wu, P.H., 1981, Catecholamines and the sodium pump in excitable cells. Prog. Neurobiol., 17:141.

Rogawski, M.A. and Aghajanian, G.K., 1980, Modulation of lateral geniculate neurone excitation by noradrenaline microiontophoresis or locus coeruleus stimulation. Nature, 287:731.

Sara, S.J., 1985, Memory disfunction: an integration of animal and human research from preclinical and clinical perspectives. D.S. Olton, E. Gamzu and S. Corkin (Eds.) Ann. New York Acad. Sci. in press.

Sasa, M., Munekiyo, K., Ikeda, H. and Takaori, S., 1974, Noradrenaline-mediated inhibition by locus coeruleus of spinal trigeminal neurons. Brain Res., 80:443.

Segal, M., 1976, Brainstem afferents to the rat medial septum. J. Physiol. (Lond), 261:617.

Segal, M., 1977, The effects of brainstem priming stimulation of interhemispheric hippocampal responses in the awake rat. Exp. Brain Res., 28:529.

Segal, M., 1981, The action of norepinephrine in the rat hippocampus: intracellular studies in the slice preparation. Brain Res., 206:107.

Segal, M., 1982, Norepinephrine modulates reactivity of hippocampal cells to chemical stimulation in-vitro. Exp. Neurol., 77:86.

Segal, M. and Bloom, F.E., 1976, The action of norepinephrine in rat hippocampus IV the effects of locus coeruleus stimulation on evoked hippocampal unit activity. Brain Res., 107:513.

Waterhouse, B.D., Moises, H.C. and Woodward, D.J., 1981, Alpha-receptor mediated facilitation of somatosensory cortical neuronal responses to excitatory synaptic inputs and iontophoretically applied acetylcholine. Neuropharmacol., 20:907.

Woodward, D.J., Moises, H.C., Waterhouse, B.D., Hoffer, B.J. and Freedman, R., 1979, Modulatory actions of norepinephrine in the central nervous system. Fed. Proc., 38:2109.

MODULATION OF SELECTIVE PROCESSES IN LEARNING BY NEOCORTICAL AND LIMBIC DOPAMINE: STUDIES OF BEHAVIOURAL STRATEGIES

Robert Oades[1*], Michael Rea[2] and Khalid Taghzouti[3]

1, Breslauerstr.9, 6102 Pfungstadt, FRG.**2, MPG Forschungsgruppe, Univ.-Frauenklinik, 4400 Münster, FRG. 3 INSERM 259, Rue Camille Saint-Saens, F-33077 Bordeaux

INTRODUCTION

The main dopaminergic (DA) systems of the mammalian CNS are the diencephalic tubero-infundibular and incerto-hypothalamic (A11-14) and the mesencephalic long-axon (A8-10) projections to the forebrain (Lindvall and Björklund, 1978). The A9 group preferentially innervates the striatum and cingulate cortex (Beckstead et al., 1979), whereas the ventral tegmental area (VTA-A10) projects to the prefrontal, cingulate, suprarhinal and entorhinal cortices (mesocortical system, M/C) and underlying limbic structures such as the amygdala, septo-hippocampal complex and the nucleus accumbens (mesolimbic system, M/L, Simon, 1981).

Some of the major projection regions of the M/C and M/L systems show an extraordinary convergence of pathways from divergent sources (Phillipson, 1979). In discussing the integrative functions that this convergence implies we shall consider the prefrontal cortex and the septum (or N. accumbens) as representative of the M/C and M/L systems respectively (cf. fig.1).

Anatomical and electrophysiological studies show that the prefrontal cortex and septum are polysensory association areas (Mercer and Remley, 1979; Rosenkilde, 1979). Brain damage in both areas leads to impairments in learning and to changes in the "flexibility" of the early attention-related stages of information processing (Numan and Lubar, 1974; Brody and Pribram, 1978; Oades, 1979; 1982). Animals with prefrontal damage have difficulty with cross-modal integration (Petrides and Iversen, 1976), but the presentation of relevant cues ameliorates their impairment on delayed spatial or timing tasks (Crowne, 1983; Rosenkilde, 1979). After septal damage rats may learn abnormally fast (shock avoidance) or slow (appetitive operant/maze tasks). In some situations a relevant cue helps (classical/operant conditioning or maze learning: Burton and Toga, 1982; Donovick et al., 1979; Ellen et al., 1977); in others it does not help (spatial or avoidance tasks: Beatty and Carbone, 1980; Numan et al.,1982).

---

* Senior NATO and Fyssen Fellow.
** Present Address: Dept. of Human Physiology, Flinders Univ. School of Medicine, Bedford Park, South Australia 5042.

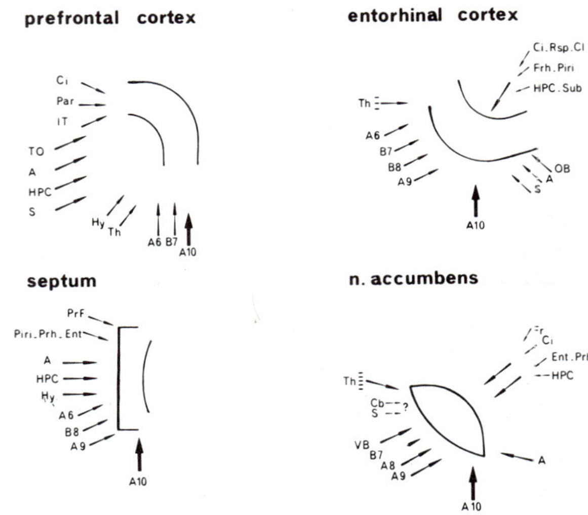

Fig.1. Convergence of afferents on M/C and M/L projection areas. A, Amygdala; Cb, Cerebellum; Ci, Cingulate cortex; Cl, Claustrum; Ent, Entorhinal cortex; Fr, Frontal cortex; HPC, Hippocampus; Hy, Hypothalamus; IT, Inferotemporal cortex; OB, Olfactory bulb; Par, Parietal cortex; Piri, Piriform cortex; PrF, Prefrontal cortex; Prh, Perirhinal cortex; S, Septum; Th, Thalamus; TO, Tuberculum olfactorium; A, B6-10, several monoaminergic nuclei.

DOPAMINE MODULATION

VTA-A10 lesions that deplete DA in the forebrain impair learning of a variety of complex appetitively and aversively reinforced tasks (Le Moal et al., 1977); but, as is usual after damage to complex association cortices, simple discriminations are not impaired (Simon, 1981). Yet appropriate levels of DA activity may not just be helpful but are probably crucial to adaptive learning. Extensive A9/A10 lesions can cause adipsia, aphagia sensory neglect and abolish food-hoarding, pup-nursing, delayed alternation and active avoidance learning (see Oades, 1985).

In general terms DA may modulate the initiation and timing of responses and the switching of input-output relations in DA-innervated regions (Robbins and Everitt, 1982; Oades, 1985). But these accounts do not attempt to interpret the specific role of this modulation in the information processing particular to one or another brain area. The differences of learning performance found after non-specific electrolytic or specific 6-hydroxydopamine (6-OHDA) damage to DA projections illustrate the need for comparing M/C and M/L function. Thus delayed alternation is impaired after non-specific damage to the VTA, prefrontal cortex or septum (Simon, 1981; Markowitsch and Pritzel, 1976; Thomas,1979). It is also impaired after 6-OHDA damage to the VTA and prefrontal cortex (Simon, 1981; Brozovski et al., 1979). On active avoidance tasks electrolytic damage of the VTA produces a mild (Le Moal et al., 1969) or of the septum a strong facilitation (Numan et al., 1982), whereas prefrontal damage results in a mild deficit (Markowitsch and Pritzel, 1976). By contrast 6-OHDA lesions of the prefrontal cortex produce no effect and 6-OHDA lesions of the

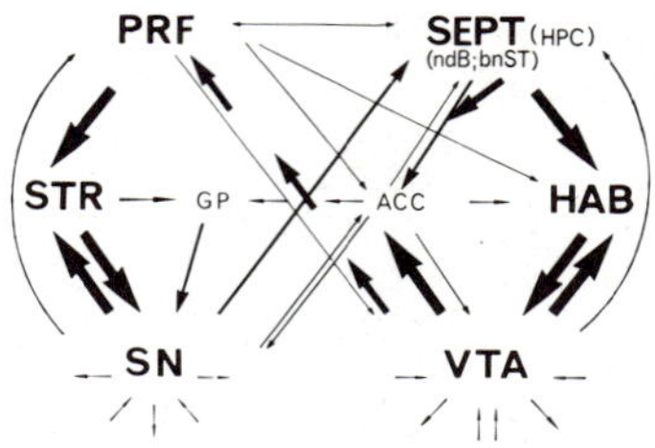

Fig.2. Frontostriate and septo(limbic)habenula axes with reference to M/C- and M/L-DA innervation from the substantia nigra (SN) and VTA. (Arrow size,- output importance as considered in text: note inclusion of closely linked limbic structures with septum (S) and integrative role of VTA shown by input/output arrows). ACC, N. Accumbens; HAB, Habenula; ndB, N. of diagonal band of Broca; PRF, Prefrontal cortex; STR, Neostriatum.

septum result in a mild facilitation of active avoidance, but 6-OHDA lesions of the VTA disrupt acquisition (cf fig.3).

STUDIES USING BEHAVIOURAL RECORDS

To improve our understanding of these region- and task-specific DA/non-DA interactions, one needs to know more than whether a given manipulation changes the latency or accuracy of performance on a task battery. One must ask how a stimulus can control or is hindered from controlling the output of a given region or system, for which behaviour is an indicator. When does a stimulus control? When is a response controlled? In the former case one finds out under what stimulus conditions an information processing channel is allocated (e.g. learned inattention). In the latter one records components of behavioural sequences as indicators of the control over information processing that allowed the particular output (e.g. task-solving strategy). In practice the two approaches overlap. A strategy of response is always required; the difference often lies with the number of response-determining events. The reasoning behind learned inattention has been elaborated elsewhere (Mackintosh, 1975). This discussion emphasizes the study of strategy (response sequences) that has only more recently attracted attention.

The importance of investigating behavioural strategies to the study of M/C and M/L function is illustrated by reference to the output of the prefrontal cortex and septum. The neostriatum receives a massive input from the frontal cortex (Beckstead,1979) and the septal (limbic) relations with the habenula and nucleus accumbens are equally impressive (fig.2; Sutherland, 1982). Neostriatal activity seems to be important for gating sequences of behaviour appropriate to external (e.g.neighbours' behaviour) and to internal contingencies (e.g. own ongoing behaviour; Van den Bercken and Cools, 1982). Neostriatal treatment with DA antagonists or agonists can reduce or facilitate, respectively, the number of strategies rats may try in order to escape from drowning in a basin of water (Cools, 1980). Similarly damage to the habenula can drastically reduce the frequency of behavioural changes and the number of categories of behaviour (Thornton and Evans, 1982).

"Sequence-dependent Strategy"

We have been studying the efficiency of acquisition and the types of strategy used in a holeboard search task. In this task rats search a 16-hole arena for the 4 holes that always contain food. After 70-100 trials (10 trials/session, 2 sessions/day) rats learn to avoid empty holes and to visit food holes in a preferred order (strategy) that differs between individuals (Oades and Isaacson, 1978; Oades, 1981a; 1982).

In a recent experiment animals received the neuroleptic spiroperidol (2 µg/0.5 µl), apomorphine (DA agonist) or vehicle (pH adjusted tartarate solution) directly into the VTA 15 min before sessions 4 and 7 (Oades, 1981c). Although a degree of diffusion was expected after this treatment, the neurochemical and behavioural sequelae of injection sites 1.5 mm outside the VTA were quite different. Neuroleptic but not other treatments increased rearing (3-4 fold), the number of errors (50-100%) and the number of switches away from the preferred strategy (Oades, 1981c). Some animals were sacrificed after session 7. After neuroleptic treatment there was an increase of DA utilization in the N. accumbens (table 1). This increase correlated with increased rearing (N, 18; $r=0.52$; $p<0.05$; Spearman rank correlation) and with the number of errors ($r=0.52$; $p<0.05$). The correlation was similar whether the errors were classified according to working or reference memory. But there was no significant correlation with the number of strategy repetitions or changes ($r=0.2$).

In view of the hippocampal input to the habenula and N.accumbens it is interesting that an earlier report showed that hippocampal damage resulted in animals making more errors and frequent changes of strategy. After haloperidol treatment (ip) such animals made fewer errors but continued to change their strategy (Oades and Isaacson, 1978). Thus it is possible that whereas the M/L system may be involved in an error evaluation process (necessary for a succesful strategy), an important contribution to the sequential organization of appropriate responses may arise elsewhere. The work of Cools and our own report that haloperidol alone can disrupt strategy development suggests that the fronto-striate axis may be important. Whereas haloperidol treatment reduced the number of strategy repetitions (quantity), cortical damage increased the range of hole-visit sequences attempted (quality)(Oades and Isaacson, 1978). Damage was more temporo-parietal than frontal, but the possibility of an interaction at the level of the prefrontal cortex was supported by the results of treating cortically-damaged animals with a neuroleptic. Working but not reference memory types of error (ie intra-trial) showed a slight improvement (Oades, 1981b). A differential impairment for working/episodic rather than for

Table 1. Levels (pmol/mg wet weight) and ratios of monoamines and metabolites in N. Accumbens (±SEM)

Group	DA	DOPAC	DOPAC/DA	NA	5HIAA/5HT
C	38.8 ±6.2(7)	7.4 ±2.1(6)	0.19 ±0.04(6)	4.74 ±1.14(5)	0.29 ±0.01(7)
SP	64.1 ±2.6(5)	18.6 ±1.5(5)	0.30§§ ±0.03(5)	4.6 (1)	0.34 ±0.03(5)
AP	23.0 ±3.1(4)	4.3 ±1.4(3)	0.18 ±0.03(3)	3.0 ±0.43(3)	0.47 ±0.17(4)

Reverse phase HPLC analysis, electrochemical detection (0.8 v)(0.1M citrate-phosphate buffer, pH 3.5, 0.1 mM EDTA, 0.4 mM sodium octylsulphate, 15% (v/v) methanol (Rea et al., 1982). C, control; SP, spiroperidol- and AP. apomorphine-treated rats (in VTA): n, number of samples: §§ $p<0.02$ Student's T test.

reference/semantic memory tasks has also been noted for Parkinsonian patients (Weingartner et al., 1984).

The implication is that frontal areas may be involved in strategy formation and the prefrontal-DA component is active in determining the adaptive quantity of a particular strategy. This is not entirely unexpected. Monkeys with frontal damage are slow to acquire a 3-key press contingent discrimination (Brody and Pribram, 1978) and frontally damaged rats are poor at relearning a rod-lever press sequence (Mogenson and Divac, 1984). Human subjects with frontal damage have difficulty in using experimenter-provided negative feedback in directing the future sequence of discrimination choices between stimuli varying on 4 binary dimensions (i.e. impaired development of an adaptive lose-shift strategy, Cicerone et al., 1983). Further definition of the role of prefrontal DA could be achieved by manipulation of the number, salience and relevance of cues available to determine the strategies chosen and to solve the tasks chosen (cf next section).

"Event-dependent Strategy"

We have started to study stimulus control in a two-way active avoidance task. The rate of learning of normal rats improved slightly after adding a light to a tone stimulus (simple/compound CS). This was clearly so for prefrontal- but not for septal-6-OHDA-treated animals (fig.3). (This recalls effects of cues on task performance after electrolytic damage; see introduction).

Animals treated with 6-OHDA in the septum were moved onto a conditioned blocking test. (Avoidance responses to separately presented non-reinforced tone and light stimuli shown by animals with shocked tone and light (ab) experience for two sessions were compared to those from animals pre-exposed to one of the stimuli on the first session (a/ab)). The conditioned avoidance blocking ratio (b/a+b) over 20 presentations of the stimuli alone for the septal-6-OHDA group was a satisfactory control value of 0.18(i.e. animals did not learn well the association between "b" and the shock). This is opposite to expectations from previous results obtained for animals with electrolytic damage of the septum

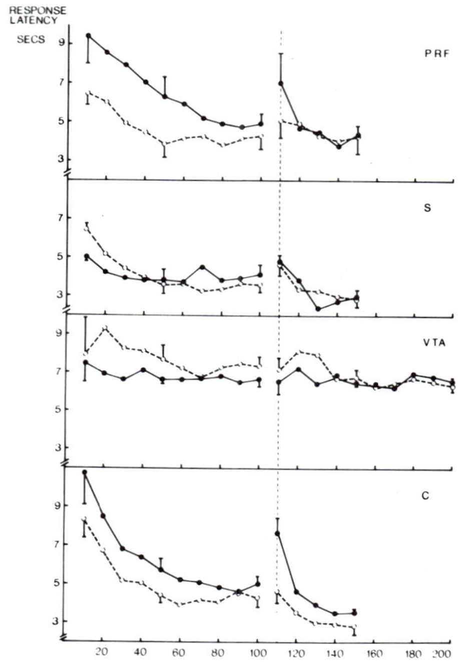

Fig.3. Latency (SEM) to avoid footshock in a shuttle box over two 100 trial sessions is shown for 4 groups: ascorbic acid (C;n,18) or 6-OHDA (4 ug/ul/side + 20 mg/kg DMI) in A10 (VTA; n,10), septum (S; n,11) or prefrontal cortex (PRF; n,13). A tone (1 KHz, 85 dB: white noise 65 dB) (——) or a tone with light (100 Lux)(---) were presented 5 sec before shock (1 mA). Training stopped after acquisition (80% on 3 successive 10 trial blocks). The VTA group did not learn to avoid: the two vs one stimulus condition facilitated acquisition for the prefrontal and control but not for the septal group (Analysis of variance, H 14.8; df 2, $p<0.001$, Kruskal Wallis).

(Weiss et al., 1974). But a closer look at the response latencies to the first and last three stimuli shows this to be an over simplified conclusion (fig.4). At first,blocking was attenuated. Finally the septal group showed better blocking than the controls. Overall the demonstration of blocking in the animals treated with 6-OHDA in the septum confirms the hypothesis of Solomon whereby an attenuation of blocking results from DA hyperactivity (Crider et al., 1982; Solomon et al., 1981). But taken trial by trial we found a "change-over" of responsiveness.

A potentially similar "change-over" from responding to non-responding depending on stimulus exposure has also been reported on a latent inhibition task. Burton and Toga (1982) pre-exposed animals to non-reinforced presentations of the to-be-discriminated stimulus. Rats with largely unilateral electrolytic damage to the septum showed an attenuation of latent inhibition after many exposures (180+) but an enhancement after fewer pre-exposures (<100). They suggested that there may have been a change of

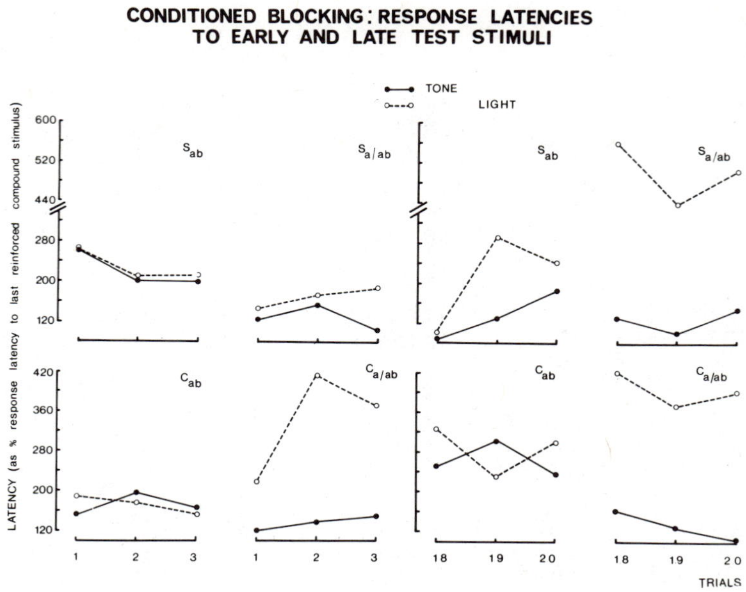

Fig.4. Response latencies to the first (left) and last (right) three presentations of a non-reinforced tone (a:——) or light (b:---) in a shuttle box after avoidance training with "ab" or "a" followed by "ab". Latencies are shown as a percentage of that for the last reinforced presentation of "ab". (Test program: a,b,a,ab,b,a,b,ab, etc). Animals received 6-OHDA (S; n,11) or vehicle (C; n,18) in the septum. Blocking to "b" followed attenuation in the S a/ab group (X_r^2 11.6, $p<0.05$, Friedman two way analysis of variance).

sensitivity or a sensitization effect. Although animals with electrolytic septal lesions are reported to be more sensitive to shock and light (Clody and Carlton, 1969; Chafetz et al., 1981), we have not found animals with 6-OHDA lesions of the septum to be especially sensitive to footshock or tone (in prep). We also looked at the latency to enter a dark then to return to a strongly lit (100 W) compartment of a two-chamber box (Bruneau et al., 1980). After septal-6-OHDA treatment rats entered the dark compartment with a latency comparable to controls (mean, S, 11.8; C, 18.4 sec) but they returned to the light chamber later (104 vs 52 sec; $p<0.05$ Mann Whitney U test). A similar result was found after 6-OHDA treatment of the N. accumbens (Taghzouti, 1983). But 9 of the septal animals in fact poked their noses back into the light chamber long before they emerged (38 vs 141 sec). The remaining septal animals were comparable to controls in that they usually emerged after poking their noses out into the light chamber.

Another example of an apparent "change-over" of responsiveness was found in the sequences of unrewarded visits shown by rats exploring the three arms of a Y-maze (cf Oades et al.,1985). Overall both lesioned (septal-6-OHDA) and control animals showed a strategy for visiting the least-recently visited arm (choice alternation; 61-70%). But whereas a control animal revisited the most-recently visited arm (perseverative choice) once in the in the first 8-12 choices, a lesioned animal made 3-4 changes of choice strategy in this time. Yet after 25-30 choices lesioned animals showed perseverative choices less often than controls (i.e. fewer strategy changes). In assessing this result it is important to note that opposite results have been reported with different techniques and animals. Improved alternation, spatial discrimination and reversal performance were reported after septal-6-OHDA damage in mice (Galey et al., 1984), but small amounts of electrolytic septal damage in rats resulted in performance deficits (Thomas, 1979).

The DA-A10 input to the septum exerts a tonic inhibition and 6-OHDA damage to the septum a disinhibitory effect on cholinergic activity in the septo-hippocampal pathway (Durkin et al., 1983; Galey et al., 1984). This lesion may decouple the coordination of information processing controlled by septo-hippocampal input from the temporal pattern of environmental factors that inform on the appropriateness of the response and ultimately the need for a change of strategy. What might these factors be? In normal rats the nature of the reinforcement schedule and the degree of exposure to the stimuli (exploration) determine the tendency to repeat or to shift response choice (Evenden and Robbins, 1984; Haig et al., 1983). Since the suppression of responding by animals with electrolytic septal damage on schedules requiring differential responding (DRL) is reported not to be solely determined by the density and distribution of reinforcement (Ellen et al., 1978), we would emphasize the important role of stimulus exposure and timing factors. In animals with 6-OHDA-septal damage the influence of important strategy determining information is displaced, not absolutely, but temporally. This "decoupling" of septo-hippocampal processing from an input that inhibits processing (i.e.disinhibition) leads at the behavioural level to a change of responsiveness temporally inappropriate to the environmental circumstance.

CONCLUSIONS

DA activity in both M/C and M/L systems is crucially involved in adaptive learning. Studies of behavioural strategies in simple-event and complex-task situations both contribute to elucidating the interdependence of these systems in organizing appropriate response sequences (nature and quantity: frontostriatal and septo(limbic)habenula axes). The M/L system appears important for the evaluation and temporal coordination of response feedback and the M/C system for the integration of succesful responses in to appropriate strategies. Both processes are dependent on timing and on input/output switching that are attributable to DA activity (Oades, 1985; Robbins and Everitt, 1982). Speculative as these working hypotheses are, they encourage further development of a combined psychological, ethological and neurobiological approach to the investigation of the M/C and M/L systems for which this report represents but an early stage.

REFERENCES

Beatty, W. and Carbone, C. 1980, Septal lesions, intramaze cues and spatial behavior in rats, Physiol. Behav. 24: 675.

Beckstead, R. M., 1979, An autoradiographic examination of corticostriatal and subcortical projections of the mediodorsal

projection (prefrontal) cortex in the rat, J. Comp. Neurol. 184: 43.
Beckstead, R. M., Domesick, V. B. and Nauta, W. J. H., 1979, Efferent connections of the substantia nigra and ventral teg. mental area in the rat, Brain Res., 175: 191.
Brody, B. A. and Pribram, K. H., 1978, The role of frontal and parietal cortex in cognitive processing: tests of spatial and sequence functions, Brain, 101: 607.
Brozovski, T. J., Brown, R. M., Rosvold, H. E. and Goldman, P. S. Cognitive deficit caused by regional depletion of dopamine in prefrontal cortex of rhesus monkey, Science, 205: 929.
Bruneau, N., Simon, H. and Le Moal, M., 1980, Dark-light emergence at two different times of the nycthemeral cycle in the four and twelve week old rat, Behav. Proc. 5: 281.
Burton, H. A. and Toga, A. W., 1982, Successive discrimination performance improves with increasing numbers of stimulus preexposures in septal rats, Behav. Neur. Biol., 34: 141.
Chafetz, M. D., Thompson, R. G., Evans, S. H. and Gage, F. H., Biochemical specificity of septal hyperreactivity: a behavioral discrimination, Behav. Brain Res., 1: 27.
Cicerone, K. D., Lazar, R. M. and Shapiro, W. R., 1983, Effects of frontal lobe lesions on hypothesis sampling during concept formation, Neuropsychologia, 21: 513.
Clody, D. E. and Carlton, P. L., 1969, Behavioral effects of lesions of the medial septum of cats, J. Comp. Physiol. Psychol., 67: 344.
Cools, A. R., 1980, The role of neostriatal dopaminergic activity in sequencing and selecting behavioral strategies: facilitation of processes involved in selecting the best strategy in a stressful situation, Behav. Brain Res., 1: 361.
Crider, A., Solomon, P. and McMahon, M., 1982, Disruption of selective attention in the rat following chronic d-amphetamine administration: relationship to schizophrenic attention disorder, Biol. Psychiat., 17: 351.
Crowne, D. P., 1983, The frontal eye field and attention, Psychol. Bull., 93: 232.
Donovick, P. J., Burright, R. G. and Fink, E. A., 1979, Discrimination behavior of rats with septal lesions in a cue-deletion paradigm, Physiol. Behav., 22: 125.
Durkin, T. P., Hashem-Zadeh, H., Mandel, P., Kempf, J. and Ebel, A., 1983, Genotypic variation in the dopaminergic inhibitory control of striatal and hippocampal cholinergic activity in mice, Pharmac. Biochem. Behav., 19: 63.
Ellen, P., Dorsett, P. G. and Richardson, W. K., 1977, The effect of cue fading on the DRL performance of septal and normal rats, Physiol. Psychol., 5: 469.
Ellen, P., Makohon, L. and Richardson, W. K., 1978, Response suppression on DRL by rats with septal damage, J. Comp. Physiol. Psychol., 92: 511.
Evenden, J. L. and Robbins, T. W., 1984, Win-stay behavior in the rat, Q. J. Exp. Psychol., 36B: 1.
Galey, D., Durkin, T., Sifakis, G. and Jaffard, R., 1984, Amélioration de conduites spatiales spontanées et acquises après lésion des afférences dopaminergiques septales chez la souris: relations possibles avec l'activité cholinergique hippocampique, C. R. Acad. Sci. Fr., in press.
Haig, K. A., Rawlins, J. N. P., Olton, D. S., Mead, A. and Taylor, B., 1983, Food searching strategies of rats: variables affecting the relative strength of stay and shift strategies, J. Exp. Psychol. (Anim. Behav. Proc.) 9: 337.
Le Moal, M., Stinus, L., Simon, H., Tassin, J-P., Thierry, A-M.,

Blanc, G., Glowinski, J. and Cardo, B., 1977, Behavioral effects of a lesion in the ventral mesencephalic tegmentum: evidence for involvement of A 10 dopaminergic neurons, Adv. Biochem. Psychopharmacol., 16: 237.

Lindvall, O. and Björklund, A., 1978, Anatomy of the dopaminergic neuron system in the rat brain, Adv. Biochem. Psychopharmacol., 19: 1.

Mackintosh, N., 1975, A theory of attention: variations in the associability of stimuli with reinforcement, Psychol. Rev., 82: 276.

Markowitsch, H. J. and Pritzel, M., 1976, Learning and the prefrontal cortex of the cat: anatomico-behavioral interrelation, Physiol. Psychol., 4: 247.

Mercer, L. and Remley, N., 1979, Mapping of sensory responsive cells in septal area of the rat, Brain Res. Bull., 4: 483.

Mogenson, J. and Divac, I., 1984, Sequential behavior after modified prefrontal lesions in rat, Physiol. Psychol., 12: 41.

Numan, R. and Lubar, J., 1974, Role of the proreal gyrus and septal area in response modulation in cat, Neuropsychologia, 12: 213.

Numan, R., Ward, C. and Clark, J., 1982, Septal lesions and active avoidance performance in rats: effects of differential intra-box cues, Physiol. Behav., 29: 489.

Oades, R. D., 1979, Search and attention: interactions of the hippocampal-septal axis, adrenocortical and gonadal hormones, Neurosci. Biobehav. Rev., 3: 31.

Oades, R. D., 1981a, Types of memory or attention? Impairments after lesions of the hippocampus and limbic ventral tegmentum, Brain Res. Bull., 7: 221.

Oades, R. D., 1981b, Impairments of search behavior in rats after haloperidol treatment or hippocampal damage suggest a mesocorticolimbic role in cognition, Biol. Psychol., 12: 77.

Oades, R. D., 1981c, Dopaminergic agonistic and antagonistic drugs in the ventral tegmentum of rats inhibit and facilitate changes of food search behavior, Neurosci. Lett., 27:77.

Oades, R. D., 1982, Attention and Schizophrenia: Neurobiological Bases, Pitman Press, London.

Oades, R. D., 1985, The role of noradrenaline in tuning and dopamine in switching between signals in the CNS, Neurosci. Biobehav. Rev., 9: in press.

Oades, R. L. and Isaacson, R. L., 1978, The development of food search behavior by rats: effects of hippocampal damage and haloperidol treatment, Behav. Biol., 24: 327.

Oades, R. D., Taghzouti, K., Simon, H. and Le Moal, M., 1985, Dopamine-sensitive alternation and collateral behaviour in a Y-maze: effects of d-amphetamine and haloperidol, Psychopharmacol., in press.

Petrides, M. and Iversen, S. D., 1976, Cross-modal matching and the primate frontal cortex, Science, 192; 1023.

Phillipson, O. T., 1979, Afferent projections to the ventral tegmental area of Tsai and interfasicular nucleus: a horseradish peroxidase study in the rat, J. Comp. Neurol., 187: 117.

Rea, M. A., Aprison, M. H. and Felten, D. L., 1982, Catecholamines and serotonin in caudal medulla of rat: a combined neurochemical and histofluorescence study, Brain Res. Bull., 9: 227.

Robbins, T. W. and Everitt, B. J., 1982, Functional studies of the central catecholamines, Internat. Rev. Neurobiol., 23: 245.

Rosenkilde, C., 1979, Functional heterogeneity of the prefrontal cortex in the monkey, Behav. Neur. Biol., 25: 301.

Simon, H., 1981, Neurones dopaminergiques A 10 et système frontal J. Physiol.,(Paris), 77: 81.

Solomon, P. R., Crider, A. R., Winkelman, J. W., Turi, A., Kamer, R. M. and Kaplan, L. J., 1981, Disrupted latent inhibition in the rat with chronic amphetamine or haloperidol-induced supersensitivity: relationship to schizophrenic attention disorder, Biol. Psychiat., 16; 519.

Sutherland, R. J., 1982, The dorsal diencephalic conduction system: a review of the anatomy and functions of the habenular complex, Neurosci. Biobehav. Rev., 6: 1.

Taghzouti, K., 1983, L'innervation dopaminergique du noyau accumbens et du septum, étude comportementale, unpubl. Thesis, University of Bordeaux II.

Thomas, G. J., 1979, Comparison of effects of small lesions in postero-dorsal septum on spontaneous and rerun correction (contingently reinforced) alternation in rats, J. Comp. Physiol. Psychol., 93: 685.

Thornton, E. W. and Evans, J. C., 1982, Role of the habenula nuclei in the selection of behavioral strategies, Physiol. Psychol., 10: 361.

Van den Bercken, J. H. L. and Cools, A. R., 1982, Evidence for a role of the caudate nucleus in the sequential organization of behaviour, Behav. Brain Res., 4: 319.

Weingartner, H., Burns, S., Diebel, R. and LeWitt, P., 1984, Cognitive impairments in Parkinson's disease: distinguishing between effort demanding and automatic cognitive processes, Psychiat. Res., 11: 223.

Weiss, K. R., Friedman, R. and McGregor, R., 1974, Effects of septal lesions on latent inhibition and habituation of the orienting response in rats, Acta Neurobiol. Exp., 34: 491.

HORMONAL INFLUENCES ON MEMORY: INTERACTION OF CENTRAL AND PERIPHERAL SYSTEMS

James L. McGaugh, K.C. Liang,*
M. Catherine Bennett and Debra B. Sternberg

Center for the Neurobiology of Learning and Memory
and Department of Psychobiology
University of California, Irvine, California

INTRODUCTION

A number of stress-related hormones, including adrenergic catecholamines as well as peptide hormones, have been shown to affect retention when administered to animals after training (de Wied, 1980; Gold and Zornetzer, 1983; McGaugh, 1983). It is well-known that many of the hormones found to affect retention are released from the adrenal medulla and pituitary when animals are given stimulation of the kind typically used in training. In view of this evidence it has seemed reasonable to propose that endogenous hormones released at the time of a learning experience may play an important role in the normal physiology of memory storage (Gold and McGaugh, 1975). According to this view, the susceptibility of memory to posttraining modulating influences serves an important adaptive function: The period of susceptibility allows hormones released by experiences to influence memories of the experiences.

ADRENERGIC INFLUENCES ON MEMORY STORAGE

Studies in our laboratory have focussed on the influence of adrenergic hormones on memory storage processes. In previous work, Gold and van Buskirk (1976) found that, in low doses, epinephrine enhanced rats' retention of an inhibitory avoidance task. The enhancing effects of epinephrine are also seen in other tasks: Epinephrine enhances retention of an active avoidance task (Liang, Bennett, and McGaugh, in press) as well as a discrimination task using appetitive motivation (Sternberg, Isaacs, Gold, and McGaugh, unpublished findings). Such generality is required, of course, if it is assumed that epinephrine has a general role as in the modulation of memory storage.

The findings of several studies indicate that the effects of a variety of treatments known to affect memory when administered after training may involve influences on adrenergic systems (Gold and Sternberg, 1978; Sternberg and Gold, 1980). For example, adrenergic antagonists block the amnestic effects of several drugs, including pentylenetetrazol, cycloheximide, and diethyldithiocarbamate, as well as the effects of electrical stimulation of the frontal cortex and amygdala. Recently we found that adrenergic antagonists also block the enhancing effects of electrical stimulation of the brain (Sternberg, Gold and McGaugh, 1983).

As Figure 1 shows, posttraining electrical stimulation of the frontal cortex enhanced retention in animals trained on an active avoidance task with low footshock. The enhancing effect of the brain stimulation was blocked in animals given the beta-antagonist, propranolol, prior to training.

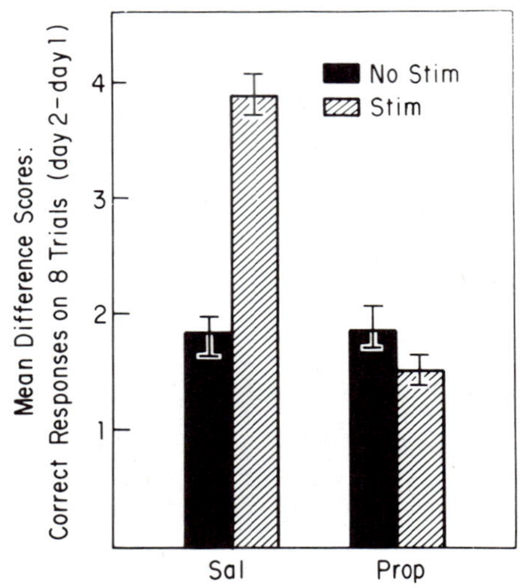

Figure 1. Effects of posttraining frontal cortex stimulation (5 mA/1.0 s) on retention of an active avoidance response. Saline or propranolol (0.5 mg/kg) was administered 30 min prior to training.

ADRENERGIC INVOLVEMENT IN EFFECTS OF AMYGDALA STIMULATION

The hypothesis that endogenous adrenergic hormones are involved in the modulation of memory storage is also supported by evidence that the effects of posttraining electrical stimulation of the amygdala are altered by adrenal demedullation or denervation (Bennett, Liang, and McGaugh, in press). In these studies, rats were implanted bilaterally with amygdala electrodes and then given sham adrenal surgery or were demedullated or denervated. They were then trained on inhibitory and active avoidance tasks. Amygdala stimulation was administered immediately after training and retention was tested 24 hours following training. Retention performance was not significantly affected by either the electrode implantation or the adrenal demedullation. However, retention was significantly impaired in the demedullated implanted controls. Further, demedullation significantly altered the effect of posttraining amygdala stimulation. Amygdala stimulation impaired retention in animals with intact adrenal medullae, in comparison with sham-stimulated controls with intact adrenals. However, amygdala stimulation enhanced retention in demedullated rats, in comparison with controls that were implanted and demedullated.

Additional recent evidence supports the interpretation that these effects of adrenal demedullation are due specifically to depletion of adrenal epinephrine (Liang, et al., in press). First, as the findings summarized in Figure 2 indicate, the retention deficit seen in demedullated rats with implanted amygdala electrodes is attenuated by posttraining administration of epinephrine. Second, posttraining amygdala stimulation

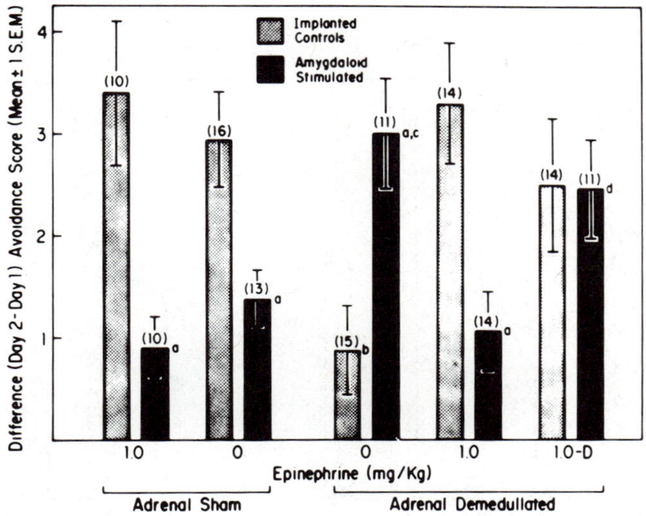

Figure 2. Effect of posttraining epinephrine and amygdala stimulation of retention of inhibitory avoidance in adrenal demedullated and sham control rats.

impairs retention in demedullated rats if epinephrine is administered following the training but immediately prior to the stimulation. However, if the administration of epinephrine is delayed until after the amygdala stimulation, retention is enhanced, as it is in demedullated rats that are not given posttraining epinephrine. These findings clearly suggest that the effect of amygdala stimulation on retention depends upon epinephrine levels present at the time of the stimulation. If a sufficient level of peripheral epinephrine is present at the time of the stimulation (either because epinephrine is released by the training stimulation or because it was administered prior to the stimulation) retention is impaired. If epinephrine is depleted, the amygdala stimulation enhances retention.

In other recent studies we have found that the effects of amygdala stimulation on memory are blocked by lesions of the stria terminalis (Liang and McGaugh, 1983a) as well as by administration of naloxone into the bed nucleus of the stria terminalis prior to the amygdala stimulation (Liang, Messing and McGaugh, 1983). These findings indicate that the effects of amygdala stimulation on memory are due to influences mediated by the stria terminalis. Overall, the findings of our recent studies of the effects of amygdala stimulation add to the growing evidence suggesting that the amygdala may be part of a brain system involved in memory consolidation (Mishkin, 1984; Squire and Zola-Morgan, 1983; Liang, McGaugh, Martinez, Jensen, Vasquez, and Messing, 1982).

Other findings from our laboratory provide additional evidence suggesting that peripheral epinephrine affects memory storage through influences involving the amygdala. For example, as is shown in Figure 3, the effects of posttraining epinephrine on memory are blocked in rats with stria terminalis lesions (Liang and McGaugh, 1983b). Posttraining

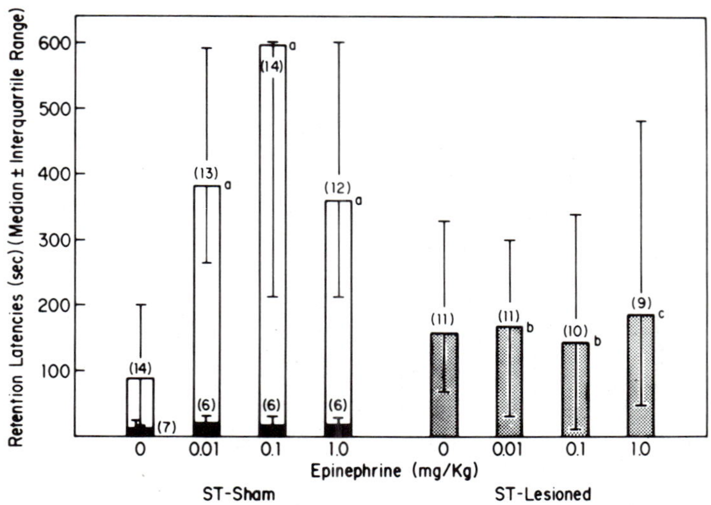

Figure 3. Effect of posttraining epinephrine on inhibitory avoidance retention in sham controls and stria terminalis lesioned rats. Black bars indicate non-shocked controls.

effects on memory are also attenuated in rats given intra-amygdala injections of propranolol immediately prior to the administration of epinephrine (Liang, Juler and McGaugh, 1984). These findings suggest that epi--nephrine affects memory either through influences on the amygdala or through influences elsewhere that require cooperative outputs from the amygdala.

In a series of studies Gallagher and her colleagues (Gallagher, Kapp, Pascoe, and Rapp, 1981) have shown that, in rats, retention can be altered by administration of adrenergic antagonists (as well as opiate agonists and antagonists) directly into the amygdala. Further, intra-amygdala administration of high doses of norepinephrine has been found to impair retention (Ellis and Kesner, 1983). We have obtained additional evidence that retention is altered by adrenergic stimulation of the amygdala. We have found (see Figure 4) that retention is enhanced by posttraining intra-amygdala administration of low doses of norepinephrine and that the enhancement is blocked if propranolol is administered with the epinephrine (Liang and McGaugh, unpublished findings). These findings are consistent with the view that peripheral epinephrine may influence the amygdala through activation of adrenergic receptors in the amygdala.

At present, little is known about the detailed mechanisms by which peripheral adrenergic hormones affect brain systems. While it is generally held that adrenergic hormones do not pass the blood-brain barrier, it may well be that they do under some conditions. It is also possible that peripheral hormones stimulate brain systems by activating visceral afferents. Figure 5 presents a simple model of the possible involvement of peripheral hormones and central systems in the modulation of memory storage. The model assumes that the strength of storage in a memory system is influenced by a memory modulation system. The modulation

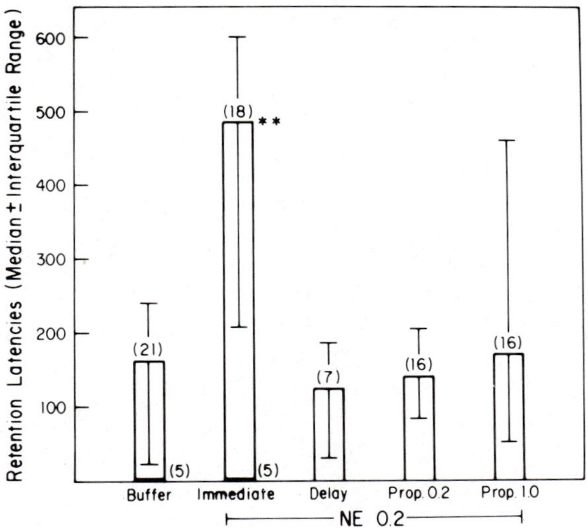

Figure 4. Effects of immediate or delayed (2 hr) post-training intra-amygdala administration of norepinephrine and propranolol (doses in micrograms) on retention of an inhibitory avoidance response. Black bars in the Buffer and Immediate norepinephrine groups indicate performance of non-shocked controls.

system is activated by sensory stimulation and, also, by influences from peripheral hormones. As we indicated above, several lines of evidence suggest that the amygdala may be part of the proposed memory modulation system. For example, if the amygdala is part of a modulating system then disruption of the stria terminalis, a major amygdala pathway, should be expected to interfere with the modulating effects of the amygdala. Further, disruption of the stria terminalis should, according to the model, also block the enhancing effect of peripheral epinephrine on memory. And, if the modulating effects of the amygdala are altered by influences of adrenergic hormones on the amygdala, the effects of the stimulation should vary with peripheral levels of epinephrine. The findings of the studies summarized above generally fit well with this simple model.

As we noted, a number of hormones are released peripherally, as well as within the brain, by stimulation of the kinds used in training. It is increasingly evident that several different types of hormones are effective in modulating memory when administered posttraining. A central question is whether the hormones released at the time of training are normally involved in the endogenous modulation of memory storage. If so, do the various hormones all act through a common modulation system in the

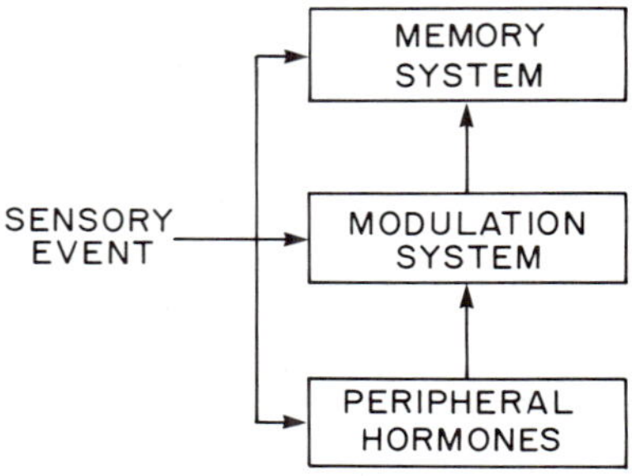

Figure 5. Interaction of peripheral hormones with central modulating and memory systems.

brain or do they act on different systems? It is clear from other recent research that hormones interact in influencing memory. For example, the effects of epinephrine are attenuated by corticosterone (Borrell, de Kloet, and Bohus, 1984) and are potentiated by vasopressin (Borrell, de Kloet, Versteeg, and Bohus, 1983). Understanding the bases of these interactions should provide additional insights into the central mechanisms underlying the modulating influences of hormones on memory storage processes.

REFERENCES

Bennett, C.B., Liang, K.C., and McGaugh, J.L., in press, Depletion of adrenal catecholamines alters the amnestic effect of amygdala stimulation, Behav. Brain Res.

Borrell, J., de Kloet, E. R., and Bohus, B., 1984, Corticosterone decreases the efficacy of adrenaline to affect passive avoidance retention of adrenalectomized rats, Life Sci., 34:99.

Borrell, J., de Kloet, E. R., Versteeg, D. H. G., and Bohus, B., 1983, The role of adrenomedullary catecholamines in the modulation of memory by vasopressin, in: "Integrative Neurohumoral Mechanisms," E. Endroczi, ed., Elsevier, Amsterdam.

DeWied, D., 1980, Pituitary neuropeptides and behavior, in: "Central Regulation of the Endocrine System," K. Fuxe, T. Hokfelt & R. Luft, eds., Plenum Press, New York.

Ellis, M.E. & Kesner, R.P., 1983, The noradrenergic system of the amygdala in aversive information processing, Behav. Neurosci., 97:399.

Gallagher, M., Kapp.,B.S., Pascoe, J.P., and Rapp, P.R., 1981, A neuropharmacology of amygdaloid systems which contribute to learning and memory, in: "The Amygdaloid Complex," Y. Ben-Ari, ed., Elsevier, North Holland.

Gold, P.E. and McGaugh, J.L., 1975, A single-trace, two-process view of memory storage processes, in: "Short-Term Memory," D. Deutsch and J.A. Deutsch, eds., Academic Press, New York.

Gold, P.E. and Sternberg, D.B., 1978, Retrograde amnesia produced by several treatments. Evidence for a common neurobiological mechanism, Science, 209:836.

Gold, P.E. and van Buskirk, R., 1976, Effects of posttrial hormone injections on memory processes, Horm. Behav., 7:509.

Gold, P.E. and Zornetzer, S.F., 1983, The mnemon and its juices: neuromodulation of memory processes, Behav. & Neural Biol., 38:151.

Liang, K.C., Bennett, C., and McGaugh, J.L., in press, Peripheral epinephrine modulates the effects of posttraining amygdala stimulation on memory, Behav. Brain Res.

Liang, K.C., Juler, R., and McGaugh, J.L., 1984, Amygdala noradrenergic system and memory modulation: Involvement in the enhancing effect of peripheral epinephrine, Neurosci. Soc. Absts., 10:254.

Liang, K.C. and McGaugh, J.L., 1983a, Lesions of the stria terminalis attenuate the amnestic effect of amygdaloid stimulation on avoidance responses, Brain Res. 274:309.

Liang, K.C. and McGaugh, J.L., 1983b, Lesions of the stria terminalis attenuate the enhancing effect of post-training epinephrine on retention of an inhibitory avoidance response. Behav.Brain Res., 9:49.

Liang, K.C., McGaugh, J.L.,Martinez, J.L., Jr., Jensen, R.A., Vasquez, B.J. and Messing, R.B., 1982, Posttraining amygdaloid lesions impair retention of an inhibitory avoidance response, Behav. Brain Res., 4:237.

Liang, K.C., Messing, R.B. and McGaugh, J.L., 1983, Naloxone attenuates amnesia caused by amygdaloid stimulation: The involvement of a central opioid system, Brain Res., 271:41.

McGaugh, J.L., 1983, Hormonal influences on memory. Ann. Rev. of Psychol., 34:297.

Mishkin, M., Malamut, B., and Bachevalier, J., 1984, Memories and Habits: Two Neural Systems, in: "Neurobiology of Learning and Memory," G. Lynch, J.L. McGaugh & N.M. Weinberger, eds., Guilford Press, New York.

Sternberg, D.B., and Gold, P.E., 1980, Effects of alpha- and beta-receptor antagonists on retrograde amnesia produced by frontal cortex stimulation, Behav. & Neural Biol., 29:289.

Sternberg, D.B., Gold, P.E. and McGaugh, J.L., 1983, Memory facilitation and impairment with supraseizure electrical brain stimulation: Attenuation with pretrial propranolol injections, Behav.& Neural Biol., 38:261.

Sternberg, D.B., Isaacs, K., Gold, P.E., and McGaugh, J.L., Epinephrine facilitation of appetitive learning: Attenuation with adrenergic receptor antagonists, Submitted.

Squire, L. and Zola-Morgan, S., 1983, The neurology of memory: The case for correspondence between the findings for human and nonhuman primate, in: "The Physiological Basis of Memory," J.A. Deutsch, ed., Academic Press. New York.

---

*Current address: Department of Psychology, National Taiwan University, Taipei, Taiwan.

The research reported in this paper was supported by Research Grants MH12526, AG00538 and Contract N00014-84-K-0391 from the Office of Naval Research (to JLMcG).

EFFECTS OF CENTRAL ADMINISTRATION OF CORTICOSTERONE OR ADRENALECTOMY IN MICE ON MEMORY AND EVOKED ACTIVITY IN THE HIPPOCAMPUS

Edith Pomet, Jacques Micheau*, Edmond Carlier
and Bernard Soumireu-Mourat

Lab. Neurobiologie des Comportements, CNRS U.A. 372,
Université de Provence, Marseilles,
and *Lab. Psychophysiologie, Univ. Bordeaux I, France

INTRODUCTION

There is now a large body of evidence for corticosteroid actions on brain and behavior. Some authors have emphasized that corticosterone can play a role in adaptive behaviors, and particularly in learning and memory, through interactions with hippocampal functions (McEwen, 1979; McEwen and Micco, 1980); the hippocampus could be the major glucocorticoid target site of the central nervous system (CNS). Relations between steroids and the hippocampus have been shown in different ways, mainly in rats. Lesions or stimulations of the hippocampal system modify the pituitary adrenocortical functions (Feldman and Conforti, 1980; Feldman et al., 1982; Dunn and Orr, 1984; and review in Pfaff et al., 1971, and Bohus, 1975). Conversely, modifications of circulating corticosterone levels change morphological aspects of hippocampal neurons (Scheff and Cotman, 1982) and induce neurochemical hippocampal modifications (Etgen et al., 1979; McEwen and Micco, 1980; Nestler et al., 1981). In aging, correlations have been observed between degeneration in the hippocampus and plasma corticosterone levels (Landfield and Lynch, 1977). Moreover, steroids can modulate electrophysiological activity of the hippocampus (review in McEwen, 1979) as measured by spontaneous cell activity (Pfaff et al., 1971) as well as by commissural evoked potentials (Dana et al., 1982). Corticosterone has mainly been tested, but other steroids have also been used (Foy and Teyler, 1983). Of particular importance, it has been shown that glucocorticoid receptors are present in the hippocampus (McEwen, 1979; Sapolsky et al., 1983). This structure has the largest number of glucocorticoid receptors in the brain (McEwen and Micco, 1980; Hall, 1982). It has been shown recently that mineralocorticoid receptors are also present in high concentration in the hippocampus (Coirini et al., 1983). Corticosterone is the main corticosteroid in rats and mice (Butte et al., 1972; Veldhuis and De Kloet, 1982). However, nature and localization of the corticosterone receptors in the hippocampus, as well as their function (De Kloet, 1983), are not yet well understood (Moguilewsky and Raynaud, 1980; Hall, 1982).

In parallel, behavioral studies have been developed in order to show the influence of steroids on learning and memory. Several steroids have been studied with various behavioral tasks (review in Bohus, 1975; McEwen, 1979; and McEwen and Micco, 1980). Most of the time, peripheral administration has been chosen. Behavioral effects have been observed, but due to contradictory results they are difficult to analyze. Most of these experiments concern pre-session administration of the drugs and describe proactive effects of the treatment. Only a few authors have delivered the drug directly to the brain: Van Wimersma Greidanus and De Wied (1969), and Bohus (1973 and 1975) put crystals of corticosterone directly in the hippocampus, and Cottrell and Nakajima (1977) injected hydrocortisone into this structure. Only these latter authors were using the post-trial paradigm for central administration of steroids.

Finally, as far as we know, no intrahippocampal injection of solubilized corticosterone has been performed to study the behavioral effects of the drug. Our experiments attempt to analyze the role of glucocorticoids on hippocampal functions and memory processes, using post-trial injections into the dorsal hippocampus through chronically implanted cannulae. We decided to use a large variety of behavioral learning tasks, with positive or negative reinforcement. Indeed, most of the literature (Van Wimersma Greidanus and De Wied, 1969; Bohus, 1973 and 1975; Cottrell and Nakajima, 1977) concerns avoidance behaviors, which also could induce strong emotional reactions. In such cases, it can be difficult to analyze the behavioral effects of steroids. We began recently to study adrenalectomized animals using positively reinforced tasks. All these experiments have been performed with male mice of the BALB/c strain.

EFFECTS OF CORTICOSTERONE ON LEARNED BEHAVIORS

Centrally administered corticosterone

Two weeks before training, the dorsal hippocampus of each mouse was bilaterally implanted with a 26-gauge guide cannula. In all the experiments, the treatment was administered one minute after the end of the session (either first learning session, retention, or extinction session). The treatment consisted of bilateral injection (0.7 µl in 140 sec on each side, one after the other) of either saline, vehicle (28% ethanol + benzylic alcohol solution), or corticosterone-21-acetate (1.0 µg by injection, i.e. 8 µg/100g of body weight).

At first, the effect of corticosterone was tested on operant conditioning, with a continuous reinforcement schedule (CRF, 6.0 mg

Table 1. Effect of intra-hippocampal injection of corticosterone on CRF conditioning.
Mean lever-presses (± S.E.M.) are presented for each 5-min block of the first and second sessions.

	First session			Second session					
(min)	5	10	15	5	10	15	20	25	30
Saline	4.7	5.1	7.0	15.0	16.7	18.9	18.6	18.0	18.0
(N=11)	0.3	0.8	0.6	1.6	1.6	1.3	1.0	1.1	0.8
Vehicle	5.1	4.9	7.0	12.6	16.6	17.5	18.7	19.8	18.9
(N=11)	0.5	0.7	0.5	1.7	2.1	2.2	1.9	2.0	1.7
Cortico.	5.3	5.3	6.5	14.6	15.7	17.6	17.9	18.1	18.4
(N=19)	0.5	0.5	0.4	1.5	1.5	1.0	1.1	1.0	1.0

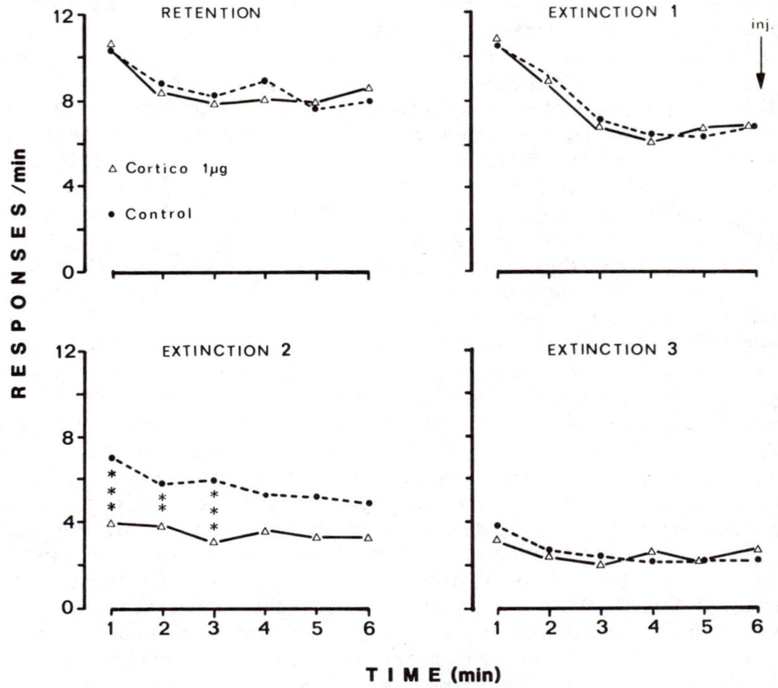

Fig. 1. Effect of intra-hippocampal injection of corticosterone on subsequent extinction of CRF conditioning.
The arrow shows time of injection.
Student's t-test: ** $p < 0.02$; *** $p < 0.01$

pellets). As described elsewhere (Destrade et al., 1973), the paradigm consisted of a 15-minute partial learning session followed, 24 hrs later, by a 30-min retention session. Control subjects were injected with either saline or vehicle. No difference in performance was observed between the two control groups. Corticosterone was administered after the initial learning session. The results are presented in Table 1: the drug had no effect on performance during the second session of the CRF conditioning.

In the second experiment, the effect of the same dose of corticosterone was tested on extinction of the same operant conditioning. After complete learning of the reinforced task, mice were submitted to a 6-min retention session. On each following day they underwent a 6-min extinction session. The injection was done immediately after the first extinction session. The control group (N=12) consisted of non-implanted, saline-injected and vehicle-injected mice, and no difference in performance was observed between them during the experiment. As can be seen on Figure 1, the corticosterone-treated animals (N=10) exhibited a facilitation of extinction at the beginning of the second extinction session ($p < 0.02$). Using extinction of CRF conditioning, the dose-effect of corticosterone was studied: a dose as low as 0.01 µg was still effective on extinction. Under the same conditions, Dexamethasone (1.0 µg in each hippocampus) had no effect.

These two first experiments could indicate that for our conditions and with a dose of 1.0 µg in each hippocampus, corticosterone does not directly facilitate memory consolidation per se, but only improves other processes such as behavioral suppression expressed in extinction. In order to verify this hypothesis, another experiment was run, using an operant discriminative conditioning task in the same Skinner box (Destrade and Jaffard, 1978). Briefly, the positive stimulus (SD) was a light and a buzzer presented simultaneously; the negative stimulus (SΔ) corresponded to the conditions of the CRF experiment. Each session consisted of two blocks of 10 min; each block included periods with a mean duration of 50 sec for a total of 5 min under SD with all lever presses reinforced, and a total of 5 min under SΔ with no reinforcement. The animals were pretrained with a continuous reinforcement until complete learning. One week later, each animal was given a 20-min session of the discriminative task every 24 hrs for 3 days, and was injected intrahippocampally at the end of the first and second sessions. Performances were individually expressed as a discrimination ratio SD/(SD+SΔ), where SD and SΔ are, respectively, the number of responses during the reinforced and the nonreinforced periods of each 10-min block. The scores of two groups injected with vehicle or saline were compared with those of a sham-operated group. As no difference was observed either in the discrimination ratio or in the absolute number of responses during the SD or SΔ periods, all these animals were pooled into a single control group (N=15). As shown on Figure 2, the discrimination ratio was significantly improved in the corticosterone-injected group (N=10) on the first 10-min block of the 2nd ($p < 0.001$) and the 3rd ($p < 0.05$) sessions. This improvement was mainly the consequence of improved suppression of the nonreinforced responses to SΔ ($p < 0.01$). In another experiment, time-dependent effects of post-trial intrahippocampal injection of corticosterone were studied on the same learning task. The drug was still effective when given 3 hrs after the first learning session, but not after 6 hrs.

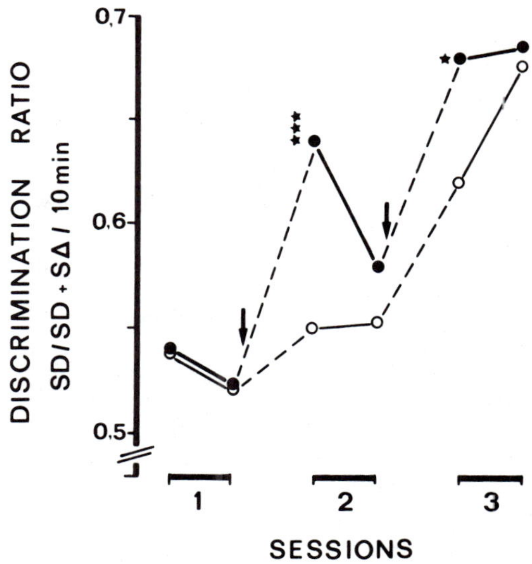

Fig. 2. Effect of intra-hippocampal injection of corticosterone on the operant discriminative conditioning, represented by the discrimination ratio for each 10-min block during 20-min daily sessions. Student's t-test: * $p<0.05$; *** $p<0.001$

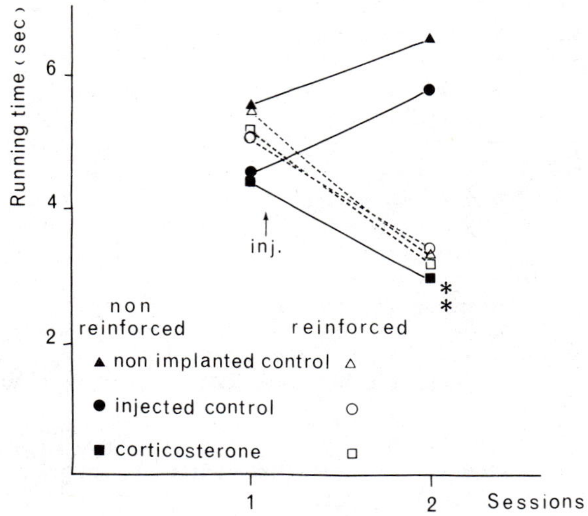

Fig. 3. Effect of intra-hippocampal injection of corticosterone on the Go-NoGo discrimination task.
Student's t-test: ** $p < 0.02$

Thus, we observed that a post-training injection of corticosterone in each dorsal hippocampus induced no effect on retention of CRF conditioning in the Skinner box, but produced a facilitation of extinction under the same conditioning and an improvement of an operant discrimination related to a reduction of non-reinforced responses. Using the intracerebroventricular route, we found similar results with operant conditioning (no effect with CRF schedule, and facilitation with the discriminative program). The present results seem to indicate that corticosterone is behaviorally active directly on the hippocampus; moreover, it could be hypothesized that corticosterone exerts its action by improving behavioral suppression.

However, in another experiment, the subjects were trained to perform a Go-NoGo discrimination task in parallel runways (one dark, one white), using food reinforcement. Each subject underwent 12 randomized trials in a daily session (6 trials on the reinforced side [S+] and 6 on the non-reinforced side [S-]). Half of the animals were reinforced in the dark alley, the other half in the white alley. Performances were quantified using the running time between the start box and the goal box: throughout learning, running time progressively decreased in S+ and increased in S-. Performances of corticosterone-injected mice ($N = 11$) were compared on the one hand with those of non-implanted mice ($N = 13$), and on the other hand with those of mice injected with saline or vehicle ($N = 16$). Corticosterone was administered under the same conditions as in the previous experiments, in the minutes following the first learning session. On the first learning session, S+ and S- running times for all the groups were not significantly different. On the second session, S+ running times decreased but there were no differences between groups. On the contrary, S- running times increased for non-implanted mice and mice injected with saline or vehicle, corresponding to the discriminative learning; for corticosterone-injected mice, S- running times decreased in such a way that they displayed no discrimination between S+ and S- (Figure 3). In other words, they revealed a deficit compared to non-implanted ($p < 0.025$) and saline- or vehicle-injected mice ($p < 0.01$).

Table 2. Effects of corticosterone on one-trial passive avoidance conditioning.

	ACQUISITION Latency (sec)	Inj. ↓ 24 hrs	RETENTION Latency (sec)	RETENTION Total time (sec)
Control (N=13)	7 (4.5-9)		22 (11-340)	141 (39-280)
Cortico. (N=10)	8.5 (7-13)		361** (197-503)	79 (23-209)

Values are medians (and interquartiles in parentheses).
Total time is the duration of time spent in the dark compartment within a maximum of 600 sec. Mann-Whitney's U-test: $**p<0.02$

However, treated subjects were able to learn the discrimination in spite of this initial deficit.

Our initial hypothesis was that corticosterone could induce modifications of hippocampal functioning as far as behavioral suppression was involved. However, considering the effects of corticosterone on a Go-NoGo discriminative task, this hypothesis did not fit. The effects of the treatment seemed to be more subtle, and may be related to procedural variables, maybe to the strength of the reinforcement and the nature of conflict due to non-reward. Two sets of experiments were performed in order to investigate this possibility.

At first we examined the effects of corticosterone on two different passive avoidance tasks. One-trial passive avoidance was investigated by placing the animals in a box divided into two compartments, a dark and a light one. A weak footshock (0.3 mA peak to peak) was delivered when the animal went from the light to the dark compartment. On the next day, the step-through latency was measured with a 600 sec ceiling. Corticosterone was injected after the learning trial on the first day. The control group was, as previously described, comprised of non-implanted, saline- and vehicle-injected mice. As shown in Table 2, corticosterone increased the step-through latency on the second day, and consequently decreased the time spent inside the dark compartment after multiple entries within the ceiling. Thus corticosterone facilitated retention of the one-trial passive avoidance conditioning when using a weak shock.

Table 3. Effects of corticosterone on retention of passive avoidance conditioning in the Skinner box.
Performances (in sec) are expressed by means ($\pm$ S.E.M.).
Student's t-test: $* \ p<0.05$; $** \ p<0.01$

	ACQUISITION 1st press latency	Inj. ↓ 24 hrs	RETENTION Entrance latency	RETENTION 1st press latency
Control (N=15)	12.5 2.8		62.6 24.9	36.7 8.5
Cortico. (N=10)	7.0 2.3		7.6* 2.8	8.6** 2.5

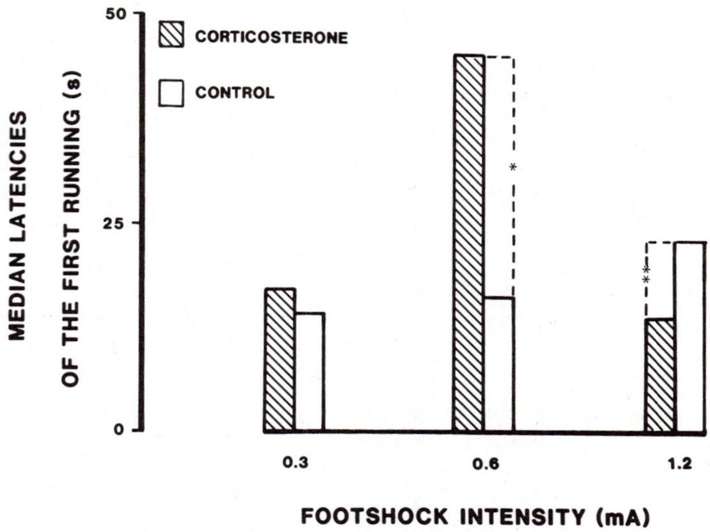

Fig. 4. Reactivity to a noxious stimulus measured by the time before crossing the whole box (median latency) as a function of the intensity of the shock (10 animals for each group). Mann-Whitney's U-test: * $p<0.05$; ** $p<0.02$

Passive avoidance behavior was also investigated in food-deprived mice in the Skinner box. The animals were first trained to learn the CRF schedule to criterion. On the next day, they received a strong electric shock (5 mA) through the lever at the end of a 10-min retention session of the appetitive task; 24 hrs later, the latency to enter the box from the waiting compartment and the latency to press the lever for the first time were measured. As shown in Table 3, the corticosterone-injected mice (N = 10) showed disrupted passive avoidance as both their retention latencies were shorter than those of controls (N = 15: 5 non-implanted, 5 vehicle- and 5 saline-injected mice).

Briefly, the same dose of corticosterone, centrally injected, induced contradictory findings which may be related to the strength of the shock used. In order to explain these data, we studied the reactions of mice to a noxious stimulus (electric shock of variable intensity), using Henderson's procedure (1968). The animal was placed in a 24x30 cm box with a metal grid floor. Activity was measured over a 30 second period by the number of interruptions of 2 photo-cell beams which divided the box into three equal parts. Then the subject received an electric shock during a 2 sec period at one of three intensities : 0.3 mA, 0.6 mA or 1.2 mA. The animal was then injected with corticosterone as in the learning experiments. On the next day, it was placed again into the box for 5 min, and the initial time to cross the whole box was measured. Results are represented on Figure 4. After the weak shock (0.3 mA), there was no difference between treated animals and controls. After the intermediate shock (0.6 mA), treated animals displayed a longer latency to cross the box than controls ($p < 0.05$). On the contrary, after a strong shock (1.2 mA), the effect was reversed : the latency was shorter for corticosterone-injected mice than for controls ($p < 0.02$). The response after corticosterone injection with respect to the shock intensity produced an inverted U-shaped curve, while for the control group this latency was proportional to the footshock intensity.

Table 4. Performances in acquisition and retention of continuously reinforced operant conditioning after adrenalectomy. Mean lever-presses (± S.E.M.) for each 5-min block.

	First session			Second session					
(min)	5	10	15	5	10	15	20	25	30
Control	5.2	4.1	5.4	6.6	11.8	21.0	24.1	24.6	22.6
(N=14)	0.6	0.4	0.8	1.4	1.6	2.1	2.0	1.8	1.4
ADX	5.1	5.1	5.8	6.7	6.6	11.9	17.7	24.2	22.3
(N=9)	0.5	0.5	0.7	2.1	2.1	2.4	2.4	1.6	1.6

Thus, the effect of the treatment may depend on the strength of the reinforcement. This could explain the disturbance for the passive avoidance in the Skinner box (using a strong shock) and the improvement induced in the one-trial passive avoidance (using a lower shock).

Peripherally administered corticosterone

For some of the preceding learning tasks, we studied the effects of peripheral administration of corticosterone. We chose a dose of 1 mg in 0.5 ml propylene glycol injected i.p. Using the same three situations (retention of the CRF operant conditioning, extinction of the same task, and Go-NoGo discrimination), peripheral treatment with this dose of corticosterone had no effect.

ADRENALECTOMY, LEARNING AND RETENTION

After bilateral adrenalectomy (ADX), animals were kept under normal diet and saline water. The first experiment consisted of the operant conditioning with the same protocol as in previous experiments (15 min partial learning session, then, 24 hrs later, a second learning session). The first learning session took place around 20 days after surgery. As shown in Table 4, no differences were observed during the initial partial learning session, but adrenalectomized mice showed a deficit during the second session ($p < 0.05$ for the second to fourth 5-min block of the session), but they were able to reach the same level of performance as the intact animals at the end of the retention session.

In the Go-NoGo discrimination three learning sessions were run, one day apart, starting 14 days after surgery. As shown in Table 5,

Table 5. Performances in the Go-NoGo discrimination task after adrenalectomy. Mean latencies (± S.E.M.) for each daily session. Student's t-test: ** $p<0.01$
S+ : reinforced running times; S- : non-reinforced running times.

	1st session		2nd session		3rd session	
	S+	S-	S+	S-	S+	S-
Control	7.10	7.23	4.94	7.40	3.04	8.07
(N=22)	0.64	0.95	0.61	0.96	0.21	0.70
ADX	9.04	8.63	6.50	12.31**	3.72	9.67
(N=14)	1.24	1.36	0.83	1.83	0.42	1.34

adrenalectomized mice were slower than control subjects (intact + sham mice). The difference was not significant for the S+ (reinforced) latencies, but was highly significant for the S- (non-reinforced) latencies ($p < 0.01$ for the whole learning curve). Thus, adrenalectomized mice seemed to show better discrimination learning compared to controls.

These preliminary results in adrenalectomized animals do not reveal whether the effects in retention of appetitive tasks are related either to memory processes (through hippocampal mechanisms) or to a reduced activity as shown primarily by the data from discrimination task.

EFFECTS OF CORTICOSTERONE OR ADRENALECTOMY ON HIPPOCAMPAL EVOKED ACTIVITY

An electrophysiological study analyzed the effects of central administration of corticosteroids on hippocampal evoked activity. Evoked responses were recorded in dorsal CA_1 of mice anesthetized with sodium penthiobarbital after homotopic commissural stimulation. The amplitude of the population spike was measured as the mean of 20 evoked responses. Intracerebroventricular corticosterone administration (1 µg) induced a decreased amplitude of the population spike during at least 90 min (Figure 5). After Dexamethasone injected under the same conditions, half of the animals did not reveal any modifications, and the other half showed less decrease in amplitude over a shorter time period (≤ 10 min). No change was observed after administration of vehicle or saline. Seven days after adrenalectomy, the effect of corticosterone under the same conditions was reversed in that the amplitude of the population spike was increased by 100% to 200% compared to adrenalectomized non-injected mice, during several hours. There was no clear effect of adrenalectomy itself on evoked activity. Dexamethasone induced only a slight and brief increase in adrenalectomized mice, and vehicle had no effect.

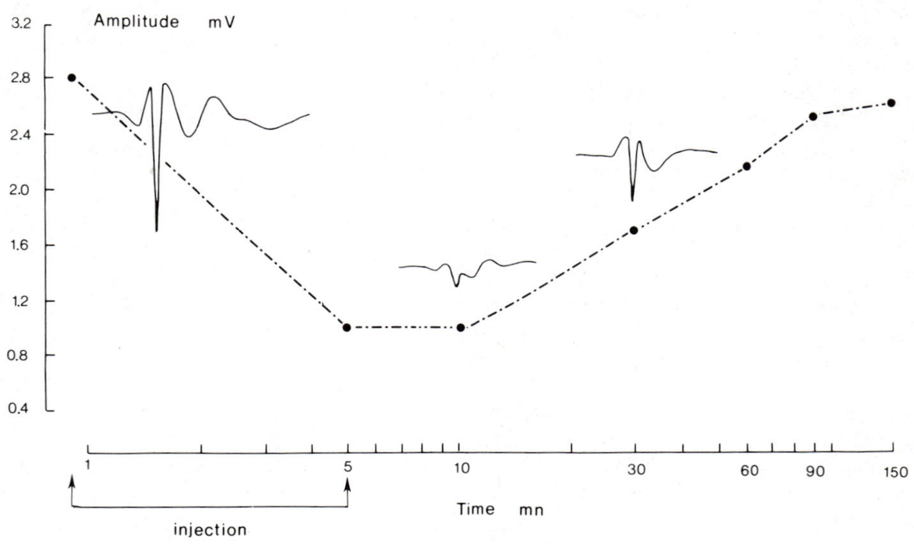

Fig. 5. Effects of corticosterone on hippocampal commissural evoked potentials: amplitude of the population spike as a function of time after intra-hippocampal injection for a 180 µA intensity stimulation.

DISCUSSION

Our results show that post-trial intrahippocampal administration of corticosterone can modify retention and extinction of learned behaviors. Some of these data reproduce the effects obtained by intracerebroventricular injection and thus indicate that the main structure responsible for the effects of corticosterone on learned behaviors could be the hippocampus. However, this hippocampal involvement of corticosterone needs further analysis. Especially, it needs to be demonstrated that an injection under similar conditions in other structures has no effect, to establish that our results are not due to a spread of the drug to surrounding tissues. However, we feel this is unprobable given the quantities of corticosterone injected.

In terms of behavior, we found faster extinction with both positively and negatively reinforced tasks, as frequently quoted in the literature on peripheral administration of corticosterone (Van Wimersma Greidanus, 1970; Garrud et al., 1974; Bohus, 1975). The fact that we found no effect in our experiments using peripheral administration could be related to the dose we used. However, Micco et al. (1979) also found no facilitation of peripheral corticosterone on appetitive extinction in the rat.

The analysis of our behavioral effects is unclear. According to most authors, corticosteroids enhance internal inhibitory processes leading to the suppression of fear (e.g. Bohus, 1973). This could be related to changes in the motivational value of the stimuli leading to elimination of the non-reinforced response (Bohus, 1973). Our experiments using the appetitive operant tasks are consistent with this hypothesis. However, Bohus (1973) suggested that corticosteroids can also suppress the inhibitory value of a previously non-reinforced signal. Consistent with this latter hypothesis, Micco et al. (1979) found facilitation of extinction after adrenalectomy and a compensatory effect of corticosterone, which suggested a disinhibitory behavioral effect of the steroid by inhibiting the hippocampus (itself involved in inhibition of responses). In other words, they proposed that corticosterone produced behavioral persistence in the face of frustrative non-reward. In fact, most of our results do not fit with the hypothesis of Micco et al., but rather with Bohus' motivational hypothesis. Considering all our data, relations between steroids and the hippocampus appear to be both subtle and complex. There does not seem to be a direct effect on memory processes, or even an unitary effect on behavioral inhibition as hypothetized above. From the behavioral point of view, the corticosterone effects are probably related to the strength of the reinforcement and to the level of learning, as suggested by our results using aversive stimuli. Thus the pretraining conditions could be critical; it must be underlined that the conditions in the two discrimination tasks we used were very different.

After adrenalectomy, we found a tendency for improved retention in the discrimination task. However, at this point of our research, we are unable to state that adrenalectomy has strictly opposite effects to the central administration of corticosterone. In the CRF task, we observed an impairment of performance as have several authors who used other tasks (Borrell et al., 1984). Behaviorally, administration of corticosterone compensates the deficit induced by adrenalectomy (Veldhuis and De Kloet, 1983; Nyakas et al., 1983), and performances in appetitive tasks vary as a function of plasma corticosterone (O'Reilly et al., 1981).

The physiological bases of the behavioral effects of corticosterone administration or adrenalectomy are unclear. Most authors hypothesize that these treatments act on steroid receptors of the limbic system, mainly in the hippocampus (De Kloet, 1983). However, the exact nature of these receptors is still controversial. There are several corticoid binding entities in the brain (Moguilewsky and Raynaud, 1980; Bohus et al., 1982). Besides the glucocorticoid receptor, the mineralocorticoid receptor is present in the hippocampus (Beaumont and Fanestil, 1983), and it is difficult to distinguish between the specific glucocorticoid or mineralocorticoid effects of corticosterone (Veldhuis et al., 1982). Moreover, dexamethasone, a synthetic glucocorticoid, binds more to pituitary receptors than corticosterone; in contrast, the highest concentration of corticosterone receptors is found in the hippocampus. This raises the question of the exact nature of the corticosterone receptor in the hippocampus and its specificity could explain the lack of effects we observed after administration of dexamethasone in this structure.

The behavioral effects of corticosterone are probably related to the saturation level of these receptors in the CNS (Micco et al., 1979). In intact animals, the lack of effect in CRF conditioning could be due to the saturation of these receptors by endogenous steroids. This does not seem to be true, however, as in discrimination tasks we observed clear effects. A lower level of occupation of these receptors after adrenalectomy could explain the reversal of corticosterone effects on the amplitude of the evoked potentials that we observed in ADX mice compared to intact mice. These preliminary electrophysiological results partially corroborate our behavioral data. Reduction of electrophysiological activity by corticosterone in intact animals was also observed by Pfaff et al. (1971), and the expression of hippocampal electrical activity is suggested to be under the control of corticosterone (Cottrell et al., 1984). However, long-term potentiation in CA_1 field of the hippocampus and in dentate gyrus does not seem related in the same way to changes in adrenal hormones (Harris and Teyler, 1983; Nowicky et al., 1983). Perhaps, several mechanisms are possible for the cellular actions of corticosterone on the hippocampal neurons (Hall, 1982; Keller et al., 1982; Bohus et al., 1982).

In conclusion, an action of corticosterone on the hippocampus in adaptive behaviors is probable, but its functional bases are unclear and its cellular mechanisms are still unknown (Bohus et al., 1982). As for behavioral analysis, this action seems to be a function of the steroid dose, of the level of free receptors, and of the task parameters (conditions of the pre-training, strength of the reinforcement and level of learning). The time-dependent effects we observed seem to indicate that corticosterone is acting during the so-called consolidation period but corticosterone is probably acting only indirectly on memory processes. All together, our data favoured a role of corticosterone in modulation of memory processes through hippocampal mechanisms during the hours following learning of new information, as suggested by Gold and McGaugh (1975) for different hormones. In fact, in a more general vein, corticosterone, and probably other steroids, may have a modulatory role in adaptive behaviors by influencing hippocampal mechanisms.

REFERENCES

Beaumont, K., and Fanestil, D.D., 1983, Characterization of rat aldosterone receptors reveals high affinity for corticosterone, Endocrinology, 113:2043.

Bohus, B., 1973, Pituitary-adrenal influences on avoidance and approach behavior of the rat, Progr. Brain Res., 9:407.
Bohus, B., 1975, Pituitary-adrenal system hormone in: "The hippocampus", R.L. Isaacson and K.H. Pribram, eds, Plenum Press, New York.
Bohus, B., De Kloet, E.R., and Veldhuis, H.D., 1982, Adrenal steroids and behavioral adaptation: relationship to brain corticoid receptors, in: "Adrenal actions on brain", Current topics in Neuroendocrinology, vol.2, D. Ganten and D. Pfaff, eds., Springer Verlag, Berlin.
Borrell, J., De Kloet, E.R., and Bohus, B., 1984, Corticosterone decreases the efficacy of adrenaline to affect passive avoidance retention of adrenalectomized rats, Life Sci., 34:99.
Butte, J.C., Kakihana, R., and Noble, E.P., 1972, Rat and mouse brain corticosterone, Endocrinology, 90:1091.
Coirini, H., Marusic, E.T., De Nicola, A.F., Rainbow, T.C., and Mc Ewen, B.S., 1983, Identification of mineralocorticoid binding sites in rat brain by competition studies and density gradient centrifugation, Neuroendocrinology, 37:354.
Cottrell, G.A., and Nakajima, S., 1977, Effect of corticosteroids in the hippocampus on passive avoidance behavior in the rat, Pharmac. Biochem. Behav., 7:277.
Cottrell, G.A., Nyakas, C., De Kloet, E.R., and Bohus, B., 1984, Hippocampal kindling: corticosterone modulation of induced seizures, Brain Res., 309:377.
Dana, R.C., Gerren, R.A., Sternberg, D.B., Martinez, J.L. Jr, Hall, J., Stansbury, N.A., and Weinberger, N.M., 1982, Long-term potentiation is impaired by adrenalectomy and restored by corticosterone, Soc. Neurosci. Abs., 8:316.
De Kloet, E.R., 1983, Function of steroid receptor systems in the central nervous system, J. Ster. Biochem., suppl. 19:162S.
Destrade, C., Soumireu-Mourat, B., and Cardo, B., 1973, Effects of post-trial hippocampal stimulation on acquisition of operant behavior in the mouse, Behav. Biol., 8:713.
Destrade, C., and Jaffard, R., 1978, Post-trial hippocampal and lateral hypothalamic electrical stimulation. Facilitation on long-term memory of appetitive and avoidance learning tasks, Behav. Biol., 22:354.
Dunn, J.D., and Orr, S.E., 1984, Differential plasma corticosterone responses to hippocampal stimulation, Exp. Brain Res., 54:1.
Etgen, A.M., Lee, K.S., and Lynch, G., 1979, Glucocorticoid modulation of specific protein metabolism in hippocampal slices maintained in vitro, Brain Res., 165:37.
Feldman, S., and Conforti, N., 1980, Participation of the dorsal hippocampus in the glucocorticoid feedback effect on adrenocortical activity, Neuroendocrinology, 30:52.
Feldman, S., Conforti, N., and Siegel, R.A., 1982, Adrenocortical responses following limbic stimulation in rats with hypothalamic deafferentations, Neuroendocrinology, 35:205.
Foy, M.R., and Teyler, T.J., 1983, 17-α-estradiol and 17-β-Estradiol in hippocampus, Brain Res. Bull., 10:735.
Garrud, P., Gray, J.A., and De Wied, D., 1974, Pituitary-adrenal hormones and extinction of rewarded behaviour in the rat, Physiol. Behav., 12:109.
Gold, P.E., and McGaugh, J.L., 1975, A single trace, two-process view of memory storage processes, in: "Short-term memory", D. Deutsch and J.A. Deutsch, eds., Academic Press, New York.
Hall, E.D., 1982, Glucocrticoid effects on central nervous excitability and synaptic transmission, Int. Rev. Neurobiol., 23:165.
Harris, K.M., and Teyler, T.J., 1983, Age differences in a circadian influence on hippocampal LTP, Brain Res., 261:69.

Henderson, N.D., 1968, Genetic analysis of acquisition and retention of a conditioned fear in mice, J. Comp. Physiol. Psychol., 65:325.

Landfield, P.W., and Lynch, G.S., 1977, Brain aging and plasma steroid levels: quantitative correlations, Soc. Neurosci. Abs., 3:111.

Keller, B.T., Landes, G.M., and Kitos, P.A., 1982, Evidence for more than one mechanism of action of the glucocorticoid hormones, Biochim. Biophys. Acta, 717:228.

McEwen, B.S., 1979, Influences of adrenocortical hormones on pituitary and brain function, in: "Glucocorticoid hormone action", Monographs on endocrinology, vol.12, J.D. Baxter and G.G. Rousseau, eds., Springer-Verlag, Berlin.

McEwen, B.S., and Micco, D.J., 1980, Toward an understanding of the multiplicity of glucocorticoid actions on brain function and behavior, in: "Hormones and the brain", D. De Wied and P.A. Van Keep, eds., MTP Press Ltd, Lancaster.

Micco, D.J., McEwen, B.S., and Shein, W., 1979, Modulation of behavioral inhibition in appetitive extinction following manipulation of adrenal steroids in rats: implications for involvement of the hippocampus, J. Comp. Physiol. Psychol., 93:323.

Moguilewsky, M., and Raynaud, J.P., 1980, Evidence for a specific mineralocorticoid receptor in rat pituitary and brain, J. Steroid Biochem., 12:309.

Nestler, E.J., Rainbow, T.C., McEwen, B.S., and Greengard, P., 1981, Corticosterone increases the amount of protein 1, a neuron-specific phosphoprotein, in rat hippocampus, Science, 212:1162.

Nowicky, A.V., Vardaris, R.M., and Teyler, T.J., 1983, Corticosterone and long-term potentiation in the hippocampus, Soc. Neurosci. Abs., 9:1211

Nyakas, CS., De Kloet, E.R., Veldhuis, H.D., and Bohus, B., 1983, Hippocampal corticosterone receptors and novelty-induced behavioral activity: effect of kainic acid lesion in the hippocampus, Brain Res., 288:219.

O'Reilly, H.M., Coleman, G.J., and Armstrong, S., 1981, Endogenous corticosterone alterations and appetitive learning in the rat, Physiol. Behav., 27:515.

Pfaff, D.W., Silva, M.T.A., and Weiss, J.M., 1971, Telemetered recording of hormone effects on hippocampal neurons, Science, 172:394.

Sapolsky, R.M., McEwen, B.S., and Rainbow, T.C., 1983, Quantitative autoradiography of [^{3}H]corticosterone receptors in rat brain, Brain Res., 271:331.

Scheff, S.W., and Cotman, C.W., 1982, Chronic glucocorticoid therapy alters axon sprouting in the hippocampal dentate gyrus, Exp. Neurol., 76:644.

Veldhuis, H.D., and De Kloet, E.R., 1982, Significance of $ACTH_{4-10}$ in the control of hippocampal corticosterone receptor capacity of hypophysectomized rats, Neuroendocrinology, 34:374.

Veldhuis, H.D., and De Kloet, E.R., 1983, Antagonistic effects of aldosterone on corticosterone-mediated changes in exploratory behavior of adrenalectomized rats, Horm. Behav., 17:225.

Veldhuis, H.D., Van Koppen, C., Van Ittersum, M., and De Kloet, E.R., 1982, Specificity of the adrenal receptor system in rat hippocampus, Endocrinology, 110:2044.

Van Wimersma Greidanus, Tj. B., and De Wied, D., 1969, Effects of intracerebral implantation of corticosteroids on extinction of an avoidance response in rats, Physiol. Behav., 4:365.

Van Wimersma Greidanus, Tj. B., 1970, Effects of steroids on extinction of an avoidance response in rats. A structure-activity relationship study, Progr. Brain Res., 32:185.

VASOPRESSIN, HIPPOCAMPAL EXCITABILITY AND PAIRED-PULSE POTENTIATION

J.E. Smithson and M.W. Brown

Department of Anatomy
University of Bristol, The Medical School
Bristol BS8 1TD, England

INTRODUCTION

The essentially simple organization of the hippocampal formation renders it well suited to electrophysiological and pharmacological studies. The basic structure comprises a trisynaptic circuit repeated in lamellar fashion along its longitudinal axis. This cellular architecture allows the extracellular recording of large evoked potentials which closely reflect events at the level of the single cell (Andersen et al., 1971). Furthermore, thin transverse sections can be prepared from the hippocampus and maintained *in vitro* whilst remaining functionally intact and thus providing a convenient model for the study of the actions of pharmacological agents (Langmoen and Andersen, 1981; Schwartzkroin, 1981).

Considerable attention has been focussed on the various forms of synaptic plasticity supported by the elements of the hippocampal trisynaptic circuit, in view of the longstanding association of this region with mnemonic processes (Scoville and Milner, 1957). The majority of studies have concentrated upon long-term potentiation, an enduring enhancement of synaptic efficacy induced by brief high frequency stimulus trains. Altogether less investigation has been directed towards the varieties of short-term potentiation seen in the hippocampal formation, phenomena similar to those described elsewhere in the central nervous system and at the neuromuscular junction (Creager et al., 1980; McNaughton, 1983).

The present experiments have investigated one form of short-term potentiation, paired-pulse potentiation (PPP), in which there is an increased response to the second of a pair of brief electrical pulses separated by a short interval. PPP decays in exponential fashion with a time constant of about 100ms (Creager et al., 1980). It has been suggested that the process is presynaptic, being due to an increased, calcium-dependent, release of transmitter quanta (Creager et al., 1980; Kandel, 1981). After receiving detailed investigation by Lømo (1971) at the perforant path - dentate granule cell synapse, PPP has since been demonstrated *in vivo* or *in vitro* at each stage of the trisynaptic circuit (Andersen, 1975; Alger and Teyler, 1976). The possible significance of PPP for hippocampal function remains to be established, though amplification mechanisms such as those responsible for PPP may well prove essential for mnemonic acquisition and/or retrieval processes.

The work of de Wied (1980) and others has suggested that the nonapeptide vasopressin has behavioural effects in addition to its classical peripheral endocrine actions. The thesis of de Wied and collaborators is that the peptide has a direct, specific, long-term effect in facilitating the acquisition and retention of avoidance behaviour in rats and mice. However, in a continuing controversy, Gash and Thomas (1984), amongst others, have argued against a direct central role for vasopressin and in favour of a visceral action giving rise indirectly to the reported behavioural effects, perhaps by influencing arousal and/or other performance factors.

Nevertheless, there is evidence that vasopressin may be released from terminals within the brain (Buijs and Van Heerikhuize, 1982). The peptide has been demonstrated immunocytochemically to have a widespread distribution throughout the brain stem and limbic system (Buijs, 1978). In the hippocampus, vasopressin-positive fibres have been observed in close proximity to pyramidal cell apical dendrites (Buijs, 1978). Vasopressin has been shown to affect the spontaneous activity of hippocampal neurones both in vivo (Joëls and Urban, 1982) and in vitro (Mühlethaler et al., 1982, 1984). Effects have been shown to be dose-dependent, reversible and blocked by specific antagonists, though it should be noted that oxytocin may be even more potent than vasopressin (Mühlethaler et al. 1982, 1984). The presence of central terminals and receptors for hormones would not necessitate a rôle which is specific to memory processes for these substances, though such receptors might provide convenient anatomical substrates by which emotional and endocrinological inputs could influence mnemonic mechanisms. Thus such hormonal influences on learning and memory processes could be direct and central without being exclusively mnemonic.

The short time course of PPP and its potentially large magnitude reflect the powerful flexibility of hippocampal synapses, flexibility allowing them to change their strength markedly in response to rapidly changing input patterns. Such flexibility could play an important rôle in mnemonic processes. Present evidence suggests that vasopressin affects processes related to memory and that it might function as a transmitter or neuromodulator in the hippocampus. The present study was therefore undertaken to investigate the effect of arginine vasopressin (AVP) on PPP, as a means of further exploring neurohormonal influences on potential memory mechanisms.

MATERIALS AND METHODS

Transverse slices (400μm thick) were cut from the middle third of the hippocampal formation of male Wistar rats (average weight 250g) and maintained at $35\pm1^{\circ}C$ in an incubation chamber perfused (1ml/min) with artificial c.s.f. (Wheal, 1981). Tungsten microelectrodes placed in the stratum pyramidale and stratum radiatum of area CA1 simultaneously recorded population spikes and field EPSPs evoked by brief (100μs) electrical pulses delivered to a stimulating electrode placed across the Schaffer collaterals (Fig. 1). Pairs of pulses were delivered every 5s. Evoked potentials were amplified, displayed and stored on magnetic tape and computer disc for subsequent averaging and analysis. Series of evoked potentials were collected before, during and after the addition of arginine vasopressin (AVP) at a concentration of 2×10^{-6}M to the artificial c.s.f. Records were not taken during the first 20min of AVP perfusion, nor during the first 20min after the end of perfusion.

Evoked potential measurements were compared using two-tailed, paired t-tests on data logarithmically transformed to improve the normality of distribution.

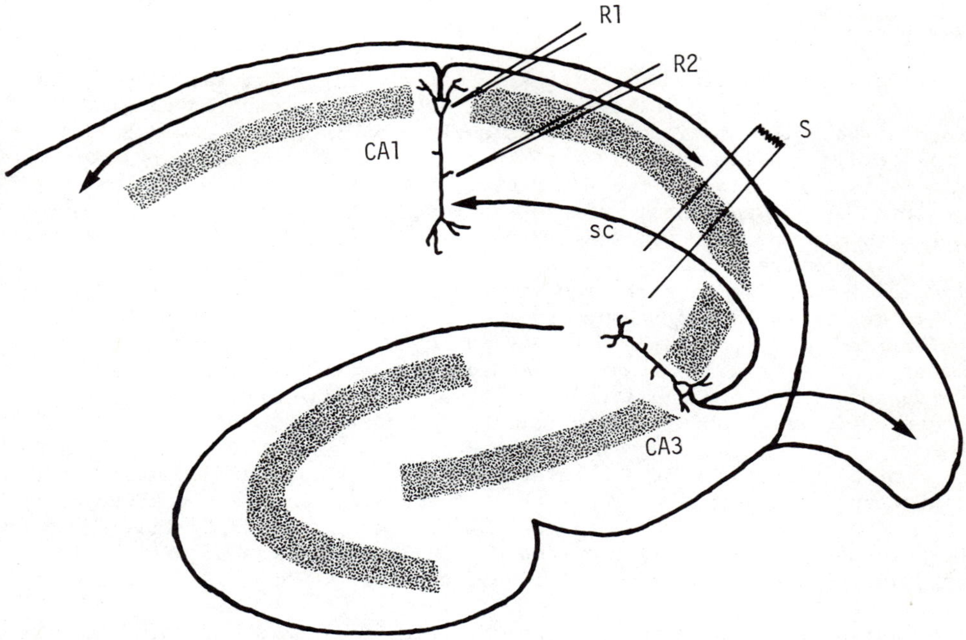

Fig.1. Diagram of hippocampal slice indicating position of stimulating electrode (S) and recording electrodes in stratum pyramidale (R1) and stratum radiatum (R2). CA1 and CA3, hippocampal subfields; sc, Schaffer collaterals.

Fig.2. (a) Evoked potential recorded in stratum radiatum to indicate measurement of slope (dotted line) of field EPSP; (b) Evoked potential recorded in stratum pyramidale indicating measurement of amplitude (h) of population spike. Scale bars for a) and b) 200μV and 2ms.

RESULTS

Data were obtained from one slice from each of 7 rats. Experiments were only performed on slices for which the maximum unpotentiated CA1 population spike amplitude in response to stimulation of the Schaffer collaterals was >1mV. The threshold current for production of a population spike was 150±30μA (mean ± s.e.m.). PPP of the field EPSP was measured as the percentage change in slope of the EPSP evoked by the second of the pulses compared to that of the first (Fig. 2). PPP of the population spike was measured as the percentage increase in amplitude of the chief negative wave recorded in the stratum pyramidale (Fig. 2). There was no consistent change in response for control values obtained before and after perfusion with AVP ($2x10^{-6}$M). Thus there was no evidence that any effect of AVP persisted 20min after the end of its perfusion. Response amplitudes during AVP perfusion have therefore been compared to the mean of values for the two control periods.

At a current giving a population spike of half maximum amplitude and at an interpulse interval of 100ms, PPP of the field EPSP was unaffected by AVP (PPP mean ± s.e.m. : AVP = 26.7 ± 8.8%; Control = 25.1 ± 4.0%). However, PPP of the simultaneously recorded population spike was significantly ($p < 0.01$) reduced (AVP = 66.5 ± 12.3%, control = 125.3 ± 27.3%), see Fig. 3.

Further analysis indicated that for the field EPSP neither the response to the first pulse nor that to the second was significantly changed compared to control values by the presence of AVP (Fig. 4). The magnitude of the population spike in response to the first pulse was significantly ($p < 0.02$) increased during AVP perfusion (mean response as a percentage of unpotentiated maximum population spike : AVP = 67.8 ± 6.6%; Control = 56.2 ± 7.7%). However, the magnitude of the population spike in response to the second pulse was not significantly altered.

Thus at an inter-pulse interval of 100ms, AVP perfusion was associated with an increase in post-synaptic excitability as measured by an increase in size of the population spike without change in the field EPSP. This increased excitability was found for pulses separated by ∿5s (the time between successive pairs of pulses), but not for the second of two pulses occurring 100ms after the first. Similar results were also found for interpulse intervals from 10-400ms.

DISCUSSION

The results show that AVP reduces PPP of the population spike by increasing the amplitude of the response to the first of a pair of stimulus pulses while leaving the second population spike unaltered. This action of AVP provides a noteworthy example of how potentiation may be modulated by an external influence. There was no effect of AVP upon either the size or PPP of the Schaffer collateral input to CA1 pyramidal cells as measured by the field EPSP. Thus the peptide does not affect the putative presynaptic mechanism of PPP (Creager et al., 1980) and could not have produced its effect through variation in the Schaffer collateral presynaptic volley. AVP's effect was therefore postsynaptic, being upon the CA1 pyramidal cell output as measured by the population spike. Furthermore, the second (and potentiated) of the responses to the paired stimulus pulses remained unaltered by AVP; it was the increase in the first (and unpotentiated) population spike that resulted in the decrease in PPP. Since PPP is defined by the ratio of responses to a pair of equal stimuli, the resulting change represents a true decrease in potentiation - however, if the mechanism of PPP is held to be presynaptic, then the effect on PPP is indirect. It should be noted that since the second population spike was

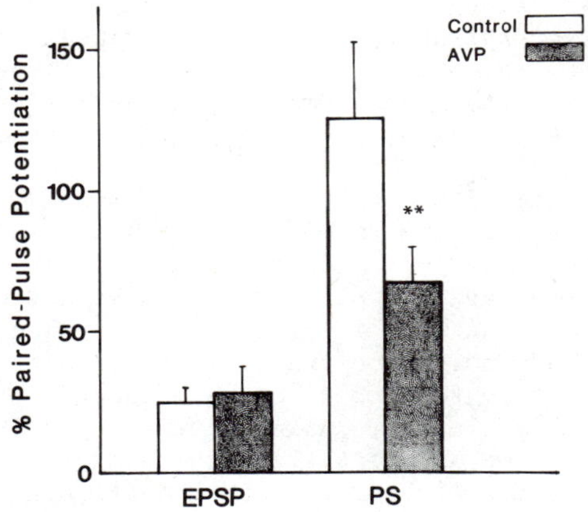

Fig.3. Mean (±s.e.m.) percentage paired-pulse potentiation of field EPSP and population spike (PS) during control and AVP ($2x10^{-6}M$) perfusion periods. **indicates significant ($p<0.01$) difference from control measure.

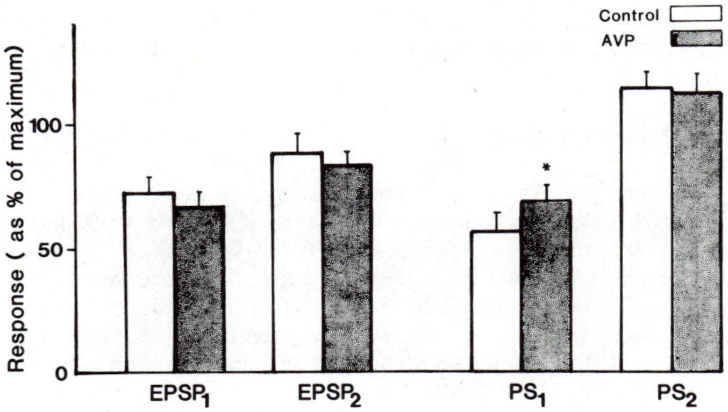

Fig.4. Mean (±s.e.m.) size of first ($EPSP_1$, PS_1) and second ($EPSP_2$, PS_2) of the paired evoked potentials expressed as a percentage of the maximum unpotentiated field EPSP and population spike (PS) as appropriate, for control and AVP ($2x10^{-6}M$) perfusion periods. *indicates significant ($p<0.02$) difference from control measure.

clearly sub-maximal, the failure of AVP to increase its size cannot be attributed to a ceiling effect.

The action of AVP was specific in that it depended on the frequency of stimulation. Population spikes in response to stimuli approximately 5s apart (the time between the pairs of pulses) were enhanced, those in response to stimuli separated by ≤400ms were not. Effects at interpulse intervals between 400ms and 5s have not yet been investigated.

Analysis of hippocampal evoked potentials has shown that the size of the population spike reflects the number of pyramidal cells synchronously activated by the stimulating pulse (Andersen et al., 1971). Assuming that AVP leaves unchanged the waveform of the individual action potentials of the pyramidal cells, the increase in size of the population spike may therefore be taken as an indication that the excitability of CA1 pyramidal cells in response to afferent stimulation is enhanced in the presence of AVP. In contrast to this finding, Mühlethaler et al. (1984) have reported that AVP increases the spontaneous activity of a class of non-pyramidal neurones while being either without effect or decreasing the spontaneous activity of CA1 pyramidal cells recorded in hippocampal slices. Preliminary results in our laboratory suggest that the effect of AVP upon the population spike may disappear if currents close to threshold for the production of a population spike are employed. Thus the action of AVP may in addition be specific in being dependent on the degree of afferent stimulation. If such is the case, AVP might not be expected to result in increases in the low levels of spontaneous pyramidal cell activity recorded by Mühlethaler et al. (1984). In further support of this suggestion is the finding by Joëls and Urban (1982) that the rate of firing of hippocampal cells _in vivo_ in response to applied glutamate (a possible transmitter in this system) is enhanced if AVP is administered at the same time, whilst AVP alone did not necessarily increase firing rate. A possible prediction from this argument, if the characteristics of AVP's action on evoked potentials are accurately reflected at the level of the single cell, is that the temporal specificity of the peptide's action when the cell is receiving sufficient afferent stimulation would result in a particular decrease in long (≥5sec, but possibly >400ms) interspike intervals in the presence of AVP. A dichotomy similar to that proposed here for the effects of AVP on spontaneous and evoked activity has been found in the action of noradrenalin in visual cortex (Segal, this volume): noradrenalin inhibits the spontaneous activity of single cells while enhancing their response to preferred afferent stimulation.

From the discussion above, the view emerges that the action of AVP on hippocampal neurons is highly specific, depending both on the strength and the temporal pattern of afferent stimulation. However, the means by which such specificity is achieved remains unclear. Recent work has demonstrated the existence of various classes of non-pyramidal cells (Schwartzkroin and Knowles, 1983) which may play an important rôle in the regulation of hippocampal output. Such cells are postulated to mediate profound feed-forward and feed-back influences on the pyramidal cell population. Indeed, Mühlethaler et al. (1984) have suggested that inhibition of the spontaneous activity of pyramidal cells by AVP is effected via the peptide's action on local interneurons. A similar mechanism has been proposed as underlying the excitatory effects of opiates in the hippocampus (Zieglgänsberger et al. 1979). Thus AVP's specificity of action may result from the properties of local neuronal circuits. Alternatively, the mechanisms involved may reside in the pyramidal cells themselves: for example, Andersen (1983) has described how pyramidal cells respond differently to sudden and to steady state activation, possibly through the recruitment of different ionic currents. Further investigation of such possibilities is presently under way in our laboratory.

In the foregoing discussion, consideration has been given only to the activity of vasopressin in a selected part of the hippocampal region. However, vasopressin has been shown to influence cellular events at other loci, both in the hippocampus (Auerbach and Lipton,1982; Church,1983), and elsewhere in the CNS (Kovacs et al., 1979; Joëls and Urban, 1982 ; Burns et al., 1983a,b). Further, other substances have been shown to influence short-term (Brown et al., 1983) and long-term (Bliss et al., 1983) potentiation, so that these processes may be subject to powerful but precise control by a number of different systems. Indeed, there is also the possibility of complex interactions between such systems, for example, interactions between the vasopressinergic and monoaminergic systems (Kovacs et al., 1979). Thus there are many possible sites of action and means by which vasopressin may exert an influence on memory processes.

In conclusion, there is mounting evidence for central actions of vasopressin, though it still remains to be shown to what extent and exactly how such actions may explain its behavioural effects (de Wied, 1980). In particular, it is of interest to determine whether the action of AVP in increasing postsynaptic excitability and thereby reducing PPP in the hippocampus is a finding capable of generalisation to other systems, so proving to be a model for specific interaction between neuroendocrine and plastic systems in the central nervous system.

SUMMARY

1. Experiments have been performed on transverse hippocampal slices from Wistar rats.
2. Paired-pulse potentiation (PPP) of the population spike recorded in CA1 after stimulation of the Schaffer collateral pathway was significantly reduced in the presence of arginine vasopressin (AVP) at a concentration of $2x10^{-6}$M. The reduction occurred for pairs of pulses delivered every 5s with interpulse intervals of 10-400ms and at a level of afferent stimulation evoking a half-maximal, unpotentiated population spike.
3. The reduction in PPP of the population spike was due to a significant increase in size of the first of the pair of population spikes, the size of the second remaining unchanged.
4. Neither the size nor PPP of the field EPSP was changed by AVP ($2x10^{-6}$M). Thus the effect of AVP appears to be due to an increase in the postsynaptic excitability of CA1 pyramidal neurones in response to the first, but not the second, of two pulses separated by a brief interval ($\leq$400ms).
5. It is concluded that within the hippocampus AVP has a direct action which is specific to the frequency and probably the strength of afferent stimulation.

ACKNOWLEDGEMENTS

The financial support of the MRC is gratefully acknowledged.

REFERENCES

Alger, B. E., and Teyler, T. J., 1976, Long term and short term plasticity in the CA1, CA3 and dentate regions of the rat hippocampal slice, Brain Res., 110:463

Andersen, P., 1975, Organisation of hippocampal neurons and their interconnections, in "The Hippocampus", Vol. 1, R. L. Isaacson and K. H. Pribram, eds., Plenum Press, New York.
Andersen, P., 1983, Operational principles of hippocampal neurons, in: "Neurobiology of the hippocampus", W. Seifert, Ed., Academic Press, London.
Andersen, P., Bliss, T. V. P., and Skrede, K. K., 1971, Unit analysis of hippocampal population spikes, Exp. Brain Res., 13:208
Auerbach, S., and Lipton, P., 1982, Vasopressin augments depolarization-induced release and synthesis of serotonin in hippocampal slices, J. Neurosci., 2:477
Bliss, T. V. P., Goddard, G. V., and Riives, M., 1983, Reduction of long-term potentiation in the dentate gyrus of the rat following selective depletion of monoamines, J. Physiol., 334:475
Brown, M. W., Rose, D., and Ahlquist, J., 1983, Amnesia-producing drugs affect hippocampal frequency potentiation, Neuroscience, 10:697
Buijs, R. M., 1978, Intra- and extrahypothalamic vasopressin and oxytocin pathways in the rat, Cell Tiss. Res., 192:423
Buijs, R. M., and Van Heerikhuize, J. J., 1982, Vasopressin and oxytocin release in the brain - synaptic event, Brain Res., 252:71
Burns, B. D., Ebenezer, I. S., Summerlee, A. J. S., and Webb, A. C. 1983a, Does arginine vasopressin alter an electrophysiological measure of arousal in unrestrained rats and rabbits? J. Physiol., 349:26P
Burns, B. D., Ebenezer, I. S., and Webb, A. C., 1983b, The effects of arginine vasopressin upon the electrical activity of the cerebral cortices of rats anaesthetized with urethane, J. Physiol., 349:25P
Church, A. C., 1983, Vasopressin potentiates the stimulation of cAMP accumulation by norepinephrine, Peptides, 4:261
Creager, R., Dunwiddie, T., and Lynch, G., 1980, Paired-pulse and frequency facilitation in the CA1 region of the in vitro rat hippocampus, J. Physiol., 299:409
De Wied, D., 1980, Behavioural actions of neurohypophysial peptides, Proc. Roy. Soc. (Lond.) B, 210:183
Gash, D. M., and Thomas, G. J., 1984, Reply to 'The importance of vasopressin in memory', Trends Neurosci., 7:64
Joëls, M., and Urban, I. J. A., 1982, The effects of microiontophoretically applied vasopressin and oxytocin on single neurones in the septum and dorsal hippocampus of the rat, Neurosci. Lett., 33:79
Kandel, E. R., 1981, Calcium and the control of synaptic strength by learning, Nature, 293:697
Kovacs, G. L., Bohus, B., and Versteeg, D. H. G., 1979, The effects of vasopressin on memory processes: the role of noradrenergic neurotransmission, Neuroscience, 4:1529
Langmoen, T. A., and Andersen, P., 1981, The hippocampal slice in vitro. A description of the technique and some examples of the opportunities it offers, in: "Electrophysiology of Isolated Mammalian CNS Preparations", G. A. Kerkut and H. V. Wheal, eds., Academic Press, London.
Lømo, T., 1971, Potentiation of monosynaptic EPSPs in the perforant path - dentate granule cell synapse, Exp. Brain Res., 12:46
McNaughton, B. L., 1983, Activity dependent modulation of hippocampal synaptic efficacy: some implications for memory processes, in: "Neurobiology of the hippocampus", W. Seifert, Ed., Academic Press, London.
Mühlethaler, M., Charpak, S., and Dreifuss, J. J., 1984, Contrasting effects of neurohypophysial peptides on pyramidal and non-pyramidal neurones in the rat hippocampus, Brain Res., 308:97
Mühlethaler, M., Dreifuss, J. J., and Gähwiler, B. H., 1982, Vasopressin excites hippocampal neurones, Nature, 296:749

Schwartzkroin, P. A., 1981, To slice or not to slice, in: "Electrophysiology of Isolated Mammalian CNS Preparations", G. A. Kerkut and H. V. Wheal, eds., Academic Press, London.

Schwartzkroin, P. A., and Knowles, W. D., 1983, Local interactions in the hippocampus, Trends Neurosci., 6:88

Scoville, W. B., and Milner, B., 1957, Loss of recent memory after bilateral hippocampal lesions, J. Neurol. Neurosurg. Psychiat., 20:11

Wheal, H. V., 1981, Characteristics of CA1 cells in the rat hippocampus in vitro, in: "Electrophysiology of Isolated Mammalian CNS Preparations", G. A. Kerkut and H. V. Wheal, eds., Academic Press, London.

Zieglgänsberger, W., French, E. D., Siggins, G. R., and Bloom, F. E. 1979, Opioid peptides may excite hippocampal pyramidal neurons by inhibiting adjacent inhibitory interneurons, Science, 205:414

EFFECTS OF LYSINE-VASOPRESSIN ON SPONTANEOUS BEHAVIOR AND LEARNING IN APPETITIVE TASKS IN THE RAT

Tatiana Alexinsky* and Josette Alliot**

*Département de Psychophysiologie
LPN-CNRS
Gif-sur Yvette-91190
and
**Laboratoire de Psychophysiologie
Université Clermont II B.P.45
63170 Aubière, France

INTRODUCTION

As a result of the demonstration by de Wied (1966) that pitressin, a neurohypophyseal extract, increased resistance to extinction of avoidance tasks in the rat, it was suggested that vasopressin may play an important role in memory processes. Several experimental findings have lent empirical support to this notion. The existing evidence can be summarized as follows :
1) Injections of vasopressin have been shown to alleviate memory impairments caused by different kinds of experimental or natural disruptive events such as hypophysectomy (de Wied, 1969, Bohus et al.,1975), lack of vasopressin release due to hereditary hypothalamic diabetes insipidus (Bohus et al., 1977) or the administration of amnestic agents such as puromycin (Walter et al.,1975), pentylenetetrazol (Bookin and Pfeifer, 1977), CO_2 or ECS (Rigter et al., 1974);
2) Post-training administration of vasopressin has been shown to facilitate retention of several aversive tasks (de Wied et al., 1984);
3) The inactivation of endogenous vasopressin by a specific anti-serum injected intraventricularly or intrahippocampally results in impaired avoidance performance (Van Wimmersma-Greidanus et al.,1976). Taken together, existing evidence is in support of a "mnemoactive" role of vasopressin (Gold and Zornetzer, 1983) and suggests that this peptide might have clinical applications in the treatment of amnesic syndroms in humans. However, Jolles (1983) has pointed out that an evaluation of clinical efficiency of vasopressin is difficult since both positive and negative results have been obtained.

The first discrepancies among the original claims were noted in the research on animals, and were due, at least in part, to the use of multiple behavioral tests (for review and discussion see Gash and Thomas, 1983, 1984; Messing and Sparber, 1984). From a theoretical point of view, two major obstacles stood in the way of accepting vasopressin as a "mnemotropin". The first obstacle was that most results were obtained in experiments that used extinction procedures. The second was that in most experiments footshocks were used and it is likely that facilitatory effects might have been due to modified reactions to the aversive nature of the task rather than to a specific effect on consolidation or retrieval processes. Therefore it is important to know whether or not vasopressin affects behavior in other tasks than shock-motivated avoidance ones, so that we can choose between one of the three following hypotheses concerning the role of vasopressin in learning and memory processes:

1) A direct influence of vasopressin on neural mechanisms underlying engrams, as was postulated by de Wied (1979) who suggested that vasopressin carries not only an endocrinological message but also a behavioral one which may be encoded into some of its amino-acid sequences.

2) An influence on memory through modulation of emotionality, motivation and/or attention.

3) An effect on performance rather than on associative learning per se (e.g.increase or decrease in activity).

The following is a brief survey of our work and that of others on the effects of sub-cutaneous vasopressin injections on spontaneous behavior and retention of non-aversive tasks in the rat.

EFFECTS OF VASOPRESSIN ON APPETITIVE TASKS

Effects of pre-trial injections

Several workers attempted to study the behavioral effects of vasopressin in non-aversive situations using pre-trial injections. Depending on the nature of the task and dosage, various investigators reported either a lack of effect in a runway (Garrud et al., 1974) or in a radial maze (Buresova and Skopkova, 1979), an impairment in a bar-pressing task (Andrews et al., 1983) or a facilitation of either performance in a visual dicrimination task (Hostetter et al., 1977, Sara et al., 1983) or in bar-pressing rates (Messing and Sparber, 1984). Thus, the results are contradictory but the debate appears, so far to be non-productive since "In general the strategy of administering a treatment prior to learning in an attempt to discover something fundamentally unambiguous regarding memory itself, is at best extremely complicated and at worst useless"(Gold and Zornetzer, 1983).

Effects of post-trial injections

Evidence concerning the effects of post-trial injections of vasopressin on appetitive tasks is scarce. Amongst reported data, Buresova and Skopkova (1979, 1982) have found either no effect or an impairment of retention of a radial maze task. Facilitation has been reported in males traversing a maze to reach a sexually receptive female (Bohus, 1977); however, vasopressin injections were found to delay the disappearance of ejaculation and intromission in the rat so that a positive effect of vasopressin on the level of sexual motivation cannot be easily dismissed. Furthermore given the high level of interest and experimental activity surrounding the problem of vasopressin and memory, it is not unlikely, that there are unpublished negative results.

In our experiments we aimed to clarify the effects of vasopressin effects on non-aversive tasks in rats. Three different situations were used in which male Sprague Dawley rats (250-300 g) were given post-trial sub-cutaneous injections of 0 .2 ug of lysine vasopressin (LVP)and their performance compared to that of control rats injected with saline.

<u>Modulation of bar pressinq rate in operant schedules</u> . Two different operant schedules were used. Under a continuous reinforcement (CRF) schedule, LVP rats reached the learning criterion significantly more

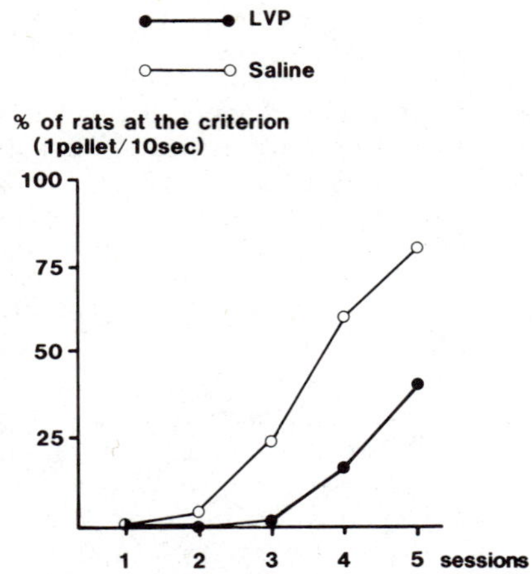

Fig. 1 Effects of post-trial LVP injections on acquisition of bar pressing under CRF schedule.

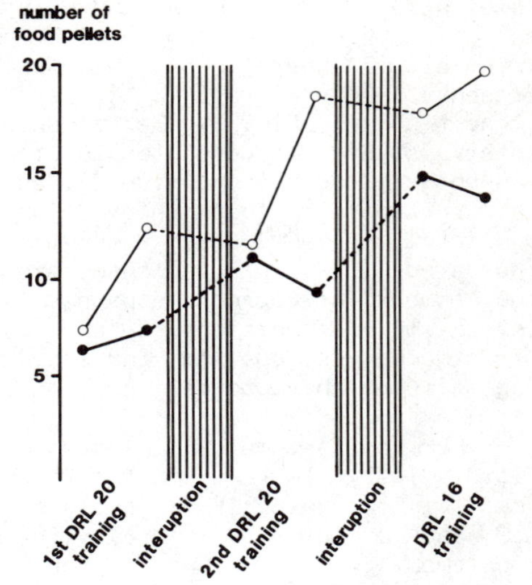

Fig. 2 Effects of post-trial injections on successible daily sessions in a DRL schedule.

slowly than did control rats (Alliot and Alexinsky, 1984; Fig. 1). Impairment was also observed under differential reinforcement of low rates (DRL) schedules; (Alliot and Alexinsky 1983; Fig. 2).It is noteworthy that performance deteriorated when training and injections were given daily. After a prolonged interruption, LVP rats reached the same level of performance as did control rats on the first day of retraining not preceded by an injection. This indicates that vasopressin impaired performance without affecting memory.

Discrimination learning . We have found (Alliot and Alexinsky, 1982) that LVP injections did not alter discrimination learning when bar-presses were reinforced only during 'light on' periods. In this " go-no go" discrimination, control and LVP rats mastered the task at comparable rates.

On another discrimination task (black/white in a T-maze), the rats injected with vasopressin after the choice of S+ (white) alley reached the learning criterion more slowly than did control rats (Alliot and Alexinsky, unpublished results).

Working memory task. Rats were taught a delayed matching-to-sample (DMS) rule in a three choice maze: correct responses consisted of an entry into the goal-box which had the same characteristics as the "sample box" previously presented. One goal box was black, the other, white and the third had black and white vertical stripes (Alexinsky and Chapouthier, 1978). Acquisition of this rule is presumed to be based on reference memory (Olton et al., 1978). Following acquisition of the reference memory, delays were interpolated between the sample box and test so that retention of the characteristics of the sample box depends on working memory. To study LVP effects on working memory, rats were alternately injected with saline or LVP immediatly after removal from the sample box. Previous extensive studies showed that performance is directly related to the duration of presentation of the sample and inversely proportional to the duration of the sample-to-test interval (Alexinsky, 1981). Four combinations of interval and duration of sample presentations were chosen to study LVP effects on working memory. As reported earlier, when a long interval (3 hours) was interposed between a brief (5 min.) sample presentation and test, the performance of animals was at chance levels (condition A, Fig. 3). Performance could be improved in two ways: 1) When intervals are reduced to 10 minutes, performance reaches 70% correct responses levels (condition B), an improvement that could have been due to the possibility that information stored in short term memory had not completely decayed. 2) When the duration of stimulus presentation is increased to 30 minutes then, even with a delay as long as 3 or 24 hours, the number of correct responses is also significantly greater than chance (65-75 %), (conditions C and D, Fig.3). In all probability, this good performance is due to the increase in the time available for processing the stimulus characteristics.

LVP post-stimulus injections did not alleviate forgetting (see A in Fig.3. left). In fact, in the three situations in which good retention was found in controls, LVP injections produced an impairment in retention (see B,C and D in Fig.3 left). It is interesting to note that the LVP impairment was seen under both conditions which normally facilitate performance in control animals, that is increasing the time for stimulus processing (C and D) and reduction in the time for stimulus decay (B). Only the working memory was affected. Reference memory remained unmodified as evidenced by the high performance of the same rats when the stimulus presentation was not followed by the injection of LVP. It can also be noted that the longer the duration of the presentation of the stimulus box, the greater the disruption induced by LVP.

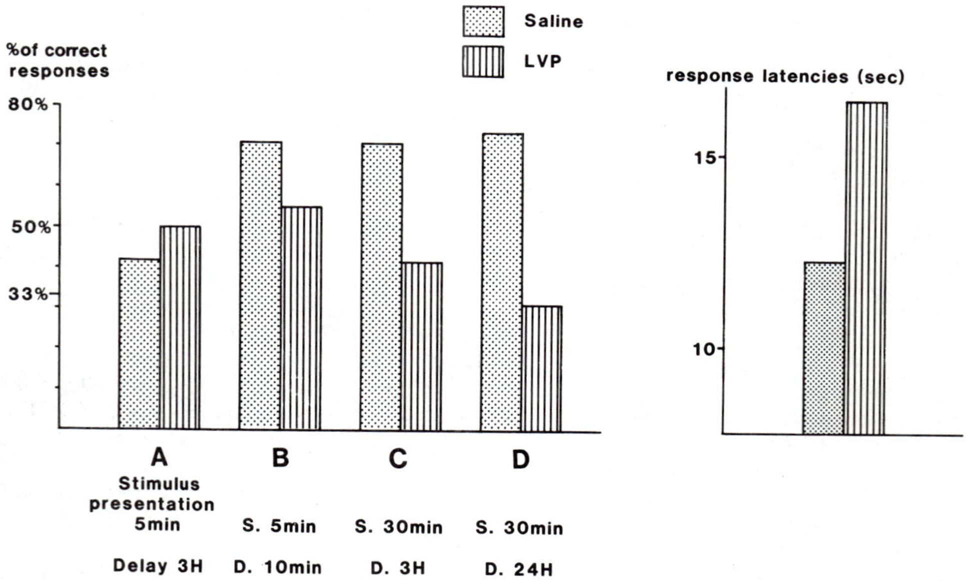

Fig. 3 Left : % of correct responses in DMS task obtained after injections of saline or LVP in different conditions of stimulus duration and interval between sample and test. Right : in all conditions the response latencies were increased after association of LVP and sample.

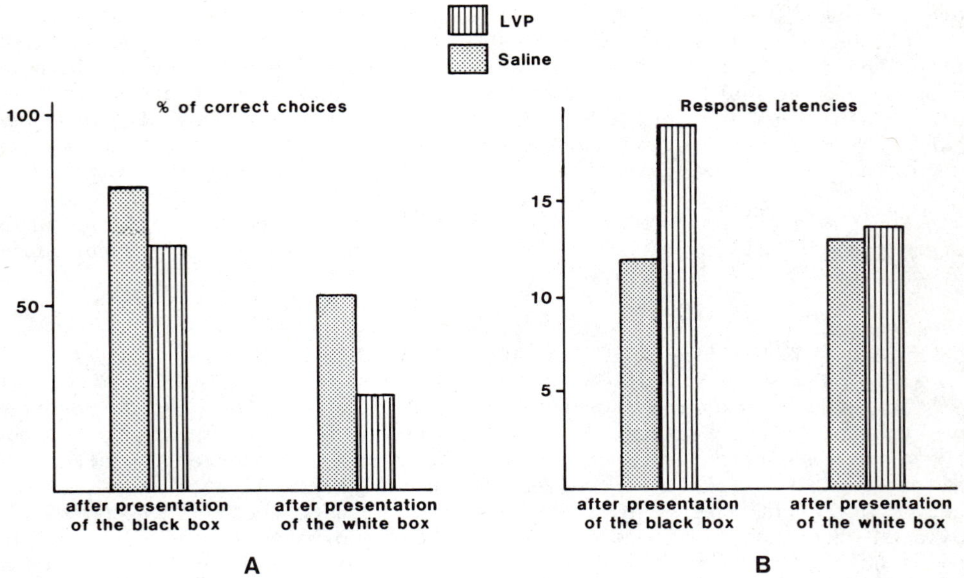

Fig. 4 Modulation of effects of LVP injections on (A)% of correct responses and (B) response latencies (sec) as a consequence of the color of sample.

Furthermore the impairment was selective to the matching-to-sample paradigm: when rats in a yoked control group were submitted to the same procedure but without matching contingencies, with reinforcement randomly distributed in the three boxes, no differences appeared between LVP and saline injected rats. We also found that response latencies increased after LVP injections compared with the latencies after saline injections (Fig. 3, right).

In summary, these experiments provided evidence that vasopressin not only does not facilitate but rather impairs working memory.

Further analysis of performance in the matching-to-sample task indicated that the impairment was, perhaps partially, due to differential reactions to the light and dark boxes. LVP injections reduced responding to the correct goal box, but response latencies were not significantly modified after the presentation of the white box (Fig.4). In contrast after the presentation of the black box, a significant increase in response latencies was seen, even when responses were correct. Moreover,such a reduction in the number of correct choices and increased latencies were enhanced when the duration of the presentation of the stimulus box was increased. Therefore, these effects seemed to be associated with a longer duration of processing of the characteristics of the sample. However in the yoked control group, LVP post-stimulus injections led to another result: increase in the number of choices of the white box (Fig.5). Thus, a modification of the spontaneous reactivity toward the light and dark boxes cannot account for the results obtained in the DMS task. Another factor in support of this conclusion was that the difference in performance after injections of saline or LVP did not appear until the third of fourth trial.

Thus, our results showed that injections of LVP, even at very low dosages, did modify behavior. Though facilitation was never observed, rats were impaired in learning and retention of bar pressing responses, of correct choices in a T-maze or in DMS task. On the other hand, injections after the learning sessions did not affect the differential adaptation of responses when rats were presented with S+ (light) or S- (dark). In a DRL task, rats treated after a learning session had less correct responses than control rats during subsequent sessions but the same rats showed an improvement after an interruption of treatment and training.

In summary, performance was impaired in several appetitive tasks after the injection of LVP but the ability to learn or store information was not affected by this compound.

EFFECTS OF VASOPRESSIN INJECTIONS ON SPONTANEOUS BEHAVIOR

In order to control for non-specific effects of the peptide, several investigators have associated the injection of vasopressin with food reward or a particular experimental situation, and observed the effect of this associated treatment on subsequent behavior. Sahgal et al. (1983) have found that response latencies were increased when injections of LVP followed footshocks. Ettenberg et al. (1983) and Dantzer (1984) found that an association of LVP injections with a particular place or taste resulted in an avoidance reaction (conditioned aversion). Similarly, Miller et al. (1968) demonstrated the possibility of an association of vasopressin with one of two arms of a T-maze after 30 associations only when their rats were previously preloaded with water.

Other experiments which attempted to assess the effects of vasopressin on different measures of spontaneous behavior have produced contradictory findings. Thus, in the open field, Ettenberg et al.(1983), and Andrews et al. (1983) observed a reduction of locomotion whereas Schulz et al. (1974) and Krejci et al.(1979) did not find any effect of LVP injections. Latencies prior to shock in a passive avoidance study were not decreased in Sahgal's (1983) experiment, in contrast to the increased latencies observed by Hostetter et al.(1980) for her control group. Messing and Sparber (1984) found an increase in the number of nose poke responses. Van Ree et al. (1977) and Dorsa et al. (1979) reported a reduction of self-stimulation and heroin self-injection. Other modifications of behavior have been noted: reduction of grooming and scratching , increased freezing (Meisenberg, 1981), decreased agoraphobia (Crine, 1981), decreased photophobia (Frucht-Celaru and Sterescu-Voranschi, 1975), acceleration of habituation to a novel environment (Sadile, 1981). While such results do not point to any consistent pattern of effects, they strongly support the notion that LVP does actually modify spontaneous behavior.

In order to study the effects of vasopressin on spontaneous behavior more systematically, rats were placed in a T-maze and LVP was injected immediatly after the choice of one specific arm or both, or injections were simply given prior to exploration of the maze.

Either one or five injections of 0.2ug of LVP were given 24 hours or one hour before testing. In some conditions, rats were food deprived, in others they were fed ad libitum. The number and latencies of choices of either black or white arms in a T-maze were compared after saline and LVP injections (Alliot and Alexinsky, 1984). It should be noted that in the same apparatus, when choices of the white arm were reinforced by food, rats injected with LVP after a white choice learned less rapidly than did control rats. The results can be summarized as follows:

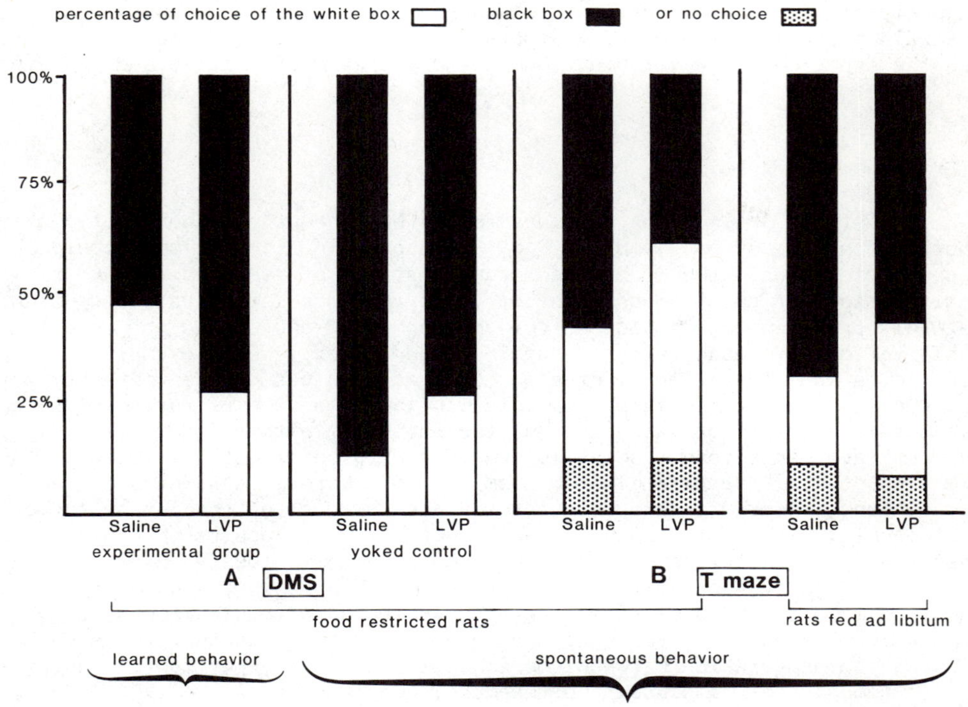

Fig. 5 Effects of LVP in different situations of choice between white and black arm.

1) In experiments without food reward associated with a specific arm, two different behaviors were observed. First, rats injected with vasopressin after a choice of the white arm, chose it significantly more frequently than did rats injected with saline. As in the delayed matching-to-sample task (for yoked control group), this effect cannot account for the results observed when vasopressin was associated with the reinforcement. Secondly, LVP injections after the choice of the black arm did not modify the proportions of choices of either white or black arms. The performance of the rats does not reflect a conditioned aversion because a conditioned aversion would be seen in a reduction of choices of the black arm after pairing of black arm entry with injection as well as in a reduction of white arm choices after the pairing of white arm entry with injection.

2) The injection of LVP, 24 hours before the test, but without a preliminary association with the testing situation, resulted in a significantly reduced preference for darkness (Fig.5 B).

3) Finally, when LVP was injected in rats that were fed ad lib., we also saw an effect on subsequent observed spontaneous behavior (Fig.5 B). Whereas control rats injected with saline showed the expected preference for the black arm, LVP treated rats made more choices of the white arm with increased start latencies. These effects were seen for injection to test intervals of 24 hours but were greater with a delay of only one hour. The difference between the effects of LVP and saline was reduced when five, instead of one injection preceded the test session.These results suggest that: Spontaneous choice behavior can be modified by LVP

injections; this modification is characterized by a reduction of the natural preference for dark, an increase in exploration of the other parts of the experimental environment and/or an increase of emotive and freezing responses.

DISCUSSION

The first point to consider is whether or not the modifications in spontaneous behavior induced by vasopressin injections could be responsible for the observed facilitation of learned behavior. A modification of spontaneous behavior (e.g. response latencies, locomotion, natural preferences) could affect performance scores and yet appear as facilitation of associative learning and memory. For example, in a discrimination task, an increased number of choices of the white arm may be due to a decrease in photophobia, an increase in exploration or to an improved capacity of associating the white arm with reinforcement. In a passive avoidance task, memory is considered to be directly related to the time spent in the illuminated compartment before rats enter the dark compartment; however, this behavior may also be due to freezing or reduced photophobia. Thus it appears that some of the results obtained after vasopressin injections might be more parsimoniously explained as modifications of spontaneous behavior rather than as enhanced learning and memory. In both DMS and T-maze tasks, however, behavioral effects obtained after injections of LVP were opposite depending on whether there was or was not an association between the injection and the experimental context, reinforcement or response. Furthermore, Garrud et al. (1974), Sahgal (1982) and Ettenberg (1983) reported that, in different experimental situations, LVP injections do not produce the same effects when they are associated with the stimulus or with reinforcement, or if it is injected independently of either one. This latter evidence is incompatible with a systematic interpretation of all vasopressin effects on learning in terms of its effects on performance only.

Has vasopressin a direct effect on learning processes and memory storage?

First, it should be emphasized that in all our experiments described above in which positive reinforcement was used, we have never seen facilitation of learning and memory as a result of LVP injections. Thus, facilitatory effects of post-trial injected LVP appeared to be strictly limited to aversive tasks.

Second, LVP injections can result in impaired performance. A careful analysis of behavior led us to conclude that such impairments were not due to a disruption of mnemonic abilities (e.g. reference memory was not impaired in the DMS task, and in the DRL paradigm, LVP rats performed within a normal range after an interruption of both training and treatment).

Third, it has been shown by Messing and Sparber (1984) that the effects of LVP might be mimicked by a saline injection, a substance which so far has not been implicated in the formation of an engram. Similarly, another substance which cannot be considered as a natural mnemonic modulator -lithium chloride- has been shown to have the same behavioral effect as vasopressin (Ettenberg, 1983).

Finally, the finding that spontaneous behavior can be modified in situations in which no demands are made to learn and remember, makes it more difficult to accept that vasopressin exerts a specific effect on learning. Taken together, the existing evidence is not in favor of the notion of vasopressin as a "mnemotropin", unless this role is restricted to negatively reinforced learning processes.

The other proposed explanations for the role of neuropeptides on memory emphasize the nonspecific modulatory effects of peptides which, in turn, would affect learning and memory processes. For example, the arousal theory (Sahgal 1983) proposes that the activation induced by the peptides would enhance attention to stimuli and, thus, accelerate information processing. Increased activation may also stimulate neuronal mechanisms which underlie consolidation of memory. This would account for the well known facilitatory effects of pre-test vasopressin injections, on appetitive as well as on negatively reinforced tasks. But how can the disrupting effects of the same dose of LVP when it is injected after a trial be explained?

Our results have stressed the importance of an associative link which may exist between vasopressin injections and either the stimulus or the response. Since recent investigations using place and taste aversion have shown that a vasopressin injection may constitute an aversive stimulus by itself (Ettenberg et al.1983 and Dantzer, 1984), we can assume that aversive properties of post-training injections of LVP become associated with the conditioned stimulus.

Such an explanation may account for the contrasting results obtained in appetitively and aversively reinforced tasks: Vasopressin may potentiate the negative value, or reduce the positive value, acquired by the stimulus during learning. However, this explanation restricts the potential associative strength of LVP with the conditioned stimulus. In fact, the results obtained with our study of spontaneous behavior lead us to emphasize the importance of an interaction between the increased level of vasopressin and behavioral reactivity.

Numerous studies have reported that vasopressin produces large and varied physiological effects which are frequently related to stressful situations. This suggests that the internal state associated with endogenous hormone release during stress provides an attribute of aversiveness which can also be induced by vasopressin injection. The behavioral consequences of such treatment would be the occurrence of fear responses added to the environmental cues during subsequent test, resulting, in turn, in a decrease of approach responses.

In conclusion, to account for the opposite effects of vasopressin on appetitively and shock-motivated tasks, we propose the following hypothesis: The physiological consequences of increased vasopressin levels induces a behavioral state which serves as an internal cue for subsequent retrieval and which can modulate the strength of learned associations. This behavioral state might increase avoidance responding so that performance may appear to reflect a facilitation of learning and memory of aversive tasks. However, at the same time, this state is not compatible with approach behavior or spontaneous preferences, so that appetitive tasks are impaired and spontaneous behavior modified.

REFERENCES

Alexinsky, T.,1981, Analyse de l'évolution de la rétention "à court terme" par la méthode d'appariement retardé, in : "Mémoire, Conditionnement, Evolution", Publ. de la Sorbonne, Paris.

Alexinsky, T.and Chapouthier, G.,1978, A new behavioral model for studying delayed response in rats, Behav.Biol. , 24:442.

Alliot, J. and Alexinsky,T., 1982, Effects of posttrial vasopressin injections on appetitively motivated learning in rats, Physiol. Behav. , 28:525.

Alliot, J. and Alexinsky, T., 1983, Repeated posttrial administration of vasopressin impairs subsequent differential reinforcement of low rates (DRL) performances, Behav. Processes, 5: 345.

Andrews, J.S., Newton, B.A. and Shagal, A.,1983, The effects of vasopressin on positively rewarded responding and on locomotor activity in rats, Neuropeptides, 4:17.

Bohus, B., Gispen, W.H. and De Wied D., 1973, Effect of lysine-vasopressin and ACTH 4-10 on conditioned behavior of hypophysectomized rats, Neuroendocrinology, 11:137

Bohus, B., Van Wimmersma-Greidanus, T.J.B. and De Wied, D., 1975, Behavior and endocrine responses of rats with hereditary hypothalamic diabetes insipidus (Brattleboro strain), Physiol. Behav., 14:609.

Bohus,B., 1977, Effect of desglycinamide-lysine vasopressin (DG-LVP) on sexually motivated T-maze behavior of the male rat, Horm. Behav., 8:52

Bookin, H.B.and Pfeifer,W.D.,1977, Effect of lysine vasopressin on pentylenetetrazol induced retrograde amnesia in rats, Pharmacol. Biochem.Behav., 7:277..

Buresova, O. and Skopkova, J.,1979, Vasopressin analogues and spatial short-term memory in rats, Peptides, 1:261.

Buresova, O. and Skopkova, J.,1982, Vasopressin analogues and spatial working memory in the 24-arm radial maze, Peptides, 3:725.

Carey, R.J. and Miller, M., 1982, Absence of learning and memory deficits in the vasopressin-deficient rat, (Brattleboro strain), Behav. Brain Res., 6:1.

Crine, A.F., 1981, Effect of vasopressin on open field behavior in rats, Physiol. psychol., 9:113.

Dantzer, R., 1984, personal communication.

De Wied, D., 1969, Effects of peptide hormones on behavior, in : "Frontiers in Neuro-endocrinology", Ganong, W.F. and Martini, L., ed., Oxford University Press.

De Wied, D., 1976, Behavioral effects of intraventricularly administered vasopressin and vasopressin fragments, Life Sci., 19:685.

De Wied, D., Gaffori, O., Van Ree, J.M. and Dejong, W., 1984, Central target for the behavioral effects of vasopressin neuropeptides, Nature, 308:276.

De Wied, D. and Versteeg, D.H.G., 1979, Neurohypophyseal principles and memory, Fed. Process., 38:2348.

Dorsa, D.M., and Van Ree, J.M., 1979, Modulation of substantia nigra self-stimulation by neuropeptides related to neurohypophyseal hormones, Brain Res., 172:367.

Ettenberg, A., Van Der Kooy, D., Le Moal, M., Koob, G.F. and Bloom, F.E., 1983, Can aversive properties of (peripherally-injected) vasopressin account for its putative role in memory?, Behav. Brain Res., 7:331.

Frucht-Celaru, M., Strerescu-Volanschi, M., 1975, The effect of lysine vasopressin on short-term recall of noxious significance, Rev. Roum. morph. embryol. physiol., 12:285.

Gash, D.M. and Thomas, G.J., 1983, What is the importance of vasopressin in memory processes?, Trends Neurosci., 6:197.

Gash, D.M. and Thomas, G.J., 1984, Reply to D. De Wied, Trends Neurosci., 7:64.

Garrud, P., Gray, J.A. and De Wied, D., 1974, Pituitary-adrenal hormones and extinction of rewarded behavior in the rat, Physiol. Behav., 12:109.

Gold, P.E. and Zornetzer, S.F., 1983, The mnemon and its juices: neuromodulation of memory processes, Behav. Neural Biol., 38:151.

Hostetter,G. Jubb, S.L. and Kozlowsky, G.P., 1977, Vasopressin affects the behavior of rats in a positively-rewarded discrimination task, Life Sci. , 21:1323.

Hostetter,G. Jubb, S.L. and Kozlowsky, G.P., 1980, An inability of subcutaneous vasopressin to affect passive avoidance behavior, Neuroendocrinology ,30:174.

Jolles, J., 1983, Vasopressin-like peptides and the treatment of memory disorders in Man, in "The Neurohypophysis:structure, function and control", Progress in Brain Research, 60:169.

Krejci, Y.B., Kupkova, B, Metys, J., Barth, T. and Jost, K., 1979, Vasopressin analogs:sedative properties and passive avoidance behavior in rats, Eur. J. Pharmacol. , 56:347.

Meisenberg, G., 1981, Short term behavioral effect of posterior pituitary peptides in mice. Peptides , 2:1.

Messing, R.B. and Sparber, S.B., 1983, Facilitation of appetitively motivated learning and memory by desglycinamide arginine vasopressin (DG-AVP), J. Pharmacol. , 89:43.

Messing, R.B. and Sparber, S.B., 1984, Does vasopressin really facilitate memory processes? Trends Pharmacol. Sci. , 5:149.

Miller, N.E., DiCara, L.V., and Wolf, G., 1968, Homeostasis and reward:T-maze learning induced by manipulating antidiuretic hormone, Amer.J.Physiol. , 215:684.

Olton, D.S., 1978, Characteristics of spatial memory. in : Cognitive processes in animal behavior, ed. S.H. Hulse, H. Fowler and W.H. Honig. Hillsdale, N.J.:Lawrence Erlbaum Associates..

Rigter, H., Van Riezen, H. and De Wied, D., 1974, The effects of ACTH and vasopressin analogues in CO2 induced retrograd amnesia in rats. Physiol. Behav. , 13:381.

Sadile, A., Keith, A.B., Wright, C. and Edwardson, J.A., 1982, Failure of vasopressin to enhance memory in a passive avoidance task in rats, Neurosci. Letters ,28:87.

Sahgal, A., 1983, Vasopressin and behavior, some arguments for an arousal hypothesis, in :"Neuropeptides and Psychosomatic Processes", Endroczi, A., ed., Budapest.

Sara, S.J., Barnett, J. and Toussaint, P., 1982, Vasopressin accelerates appetitive discrimination learning and impairs its reversal, Behav. Processes , 7:157.

Schulz, H., Kovacs, G.L. and Telegdy, G., 1974, Effect of physiological doses of vasopressin and oxytocin on avoidance and exploratory behavior in rats, Acta physiol.Acad. Sci. Hung. , 45:211.

Van Ree, J.M. and De Wied, D., 1977, Modulation of heroin self administration by neurohypophyseal principles, Eur.J. Pharmacol. 43:199.

Van Wimersma Greidanus, T.J.B. and De Wied, D.,1976, Dorsal hippocampus: a site of action of neuropeptide on avoidance behavior?, Pharmacol. Biochem. Behav. ,5:29.

Walter,R., Hoffman, P.L., Flexner, J.B. and Flexner, L.B., 1975, Neurohypophyseal hormones, analogs and fragments, their effect on puromycin-induced amnesia, Proc. Nat. Acad. Sci. , 72: 4180.

OXYTOCIN AND VASOPRESSIN IN MEMORY AND AMNESIA

Gábor L. Kovács[+], Gyula Telegdy[+], Ferenc Laczi[++] and Ferenc A. László[++]

Institute of Pathophysiology[+] and Endocrine Unit and Research Laboratory[++], University Medical School, Szeged, Hungary

INTRODUCTION

The pituitary gland plays a role in the acquisition and maintenance of learned behavior. Treatment of rats with lysine-vasopressin (LVP) or arginine-vasopressin (AVP) has been shown to produce a long-lasting delay in the extinction of pole-jumping (De Wied et al., 1971) and bench-jumping (Schulz et al., 1974) avoidance reactions. LVP and AVP treatments also improved passive avoidance behavior (Bohus et al., 1972, Kovács et al., 1977). De Wied et al. (De Wied 1976) suggested that memory processes might be facilitated by vasopressin. Additional evidence for effects of AVP and LVP on memory processes was provided by experiments on amnesia which indicated that these peptides prevented or reversed puromycine-, CO_2- or pentylenetetrazole-induced retrograde amnesia in rats and mice (Lande et al., 1972, Rigter et al., 1974, Bohus et al., 1982).

EFFECTS OF DG-AVP ON MEMORY IN DIABETES INSIPIDUS PATIENTS

Structure-activity studies have revealed that the effect of vasopressin on memory processes was not related to the "classical" endocrine activities of these hormones (De Wied et al., 1972; Burbach et al., 1983). For example, des-glycinamide-LVP (DG-LVP) and des-glycinamide-AVP (DG-AVP), which are practically devoid of antidiuretic activity, facilitated avoidance behavior in rats (De Wied et al., 1972). The same results have also been found for the action of vasopressin on memory processes in human patients. It has been shown that LVP improves memory processes, attention and concentration in human volunteers (Legros et al., 1978, Laczi et al., 1982). Consistent with the animal studies, Laczi et al. (1983) found that acute i.m. injection of DG-AVP improved some aspects of short-term memory in patients with central diabetes insipidus and in non-diabetic controls. Also, subchronic intra-nasal administration of DG-AVP facilitated short-term memory more consistently and in addition improved long-term memory (Laczi et al., 1983). In a placebo controlled follow-up study, we found that DG-AVP treatment

TABLE 1.

INFLUENCE OF INTRA-NASAL DG-AVP TREATMENT FOR 7 DAYS ON THE PERFORMANCE OF ACQUIRED DIABETES INSIPIDUS PATIENTS AS ASSESSED IN DIFFERENT PSYCHOLOGICAL TESTS.

TREATMENT	PSYCHOLOGICAL TESTS[x]				
	ATTENTION	SHORT-TERM MEMORY			LONG-TERM MEMORY
	Bourdon-test (%)	Acoustic (scores)	Optical (scores)	Word-pair short-term (scores)	Word-pair long-term (scores)
PLACEBO TREATMENT (N = 9)					
Baseline	91.14±1.86[+]	6.11±0.63	11.56±0.67	79.63±1.88	70.38±3.35
Placebo	91.97±1.47	6.33±0.37	10.89±0.73	81.85±2.09	69.27±2.71
Placebo (14 days later)	92.25±1.82	7.11±0.51	12.88±0.61	77.78±1.84	73.33±3.14
PEPTIDE TREATMENT (N = 10)					
Baseline	89.15±1.43	6.20±0.48	10.10±0.80	80.00±1.48	68.94±2.73
DG-AVP (80 ug/day)	93.25±1.08	9.60±0.22*	11.90±0.77	94.01±1.29*	88.34±3.38**
DG-AVP	94.28±0.87	7.60±0.45	12.60±0.52	90.97±1.73*	85.67±2.00*

+ Mean ± SEM
* $P<0.05$; ** $P<0.02$ (ANOVA, followed by paired nonparametric comparisons)
x The detailed methods of scoring have been described elsewhere (Laczi et al. 1982, 1983).

induced some improvement in short-term and long-term memory in patients with acquired diabetes insipidus; these effects persisted for a relatively long period (two weeks) after the discontinuation of the treatment (Table 1).

EFFECTS OF OXYTOCIN ON RETROGRADE AMNESIA

Oxytocin (OXT) affects behavior in a direction opposite to that of AVP or LVP (Telegdy and Kovács 1979a,b, 1982). OXT treatment has been shown to facilitate the extinction of an active avoidance bench-jumping reaction (Schulz et al., 1974) and to attenuate passive avoidance behavior (Kovács et al., 1978; Bohus et al., 1978). It was tempting to speculate that OXT attenuated memory processes in rats. Bohus et al. (1978) demonstrated that OXT suppressed both consolidation and retrieval of memory processes in rats and concluded that OXT might be a naturally-occurring amnestic neuropeptide. We have found further evidence of the action of OXT on memory processes by studying its effects on retrograde amnesia in rats (Fig. 1).

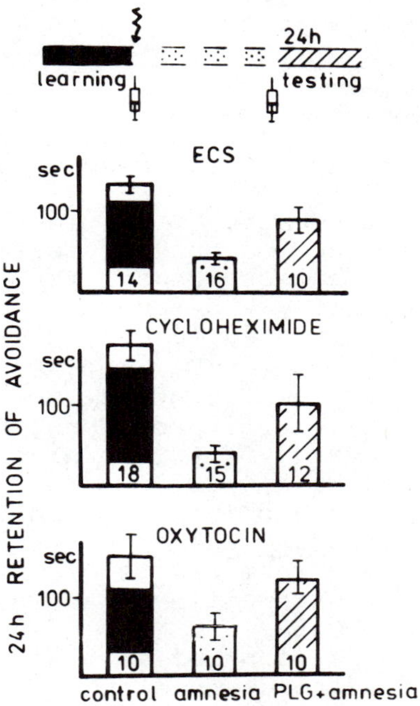

Fig. 1. Oxytocin induced retrograde amnesia comparable to that evoked by electroconvulsive shock or protein synthesis inhibition. Rats were trained in a passive avoidance apparatus. Immediately after the learning trial, one of three amnestic treatments were given. Electroconvulsive shock (ECS) was administered via ear-clip electrodes. Cycloheximide was injected i.p. in a dose of 2.5 mg/kg. OXT was administered s.c. in an amount of 1 ug/animal. Pro-Leu-Gly-NH_2 (PLG) was injected in an amount of 100 ug/rat 1 hr before the 24 hr retention trial.

Treatments such as electroconvulsive shock (ECS) or protein synthesis inhibitors (cycloheximide) disrupt passive avoidance in rats provided that these treatments are given shortly after the learning trial. All these treatments induce retrograde amnesia by interfering with either consolidation or retrieval processes (McGaugh et al., 1979). OXT resulted in a deficit of passive avoidance behavior, an effect comparable with those of ECS or protein synthesis inhibition. We also found that the retrograde amnesia induced by ECS, protein synthesis inhibition and OXT treatment was alleviated by peripheral treatment with prolyl-leucyl-glycinamide (PLG, MIF-1, the C-terminal tripeptide of OXT), suggesting that common or similar brain mechanisms were affected by the three amnestic treatments used. Consistent with these animal experiments, recent data (Ferrier et al., 1980, Kennett et al., 1982) suggest that the recall of learned information was impaired by OXT in human subjects too. These findings substantiate the idea that OXT is an amnestic neuropeptide in humans.

In conclusion, the data indicate that vasopressin and oxytocin influence behavioral processes related to information processing, learning and memory in a manner opposite to each other. The beneficial effect of vasopressin might be of therapeutic significance in human mental disorders.

REFERENCES

Bohus, B., Ader, R., and De Wied, D., 1972, Effects of vasopressin on active and passive avoidance behavior, Horm. Behav. 3:191.

Bohus, B., Conti, L., Kovács, G. L., and Versteeg, D. H. G., 1982, Modulation of memory processes by neuropeptides: interaction with neurotransmitter systems. In: Neuronal Plasticity and Memory Formation. C. Ajmone Marsan and H. Matthies (Eds.), Raven Press, New York.

Bohus, B., Kovács, G. L., and De Wied, D., 1978, Oxytocin, vasopressin and memory: opposite effects on consolidation and retrieval processes, Brain Res. 157:414.

Burbach, J. P. H., Kovács, G. L., De Wied, D., Van Nispen, J. W., Greven, H. M., 1983, A major metabolite of arginine vasopressin in the brain is a highly potent neuropeptide, Nature 221:1310.

De Wied, D., 1971, Long term effect of vasopressin on the maintenance of a conditioned avoidance response in rats, Nature 232:58.

De Wied, D., 1976, Behavioral effects of intraventricularly administered vasopressin and vasopressin fragments. Life Sci. 19:685.

De Wied, D., Greven, H. M., Lande, S., and Witter, A., 1972, Dissociation of the behavioral and endocrine effects of lysine vasopressin by tryptic digestion, Br. J. Pharmacol. 45:118.

Ferrier, B. M., Kennett, D. J., and Devlin, M. C., 1980, Influence of oxytocin on human memory processes, Life Sci. 27:2311.

Kennett, D. J., Devlin M. C., and Ferrier, B. M., 1982, Influence of oxytocin on human memory processes: validation by a control study. Life Sci. 31:273.

Kovács, G. L., and Telegdy, G., 1982, Role of oxytocin in memory and amnesia, Pharmacol. Ther. 18:375.

Kovács, G. L., Vécsei, L., and Telegdy, G., 1978, Opposite action of oxytocin to vasopressin in passive avoidance behavior in rats, Physiol. Behav. 20:801

Kovács, G. L., Vécsei, L., Szabó, G., and Telegdy, G., 1977, The involvement of catecholaminergic mechanisms in the behavioral action of vasopressin, Neurosci, Lett. 5:337

Laczi, F., Valkusz, Zs., László, F. A., Wagner, A., Járdánházy, T., Szász, A., Szilárd, J., and Telegdy, G., 1982, Effects of lysine-vasopressin and 1-deamino-8-D--vasopressin on memory in healthy individuals and diabetes insipidus patients, Psychoneuroendocrin, 7:185

Laczi, F., Van Ree, J. M., Wagner, A., Valkusz, Zs., Járdánházy, T., Kovács, G. L., Telegdy, G., Szilárd, J., László, F. A., and De Wied, D., 1983, Effects of desglycinamide-arginine-vasopressin (DG-AVP) on memory processes in diabetes insipidus patients and non--diabetic subjects. Acta Endocrinol. 102:205

Lande, S., Flexner, J. B., and Flexner, L. B., 1972, Effect of corticotropin and desglycinamide-9-lysine vasopressin on suppression of memory by puromycin, Proc.nat.Acad. Sci. (Wash.) 69:558

Legros, J. L., Gilot, P., Seron, X., Claessens, J., Adams, A., Moeglen A., Audibert, J. M., and Berchier, P., 1978, Influence of vasopressin on learning and memory, Lancet 1:41

McGaugh, J. L., Gold, P. E., Handwerker, M. J., Jensen, R. A., Martinez, J. L., Meligeni, J. A., and Vasquez, B. J., 1979, Altering memory by electrical and chemical stimulation of the brain, In: Brain Mechanisms in Memory and Learning: From Single Neuron to Man. IBRO Monograph Series. Vol.4. M. A. B. Brazier (Ed.) Raven Press, New York.

Rigter, H., van Riezen, H., and De Wied, D., 1974, The effects of ACTH- and vasopressin-analogues on CO_2-induced retrograde amnesia in rats, Physiol. Behav. 13:381.

Schulz, H., Kovács, G. L., and Telegdy, G., 1974, Effect of physiological doses of vasopressin and oxytocin on avoidance and exploratory behaviour in rats. Acta Physiol. Hung. 45:211.

Telegdy, G., and Kovács, G. L., 1979a, Role of monoamines in mediating the action of hormones on learning and memory, In: Brain Mechanisms in Memory and Learning: From Single Neuron to Man. IBRO Monograph Series. Vol.4. M. A. B. Brazier, (Ed.) Raven Press, New York, 249

Telegdy, G., and Kovács, G. L., 1979b, Role of monoamines in mediating the action of ACTH, vasopressin and oxytocin, In: Central Nervous System Effects of Hypothalamic Hormones and Other Peptides, Collu, R., A. Barbeau, J. R. Ducharme and J. G. Rochefort (Eds.) Raven Press, New York, 189.

CHOLECYSTOKININ, LEARNING AND MEMORY

Gyula Telegdy, Tibor Kádár and Mátyás Fekete

Inst. of Pathophysiology, Univ. Med. School
Szeged, POB 531, Hungary

INTRODUCTION

The presence of cholecystokinins (CCK-s) in the central nervous system has been well established (Rehfeld 1977, Dockray et al., 1978, Müller et al., 1978). In recent years, a large number of data have accumulated demonstrating many different actions of CCK-s on brain functions. Nevertheless the physiological significance of brain CCK-s is still poorly understood.

During the past few years, intensive research has been carried out in our Department to elucidate the possible action of CCK-s on different aspects of brain function in rat (Telegdy et al., 1979, Telegdy 1981, 1982). We have shown that CCK octapeptide (CCK-8), in its sulphated form (CCK-8-SE) can influence the brain dopamine (DA), norepinephrine (NE) and to some extent serotonin levels in different brain regions (Telegdy et al., 1980, Fekete et al., 1981c,f). Turnover studies revealed that mainly DA and NE are affected (Fekete et al., 1981c). DA release is inhibited in most of the limbic brain regions, with the exception of the striatum in which DA release is facilitated. In contrast, CCK in desulphated form (CCK-8-NS) was practically without effect on these transmitters (Fekete et al., 1981b). Structure - activity studies revealed that mainly the N-terminal portion of CCK-8-SE is important for the effects of this sulphated peptide on neurotransmitter turnover. The fragments of CCK-8-SE that are active in modulating transmitters were also active in stimulating pituitary-adrenal secretion (Fekete et al., 1981b). In contrast to their differential effects on neurochemistry, both CCK-8-SE and CCK-8-NS were found to affect behavior (Telegdy et al., 1982).

In the present work, we report the effects of CCK-8-SE and CCK-8-NS on the learning and memory of different types of conditioned behaviour. The following behavioural paradigms were used: active and passive avoidance behaviour, conditioned feeding and electroshock (ECS) -induced amnesia.

EFFECTS OF CCK-S ON ACTIVE AVOIDANCE CONDITIONING AND CONDITIONED FEEDING BEHAVIOUR

Active avoidance conditioning was performed as described in detail elsewhere (Telegdy et al., 1968). Briefly, the animals were trained to avoid an electric foot-shock by jumping onto a platform fixed to the wall of the box using light as the conditional signal; ten conditioning trials were given daily. CCK was administered intraperitoneally (ip) or intraventricularly (icv) via a chronically implanted stainless-steel cannula in the lateral brain ventricle. For icv administration, the dose was from 80 fmoles to 800 pmoles, whereas for ip administration the dose ranged up to 400 nmoles/kg. In general, to obtain comparable results, doses about 1000 times higher were needed for ip than for icv administration (Fekete et al., 1981e, 1982, Telegdy et al., 1982). The CCK and the fragments used in this study were synthetized by Penke et al. in the Dept. of Medical Chemistry, Szeged.

Both CCK-8-SE and CCK-8-NS were found to attenuate the acquisition of the conditioned avoidance reflex (Fekete et al., 1982).

The study of conditioned feeding behavior was carried out by the method we have described earlier (Fekete et al., 1981e). The conditional signal was a light, and the reinforcement was a piece of grain. Ten conditioning trials were given daily. CCK was administered as in the previous experiments. The results were similar to those of the previous experiments in that both CCK-8-SE and CCK-8-NS impaired the acquisition of conditioned food intake (Fekete et al., 1981a).

The effects of CCK-8 on extinction were studied in other series of experiments. In both active avoidance and conditioned food intake paradigms the animals were trained without any treatment until they had reached the criterion of learning, which was set at 80 % positive response for three consecutive days without reinforcement. In both experimental paradigms the extinctions were facilitated by CCK-8 administration (Fekete et al., 1981a,e).

From the experiments described here the following alternative conclusions can be drawn: CCK given either ip or icv can delay the acquisition of fear and hunger-motivated behaviour and facilitate extinction. These findings may indicate that CCK impairs learning by its effects on memory consolidation, memory storage, or other components of learning, such as arousal, motivation, etc. As the effects were similar in the two experimental designs, the action is probably not specific to basic motivational processes. Although it has been shown that CCK can act as a sedative too (Itoh et al., 1981), such an effect does not seem to be responsible for the action observed, as at the doses used CCK did not affect open field activity (Fekete et al., 1982). However, using different doses other workers have found that CCK decreases open-field activity (Crawley et al., 1981). Other data demonstrate a neuroleptic-like effect of CCK (Van Ree et al., 1983, Fekete et al., 1984) and our data on transmitter effects favour this suggestion (Telegdy et al., 1979, Fekete et al., 1981b, 1981f). It is interesting to note that neuroleptics can decrease avoidance responding (Worms et al., 1983).

EFFECT OF CCK-S ON PASSIVE AVOIDANCE CONDITIONING

From the experiments described in the last section, we cannot conclude whether or not CCK impairs learning and/or memory. To answer this question, the following experiments were designed: Firstly, a passive avoidance task was used and the effects of CCK given either prior to electric foot-shock, immediately after electric foot-shock or 23 h later were studied. Secondly, we studied the action of CCK on ECS-induced amnesia in a passive avoidance paradigm.

The passive avoidance paradigm was the same as that employed by Ader and De Wied (1972). Briefly, the animals were trained on the first day to enter a dark box from an illuminated platform. On day 2 the animals were allowed to enter the dark compartment and electric foot-shock was given (3 mA, 2 s). Retention of the passive avoidance response was studied on the third day by measuring the latency of entry of the animal up to a maximum of 300 s (Fekete et al., 1981d, Kádár et al., 1981). CCK-s were administered icv in equimolar doses in a volume of 5 ul. When treatment was carried out 20 min before the learning trial, some of the larger doses of CCK-8-SE and CCK-8-NS were effective in delaying latency (Fig. 1).

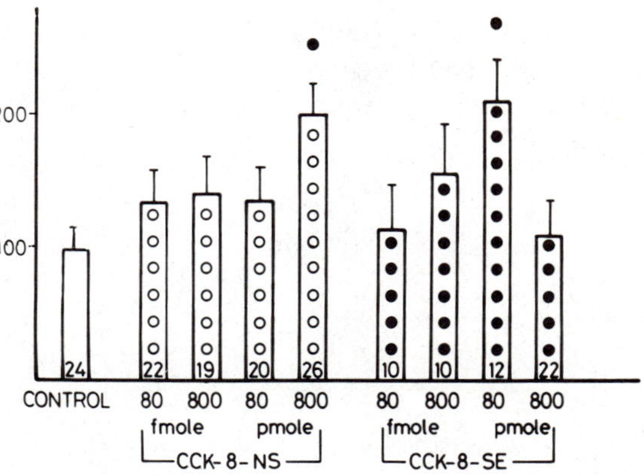

Fig. 1. The effects of CCK-s on latency of passive avoidance task following pre-trial administration. (o : $p<0.05$. Figures at the bottom of the bars represent the number of animals used).

When CCK-8 was given immediately after the learning trial and the animals were tested 24 h later, the passive avoidance latency was lenghthened in a dose-dependent manner. This might indicate that CCK-8 can influence memory consolidation (Fig. 2).

When CCK-8 was given 20 min before testing on day 3, mainly the smaller doses facilitated the passive avoidance response. This suggests that CCK-8 might influence retrieval processes too (Fig. 3).

EFFECTS OF CCK-S ON ECS-INDUCED AMNESIA

The effect of CCK-8 on ECS-induced amnesia was also studied in the passive avoidance task described above. The ECS was given through ear-clip electrodes (190 V, 0.5 s). Immediately after the learning trial. The animals were tested of retention 24 h later.

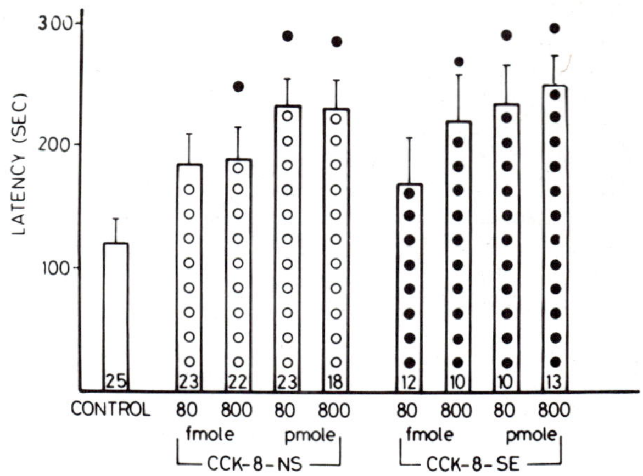

Fig. 2. The effects of CCK-s on latency of passive avoidance task following post-shock administration. o: $p < 0.05$ (For abbreviations see Fig. 1)

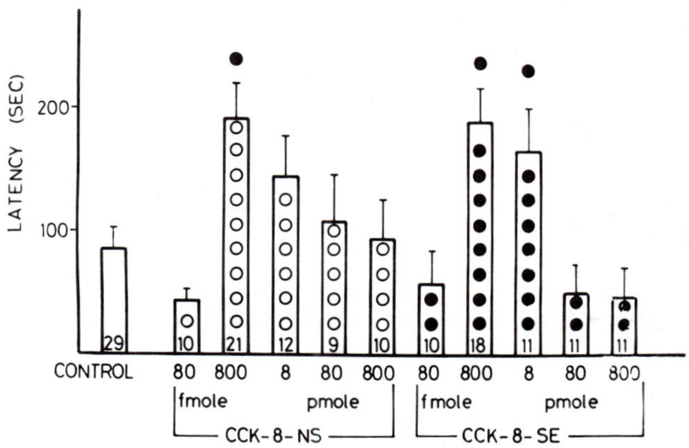

Fig. 3. The effects of CCK-s on latency of passive avoidance task 23 h following post-shock treatment. o: $p < 0.05$ (For abbreviations see Fig. 1)

When CCK treatment (icv) was given 10 min before the learning trial, both CCK-8-SE and CCK-8-NS attenuated the amnesia but this result may have reflected the fact that CCK-8 reduces ECS-induced seizure (Kádár et al., 1964). When CCK treatment was given immediately after ECS, one

Table 1. Effects of CCK octapeptides on electroconvulsive shock-induced retrograde amnesia in a passive avoidance test after icv injections

Treatment		Latency (sec) I.	II.	III.[a]
Saline	-ECS	126 ± 30	161 ± 24	112 ± 23[b]
Saline	+ECS	10 ± 2 ooo	26 ± 7 ooo	32 ± 9 oo
CCK-8-SE				
0.8 pmole	"	26 ± 9	27 ± 7	26 ± 11
8.0 pmole	"	94 ± 25 xx	13 ± 3	32 ± 13
80 pmole	"	49 ± 26	19 ± 4	40 ± 9
CCK-8-NS				
0.8 pmole	"	72 ± 32	95 ± 23 xx	19 ± 5
8.0 pmole	"	73 ± 37	45 ± 16	12 ± 3
80 pmole	"	60 ± 22 x	46 ± 16	29 ± 18

a: I: icv treatments 10 min prior to learning trial and ECS; II.: icv. treatments immediately after ECS (or after the learning trial in Saline - ECS group); III: icv treatments 20 min prior to testing 24-hr retentions
b: means ± SEM
o: statistically significant vs Saline - ECS group
oo: $p<0.01$; ooo: $p<0.001$.
x: statistically significant vs Saline + ECS group
x: $p<0.05$; xx: $p<0.01$.

dose (0.8 pmole) of CCK-8-NS prevented ECS-induced amnesia. However, when CCK-8 was given 20 min before the 24-h retention test, no effect was observed. (Table 1).

These data show that CCK-8-S and CCK-8-NS, though in different doses, prevent or modify the ECS-induced amnesia either before or immediately after ECS treatment. Although the mechanism may be different, the results indicate that CCK-8 improves memory processes too.

SUMMARY

Our studies with active avoidance conditioning and conditioned feeding behavior showed that CCK-8-SE and CCK-8-NS delayed acquisition and facilitated extinction. These effects may be related to the sedative, tranquillizing or neuroleptic effects of CCK. Thus, such effects might have masked any basic learning or memory processes. In passive avoidance conditioning, CCK at high doses improved later retention scores when the drug was given immediately after the learning trial and retention tested 24 h later, but only a lower dose of CCK was effective when given before the 24-h test. These latter results indicate the participation of CCK in memory consolidation and perhaps in memory retrieval also. One low dose of CCK also prevented ECS-induced amnesia, which stregthens these suppositions.

REFERENCES

Ader, R., and De Wied, D., 1972, Effects of vasopressin on active and passive avoidance learning, Psychon. Sci., 29:46.

Crawley, J. N., Hays, S. E., Paul, M. S., and Goodwin, F. K., 1981, Cholecystokinin reduces exploratory behaviour in mice, Physiol. Behav. 27:407.

Dockray, G. J., Gregory, R. A., Hutchison, J. B., Harris, J. I., and Ruswick, M. J., 1978, Isolation, structure and biological activity of two cholecystokinin octapeptides from sheep brain, Nature 274:711.

Fekete, M., Balázs, M., Penke, B., and Telegdy, G., 1981a, Influence of sulfated and unsulfated cholecystokinin octapeptide on conditioned feeding behaviour in rats. Peptides. 2:385.

Fekete, M., Bokor, M., Penke, B., Kovács, K., and Telegdy, G., 1981b, Effects of cholecystokinin octapeptide and its fragments on brain monoamines and plasma corticosterone. Neurochem. Intern., 3:165.

Fekete, M., Kádár, T., Penke, B., Kovács, K., and Telegdy, G., 1981c, Influence of cholecystokinin octapeptide sulfate ester on brain monoamine metabolism in rats. J. Neural Transmission, 50:81.

Fekete, M., Kádár, T., Penke, B., and Telegdy, G., 1981d, Modulation of passive avoidance behaviour by cholecystokinin octapeptides in rats, Neuropeptides, 1:301.

Fekete, M., Szabó, A., Balázs, M., Penke, B., and Telegdy, G., 1981e, Effects of intraventricular administration of cholecystokinin octapeptide sulfate ester and unsulfated cholecystokinin octapeptide on active avoidance and conditioned feeding behaviour of rats. Acta Physiol. Acad.Sci.Hung. 58:39.

Fekete, M., Várszegi, M., Kádár, T., Penke, B., Kovács, K., and Telegdy, G., 1981f, Effect of cholecystokinin octapeptide sulphate ester on brain monoamines in the rat. Acta Physiol, Acad, Sci. Hung. 57:37.

Fekete, M., Bokor, M., Penke, B., and Telegdy, G., 1982, Effects of cholecystokinin octapeptide sulphate ester and unsulphated cholecystokinin octapeptide on active avoidance behaviour in rats, Acta Physiol. Acad.Sci. Hung. 60:57.

Fekete, M., Lengyel, A., Hegedüs, B., Penke, B., Zarándi, M., Tóth, G., Telegdy, G., 1984. Further analysis of the effects of cholecystokinin octapeptides on avoidance behaviour in rats, Eur. J. Pharmacol. 98:79.

Itoh, S., Katsuura, G., 1981, Sedative action of intracerebroventricularly administered cholecystokinin in the rat, Neurosci Lett, Suppl. 7. S323.

Kádár, T., Fekete, M., and Telegdy, G., 1981, Modulation of passive avoidance behavior of rats by intracerebroventricular administration of cholecystokinin octapeptide sulfate ester and nonsulfated cholecystokinin octapeptide. Acta Physiol. Acad.Sci.Hung. 58:269.

Kádár, T., Pesti, A., Penke, B., Telegdy, G., 1984, Inhibition of picrotoxin and electroshock-induced seizures by cholecystokinin octapeptides and their fragments in Neuropharmacology, 23:955.

Muller, J. E., Strauss, E., and Yalow, R. S., 1978, Cholecystokinin and its COOH-terminal octapeptide in the pig brain, Proc.Natl. Acad.Sci. USA 75:486.

Rehfeld, J. F., 1977, Cholecystokinins and gastrin in brain and gut, Acta pharmac, toxicol., 41:24.

Telegdy, G., 1981, The effect of neurohormones on the brain and endocrine system. Acta Phys.Acad.Sci.Hung. 55:273.

Telegdy, G., 1982, Hormone-induced alterations in brain transmitter metabolism and brain function. Endocr. Exper. 16:217.

Telegdy, G., Fekete, M., Kádár, T., 1982, Cholecystokinin in the brain, Intern. Med. 2:2.

Telegdy, G., Hadnagy, J., Lissák, K., 1968, The effect of gonads on conditioned avoidance behavior of rats. Acta Physiol. Acad.Sci.Hung. 33:439.

Telegdy, G., Fekete, M., and Várszegi, M., 1979, Effects of peptide hormone on the neurotransmitter metabolism of the central nervous system. In: "Recent Results in Peptide Hormone and Androgenic Steroid Research", F. A. László, ed., Akadémiai Kiadó, Budapest.

Telegdy, G., Fekete, M., Várszegi, M., and Kádár, T., 1980, Effect of peptide hormones on biogenic amines of the central nervous system. In: Advances in Pharmacological Research and Practice. Gen.ed., J.Knoll, Vol.VII. Aminergic and peptidergic receptors eds.: E. S. Vizi and M. Wolleman, Pergamon Press.Akadémiai Kiadó, Budapest.

Van Ree, J. M., Gaffori, O., and De Wied, D., 1983, In rats the behavioural profile of CCK-8 related peptides resembles that of antipsychotic agents, Eur. J. Pharmacol. 93:63.

Worms, P., Broekkamp, L. E., and Lloyd, K. G., 1983, Behavioral effects of neuroleptics. In: "Neuroleptics: Neurochemical, Behavioral and Clinical Perspectives". J. T. Coyle and S. J. Enna, eds., Raven Press, New York.

POSSIBLE INVOLVEMENT OF NEUROEXCITATORY AMINO ACIDS AND RELATED PEPTIDES IN LEARNING AND MEMORY PROCESSES

Arielle Ungerer, Monique Bourgeois, Joseph Reinbolt* and Yves Boulanger*

Laboratoire de Psychophysiologie, Université Louis Pasteur
7 rue de l'Université, 67000-Strasbourg, France
*Institut de Biologie Moléculaire et Cellulaire, C.N.R.S., rue Descartes, 67000-Strasbourg, France

INTRODUCTION

Aspartate and glutamate were often considered during the 1950s and 1960s as facilitating learning (Zimmerman and Ross, 1944; Hughes and Zubek, 1956; Wincze and Vogel, 1969). However, no clear explanation on the mechanisms involved was provided. More recently, the hypothesis that glutamate has a role in memory processes was first formulated by Van Harreveld and Fifkova (1974). These authors proposed that neural activity associated with learning experiences provokes glutamate release into the extracellular space which leads to modifications in neuronal permeability to sodium and, as a consequence, to swelling of dendritic spines and facilitation of synaptic transmission. These specific morphological changes would constitute the basis of facilitated neuronal pathways which are supposed to underlie memory storage. Although this hypothesis remains to be tested, there is growing evidence that excitatory amino acid systems could be involved in neuronal synaptic plasticity and thereby might have a predominant role in learning and memory processes.

CHARACTERISTICS AND DISTRIBUTION OF ACIDIC AMINO ACID RECEPTORS

A great deal of experimental data indicates that glutamic acid (Glu) and aspartic acid (Asp) satisfy to a large extent the criteria for classification as neurotransmitters and may be considered as the major excitatory mediators in the mammalian central nervous system (for a review, see Fonnum, 1984). Their action as neurotransmitters is mediated by various types of receptors which have been characterized on the basis of their pharmacological profile, their ionic requirement and their electrophysiological characteristics ; however, the complexity in this area has considerably increased during recent years as studies have developed and no clear classification of glutamate receptors can be proposed yet (for a review, see Mc Lennan, 1983).

At present, electrophysiological experiments suggest the existence of four classes of receptors characterized by their selective sensitivity to N-methyl-D-aspartate (NMDA), quisqualic acid (QA), kainic acid (KA) and 2-amino-4-phosphono butyric acid (APB). This classification only partially overlaps that which can be proposed according to ion sensitivities of glutamate receptors. Neurochemical data and binding studies have identified both sodium-dependent binding sites which present characteristics of high-affinity glutamate uptake sites, and sodium-independent binding sites. Among the latter, one can distinguish Cl-/Ca2+ -dependent binding sites sensitive to APB and Cl-/Ca2+ -independent binding sites that present some pharmacological similarities with NMDA and quisqualate receptors (Whittemore et al. 1983 ; Fagg et al., 1984).

Glutamate receptors are largely distributed throughout the brain, with the highest concentrations in cerebral cortex, striatum, hippocampus and cerebellum. Many pathways related to these structures have been shown to use acidic amino acids as their neurotransmitter. Fagg and Foster (1983) described some of these pathways in the mammalian brain, principally, the corticofugal pathways projecting to subjacent structures and the hippocampal fibres which project via the fimbria to the lateral septum, nucleus accumbens, diagonal band of Broca, bed nucleus of the stria terminalis, hypothalamus and mammillary bodies (Fig. 1).

Likewise, excitatory amino acids appear at present to be the main candidates for transmitters in intrinsic hippocampal circuitry. Several studies have demonstrated that the perforant path which originates from the entorhinal cortex and projects to dentate granule cells, the mossy fibres which originate from granule cells and project to CA3 pyramidal cells and the Schaffer collaterals-commissural fibres which connect CA3 to CA1 areas respectively ipsi- and contralaterally, all use Asp and/or Glu as their transmitters.

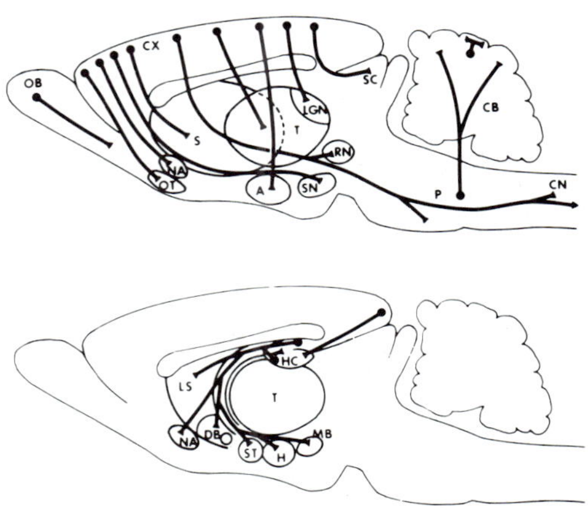

Fig.1 : Some pathways proposed to use acidic amino acid transmitters in the mammalian brain. Abbreviations:A,amygdala; CB,cerebellum; CN,cuneate nucleus; CX,cerebellar cortex; DB,nucleus of the diagonal band; GP,globus pallidus; HC,hippocampus; H,hypothalamus; LGN,lateral geniculate nucleus; LS,lateral septum; LVN,lateral vestibular nuclei; MB,mammillary body; NA,nucleus accumbens; OB,olfactory bulb; OT,olfactory tubercle; P,pons; RN,red nucleus; S,striatum; SC,superior colliculus; SN,substantia nigra; ST,bed nucleus of the stria terminalis; T,thalamus.

(From Fagg and Foster, 1983)

Moreover, all types of receptors cited above seem to be involved in hippocampal circuitry, where their distribution appears highly specific (Monaghan et al. 1983). For example, there is evidence that mossy fibres use kainate-type receptors (Forster et al., 1981 ; Berger and Ben-Ari, 1983 ; Coyle, 1983) ; on the other hand, the lateral and medial entorhinal cortex perforant paths use different types of receptors, the former APB-type receptors and the latter quisqualate and/or kainate-type receptors (Koerner and Cotman, 1981 ; Crunelli et al., 1983).

In summary, neuroexcitatory amino acid systems appear to be a major substrate of neuronal activity in those structures such as cerebral cortex, hippocampus and other regions of the limbic system which are known to have an essential role in learning and memory. Such a functional role and anatomical distribution are a first argument to suggest their possible involvement in memory processes. A second argument is provided by recent data which suggest that acidic amino acids could be involved in synaptic plasticity of the hippocampal formation.

EXCITATORY AMINO ACIDS AND SYNAPTIC PLASTICITY IN THE HIPPOCAMPUS

Brief high frequency stimulation of either the perforant path, the mossy fibres or the Schaffer collaterals-commissural fibres induce a long-lasting enhancement of synaptic efficiency at monosynaptic junctions (for reviews, see Eccles, 1983 ; Voronin, 1983 ; Teyler and Discenna, 1983). This enhancement, currently called "long-term potentiation" (LTP), may persist for days or weeks. Owing to its properties, the form of synaptic plasticity exhibited by LTP is generally considered as being a most valuable model for enduring synaptic changes that are supposed to underlie memory encoding. As this argument is expressed elsewhere in this volume (see the chapters by Buzsaki and Laroche), we shall not consider it further here.

It must be emphasized that LTP seems relatively specific to those hippocampal pathways that use excitatory amino acids as transmitters, although LTP with somewhat different characteristics has recently been observed in the neocortex by Lee (1983) and in various pathways of the limbic system by Racine et al. (1983). This distribution led Storm-Matthisen (1977) to suggest that acidic amino acids may be involved in LTP and both pre- and post-synaptic mechanisms of mediation have been proposed (Bliss and Dolphin, 1982).

Presynaptic mechanisms . LTP induction is accompanied by a prolonged increase in glutamate release at the terminal fields of the stimulated pathway. This was first demonstrated by Skrede and Malthe-Sorenssen (1981) using hippocampal slices : LTP induction in afferent fibres of CA1 pyramidal cells produced a long-lasting increase in both resting and evoked release of the glutamate analog [3H]-D-Asp. This increase appeared to be correlated to LTP induction, since in the absence of Ca^{2+} in the medium, no LTP can be induced and no increase in D-Asp release was observed. Similarly, Dolphin et al. (1982) using an *in vivo* technique found that LTP induction in the perforant path was correlated with a prolonged increase of newly synthetized glutamate in the dentate gyrus.

Post-synaptic mechanisms have been proposed by Baudry, Lynch and their colleagues. These authors found that high-frequency stimulation of hippocampal slices produced a marked and lasting increase in the number of [3H]-glutamate binding sites in hippocampal membranes correlated with LTP induction (Baudry et al., 1980 ; Lynch et al. 1982). The same increase was observed when low concentrations of calcium were added to the medium bath (Baudry et al. 1981). In light of these data, these workers proposed the intriguing hypothesis that high-frequency stimulation leads to an influx of calcium ions into dendrites, activating a calcium-dependent protease which unmasks additional post-synaptic glutamate receptors ; they postulated that such a mechanism may be involved in memory storage (Lynch and Baudry, 1984). In agreement with their hypothesis, they found that calcium degrades fodrin (a membrane associated protein) in a similar way to processes postulated to lead to uncovering of glutamate binding sites. However, Lynch and Baudry's (1984) hypothesis has recently been challenged by Sastry and Goh (1984) who observed that low-frequency stimulation of Schaffer collaterals produced a depression of CA1 population spikes associated with an increase in [3H]-glutamate binding, whereas LTP induction was without effect on [3H]-glutamate binding. According to these authors, LTP would be presynaptic and the increase in the number of receptors observed by Baudry and Lynch might in fact concern only the extrasynaptic glutamate receptors.

Pharmacological data provide arguments for both pre- and postsynaptic hypotheses. On the one hand, exogenous application of glutamate or kainic acid to hippocampal slices induced long-term potentiation after stimulation of Schaffer Collaterals (Collingridge et al., 1983 ; Goh and Sastry, 1983). On the other hand, the antagonists of glutamate receptors, APB, 2-amino-5 phosphono-valeric acid (APV, a NMDA-receptor antagonist) and Gamma-D glutamyl-glycine (γ-DGG, an antagonist of NMDA-, KA- and QA-receptors) blocked LTP although APV and γ-DGG seemed to mask rather than prevent LTP (Dunwiddie et al. 1978 ; Koerner and Cottman, 1981 ; Dolphin, 1983 ; Sastry et al. 1984).

All these results strongly suggest a participation of excitatory amino acids in LTP phenomena. Their mechanisms and sites of action, pre- and/or post- synaptic, remain to be determined more precisely; but, it can be outlined that both pre- and post-synaptic hypotheses have been proposed for LTP and the genesis and maintenance of LTP may well involve the two types of mechanisms (Bliss and Dolphin, 1982).

EXCITATORY AMINO ACIDS AND LEARNING AND MEMORY

In light of these findings, the question arises whether glutamate and aspartate systems are involved in learning and memory processes but little work has been done in this area. This is essentially due to three kinds of difficulty which strongly limit behavioral studies : 1) The neurotoxic effects produced by most excitatory amino acids and analogs or antagonists ; 2) the lack of highly selective agonists and antagonists for the various types of receptors ; 3) the fact that most of these substances do not cross the blood brain barrier except at the circumventricular level.

The most extensive studies during the last ten years have been carried out in chicks, first by Van Harreveld and Fifkova (1974) and then by Cherkin et al. (1976). They found that L-proline and L-glutamine, two amino acids which inhibit endogenous glutamate release, provoked a retrograde amnesia in an one-trial avoidance task. These amnestic effects were time-dependent and did not seem to be due to electrophysiological disturbances or behavioral deficits. N-acetyl-L-glutamate which primarily affects glutamate uptake in synaptosomes produced similar memory deficits (Davis et al. 1981). Moreover, amnestic potency of L-proline analogs was correlated with the degree of glutamate inhibition and appeared highly stereospecific since D-proline had no effect on learning retention (Pico et al., 1983). Gibbs et al. (1977) confirmed amnestic action of L-proline, but suggest that it could be due to interference of L-proline with amino-acid transport into the brain, leading to impairment of cerebral protein synthesis. However, no confirmation of this hypothesis has emerged yet (Cherkin et al. 1981).

Freed and Wyatt (1981) found that glutamate-diethyl ester (GDEE, an antagonist of quisqualate receptors) deeply impaired acquisition of a food-motivated instrumental task in rats; in contrast, no performance impairment was observed in rats that had previously learned the task. These results were interpreted by the authors as demonstrating that excitatory processes would play an important role in learning, particularly when learning involves acquisition of new behavioral patterns. In addition, it can be noted that low doses of glutamic acid into the hippocampus (Smialowski, 1983) or into the reticular formation (Bernardi, 1982) provoked behavioral arousal and activation of electrocorticogram suggesting that glutamate systems may be involved in the control of electrocorcortical activity.

More recently, it was found that classical conditioning of the eyelid response in rabbits was accompanied by a long-lasting and learning-dependent increase in the number of [3H]-glutamate binding sites in the hippocampal membranes (Mamounas et al., 1984). Furthermore, intra ventricular perfusion of leupeptin (an inhibitor of the calcium-activated enzymes) during several days, deeply impaired spatial memory either in an 8-arm maze or in a delayed alternation task (Staubli et al., 1984). Leupeptin also deeply impaired performance on an olfactive discrimination task, but had no effect on inhibitory avoidance behavior or on active avoidance conditioning. From these results and according to the fact that calcium is quite ineffective in unmasking glutamate receptors in several brain stem regions, Lynch and Baudry (1984) concluded that calcium-dependent mechanisms postulated to be involved in memory storage would be operative only in telencephalic brain stuctures and that some forms of learning and memory may be mediated by other cellular processes.

ISOLATION OF AN ACIDIC PEPTIDE THAT IMPAIRS LEARNING RETENTION

We recently isolated from Datura stramonium (solanaceae), a plant known since Antiquity for its amnestic effects, a peptide composed of aspartate and glutamate which deeply impairs learning retention in mice. In our first studies we showed that the aqueous extract of Datura produced a significant

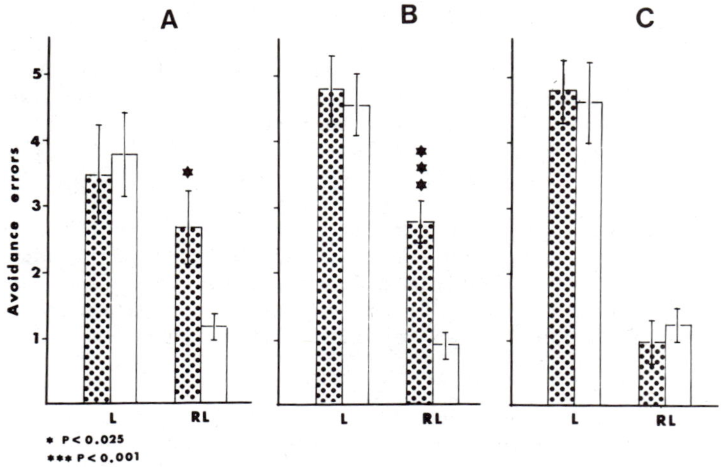

Fig. 2 : Effects of Datura extract on the retention of a Y-maze avoidance learning. The animals received systemic injection of either Datura extract (dotted columns) or 9°/oo NaCl (white columns), 1h. (A), 24h. (B) or 7 days (C) after a first learning session and were submitted to a relearning session 48h. following the treatment.

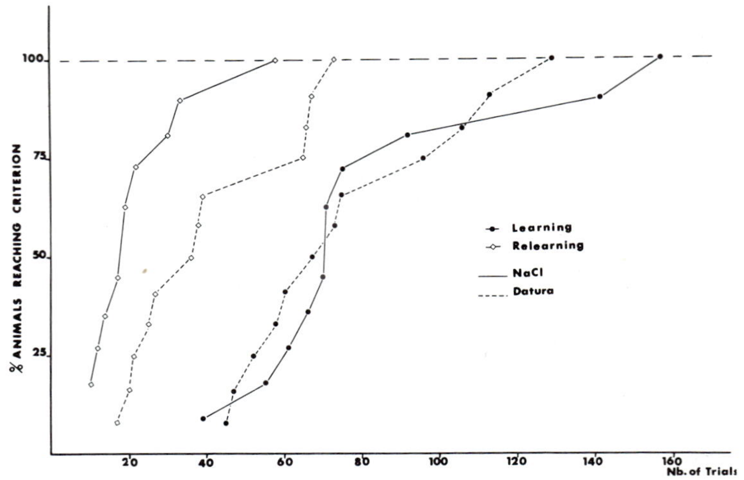

Fig. 3 : Effects of Datura extract on the relearning of a two-way active avoidance. Cumulative records of animals that reached the criterion of 9 correct responses on 10 consecutive trials during learning session, 24h. before treatment (dark symbols) and during relearning session, 3 days following the treatment (open symbols).

retention deficit in various learning tasks (Ungerer et al., 1978). Systemic injection of the extract 1h. or 24h. after acquisition of an Y-maze avoidance task severely impaired reacquisition of the task 48h. later, as expressed by the number of avoidance errors made during the relearning session by treated animals as compared to control animals (Fig. 2). No effects were observed when the extract was administered seven days after learning session, thus indicating that Datura effects on learning retention were time dependent. Similarly, the aqueous extract administered 24h. after acquisition of two-way active avoidance learning impaired reacquisition on the task three days later (Fig. 3). The Datura extract also induced a significant retention deficit on an appetitively reinforced CRF task : the animals which received systemic injection of the extract 24h. after a single 45 min. learning session made few lever-pressing responses during retention testing (N = 20 ; M = 54.15 ± 4.79) when compared to control animals (N = 25 ; M = 78.36 ± 3.78)(F 1/43 = 16.51 ; P<0.001).

We demonstrated that these deficits could not be ascribed to performance impairments nor to an action on locomotor activity or emotional reactivity of the animals as shown by open-field and actographic measures. Moreover, similar retention deficits were observed whether the extract was injected systemically or into the lateral ventricle, which suggests a direct action on the central nervous system.

Chemical analysis of the Datura extract led us to isolate a new peptide. Isolation was obtained by successive fractionning on DEAE-trisacryl, Sephadex G-25, SP-Sephadex G-25 and in the last step by electrophoresis at pH 4.4 on silicium plates. Analysis of the isolated peptide on amino acid analyzer before and after hydrolysis (Fig. 4, left graph) revealed that the active principle of Datura was a small acidic peptide composed of aspartic acid and glutamic acid.

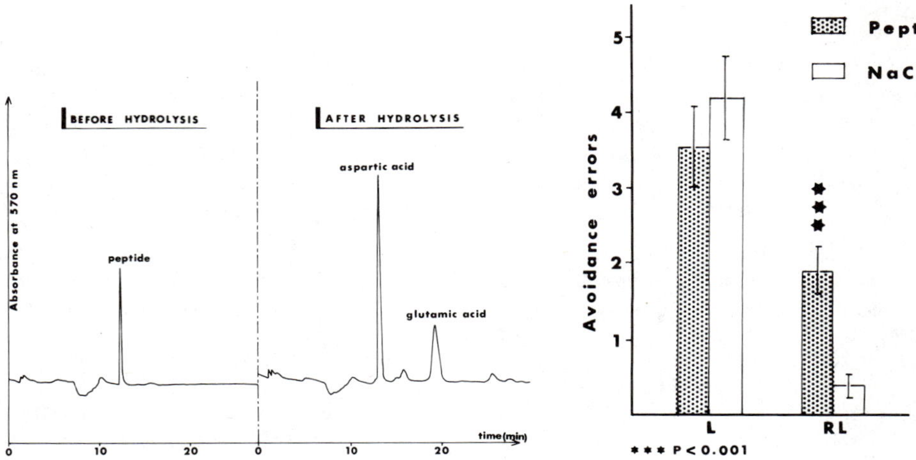

Fig. 4 : The purified peptide analyzed before and after hydrolysis on amino acid analyzer (left graph). Effects of the purified peptide on the retention of Y-maze avoidance learning when administered 24h. after the first learning session (L), the retention testing (RL) being carried out 48h. following the treatment (right graph).

Behavioral studies showed that the purified peptide systemically administered at nanomolar doses 24h. after acquisition of Y-maze avoidance learning impaired retention in the same way as the Datura extract (Fig. 4, right graph). At these doses, the peptide is devoid of toxicity and has no apparent effect on the animal's general behavior.

From these results, it can be concluded that the new peptide differs from various peptides known to affect memory processes, 1) by virtue of its long delays of effectiveness and its lasting effects on learning retention still apparent 48h. after treatment, and 2) by its particular molecular composition. Such charateristics lead us to suggest that the isolated peptide may affect some slow memory processes such as those evoked by Mc Gaugh and Gold (1976) and more recently by Squire and Spanis (1984) and that this action could be mediated by neuroexcitatory amino acid systems.

Preliminary studies have been carried out in order to test this hypothesis : it has been found that the peptide at nanomolar concentrations partially inhibited [3H]-glutamate binding to striatal and hippocampal membranes but had no effect on glutamate binding in cerebellar membranes (De Barry and Ungerer, unpublished data). It must be noted that in contrast to the hippocampal and striatal glutamate binding sites, the cerebellar sites are not calcium-dependent and differ from the former by their pharmacological charateristics (Baudry and Lynch, 1980; Whittemore et al., 1983). According to these data, we suggest that the isolated peptide might have a specific activity on excitatory amino acid systems and thus provide support for the hypothesis of a role of those systems in memory processes. Research is underway to determine more precisely the site of action of the isolated peptide. If specificity were demonstrated, owing to its action on learning retention and its lack of toxicity, the peptide would constitute a valuable tool for studying excitatory amino acids systems and their role in learning and memory processes.

ACKNOWLEDGMENT

We thank Dr. I. Arimi for his assistance in chemical studies of the isolated peptide and Dr. J. Anderson for his help in translating the manuscript.

REFERENCES

Baudry, M., Bundman, M. C., Smith, E. K., and Lynch, G., 1981, Micromolar calcium stimulates proteolysis and glutamate binding in rat brain synaptic membranes. Science, 212:937.

Baudry, M., Lynch, G., 1980, Regulation of hippocampal glutamate receptors : evidence for the involvement of calcium activated protease. Proc. Nat. Acad. Sci. US, 77,2298.

Baudry, M., Oliver, M., Creager, R., Wieraszko, A., and Lynch, G., 1980, Increase in glutamate receptors following repetitive electrical stimulation in hippocampal slices. Life Sci., 27:325.

Berger, M., and Ben-Ari, Y., 1983, Autoradiographic visualization of [3H]-kainic acid receptor subtypes in the rat hippocampus. Neurosci. Lett., 39:237.

Bernardi, N., 1982, Possible involvement of glutamate in electrocortical activity. Neuropharmacol., 21:937.

Bliss, T. V. P., and Dolphin, A. C., 1982, What is the mechanism of long-term potentiation in the hippocampus. Trends Neurosci., 5:289.

Cherkin, A., Bennett, E. L., and Davis, J. L., 1981, Amnestic effect of L-proline does not depend upon inhibition of brain protein synthesis. Brain Res., 223:455.

Cherkin, A., Eckhardt, M. J., and Gerbrandt, L. K., 1976, Memory : proline induces retrograde amnesia in chicks. Science, 193:242.

Collingridge, G. L., Kehl, S. J., Loo, R., and Mc Lennan, H., 1983, Effects of kainic and other amino acids on synaptic excitation in rat hippocampal slices : I- Extracellular analysis. Exp. Brain Res., 52:170.

Coyle, J. T., 1983, Neurotoxic action of kainic acid. J. Neurochem., 41:1.

Crunelli, V., Forda, S., and Kelly, J. S., 1983, Blockade of amino acid-induced depolarizations and inhibition of excitatory post-synaptic potentials in rat dentate gyrus. J. Physiol.(London), 341:627.

Davis, J. L., Cherkin, A., and Hewitt, M. D., 1981, Intraventricular N-acetyl-L-glutamate induces retrograde amnesia in chicks. Pharmacol. Biochem. Behav., 14:919.

Dolphin, A. C., 1983, The excitatory amino acid antagonist γ-D-glutamylglycine masks rather than prevents long-term potentiation of the perforant path. Neuroscience, 10:377.

Dolphin, A. C., Errington, M. L., and Bliss, T. V. P., 1982, Long-term potentiation of the perforant path in vivo is associated with increased glutamate release. Nature, 297:496.

Dunwiddie, T. V., Madison, D., and Lynch, G., 1978, Synaptic transmission is required for initiation of long-term potentiation. Brain Res., 150:413.

Eccles, J. C., 1983, Calcium in long-term potentiation as a model for memory. Neuroscience, 10:1071.

Fagg, G. E., and Foster, A. C., 1983, Amino acid neurotransmitters and their pathways in the mammalian central nervous system. Neuroscience, 9:701.

Fagg, G. E., Riederer, B., and Matus, A., 1984, Sodium ions regulate a specific population of acidic amino acid receptors in synaptic membranes. Life Sci., 34:1739.

Fonnum, F., 1984, Glutamate : a neurotransmitter in mammalian brain. J. Neurochem., 42:1.

Foster, A. C., Mena, E. E., Monaghan, D. T., and Cotman, C. W., 1981, Synaptic localization of kainic acid binding sites. Nature, 289:73.
Freed, W. J., and Wyatt, R. J., 1981, Impairment of instrumental learning in rats by glutamic acid diethyl ester. Pharmacol. Biochem. Behav., 14:223.
Gibbs, M. E., Ng, K. T., and Richdale, A. L., 1977, L-proline inhibition of long-term memory formation. Neurosci. Lett., 6:355.
Goh, J. W., and Sastry, B. R., 1983, Effects of localized applications of L-glutamate on the population spike in the hippocampal slice preparation. Life Sci., 33:1673.
Hughes, K. R., and Zubek, J. P., 1956, Effect of glutamic acid on the learning ability of bright and dull rats : I. Administration during infancy. Canad. J. Psychol., 10:132.
Koerner, J. F., and Cotman, C. W., 1981, Micromolar L-2-amino-4-phosphono butyric acid selectively inhibits perforant path synapses from lateral entorhinal cortex. Brain Res., 216:192.
Lee, K., 1983. Sustained modification of neuronal activity in the hippocampus and cerebral cortex. In : "Molecular Cellular and Behavioral Neurobiology of the Hippocampus", W. Seifert, ed., Academic Press, New-York.
Lynch, G., and Baudry, M., 1984, The biochemistry of memory : a new and specific hypothesis. Science, 224:1057.
Lynch, G., Halpain, S., and Baudry, M., 1982, Effects of high frequency synaptic stimulation on glutamate binding studies with a modified in vitro hippocampal slices. Brain Res., 244:101.
Mc Gaugh, J. L., and Gold, P. E., 1976, Modulation of memory by electrical stimulation of the brain. In : "Neural Mechanisms of Learning and Memory", M. R. Rosenzweig and E. L. Bennett, ed., MIT Press, Cambridge, Massachusetts.
Mc Lennan, H., 1983, Receptors for the excitatory amino acids in the mammalian central nervous system. Progress Neurobiol., 20:251.
Mamounas, L. A., Thompson, R. F., Lynch, G., and Baudry, M., 1984, Classical conditioning of the rabbit eyelid response increase glutamate receptor binding in hippocampal synaptic membranes. Proc. Nat. Acad. Sci. US, 81:2548.
Monaghan, D. T., Holets, V. R., Toy, D. W., and Cotman, C. W., 1983, Anatomical distribution of four pharmacologically distinct [3H]-L-glutamate binding sites. Nature, 306:176.
Pico, R. M., Keller, E., Cherkin, A., and Davis, J. L., 1983, Brain glutamate inhibition and amnesia : evidence provided by proline analog action. Develop. Brain Res., 9:223.
Racine, R. J., Milgram, N. W., and Hafner, S., 1983, Long-term potentiation phenomena in the rat limbic forebrain. Brain Res., 260:217.
Sastry, B. R., and Goh, J. W., 1984, Long-lasting potentiation in hippocampus is not due to an increase in glutamate receptors. Life Sci., 34:1497.
Sastry, B. R., Goh, J. W., and Pandanaboina, M. M., 1984, Verapramil counteracts the masking of long-lasting potentiation of hippocampal population spike produced by 2-amino-5-phosphonovalerate. Life Sci., 34:323.
Skrede, K. K., and Malthe-Sorenssen, D., 1981, Increased resting and evoked released of transmitter following repetitive electrical tetanization in hippocampus : a biochemical correlate to long-lasting synaptic potentiation. Brain Res., 208:436.
Smialowski, A., 1983, Excitatory effect of intrahippocampal injection of glutamic acid on rabbit EEG. J. Neural Transmission, 58:205.
Squire, L. R., and Spanis, C. W., 1984, Long gradient of retrograde amnesia in mice : continuity with the finding in humans. Behav. Neurosci., 98:345.
Staubli, U., Baudry, M., and Lynch, G., 1984, Leupeptin, a thiol proteinase inhibitor, causes a selective impairment of spatial maze performance in rats. Behav. Neural Biol., 40:58.

Storm-Mathisen, J., 1977, Localization of transmitter candidates in the brain : the hippocampal formation as a model. <u>Progress Neurobiol.,</u> 8:119.

Teyler, T. J., and Discenna, P., 1984, Long-term potentiation as a candidate mnemonic device. <u>Brain Res. Rev.,</u> 7:15.

Ungerer, A., Schuber, F., Chauvin, R., 1978, A factor isolated from Datura stramonium that affects learning retention in mice. <u>Behav. Biol.,</u> 24:349.

Van Harreveld, A., Fifkova, E., 1974, Involvement of glutamate in memory formation. <u>Brain Res.,</u> 81:455.

Voronin, L. L., 1983, Long-term potentiation in the hippocampus. <u>Neuroscience,</u> 10:1051.

Whittemore, S. R., Mena, E. E., Monaghan, D. T., and Cotman, C. W., 1983, Regional distribution and ionic requirement of Cl-/Ca2+ activated and Cl-/Ca2+ independent glutamate receptors in rats. <u>Brain Res.,</u> 277:99.

Wincze, J. P., and Vogel, W., 1969, The effects of glutamic acid upon operant conditioning in rats. <u>J. gen. Psychol.,</u> 115:97.

Zimmerman, F. T., and Ross, S., 1944, Effect of glutamic acid and other amino acids on maze learning in the white rat. <u>Arch. Neurol. Psychiat. (Chic.),</u> 51:446.

FUNCTIONAL DEFICITS AND ANATOMICAL SPARING AFTER PRENATAL BRAIN DAMAGE IN THE RAT

Sandra C. Pereira, Charles R. Legg*, Ian Steele Russell
and Mitchell Glickstein

MRC Unit on Neural Mechanisms of Behaviour
Department of Anatomy & Embryology
University College London
London WC1E 7JG
England

*
The Psychology Unit
Department of Social Science and Humanities
The City University
London EC1V 0HB
England

One important aspect of plasticity is the extent to which animals can develop normal behaviour despite interference with the structural development of the brain. A teratogen derived from the glucoside cycasin, called methylazoxymethanol acetate (MAMa), selectively destroys dividing cells. When administered to a pregnant rat MAMa will inhibit the development of structures forming in the foetus at that time. Treatment with MAMa on embryonic day 15 produces offspring whose brain is the only organ revealing gross malformation. The brain is microencephalic, a phenomenon first described by Spatz and Laqueur (1968). This reduction in brain size in MAMa treated rats is mainly due to a failure of neocortical development (see Fig. 1).

We have been using MAMa to induce microencephaly to investigate the following questions: (1) what sorts of behavioural deficits are associated with microencephaly? (2) what are the connections of the surviving cortex? (3) is there functional compensation after such an early insult to the developing brain?

From casual observation of behaviour, such rats appear to be entirely normal. Rabe and Haddad (1972) reported that they could learn to perform a variety of simple operant conditioning tasks as well as normal animals. However, they found that microencephalic rats showed deficits in tasks involving spatial learning, i.e. the Hebb-Williams maze and the Maier reasoning test. We have found (Pereira and Russell, 1981) that microencephalic rats were identical in their motor abilities to normal animals. We could detect no deficits of any kind in their locomotion. They were indistinguishable from normal rats when tested on a comprehensive battery of tasks ranging from

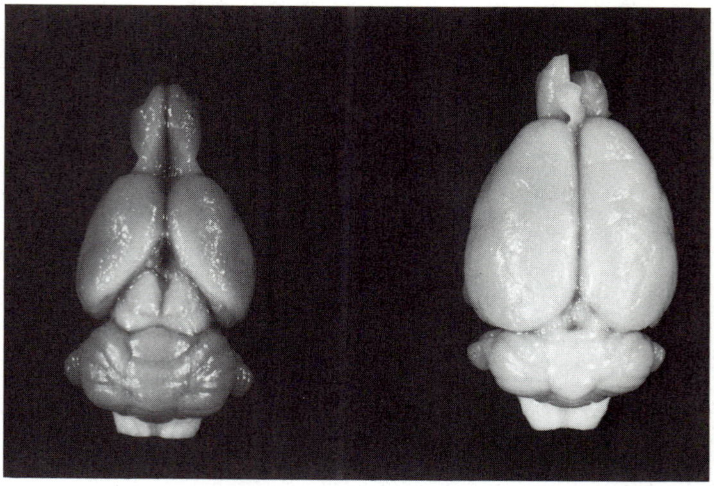

Fig. 1. Microencephalic (left) and normal (right) rat brains. in the microencephalic brain the caudal cortex is absent and the colliculi are exposed. Both the cerebellum and brain stem appear normal.

simple reflexes to negotiating a 60 cm. long 6mm. wide flexible elevated track for food reward.They also showed normal performance when tested with a brightness discrimination task. In contrast, we found that microencephalic rats were greatly impaired on visual pattern discrimination problems.

The present study was designed both to confirm and extend these earlier behavioural findings of a visual discrimination deficit in microencephalic rats. In addition we have begun to study some of the connections of microencephalic brains in the hope of eventually relating the anatomical changes produced by MAMa to the observed behavioural deficits.

BEHAVIOURAL DEFICITS

Rabe and Haddad (1972) reported that microencephalic rats are hyperreactive, an observation which we have also confirmed. Our work on visual discrimination used shock motivation where the microencephalic rats were found to be more emotionally responsive than normal animals to the level of shock applied (1 ma). It was therefore possible that their visual impairment was due to this motivational difference rather than to a lack of cortex. Accordingly, we investigated their visual abilities using lower shock levels to avoid sensitizing the microencephalic animals and to determine whether or not there was a true visual deficit. Both normal and microencephalic rats were trained on a series of visual tests using a low shock level of 0.2 ma. This mild shock level was sufficient to motivate performance of the task by both normal and microencephalic rats without any signs of emotional disturbance.

Microencephalic rats were produced as previously described (Pereira and Russell, 1981). Pregnant Wistar (albino) females were injected intra-peritoneally with a dose of 20 mg per Kg of MAMa on day 15 of gestation (day 0 was taken as the day when a sperm positive vaginal smear was obtained). A comparable volume of 0.9% saline was used to inject control rats.

Five microencephalic and four control rats were examined in this study. All rats were approximately 4 months old when behavioural testing was begun. The apparatus and training regime have been described in detail previously (Steele Russell, Bookman and Mohn, 1979). Briefly, the two-choice discrimination apparatus consisted of a start box from which the animal passed into a choice area where he was confronted with two doors on which were the two targets to be discriminated. The animal could exit to avoid the shock only through the door which displayed the correct target. Shocks of 0·5 second duration were given every two seconds after an initial five second avoidance period. A correct choice was defined as the entry into the goal box behind the correct target without any prior responses to the incorrect display. A correction procedure was used where each trial was continued until a correct choice was made and the rat entered the goal box. All animals were trained for 30 trials per day until 420 trials were given on each discrimination. The visual displays attached to each door consisted of 8cm. diameter circular cards. A black versus a white circle was used for the brightness discrimination and circles composed of 6mm. wide alternating black and white stripes were used for the pattern discriminations (either horizontal versus vertical or 45 degrees versus 135 degrees). After the training period for each discrimination, stimulus control tests were made to determine whether the animals behaviour might be due to non-visual strategies. During these tests both goal box doors displayed identical targets but one was locked and the other open as in the visual discrimination phase. With identical targets the animals performance should have fallen to chance if they had been using vision during the discrimination. If there were spurious cues from the apparatus, performance should still have been above chance.

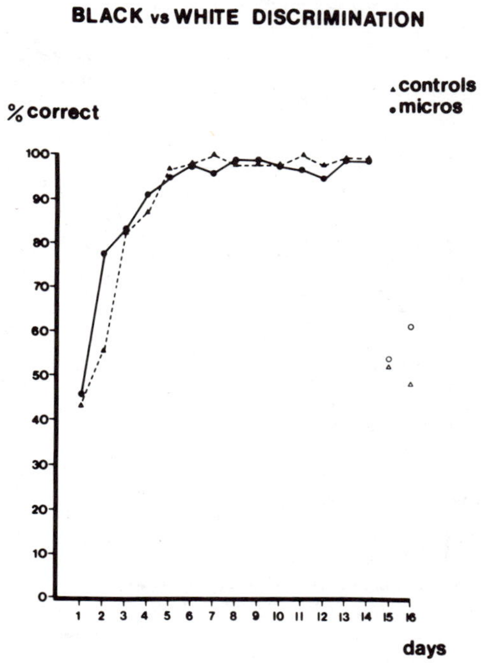

Fig. 2. Performance of normal (dotted line) and microencephalic rats (solid line) on the brightness discrimination. O and Δ represent performance when two black (day 15) or two white (day 16) targets were given (stimulus control tests).

The rats were first trained on a brightness task and their performance is shown in figure 2. It is clear that microencephalic rats performed as well as normals and had no deficit in brightness vision. Both normals and microencephalics learned the task rapidly and in both groups the stimulus control tests reduced performance levels to chance, illustrating that the animals had been using vision to perform the task. The same subjects were then tested for pattern vision using a horizontal versus vertical discrimination task. The results are shown in figure 3. The normal animals reached the criterion of 90% correct after 90 trials and remained at this level during the 420 trial period; the stimulus control tests reduced their performance to chance. The microencephalic group showed a clear deficit and never reached criterion during the 420 training trials. However, their performance did improve with time and the stimulus control tests reduced it to chance. This suggests that they were using vision to perform the task. Finally, all subjects were presented with the tilt discrimination consisting of 6mm. stripes orientated at 45 or 135 degrees. Figure 4 shows that the normals were able to learn the task in 420 trials. The stimulus control tests again illustrate that they were using vision to solve the task. The microencephalic rats were unable to perform this discrimination above chance.

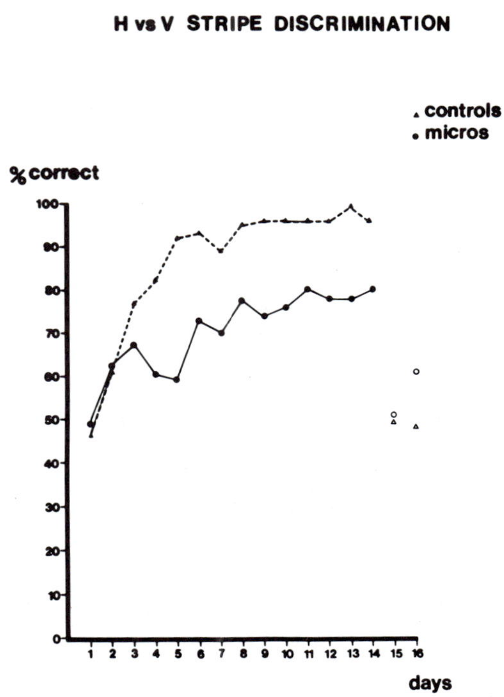

Fig. 3. Performance of normal(dotted line) and microencephalic rats (solid line) on the horizontal/vertical stripe discrimination. O and Δ represent performance when two horizontal or two vertical targets were given (stimulus control tests).

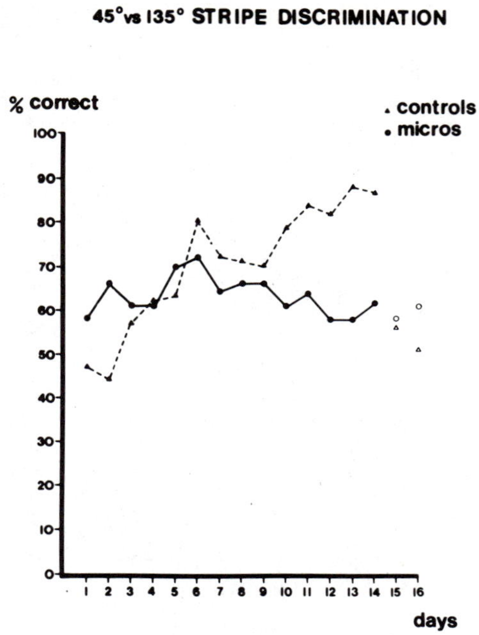

Fig. 4. Performance of normal (dotted line) and microencephalic rats (solid line) on the tilt discrimination. O and Δ represent performance when two 45 degree or two 135 degree striped targets were given (stimulus control tests).

From these results it would appear that the visual capacities of the microencephalic rat are as follows: (a) these animals can discriminate brightness normally, (b) they are deficient on a horizontal versus vertical stripe discrimination but may be able to make some visual distinction on the basis of local flux differences(e.g. they could simply be looking at the bottom line of the horizontal display and making a grey level analysis with an equivalent area of the vertical display) which could mediate the improvement of performance over time, (c) on a discrimination task which relies solely on pattern detection(oblique stripes) they show no improvement over 420 trials, (d) microencephalic rats both ran normally through the visual testing apparatus and they performed the brightness task successfully. This suggests that the deficit in pattern vision is not simply one of spatial learning as reported by Rabe and Haddad(1972).

ANATOMICAL SPARING

Microencephalic rats have a significant visual deficit and yet we have observed no comparable motor deficit. We have been studying some connections within the microencephalic brain in an attempt to correlate the behavioural sparing and loss with the structural defects produced by MAMa. On day 15 of gestation the embryonic cerebral cortex is still actively developing. The

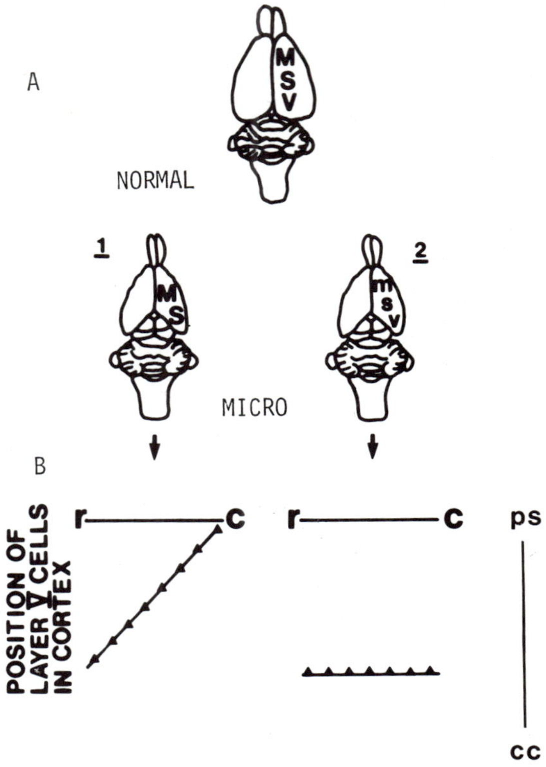

Fig. 5.

A. Schematic representation of the normal rat brain showing main functional areas; m = motor, s = somatosensory, v = visual.

(1) Microencephalic brain with rostral cortical areas only.
(2) Microencephalic brain with all functional areas represented.

B. Predicted appearance of layer V pyramidal cells in the two hypothetical cortices (1) and (2) above.
r = rostral, c = caudal, ps = pial surface, cc = corpus callosum.

normal sequence of cortical development is from rostral to caudal (Caviness and Sidman, 1973: Hicks and D'Amato, 1968). Within each region of cortex the cellular laminae are laid down in an orderly sequence from layer 6 to layer 2 (Berry, 1974). This raises the question of whether these principles of normal brain development can be applied to explain the microencephalic developmental deficit. Two possible explanations of the development of the microencephalic cortex are schematized in figure 5. If the cortex were simply reduced in size overall, there might just be a systematic and proportional reduction in length, width and thickness of cortex at all levels, and all areas (e.g. motor,somatosensory and visual cortex) would be present but smaller (figure 5A(2)). However, since MAMa affects dividing cells, when administered during cortical development, it might interfere maximally with more caudal regions of cortex leaving the most rostral parts essentially normal (figure 5A(1)). We could get some idea of the nature of MAMa's effects by simply looking at Nissl stained sections through the cortex and observing the relative thickness in different regions and the kinds of cells that are present. But such studies are limited, as if, for example, the cortex is thinner in some regions, it would be difficult to establish which cortical laminae were affected by MAMa, with cell staining alone.

Another, more efficient way to learn more about the effects of MAMa on brain structures is to study the connections of the surviving cortex using horseradish peroxidase (HRP) as a retrograde tracer. One of the principal efferent targets of the cerebral cortex is the pontine nuclei. In all mammals these pontine projections arise from pyramidal cells in layer V of the cortex (Albus and Donate-Oliver, 1977; Glickstein et al, 1980) and the cortico-pontine system can thus serve as an indicator of cortical development. By injecting horseradish peroxidase into the pontine nuclei and plotting labelled cortical cells we can compare the locus and distribution of cortico-pontine cells in normal and microencephalic brains. If the cortex were normal in laminar structure but simply reduced in length and width, one would expect to see layer V cells at the normal depth throughout the cortex (figure 5B(2)). However, if the caudal cortex were more affected by the MAMa than the rostral cortex then one would expect to see the cells at the normal depth rostrally, becoming more superficial in a thinner cortex as one moves caudally (figure 5B(1)).

A 20% solution of Sigma type 6 HRP in water was injected via 1μl Hamilton syringe under stereotaxic guidance. Two 0·125μl injections were made at different loci to fill the entire pons. The animals were allowed to survive for two days and then were deeply anaesthetized and perfused via the aorta according to the procedures outlined by Mesulam (1982). The frozen brains were sectioned at 60μ and the tissue processed with tetramethylbenzidine to reveal the presence of retrograde filled cortical cells. In normal animals there was filling in layer V pyramidal cells throughout almost all the cortex. Figure 6 shows a 60μ section through normal rat cortex to illustrate the depth and uniformity of cells in the normal brain. We found a completely different arrangement in the microencephalic animals in that labelled pyramidal cells were found at the normal depth in layer V rostrally and became more superficial caudally (figure 6). In the normal section one can see an orderly lamina of cells deep in cortex. In the microencephalic caudal section the cortical cells are not only more superficial but also lack the organisation of the normal cells. In the sections of the microencephalic brain which contained cortex, the depth of each labelled pyramidal cell was measured. It was then possible to systematically compare the depths of the cells in the rostral cortex to those in the caudal cortex(figure 7).

CONCLUDING REMARKS

From our results it would appear as though the rostral cortex of the

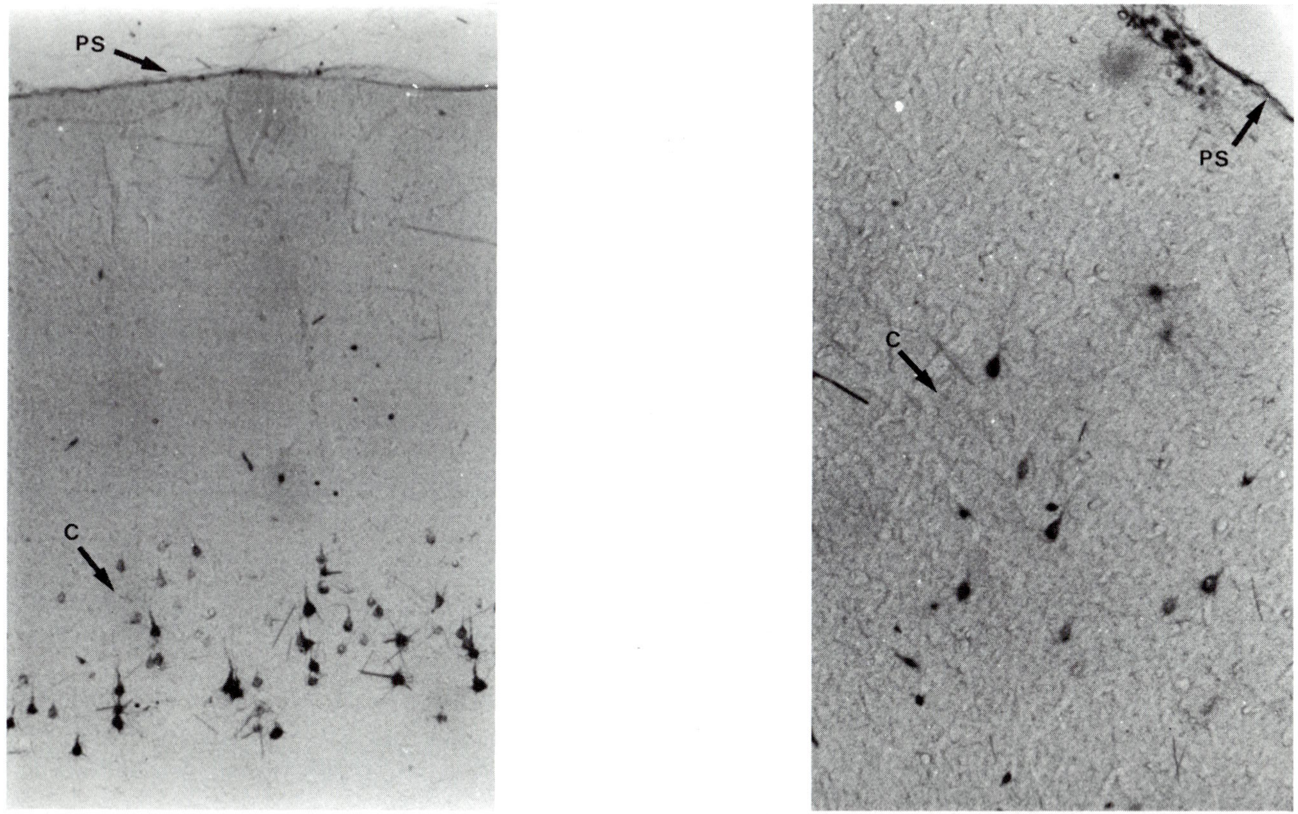

Fig. 6. Photomicrograph on the left is from a normal rat cortex after HRP injection into the pons. The layer V pyramidal cells form an orderly lamina deep within the cortex. The photomicrograph on the right is from the caudal region of microencephalic cortex. The cells here are more superficially located and less orderly. ps = pial surface, c = layer V labelled cells.

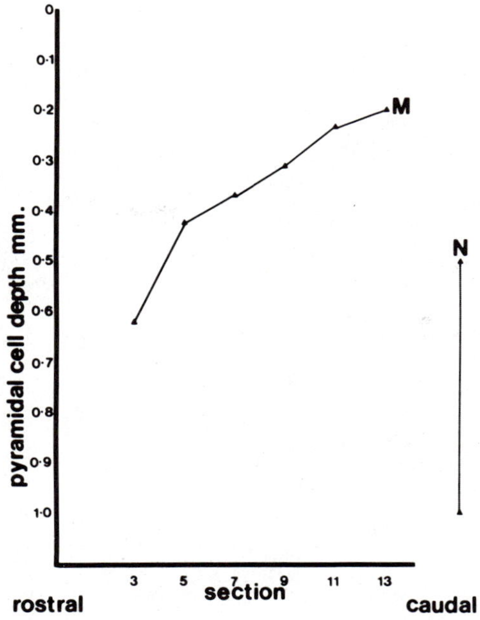

Fig. 7. Layer V pyramidal cell depths from the cortical surface in the microencephalic brain (M). Bar labelled N represents the normal range of depths at which layer V cells are found after injection of HRP into the normal rat pons.

microencephalic rat has developed in the normal way. The main effect of the injection of MAMa on embryonic day 15 is on the development of the caudal region, where the cortex is either reduced in size or absent. It is likely that the spared rostral cortex includes an intact sensory-motor cortex, which would explain our earlier finding of normal motor control and locomotor function in microencephalic rats. Further, the restriction of the developmental impairment to the caudal regions of cortex, including the visual projection areas, is entirely consistent with the present report of visual deficits. This finding, that the developmentally induced impairment of visual cortex results in a specific loss of pattern but not brightness vision, is the same as the effect of lesions of visual cortex in adult rats and rabbits (Lavond and Dewberry, 1980: de Vos and van Hof, 1981). The present results provide no evidence of developmental plasticity in the microencephalic rat visual system. It appears that the loss of visual cortical areas during development produces a non-compensated functional deficit.

REFERENCES

Albus, K., and Donate-Oliver, F., 1977, Cells of origin of the occipito-pontine projections in the cat: functional properties and intracortical location, Exp. Brain Res. ,28:167.

Berry, M. 1974, Development of the cerebral neocortex of the rat, in: "Aspects of Neurogenesis Vol. 2; Studies on the development of behaviour and the nervous system", G.Gottlieb, ed., Academic Press, New York.

Caviness Jr., V.S., and Sidman, R.L., 1973, Time of origin of corresponding cell classes in the cerebral cortex of normal and reeler mutant mice; an autoradiographic analysis, J.Comp. Neurol.,148:141.

de Vos Korthals, W.H., and van Hof, M.W., 1981, Pattern discrimination after unilateral and bilateral ablation of the occipital lobe in the rabbit, Behav. Brain. Res., 2:219.

Glickstein, M., Cohen, J.L., Dixon, B., Gibson, A., Hollius, M., La Bossiere, E. and Robinson, F., 1980, Cortico-pontine visual projections in Macaque monkeys, J. Comp. Neurol., 190:209.

Hicks, S.P., and D'Amato, C.J., 1968, Cell migrations to the isocortex in the rat, Anat. Rec., 160:619.

Lavond, D.G., and Dewberry, R.G., 1980, Visual form perception is a function of the visual cortex: II The rotated horizontal-vertical and oblique stripes pattern problems, Physiol. Psych., 8(1):1.

Mesulam, M.M., 1982, " Tracing neural connections with horseradish peroxidase", IBRO Handbook Series: Methods in the Neurosciences, John Wiley and Sons, Chichester.

Pereira, S.C., and Russell, I.Steele, 1981, Learning in normal and microcephalic rats, in:"Functional Recovery From Brain Damage", M.W. van Hof and G.Mohn, eds., Elsevier/North Holland Biomedical, Amsterdam.

Rabe, A., and Haddad, R.K., 1972, Methylazoxymethanol induced microencephaly in rats: behavioural studies., Fed. Proc., 31:1536.

Russell, I Steele, Bookman, J.F., and Mohn, G., 1979, Interocular transfer of visual learning in the rat, in: "Structure and Function of the Cerebral Commissures", I. Steele Russell, M.W. van Hof and G. Berlucchi eds., Macmillan Press Ltd., London.

Spatz, M., and Laqueur, G.L., 1968, Transplacental chemical induction of microencephaly in two strains of rat, 1., Proc. Soc. Exp. Biol. Med., 120:705.

VISUAL DISCRIMINATION LEARNING IN RATS FOLLOWING EARLY LIFE UNDERNUTRITION: RECENT FINDINGS AND REVIEW

James L. Smart and John Tonkiss

Department of Child Health
The Medical School
Manchester M13 9PT, England

INTRODUCTION

Smart (1977) made an exhaustive review of studies on the effects of early life undernutrition on learning ability in experimental animals. This was confined to experiments in which undernutrition was imposed in early life (gestation, suckling period, early post-weaning period) and was followed by a lengthy period of nutritional rehabilitation before behaviour was tested in adulthood. The results were found to be equivocal. In the rat, by far the most frequently-used species in this research, there were 37 experiments showing no significant difference between groups, 18 instances in which the well-nourished controls (C) performed better and 5 instances of superior performance by the previously undernourished (PU) animals. One of the several possible reasons for disagreement among the published results is that different genetic stocks have been used in different laboratories and the effects of early undernutrition may depend to some extent on the genetic constitution of the animals concerned (genotype x environment interaction). Indeed this possibility has been substantiated for behaviour in an open field (Blizard and Randt, 1974; Tonkiss and Smart, 1983a), social behaviour (Randt et al. 1975; Tonkiss and Smart, 1983a) and sexual behaviour (Tonkiss et al., 1984). The possibility of genotype x early nutrition interaction in the development of visual discrimination ability was investigated in the present study using a pigmented and an albino stock of rats (Lister and Wistar respectively).

A brief account of our own findings is presented and these are described in conjunction with a comprehensive review of studies of visual discrimination in PU rats. We shall be concerned throughout with investigations of general undernutrition or protein deficiency and not with specific mineral or vitamin deficiencies.

METHODS

Methods and apparatus are described more fully elsewhere (Tonkiss and Smart, 1983b). Briefly, rats of two outbred stocks were used: Lister (black-and-white hooded) and Wistar (albino). These were undernourished during development by feeding their mothers a restricted quantity of a good quality diet throughout pregnancy and lactation. This was about half

Table 1. Mean Body Weight (g) ± SEM of Previously Undernourished (PU) and Control (C) Rats of the Lister (L) and Wistar (W) Stocks at Three Ages.

Group	Birth	25 d	90 d
L-C	6.0 ± 0.2	63.2 ± 1.2	370 ± 10
L-PU	5.0 ± 0.1	19.8 ± 1.1	260 ± 4
W-C	5.5 ± 0.2	59.0 ± 1.7	276 ± 5
W-PU	5.2 ± 0.2	24.6 ± 1.9	222 ± 7

Results of two-way analysis of variance

Component	F values		
of variance	Birth	25 d	90 d
Nutrition	26.47***	739.38***	139.98***
Stock	1.03	0.14	91.66***
Interaction	9.22**	10.75**	16.60***

**P < 0.01;   ***P < 0.001;   df = 1/36 in each case.

that taken by control mothers fed ad libitum at these times. All offspring were fed ad libitum from 25 days. The subjects for behavioural testing were 40 male rats, 10 per treatment group, and all drawn from different litters.

Training commenced when rats were 13 weeks old. They were tested on the Lashley jumping stand (Lashley, 1930) using the procedure devised by Turkewitz (1975) and described in detail by Fleischer and Turkewitz (1979a). This procedure does not employ the usual reinforcers for tests of animal learning (food, water or electric shock) and is, therefore, particularly good for studies of PU rats, which are known to show heightened responsiveness to these reinforcers (see Smart, 1979). The rat is required to jump from an elevated platform of small surface area through a door bearing the positive visual stimulus to attain a 'safe', enclosed shelf. Responding to the negative stimulus or failing to respond at all is punished by a fall of 64 cm on to a foam-rubber pad. After preliminary training rats were given 10 trials per day till they reached a criterion of 9/10 correct on any one day on each of three discriminations: black vs white (B/W), horizontal vs vertical stripes (H/V), circle vs triangle (C/T). The positive stimuli were white, vertical stripes and triangle, in that order.

RESULTS

Body Weight

Two-way ANOVA revealed significant effects of nutrition and significant nutrition x stock interactions at all ages (Table 1). The interactions reflected the greater effect of undernutrition on the Lister rats.

Visual Discrimination

PU rats performed worse than controls on the H/V and C/T discriminations and Wistar rats were worse than Listers on these same tasks (Table 2, Fig.1). The significant nutrition x stock interactions for the B/W and C/T discriminations prompted analysis of the effects of undernutrition in the two stocks separately. PU rats of the Lister stock showed impaired performance on all three discriminations (all P < 0.05, t-tests) whereas PU Wistars performed worse than their controls only on the H/V task (Fig.1,P < 0.01).

Table 2. Results of a Two-way Analysis of Variance of Number of Days to Criterion.

Component of variance	F values B/W	H/V	C/T
Nutrition	2.27	25.85***	35.83***
Stock	0.09	15.44***	26.74***
Interaction	4.45*	0.21	16.67***

Abbreviations: B/W, black vs. white discrimination problem; H/V, horizontal vs. vertical stripes discrimination problem; C/T, circle vs. triangle discrimination problem.
$*P < 0.05$; $***P < 0.001$; df = 1/36 in each case

Table 3. Mean ± SEM Percentage of Days on Which Rats Showed Response Fixation.

Group	B/W	H/V	C/T
L-C	0.0 ± 0.0	7.8 ± 4.1	11.5 ± 5.4
L-PU	20.0 ± 7.4	27.2 ± 5.6	44.9 ± 8.7
W-C	0.0 ± 0.0	21.1 ± 6.4	36.1 ± 8.0
W-PU	6.7 ± 6.7	19.0 ± 6.4	38.0 ± 6.9

Results of two-way analysis of variance

Component of variance	F values B/W	H/V	C/T
Nutrition	7.26*	2.31	5.76*
Stock	1.82	0.20	1.44
Interaction	1.82	3.57	4.59*

Abbreviations as in Tables 1 and 2. $*P < 0.05$; df = 1/36 in each case

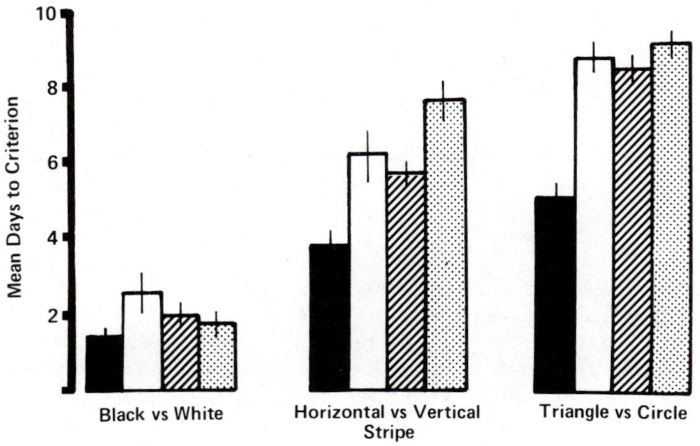

Fig.1. Mean days (± SEM) to criterion for previously undernourished (PU) and control (C) rats of the Lister (L) and Wistar (W) stocks on three visual discrimination problems.
L-C (■), L-PU (□), W-C (▨), W-PU (⬚).

Many rats displayed 'response fixation', which clearly interfered with efficient performance of the discriminations. That is, they jumped only to the door on one side of the stand, irrespective of the visual stimulus. An analysis of the percentage of days on which rats showed response fixation (jumped to the same side on all 10 trials) revealed significant effects of nutrition on response fixation for two of the discriminations (Table 3), though the significant nutrition x stock interaction for the C/T discrimination reflected the finding that undernutrition significantly affected response fixation only in Lister rats on that task ($P < 0.01$).

DISCUSSION

Genotype x Environment Interaction

The results presented above are a clear demonstration of genotype x nutrition interaction in the development of visual discrimination ability with respect at least to some visual discrimination tasks. Statistically significant interaction components of variance were found for two of the discriminations (Table 2). Put another way, there were significant effects of early undernutrition on the performance of Lister rats on all three tasks, whereas Wistar rats were affected thus on only one discrimination. Had one used only Wistar rats and tested them only on the B/W or C/T discriminations, one might have concluded that there were no effects of early life undernutrition on visual discrimination.

Genotype x environment interactions are often treated at the level of phenomena which either do or do not occur, without much attempt at insight into how they might come about. In fact, there may be some clue as to how the present interaction may have arisen in the body weight results. The stock x nutrition interactions for body weight (Table 1) reflect the lesser effect of maternal undernutrition on the growth of Wistar offspring and this parallels the findings for visual discrimination performance. It would not be surprising if the brains of the Wistars, like their bodies, had been rather less growth-retarded by the undernutrition than those of the Lister rats, and this might have resulted in less impairment of discrimination learning.

It should be emphasised that the body weight interaction occurred in spite of the fact that equally severe undernutrition was imposed on each stock (the ad libitum food intake of the two stocks was the same and both underfed groups received 50% of the ad libitum amounts). One factor that might have stood the Wistars in better stead, was their tendency to produce fewer young at birth which may have been relatively spared up to that stage through lesser 'competition' for available resources in utero. Even though litter size was standardised at birth, this better start to postnatal life may have had lasting benefit.

Review

We have made a comprehensive review of the literature on lasting effects of early life undernutrition on visual discrimination to enable us to attempt to draw general conclusions. The review is confined to studies of PU (nutritionally rehabilitated) rats and adheres to the criteria and definitions adopted by Smart (1977). Briefly, experiments are categorised as showing (i) significantly superior performance by C rats or (ii) no significant difference (there were no experiments in which PU rats performed better). An experiment is defined as a comparison between one C and one PU group. The findings from three apparently relevant papers

Table 4. Analysis of Studies on Black vs White and Light vs Dark Discrimination in Well-fed Control (C) and Previously Undernourished (PU) Rats. Numbers of Experiments*.

	C group performed better	No significant difference
Albino	5	22
Pigmented	3	3
TOTAL	8	25

There were no experiments in which the PU group performed better.
* From Barnes et al. (1966), Celedon et al. (1982), Crnic (1983), D'Amato (1960), Fleischer and Turkewitz (1979a, b), Galler (1981), Halas and Sandstead (1980), Slob et al (1973), Tachibana (1979), Tonkiss and Smart (1983b), Tsukada et al. (1979), Wetzel et al. (1979).

are not included in the tables for the following reasons. Vore and Ottinger (1970) report their results as total scores such that one cannot separate original learning of a brightness discrimination from serial reversal learning, and Stewart et al. (1980) likewise give total scores, this time for performance of the three discrimination problems described in the present paper. Tikal et al. (1972) give no information on the stock of rat used.

B/W and light vs dark (L/D) discrimination has been analysed separately from pattern discrimination, since it would appear to be qualitatively different, requiring no visual acuity whatsoever. Such experiments are summarised in Table 4 according to whether the subjects were albino or pigmented rats. There have been far fewer investigations of pigmented animals, but a rather higher proportion have yielded positive results. The same is true for (H/V) discrimination on which 3/4 experiments with pigmented rats gave positive findings, but only 2/9 with albinos (Table 5).

Parenthetically, there are reasons for not treating the results of investigations of B/W and L/D discrimination very seriously, apart from the simplicity of the task. The most important is that the animals may have a marked preference for one or the other stimulus. Unknown to us at the time our animals were tested, rats jump preferentially to a white rather than a black stimulus, at least when the background is dark (Lashley, 1930). White was positive in our test and the rats performed the task remarkably efficiently (Fig.1), demonstrating that they were discriminating but telling us little about learning. This kind of consideration seldom seems to be taken into account, to the extent that few investigators report which stimulus was designated positive. Pattern discrimination is less likely to be susceptible to this kind of problem.

Two pattern discriminations have been used frequently, H/V stripes and C/T, and when both have been tested in the same study it has always been in the order stated, probably because the C/T problem is said to be the more difficult (Lashley, 1930). Inspection of Table 5 reveals that differences in performance have shown up more frequently on the H/V task. The order in which the discriminations were presented might be a factor, in that the animals may have mastered the generalities of the type of problem before they encountered the C/T discrimination. However, the absence of an effect of nutrition in the one study in which the C/T task was presented first (Rogers et al., in press) does not support this suggestion. A highly speculative interpretation derives from the fact

Table 5. Synopsis of Studies on Pattern Discrimination in Previously Undernourished rats

Reference	Stock	Sex	Nutrition	Task	Reinforcement	C or PU group performed better	
						H/V	Δ/O
Galler (1981)	hooded	M	G^-	Lashley stand	fall	C	ns
		F				ns	ns
Celedon et al.(1982)	Lister	M	G^-L^-	Sutherland box	water	C	-
Tonkiss & Smart (1983)	Lister	M	G^-L^-	Lashley stand	fall	C	C
	Wistar					C	ns
Rajalakshmi et al.(1967)	albino	?	L^-	T-maze	food	ns	-
Slob et al.(1973)	Wistar	M	L^-	Thompson box	shock	ns	-
		F				ns	-
Fleischer & Turkewitz (1979a)	Sprague-Dawley	M	L^-	Lashley stand	fall	C	ns
		F				ns	ns
Fleischer & Turkewitz (1979b)	Sprague-Dawley	M	L^-	Lashley stand	fall	ns	ns
		F				ns	ns
Crnic (1983)	Sprague-Dawley	M	L^-	Lashley stand	fall	ns	ns
Rogers et al.(in press)	Lister	M	L^-W^-	Lashley stand	fall	-	ns

C = well-fed control, PU = previously undernourished, M = male, F = female, G^- = undernourished during gestation, L^- = undernourished during lactation, W^- = undernourished post-weaning, H/V = horizontal vs vertical stripes discrimination, Δ/O = triangle vs circle discrimination, ns = not significant.

that the rats will have had experience of horizontal and vertical visual stimuli, in the form of cage bars, cage racks etc, throughout their lives, and from the finding that PU rats show little evidence of 'exposure learning' (Rogers et al., in press). Perhaps differences show up on the H/V discrimination because well-fed control rats have benefitted from exposure to H and V stimuli whereas PU rats have not.

Two other tentative conclusions emerge from Table 5. The first relates to the timing and duration of the period of undernutrition. Lasting effects appear to be more likely to result from undernutrition during either gestation alone or gestation plus the suckling period than from undernutrition during the suckling period alone or in combination with post-weaning restriction. This conclusion receives further support from the findings of Stewart et al. (1980) for rats undernourished during gestation or during gestation and the suckling period, which showed impaired visual discrimination (not tabulated for reasons given above). Hence, what evidence there is suggests that gestation is the most vulnerable period for any influence of undernutrition on the development of visual discrimination. The second general conclusion is that female rats are less susceptible to effects of early undernutrition than males, in that none of the four investigations of PU females revealed impaired performance (Table 5).

By integrating these conclusions one can put forward a 'most likely to fail' group; that is, the experimental group most likely to show impaired visual discrimination performance. This group would be male, pigmented rats, undernourished during gestation (and the suckling period?) and tested for discrimination of H vs V stripes.

Mechanisms

We have tended to imply up to now that deficient performance of a visual discrimination task indicates impaired learning, and, of course, this may not be the case. For instance, many of the rats in the present study exhibited the trait of 'response fixation', and this suggests the possibility that the differences in visual discrimination performance may be explicable just as readily in terms of differences in 'emotionality' as differences in cognition. Given that rats are more prone to response fixation under anxiety-provoking conditions than under non-stressful conditions (Farber, 1948), it is reasonable to suggest that, in a particular situation more 'anxious' rats will show more response fixation than less anxious rats. PU rats certainly seem to be more responsive than controls in a variety of unpleasant situations (e.g. Levitsky and Barnes, 1970), and so it may be permissible to say that they display heightened 'anxiety'. If so, PU rats may be more anxious in the somewhat stressful test situation and consequently show more response fixation, resulting in poorer performance of the discrimination task. This causal chain, together with some other possibilities, is depicted in Table 6 (middle chain). In this particular case, heightened anxiety is made the first link in the chain, and in the other chains other possible differences (perceptual, cognitive etc) are proposed as the first link. In our present state of knowledge, or ignorance, it is impossible to choose between these various hypotheses.

A further possibility is that the deficient performance may be a consequence of visual rather than psychological impairment (i.e. sensory rather than cognitive or emotional). Indeed, a deficiency might reside anywhere in the visual pathways from eye to visual cortex, or even outside the visual pathways altogether. There is both histological and electrophysiological evidence that the visual pathways are affected by early undernutrition. Though postnatal undernutrition does not appear to have lasting effects on the structure of the optic nerve in terms of fibre

Table 6. Hypothetical Routes by Which Early Life Undernutrition May Have Given Rise to Poor Discrimination Learning Performance

Early life undernutrition → Impaired discrimination learning ability and/or ↑Response fixation → Poor discrimination learning performance

Early life undernutrition → ↑Anxiety → ↑Response fixation → Poor discrimination learning performance

Early life undernutrition → Impaired discrimination learning ability → ↑Anxiety → ↑Response fixation → Poor discrimination learning performance

↑= "increase in".

number, size or size distribution (Bedi and Warren, 1983), there is some evidence that undernutrition during gestation only results in a permanently increased proportion of smaller-diameter fibres within the optic nerve (Sima and Sourander, 1974). There are further indications of gestational vulnerability from studies of cortical thickness in PU rats. Of a number of groups of rats which had been undernourished during various periods or combinations of periods in early life, only those whose period of undernutrition included gestation had thinner visual cortices than well-fed controls (Katz et al., 1982). The apparently greater vulnerability of the visual pathways to undernutrition during gestation may be associated with the fact that that is the time of neuronal proliferation in those regions (Rodier, 1980). It is noteworthy that this tentative conclusion of gestational vulnerability for anatomical measures, parallels that for visual discriminaton performance. This suggestion ought to be followed up with definitive studies in both disciplines, preferably on the same rats.

The electrophysiological evidence does not parallel the histological findings closely, in that there appears to be fairly convincing evidence of vulnerability to postnatal undernutrition. (In fact, there has been very little electrophysiological investigation of prenatally undernourished animals.) For instance, peak latencies of visual evoked responses are still abnormally long 35 days after a 5 day episode of undernutrition during the suckling period (Mourek et al., 1967), though other evidence suggests that further recuperation to normal occurs with age (Morgane et al., 1978). Most convincing are the studies of electroretinograms (ERG's) in adult rats which had been undernourished during the suckling period only (Wiggins et al., 1980). The response amplitudes of the A and B waves are very substantially lower in the PU rats.

The disclaimer was made in the Introduction that the paper was concerned with deprivation of major dietary constituents and not with specific vitamin or mineral deficiences. Vitamin A is, of course, essential for the development of normal vision (McLaren, 1980) and it might be prudent to consider whether 'general undernutrition' might have resulted in deficiency, inter alia, of vitamin A and whether this might have been the source of the various effects on the visual pathways and on visual discrimination. This is probably unlikely, though not out of the question, since (i) low protein diets are always supplemented with adequate amounts of vitamins, and (ii) proprietary rat chow usually contains a superabundance of vitamins, such that even when the diet is fed in restricted quantities the amounts of vitamins ingested are fairly adequate. Perhaps more likely is the possibility described by Wiggins et al. (1980) that protein deprivation may result in reduced levels of serum proteins, such as the retinol-binding protein. Lack of this normal transport mechanism might result in depletion of vitamin A in the retina, in spite of an adequate dietary supply and liver store of the vitamin.

Whether any of the above histological and electrophysiological differences have functional consequences is not known. Clearly, the possibility that they may do should be borne in mind when interpreting findings of impaired visual discrimination in PU rats. The impairment may be wholly or partly visual or perceptual rather than, or as well as cognitive.

CONCLUSIONS

1. Genotype x nutrition interaction has been demonstrated with respect to the effects of early life undernutrition on the learning of visual discrimination tasks in adulthood. The performance of pigmented, Lister, rats was more consistently impaired than was that of albino, Wistar, rats.

2. A comprehensive review of the literature on visual discrimination in previously undernourished rats led to the following tentative conclusions. Performance is more likely to be impaired in male rats than females, and in pigmented rats rather than albinos. Gestation appears to be the most vulnerable period for the effect. The discrimination of horizontal vs vertical stripes is the most likely to be affected of those that have been used.

3. There is histological and electrophysiological evidence of lasting effects of early undernutrition on the visual pathways. Hence, it is prudent to bear in mind the possibility that a variety of different mechanisms could account for the observed impairments in the performance of visual discrimination tasks, from sensory through to cognitive or even emotional.

ACKNOWLEDGEMENTS

This research was funded by the Medical Research Council and the National Fund for Research into Crippling Diseases. We are grateful to Dr. Peter Rogers for his help in the preparation of the manuscript.

REFERENCES

Barnes, R.H., Cunnold, S.R., Zimmermann, R.R., Simmons, H., MacLeod, R.B., and Krook, L., 1966, Influence of nutritional deprivations in early life on learning behavior of rats as measured by performance in a water maze, J. Nutr., 89: 399.

Bedi, K.S., and Warren, M.A., 1983, The effects of undernutrition during early life on the rat optic nerve fibre number and size-frequency distribution, J. comp. Neurol., 219: 125

Blizard, D.A., and Randt, C.T., 1974, Genotype interaction with undernutrition and external environment in early life, Nature, 251: 705.

Celedon, J.M., Smart, J.L., and Dobbing, J., 1982, Effects of level of motivation on visual discrimination, transfer of learning, and long-term memory in previously undernourished and control rats, Nutr. Behav., 1: 89.

Crnic, L.S., 1983, Effects of nutrition and environment on brain biochemistry and behavior, Develop. Psychobiol., 16: 129.

D'Amato, M.R., 1960, Effect of litter size on brightness discrimination and reversal, Psychol. Rep., 7: 91.

Farber, I.E., 1948, Response fixation under anxiety and non-anxiety conditions, J. exp. Psychol., 38: 111.

Fleischer, S.F., and Turkewitz, G., 1979a, Effect of neonatal stunting on development of rats : large litter rearing, Develop. Psychobiol., 12: 137.

Fleischer, S.F., and Turkewitz, G., 1979b, Behavioral effects of rotation between lactating and nonlactating females, Develop. Psychobiol.,12: 245.

Galler, J.R., 1981, Visual discrimination in rats : the effects of rehabilitation following intergenerational malnutrition, Develop. Psychobiol., 14: 229.

Halas, E., and Sandstead, H.H., 1980, Malnutrition and behavior : the performance versus learning problem revisited, J. Nutr., 110: 1858.

Katz, H.B., Davies, C.A., and Dobbing, J., 1982, Effects of undernutrition at different ages early in life and later environmental complexity on parameters of the cerebrum and hippocampus in rats, J. Nutr., 112: 1362.

Lashley, K.S., 1930, The mechanism of vision : I. A method for rapid analysis of pattern vision in the rat, J. genet. Psychol., 37: 453.

Levitsky, D.A. and Barnes, R.H., 1970, Effect of early malnutrition on the reaction of adult rats to aversive stimuli, Nature, 225: 468

McLaren, D.S., 1980, "Nutritional Ophthalmology", Academic Press, London.

Morgane, P.J., Miller, M. Kemper, T., Stern, W., Forbes, W., Hall, R., Bronzino, J., Kissane, J., Hawrylewicz, E., and Resnick, O., 1978, The effects of protein malnutrition on the developing central nervous system in the rat, Neurosci. Biobehav. Rev., 2: 137.

Mourek, J., Himwich, W.A., Myslivecek, J., and Callison, D.A., 1967, The role of nutrition in the development of evoked cortical responses in the rat, Brain Res., 6: 241.

Rajalakshmi, R., Ali, S.Z., and Ramakrishnan, C.V., 1967, Effect of inanition during the neonatal period on discrimination learning and brain biochemistry in the albino rat, J. Neurochem., 14: 29.

Randt, C.T., Blizard, D.A., and Friedman, E., 1975, Early life undernutrition and aggression in two mouse strains, Develop. Psychobiol., 8: 275.

Rodier, P.M., 1980, Chronology of neuron development : animal studies and their clinical implications, Dev. Med. Child Neurol., 22: 525.

Rogers, P.J., Tonkiss, J., and Smart, J.L., in press, (abstract), Discrimination learning in rats following early-life undernutrition and stimuli pre-exposure, Behav. Brain Res.

Sima, A., and Sourander, P., 1974,The effect of early undernutrition on the calibre spectrum of the rat optic nerve : A morphometric ultra-structural study, Acta Neuropathol., 28:1.

Slob, A.K., Snow, C.E., and de Natris-Mathot, E., 1973, Absence of behavioral deficits following neonatal undernutrition in the rat, Develop. Psychobiol., 6: 177.

Smart, J.L., 1977, Early life malnutrition and later learning ability : a critical analysis, in: "Genetics, Environment and Intelligence", A. Oliverio, ed., Elsevier/North Holland, Amsterdam.

Smart, J.L., 1979, Undernutrition and the development of brain and behaviour, in: "Chemical Influences on Behaviour", K. Brown and S.J. Cooper, eds., Academic Press, London.

Stewart, R.J.C., Sheppard, H., Preece, R. and Waterlow, J.C., 1980, The effect of rehabilitation at different stages of development of rats marginally malnourished for ten to twelve generations, Br. J. Nutr., 43: 403.

Tachibana, T., 1979, Effects of early undernutrition on motivation and discrimination learning in rats, Develop. Psychobiol., 12: 539.

Tikal, K., Benesova, O., and Frankova, S., 1972, The effect of pyrithioxine (Encephabol Merck) on behavior, learning and biochemical variables of brain in rats malnourished in early life. II. Active avoidance and discrimination learning, Activ. Nerv. Sup., Praha, 14: 174.

Tonkiss, J., and Smart, J.L., 1983a, Interactive effects of genotype and early life undernutrition on the development of behavior in rats, Develop. Psychobiol., 16: 287.

Tonkiss, J., and Smart, J.L., 1983b, Discrimination learning in pigmented and albino rats following early life undernutrition, Nutr. Behav., 1: 207.

Tonkiss, J., Smart, J.L., and Griffiths, E.C., 1984, Mating behaviour of male rats following pre- and early postnatal undernutrition : a comparison of two outbred stocks, Physiol. Behav., 32: 397.

Tsukada, Y., Kohsake, S., and Nagai, K., 1979, Effect of mild protein restriction during prenatal and postnatal development of rats on discriminative learning ability, in: "Behavioral Effects of Energy and Protein Deficits", J. Brozek, ed., USDHEW, NIH Publication No. 79 - 1906, Washington.

Turkewitz, G., 1975, Learning in chronically protein deprived rats, in: "Nutrition and Mental Functions", G. Serban, ed., Plenum, New York.

Vore, D.A., and Ottinger, D.R., 1970, Maternal food restriction : effects on offspring development, learning, and a program of therapy, Develop. Psychobiol., 3: 337.

Wetzel, W., Ruthrich, H.L., and Matthies, H., 1979, Memory impairment in adult rats after postnatal undernutrition, Behav. neural Biol., 25: 157.

Wiggins, R.C., Fuller, G.N., and Dafny, N., 1980, The electroretinogram of adult rats following a period of postnatal undernutrition and prolonged nutritional rehabilitation, Life Sci., 26: 1169.

ONTOGENY OF ACQUISITION AND RETENTION OF TWO-WAY ACTIVE AVOIDANCE IN THE RAT : EFFECTS OF EARLY SEPTAL DAMAGE

Françoise Eclancher

Laboratoire de Neurophysiologie - Centre de Neurochimie
du CNRS, 5 rue Blaise Pascal
67084 Strasbourg Cédex, France

INTRODUCTION

It has been reported that young rats can perform similarly to adults in an active avoidance task (Kirby, 1963 ; Campbell, 1967 ; Denenberg and Kline, 1958 ; Feigley and Spear, 1970 ; Riccio et al., 1968). However, other authors have reported that active avoidance performances appear to improve as a function of age. Egger and Livesey (1972) found that 24 day-old rats required more training to learn the active avoidance task than did 100 day-old rats, and Izquierdo et al. (1975) observed that 20-21 day-old rats made fewer conditioned avoidance responses than did 55-70 day-old adult rats in an acquisition session and showed no retention of these responses.

Generally, authors have been interested in the comparison of performances between young and old animals and only a few of them have been interested in the development of behavior of animals throughout ontogenesis. In studies on early versus late acquisition of conditioned avoidance by rats, a one-way active avoidance task was generally used, except in the studies of Izquierdo et al. (1975) and Bauer (1978) who used a 2-way active avoidance task. The one-way task is easier to learn for the rat than the 2-way task, and requires that an animal respond to a conditioned stimulus (CS) e.g. a tone, by jumping or running from a start compartment (which is always the source of the aversive stimulation) into a distinct safe compartment (which may acquire secondary reinforcing properties). The 2-way active avoidance task involves a place conflict, particularly during early training trials during which subjects show obvious indecision in crossing to the opposite end of the box where they were just shocked on the previous trial. In this task, the rats are shocked for failing to shuttle between compartments in response to the CS.

The first purpose of our study was to determine if there are developmental changes in the learning capability of a 2-way active avoidance in the rat. The developmental stages studied were 20, 25, 30, 40, 50 and 90 days of age.

No research has been directed at measuring the relative persistence of retention of early acquired responses until adult age. In the studies of retention of an avoidance response in rats as a function of age, only short-term retention was considered (Klein and Spear, 1969).

Another question was to know whether retention of the acquired properties of the CS that occur early in life is more fragile than retention of similar stimuli given at a later stage in ontogeny. We also wanted to know whether changes in long-term retention exist during ontogeny. Thus the rats that had received prelearning at 20, 30, 40 and 50 days, were retested at 90 days of age.

The septal area has been shown to mediate inhibitory responding such as that observed in the 2-way active avoidance behavior. Lesions of this telencephalic area performed at 7 days or 90 days of age have been shown to induce high levels of reactivity as well as a facilitation of the 2-way active avoidance in the adult rat (Eclancher and Karli, 1979, 1981; Feigley and Hamilton, 1971; Johnson, 1972; Molino, 1975). Ontogeny of arousal and inhibitory functions in the brain is possibly related to the caudal-rostral sequence of central nervous system maturation (Campbell and Spear, 1972; Moorcroft et al., 1971). As stated by Jackson (1931), brainstem and midbrain structures mature first and are followed by phylogenetically newer forebrain systems. These late maturing forebrain structures have been hypothesized to acquire inhibitory control over the original primitive systems, including those in the midbrain that mediate behavioral arousal. It follows that the developing organism might go through a brief period during which arousal is unchecked because of immaturity of forebrain inhibitors. Spontaneous activity of the rat has been shown by Moorcroft et al. (1971) and Campbell and Mabry (1972) to increase sharply between 15 and 20 days and to decrease to adult levels by 25 and 30 days of age. The septum being part of these inhibitory forebrain structures, we may suppose that a lesion of this area performed at 6 days, i.e. before its functional maturation, will induce behavioral changes such as hyperreactivity to stimuli received in active avoidance tasks. The age at which these behavioral changes appear should reflect the time at which the septal area has reached functional maturation.

The second purpose of the present study was thus to determine the age at which early septal-induced changes first appear, by submitting early septal-damaged rats to the 2-way active avoidance task at 20, 25, 30, 40, 50 or 90 days of age. Another question was whether the early septal lesion disrupts retention in adult rats of responses acquired at an early age.

METHOD

Subjects

Pregnant female Wistar rats bred in our laboratory were housed singly in 42 x 26 x 15 cm translucent Makrolon cages. The cage floors were covered with pine shavings. These animals were maintained on a free feeding schedule and kept upon a 12 hr : 12 hr light-day cycle, with lights on at 8 o'clock. Within 2 days after parturition, each litter was culled to 5-8 male pups that were used in the present study. The split-litter technique was used in selecting animals for the septal and control groups. Half of the pups in each litter were given a septal lesion, the other half were given a sham lesion (control animals).

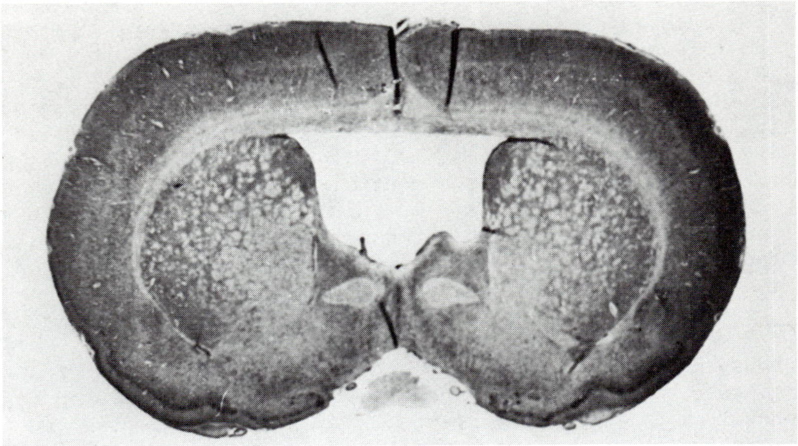

Fig. 1. Photomicrograph of a frontal section through the septum from a rat that sustained a septal lesion at 6 days of age.

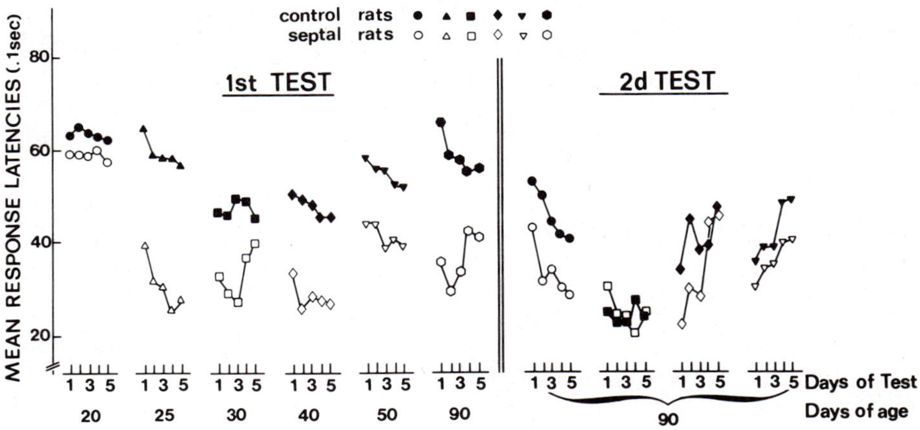

Fig. 2. Mean response latencies in the 2-way active avoidance of the rats, at various ages and for each day of test.

Surgery

Septal lesions (Fig. 1) were produced in 6 day-old rat pups. The surgery was performed stereotaxically under sodium pentobarbital (Nembutal) anesthesia (0.22 mg/10 g, IP) (for the details of surgery, see Eclancher and Karli, 1981).

The rats with lesions and the control rats were housed in groups of 5-7 rats per cage (42x28x27 cm). The following groups were formed according to the treatment received (sham lesion : C, or septal lesion : S) and to the age (20, 25, 30, 40, 50 and 90 days) at which the rats were submitted (in the morning) to the open-field and 2-way active avoidance tests, with the number (n) of subjects in each group that survived surgery : 20 C (n =12), 25 C (n = 18), 30 C (n = 19), 40 C (n = 23), 50 C (n = 23) and 90 C (n = 20) for the control rats and 20 S (n = 15), 25 S (n = 13), 30 S (n = 14), 40 S (n = 18), 50 S (n = 13) and 90 S (n = 20) for the septal rats. The rats that had been tested before 90 days were tested again at 90 days of age.

Shuttle-box apparatus

The 48x21x22 cm shuttle-box made of plexiglas was placed in a sound-proof enclosure. It was divided into two compartments separated by a partition with a 9x9 cm square opening. The grid floor consisting of 0.3 cm stainless steel rods, mounted 1.5 cm apart, pivoted at the center so that it tilted when the rat's total body weight passed the center. A 4000 Hz, 85 dB buzz, 7 sec in duration, served as the CS. The unconditioned stimulus (US) was a constant current of 0.3 mA scrambled footshock.

Avoidance training procedure

Each rat was placed in the shuttle-box and was allowed a 5 min free exploration before receiving a session of 50 trials. If the animal had not responded by the end of the 2nd second of the buzzer presentation, continuous shock was administered along with the buzzer. During each trial, termination of both CS and footshock was contingent upon crossing over to the opposite compartment or the end of the 7 sec buzzer presentation. Responding with a latency of less than 2 sec from CS onset enabled the rat to avoid shock. Crossings occurring during the 13 sec intertrial interval (ITI) were not punished. The rats were trained for 5 days with 50 trials per day.

Data analysis

With respect to acquisition of the 2-way active avoidance task, the following data were calculated for each group of rats : mean escape or avoidance latency, mean number of crossings from one side of the shuttle box to the other one and mean number of shocks received over 50 trials.

A 2-way analysis of variance (lesion x age) for unequal numbers completed by Scheffé tests for multiple two by two comparisons (at the level of .05 was performed to compare the daily performance of the rats. The same analysis was carried on the performance of the rats during the retesting period at 90 days of age.

RESULTS

For the five days of testing, the 30 and 40 day-old control rats performed significantly better than all other control rats. Their response latency was shorter (Fig. 2) and the number of their crossings was higher than that of the other age groups (Fig. 3). However, the number of shocks received by the 30 and 40 day-old rats was not statistically different from that of other groups of rats (Fig. 4).

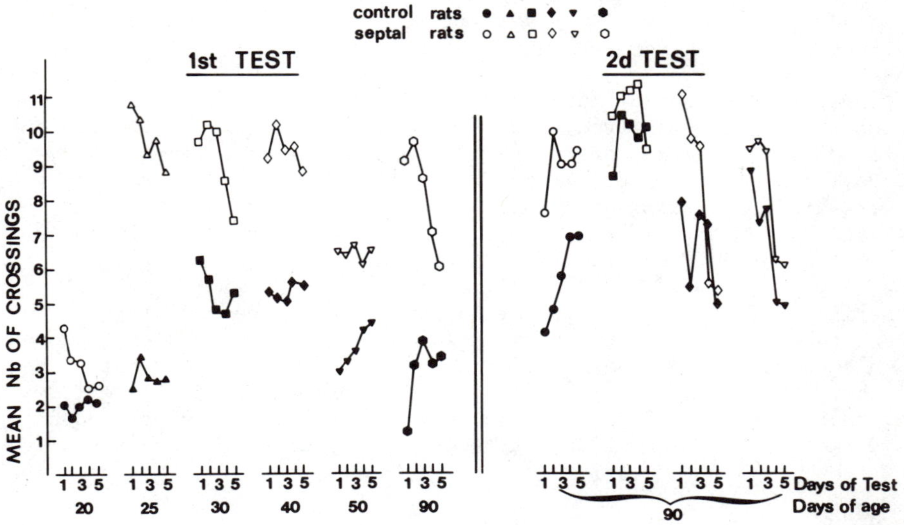

Fig. 3. Mean number of crossings between the two compartments of the shuttle box for the rats of various ages and for each day of test.

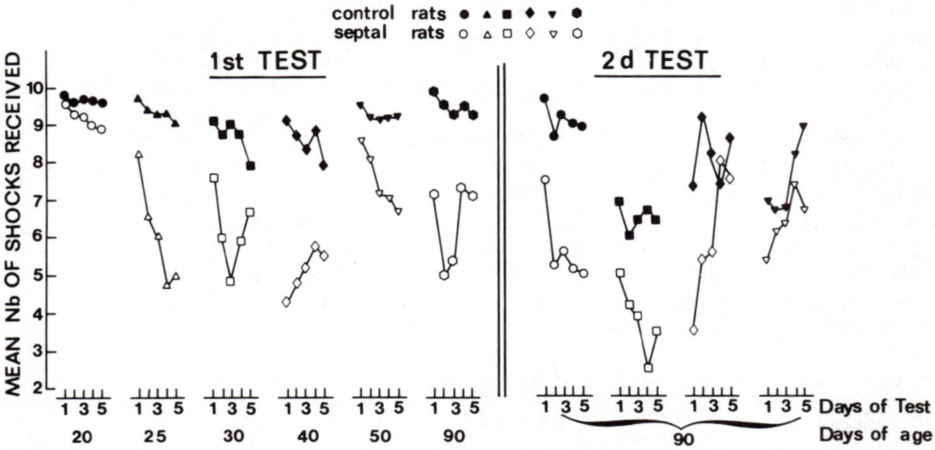

Fig. 4. Mean number of shocks received in the 2 way active avoidance task by the rats of various ages and for each day of test.

During the retesting period, and in comparison to the first testing, the 30 C rats were significantly improved on every day for each variable considered. The 20 C and 50 C rats were improved only on some days of retesting, while the 40 C rats were not improved at all.

Concerning the septal rats, their performance (lower response latency and number of shocks received, and higher number of crossings) was significantly better than that of the respective control rats, from 25 days on. At 20 days of age, the septal rats differed from the control rats only by a higher number of crossings on the first days of testing. It was at 25, 30, 40 and 90 days of age that the septal rats had better performances, whatever the variable considered.

During the retesting period at 90 days of age, the septal rats that had been tested for the first time at 20 days of age, showed an improvement on each measure of their performance and for each day of retesting. The other groups of septal rats showed no improvement of their performance during their retesting at 90 days, except for the rats first tested at 30 days that received at 90 days a significantly fewer number of shocks.

DISCUSSION

During the course of ontogenesis of the rat, there is a period from 30 to 40 days of high levels of performance in the 2-way active avoidance task in intact animals (in terms of response latency and crossings, but only slightly in terms of fewer shocks received). These levels are higher than those shown at both younger (20 and 25 days) and older ages (50 and 90 days). However, these changes in reactivity of the rat to stimuli received in the shuttle box are only weakly associated with changes in ability to avoid shocks. It is obvious that the number of trials given to the rats at either age was not sufficient for the rats to get conditioned. But the training procedure had to last for a short period, otherwise the animals tested at later stages of their development would have reached adult age by the time they were conditioned. In that case, comparison of performances between the various ages would have been difficult.

The shorter response latency and higher number of crossings displayed by the intact rats at 30 and 40 days of age may be related to a decrease of sensitivity to the approach-avoidance conflict induced by the necessity to cross to the other compartment where they had just been shocked on the previous trial. But the high level of reactivity of these rats in the shuttle box should also be related to the particularly high level of reactivity rats display at 30 and 40 days when confronted to a novel environment such as the open-field (Eclancher, 1984). Indeed Bauer (1978), who found significant positive correlations between activity and 2-way active avoidance acquisition in immature rats, reported that avoidance ability increases from 21 to 51 days of age and decreases to 90 and 200 days of age. However, Izquierdo et al. (1975) found that younger rats (20-21 day old) made more intertrial crossings than the adult ones (55-70 day-old) during pseudoconditioning (footshocks interspersed among the buzzers) and during conditioning of the 2-way task, the weanling rats learned this task to a much lesser extent than adults. Thus high levels of activity are not necessarily related to high levels of avoidances.

The present findings are consistent with the suggestion of Altman et al. (1973) that immature rats should acquire 2-way active avoidance faster than adults. Nevertheless it is worth adding that this is true only for the early ages of 30 and 40 days. The ages of 30 and 40 days correspond to the period of puberty with release of sexual hormones that are probably responsible for some behavioral changes observed at these two ages. The peripubertal stage in male rats starts around Day 30 as mature Leydig cell function begins and testosterone becomes the dominant androgen. Puberty terminates as the first sperm are released from the germinal epithelium around 40-45 days of age (Ramaley, 1979).

Thirty days of age represents a critical age at which the initial learning (50 trials a day for 5 days) of a 2-way task, that is not sufficient for the rat to get conditioned results in the facilitation of relearning of the same task when adult with a significant decrease of shocks avoided. Thus, avoidance conditioning in the rat at this early age of 30 days is particularly well memorized until adult age. Although the 40 day-old rats have reached the same level of performances as the 30 day-olds in the initial test sessions, they showed no clear savings during retention testing when adult.

Following an early septal lesion, the rats showed a higher level of reactivity to the stimuli presented during the 2-way active avoidance acquisition from 25 days of age onwards. Although the number of crossings of septal rats was significantly higher than that of control rats on the first day, at the age of 20 days, they were nevertheless able only to escape the shocks, as did the control rats. It is possible that the septal rats failed to avoid shocks at 20 days because of their inability to make the required locomotor response at a sufficiently high speed. Thus control rats showed generally poor active avoidance learning after the 250 trials given at any age, whereas the septal rats were able to avoid half of the shocks given, but only at the ages of 25, 30, 40 or 90 days. The 50 day-old septal rats were not so significantly improved in comparison of control rats of same age.

Our present findings corroborate our previous results (Eclancher and Karli, 1980; 1981) and those of other authors (Johnson, 1972; Johnson et al., 1972; Molino, 1975) concerning the hyperreactivity and the facilitation of a 2-way active avoidance acquisition by adult rats that had sustained a septal lesion at a very early age, and generally no recovery of function has been found. Schoenfeld et al. (1974) also reported a significant facilitation of 2-way active avoidance in 25 day-old rats, after septal lesions inflicted at 15 or 20 days of age.

The fact that high responsiveness and facilitation of the 2-way task does not appear in the early septal-damaged rat at 20 days of age but only at 25 days of age suggests that, in the intact rat, the septum has reached its functional maturity at the age of 25 days. The lower level of reactivity in the 50 day-old septal rats cannot be put into account of recovery during the long period of time between the day 6 (septal lesion) and the day of testing, since the 90 day-old rats tested after a longer period, displayed an even higher level of reactivity.

Some authors (Douglas et al., 1973; Altman et al., 1973) have reported that behavior of adult rats with hippocampal lesions was similar to the behavior of weanling rats. As both the septum and the hippocampus have been implicated in the inhibitory control of behavior, one can be tempted to compare the behavior of adult rats with septal damage to weanling intact rats such as 20 day-old rats, having a septum that has not yet reached functional maturity for the inhibition of behavior displayed in the 2-way task. It appears that these 20 day-old intact rats have a very lower level of reactivity than that of septal rats of 25 days of age and more. Consequently, the hyperreactivity induced by the early septal lesion is far higher than the possible hyperreactivity that would result from the lack of maturation of structures like the septum, responsible for behavioral inhibition.

The lasting functional disruption induced by early septal damage proves that no apparent compensation occurred for the major role played by the septum in the inhibitory control of reactivity as well as in the utilization of spatial and/or other discriminative cues to get oriented (SREBRO et al., 1976) and thus develop an approach-avoidance conflict that prevents the control rat to get easily conditioned.

Our results also show that in septal rats,as we have seen for control rats, retention of escape and avoidance experienced at 30 days of age was stronger than that of similar events experienced at other ages. This improved retention of training that occurred at 30 days does not seem to be due to the high level of acquisition obtained at that age because the 40 day-old rats that had reached the same level of acquisition as the 30 day-old rats had a much lower level of reacquisition of the 2-way active avoidance task when tested at 90 days of age. The similar facilitation to reacquire a 2-way active avoidance task by an adult septal rat when a pre-learning of this task has been experienced at the age of 30 days shows that the early septal lesion did not affect retention in these rats. The septum does not seem to be critical for processes underlying long term retention of events experienced at puberty.

ACKNOWLEDGEMENTS

The author thanks Mrs Marie-Josée Angst and Ms. Gaby Rudolf for their skillful technical assistance and also Mrs Barbara Fritsch for typing the manuscript.

REFERENCES

Altman, J., Brunner, R.L., and Bayer, S., 1973, The hippocampus and behavioral maturation, Behav. Biol., 8:557.

Bauer, R.H., 1978, Ontogeny of two-way avoidance in male and female rats, Develop. Psychobiol., 11:103.

Campbell, B.A., 1967, Developmental studies of learning and motivation in infraprimate mammals, in: "Early behavior : comparative and developmental approaches", H.W. Stevenson, E.H. Hess and H.L. Rheingold, eds., Wiley, New York.

Campbell, B.A., and Mabry, P.D., 1972, Ontogeny of behavioral arousal : A comparative study, J. comp. physiol. Psychol., 21:371.

Campbell, B.A.,and Spear, N.E., 1972, The ontogeny of memory, Psychol. Rev., 79:215.

Denenberg, V.H., and Kline, N.J., 1958, The relationship between age and avoidance learning in the hooded rat, J. comp. physiol. Psychol., 51:488.

Douglas, R.J., Peterson, J.J., and Douglas, D., 1973, The ontogeny of a hippocampus-dependent behavior in two rodent species, Behav. Biol., 8:27.

Eclancher, F., 1984, Eveil comportemental et acquisition d'un conditionnement au cours de l'ontogenèse : Rôle joué par le septum, SFECA Bulletin intérieur, n°1:27.

Eclancher, F., and Karli, P., 1979, Septal damage in infant and adult rats: effects on activity, emotionality and muricide, Aggres. Behav., 5:389.

Eclancher, F. and Karli, P., 1981, Influence of rearing conditions on the acquisition of the two-way active avoidance responses by rats septalectomized at an early age, Behav. Brain Res., 3:83.

Egger, G.J., and Livesey, P.J., 1972, Age effects in the acquisition and retention of active and passive avoidance learning by rats, Develop. Psychobiol., 5:343.

Feigley, D.A., and Hamilton, L.W., 1971, Response to novel environment following septal lesions or cholinergic blockade in rats, J. comp. physiol. Psychol., 76:496.

Feigley, D.A., and Spear, N.E., 1970, Effect of age and punishment condition on long-term retention by the rat of active- and passive-avoidance learning, J. comp. physiol. Psychol., 73:515.

Izquierdo, I., Salzano, F., Thome, F.S., and Thaddéu , R., 1975, Shuttle behavior in weanling and in adult rats, Behav. Biol., 14:361.

Jackson, J.H., 1931, Selected writings of John Hughlings Jackson, vol. I and II, Hodder and Stoughton, London.

Johnson, D.A., 1972, Developmental aspects of recovery of function following septal lesions in the infant rat, J. comp. physiol. Psychol., 78:331.

Johnson, D.A., Poplawsky, A., Bieliauskas, L., and Liebert, D., 1972, Recovery of function on a two-way conditioned avoidance task following septal lesions in infancy : effects of early handling, Brain Res., 45:282.

Kirby, R.H., 1963, Acquisition, extinction, and retention of an avoidance response in rats as a function of age, J. comp. physiol. Psychol. , 56:158.

Klein, S.B., and Spear, N.E., 1969, Influence of age on short-term retention of active-avoidance learning in rats, J. comp. physiol. Psychol., 69:583.

Molino, A., 1975, Sparing of function after infant lesions of selected limbic structures in the rat, J. comp. physiol. Psychol., 89:868.

Moorcroft, W.H., Lytle, L.D., and Campbell, B.A., 1971, Ontogeny of starvation-induced behavioral arousal in the rat, J. comp. physiol. Psychol., 75:59.

Ramaley, J.A., 1979, Development of gonadotropin regulation in the prepubertal mammal, Biol. Reprod., 20:1.

Riccio, D.C., Rohrbaugh, M., and Hodges, L.A., 1968, Developmental aspects of passive and active avoidance learning in rats, Develop. Psychobiol., 1:108.

Schoenfeld, T.A., Hamilton, L.W., and Gandelman, R., 1974, Septal damage during the maturation of inhibitory responding : Effects in juvenile and adult rats, Develop. Psychobiol., 7:195.

Srebro, B., Ellertsen, B., and Ursin, H., 1976, Deficits in avoidance learning following septal lesions in the albino rat, Physiol. Behav., 16:589.

DISSOCIATION OF TWO BEHAVIORAL FUNCTIONS IN THE MONKEY AFTER EARLY HIPPOCAMPAL ABLATIONS

Helen Mahut

Psychology Department, Northeastern University
Boston, Massachusetts, 02155

INTRODUCTION

Bilateral excisions of medial temporal lobe structures (uncinate-amygdaloid region, hippocampus, hippocampal gyrus) in patients suffering from long-standing psychotic, or epileptic, symptoms resulted in a well-known severe anterograde amnesia (Scoville, 1954; Scoville and Milner, 1957). Subsequent clinical evidence pointed to the hippocampus as the critical structure in the memory loss (Penfield and Milner, 1958; Milner and Teuber, 1968; Milner, Corkin and Teuber, 1968).

Follow-up examinations of one of Dr. Scoville's patients (H.M.), operated upon in 1953, with permanent and severe anterograde amnesia, confirmed that the salient deficit remains in the transmodal establishment of new memories after delays as short as two minutes (Prisko, 1963; Sidman, Stoddard and Mohr, 1968). However, in spite of his global amnesia, patient H.M. performs normally on tests of motor and perceptual skills though he cannot remember having encountered the tasks before (Milner, 1962). More recently, he has been able to master a complex visual puzzle (the Tower of Hanoi) within a normal range of trials and to retain it with savings one year later (Cohen, Eichenbaum, DeAcado and Corkin, 1985). Instances of preserved learning co-existing with amnesia have been repeatedly noted in patients with brain damage other than that of medial temporal lobe structures (Warrington and Weiskrantz, 1968; Lhermitte and Signoret, 1972; Cermak and Butters, 1972; Brooks and Baddeley, 1976; Weiskrantz and Warrington, 1979; Cohen and Squire, 1980). The finding that the hippocampus is not essential for all learning is paralleled by extensive evidence obtained with rats (for reviews, see Hirsh, 1974; O'Keefe and Nadel, 1978; Olton, Becker and Handelman, 1979).

Though, in the clinic, the critical structure for memory was presumed to be the hippocampus, in all cases of amnesia there was either bilateral excision of both hippocampus and amygdaloid complex, as in patient H.M., or bilateral excision of hippocampus with unilateral excision of amygdala. No comparable memory loss was found after bilateral ablations limited to the uncinate-amygdaloid region (Scoville and Milner, 1957). However, in the absence of cases with selective damage to the hippocampus, its role in memory remained to be established.

To assess the behavioral functions of the hippocampus and to obtain tests. When appropriate, age x lesion interactions were examined with the use of two-way analyses of variance, followed by tests of simple main effects and by *a posteriori* between-group comparisons using the Newman-Keuls test (Kirk, 1968). Since, in most cases, the number of trials to learning criterion paralleled the number of errors, statistical comparisons were based on the number of errors. It should be noted that statistical analyses were carried out for all groups of infants and juveniles, including those with fornix sections in both age groups. However, results reported here concern only groups of normal, control and hippocampectomized monkeys. All levels of confidence were 2-tailed.

Results - Spatial reversals

On the first post-operative tests, groups of hippocampectomized infants and juveniles were significantly impaired (p's $<.032-.016$). The deficit persisted longer in infant, than it did in juvenile, monkeys: on the second round of tests, one year after surgery, infant hippocampectomized monkeys, unlike hippocampectomized juveniles, were still significantly impaired. Two years following surgery, significant impairment was no longer found.

- Object reversals

Since adult monkeys with hippocampal ablations obtain normal scores on this task, we did not expect, nor did we find, any post-operative abnormalities in the hippocampal group of infants or juveniles.

- Successive retentions - 24-hr and 1-hr intervals

In contrast with the performance of adult operated monkeys, no significant differences were found between normal and operated groups in the two age groups in either of the two conditions. This finding did not contribute, as yet, to the understanding of the nature of preserved capacities after hippocampectomy. Or so it seemed, as long as the spared performance on the successive retentions task was viewed in isolation. However, as mentioned in the Introduction, assessment of chronic effects of early hippocampectomy was to provide another example of spared learning capacity. Taken together, the two instances became easier to interpret in the context of a co-existing memory deficit, uncovered shortly afterwards. The differential effects of early hippocampal damage on associative learning, as contrasted with recognition memory, are described below.

Associative learning spared, recognition and association memory not spared after early resections of hippocampus.

- Concurrent object discrimination task

Though infants and juveniles with hippocampal ablations were not impaired, as a group, on the successive retentions task, individual monkeys had obtained elevated learning scores. Thus, the sparing of the capacity that mediates retention, and re-learning, of simple object discriminations was not clear-cut. Accordingly, the concurrent object discrimination task, known to be sensitive to the effects of hippocampal ablations in adult monkeys was chosen to serve as an unfamiliar test of associative learning (see Introduction, for descriptions of the task). At the time when it was administered, monkeys that had been operated upon in infancy were approximately 4 to 5 yrs of age, and those that had been operated on as 2 yr old juveniles, were approximately 7 yrs of age.

The results were clear-cut: Only older, but not younger hippocampectomized monkeys were impaired on this task. Two-way analysis of variance precise verification of surgical damage, we first studied anterograde effects of damage to the hippocampal formation in young adult non-human primates (rhesus macaque monkeys). This work revealed a complex post-operative behavioral profile that included both impaired and unimpaired performance on different tasks (Mahut, 1971; Mahut and Zola, 1973; Zola and Mahut, 1973; Mahut, Moss and Zola-Morgan, 1981; Moss, Mahut and Zola-Morgan, 1981). It was not easy, however, to make a meaningful distinction between the nature of impaired and preserved underlying capacities. Also, though impairment in the retention of simple visual, or tactual, object discriminations after 1-hr, or 24-hr, intervals was included in the profile (Mahut, et al., 1981), the quality and degree of this defect did not approximate memory losses found in the clinic. In retrospect, it is easier to see that the failure to find memory losses was due to our use of standard laboratory tasks that were more suited to the study of discrimination learning than they were to the study of short-term memory processes. In a subsequent project, we investigated the effects of hippocampal damage as a function of age with ablations of hippocampus performed either in 2 mos old infant, or 2 yr old juvenile, monkeys. In the first phase of the project, lesion effects were assessed shortly after the animals recovered from surgery, with the use of three discrimination tasks. To investigate possible recovery of initially impaired functions, monkeys were re-tested on the same three tasks twice, at one year intervals. The second phase of the project consisted of an assessment of chronic effects of hippocampectomy, five to seven years following surgery. During this phase, monkeys were tested on a new, previously unencountered discrimination task. Also, we had at our disposal a task contributed by Gaffan (1974) which closely resembled tests of recognition memory given to patients. The results of the longitudinal examination of the effects of early and late hippocampal ablations revealed a dissociation of performance on two classes of tasks and constitute the subject of this report. However, to provide a background against which the effects of hippocampal damage in infants and juveniles can be appreciated, a brief description will be given of the post-operative profile of adult monkeys, together with the description of the tasks that were used.

1. Discrimination reversal learning: Selective deficit on spatial tasks. Adult hippocampectomized monkeys are impaired on tasks that require them to displace first the left, then the right, of two identical plaques in order to find food on every trial (Mahut and Cordeau, 1964; Mahut, 1971; Mahut and Zola, 1973). In parallel with clinical findings (Corkin, 1965), the deficit on the spatial task is even more severe in the tactual modality (Mahut and Zola, 1973). However, the same monkeys are not impaired on an object discrimination reversal task in which one of two different objects is the positive stimulus on alternate sessions, regardless of its R-L position (Mahut, 1971; Mahut and Zola, 1973).

2. Deficit in retention of object discriminations: Successive retentions task. Adult hippocampectomized monkeys were impaired on a task in which initial acquisition of easy two-choice object discriminations was followed by five successive retention, and relearning, sessions, with the same member of a given pair of objects always the positive stimulus (Zola, 1973; Mahut, et al., 1981). The time intervals between acquisition and repeated retention tests were either 1-hr or 24-hrs. A trial by trial examination of performance revealed that retention of past events on the first trials of all sessions by operated monkeys equalled that of normal monkeys. It was the capacity to relearn object-reward attachments within the session that was impaired (Mahut and Moss, 1984).

3. Deficit in concurrent object discrimination learning. Operated monkeys are not impaired, as a rule, in learning individual easy object discriminations. However, they are impaired in learning several discrim-

inations simultaneously: A deficit is found on a task in which eight pairs of objects are presented in an intermingled fashion during each testing session, with one member of each pair always the positive stimulus, until all eight discriminations are mastered. The finding of the deficit in both visual and tactual modalities (Moss, *et al.*, 1981) emphasizes the supramodal involvement of the hippocampus in the formation and retention of object-reward associations in adult monkeys. Day by day analysis of performance revealed an abnormal susceptibility of operated monkeys to interpair interference (Moss, *et al.*, 1981).

EFFECTS OF HIPPOCAMPAL DAMAGE AS A FUNCTION OF AGE

Detailed descriptions of the subjects, surgical procedures and histological verifications of legions are given elsewhere (Rehbein, Mahut and Moss, 1985) and will only be summarized here.

Findings will be described for 22 rhesus macaque monkeys (Macaca mulatta). Of these, 12 were infants and 10 were juveniles. Infants , born at the California Primate Research Centre in Davis, were between 37 and 71 days of age on arrival and were bottled-fed until weaning time. With the exception of 3 juveniles born and raised in the California Primate Research Centre, all juveniles were fertile. They were approximately 2 yrs of age upon arrival, with weights ranging from 2,071-3,336 gr.

Infants were operated on at approximately 2 mos of age and all aseptic surgery was performed in one stage. The hippocampus, approached either through a small incision in the cortex at the caudal tip of the rhinal sulcus, or through the cortex medial to the occipito-temporal sulcus, was aspirated with the roof of the lateral ventricles used as a guide. Paraffin or frozen coronal sections were cut at 10, with every 20th section stained with cresyl violet for cell-body Nissl substance, and every 40th section stained with luxol blue for fibers. Lesions of hippocampus were reconstructed in coronal planes as well as in lateral and ventral views of the brain.

Four of the seven monkeys with intended resections of the hippocampus sustained radical bilateral removals of Ammon's horn, dentate gyrus and subicular complex. Three monkeys sustained near-total removals of hippocampus in one hemisphere, with one third of the hippocampus spared in the contralateral hemisphere. However, in all three animals fimbria fibers were damaged bilaterally in the most caudal aspect of the hippocampus. All seven monkeys in the group had additional damage to areas TE and TF-TH, and all but two had damage to the entorhinal cortex. The extent of extra-hippocampal damage varied from small unilateral to extensive bilateral.

Juveniles were operated on at approximately 2 yrs of age. Surgical and histological procedures, together with detailed descriptions and illustrations of representative lesions were given earlier (Zola-Morgan, Dabrowska, Moss and Mahut, 1983) and will be only summarized here. All bilateral surgery was done in one stage. The hippocampus was removed by suction after retraction of the occipito-temporal sulcus. Of the six monkeys with intended hippocampal removals, four sustained total, or near-total, bilateral removals of Ammon's horn, dentate gyrus and subicular complex. In the other two, the anterior third of the hippocampus was spared in the right hemisphere.

Five infants and four juveniles served as normal, control groups. Groups of infants with either bilateral, or unilateral, fornix sections, and a group of juveniles with bilateral fornix sections also participated in the study and are described elsewhere (Mahut and Moss, 1985; Rehbein, Mahut and Moss, 1985).

Methods

Order of task presentation. Upon recovery from surgery, all groups of infants and juveniles, together with corresponding groups of normal monkeys, were tested on three tasks: i. Spatial reversals, ii. Object reversals and iii. Successive retentions task, with 24-hr intervals. Testing with 1-hr intervals had to be delayed by one year because many 3-mos old infants refused to work when several testing sessions were massed in one day.

To explore possible age, or task, specific rates of functional recovery, all monkeys were re-tested on the same tasks one year following surgery. In addition, they were tested for the first time on the successive retentions task with 1-hr intervals. Two years following surgery, all groups of monkeys were tested for the third, and last, time on the spatial and object reversal tasks and, for the second time on the retentions task with 1-hr intervals. All monkeys in infant and juvenile N group were also tested on the successive retentions task with 24-hr intervals, but only individual hippocampectomized monkeys in the two age groups were tested for the third time in this condition.

Tasks. All testing was carried out in a modified Wisconsin General Test apparatus. An opaque screen separated the monkeys from the tray and a one-way vision screen concealed the experimenter from the monkeys. Banana pellets, raisins or small pieces of grapes, were used as bait.

i. Spatial reversals. Two identical black plaques covered the two food wells and, during initial position learning, the bait was under the left plaque for all animals. Twenty trials a day were given, using a non-correction method, until a learning criterion of 18 correct responses in 20 consecutive trials was reached. This was followed, next day, by a reversal to the opposite side and, at 20 trials a day, training was continued until learning criterion was met. Five successive reversals were administered in all.

ii. Object reversals. The procedure was the same as that followed for the spatial reversal task but, this time, two junk objects were used. Their position varied in a pre-determined order between right and left wells within each session and monkeys had to choose a previously incorrect object on each consecutive reversal.

iii. Successive retentions task. A discrimination between two objects was taught on one day, to a lenient learning criterion of 9 correct in 10 consecutive trials. Initial learning was followed by five retention tests, each to the same learning criterion, with the choice of the object rewarded during initial acquisition remaining the correct one throughout all sessions. Monkeys were thus tested with the same pair of objects for six consecutive days in the 24-hr condition. In the 1-hr condition, there were six consecutive sessions in one day, one hour apart. Two different pairs of objects were used in each of the two conditions so that two measures of initial learning and retention were obtained for each length of interval. In order to have initial discrimination problems that were neither too easy, nor too difficult, to learn in one session, we rejected pairs that could be learned under 20 trials and those that were not learned within 60 trials. As a result, pairs of objects that were used required uniformly between 31 and 59 trials, with 9 to 20 errors, to achieve learning criterion; means ranged between 33.3-41.5 for trials, and 14.5-17.3 for errors, across all groups and all conditions.

Statistical comparisons. The data obtained in the first phase of the project were subjected to Kruskal-Wallis one-way analyses of variance, followed by individual comparisons between groups, using Mann-Whitney *U*-

revealed a significant interaction between age and lesion ($p < .0005$). Between group comparisons revealed that older monkeys with hippocampal ablations made significantly more errors than did younger monkeys with comparable damage (p's<.05). Main effects. The detailed results obtained with the Newman-Keuls tests account for the significant main effect for lesion ($p<.0001$). It is worth noting that the extent of impairment shown by juvenile operated monkeys paralleled that seen in adult monkeys with hippocampal ablations. Also worth noting is the finding that younger animals with hippocampal ablations were not impaired in spite of the fact that all of them had sustained additional damage of the inferotemporal neocortex which, in itself, is sufficient to produce a deficit on the concurrent object discrimination task (Iwai and Mishkin, 1968; Cowey and Gross, 1970). The only meaningful correspondence between learning scores and extent, and site, of extrahippocampal damage was seen in the three best error scores obtained by monkeys with incomplete ablations of hippocampus.

- Recognition memory

As mentioned in the Introduction, a novel and appropriate method of assessing memory in nonhuman primates was contributed by Gaffan (1974). With a matching-to-sample technique, monkeys were trained to indicate whether they were able to discern, after a single previous presentation a familiar (i.e. previously presented) from a novel (i.e. not previously presented) test object. In one version of the task, delays between familiarization with the sample object and its re-presentation with a novel one were increased, in steps, from 10 to 30, 70 and 130 secs. In another version, several sample objects were presented serially and each 'list' was followed by matching trials of each list member with a novel object. Since, given a large pool of objects, several days would elapse before the same object was used again, but never in the same combination with another object, the task approximated a trial-unique test of recognition memory. In many respects, Gaffan's task resembled, more than did other standard laboratory tasks, some of the clinical tests that reveal memory deficits in amnesic patients.

In our experiments, we used a modified version of the task based on delayed non-matching since this procedure exploits the tendency of rhesus macaque monkeys to respond to novel stimuli and results in much more rapid learning (Mishkin and Delacour, 1975). Detailed procedures were described earlier, together with the results obtained with groups of older monkeys, operated on at 2 yrs of age (Mahut, Zola-Morgan and Moss, 1982). Preliminary results obtained with groups of younger animals, operated upon in infancy, were reported earlier (Rehbein, Zola-Morgan, Mahut and Moss, 1980; Rehbein and Mahut, 1983). A full description will be given elsewhere (Rehbein, *et al.*, 1985). However, salient findings will be summarized here.

- Object recognition task - Delays

Two monkeys in group H learned the basic task, with 10-sec delays, as readily as did normal, control monkeys. However, the remaining five monkeys in the group were only able to attain an average of 76% correct responses within the limits of testing. A significant interaction effect was found for groups and delays ($p<.0001$). Tests of simple main effects revealed significant group differences at each delay (p's<.01). Even the two monkeys, unimpaired on the basic task, were significantly impaired with longer intervals (p's<.05). The deficit shown by younger monkeys was comparable to that found with older hippocampectomized monkeys.

- Object recognition task - Lists

Though the group x list length interaction effects only approached significance(p=.06), tests of simple main effects revealed significant group differences at each list length (p's<.01). The level of accuracy with lists of 1 item reached by monkeys in group H was significantly lower than that reached by monkeys in the normal control group. It was also significantly lower than that of normal control monkeys with all other list lengths (p's<.05).

The performance of normal monkeys remained at consistently stable levels of accuracy in the delays version of the task and ranged from 90% to 97% correct responses throughout all delays. However, in the list version, a more difficult task, the group mean level of accuracy decreased as the list length increased. Detailed comparisons revealed a significantly greater decrease in accuracy of performance of monkeys in the hippocampal group than that of monkeys in the normal, control group (p's<.05). These results paralleled those found with older monkeys (Mahut, et al., 1982).

It should be noted that all animals with hippocampal ablations sustained bilateral damage of area TF-TH. In addition, in all but one monkey, there was small, to extensive, bilateral damage of the inferotemporal cortex and radical, bilateral ablations of this structure are sufficient to impair performance on the delayed recognition task (Mishkin and Oubre, 1976). The small number of animals and the variability in the site, and extent, of extrahippocampal damage make it difficult to assess possible interactive effects of unilateral lesions of area TE and entorhinal area with those of bilateral, radical ablations of hippocampus. However, the performance of monkeys in the hippocampal group on either version of the task did not appear to reflect individual differences in the size and site of lesion.

- Position recognition memory

Spatial function was not spared by early hippocampal ablations, but the initial post-operative deficit was no longer present on the second retest, 2 yrs following surgery. It was difficult to decide, however, whether the improved performance was due to the effects of practice with the task, or whether it reflected a recovery of spatial reversal function. Therefore, approximately three years later, monkeys were tested on the delayed alternation task (DA) which required choices of either the left, or the right, of two identical plaques on alternate trials within each session of 30 trials. Normal, control monkeys find this task to be more difficult than consecutive spatial reversals that were used post-operatively. Nevertheless, it was found that monkeys with hippocampal ablations were not significantly impaired, as a group, on the delayed alternation task (Rehbein and Mahut, 1983).

Since examination of chronic effects of hippocampectomy, 5 or more years after surgery, revealed efficient learning of operated monkeys of tasks administered with repeated trials (successive retentions, concurrent object discrimination), and impaired learning of trial-unique recognition memory tasks, we could expect them to be impaired in learning a trial-unique spatialrecognition task. Accordingly, all younger monkeys were given a delayed, non-matching-to-sample position recognition task. In this task, monkeys had to discern on every trial, an unfamiliar location of one of two identical small plaques (i.e., one previously not presented) from a familiar one (i.e., one previously presented), on a tray containing 18 food wells. As in the object recognition memory task, delays of 10 secs to 130 secs were introduced between the presentation of the position sample and its re-presentation with a plaque in a novel position (Rehbein and Mahut, 1983).

A significant interaction effect was found between delays and groups

($p<.0002$). Tests of simple main effects revealed significant group effects with each delay (p's$<.01$). All seven monkeys with hippocampal ablations were severely impaired and the accuracy of their responses was close to chance levels.

In summary: When the effects of early and late hippocampal damage were assessed five to seven years following surgery, it was found that learning of mutiple object-reward associations was spared (concurrent object discriminations) after early, but not after late, lesions. However, operated monkeys in both age groups were impaired on tests of trial-unique object recognition memory and on tests of delayed position recognition.

CONCLUSIONS

Whether they were inflicted in infancy, or at 2 yrs of age, selective ablations of hippocampus spared performance on a task in which two-choice object-reward associations had to be retained, and re-learned with repeated trials (successive retentions task, with 1-hr and 24-hr intervals). Nor were monkeys with early lesions impaired in learning multiple object-reward attachments when one member of each pair of objects retained the same reward value for several hundred trials (concurrent object discrimination task).

In marked contrast, both early and late ablations of hippocampus impaired memory processes. Thus, operated monkeys experienced difficulty in recognizing the quality of objects seen only once before after delays as short as 30 secs. Recognition was even less accurate when 'lists' of as few as three objects had to be memorized in only one presentation (trial-unique delayed recognition task). In addition, monkeys with early lesions were significantly impaired in discerning, after a single presentation, a familiar, from an unfamiliar, location of an object (trial-unique delayed position recognition task).

The dissociation of effect on two classes of tasks found after early hippocampal ablations is best understood in the light of a model of learning proposed by Hirsh (1974) and by Hirsh and Krajden (1982). The hippocampus is considered to be a cognitive, Tolmanian, system which mediates selective processes that determine which limited subset of memories, out of many potentially relevant memories, will guide behavior at any given time. However, the brain is also assumed to contain an associative learning system which operates, literally, in the Hullian sense of a summating build-up of stimulus-response associations through reinforcement. On this view "a strong association involving a given stimulus can exist only when the representation of that stimulus is incorporated into no more than one association". This subsidiary brain system relies, therefore, on 'habit formation' (Hirsh, 1974, p.439).

By choice, Hirsh's theory of dual brain mechanisms remains at a 'metaphysiological' level. In an earlier attempt to account for instances of spared learning in adult hippocampectomized monkeys, Mahut and Moss (1984) suggested that the two learning systems may have distinct neural correlates, with the striatum identified as an important component of the S-R learning system. The neural distinctiveness of the two systems would account for the results obtained with early lesions: the learning of object discrimination tasks with repeated trials was spared because it was mediated by an intact S-R system. However, the learning of trial-unique tasks was impaired because the memory system was disrupted by ablations of hippocampus.

The finding that dissociable effects of hippocampectomy are more marked the earlier the age of lesion reveals possible interactive effects between the two brain systems: the S-R system may be able to mediate learning most effectively before it has a chance to function in the presence of an intact hippocampal system. Once the two systems are allowed to co-exist, at least for the first 2 yrs of life, disruption of one interferes with the normal function of the other.

REFERENCES

Brooks, D.N. and Baddeley, A., 1976, What can amnesic patients learn? Neuropsychologia, 14:111.

Cermak, L.S. and Butters, N., 1972, The role of interference and encoding in the short-term deficits of Korsakoff patients. Neuropsychologia, 10: 89.

Cohen, N.J. and Squire, L.R., 1980, Preserved learning and retention of pattern analyzing skill in amnesia: Dissociation of knowing how and knowing that. Science, 210-207.

Cohen, N.J., Eichenbaum, H., DeAcado, B. and Corkin, S. Different memory systems underlying acquisition of procedural and declarative knowledge. Ann. N.Y. Acad. Sci., in press.

Corkin, S., 1965, Tactually-guided maze learning in man: Effects of unilateral cortical excisions and bilateral hippocampal lesions. Neuropsychologia, 3:339.

Cowey, A. and Gross, C.G., 1970, Effects of foveal prestriate and inferotemporal lesions on visual discrimination by rhesus monkeys. Exper. Brain Res., 11:128.

Gaffan, D., 1974, Recognition impaired and association intact in the memory of monkeys after transection of the fornix, J. Comp. Physiol. Psychol., 86:1100.

Hirsh, R., 1974, The hippocampus and contextual retrieval of information from memory: A theory, Behav. Biol., 12:421.

Hirsh, R. and Krajden, J., 1982, The hippocampus and the expression of knowledge, in: "The expression of knowledge", R.L. Isaacson and N.E. Spear, eds., Plenum Press, New York.

Iwai, E. and Mishkin, M., 1968, Two visual foci in the temporal lobe of monkeys. Paper presented at the Japan-US joint seminar on neurophysiology of memory, Kyoto, Japan.

Kirk, R.E., 1968, Experimental design: Procedures for the behavioral sciences, Brooks-Cole, Belmont, CA.

Lhermitte, F. and Signoret, J.L., 1972, Analyse neuropsychologique et differenciation des syndromes amnesiques, Rev. Neurol., 126:161.

Mahut, H., 1971, Spatial and object reversal learning in monkeys with partial temporal lobe ablations, Neuropsychologia, 9:409.

Mahut, H. and Cordeau, J.P., 1964, Some long-term effects of temporal lobe resections on auditory and visual discrimination in monkeys, Brain, 87:177.

Mahut, H. and Moss, M., 1984, Consolidation of memory: the hippocampus revisited, in: "Neuropsychology of memory", L.R. Squire and N. Butters, eds., Guilford Press, New York.

Mahut, H. and Moss, M., The Monkey and the Sea Horse, in: "The hippocampus", R.L. Isaacson and K.H. Pribram, eds., Plenum Press, New York. In press.

Mahut, H., Moss, M. and Zola-Morgan, S., 1981, Retention deficits after combined amygdalo-hippocampal and selective hippocampal resections in the monkey, Neuropsychologia, 19:201.

Mahut, H. and Zola, S., 1973, A non-modality specific impairment in spatial learning after fornix lesions in monkeys, Neuropsychologia, 11:255.

Mahut, H. and Zola, S., 1977, Ontogenetic time-table for the development of three functions in infant macaques and the effect of early hippocampal damage upon them, Abst. Soc. Neurosci., 3:428.

Mahut, H., Zola-Morgan, S. and Moss, M., 1982, Hippocampal resections impair associative learning and recognition memory in the monkey. J. Neurosci., 2:1214.
Milner, B., 1962, Les troubles de la mémoire accompagnant des lésions hippocampiques bilatérales, in: Physiologie de l'hippocampe, P. Passouant, ed., CNRS, Paris.
Milner, B. and Teuber, H.-L., 1968, Alteration of perception and memory in man: Reflections on methods, in:"Analysis of behavioral change", L. Weiskrantz, ed., Harper and Row, New York.
Milner, B., Corkin, S. and Teuber, H.-L. 1968, Further analysis of the hippocampal amnesic syndrome: 14-year follow-up study of H.M. Neuropsychologia, 6:215.
Mishkin, M. and Delacour, J., 1975, An analysis of short-term visual memory in the monkey. J. Exper. Psychol.: Animal Behav. Processes, 1:326.
Mishkin, M. and Oubre, J.L., 1976, Dissociation of deficits on visual memory tasks after inferior temporal and amygdala lesions in monkeys, Abst., Soc. Neurosci, 2:1127.
Moss, M., Mahut, H. and Zola-Morgan, S., 1981, Concurrent discrimination learning of monkeys after hippocampal, entorhinal or fornix lesions, J. Neurosci., 1:227.
O'Keefe, J. and Nadel, L., 1978, The hippocampus as a cognitive map, Clarendon Press, Oxford.
Olton, D.S., Becker, J.T. and Handelman, G.E., 1979, Hippocampus, space and memory, Behav.Brain Sci., 2:313.
Penfield, W. and Milner, B., 1958, Memory deficit produced by bilateral lesions in the hippocampal zone. A.M.A. Arch. Neurol. Psychiat., 79: 475.
Prisko, L.H., 1963, Short-term memory in focal cerebral damage. Unpub. Doct. Dissert., McGill University, Montreal, Canada.
Rehbein, L. and Mahut, H., 1983, Long-term deficits in association and spatial recognition memory after early hippocampal damage in monkeys. Abst., Soc. Neurosci., 9:639.
Rehbein, L., Mahut, H., and Moss, M., Hippocampal ablations in infant monkeys: A long-term assessment of behavioral change. In preparation, to be submitted to J. Neurosci.
Rehbein, L., Zola-Morgan, S., Mahut, H. and Moss, M., 1980, Failure of sparing, or recovery, of recognition memory after early hippocampal resections in the rhesus macaque, Abst., Soc. Neurosci., 6:88.
Scoville, W.B., 1954, The limbic lobe in man., J.Neurosurg., 11:64.
Scoville, W.B. and Milner, B., 1957, Loss of recent memory after bilateral hippocampal lesions. J. Neurol. Neurosurg. Psychiat., 20:11.
Sidman, M., Stoddard, L.T. and Mohr, J.P., 1968, Some additional quantitative observations of immediate memory in a patient with bilateral hippocampal lesions, Neuropsychologia, 6:245.
Warrington, E.K. and Weiskrantz, L., 1968, New method of testing long-term retention with special reference to amnesic patients, Nature, 217:972.
Weiskrantz, L. and Warrington, E.K., 1979, Conditioning in amnesic patients, Neuropsychologia, 17:187.
Zola, S., 1973, The effects of fornix transection on reversal learning in the monkey. Unpub. Doct. Dissert., Northeastern Univer, Boston, MA
Zola, S. and Mahut, J., 1973, Paradoxical facilitation of object reversal learning after fornix lesions in monkeys, Neuropsychologia, 11:271.
Zola-Morgan, S., Dabrowska, J., Moss, M. and Mahut, H., 1983, Enhanced preference for perceptual novelty in the monkey after section of the fornix but not after ablation of the hippocampus, Neuropsychologia, 21:433.

IS THE HIPPOCAMPUS REALLY INVOLVED IN MEMORY?

Leonard E. Jarrard

Department of Psychology
Washington and Lee University
Lexington, Virginia

INTRODUCTION

When asked about the functions of the hippocampus, most investigators in the area agree that the hippocampus is involved in some way in memory. A good summary of how many feel was provided by John O'Keefe (1983) in a panel discussion at a recent conference. O'Keefe said, "...there is general agreement or consensus that it (the hippocampus) has something to do with memory and a disagreement as to what kinds of memories" (p. 614).

The idea that the hippocampus is involved in memory has been around for many years, but received considerable support from the work of Scoville and Milner (1957). These investigators described the (now) well known patient, H. M., who was operated on for intractable epilepsy, and subsequently experienced a severe and permanent recent memory impairment following neurosurgical removal of the medial temporal cortex (which included the hippocampus). Subsequent attempts to study amnesia by subjecting monkeys to the same hippocampal ablations that had caused amnesia in man, resulted in little evidence for an involvement of the hippocampus in memory (Orbach et al., 1960). Since these early studies, there have been numerous experiments in which it is reported that ablation of the hippocampus interferes with performance of a variety of learning and memory tasks (see Gray and McNaughton, 1983).

From time to time, there have been investigators who have questioned the involvement of the hippocampus in memory. Jim Horel (1978), in a thoughtful analysis of the literature and a careful series of experiments, concluded that the memory impairments frequently reported following removal of the temporal lobe are not caused by damage to the hippocampus, but rather by damage to the inferior-medial temporal lobe with its' connecting stem. (However, a report published in 1982 by Zola-Morgan et al., does not agree with this conclusion.) In a series of experiments, Stein and his colleagues (Isseroff et al., 1976; Stein et al., 1969) showed that rats with large bilateral lesions of the hippocampus are still able to learn and remember complex maze and memory tasks as well as nondamaged controls if the lesions are carried out in stages. More recently, Vanderwolf and Leung (1983) have stated that the results of their recording studies have forced them to question whether the hippocampus is involved in learning.

Much of the evidence that has been employed to implicate the hippocampus in learning and memory comes from lesion experiments. While there are conceptual problems of interpretation that arise from the use of lesions (see Gregory, 1961; Webster, 1973), there are also the practical problems associated with limiting the damage to the intended structure. In this regard, it is interesting that Green, in 1960, stated that "...ablation of the hippocampus without serious damage to large areas of the brain has not yet proved possible" (p. 1381). Certainly, conventional lesion techniques (ablation, thermocoagulation) invariably interrupt fibers-of-passage, and it is frequently the case that areas adjacent to the structure of interest are damaged. Interpretation of lesion studies is further complicated by the structural reorganization that is known to occur following damage to some brain areas, and changes in underlying behavioral processes (attention, sensory-motor functions, motivation) that may serve to confound the behaviors being studied. While there are these potential complicating factors, only with a lesion approach can one determine which behaviors can and which cannot be performed in the absence of a particular brain area (Olton et al., 1979).

We have recently completed a series of experiments that involved using neurotoxins to selectively damage cells in the hippocampus (Jarrard, 1983; Jarrard, in press). The advantage of using neurotoxins in lesion studies is that cells in a region can be removed without damaging either fibers-of-passage or afferents that terminate in the area. To be selective and yet limited, we find that the neurotoxin must be injected in the least concentration and volume possible, injected at a slow rate, and the cannula must be left in place for several minutes to prevent chemical flow up the needle track. Further, careful histological analysis of the lesion with both cell and silver stains is important.

In our studies to date, we find that performance of complex place and cue memory tasks is minimally affected following interruption of different components of the intrahippocampal circuit. Specifically, bilateral removal of cells in the dentate gyrus with colchicine, the CA3 cell field with kainic acid, or more extensive removal of the hippocampus with ibotenic acid (IBO), all have only a small, temporary effect on the performance of our complex memory tasks. In contrast with these results, removal of the hippocampus with aspiration, and interruption of the main connections of the hippocampus with other structures (fimbria-fornix, entorhinal cortex), all have major effects on performance (for a review see Jarrard, in press).

Since much of the research that is used to implicate the hippocampus in memory has been obtained using the more conventional lesion techniques that inadvertently cause extra damage, it seemed that a direct comparison of the effects of aspiration, radio-frequency, and the more specific neurotoxin (IBO) lesions of the hippocampus on memory was in order. Thus, the purpose of this experiment was to see if removing the cells in the hippocampus (CA1 - CA3 pyramidal cells, and cells in the dentate gyrus) using a technique that spares fibers-of-passage, would have a similar effect on memory as removing the cells with more conventional lesion techniques.

METHOD

A radial arm maze and a within-subjects design were used, together with a procedure that permits determining two kinds of learning (place and cue) and two memory functions (reference and working memory). The radial arm maze was similar to the one described elsewhere (Jarrard, 1983).

In the place task, the rats learned to choose four out of eight similar arms where all arms remained in the same spatial location from trial to trial. There were many extramaze cues readily visible (table, door, blackboard, panel of fluorescent lights, rack of cages). In the cue task, different textured floor inserts (hardware cloth, carpet, cloth, aluminum, ceiling tiles, screen, sandpaper) in the eight arms were moved in a random order from trial to trial, and the rat was rewarded for choosing the same four cues. The lighting in the cue task consisted of a single light bulb with a reflector located directly over the maze, and this served to maximize the visability of the intramaze cues and minimize the room cues. Thus, performance in the place task required the use of distal, spatial cues, whereas, proximal intramaze cues were important for performance in the cue task.

On a single trial the rat was placed on the maze where it remained until all four reinforcements were received, until 16 choices were made, or until 5 min had elapsed, whichever occurred first.

Since only four out of eight arms (or cues) were baited, the types of memory errors that were made can be analyzed. Reference memory errors are operationally defined as choices of arms (or cues) that are never baited; working memory errors are repeated entries into baited and unbaited arms (or cues) that have already been visited within the trial. Working memory errors can be further divided into reentries into baited arms (or cues) (working memory-correct), and reentries into arms (or cues) that have never been baited (working memory-incorrect). Since a within-subjects design was used, information is obtained from each animal concerning his performance on both the place and cue task, as well as working memory and reference memory within each task (see Jarrard, in press).

In the experiment, the rats were trained before the operations to a criterion of performance on each task. The preoperative training required approximately 60 trials on each task, with the rats receiving two trials each day on each task. The operations were next carried out, and after recovery the animals were retested for 50 trials on each task using the same conditions as those that had been used before the operations.

There were four groups of subjects. In three of the groups an attempt was made to maximize the amount of damage to the hippocampus and minimize the damage to adjacent structures. Animals in the Complete Hippocampal Aspiration group (CHasp; N = 9) had the hippocampus removed bilaterally by suction (see Jarrard, 1976). After removing the bone over the dorsal hippocampus, and with the aid of a dissecting microscope, the neocortex overlying the hippocampus was suctioned, and then as much hippocampus as possible was removed including the fimbria. In the Complete Hippocampal Radio-frequency group (CHrf; N = 5) lesions were made at four different sites on each side. Thus, each rat had a total of eight electrode penetrations. In animals in the Complete Hippocampal Chemical group (CH-C; N = 16), the lesions were made by injecting .10 μl of a 10 μg/μl solution at each of seven different sites on each side. These animals had 14 injections per animal (see Jarrard, in press). The fourth group of rats consisted of operated and unoperated controls (N = 17), whose similar data was subsequently combined into a single control group. The rats were allowed a 14-day recovery period before testing was resumed.

RESULTS

Anatomical

Histological preparation of the brains included a cresyl violet stain for cells, and a silver stain (Fink and Heimer, 1967) to help identify degenerating and damaged brain areas.

Previous experience indicated that it is extremely difficult to selectively (and yet completely) lesion the hippocampus with neurotoxins (Jarrard, in press). Even with great care, the damage in some animals is either too limited, or is so extensive as to involve other structures (especially subiculum). Of the 16 animals in the CH-C group, 9 were judged as having selective lesions that included most of the cells in the hippocampus (pyramidal cells and cells in the dentate gyrus) and only minimal damage to other structures. In two rats the damage to hippocampus was limited mostly to bilateral damage of the CA3 cell field, and these rats were dropped from the study. In five rats there was complete loss of cells in the hippocampus, but there was also considerable bilateral damage to the subiculum. The data for these animals were not included in the analyses. Decisions concerning bilateral damage to subiculum were greatly aided by the use of silver stained brain sections. While it is often difficult with a cell stain to determine the existence of cell loss in a structure like the subiculum, with the silver stain one can examine the terminal fields and fiber bundles of subicular projections for the presence of degeneration. For example, a good estimate of the amount of subicular involvement can be obtained by looking in the descending columns of the fornix in material stained with the silver stain.

The loss of hippocampal cells in the 9 rats that remained in the CH-C group was extensive. In Figure 1 are cell stained brain sections from a control animal (left) and from a representative animal in the CH-C group (right). Any sparing in these animals was limited for the most part to CA1 cells in the dorsal and posterior areas. In most animals the sparing was unilateral and consisted of 15 to 20% of the CA1 cells. In addition, there were often some cells that survived in the dentate gyrus but these were few in number. Any degeneration in postcommissural fornix in these animals was minimal.

The nature and extent of the damage in the CHasp and CHrf groups can be seen in the microphotographs shown in Figure 2. The aspiration lesions were generally similar to those previously described (Jarrard, 1976). As is usually the case with this approach, 80 to 85% of the hippocampus was removed. In most cases the fimbria was bisectedlaterally, and the only sparing of hippocampus was found in the posterior-ventral area. There was generally a greater amount of extrahippocampal damage in the CHrf animals than in the other lesion groups. As can be seen in the representative brain sections in Figure 2 on the right, this damage included deep layers of the neocortex, and in some cases involved bilateral damage to the thalamus.

In terms of amount of damage to the hippocampus, the loss of cells in the structure was generally quite similar in the three operated groups. In fact, it would appear that a slightly greater loss of pyramidal and dentate gyrus granule cells was found in animals in the CH-C group.

Behavioral

The main behavioral results of the experiment are presented in Figures 3 and 4. Statements of differences between groups that follow are based on

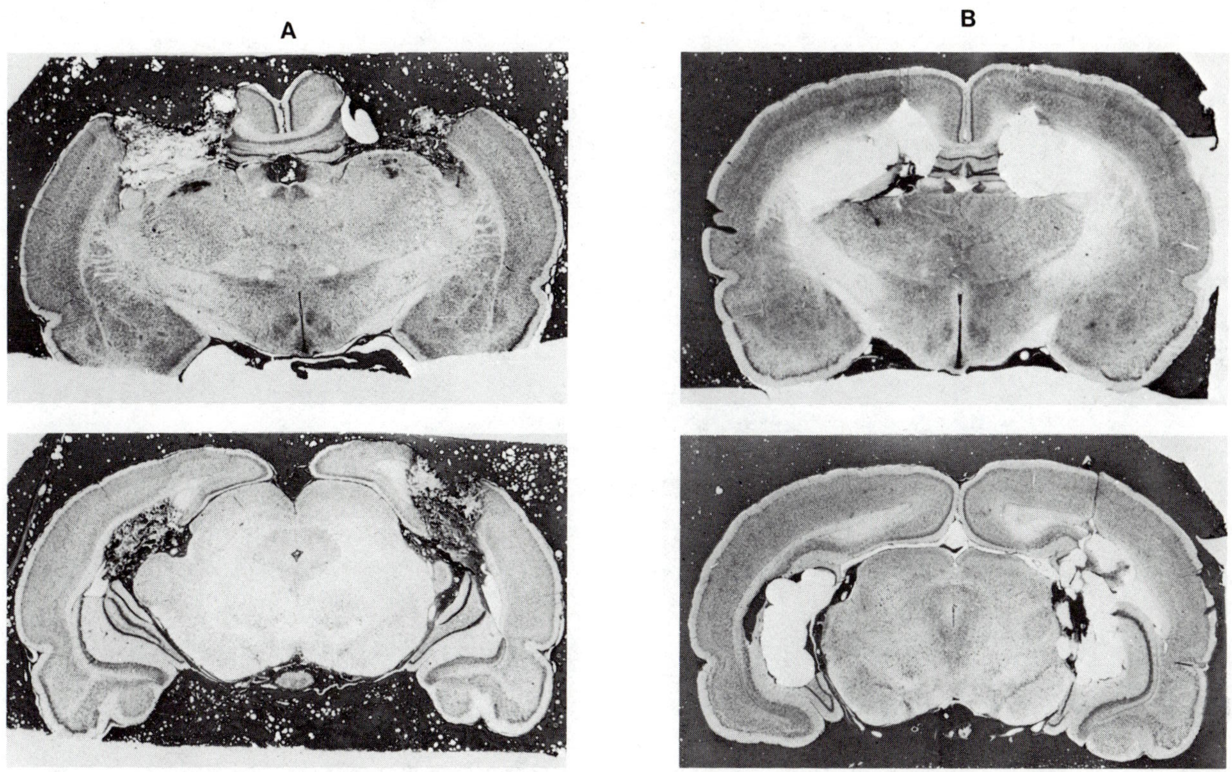

Fig. 1. Photomicrographs of coronal, cresyl-violet stained sections at anterior (top) and posterior (bottom) levels of the hippocampus for an unoperated animal (left), and a rat with a complete hippocampal ibotenate lesion (CH-C group; right). In the ibotenate injected animal the pyramidal cells in the hippocampus and the granule cells in the dentate gyrus have been replaced by glial cells.

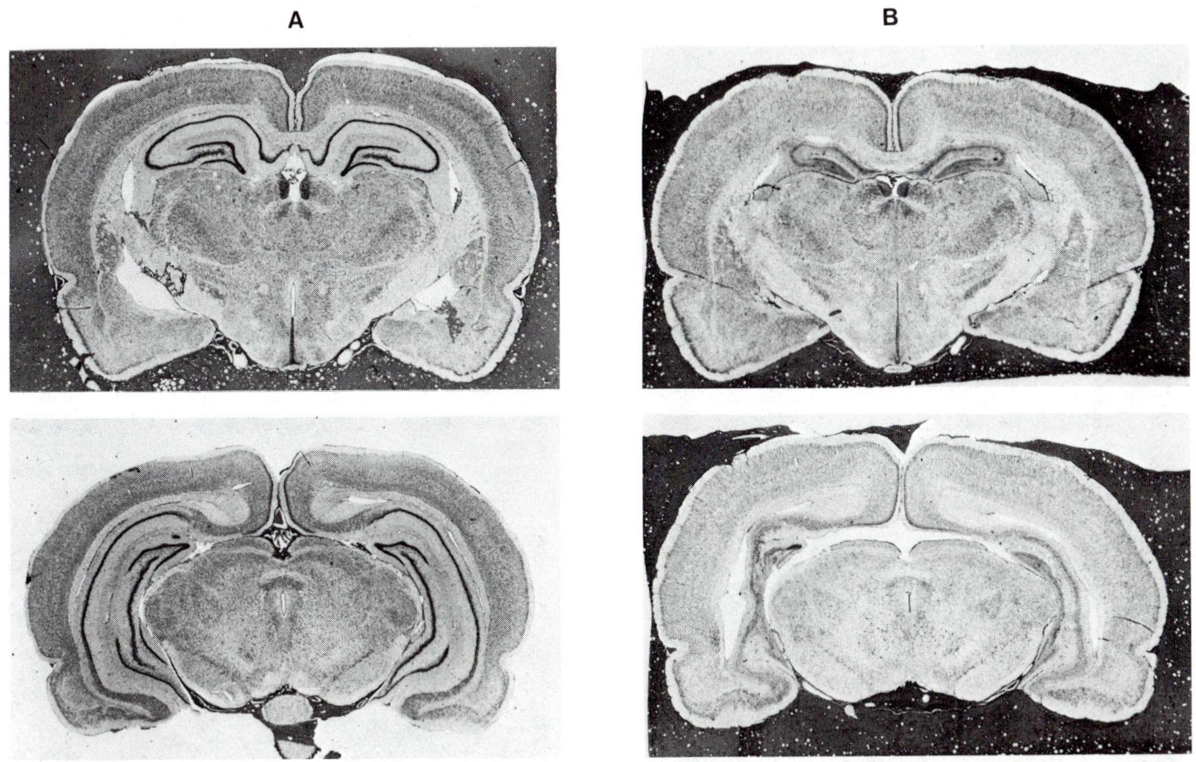

Fig. 2. Photomicrographs of coronal sections at anterior (top) and posterior (bottom) levels of the hippocampus for an animal with a complete hippocampal aspiration lesion (CHasp group; left), and a rat with a complete hippocampal radio-frequency lesion (CHrf group; right) (see text).

statistical analyses of the data, in most cases repeated-measures analysis of variance.

In Figure 3 is plotted the mean number of errors made by the groups on the place and cue tasks as a function of blocks of trials. One obvious difference is that all three groups with hippocampal lesions generally made more errors on the place than on the cue task. The most impaired animals were those that had radio-frequency lesions (CHrf group), and the next most impaired were the animals with aspiration lesions of the hippocampus (CHasp group). Perhaps the most surprising finding was that removal of the cells in the hippocampus with IBO (the CH-C group) had only a slight, temporary effect on performance on the place task, and no effect on the cue task. Rats with IBO lesions differed from controls on the first 20 trials on the place task, but on subsequent trials performance was like that of controls.

More specific information concerning the nature of the impairments in the operated animals is provided by analyses of the types of errors that were made (see Figure 4). Reference memory errors are original choices of arms (or cues) that were never baited; working memory-correct errors are repeated entries into correct, baited arms that had already been visited; working memory-incorrect are repeated entries into arms that were never baited. On the place task, animals in the CHrf and the CHasp groups made many errors of all three types. Observation of these animals during testing made one think the rats were almost choosing arms at random on the place

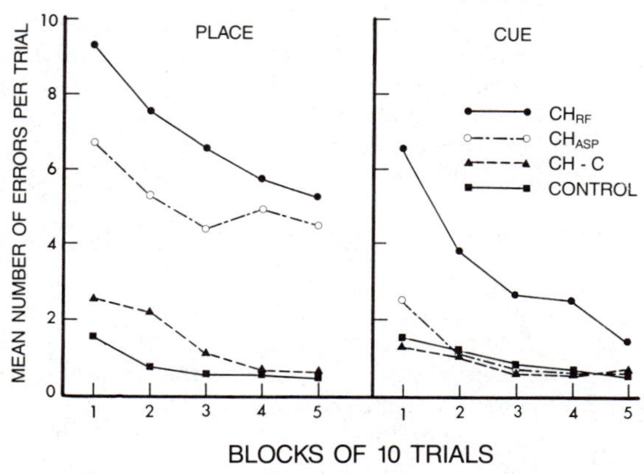

Fig. 3. Mean number of errors for the groups on the place and cue tasks are plotted as a function of blocks of trials during testing for retention. (Abbreviations: CHrf = complete hippocampal, radio-frequency lesion; CHasp = complete hippocampal, aspiration lesion; CH-C = complete hippocampal, ibotenic acid lesion; Control = operated and unoperated control.)

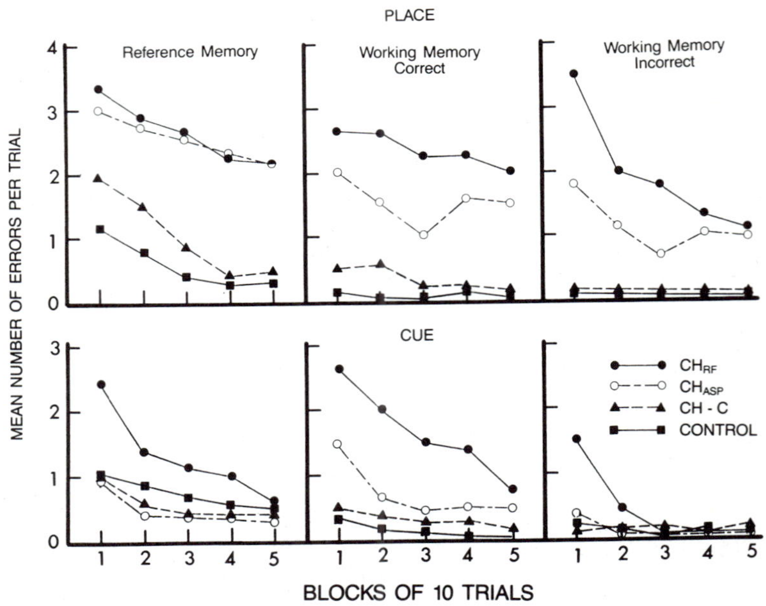

Fig. 4. Mean number of different types of errors are shown for the various groups on the place (top) and cue (bottom) tasks. (See caption to Figure 3 for identification of the groups.)

task; however, this was not the case on the cue task. Both CHasp and CHrf rats avoided incorrect arms (and cues) they had already visited (working memory-incorrect errors) on the cue task. Rats in the CHasp group tended to reenter correct arms (cues) they had already visited for the first 10 trials (working memory-correct errors), but the difference was not statistically significant. Inspection of Figure 4 shows that errors made by CHrf rats on the cue task were primarily reference and working memory-correct errors. Animals with the hippocampus removed by injection of IBO (CH-C group) made more reference memory errors on the place task than controls early during testing, but did not differ from controls after 20 trials.

CONCLUSIONS

Even though the hippocampus is usually thought to be intimately involved in memory, our results show that the structure can be removed with little effect on the complex memory tasks we used. Specifically, there was only a temporary effect on performance in the place task following removal of hippocampus with ibotenate, but removal with more conventional lesion techniques (aspiration, radio-frequency) resulted in a long-lasting impairment. The total amount of cell loss in the hippocampus was generally similar with all three lesions. However, the IBO lesion did not interrupt fibers-of-passage, and there was less incidental damage to adjacent areas.

The damage to fibers-of-passage may be especially important. Cells in the subiculum send axons in a rostral direction through the alveus to the septum, nucleus accumbens, and to the hypothalamus (Kelley and Domesick, 1982; Swanson, 1979). Then, too, there are entorhinal-accumbens projections (Krayniak et al., 1981), as well as septo-subicular fibers (Swanson and Cowan, 1977). Most of these projections were interrupted with our aspiration and radio-frequency lesions, but were not damaged with the ibotenate lesions. Further, it is difficult not to inadvertently damage thalamus and the subiculum with the aspiration and radio-frequency lesions, and there is unavoidable damage to neocortex. Our other research seems to indicate that the subiculum and its projections through the alveus may be especially involved (Jarrard, in press). Thus, with the more conventional lesion techniques there is extrahippocampal damage, whereas, with the neurotoxin lesions this additional damage is minimized.

It may be important that some animals in the CHasp and CHrf groups seemed to have difficulty recovering from the operations. A number of animals were hyperreactive and difficult to handle for several days. Further, several animals in these groups were observed to have spontaneous seizures. These observations suggest that brain regions in addition to hippocampus may have been affected.

So, is the hippocampus involved in memory? Of course, our findings do not provide a definite answer to this important question. However, they do show that if one is careful, most of the cells in the hippocampus can be removed with minimal effects on the complex memory tasks used in our experiments -- tasks shown to be sensitive to the effects of other lesions. Considering the amount of evidence for an involvement of the hippocampus in memory obtained with other experimental techniques (e.g., recording, stimulation), it is difficult not to believe that the hippocampus is normally involved in memory in some way. It should now be possible, with the use of axon-sparing neurotoxin lesions, to determine if the hippocampus really does play a critical role in memory.

ACKNOWLEDGMENTS

The author is indebted to Beverly Bowring for assistance with the histology and to Paul Chapman and Hugh Jarrard for testing the animals. This research was supported by grants from the National Science Foundation (BNS 80-05735, 83-06840).

REFERENCES

Fink, R. P., and Heimer, L., 1967, Two methods for selective silver impregnation of degenerating axons and their synaptic endings in the CNS, Brain Res., 4:369.

Gray, J. A., and McNaughton, N., 1983, Comparison between the behavioral effects of septal and hippocampal lesions: a review, Neurosci. Biobehav. Rev., 7:119.

Green, J. D., 1960, The hippocampus, in: "The Handbook of Physiology," Vol. II, J. Field, ed., Williams and Wilkins Co., Baltimore.

Gregory, R. L., 1961, The brain as an engineering problem, in: "Current Problems in Animal Behavior," W. H. Thorpe and O. L. Zangwill, eds., Cambridge University Press, Cambridge.

Horel, J. A., 1978, The neuroanatomy of memory, Brain, 101:403.

Isseroff, A., Leveton, L., Freeman, G., Lewis, M., and Stein, D. G., 1976, Differences in the behavioral effects of single-stage and serial lesions of the hippocampus, Exptl. Neurol., 53:339.

Jarrard, L. E., 1976, Anatomical and behavioral analysis of hippocampal cell fields, J. comp. physiol. Psychol., 90:1035.

Jarrard, L. E., 1983, Selective hippocampal lesions and behavior: effects of kainic acid lesions on performance of place and cue tasks, Behav. Neurosci., 97:873.

Jarrard, L. E., 1984, Selective hippocampal lesions and behavior: implications for current research and theorizing, in: "The Hippocampus," R. L. Isaacson and K. Pribram, eds., Plenum Press, New York, in press.

Jarrard, L. E., Okaichi, H., Steward, O., and Goldschmidt, R., 1984, On the role of the hippocampal connections in the performance of place and cue tasks: comparisons with damage to hippocampus, Behav. Neurosci., in press.

Kelley, A. E., and Domesick, V. B., 1982, The distribution of the projection from the hippocampal formation to the nucleus accumbens in the rat: an anterograde- and retrograde-horseradish peroxidase study, Neurosci., 7:2321.

Krayniak, P. F., Meibach, R. C., and Siegel, A., 1981, A projection from the entorhinal cortex to the nucleus accumbens in the rat, Brain Res., 209:427.

O'Keefe, J., 1983, Spatial memory within and without the hippocampal system, in: "Neurobiology of the Hippocampus," W. Seifert, ed., Academic Press, New York.

Olton, D. S., Becker, J. T., and Handelmann, G. E., 1979, Hippocampus, space, and memory, Behav. Brain Sci., 2:313.

Orbach, J., Milner, B., and Rasmussen, T., 1960, Learning and retention in monkeys after amygdala-hippocampus resection, Arch. Neurol., 3:230.

Scoville, W. B., and Milner, B., 1957, Loss of recent memory after bilateral hippocampal lesions, J. Neurol., Neurosurg., Psychiat., 20:11.

Stein, D. G., Rose, J. J., Graziadec, J., Mishkin, D., and Brink, J., 1969, Central nervous system: recovery of function, Science, 166:528.

Swanson, L. W., 1979, The hippocampus: new anatomical insights, Trends in Neurosci., 2:9.

Swanson, L. W., and Cowan, W. M., 1977, An autoradiographic study of the organization of the efferent connections of hippocampal formation in the rat, J. Comp. Neurol., 172:49.

Vanderwolf, C., and Leung, L. -W. S., 1983, Hippocampal rhythmical slow activity: a brief history and the effects of entorhinal lesions and phencyclidine, in: "Neurobiology of the Hippocampus," W. Seifert, ed., Academic Press, New York.

Zola-Morgan, S., Squire, L. R., and Mishkin, M., 1982, The neuroanatomy of amnesia: amygdala-hippocampus versus temporal stem, Science, 218:1337.

NEOCORTEX, HIPPOCAMPUS AND PERFORMANCE IN LASHLEY'S MAZE III

Lesley C. Eames and David A. Oakley

Department of Psychology and Centre for Neuroscience
University College London
Gower Street, London WC1E 6BT, G.B.

INTRODUCTION

There is now a long history of ideas and experimentation on the role of neocortex in associative learning (see reviews by Oakley, 1979a,b,; Russell, 1980). So far as Pavlovian conditioning is concerned, Pavlov himself proposed neocortex as the primary substrate of learning. Subsequent work, however, has shown that Pavlovian conditioning of heart-rate and gastrointestinal responses (DiCara, Braun and Pappas, 1970), nictitating membrane responses (Oakley and Russell, 1977) and autoshaped bar-presses (Oakley et al. 1981) is not impaired, and may even be enhanced, after removal of most or all of neocortex (neodecortication) in rats and rabbits. In the case of the conditional nictitating membrane response interest has recently centred on the cerebellum (Thompson, 1983) and in particular on cerebellar cortex (e.g. Yeo and Hardiman in this volume) as the essential substrate of the learned association. Similarly, both long-term and short-term habituation of distraction to a novel tone survive total removal of neocortex in rats (Yeo and Oakley, 1983).

Evidence that neocortex is not essential for instrumental learning has appeared more recently than that for Pavlovian conditioning. DiCara, Braun and Pappas (1970) confirmed existing views in reporting that neodecorticates failed to learn instrumental shock-avoidance responses even though the sensory and response characteristics were identical to those used in their Pavlovian conditioning study. Learning had also been reported to be very poor or absent after removal of neocortex in studies using leg-flexion (Bromiley, 1948) and head-turn responses (Bloch and Lagarrigue, 1968) with a shock-avoidance paradigm. It is only with the introduction in recent years of food reward when training neodecorticates that clear instrumental learning has been demonstrated. Bar pressing for food reward is not only possible in neodecorticated rabbits (Oakley, 1971; Oakley and Russell, 1977a,b) and rats (Oakley and Russell, 1979) but may be performed very efficiently on response:reinforcement ratios as high as 60:1 (Fixed Ratio 60) if appropriate pretraining is given (Oakley, 1979c, 1980). Comparing the performance of neodecorticated rats under similar conditions in a straight alleyway for food reward or shock avoidance confirmed that whilst shock motivation produced little or no avoidance learning in these preparations, food reward was compatible with acquisition performances which were as good, or in some instances better, than those of normal animals (Oakley, 1979d).

Subsequent studies have successfully used food reward with neodecorticates in a two-choice visual discrimination apparatus (Oakley, 1981) and water reward in a chicane alleyway, which requires the animal to perform alternating right and left turns (Herbert and Oakley, 1981, Eames and Oakley, 1985). The lesioned animals in the chicane alleyway were in fact found to acquire the learned response more readily and to run more rapidly than the normal animals. It was this particular outcome which led us to reconsider some much earlier experiments by Lashley (1929) in which he claimed to show a very powerful effect of subtotal neocortical lesions on performance on the most complex of his mazes (Maze III).

Lashley's Maze III (see insert in Figure 2) required the rat to perform a series of five alternating right/left turns upon leaving the start box in order to reach food in the goal box. Making an incorrect turn at any of these points led the animal into a short cul-de-sac, requiring it to retrace its steps back to the choice point. Lashley was able to demonstrate a strong positive correlation between the amount of neocortex destroyed, irrespective of its location, and errors during training on Maze III. When lesions removed 50% or more of neocortex the majority of his rats in fact were never able to learn to run the maze without making errors. Considerable theoretical significance was attached to the apparent correlation between cortical mass and learning capacity both by Lashley himself and by more recent reviewers. In particular his results led to a widespread and persistent belief that for some aspects of behavior, particularly more complex functions, an interpretation of neocortical capacity in terms of equipotentiality was more appropriate than a localizationist approach. If, in Lashley's terms, the entire neocortex serves as the functional area subserving learning in complex, walled mazes, any intact area of neocortex remains capable of carrying out the functions which are lost by destruction of the whole. The efficiency with which the remaining neocortical zones carry out their generalized function would then depend simply on the amount (but not the location) of residual tissue. The latter relationship was embodied in his law of mass function (or mass action) which might apply to the entire cortex, as in the present case, or more limited localized areas in the case of specific sensory or motor functions. Lashley's correlation between amount of neocortical loss and performance on Maze III is still regularly quoted by modern authors in their discussions of equipotentiality vs localization. For example, the following recent textbooks all present Lashley's data, often in the form of his original figure relating errors to cortical lesion size, as an example of neocortical equipotentiality: Bennett (1982), Kalat (1980), Levitt (1981), Schwartz (1978).

Whilst we had never tested our own neodecorticates in a maze identical to Maze III, our experience of their performance in other complex learning situations would not have led us to expect such marked learning deficits with subtotal neocortical lesions as those reported by Lashley. Fortunately Lashley left detailed descriptions of the effects of his thermocautery lesions and it is clear that the animals with larger amounts of cortical loss had also suffered damage to the hippocampus and in many the thalamus was injured as well. There is thus the possibility that the correlation which he obtained between lesion size and errors was not due to the amount of neocortex destroyed but to an increasing probability of damaging subcortical structures as the area of neocortex subjected to cautery became larger. The experiments reported in the next section were intended to investigate the replicability of Lashley's earlier findings with variable-size lesions restricted to neocortex and to explore systematically the effects of adding specific subcortical lesions to those of neocortex.

LESIONS AND LEARNING IN MAZE III

An experimental programme was initiated in which 37 male, hooded rats were prepared with neocortical lesions matching as exactly as possible the variable, and often asymmetrical ablations which Lashley reported (Lashley, 1929). The lesions in our study were produced by a process of pial stripping (see Oakley, 1979d) which causes cortical necrosis as a consequence of local devascularization. This technique avoids the possibility of damaging subcortical structures directly. As in Lashley's original study the lesions ranged from less than 10% of neocortex removed to over 80% and were distributed over all neocortical zones. A further group of 12 rats had a representative selection of similar neocortical lesions but this time accompanied by a standard aspiration removal of the dorsal two-thirds of each hippocampus. As this type of hippocampal lesion is always associated with removal of up to 15% of neocortex, this cortex-hippocampus group was produced as a parallel to Lashley's animals only for those animals which had neocortical lesions in excess of 20%. One animal in the cortex-hippocampus group was subsequently found to have accidental bilateral damage to the dorsal thalamus and the data from this animal are treated separately. No other unintended subcortical injury was present in the animals reported here. In addition to these two main groups, five sham-operated animals (trephine holes), eight animals with standard hippocampal lesions and seven with small neocortical lesions, whose size and location matched the neocortical removal required to aspirate hippocampus, were used as controls.

All animals were trained (5 trials per day) in Maze III for 150 trials or until they reached a criterion of ten consecutive errorless trials. Errors (cul-de-sac entries), the time taken to traverse the maze from start box to goal box, and trials to criterion were recorded. For convenience the data from Lashley's animals, our own neocortical lesion group and the cortex-hippocampus group were analysed on the basis of subgroups of animals having lesions of 0-10%, 10-20%, 20-30%, 30-40%, 40-50% and more than 50% of neocortex. Lashley found a clear relationship between the number of trials, taken to reach criterion and the size, but not location, of the neocortical lesion. In all, eleven of his animals failed to reach criterion within 150 training trials and, in particular, five out of the seven animals with lesions greater than 50% of neocortex failed to reach criterion. In contrast, none of the neocortically lesioned animals in our study failed to reach criterion even though our largest neocortical lesion at 89% was slightly larger than that of Lashley (81%).

Figure 1 shows representative learning curves in terms of errors (5-trial blocks) for three of our neocortically lesioned animals in comparison with the performance of the sham-operated group. The most important feature of this figure is the presence of clear learning curves, reaching the zero-error criterion, in all three neocortically lesioned animals. It is also evident that all three reach that criterion in fewer training trials than the worst of the sham-operated animals. There was no sign in any of our neocortically lesioned animals of the asymptotic high error performance which Lashley described in the more impaired of his lesioned animals. There is, nevertheless, an apparent tendency for the neocortically lesioned animals in Figure 1 to produce more errors, particularly in the first five trials, than the sham-operated animals, and this does seem to correlate with lesion size.

Total errors during training was adopted by Lashley as the most reliable index of performance and the results for his animals and all the present groups of animals on this measure are presented in Figure 2. It is evident that the neocortically lesioned animals in our study made far fewer errors than Lashley's animals. As our sham-operated controls made slightly more errors than Lashley's normal animals, this outcome cannot be accounted for by assuming that our version of Maze III is simpler in behavioural terms than Lashley's original. Looking at the animals with neocortical plus hippocampal lesions it is evident that they made far more errors than the corresponding animals with neocortical lesions alone. As noted above the standard hippocampal ablation involves damage to up to 15% of neocortex (in the present

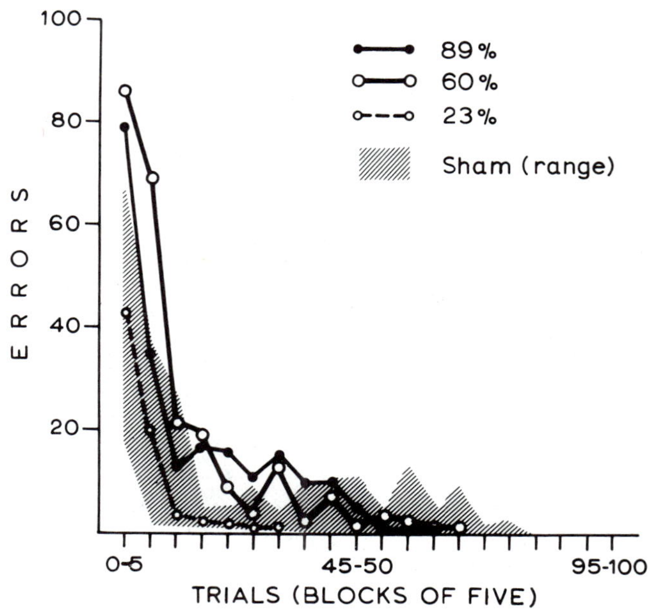

Fig. 1. Mean errors in 5-trial blocks (days) during Maze III acquisition in three animals with neocortical lesions (89%, 60% and 25% of neocortex removed) compared with the range of acquisition performance in the sham-operated control group. The hatched area is bounded by the acquisition curves of the sham-operated animals making the least and the most errors respectively during acquisition.

hippocampal group the neocortical damage ranged from 8 to 13%). The hippocampal group (H) shown in Figure 2 thus corresponds approximately to a cortex-hippocampal group with 10-20% neocortical lesions. The cortical lesion control group (CC) have neocortical lesions matching those of group H and error scores which are not different from the sham-operated group (S). The entire deficit shown by group H would thus seem to be attributable to the hippocampal lesions with no significant contribution from the small neocortical loss. It is also significant that group H has an error score which is almost identical to that of Lashley's 10-20% lesion group. At least two of the four animals in the latter group are known to have had unintentional hippocampal lesions, though no other subcortical damage is reported. In contrast none of the animals in Lashley's group 0-10% had hippocampal damage and their errors were similar to those of our 0-10% cortical lesion group as well as to normal and sham-operated animals. The effects of neocortical loss and hippocampal damage in our series bear a simple additive relationship to one another. If the mean error score of group H is added to the mean error scores for the 20-30%, 30-40%, 40-50% and 50%+ subgroups of cortically lesioned animals the result matches almost exactly the errors produced by the corresponding cortex-hippocampus subgroups. This relationship is shown in Table 1.

Comparing groups of animals with 20% or more of neocortex removed (Figure 2), the cortex-hippocampus groups still do not make as many errors in learning Maze III as the corresponding Lashley animals. A possible explanation for this discrepancy is the progressive involvement of subcortical areas in addition to hippocampus as Lashley increased his cortical lesion size. This is particularly the case in his 50%+ lesion group where five of the seven animals had thalamic damage in addition to hippocampal injury but is also true of the 20-30%, 30-40% and 40-50% groups. As noted earlier, one of our cortex-hippocampus animals had suffered unintentional bilateral damage to the dorsal half of the thalamus. This animal had a cortical lesion of 24.6% and a total error score of 777. Lashley's 20-30% lesion group had error scores in the range 193 to 809. Interestingly, the Lashley animal with 809 errors also had thalamic and hippocampal damage.

We have pointed out the contrasts between our results and those of Lashley and have noted in particular the very much lower levels of errors which we find when lesions are restricted to neocortex There is, nevertheless, in both Lashley's animals and, less strongly, in our own, a tendency for the error scores to become larger as lesion size increases in the cortical lesion groups. There existed, according to Lashley, a correlation of $r = 0.86$ between lesion size and errors in his original data and the slope (rate) which describes this relationship is 18.7. The lesion size/error correlation in our own cortical lesion group is lower at $r = 0.61$ and the slope (3.7) is very much less. The significance of this similarity between the two sets of data is considered later.

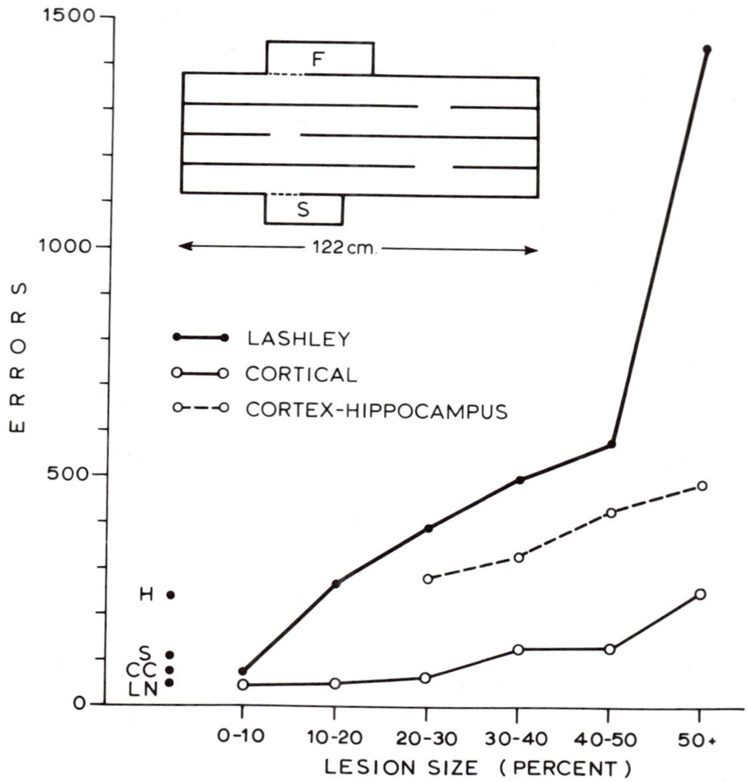

Fig. 2. Inset. Plan view of Lashley's Maze III, showing the start box (S) and goal box with food reward (F). The dashed lines indicate doors used to prevent re-entry to the start box and exit from the goal box. The openings between alleyways are shown as gaps in the partition walls.
Main Graph. Mean total errors throughout training for Lashley's Maze III animals and all the animals in the present study. H = hippocampal lesions, S = sham-operated controls, CC = cortical controls, LN = Lashley's normal animals. Lashley's cortically lesioned animals (Lashley), as well as the neocortically lesioned animals (cortical) and the animals with combined neocortical and hippocampal lesions (cortex-hippocampus) of the present study have been divided into subgroups according to lesion size (percent of neocortex removed) in the ranges 0-10%, 10-20%, 20%-30%, 30%-40%, 40%-50% and greater than 50%.

TABLE 1. THE ADDITIVE EFFECT OF NEOCORTICAL AND HIPPOCAMPAL LESIONS ON ERRORS IN LEARNING MAZE III.[a]

		20-30%	30-40%	40-50%	50%+
CORTICAL LESION (C)	-	64	135	137	258
HIPPOCAMPAL LESION (H)	219	-	-	-	-
H + C	-	283	354	356	477
CORTEX-HIPPOCAMPUS	-	280	333	434	488

a: The mean number of errors to criterion are shown for four Cortical lesion (C) subgroups and for the Hippocampal lesion (H) group. Summing these two scores gives a predicted mean error score (H+C) for combined hippocampal and neocortical lesions. Finally the actual mean error scores for the corresponding subgroups of the Cortex-hippocampus group are shown.

We are at present analysing the strategies which the various groups of animals adopted in solving the maze and the types of errors they made. So far it seems clear that the animals with neocortical lesions made their errors as a result of difficulties in identifying the openings in alleyway walls. Their errors were produced by a tendency to shuttle back and forth in an alleyway missing the doorway but eventually exiting in the right direction. The animals with hippocampal lesions, on the other hand, moved readily through the communicating openings but scored errors as a result of a tendency to retrace parts of the maze, moving back into previously explored alleys. Combining the two lesions introduced a combination of these two behaviours.

DISCUSSION AND CONCLUSIONS

We have shown that neocortical lesions alone do not produce the large numbers of errors claimed by Lashley on Maze III. In common with Lashley, on the other hand, we do find a tendency for errors to increase as neocortical lesion size increases. We would claim, however, that the more modest numbers of errors involved and the much less marked correlation between lesion size and errors which we find when lesions are restricted to neocortex makes an explanation in terms of progressive sensory and motor impairment a plausible one. It was important to Lashley's case that peripheral sensory or motor restrictions did not produce the same degree of impairment as damage to neocortical sensory and motor zones. The apparent "additional" deficit consequent upon neocortical lesions compared to peripheral sensory or motor limitations was evidence for Lashley of a common general processing capacity shared by all neocortical areas and which was in addition to their sensory and motor functions. We believe that Lashley's general capacity, which underlay both mass action and the apparent equipotentiality of neocortex for some tasks, was in reality an artefact created by the extension of his lesions into common areas of hippocampus and in some instances thalamus also. It is now a frequent finding that aspiration lesions of hippocampus disrupt spatial learning (Morris, 1983; Sutherland, Kolb and Whishaw, 1982). It is plausibly argued that this is because hippocampus acts in some way as a mapping system for the animal (e.g. O'Keefe, 1983) though alternative explanations in terms of disruptions to working memory (Olton, 1983) suggest the hippocampus should not be viewed in terms of a single function.

Our own behavioural observations are compatible with an interpretation in terms of a spatial memory deficit in the hippocampal animals and a sensory/motor deficit in the neocortically lesioned animals. The hippocampals appeared to have difficulty in maintaining a directional orientation to the goal box (or possibly in remembering where they had just been) whereas the neodecorticates, initially at least, had difficulty in identifying the salient parts of the apparatus and in executing the correct response. The neocortically lesioned animals, however, did not appear to have a reduced capacity for learning and rapidly overcame their initial difficulties to reach the level of errorless performance attained by the normal rats. This latter aspect of our data is of course consistent with the large numbers of studies mentioned above which suggest intact associative learning in the absence of neocortex.

Unlike Lashley, then, our experience of neocortically lesioned animals in Maze III does not lead us to propose general functions in addition to the established localized capacities of neocortical zones. The original Maze III findings were, and still are, taken as support for non-localizationist views as enshrined in the principles of equipotentiality and mass action. These views have been important in the study of plasticity and recovery from brain damage because they imply that, for some functions at least, undamaged regions of the brain are available to take over from damaged areas and to function in an equivalent way. The localizationist view, on the other hand, implies that once an area of the brain is destroyed its unique function is permanently lost. Recovery of function thus becomes a matter of deploying spared localized systems to tackle the behavioural problems in new ways. The two views clearly have different implications for attempts to rehabilitate individuals after brain injury. It is of practical importance, therefore, to continue to question the evidence for both of them. In the present case it would appear that a most influential set of experimental data supporting the equipotentialist view is not replicable when due care is taken to restrict lesions to their intended targets.

REFERENCES

Bennett, T.L., 1982, "Introduction to Physiological Psychology", Brooks/Cole, Monterey, Calif.

Bloch, S., and Lagarrigue, I., 1968, Cardiac and simple avoidance learning in neodecorticate rats, Physiol. Behav., 3:305.

Bromiley, R.B., 1948, Conditioned responses in a dog after removal of neocortex, J. Comp. Physiol. Psychol., 4:102.

DiCara, L.V., Braun, J.J., and Pappas, B.A., 1970, Classical conditioning and instrumental learning of cardiac and gastrointestinal responses following removal of neocortex in the rat, J. Comp. Physiol. Psychol., 73:208.

Eames, L.C., and Oakley, D.A., 1985, Learning without neocortex in a chicane alleyway. In preparation.

Herbert, B.A., and Oakley, D.A., 1981, Alleyway performance and behavioural flexibility with water reinforcement in rats without neocortex, IRCS: Med. Sci., 9:415.

Kalat, J.W., 1980, "Biological Psychology", Wadsworth, Belmont, Calif.

Lashley, K.S., 1929, "Brain Mechanisms and Intelligence", University of Chicago Press: Chicago.

Levitt, R.A., 1981, "Physiological Psychology", Holt, Rinehart and Winston, New York.

Morris, R.G.M., 1983, An attempt to dissociate "spatial mapping" and "working memory" theories of hippocampal function, in: "Neurobiology of the Hippocampus", W. Seifert, ed., Academic Press, London.

O'Keefe, J., 1983, Spatial memory within and without the hippocampal system, in: "Neurobiology of the Hippocampus", W. Seifert, ed., Academic Press, London.
Oakley, D.A., 1971, Instrumental learning in neodecorticate rabbits. Nature (New Biol.), 233:185.
Oakley, D.A., 1979a, Cerebral cortex and adaptive behaviour, in: "Brain, Behaviour and Evolution", D.A. Oakley and H.C. Plotkin, eds., Methuen, London.
Oakley, D.A. 1979b, Neocortex and learning, Trends Neurosci., 2:149.
Oakley, D.A., 1979c, Instrumental reversal learning and subsequent fixed ratio performance on simple and go/nogo schedules in neodecorticate rabbits, Physiol. Psychol., 7:29.
Oakley, D.A., 1979d, Learning with food reward and shock avoidance in neodecorticate rats, Exper. Neurol., 63:627.
Oakley, D.A., 1980, Improved instrumental learning in neodecorticate rats, Physiol. Behav., 24:357.
Oakley, D.A., 1981, Performance of decorticated rats in a two-choice visual discrimination apparatus, Behav. Brain Res., 3:55.
Oakley, D.A., Eames, L.C., Jacobs, J.L., Davey, G.C.L., and Cleland, G.C., 1981, Signal-centered action patterns in rats without neocortex in a Pavlovian conditioning situation, Physiol. Psychol., 9:135.
Oakley, D.A., and Russell, I.S., 1977, Subcortical storage of Pavlovian conditioning in the rabbit, Physiol. Behav., 18:931.
Oakley, D.A., and Russell, I.S., 1978a, Performance of neodecorticated rabbits in a free-operant situation, Physiol. Behav., 20:157.
Oakley, D.A., and Russell, I.S., 1978b, Manipulandum identification in operant behaviour in neodecorticate rabbits. Physiol. Behav., 21:943.
Oakley, D.A., and Russell, I.S., 1979, Instrumental learning on Fixed Ratio and GO-NOGO schedules in neodecorticate rats, Brain Res., 161:356.
Olton, D.S., 1983, Memory functions and the hippocampus, in: "Neurobiology of the Hippocampus", W. Seifert, ed., Academic Press, London.
Russell, I.S., 1980, Encephalization and neural mechanisms of learning, in: "Psychology Survey No.3", M.A. Jeeves, ed., George Allen and Unwin, London.
Schwartz, M.D., 1978, "Physiological Psychology", Prentice Hall, London.
Sutherland, R.J., Kolb, B., and Whishaw, I.Q., 1982, Spatial mapping: Definitive disruption by hippocampal or medial frontal cortical damage in the rat, Neurosci. Lett., 31:271.
Thompson, R.F., 1983, Neuronal substrates of simple associative learning: classical conditioning, Trends Neurosci., 6:270.
Yeo, A.G., and Oakley, D.A., 1983, Habituation of distraction to a tone in the absence of neocortex in rats, Behav. Brain Res., 8:403.

BEHAVIORAL CORRELATES OF PLASTICITY IN SUBSTANTIA NIGRA EFFERENTS

Joseph P. Huston, Sarah Morgan, and Heinz Steiner

Institute of Psychology III
University of Düsseldorf
D-4000 Düsseldorf (F.R.G.)

ABSTRACT

The present chapter describes a series of experiments which deal with the relationship between recovery from lesion-induced asymmetries in behavior and changes in crossed efferents emanating from the substantia nigra. In all of these experiments rotational behavior, which subsided with time, was induced in rats. This was usually done by means of unilateral lesions to the nervous system. Connections from the substantia nigra (SN) to the contralateral nucleus caudatus (NC) and thalamus (Th) were then examined using tract tracing techniques. Horseradish peroxidase (HRP) was usually employed. It was found that the cessation of the lesion-induced turning behavior (behavioral plasticity) was associated, in time, with an increase in the number of cells in the SN that were labeled by a tract tracer injected into the contralateral NC or Th (anatomical plasticity). It was also found that prevention of the rotational behavior suppressed both the return to symmetry and the apparent increase in the crossed nigro-thalamic projections. Thus, it is possible that changes in crossed SN efferents form part of a morphological basis of recovery from asymmetrical behavior induced by unilateral nervous system lesions.

INTRODUCTION

The experiments described here were designed to relate apparent changes in the central nervous system to alterations in behavior indicative of 'learning', i.e.,to changes in behavior resulting from experience.

Considerable evidence shows that the brain changes in response to various interventions, especially injury (Björklund and Stenevi, 1979; Cotman and Nieto-Sampedro, 1982; Cotman et al., 1981; Tsukahara, 1981). For example, removal of a portion of the afferents to an area can result in rearrangement of the remaining terminals. These terminals can sprout and replace lost synapses. Denervation can also lead to synaptic turnover in closely related regions which are not directly affected by the lesion (Hoff et al., 1981). It has additionally been found that a cell can begin to respond to stimuli,which did not previously affect it, after losing its normal input. Thus, there is some evidence that ineffective connections can become effective in response to injury (Merrill and Wall, 1978).

Although there are anatomical and electrophysiological data to validate the existence of these phenomena, there is not much evidence to link them to behavior, especially to learning. Therefore, a series of experiments was designed to systematically look at whether such physical changes can be related to behavioral plasticity.

EXPERIMENTS

Changes in Crossed Thalamic Afferents Associated with Behavioral Plasticity, Subsequent to Unilateral Removal of the Telencephalon

This series of studies was initiated by a chance finding. This first experiment was carried out in order to look for possible changes in the neural organization of the thalamic rat (Pritzel and Huston, 1981a),i.e., a rat devoid of the telencephalon (Huston and Borbély, 1974). In order to compare possible reorganization in the denervated thalamus with the intact side, a unilaterally detelencephalized preparation was used. For this the neocortex, striatum, hippocampus and basal forebrain nuclei were aspirated on one side. This lesion resulted in transient turning behavior directed towards the side of the lesion. To look at residual afferents to the thalamus the tract tracer HRP was used. To compare the remaining connections to the damaged side with those to the side that had retained its telencephalon, the HRP was injected into the thalamus either ipsilateral or contralateral to the lesion. When the HRP was injected ipsilateral to the lesion, labeled cells were found in contralateral areas, which normally show only minimal projections to the thalamus (figure 1b). Such cells were not found when the HRP was injected into the thalamus in the intact hemisphere. The areas in which an increased number of labeled cells was found included the SN - a region upon which we subsequently focussed our attention (figure 1b). Interestingly, the increase in number of labeled cells appeared at the point in time when the animal seemed to have compensated for the lesion-induced turning. They were not found in the brains of animals which had not had time to learn to compensate, i.e., those which received an injection of HRP within 24 hours of removal of the telencephalon. The appearance of these cells in animals in which rotational behavior had subsided, suggested that compensation for lesion-induced asymmetry is associated with an increase in connectivity between the thalamus and contralateral structures.

The next experiment (Neumann et al., 1982) was carried out to test the generality of the initial finding. It is well known that young animals can have greater capacity for neuronal plasticity than adults (Johnson and Almli, 1978). Therefore, this study looked at the relationship between a decrease in lesion-induced turning and changes in crossed afferents to the thalamus of the infant rat.

The telencephalon was unilaterally removed in 7-day old rats and the resulting rotational asymmetry was observed. One week after the lesion, when the spontaneous turning behavior had largely ceased, HRP was injected into the Th on the side of the lesion and the brains were examined for HRP-labeled cells. Again certain areas in the contralateral hemisphere were found to contain more HRP-labeled cells than those of control animals. Control animals were either normal infant rats (3,7, and 14 days old), or unilaterally detelencephalized infants which had received their HRP injection immediately after the lesion. As in the experiment with the adult rats, these results indicated a relationship between behavioral and neuronal plasticity. In accordance with these findings McAllister et al. (1982) have also given evidence, in the adult cat, for anatomical reorganization in the Th subsequent to hemidecortication. Unfortunately, these authors gave no behavioral correlates of this change.

Turning Behavior and Anatomical Changes after Unilateral Lesions in the Substantia Nigra

In the first two studies described above turning behavior and changes in HRP labeling were examined after an extremely large lesion, i.e., after removal of one entire telencephalon. In a subsequent experiment a more discrete lesion was employed to elicit turning behavior. Possible changes in crossed afferents to the NC, as well as those to the Th, were examined. In addition to HRP, histofluorescent tracers were used to look at such possible changes in connectivity.

It has frequently been found that unilateral destruction of the SN results in asymmetrical behavior which usually subsides over the next week (Ungerstedt, 1971). Thus, such a lesion provides an excellent preparation to examine the relationship between recovery from lesion-induced turning and possible changes in neural connectivity. The SN sends efferents to various regions, which include the Th and the NC (Gerfen et al., 1982) (see figure 1a). The SN projections to these two regions are predominantly homolateral in the normal rat, although a few crossed connections exist (Gerfen et al., 1982). We investigated possible changes in connectivity from the SN to these two regions.

Initially, changes in nigro-thalamic projections in response to a unilateral neurotoxic lesion of the SN were investigated (Pritzel and Huston, 1981b). The lesion was made by the injection of 6-OHDA or kainic acid, or a combination of both these neurotoxins, into the SN on one side. The resulting rotational behavior was recorded, and it was found to subside about 7 days after the production of the lesion. One week after the destruction of the SN, HRP was injected into the ventromedial Th ipsilateral to the damaged SN. An examination of the SN contralateral to the HRP injection side and lesion, revealed that it contained more HRP-labeled cells. Such labeled cells were not found in the SN contralateral to a thalamic HRP injection in the normal animal. However, as they have been found with more sensitive HRP methods, it can be assumed that the SN does send a sparse projection to the contralateral Th (Gerfen et al., 1982). The appearance of these cells in the recovered animal in our experiment indicated that a decrease in lesion-induced asymmetry is associated with a change in HRP uptake by terminals of crossed nigro-thalamic connections.

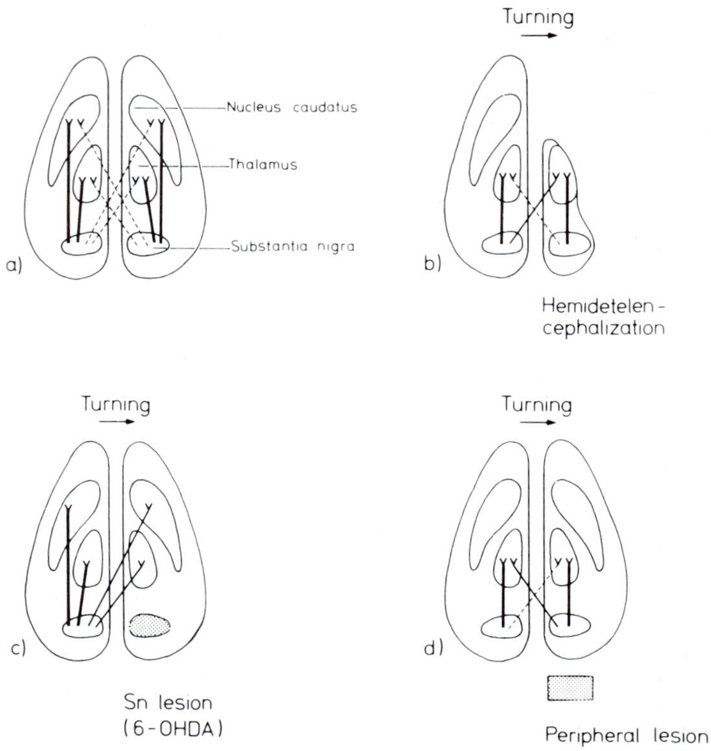

Figure 1 gives a schematic diagram depicting the results of the series of experiments. (a) The normal brain with a predominantly homolateral projection from the substantia nigra (SN) to the nucleus caudatus (NC) and thalamus (Th). There are sparse crossed projections to these two structures. (b) The hemidetelencephalized animal 7 days after the lesion with an apparent increase in connectivity between the SN (among other structures) in the intact hemisphere and the Th on the damaged side (c) The apparent changes in connectivity seen 7 days after a unilateral SN lesion. There is an enhancement of the projection from the intact SN to the Th and NC ipsilateral to the damaged SN. (d) Changes in crossed connectivity seen 16 days after the unilateral peripheral lesion. There is an apparent increase in the crossed projections from the SN on the lesion side to the Th in the hemisphere contralateral to the peripheral lesion.

Thus, alterations in crossed projections were seen irrespective of whether the rotational behavior was induced by a massive lesion (Pritzel and Huston, 1981a) (figure 1b) or a relatively discrete lesion (Pritzel and Huston, 1981b) (figure 1c).

To investigate the possibility that the alterations in crossed connections were not confined to interhemispheric nigro-thalamic projections, the effect of lesion-induced turning on crossed nigro-caudate connections was examined (Pritzel et al., 1983). Again the animals had the SN damaged by unilateral injection of neurotoxins. The animals were then injected with HRP into the NC ipsilateral to the damaged SN 7,21, or 90 days after the lesion. It was found that 7 days after the lesion the SN contralateral to the side of HRP injection contained more HRP-labeled cells than the SN contralateral to the side of HRP injection in the normal controls (see figure 1c). This finding demonstrated that changes in crossed efferents from the SN, that coincided with a decrease in asymmetry were not restricted to nigro-thalamic connections. There was no difference between the number of labeled cells found at 7 and 21 days of survival. At 90 days, however, there was no difference between the animals sustaining the SN damage and the normal controls in the number of labeled cells found in the SN contralateral to the NC containing the HRP. This latter finding suggested that these changes in crossed efferents were not permanent. Sabel et al. (in this volume) describe results which seem to confirm these findings. They observed that treatment with GM1 gangliosides was associated with enhanced behavioral recovery from unilateral partial knife transections of the nigro-striato-nigral pathway. They also found that, 15 days after the lesion, the GM1-treated animals had more HRP-labeled cells in the SN contralateral to a tracer injected into the denervated NC than did controls. This increase in cell labeling was not seen 45 days after the lesion.

Finally, to see if the results found above were an artifact of the HRP tract tracing technique, the relationship between recovery from asymmetry and the number of cells labeled by fluorescent tracers was tested (Pritzel et al., 1983). Animals were injected into the NC with Fast Blue (FB). Twenty-four hours later the SN ipsilateral to the injection site was destroyed by an injection of 6-OHDA. This resulted in turning behavior which subsided within the next week. At this point in time Nuclear Yellow (NY) was injected into the site of the FB injection. The contralateral, undamaged SN was then examined for labeled cells. It was found that there were more cells labeled with NY than with FB. Normal control animals, which had an injection of FB followed by NY 7 days later, showed no difference in the number of cells labeled by the two tracers. Thus, it was shown that the increase in crossed nigro-caudate connections, which coincided with recovery from lesion-induced turning, could be demonstrated with fluorescent techniques as well as with HRP.

Changes in Nigro-Thalamic Connections after Unilateral Peripheral Lesions

In the studies described above unilateral brain lesions were used to investigate the relationship between turning behavior and possible changes in crossed projections from the SN. In a subsequent study we considered whether turning behavior, in the absence of any central lesion, could also result in such changes (Sabel et al., 1984). To induce turning behavior the sensory and motor nerves to the hind - and fore-limbs were cut on one side. Additionally, the vibrissae and the visual input to that side were removed. The animals' turning behavior was recorded on the 1st, 5th, and 16th postoperative days. It was found that the animals' asymmetry decreased between the 1st and the 5th postoperative days and then asymptoted. HRP was injected into the ventrobasal thalamus, either ipsilateral or contralateral to the denervation, on the 16th postoperative day. It was found that when the tracer was injected into the Th on the side contralateral to the peripheral lesion, i.e., the Th presumably denervated by the lesion, there was an apparent increase in the number of labeled cells found in the contralateral SN (see figure 1d). Labeled cells were not found in the SN contralateral to a HRP injection into the Th of the normal animal, nor were they found in the contralateral SN when the HRP was injected into the Th in the hemisphere ipsilateral to the side of the peripheral lesion. In addition, the results of the anatomical data indicated an apparent decrease in projections to the Th on the lesion side from the ipsilateral SN.

Thus, the results of this experiment show that a central lesion is not a prerequisite for apparent changes in connectivity between the SN and the contralateral Th which can be related to turning. They support the hypothesis that compensation for experimentally-induced asymmetry (which presumably involves learning) is critically related to changes in crossed projections from the SN. There is additional evidence in the literature to show that cells in the CNS can change their receptive fields subsequent to different peripheral lesions (Kaas et al., 1983). Moreover, it has been demonstrated that behavioral recovery from crossed union flexor extensor nerves is associated with central changes (Fujito et al., 1982; Tsukahara et al., 1982). In our studies we have also shown that central changes are associated with compensation for the peripheral injury, suggesting that they have adaptive value.

The Effect of Prevention of Turning Behavior on Plasticity Associated with Unilateral SN Lesions

The coincidence between recovery from lesion-induced turning and the connectivity changes argues for the hypothesis that the alterations in interhemispheric connections form part of a morphological correlate of the animals' learning to compensate for behavioral asymmetry. To validate this idea further, however, it was necessary to find evidence for a possible causal relationship between the appearance of these labeled cells and the recovery from the turning behavior. If, as we have argued, the changes in HRP labeling represent a morphological correlate of learning to compensate, preventing the animals from learning should (a) suppress the behavioral recovery and (b) suppress the appearance of these cells. The next experiment was designed to test this prediction (Morgan et al., 1983).

We firstly injected the neurotoxin 6-OHDA into the SN on one side. Thereafter, the rat was suspended in a hammock and thereby prevented from turning for the next 7 days. At the end of this time the animal was either injected with HRP into the

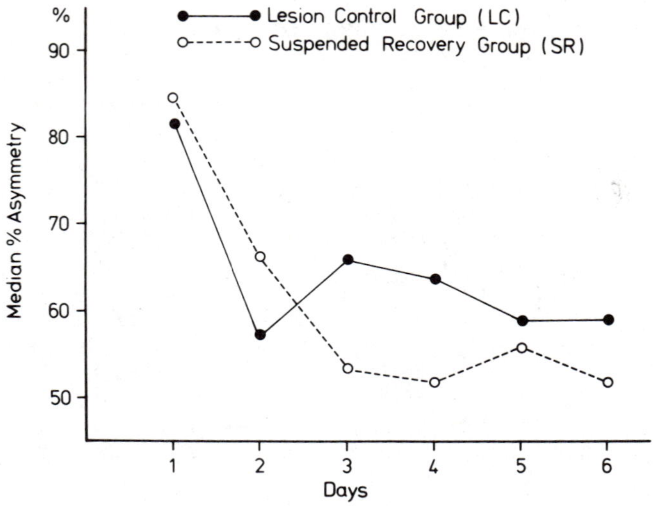

Figure 2 gives the median % asymmetry over the 6-day period after an injection of 6-OHDA into one substantia nigra (group LC) or after 7 days of prevention of turning subsequent to the injection of neurotoxin (group SR).

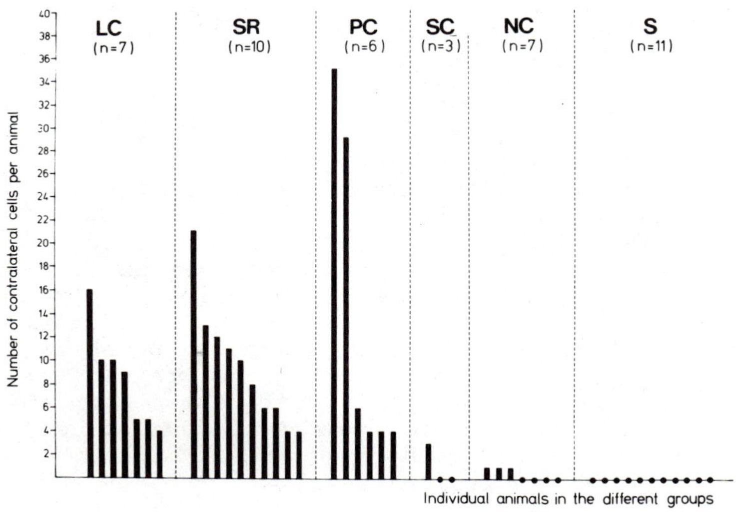

Figure 3 gives the number of labeled cells found in the substantia nigra contralateral to a thalamic horseradish peroxidase (HRP) injection for individual animals in the different groups. Animals in group LC had the substantia nigra destroyed on one side by 6-OHDA and were then allowed to turn for 6 days. Subjects in group SR had a 7-day period of prevention of turning between the injection of 6-OHDA and the 6 days of turning. Group PC was a control group with lesion (not discussed in the text). Group SC consisted of normal animals which were deprived of locomotion. NC were normal rats. Animals in group S were not allowed to turn within the 7-day period between the injection of the neurotoxin and that of HRP.

Th, or it was released and allowed one week of survival. The results of the experiment were as follows. Firstly, if the animal was released from the hammock and its turning behavior recorded for the next 6 days, then its return to symmetry was similar to that of the rat in which the SN had just been destroyed. Both groups had the same degree of asymmetry on the first day after the lesion as after release, and showed a similar return to symmetry over the next few days (see figure 2). This result argues that information feedback about motor behavior is essential for behavioral recovery. Secondly, we found no labeled cells in the animals which were injected with HRP into the Th ipsilateral to the lesion immediately after the period of motor restriction. Labeled cells were found, however, in the undamaged SN of the animals which were allowed feedback from their behavior, i.e., the animals which were allowed to turn for 7 days after the period of prevention of turning, and the animals which were allowed to turn for 7 days immediately after the lesion (see figure 3).

The finding that prevention of turning behavior suppressed behavioral recovery and prevented the increase in labeling of cells in the SN supports the hypothesis of a functional relationship between the two phenomena. Further support for this hypothesis is given by work of Held (1968). He showed that behavioral adaptation is prevented by a decrease in information feedback. Our results suggest that such adaptive changes can be related to alterations in connectivity between the two hemispheres, and that these changes could form part of a physical basis of behavioral change.

DISCUSSION

There are various explanations for the change in uptake of tracers by the crossed substantia nigra efferents. Horseradish peroxidase uptake has been found to be affected by several different factors. The number of HRP-labeled cells found at the source of an afferent projection is affected by neuronal firing rates as well as by the number of terminals present within the injection site. Moreover, uptake and transport of HRP seem to be directly related to transmitter release (Kristensson, 1978). Thus, several different mechanisms may be responsible for the increase in number of labeled cells found in our experiments. These mechanisms may act in combination or individually. Certainly the time span excludes such phenomena as the outgrowth of axons to the contralateral side. The apparent increase in labeling can, however, be attributed to changes in the rate of transmitter release, an increase in the firing rate of neurons (Eisenman and Azmitia, 1982; Nishino et al., 1979), the activation of silent synapses, or to terminal sprouting, or to a combination of several of these. All these phenomena have been shown to occur after brain injury, and HRP uptake and transport is affected by all these factors.

Zigmond et al. (1984) found evidence for an increased tyrosine hydroxylase activity in surviving terminals following a partial destruction of the dopaminergic innervation to the striatum. This resulted in an enhanced dopamine synthesis and release in these neurons and seemed to appear within a few days after the lesion. Another hypothetical explanation for the increased uptake and transport of HRP would be an increase in firing rate in the crossed substantia nigra efferents (Eisenman and Azmitia, 1982; Nishino et al., 1979). Chéramy et al. (1981) found evidence for a reciprocally organized interregulation of the two nigro-striatal dopaminergic pathways in the cat, involving interhemispheric connections.

There is some evidence to suggest that not all connections within a given region in the brain are effective (Merrill and Wall, 1978). Such ineffective connections have been shown to become active in response to injury under certain circumstances Considering the relation between transmitter release and HRP labeling, one can envisage that activity in normally silent synapses could result in more uptake of HRP.

The short term changes described above could be accompanied by, or could lead to structural changes at the synaptic level, such as terminal sprouting. Literature pertaining to synaptogenesis subsequent to brain injury is extensive, and evidence for this phenomenon has been obtained in several different brain regions, utilizing many anatomical and electrophysiological techniques, including HRP. For details of brain regions in which sprouting has been found see Cotman et al.(1981). The new terminals have been shown to be electrophysiologically active by several different experiments (Cotman et al., 1981; Tsukahara, 1981).

There are few experiments, other than ours, which correlate functional change with neural plasticity. Loesche and Steward (1977) found that after a unilateral entorhinal lesion, the postoperative performance on an alternation response, acquired prior to the lesion, recovered with the same time course as sprouting of crossed entorhinal cortex fibers in the hippocampus. However, postoperative training did not affect the rate of performance recovery. These results, like ours, showed changes in crossed connections which were related to behavioral recovery. They also suggest that the sprouted terminals can perform the same function as the lost input. Similar conclusions can be drawn from experiments by Wilson et al. (1979) and Messenheimer et al. (1979).

In a series of experiments Goldberger and Murray (1978) have looked at the relationship between axonal sprouting and recovery of movement after damage to the spinal cord or deafferentation of limbs. Although, as they pointed out, their methods do not allow a direct correlation between axonal sprouting and recovery of motor function, they have produced evidence for a dependency between the two.

Using an experimental paradigm similar to ours, Flohr and his coworkers (1981) have found a relationship between neuronal plasticity and behavioral recovery subsequent to peripheral damage. These authors looked at the mechanisms underlying recovery from head deviation resulting from unilateral labyrinthectomy in the frog. They have provided physiological evidence that compensation from asymmetrical body posture is associated with reorganization of crossed intervestibular connections. These results substantiate our hypothesis that behavioral adaptation to asymmetry can be associated with changes in crossed projections (Pritzel and Huston, 1981a, b).

An experiment which directly links recovery of function to synaptogenesis was carried out by Tsukahara and his co-workers (Fujito et al., 1982; Tsukahara et al., 1982). They examined electrophysiological changes in the red nucleus which occurred when cats had to adjust to crossed union of flexor and extensor nerves in the forelimb. They found that about two months after this operation there was a change in corticorubral EPSPs, which indicated the existence of new synapses.

It has been frequently suggested that synaptogenesis could form part of a physical basis of memory (Cotman and Nieto-Sampedro, 1982; Tsukahara, 1981). There is some evidence relating adaptation to the environment, including learning, to anatomical change in the normal animal. It has been well established that the environment can have profound influences on various physical features of the cortex, including the degree of dendritic branching (Green et al., 1983) and dendritic spine counts (Globus et al., 1973). There are very few experiments which show anatomical changes that can be directly linked to learning. Chang and Greenough (1982) have found that prolonged maze training is associated with an increase in length of oblique branches from apical dendrites in layer V pyramidal cells. They suggested that the increase in dendritic length could be due to the formation of new synapses, which were involved in the circuitry related to the capacity to perform the task. Also Vrensen and Cardozo (1981) observed that visual training was associated with both an increase in the number of "complex synaptic grids" and an increase in spine apparatus in the visual cortex of the rabbit. They did not observe any increase in synaptic density. However, since Chang and Greenough (1982) have demonstrated a change in dendritic length with training, changes in synaptic density would not necessarily be expected to occur with increases in synaptic number.

CONCLUSION

The results of the experiments described in this chapter indicate that changes occur in crossed efferents from the SN, together in time with recovery from lesion-induced asymmetry. Additionally, we consider it highly probable that there are alterations in other crossed and uncrossed systems which we did not investigate. The coincidence between the appearance of the labeled cells and the decrease in turning behavior suggested to us that changes in the crossed efferents of the SN could be related to compensation for asymmetries. A logical next step would be to determine whether turning behavior, induced by means other than injury to the nervous system, can be associated with a particular pattern of crossed substantia nigra efferents. Our preliminary results have shown that there is a relationship between the direction of amphetamine-induced turning and crossed nigro-caudate projections. The effects were not as dramatic as those found after CNS lesion. However, they do suggest that a lesion may not be necessary to demonstrate a relationship between nigro-caudate projections and asymmetrical behavior.

Implicit in our hypothesis is the assumption that recovery from lesion-induced turning required adjustments in behavior resulting from experience, i.e., learning. Therefore, in addition to providing information regarding anatomical plasticity related to recovery from unilateral brain lesions, we also consider our results to have relevance towards understanding physical changes underlying certain forms of learning.

ACKNOWLEDGEMENTS

This project was supported by grant Hu 306/3-1 from the Deutsche Forschungsgemeinschaft. We thank G. Abel for making the drawings and E. Reissner and P. Wingartz for typing this manuscript.

REFERENCES

Björklund, A., and Stenevi, U., 1979, Regeneration of monoaminergic and cholinergic neurons in the mammalian central nervous system, Physiol. Rev., 59: 62.

Chang, F.-L. F., and Greenough, W. T., 1982, Lateralized effects of monocular training on dendritic branching in adult split-brain rats, Brain Res., 232: 283.

Chéramy, A., Leviel, V., Daudet, F., Guibert, B., Chesselet, M. F., and Glowinski, J., 1981, Involvement of the thalamus in the reciprocal regulation of the two nigrostriatal dopaminergic pathways, Neurosci., 6: 2657.

Cotman, C. W., and Nieto-Sampedro, M., 1982, Brain function, synapse renewal and plasticity, Ann. Rev. Psychol., 33:371.

Cotman, C. W., Nieto-Sampedro, M., and Harris, E. W., 1981, Synapse replacement in the nervous system of adult vertebrates, Physiol. Rev., 61:684.

Eisenman, J. S., and Azmitia, E. C., 1982, Physiological stimulation enhances HRP marking of salivary neurons in rats, Brain Res. Bull., 8:73.

Flohr, H., Bienhold, H., Abeln, W., and Macskovics, I., 1981, Concepts of vestibular compensation, in:"Lesion-Induced Neuronal Plasticity in Sensorimotor Systems", H. Flohr and W. Precht, eds., Springer Verlag, New York.

Fujito, Y., Tsukahara, N., Oda, Y., and Yoshida, M., 1982, Formation of functional synapses in the adult cat red nucleus from the cerebrum following cross-innervation of forelimb flexor and extensor nerves. II. analysis of newly appeared synaptic potentials, Exp. Brain Res., 45:13.

Gerfen, C. R., Staines, W. A., Arbuthnott, G. W., and Fibiger, H. C., 1982, Crossed connections of the substantia nigra in the rat, J. Comp. Neurol., 207:283.

Globus, A., Rosenzweig, M. R., Bennet, E. L., and Diamond, M. C., 1973, Effects of differential experience on dendritic spine counts in rat cerebral cortex, J. Comp. Physiol. Psychol. 82:175.

Goldberger, M. E., and Murray, M., 1978, Recovery of movement and axonal sprouting may obey some of the same laws, in: "Neuronal Plasticity", C. W. Cotman, ed., Raven Press, New York.

Green, E. J., Greenough, W. T., and Schlumpf, B. E., 1983, Effects of complex or isolated environments on cortical dendrites of middle-aged rats, Brain Res., 264:233.

Held, R., 1968, Plasticity in sensorimotor coordination, in: "The Neuropsychology of Spatially Oriented Behavior", S. J. Freedman, ed., The Dorsey Press, Homewood, Ill.

Hoff, S. F., Scheff, S. W., Kwan, A. Y., and Cotman, C. W., 1981, A new type of lesion-induced synaptogenesis: I. synaptic turnover in non-denervated zones of the dentate gyrus in young adult rats, Brain Res., 222:1.

Huston, J. P., and Borbély, A. A., 1974, The thalamic rat: general behavior, operant learning with rewarding hypothalamic stimulation, and effects of amphetamine, Physiol. Behav., 12:433.

Johnson, D., and Almli, C. R., 1978, Age, brain damage, and performance, in: "Recovery from Brain Damage", S. Finger, ed., Plenum Press, New York.
Kaas, J. H., Merzenich, M. M., and Killackey, H. P., 1983, The reorganization of somatosensory cortex following peripheral nerve damage in adult and developing mammals, Ann. Rev. Neurosci., 6:325.
Kristensson, K., 1978, Retrograde transport of macromolecules in axons, Ann. Rev. Pharmacol. Toxicol. 18:97.
Loesche, J., and Steward, O., 1977, Behavioral correlates of denervation and reinnervation of the hippocampal formation of the rat: recovery of alternation performance following unilateral entorhinal cortex lesions, Brain Res. Bull., 2:31.
McAllister, J. P., Olmstead, Ch. E., Villablanca, J. R., and Goméz, F., 1982, Residual cerebellothalamic terminal fields following hemispherectomy in the cat, Neurosci. lett., 29:25.
Merrill, E. G., and Wall, P. D., 1978, Plasticity of connection in the adult nervous system, in: "Neuronal Plasticity", C. W. Cotman, ed., Raven Press, New York.
Messenheimer, J. A., Harris, E. W., and Steward, O., 1979, Sprouting fibers gain access to circuitry transsynaptically altered by kindling, Exp. Neurol., 64:469.
Morgan, S., Huston, J. P., and Pritzel, M. ,1983, Effects of reducing sensory-motor feedback on the appearance of crossed nigro-thalamic projections and recovery from turning induced by unilateral substantia nigra lesions, Brain Res. Bull., 11:721.
Neumann, S., Pritzel, M., and Huston, J. P., 1982, Plasticity of cortico-thalamic projections and functional recovery in the unilateral detelencephalized infant rat, Behav. Brain Res., 4:377.
Nishino, H., Ono, T., Sasaki, K., Nishino, A., and Muramoto, K., 1979, Retrograde transport of horseradish peroxidase in sciatic nerve of rats and dystrophy mice, Neurosci. Lett., 14:1.
Pritzel, M., and Huston, J. P., 1981a, Unilateral ablation of telencephalon induces appearance of contralateral cortical and subcortical projections to thalamic nuclei, Behav. Brain Res., 3:43.
Pritzel, M., and Huston, J. P., 1981b, Neural and behavioral plasticity: crossed nigro-thalamic projections following unilateral substantia nigra lesions, Behav. Brain Res., 3:393.
Pritzel, M., Huston, J. P., and Sarter, M., 1983, Behavioral and neuronal reorganization after unilateral substantia nigra lesions: evidence for increased interhemispheric nigrostriatal projections, Neurosci., 9:879.
Sabel, B. A., Pritzel, M., Morgan, S., and Huston, J. P., 1984, Interhemispheric nigrothalamic projections and behavioral recovery following turning behavior induced by unilateral peripheral sensory and motor restriction, Exp. Neurol., 83:49.
Tsukahara, N., 1981, Synaptic plasticity in the mammalian central nervous system, Ann. Rev. Neurosci., 4:351.
Tsukahara, N., Fujito, Y., Oda, Y., and Maeda, J., 1982, Formation of functional synapses in the adult cat red nucleus from the cerebrum following cross-innervation of forelimb flexor and extensor nerves, Exp. Brain Res., 45:1.
Ungerstedt, U., 1971, Postsynaptic supersensitivity after 6-hydroxydopamine induced degeneration of the nigro-striatal dopamine system, Acta physiol. Scand. Suppl., 367:69.

Vrensen, G., and Cardozo, J. N., 1981, Changes in size and shape of synaptic connections after visual training: an ultrastructural approach of synaptic plasticity, Brain Res., 218:79.
Wilson, R. C., Levy, W. B., and Steward, O., 1979, Functional effects of lesion-induced plasticity: Long term potentiation in normal and lesion-induced temporodentate connections, Brain Res., 176:65.
Zigmond, M. J., Acheson, A. L., Stachowiak, M. K., and Stricker, E. M., 1984, Neurochemical compensation after nigro-striatal bundle injury in an animal model of preclinical parkinsonism, Arch. Neurol., 41:856.

LEARNING AND MEMORY PERFORMANCE BEFORE AND AFTER UNILATERAL SELECTIVE AMYGDALOHIPPOCAMPECTOMY

T. Nadig, H.G. Wieser, and E. Perret

Neuropsychology Unit
Departments of Neurology and Neurosurgery
University Hospital, Zurich

INTRODUCTION

The detrimental effect of lesions in the amygdala and hippocampus (A+H) on the performance in tests of learning and memory has been reported for human subjects (Milner, 1974) and for monkeys (Mishkin, 1982). There are several reasons, however, to further analyze in patients the structural-functional correlation between A+H, and learning and memory. Firstly, the unilateral surgical removals in the patients studied by Milner (1980) always encompassed temporal neocortical structures in addition to A+H. The role of the latter in memory was mainly deduced from the difference between neocortical plus A+H lesions, which hampered memory performance, and only neocortical lesions, which did not - or to a lesser degree. Secondly, the selective surgical removal of A+H alone (Wieser and Yasargil, 1982) did not affect the performance of the patients in memory tests that had been shown by Corsi (1972) to be sensitive to combined temporal neocortical and A+H removals (Birri et al., 1982 and Gonser, 1983). These entirely negative findings could be discarded on the assumption of a methodological artifact, as Corsi's tests had been used in a shortened form, thus probably obscuring at least moderate memory deficits. Nevertheless, the disagreement with a large body of neuropsychological evidence was sufficiently challenging to foster further research.

The functional asymmetry of the cerebral hemispheres that has become a commonplace in neuropsychology (e.g. Walsh, 1978) has also been found relatively to A+H in learning and memory. Corsi (1972) described an impairment of verbal memory after left-sided temporal neocortical plus A+H lesions and deficient spatial memory performances after lesions in the same structures on the right side. In addition, he demonstrated a 'mass action' limited to A+H, in that the extent of the memory deficit increased with larger A+H involvement in the surgical removals. Asymmetric deficits were found by Nadig et al.(1984) after selective amygdalohippocampectomy (SAH) in learning performance , but not in the delayed recall of the verbal or figurative informations that had been previously learned to criterion. These results, however, were not specifically related to A+H, as they were obtained similarly in patients with lesions in the temporal neocortex.

While the neurosurgical removal of epileptogenic tissue drastically improves the general life conditions of patients if it relieves them from their epileptic seizures, the beneficial effects of an operation on cognitive functions in general and on learning and memory in particular is less clear. The only available information relatively to A+H has been published by Birri et al.(1982) and Gonser (1983), who found no seizable changes in the comparison of test results before and after SAH. On the contrary, Milner (1963) has clearly stated that surgical removals of temporal neocortical tissue, sparing or encroaching upon A+H, tend to slightly accentuate the preexisting functional deficit of the diseased hemisphere, while definitely improving the functional capacity of the healthy hemisphere. A similar trend has been mentioned by Nillson et al.(1984).

The purpose of the present study was to compare the learning and memory performance of epileptic patients before and after a unilateral SAH, in order to assess the behavioural consequences of the surgical procedure in itself and relatively to the detrimental functional effects of a longstanding epileptogenic focus.

METHOD

Patients

To date we have examined ten epileptic patients a few days before SAH, and again at the time of the first postoperative control on an outpatient basis.

Six patients formed the causal group: In all of them, the primary epileptogenic focus had been localized in the A+H complex (on the left side in three, on the right side in the other three) by routine EEG and by stereoelectroencephalography (SEEG, see Wieser, 1983) with implanted electrodes. The mean age was 29.3 years. The postoperative examination took place 15 weeks after SAH on the average, at which time they were free of seizures (Aebersold et al., 1983).

The four remaining patients made up the palliative group: In these cases, the A+H complex was heavily involved in the epileptic disturbances (in two each on the left, viz. on the right side), but the primary focus had been localized in neighbouring structures which could not be removed neurosurgically on the basis of previous experience of iatrogenic aphasic and psychotic disorders. The mean age of these patients was 28.5 years. The postoperative examination took place 27 weeks after SAH on the average. All four patients were still having epileptic seizures, though less frequently than before SAH and only of the clinical type caused by the primary focus.

Twelve non-patient control subjects (mean age 33.2 years) who were tested only once supplied the baseline data.

Material

Learning performance was tested with three series of 15 items each drawn or written in black ink on white cards 15 by 10 cm in size. The items were presented one at a time to the subjects. The items of the first series were nonsense designs (thereafter called "designs") (Rey, 1966 and 1968). The second series consisted of drawings of common objects (thereafter called "drawings"), and concrete nouns (shortened to "nouns" in the following) were used in the third series.

The recall of as many items as possible of each series after a filled distraction interval of 30 minutes made up the memory performance.

To avoid memory savings from pre- to postoperative testing, two parallel versions A and B of the three item series were set up. Half the subjects of each patient group were tested preoperatively with version A and postoperatively with version B, the other patients being tested first with version B and later with version A.

Procedure

In the learning of designs, the 15 items were presented to the subject at a rate of approximately one item every 2-4 seconds. The subjects were instructed to evaluate each design as "pleasant" or "unpleasant" in order to prevent verbalization. They were asked to watch the designs carefully and to take good notice of each of them, because they would be asked later on to reproduce them on paper from memory in any order. In the following presentations, only those items were presented anew which had not been reproduced in the immediately preceding trial (Buschke 1973). This procedure was repeated until the subjects were able to reproduce at least 12 out of 15 designs, but not more than 5 times. The same procedure was applied for the learning of drawings, with the difference that the drawn objects had to be named (and not rated as to their being pleasant or unpleasant), and again with the nouns, which had to be read aloud by the subjects.

The distraction interval was filled for 30 minutes with the administration of psychometric tests which did not involve learning of, or memory for new informations.

Memory performance was then assessed by asking the subjects for the delayed recall of the three types of materials, i.e., designs, drawings and nouns.

The most significant aspect of this testing procedure was that both learning and memory performances were measured with the same designs, drawing and nouns.

Analysis of Results

The causal and palliative patient groups were small, and it was not meaningful to split them according to the side of SAH. Nadig et al.(1984) have demonstrated quantitatively symmetrical relations, postoperatively, between left SAH and learning of nouns, and right SAH and learning of designs, while memory performance was not influenced whatsoever.

For the analysis of results, therefore, the side of SAH was neglected: The scores for nouns of patients operated on the left side were combined with the scores for designs of patients operated on the right side into "operated" scores (OP). Conversely, the scores for designs relatively to left SAH and the scores for nouns relatively to right SAH made up the "non-operated" scores (NOP). Given the small size of the patient groups, the individual data were averaged for the sake of clarity in the presentation of results, but no statistical tests were used to compare the mean scores.

Finally, the learning and memory performances for drawings were dropped from the analysis because they were not affected by either a left or right primary epileptogenic focus, nor by a left or right SAH.

RESULTS

Learning

Learning performance is expressed by the number of trials to reach learning criterion (Table 1).

In the causal group the preoperative learning performance is deficient in both OP and NOP compared to the combined score of the control group. From pre- to postoperative testing OP deteriorates to almost the possible maximum of 5 learning trials while NOP improves and reaches the control score. In the palliative group, OP is poor preoperatively, but NOP is close to the score of the control group. From pre- to postoperative testing OP worsens while NOP does not change.

Table 1. Learning: Mean number of trials to reach criterion (max. 5) in pre- and postoperative testing

Material	Causal group pre	Causal group post	Palliative group pre	Palliative group post	Controls
"operated"	4.3	4.5	4.0	4.3	3.1a
"non-operated"	4.5	3.0	3.5	3.5	

a - Learning scores for nouns and designs combined.

Memory

Memory performance is represented by the number of items recalled after the distraction interval (Table 2).

In the causal group the preoperative OP score is clearly deficient and it improves strongly when assessed again postoperatively. NOP is slightly higher than OP preoperatively and does not improve to the same extent at postoperative testing. Thus, neither OP nor NOP really approach the combined score of the control group at any time. The preoperative performance level of the palliative group is higher than in the causal group for OP and to a lesser degree for NOP. Both OP and NOP improve at postoperative testing. OP reaches the level of the control group and NOP still lags somewhat below it.

Table 2. Memory: Mean number of items recalled (max. 15) in pre- and postoperative testing

Material	Causal group pre	Causal group post	Palliative group pre	Palliative group post	Controls
"operated"	7.8	9.5	10.0	11.0	11.0a
"non-operated"	8.2	9.6	9.5	10.3	

a - Memory scores for nouns and designs combined.

The pre- to postoperative performance changes appear to be smaller in the palliative group than in the causal group. This difference, however, may be attributed to a ceiling effect providing that the control score is the highest attainable in the present memory test.

DISCUSSION

A primary epileptogenic focus in A+H appears to hamper learning and memory performance not only for information specific for the affected hemisphere (nouns in the left, designs in the right hemisphere) but also for the information dominantly processed by the contralateral hemisphere. This contrasts Milner's (1963) findings of only verbal deficits of memory in the presence of an epileptogenic lesion in the left temporal lobe and of only nonverbal memory deficits when the lesion is localized in the right temporal lobe. The published data, however, do not provide sufficient information as to the primacy of A+H in the epileptogenesis of Milner's patients and some differences cannot be ruled out at present. The negative findings of Birri et al. (1982) and Gonser (1983) partly with the same tests as used by Milner point to some basic difference at least between the patient groups studied in Montreal by Milner and in Zurich by Birri et al., Gonser, and Nadig et al. in the present work.

Secondary epileptogenic disturbances in A+H as recorded in the patients of the palliative group entail less severe learning and memory disorders than primary foci. Thus the difference in the primacy of the epileptogenic dysfunction in A+H is mirrored by a difference in severity of learning and memory deficits. In a further step of reasoning this is an additional confirmation of the role of A+H in learning and memory.

The outcome of a unilateral SAH on learning sheds some light on the nature of performance inhibition in the presence of an epileptogenic focus. When SAH is carried out on the primary focus it enhances the preoperative learning deficit for information for which the operated hemisphere is dominantly responsible. Phenomenally the inhibitions by a primary focus in A+H and by the excision thereof differ quantitatively more than qualitatively. In both cases inhibition seems to be passive in the sense of an absence of activation. The situation is different in the contralateral A+H where SAH allows for a strong improvement of learning performance. Therefore, inhibition of learning seems to have been active in this case and it is abolished by the excision of the focus. Milner's (1963) comparison of memory performance before and after a unilateral temporal lobectomy involving more neocortical than A+H substance is close to the present results. Such a similarity of effects of two types of surgical procedures indicate that A+H is specifically but not uniquely involved in learning and memory (Nadig, in preparation).

The effects on learning of the surgical removal of A+H when it is not primarily affected by an epileptogenic focus extend the preceding conclusions. In the hemisphere of the primary focus learning seems to have suffered from insufficient activation of A+H and is further lowered by its surgical removal. The absence of any learning change in the contralateral A+H following SAH suggests that its connections with the primary focus are not - or not only - passing through A+H of the affected hemisphere, a hypothesis at variance with, but not necessarily contradicting the situation of the A+H contralateral to a primarily affected A+H.

The relation of changes in memory performance with SAH must be left aside. There is a general improvement of performance in the causal and palliative groups, for OP as well as for NOP. The smaller improvement in the palliative group may be regarded as a ceiling effect. This merely argues against the procedure used here to assess memory performance insofar as it does not allow for broader changes, including positive ones. In addition, having the subjects reach a learning criterion of 12 items out of 15 and thus forcing most of them to recall the same items repeatedly may have produced overlearning of at least a few items. Neither neuropsychology nor experimental psychology have provided us with sufficient knowledge on overlearning to have predicted and now to explain the present memory data relatively to SAH.

CONCLUSION

The present study gives additional support to the assumption of a direct implication of A+H in learning and memory. Yet it also indicates the relativity of this implication. A+H is obviously not uniquely responsible for learning and memory and it must be considered in the framework of a complex learning and memory "circuit" or "system" (Mishkin, 1982, Markowitsch, 1984) which certainly also includes the temporal neocortex - among other structures. Moreover, A+H plays a different role when directly or indirectly implicated in the pathological epileptic process and the consequences of its surgical removal are still different.

Thus more doors to necessary future research have been opened than problems solved. Fortunately for the epileptic patients at least, the surgical removal of a primary epileptogenic focus in the amygdala and hippocampus has beneficial effects not only on their seizures, but partly also on their learning and memory performances.

ACKNOWLEDGEMENTS

The authors are grateful to Prof. M.G. Yasargil for the permission to test his patients before and after a selective amygdalohippocampectomy.

This work is supported by the Swiss National Science Foundation, Grants 3.816.79 and 3.771.80 to E. Perret.

REFERENCES

Aebersold, H., Albani, C., and Wieser, H.G., 1983, Prognostic value of postoperative EEG on the outcome of patients undergoing selective Amygdalohippocampectomy (SAH), Paper presented at the 15th Epilepsy International Symposium, Washington, D.C.

Birri, R., Perret, E., and Wieser, H.G., 1982, Der Einfluss verschiedener Temporallappenoperationen auf das Gedaechtnis bei Epileptikern, Nervenarzt, 53:144.

Buschke, H., 1973, Selective reminding for analysis of memory and and learning, J. Verb. Learn. Verb. Behav., 12:543.

Corsi, P.M., 1972, "Human Memory and the Medial Temporal Region of the Brain", Unpublished Ph.D. thesis, McGill University.

Gonser, A., 1983, "Gedaechtnisfunktionen bei hirngeschaedigten Patienten unter besonderer Beruecksichtigung des Hippokampus", Dissertation der Universitaet Zuerich, Zentralstelle der Studentenschaft.

Markowitsch, H.J., 1984, Can amnesia be caused by damage of a single brain structure? Cortex, 20:27.

Milner, B., 1963, Cognitive defects in focal and nonfocal epilepsy, Paper pesented at the 71st Ann. Conv. of the A.P.A., Philadelphia.

Milner, B., 1974, Hemispheric specialization: Scope and limits, in: "The Neurosciences. Third Study Program", F.O. Schmitt and F.G. Worden, eds., MIT Press, Cambridge, Mass.

Milner, B., 1980, Complementary functional specializations of the human cerebral hemispheres, in: "Nerve Cells, Transmitters and Behaviour", R. Levi-Montalcini, ed., Pontificia Academia Scientiarum, Vatican City.

Mishkin, M., 1982, A memory system in the monkey, Phil. Trans. R. Soc. Lond., B298:85.

Nadig, T., Perret, E., and Wieser, H.G., 1984, Learning and memory performance after unilateral selective amygdalohippocampectomy, Paper presented at the 7th Europ. Conf. of the I.N.S., Aachen.

Nillson, L.G., Christianson, S.-A., Silfvenius, H., and Blom, S., 1984, Preoperative and postoperative memory testing of epileptic patients, Acta Neurol. Scand., 69: Suppl. 99, 43.

Rey, A., 1966, "Les troubles de la memoire et leur examen psychometrique", Dessart, Bruxelles.

Rey. A., 1968, "Techniques inedites pour l'examen psychologique. Fasc. 3: Epreuves mnesiques et d'apprentissage", Delachaux & Niestle, Neuchatel.

Walsh, K.W., 1978, "Neuropsychology. A Clinical Approach", Churchill Livingstone, Edinburgh.

Wieser, H.G., 1983, "Electroclinical Features of the Psychomotor Seizure", Gustav Fischer - Butterworths, Stuttgart and London.

Wieser, H.G., and Yasargil, M.G., 1982, Selective amygdalo-hippocampectomy as a surgical treatment of mesiobasal limbic epilepsy, Surg. Neurol., 17:445.

READING VIA A NEW FUNCTIONAL PATHWAY IN AN ACQUIRED DYSLEXIA

Gregory V. Jones

Department of Psychology
University of Bristol
Bristol, England

INTRODUCTION

Studies of brain plasticity in infrahuman animals have the advantage over corresponding work with human subjects of affording in general a greater degree of experimental control. Conversely, however, work with human subjects allows analysis of behavioural manifestations of the cognitive processes involved in learning and memory to be conducted in considerably greater detail than would otherwise be possible. The potential elaborateness of such analyses perhaps reaches its apogee in the case of language. Language behaviour therefore provides a particularly appropriate domain within which to investigate mechanisms of functional recovery from the effects of damage to the human brain.

One important aspect of language behaviour that is frequently disrupted by brain damage is that of reading. Such disorders may be classified into a number of different types of acquired dyslexia (see Marshall & Newcombe, 1973; Patterson, 1981). Of the acquired dyslexias, that which has been probably the most intensively investigated in recent years is the syndrome of deep dyslexia (in particular, see Coltheart, Patterson, & Marshall, 1980). Reading behaviour in deep dyslexia is characterised by difficulties in reading several different categories of word, and also by the production of several different categories of response error, as will be described later in this chapter.

Subtractional and Substitutional Processes

Because of the diversity of the deficits that it exhibits, it is tempting to hypothesise that deep dyslexia represents a form of residual reading that remains when multiple components of the normal reading function are destroyed. Thus this view could be said to assert that the reading function in deep dyslexia arises by subtraction, in that the dyslexic patient's reading processes are merely impaired versions of normal ones. It is proposed here, on the other hand, that the reading function in deep dyslexia is not subtractional but instead substitutional. By this, it is meant that reading in deep dyslexia is supported by a mechanism that is not involved in corresponding normal behaviour.

The question thus arises as to whether the substitutional reading mechanism is one which is constructed only after brain damage has occurred, or alternatively whether it is latent within the intact organism. The present proposal implies that it is the latter that is true. That is, the new reading mechanism appears to be activated by release from inhibition or from occlusion rather than to be constructed from scratch. Moreover, the evidence suggests that this substitutional reading mechanism is mediated by the right cerebral hemisphere and takes over the reading function only when inhibition or occlusion normally provided by the left hemisphere is terminated as a consequence of brain damage.

Before turning to the substitutional hypothesis and the experimental evidence which supports it, the major features of deep dyslexia are briefly reviewed next.

Characteristics of Deep Dyslexia

At first sight, the distinguishing features of deep dyslexia are quite heterogeneous. First, patients experience greater difficulty in reading isolated words for which it is difficult to form a mental image (e.g., "idea") than in reading those for which this is easy (e.g., "apple"). Second, analyzed by syntactic category, the number of function words such as auxiliaries and prepositions (e.g., "shall", "out") read correctly is smaller than that of verbs, which is smaller than that of adjectives, which is smaller than that of nouns. Third, nonwords (e.g., "trab") are not read at all. Fourth, a considerable number of semantic paralexias occur - that is, a reading error may consist of another word which is similar in meaning to the stimulus word, but dissimilar in physical form (e.g., "town" might be read as "city"). Fifth, a number of visual paralexias occur - errors which are similar only in physical form to their stimuli (e.g., "such" might be read as "touch"). Sixth, a number of derivational paralexias occur (e.g., "defensive" might be read as "defended").

In general, the disparate nature of deep dyslexia's characteristics has been attributed to the effects of multicomponent damage to normal reading mechanisms. For example, the influential theory of Morton and Patterson (1980) attributes difficulty in the reading of low-imageability words to the effects of selective damage to the semantic representations of this class of words, and attributes both the differences in difficulty of reading among different parts of speech and the occurrence of derivational paralexias to the effects of damage to a linguistic processor. It is proposed here, however, that these characteristics are a positive manifestation of the single new pathway which has taken over the function performed prior to brain damage by normal reading processes. This pathway may be termed the predicational route to reading.

PREDICATIONAL READING

Uncertainty still remains over the identity of the information-processing mechanisms that allow one to read aloud an isolated word (see Henderson, 1982). One possibility is that the string of letters comprising a word is directly recognized as a whole, leading (again directly) to the excitation of stored information specifying appropriate phonological output. A second possibility is that individual letters or groups of letters within the stimulus are translated into phonemic representations (either by using explicit rules or else by means of analogies with other words), which then either can again lead to word recognition or can lead directly to a spoken response.

Each of the reading mechanisms described thus far is nonsemantic in that its operation does not entail access to the meaning of the word that is being read. It is postulated that in deep dyslexia brain damage results in none of these nonsemantic routes remaining intact. Instead, reading proceeds (albeit imperfectly) via a new semantic pathway, as follows. First, the string of letters comprising a stimulus is recognized as corresponding to an entry in the mental lexicon. Normal access to associated phonology at this stage, however, is not possible. Instead, the lexical entry can only summon (from long-term memory) predicates which normally serve to delimit the word's meaning (e.g., "town" might summon "has inhabitants"). Finally, it is on the basis of this predicational information alone that a response can be produced.

Application to Deep Dyslexia

It is immediately apparent that the predicational pathway provides a simple explanation of several features of deep dyslexia. Nonwords cannot be read because they do not provide access to any pre-existing entry in the mental lexicon. Semantic paralexias arise because predicational information (from which each reading response derives) is not necessarily uniquely attached to the word which originally summoned it from the long-term store (e.g., "has inhabitants" might yield "city" rather than the original "town" as response). Visual paralexias arise if a lexical entry is unsuccessful in summoning any predicates, in which case an orthographically adjacent entry may be accessed instead. Derivational paralexias can arise in the same ways as either semantic or visual paralexias.

It has been noted that the predicational hypothesis provides accounts of the generation of different types of paralexia that are highly straightforward. Although this circumstance serves to enhance the plausibility of the hypothesis, it fails to provide an opportunity for decisive experimental testing. The two remaining characteristics of deep dyslexia - the imageability effect and the syntactic category effect - are, however, provided with explanations by the predicational hypothesis whose validities are more problematic, and thus are usefully susceptible to experimental testing.

Imageability Effect

As previously noted, deep dyslexic patients enjoy greater success at reading a word for which it is easy to form a mental image of the word's referent than they do with a word for which this is difficult. Shallice and Warrington (1980, p. 136) have attributed this result to "neurological differentiations" between the representations of the two types of word, with damage only to low-imageability representations in deep dyslexia. Similar accounts have been provided by Morton and Patterson (1980, p. 115) and Coltheart (1980, p. 363). In contrast, it is suggested here that representations of low-imageability words do not suffer differential damage. Instead, words that are relatively difficult to image provide a problem in deep dyslexia only because it is generally also relatively difficult to summon predicates for them, rendering unavailable the new functional pathway to reading. That is, the factor governing reading in deep dyslexia is not word imageability as such, but a closely related one that may be termed ease of predication. The relation between ease of imaging and of predication is postulated to be general in that it applies not only to the deep dyslexic population but to the normal population also.

Extensive norms giving subjects' mean estimates of word imageability have been provided previously by Paivio, Yuille, and Madigan (1968). Their subjects were asked to judge on a seven-point scale the ease with which each word "arouses a mental image (i.e., a mental picture, or sound, or other sensory experience)" (Paivio et al., 1968, p. 4). Parenthetically, it may be noted that very similar judgements were made if instructions referred instead to a conceptually similar scale extending from abstractness (i.e., "any word that refers to an abstract concept that cannot be experienced by the senses") to concreteness (i.e., "any word that refers to objects, materials, or persons"): Paivio et al. (1968, p. 5).

Estimates of ease of predication, on the other hand, have not previously been tabulated. Accordingly, an experiment was carried out to obtain mean estimates from 30 Bristol University students. Each was provided with a list of 125 of the words studied by Paivio et al., and "ease of predication" was operationalized by asking the subject to judge "the ease with which each word can be put into simple factual statements". It was pointed out as an example that the word "dog" could be put into simple factual statements such as "A dog is a type of animal" and "A dog often lives in a kennel" (for further details, see Jones, in press).

The crucial result from this experiment was that the agreement between judgements of imageability and of ease of predication was extremely close, even though more than a dozen years - together with the Atlantic Ocean - intervened between the two sets of measurements. For the sake of compactness, Table 1 displays only the first half-dozen stimulus items. These illustrate the general finding that ease-of-predication estimates were lower in absolute terms than imageability estimates, but in relative terms were very similarly distributed; overall, $r(123) = 0.88$.

This study provides evidence therefore in favour of the parsimonious conclusion that low and high imageability words are not neurologically differentiated in terms of their susceptibility to damage. Deep dyslexic patients find words of low imageability difficult to read only because the new predicational reading route on which they rely is less likely to operate successfully for such words.

Syntactic Category Effect

The predicational hypothesis suggests that the explanation underlying this effect is similar to that for the imageability effect. That is, the increasing success with which deep dyslexic patients can read particular parts of speech (in the order function words, verbs, adjectives, and nouns) is attributed to increasing ease of predication for these categories.

Table 1. Examples of mean ease of predication and imageability

Word	Ease of Predication	Imageability[a]
agility	2.67	4.57
ambulance	5.87	6.67
anger	2.70	4.87
anxiety	2.30	4.03
army	5.63	6.53
baby	6.40	6.70

[a]These values are from Paivio et al. (1968).

A second experiment was conducted with 14 subjects to obtain mean estimates of ease of predication for these categories, using 20 instances in each case (I am grateful to Brian Alner for help in carrying out this experiment). The estimates were found to occur in the predicted configuration, with all pairwise differences between categories reaching significance (except that between adjectives and verbs). Thus the syntactic category effect, the final characteristic of the deep dyslexia syndrome to be considered, appears like the other characteristics to be usefully elucidated by the predicational hypothesis.

NEUROLOGICAL IMPLICATIONS

The evidence is consistent with the hypothesis that reading in deep dyslexia occurs by a new functional pathway which is substituted for that rendered unavailable by brain damage. The new predicational mechanism for reading aloud is presumably latent within the intact brain (with summoning of predicates acting as a basic element of the normal comprehension process), but its operation would not be generally advantageous because of the pathway's relative inaccuracy. One possibility is therefore that the operation of the predicational pathway is usually subject to inhibition from normal reading mechanisms. Further, the new pathway would come into action only as a consequence of disinhibition - damage to that part of the brain which supports the normal mechanisms. Consistent with this, a number of studies in the area of vision appear to demonstrate that, following a period of deprivation of input to one eye, the experienced eye exerts inhibition on cortical cell stimulation deriving from the deprived eye - inhibition which is removed by damage to the experienced eye (see Movshon & Van Sluyters, 1981). Geschwind (1974) has previously proposed, most relevantly, that disinhibition can play an important role in the recovery of language function in humans following damage to the dominant hemisphere. According to this hypothesis, language capabilities are present in the subordinate hemisphere, but released from inhibition only by damage to the dominant hemisphere (Geschwind proposed that this mechanism may be particularly salient in the case of recovery from childhood aphasia).

The notion of interhemispheric disinhibition provides a theoretically attractive link between the functional and neural levels of analysis of deep dyslexia. It is supported by two sets of findings. First, the existing evidence suggests that the syndrome is associated with large lesions to the left hemisphere. On the basis of computer assisted tomograms (CAT) obtained by X-ray examination of five deep dyslexia patients, Marin (1980) reported lesions to Broca's area in all cases, together with a considerable range of other left-hemisphere damage, and indeed concluded that "The cortical damage is in all cases quite extensive, and, in some intuitive way, out of proportion to the overall degree of language impairment observed in the patients" (p. 407). Further, Landis, Regard, Graves, and Goodglass (1983) have recently reported that the occurrence of semantic paralexia for unselected aphasic patients is a function of the extent of the lesion (although this interpretation has been questioned by Marshall & Patterson, 1983). The second set of findings consists of a considerable volume of recent work which has demonstrated linguistic abilities within the right hemisphere (see Coltheart, 1980, 1983; Martin, 1978; Zaidel, 1982). It is thus plausible to conclude that the predicational pathway to reading employed by the deep dyslexic patient is mediated by the right hemisphere, and that the substitution of this pathway for normal reading mechanisms is dependent upon damage-induced release from left-hemisphere inhibition.

Inhibition or Occlusion?

In objection to the foregoing conclusion, it might be argued that the results of the present analysis appear compatible also with an alternative type of neurological process distinguished by Geschwind (1974), that of occlusion.

Termination of occlusion, like that of inhibition, provides a mechanism which could underlie observed instances of functional substitution (which may also be termed "vicarious functioning" - see Marshall, 1984). In the present context, it could be argued that the linguistic ability of the right hemisphere is manifested not when it is no longer inhibited by the left hemisphere, but instead when it is no longer occluded by it. Alternatively put, the dominance of the left hemisphere could in principle be passive rather than active. The right-hemisphere predicational pathway to reading might come into play not because active inhibition from the dominant hemisphere ceases, but rather because competition from independent and more effective (e.g., faster) left-hemisphere mechanisms is terminated. Geschwind points out that an occlusive relation of this kind has been shown to hold in the cardiac conduction system (Hoffman & Cranefield, 1960). Heartbeat rhythm is maintained by whichever of a series of sets of cells possesses the fastest rate of rising to fire (normally the sinus node). Successive destruction of the fastest cells leads to the observation of increasingly slow (and previously occluded) heart rhythms.

The evidence discussed thus far, drawn from deep dyslexia, does not appear to provide a basis for choosing between inhibitory and occlusive views of cerebral dominance. However, a contrast in the predictions of the two accounts becomes apparent if we consider the consequences of brain damage which is limited to destruction of the inhibiting or occluding area only.

According to the occlusive theory, the occluding mechanism that usually prevents the expression of the predicational pathway consists simply of reading being conducted by more efficient normal pathways. Thus, according to this view, the removal of occlusion cannot be separated from the removal of normal reading pathways (and thus the genesis of deep dyslexia). According to the inhibitory theory, on the other hand, the inhibiting mechanism and the normal reading pathways are logically separable. It may therefore be possible for damage to the brain to occur in such a way that the inhibiting mechanism is impaired while the normal reading pathways are preserved. What would be the behavioural manifestation of such a pattern?

Concrete-Word Dyslexia

The major characteristic of the hypothetical disorder caused by damage to only the inhibiting mechanism in the left hemisphere should be that inaccurate information mediated by the disinhibited predicational route will pose particular problems in reading words that easily summon predicates. Because the output of the predicational pathway is often not identical with the original stimulus word (as described earlier), it will frequently conflict with the output of normal pathways. Thus the prediction is that attempts to read words with high ease of predication (or, equivalently, words with high imageability) will on occasion produce a conflict resulting in no response being made. In practice, the occurrence of a particular patient who exhibited a pronounced selective difficulty in reading highly imageable words (to which he indeed largely produced no response) has recently been described by Warrington (1981); CAT-scan showed a large area of mixed attenuation in the left temporal-parietal-occipital region. According to the present analysis, therefore, Warrington's observation provides some support for the inhibitory (rather than the occlusive) theory proposed to explain the lack of use of the right-hemisphere predicational pathway in normal reading.

As a footnote, it should be pointed out that the logic of the present reasoning is consistent not only with the inhibitory theory of hemispheric dominance, but also with any other two-factor theory which (unlike occlusive theory) separates the operations of substantive function and of hemispheric interaction. In particular, it is compatible also with the view that normal right hemisphere participation in reading is controlled by a central mechanism which regulates interhemispheric interaction, and which is localized in the language area of the left hemisphere (Zaidel & Schweiger, in press; see also Levy & Trevarthen, 1976). Indeed, it is clearly possible to regard the posited control mechanism as a simple generalization of the notion of inhibition.

RELATED DYSFUNCTIONS

Finally, it may be noted that the account of the new functional pathway observed in deep dyslexia which has been put forward here has implications also for our understanding of a number of other disorders of function caused by damage to the brain.

Phonological Dyslexia

The disorder perhaps most similar to that of deep dyslexia has been termed phonological dyslexia (or phonological alexia). Phonological dyslexic patients, like deep dyslexic patients, have great difficulty in reading nonwords. They differ, however, in that they do not produce semantic paralexias when reading words, although they do produce a proportion of derivational and of visual errors (Beauvois, Dérouesné, & Saillant, 1980). The present framework suggests that phonological dyslexia is the consequence of a lesser degree of damage than that which produces deep dyslexia. Although direct translation from grapheme to phoneme is not possible, and reading of nonwords is thereby prevented, a form of normal reading of words is preserved substantially intact. A word can still be recognized as a whole, and its phonology accessed via the relevant lexical entry (see Henderson, 1982). In consequence, the predicational pathway to reading is not called into play, and the generation of semantic paralexias is obviated.

Anomic Dementia

One may conversely envisage a syndrome which is essentially the complement of deep dyslexia. Consider the case in which the semantic structures that allow the formation of a predicational path to reading are damaged, while normal reading mechanisms remain intact. Then it would be expected that the behaviour of a patient with such an affliction should be characterised by severe difficulty in carrying out a semantic task such as classifying words into different groups (e.g., "animals" or "colours"), linked with a paradoxical absence of difficulty in reading the words aloud.

Schwartz, Saffran, and Marin (1980) have reported that just such a pattern of behaviour can be observed in the case of anomic dementia, providing further corroboration for the present framework. In terms of localisation of function, the predicational account predicts the occurrence of some right-hemisphere damage here, and the empirical evidence is consistent (though not compelling). Schwartz et al. reported that neurological examination produced evidence not of focal pathology but instead of bilateral, diffuse involvement of the cerebral hemispheres, with computerized brain tomography similarly revealing a moderate degree of generalized cortical atrophy.

Alexia without Agraphia

In this syndrome, patients typically are able to read only by the laborious strategy of individually naming each letter of a word, and then recognizing the complete series thus enunciated; for this reason, the syndrome has also been named letter-by-letter reading (e.g., Patterson, 1981). It appears to result from damage (e.g., to the posterior left hemisphere) which results in direct visual access remaining possible for the right hemisphere alone.

The disorder is of particular theoretical interest because it has been taken to constitute a problem for right-hemisphere accounts of linguistic function both by adherents of such an approach (e.g., Coltheart, 1983) and by its critics (e.g., Patterson & Besner, in press). The reasoning is that the reading skills of the right hemisphere which are supposed to be operative in deep dyslexia should function also in letter-by-letter reading. Thus semantic processing of words (at least, high-imagery ones) might be expected to be preserved, even if naming is severely impaired. In practice, however, comprehension also is greatly slowed in this syndrome (e.g., Patterson & Kay, 1982).

The observed similarity in impairment between naming and comprehension for alexia without agraphia is readily explicable within the present framework, however. This is possible because of the framework's distinction between visually-based access to the lexicon on the one hand, and subsequent access to relevant predicates on the other hand. In deep dyslexia both remain intact, leading to successful comprehension (and, ultimately, to some success in naming via the predicational pathway). In letter-by-letter reading, however, it may be hypothesised that the left-hemisphere damage, differing from that occurring in deep dyslexia, disrupts initial access to the lexicon. The important consequence is that the relevant predicates necessarily also fail to be accessed, even though their representations may be sited contralaterally and remain undamaged, and thus the observed impairment in naming is here (unlike in deep dyslexia) echoed by one in comprehension also.

SUMMARY

It has been hypothesised that reading in the acquired disorder of deep dyslexia is subserved by a new functional pathway. Although normal reading processes are damaged, it is still possible to achieve some success in reading an isolated word by responding on the basis of the predicational information which the word summons from long-term memory. It was shown that this mechanism accounts for a wide range of characteristics of deep dyslexia (e.g., the occurrence of the semantic paralexia error) and in particular elucidates the otherwise puzzling effects of a word's imageability and of its syntactic category. The present analysis also appears to assist understanding of a number of other acquired dysfunctions, such as that of phonological dyslexia.

The evidence reviewed here suggests also that the new functional pathway in deep dyslexia is mediated via the right hemisphere, and that its use is latent within the intact brain. Further, the brain damage which leads to deep dyslexia appears to do so by ending a normal state of active inhibition of the pathway, rather than by ending a state of its passive occlusion.

REFERENCES

Beauvois, M.-F., Dérouesné, J., and Saillant, B., 1980, Syndromes neuropsychologiques et psychologie cognitive. Trois exemples: Aphasie tactile, alexie phonologique et agraphie lexicale, Cahiers Psychol., 23: 211.

Coltheart, M., 1980, Deep dyslexia: A right-hemisphere hypothesis, in: "Deep Dyslexia," M. Coltheart, K. Patterson, and J. C. Marshall, eds., Routledge & Kegan Paul, London.

Coltheart, M., 1983, The right hemisphere and disorders of reading, in: "Functions of the Right Cerebral Hemisphere," A. Young, ed., Academic Press, London.

Coltheart, M., Patterson, K., and Marshall, J. C., 1980, eds., "Deep Dyslexia," Routledge & Kegan Paul, London.

Geschwind, N., 1974, Late changes in the nervous system: An overview, in: "Plasticity and Recovery of Function in the Central Nervous System," D. G. Stein, J. J. Rosen, and N. Butters, eds., Academic Press, New York.

Henderson, L., 1982, "Orthography and Word Recognition in Reading," Academic Press, London.

Hoffman, B. F., and Cranefield, P. F., 1960, "Electrophysiology of the Heart," Blakiston, New York.

Jones, G. V., in press, Deep Dyslexia, imageability, and ease of predication, Brain Lang.

Landis, T., Regard, M., Graves, R., and Goodglass, H., 1983, Semantic paralexia: A release of right hemispheric function from left hemispheric control?, Neuropsychologia, 21: 359.

Levy, J., & Trevarthen, C., 1976, Metacontrol of hemispheric function in human split-brain patients, J. exp. Psychol.: hum. Percep. Perform., 2: 299.

Marin, O. S. M., 1980, CAT scans of five deep dyslexic patients, in: "Deep Dyslexia," M. Coltheart, K. Patterson, and J. C. Marshall, eds.; Routledge & Kegan Paul, London.

Marshall, J. C., and Newcombe, F., 1973, Patterns of paralexia: A psycholinguistic approach, J. psycholing. Res., 2: 175.

Marshall, J. C., and Patterson, K., 1983, Semantic paralexia and the wrong hemisphere: A note on Landis, Regard, Graves and Goodglass (1983), Neuropsychologia, 21: 425.

Marshall, J. F., 1984, Brain function: Neural adaptations and recovery from injury, A. Rev. Psychol., 35: 277.

Martin, M., 1978, Hemispheric asymmetries for physical and semantic selection of visually presented words, Neuropsychologia, 16: 717.

Morton, J., and Patterson, K., 1980, A new attempt at an interpretation, or, an attempt at a new interpretation, in: "Deep Dyslexia," M. Coltheart, K. Patterson, and J. C. Marshall, eds., Routledge & Kegan Paul, London.

Movshon, J. A., and Van Sluyters, R. C., 1981, Visual neural development, A. Rev. Psychol., 32: 477.

Paivio, A., Yuille, J. C., and Madigan, S. A., 1968, Concreteness, imagery, and meaningfulness values for 925 nouns, J. exp. Psychol. Mon., 76: No. 1, Part 2.

Patterson, K., 1981, Neuropsychological approaches to the study of reading, Br. J. Psychol., 72: 151.

Patterson, L., and Besner, D., in press, Is the right hemisphere literate? Cogn. Neuropsychol.

Patterson, K., and Kay, J., 1982, Letter-by-letter reading: Psychological descriptions of a neurological syndrome, Q. Jl exp. Psychol., 34A: 411.

Schwartz, M.F., Saffran, E. M., and Marin, O. S. M., 1980, Fractionating the reading process in dementia: Evidence for word-specific print-to-sound associations, in: "Deep Dyslexia," M. Coltheart,

K. Patterson, and J. C. Marshall, eds., Routledge & Kegan Paul, London.

Shallice, T., and Warrington, E.K., 1980, Single and multiple component central dyslexic syndromes, in: "Deep Dyslexia," M. Coltheart, K. Patterson, and J. C. Marshall, eds., Routledge & Kegan Paul. London.

Warrington, E. K., 1981, Concrete word dyslexia, Br. J. Psychol., 72: 175.

Zaidel, E., 1982, Reading in the disconnected right hemisphere: An aphasiological perspective, in: "Dyslexia: Neuronal, Cognitive and Linguistic Aspects," Y. Zotterman, ed., Pergamon Press, Oxford.

Zaidel, E., and Schweiger, A., in press, On wrong hypotheses about the right hemisphere, Cogn. Neuropsychol.

MEMORY PROCESSES AND AGING IN RODENTS

Bernard Soumireu-Mourat

Lab. Neurobiologie des Comportements, CNRS U.A. 372
Université de Provence, Marseilles, France

INTRODUCTION

It is generally asserted that memory diminishes progressively with age. However, it is not clear to what extent these modifications are related to changes in learning abilities, memory processes, or non-associative influences such as motor, motivational or attentional factors, which can impair performance. Moreover, the neurobiological mechanisms of aging remain unknown, and it is difficult to distinguish between normal aging processes and pathology. Therefore, animal models are useful in order to better understand cognitive deficits in humans. In this chapter, some of the neurobiological correlates of aging are summarized as these could explain some of the cognitive deficits observed in aged subjects. Thereafter, data including some from our own experiments on learning and memory in aged animals are reviewed. Learning deficits, retention deficits and non-associative factors are considered successively. Finally, I will summarize the possible basis of memory deficits in aging.

BRAIN CHANGES IN AGING

Many changes in brain associated with age increases have been described. Many correlations between cerebral and behavioral modifications can be cited, but it is difficult to know what mechanisms are specifically responsible for cognitive deficits. Interestingly, several of these changes concern the hippocampus, reputed to be involved in memory processes.

First, morphological modifications appear during aging. They include loss of nerve cells in different parts of the brain, intracellular modifications, and even a reduction of the weight of the whole brain (Frolkis and Bezrukof, 1979). For instance, in mice there is a decline in neuronal density with age, which begins early compared to the survival curve of subjects (Comfort, 1979). However, some investigators observed no changes in older subjects in either density of neurons (Curcio and Coleman, 1982), synaptic density, or mean spine volume (Curcio and Hinds, 1983). Cell loss in area 17 of the cortex has been correlated with age-related differences in passive avoidance retention (Ordy et al., 1978). Morphological changes have been observed in the hippocampus of aged rats, and several authors consider these could explain memory deficits in aging (Ordy et al., 1978; Brizzee et al., 1979; Landfield et al., 1981; Stein, 1983). Moreover, plasticity is reduced in aged rats, especially in the hippocampus, as shown by dentate reinnervation induced by lesion of the hippocampal afferents (Scheff et al., 1978). However, the aged brain still has a "reserve capacity", i.e. some remaining "plasticity of function" (Stein, 1983). This is confirmed by behavioral studies. Using an active avoidance task, we found that a post-trial hippocampal electrical stimulation attenuated natural retention deficits in old F344 rats (Soumireu-Mourat et al., 1980), in the same way as it improved memory retention of different tasks in young rats (Landfield et al., 1973) and mice (Soumireu-Mourat et al., 1975).

Changes in brain physiology with aging have also been described. In some cases, the possibility of a relationship with behavioral deficits has been hypothetized. From a neurophysiological point of view, monosynaptic potentiation is impaired in hippocampal slices from aged rats (Landfield and Lynch, 1977). Changes in both spatial specificity and reliability of firing patterns during learning have also been reported (Barnes et al., 1983). Additionally, neurochemical studies have shown effects of age on neurotransmitter systems. Clear age-related decreases have been demonstrated in dopaminergic activity, and also in adrenergic, GABAergic and cholinergic systems (McGeer, 1981). However, a causal link with cognitive impairments is still discussed, and probably none of these systems is exclusively involved. Peptides could also be concerned more or less specifically with behavioral changes during aging (De Wied and Van Ree, 1982). Hormonal modifications can also influence memory performance (Gold and McGaugh, 1975). Aging results in changes in the adrenal glands. Modifications of hippocampal functioning, as those observed in old subjects, can influence hormonal levels in the blood. Indeed, there are glucocorticoid binding entities in the hippocampus participating in the regulation of adrenal hormones (McEwen and Micco, 1980). Plasticity in the hippocampus depends on plasma corticosteroid levels which are increased in aged animals (Landfield and Lynch, 1977; Hoff et al., 1982). Steroids influence hippocampal functions expressed through behavior as well as through neurophysiological activity (see Pomet et al., this volume).

LEARNING DEFICITS

Numerous animal studies show that there are age-related changes in the rate of acquisition. These have been studied mainly using shock-motivated tasks with multiple trials per session including shuttle-box avoidance conditioning in young (3 to 5 months) and old (9 to 15 months) mice (Freund and Walker, 1971), and visual discrimination avoidance in young (2 months) and old (23 months) rats (Ray and Barrett, 1973). Using a one-way active avoidance task, we found an impaired acquisition when young (3 months) and old (24 months) rats (Soumireu-Mourat et al., 1980) or mice (Soumireu- Mourat, 1981) were compared. We have also found a decline in the rate of acquisition of an appetitive task in aged mice. It has been shown that young mice of the BALB/c strain learn a continuously reinforced (CRF) operant task within a 30-minute session without any pretraining (Jaffard et al., 1974), but 24-month-old mice exhibited no such learning in the same task (Soumireu-Mourat, 1981). Such deficits in operant responding within a single learning session could reflect a deficit in short-term memory (STM) processes, which is also suggested by data obtained in monkeys. For example, the performance on a delayed-response task of aged monkeys at the shortest delay interval is comparable to that of young monkeys, but increasing the time interval to 15 seconds demonstrates a relative impairment in aged animals (Bartus et al., 1978; Marriott and Abelson, 1980). However, the apparent STM deficit could also be related to the complexity of the task, and aged animals could suffer from a temporary response bias rather than from a deficit in STM (Hamm et al., 1983).

RETENTION DEFICITS

Results obtained in different species have suggested an age-related decline in retention performance. Most of the data obtained in animals are interpreted as a deficit in long-term memory (LTM). However, such deficits raise several questions. Is there really a memory deficit in old subjects? Is this deficit related to changes in LTM or in retrieval processes? Can the observed impairments be related to non-memory alterations?

Several studies appear to demonstrate an age-related decline in retention performance in rats (Brizzee and Ordy, 1979; Lippa et al., 1980) and mice (Dean et al., 1981; Kubanis et al., 1981). Using a passive avoidance paradigm, Gold and McGaugh (1975) compared retention for different rat age groups (60 days, 1 year and 2 years) depending on the training-test interval (one day, 1 week and 6 weeks). They observed that one-year-old rats had poorer retention after a one-week interval, and two-year-old rats were deficient after all intervals. However, the older subjects did acquire the passive avoidance response, displaying correct retention performance when tested only two hours after training. The most usual paradigm in animals consists of retention testing one day after an initial learning session. Some authors did not observe any age-related change in this case. For instance, Gold et al. (1981) found no difference in retention of a one-trial passive avoidance conditioning between 70-day-old rats and one-year-old rats for a training-test interval of one day. However, factors other than the time interval before testing are probably involved in this lack of influence of age on retention. The level of the initial learning might be an important factor. A few authors found no deficit even for intervals as long as 6 months in mice (Freund and Walker, 1971). In our own experiments, comparing 3-month-old and 24-month-old mice, we found no significant difference in retention either one day after complete learning of a one-way active avoidance task or 15 days after over-training in a CRF operant task (Soumireu-Mourat, 1981).

If most authors claim the existence of retention deficits in old animals, the question remains as to the nature of this deficit. One hypothesis is that the observed retention deficits in old animals are due to differences in initial learning, a point emphasised by Ray and Barrett (1973). We noted previously that such differences do exist. In order to eliminate this hypothesis, similar performances at the end of acquisition are required whatever the age, but such a procedure is not frequently used. Retention deficits during aging could also be due to deficits in LTM processes, or to an increase of the rate of forgetting, perhaps due to increased interferences. This is supported by the lack of a deficit found at the shortest training-retention intervals and by the progressive time-dependent decrease in performance found in rats (Gold et al., 1981) and mice (Dean et al., 1981) for the aged animals compared to the young ones. Moreover, other data could be explained by slowed LTM processes. Distributed practice is more beneficial to old rats than to young rats: this seems to indicate that consolidation processes are impaired in old animals in comparison to young adults (Gold and McGaugh, 1975). Such an hypothesis is also suggested by the data of Ray and Barrett (1973) who showed that the transitory decline of avoidance performance corresponding to the Kamin effect is postponed in old animals compared to young ones. Also, the increased susceptibility of aged mice to the amnesic effect of

ECS suggests age-related changes in the rate, or the duration of memory consolidation (Essman, 1982).

Alternatively, the observed impairments in aged animals could be due to retrieval deficits. This hypothesis has been extensively studied in humans (Burke and Light, 1981). However, most of the data do not permit precise analyse of the specific mechanisms responsible for the retention impairment found in aged subjects. In recent years, studies on retrieval processes have been developed for animals (Spear, 1978) but most of these concern only adult subjects. Specific studies on recall and retrieval in old animals are needed.

NON-ASSOCIATIVE FACTORS

The main difficulty in all the studies on memory and aging is to distinguish between alterations in specific learning and memory processes and changes in non-associative factors that modify performance but not memory itself. Modifications of non-associative processes have been suggested as the main factor in performance impairment during aging by Ray and Barrett (1973). In their experiments, rats were trained in a Y-maze at the age of seventy days, then tested between one and 270 days later. They observed reduced retention in the older rats. To test only the specific effects of age, these authors studied groups for which training was done at the age when the subjects in the first experiment were retested, and these new groups were all retested one day after training. There was a decline in acquisition for the oldest rats, but also retest performance was clearly age-related. Ray and Barrett concluded that the decline in avoidance responses was not related to a memory loss, but to impaired performances.

Such a conclusion is a controversial point and probably constitutes the main question in studies on memory and aging. Several processes are modified with age. Changes in performance could be due to modifications in motor activity (Peng et al., 1980; Ingram et al., 1981) or to a reduction of arousal and attention. Both could explain a reduced rate of acquisition and impaired retrieval. It has also been suggested that motivational changes are an important factor. For example, swimming performance of mice was affected in a water maze by manipulating both age and motivation (by varying water temperature)(Werboff and Anderson, 1977). Motivational factors could explain also a paradoxical result that we observed with mice. Three-month-old and 24-month-old mice were submitted to a partial learning of an appetitive operant conditioning. They were tested for extinction nine days later: under these circumstances, old mice presented a lower level of performance than young mice. However, if tested with reinforcement two hours after extinction, there was no difference in retention between the groups. This could mean that there was no deficit in motor activity or in memory during the extinction session, but just a change in the motivational value of the situation. However, according to Bartus et al. (1978), motivational changes would not be involved in the deficits observed in monkey. Moreover, shock detection thresholds would decrease as a function of age in rats (Gordon et al., 1978). This is an important point, if true, as shock-motivated tasks are frequently used in animals. These tasks result also in stress and there are several changes in the response to stress as age increases. Particularly, urinary and blood plasma levels of substances liberated in response to stress increase with age, as do the size and functioning of the adrenal medulla (Beck, 1978). Finally, each of these changes could play a role in modifications of performance in old subjects.

DISCUSSION

Considering the behavioral data from several species, it appears that there is no single explanation for the memory deficits observed in old subjects. Perhaps the storage processes (i.e. LTM) can be slowed, at least in those circumstances which relate to the complexity of the learned information (Messing et al., 1982). This would result both in a storage deficit and in an acceleration of forgetting, mainly because of an increase of interferences. Increased interference could also be responsible for a change in retrieval. Such an analysis could explain the discrepancy among results concerning memory retention in aging: the existence of retention deficits would be negatively correlated to the level of the initial training, since the memory trace could be, in case of overtraining (as we observed in some of our experiments), quickly consolidated. However, an important part of the long-term deficit is probably due to a deficit in learning itself. In turn, this seems to be related both to a short-term memory impairment in some cases, and to a deficit in non-associative (i.e. non-specific) factors resulting in changes in performance.

Finally, there is evidence for changes in learning and memory during aging. Numerous physiological changes have been recently described in relation to age. The studies on memory and aging have to take into account all the factors discussed in this chapter. The present knowledge describes mainly correlations. Considering the complexity of cognitive operations in memory, it is no surprise that we do not know exactly the nature of the different mechanisms involved in aged-related changes in memory (Burke and Light, 1981). More studies are needed in order to understand better the neurolobiological bases of cognitive impairments during aging. Animal models present several advantages (Dean et al., 1981; Ingram, 1983). In animals, well-defined strains can be used and longitudinal experiments can be run with control over life experiences, considering the lifespan of species. Moreover, correlations between factors are easier to study than in humans. Behavioral tests have to be developed to assess more accurately the mechanisms of memory deficits in aging (Ingram, 1983). Comparison with human data must be done carefully, due to the complexity of aging processes and brain functions (Browning and Spilich, 1981).

REFERENCES

Barnes, C.A., McNaughton, B.L., and O'Keefe, J., 1983, Loss of place specificity in hippocampal complex spike cells of senescent rat, Neurobiol. Aging, 4:113.

Bartus, R.T., Fleming, D., and Johnson, H.R., 1978, Aging in the rhesus monkey: debilitating effects on short-term memory, J. Gerontol., 33:858.

Beck, C.H., 1978, Functional implications of changes in the senescent brain: a review, J. Can. Sci. Neurol., 5:417.

Brizzee, K.R., Bartus, R.T., Ordy, J.M., and Beavers, T., 1979, Short-term memory and aging: localization of cellular changes within multimodal sensory regions of the cerebral cortex and limbic system in the rhesus monkey, Soc. Neurosci. Abs., 5:3.

Brizzee, K.R., and Ordy, J.M., 1979, Age pigments, cell loss and hippocampal function, Mech. Aging Dev., 9:143.

Browning, G.B., and Spilich, G.J., 1981, Some important methodological issues in the study of aging and cognition, Exp. Aging Res., 7:1.

Burke, D.M., and Light, L.L., 1981, Memory and aging: the role of retrieval processes, Psychol. Bull., 90:513.

Comfort, A., 1979, "The biology of senescence", Churchill Livingstone, New York.
Curcio, C.A., and Coleman, P.D., 1982, Stability of neuron number in cortical barrels of aging mice, J. Comp. Neurol., 212:158.
Curcio, C.A., and Hinds, J.W., 1983, Stability of synaptic density and spine volume in dentate gyrus of aged rats, Neurobiol. Aging, 4:77.
Dean, R.L., Scozzafava, J., Goas, J.A., Regan, B., Beer, B., and Bartus, R.T., 1981, Age-related differences in behavior across the life span of the C57BL/6J mouse, Exp. Aging Res., 4:427.
De Wied, D., and Van Ree, J.M., 1982, Neuropeptides, mental performance and aging, Life Sci., 31:707.
Essman, W.B., 1982, Aging-related changes in retrograde amnesia for mice, Gerontology, 28:303.
Freund, G., and Walker, D.W., 1971, The effect of aging on acquisition and retention of shuttle-box avoidance in mice, Life Sci., 10:1343.
Frolkis, V.V., and Bezrukof, V.V., 1979, "Aging of the central nervous system", Karger Pub., Basel.
Gold, P.E., and McGaugh, J.L., 1975, Changes in learning and memory during aging, in: "Neurobiology of Aging", J.M. Ordy and K.R. Brizzee, eds., Plenum Press, New York.
Gold, P.E., McGaugh, J.L., Hankins, L.L., Rose, R.P., and Vasquez, B.J., 1981, Age dependent changes in retention in rats, Exp. Aging Res., 8:53.
Gordon, W.C., Scobie, S.R., and Frankl, S.E., 1978, Age-related differences in electric shock detection and escape thresholds in Sprague-Dawley albino rats, Exp. Aging Res., 4:23.
Hamm, R.J., Knisely, J.S., and Dixon, C.E., 1983, An animal model of age changes in short-term memory: the DRL schedule, Exp. Aging Res., 9:23.
Hoff, S.F., Scheff, S.W., and Cotman, C.W., 1982, Lesion-induced synaptogenesis in the dentate gyrus of aged rats: II. Demonstration of an impaired degeneration clearing response, J. Comp. Neurol., 205:253.
Ingram, D.K., 1983, Toward the behavioral assessment of biological aging in the laboratory mouse: concepts, terminology, and objectives, Exp. Aging Res., 9:225.
Ingram, D.K., London, E.D., Reynolds, M.A., Waller, S.B., and Goodrick, C.L., 1981, Differential effects of age on motor performance in two mouse strains, Neurobiol. Aging, 2:221.
Jaffard, R., Destrade, C., Soumireu-Mourat, B., and Cardo, B., 1974, Time-dependent improvement of performance on appetitive tasks in mice, Behav. Biol., 11:89.
Kubanis, P., Gobbel, G., and Zornetzer, S.F., 1981, Age-related memory deficits in swiss mice, Behav. Neural Biol., 32:241.
Landfield, P.W., Braun, L.D., Pitler, T.A., Lindsey, J.D., and Lynch G., 1981, Hippocampal aging in rats: a morphometric study of multiple variables in semithin sections, Neurobiol. Aging, 2:265.
Landfield, P.W., and Lynch, G., 1977, Impaired monosynaptic potentiation in in vitro hippocampal slices from aged, memory-deficient rats, J. Gerontol., 32:523.
Landfield, P.W., Tusa, R.J., and McGaugh, J.L., 1973, Effects of posttrial hippocampal stimulation on memory storage and EEG activity, Behav. Biol., 8:485.
Lippa, A.S., Pelham, R.W., Beer, B., Critchett, D.J., Dean, R.L., and Bartus, R.T., 1980, Brain cholinergic function and memory in aged rats, Neurobiol. Aging, 1:10.
Marriott, J.G., and Abelson, J.S., 1980, Age differences in short-term memory of test-sophisticated rhesus monkeys, Age, 3:7.

McEwen, B.S., and Micco, D.J., 1980, Toward an understanding of the multiplicity of glucocorticoid actions on brain function and behavior, in: "Hormones and the brain", D. De Wied and P.A. Van Keep, eds., MTP Press Ltd, Lancaster.

McGeer, E.G., 1981, Neurotransmitter systems in aging and senile dementia, Prog. Neuro-Psychopharmacol., 5:433.

Messing, R.B., Rigter, H., and Nickolson, V.J., 1982, Memory consolidation in senescence: effects of CO_2, amphetamine and morphine, Neurobiol. Aging, 3:133.

Ordy, J.M., Brizzee, K.R., Kaach, B., and Hansche, J., 1978, Age differences in short-term memory and cell loss in the cortex of the rat, Gerontol., 24:276.

Peng, M.T., Jiang, M.J., and Hsü, H.K., 1980, Changes in running-wheel activity, eating and drinking and their day/night distributions throughout the life span of the rat, J. Gerontol., 35:339.

Ray, O.S., and Barrett, R.J., 1973, Interaction of learning and memory with age in the rat, in: "Psychopharmacology and Aging", C. Eisdorfer and W.E. Fann, eds., Plenum Press, New York.

Scheff, S.W., Bernardo, L.S., and Cotman, C.W., 1978, Decrease in adrenergic axon sprouting in the senescent rat, Science, 202:775.

Soumireu-Mourat, B., 1981, Neurobiologie de la mémoire du sujet vieillissant, in: "Le cerveau âgé", J.P. Emeriau and J.M. Orgogozo, eds., Dausse Pub., Paris.

Soumireu-Mourat, B., Destrade, C., and Cardo, B., 1975, Effects of seizure and subseizure post-trial hippocampal stimulation on appetitive operant behavior in mice, Behav. Biol., 15:303.

Soumireu-Mourat, B., Martinez, J.L.Jr, Jensen, R.A., and McGaugh, J.L., 1980, Facilitation of memory processes in aged rats by subseizure hippocampal stimulation, Physiol. Behav., 25:263.

Spear, N., 1978, "The Processing of memories: forgetting and retention", Laurence Erlbaum Ass. Pub., Hillsdale.

Stein, D.G., 1983, Neurobehavioral plasticity in the aging brain, in: "Aging of the Brain", D. Samuel et al., eds., Raven Press, New York.

Werboff, J., and Anderson, A., 1977, Age and motivation in mice: performance and learning in a spatial water maze, Perceptual Motor Skills, 44:1239.

ANATOMICAL AND BEHAVIORAL STUDIES FOLLOWING LESIONS ON THE BASAL MAGNOCELLULAR NUCLEUS IN THE RAT

Hervé Simon, Willy Mayo and Michel Le Moal

Inserm U.259, Psychobiologie des Comportements Adaptatifs
Rue Camille Saint-Saëns, 33077 Bordeaux Cedex France

Increased life expectancy in the developed countries has highlighted the problems raised by both the normal and especially the pathological ageing processes of the brain. One problem in particular in terms of its socio-economic impact is represented by the senile dementia of the Alzheimer type (SDAT). It has been estimated that this disease represents the fourth or fifth main cause of death in the developed countries. The SDAT syndrome is characterized initially by temporo-spatial disturbances and impairment of recent memory followed by language disorders (aphasia), dysfunction of motor skill (apraxia) and perceptual difficulties (agnosia).

SDAT has become the subject of renewed interest over the last few years in the light of the consistent findings from either post-mortem or from biopsies of a deficiency of acetylcholine in certain brain regions in patients with this disease. Biochemical markers of the central cholinergic activity such as, choline-acetyltransferase (CAT, the synthesizing enzyme of acetylcholine), acetylcholine-esterase (the hydrolysing enzyme of acetylcholine which is also present in non-cholinergic neurons), choline uptake, and measurement of acetylcholine synthesis are found to be decreased from 40% to 90% in all neocortical areas assayed as well as in the hippocampus (Bowen et al., 1976 ; Perry et al., 1977 ; Sims et al., 1983). The reduction of CAT activity has been found to be related to the severity of the dementia (Perry et al., 1978) as well as to the number of neuritic plaques (Bowen et al., 1976 ; Perry et al., 1978). No alterations of muscarinic receptors have been observed in the various cortical areas studied (Davies and Verth, 1978). This indicates that the main lesion in SDAT probably only affects the presynaptic part of the cholinergic terminal. Interestingly, the neuritic plaques, which form with the neurofibrillary tangles the major features of the morphological alterations observed in most if not all neocortical areas of patients with SDAT, are very reactive to cholinesterase staining. The axonal debris which are present in neuritic plaques could indicate degeneration of cholinergic terminals.

As it has been demonstrated that subcortical ascending projections originating in the magnocellular nucleus of the basal forebrain (NBM) provide the major part of cortical acetylcholine (Johnston et al., 1979), it has been suggested that the neocortical cholinergic abnormality in SDAT may be the expression of pathological changes in the basal nucleus. This hypothesis is supported by the following concordant observations on this structure :

- the presence of neuritic plaques (Pilleri, 1966)
- a significant reduction in CAT activity (Rossor et al., 1982)
- a severe neuronal loss (Whitehouse et al., 1981).

While the functional significance of changes in the NBM following SDAT is unknown, dysfunction of these cholinergic neurons would seem to be at the core of dementia. Supporting this view are the results of behavioral experiments in animals and humans showing that cholinergic antagonists impair cognitive function (Bartus et al., 1982). These considerations have prompted us to study the anatomical and functional characteristics of the NBM in the rat (Dubois et al., 1985 ; Mayo et al., 1984).

EFFECTS OF NBM LESIONS ON CORTICAL CHOLINERGIC INNERVATION

Over the last few years a number of studies have thrown some light on the anatomical organization of the NBM. However, the data obtained are still incomplete. We were particularly interested in furthering investigations on the anatomical connections of the NBM in the rat as a basis for future behavioral studies. It is clear that better knowledge of the nervous structures innervated by the NBM is of crucial importance for subsequent functional studies. We therefore chose among anatomical approaches the technique of lesioning as a first step before investigating the behavioral effects of such NBM lesions. The cortical cholinergic projections from the NBM were studied by measuring the activity of CAT in three cortical areas (the antero-medial prefrontal cortex, the sensorimotor cortex and the parieto-temporal cortex) following ibotenic acid lesion of the NBM cell bodies. In the rat, the cells of the NBM are diffusely distributed along a rostro-caudal axis from the diagonal band to the sub-pallidal region. Thus we made ibotenic acid injections (7.5 µg in 0.5 µl of phosphate buffer) at several points in the NBM. We observed that the CAT depletion in the various cortical areas correlated with the depth of the lesion site. The more ventral the lesion, the greater was the reduction in CAT in the antero-medial prefrontal cortex ($r = -0.57$; $p < 0.02$). For the sensori-motor cortex and the parieto-temporal cortex, the converse was found : the more dorsal the lesion, the greater was the reduction in CAT in both these two cortical areas ($r = 0.80$ and $r = 0.59$; $p < 0.001$, respectively). Thus cortical cholinergic innervation seemed in line with a dorso-ventral organization. This led us to consider two sub-groups of cells, as shown in figure 1. One group of cells was located in the sub-pallidal region and projects preferentially towards the antero-medial prefrontal cortex, and the other group was located more dorsally in the internal border of the pallidum, innervating mainly the sensorimotor and parieto-temporal cortices.

EFFECTS OF NBM LESIONS ON BEHAVIOR

Hardly anything is known about the role of the NBM in behavior. In 1983, Coyle et al. wrote : "... there seem to be no published studies on the behavioral consequences in experimental animals that result from a discrete lesion of the NBM". Since this time, to our knowledge only four papers have been published on this subject (Lo Conte et al., 1982 ; Fliker et al., 1983 ; Friedman et al., 1983 ; Salamone et al., 1984). This is in marked contrast with the abundant literature devoted to the behavioral role of the septo-hippocampal cholinergic pathway originating in cell bodies located at the rostral pole of the NBM, in the septum and in the vertical limb of the diagonal band.

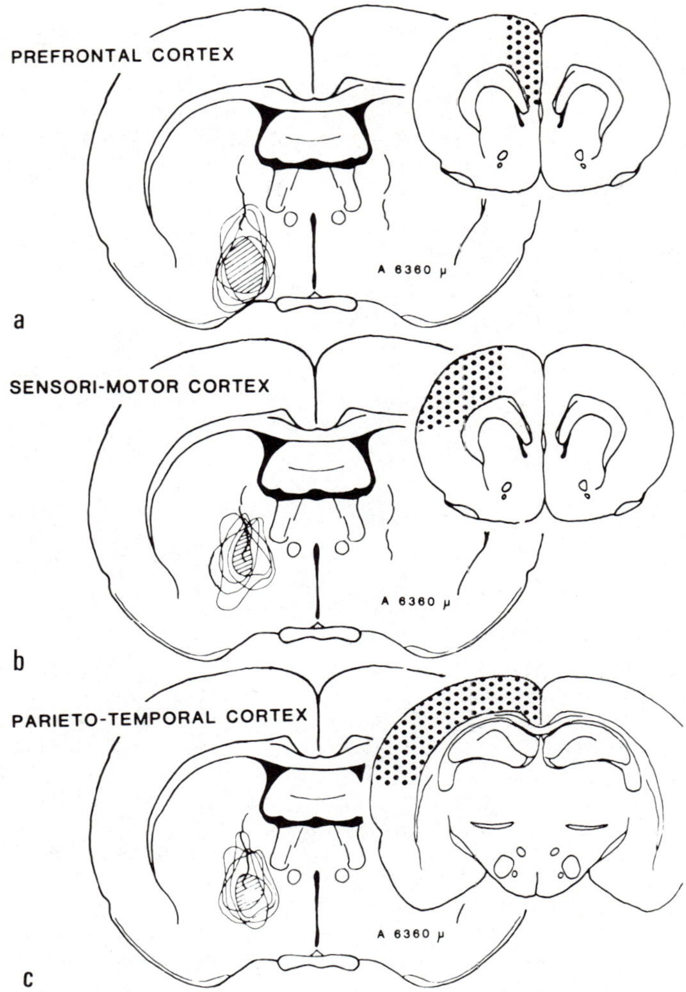

Fig.1. Sites of ibotenic acid lesions that give rise to the greatest decrease of CAT activity in the following cortical areas : a, the antero-medial prefrontal cortex ; b, the sensori-motor cortex and c, the parieto-temporal cortex. (From Mayo et al., 1984, with publisher permission).

Behavioral tests were carried out in rats with bilateral ibotenic acid lesions, of the ventral part of the NBM, i.e. lesions destroying preferentially, as we had previously demonstrated the cholinergic innervation of the prefrontal cortex.

<u>Spontaneous behaviors</u>

As essential step in a first approach to the study of the functional role of a defined anatomical pathway is an analysis of spontaneous behavior. Some changes in learned behaviors can in fact be due to disturbances of spontaneous responses.

<u>Feeding behavior</u> : Post-operatively the lesioned animals displayed a complete aphagia. Forced feeding was given by intra-gastric tubing until total recovery of the pre-operative body weight which was

reached after 7 to 10 days. At this time, while the amount of food eaten was the same in the two groups of rats, their feeding behavior measured in the home cage was quite different (Table I). By placing a paper divided into 64 squares on the tray under the wire mesh of the cage we were able to show that food spillage was more widely spread for experimental animals. Moreover, while the control animals ate virtually all of each food pellet, rats with lesions left many food pellets partially eaten.

Table I : Analysis of various parameters of home cage alimentary behavior of the rat during 24 hours (mean ± SEM)

Groups (n)	Food intake (g)	Partially eaten pellets	Food (g) spillage	Squares with food spillage
Controls (7)	20.0 ± 1.2	3.0 ± 0.7	24.3 ± 3.7	28.0 ± 3.8
Ibotenic (6) acid lesions	21.7 ± 1.1	29.3 ± 3.0 ooo	25.1 ± 2.7	54.5 ± 3.4***

***P < 0.001 Student's t test ; oooP < 0.001 Student's t test/Welch's correction.

Hoarding behavior : Hoarding is an innate behavior which has a considerable survival value for rodents. Rats were deprived of food for a short period and were then placed in their home cage at the corner of an open-field containing 61 food pellets. They were allowed to visit the open-field for 90 minutes after opening their cage. NBM lesions were found to impair hoarding behavior (Table II). In comparison to controls, the animals with lesions hoarded fewer pellets (P < 0.01) even though experimental animals and controls were equally active. The latency to leave the home cage and explore the open-field was the same for both groups (t = 0.38 ; df = 11, n.s.). It could be said that the behavior of rats with lesions was inefficient in the sense that, in comparison to controls, they returned more frequently from the open-field to the home cage without carrying a pellet (P < 0.001). In addition, although not statistically significant, the rats with lesions spent more time in the open-field. This argues against the possibility that the impairment of hoarding behavior was due to an enhanced sensitivity to stress.

Table II : Hoarding behavior (medians)

Groups (n)	Number of pellets hoarded	Time spent (sec.) in open-field	Returns in cage without pellets
Controls (7)	61	230	2
Ibotenic (6) acid lesions	20.5**	618	14.5***

**P < 0.01 ; ***P < 0.001 Mann & Whitney U test.

Locomotor activity : The locomotor activity was measured by recording every time a rat interrupted one of four light beams (located at 45° from each other) in a circular corridor. The rats with lesions exhibited a locomotor hyperactivity measured during 23 h (P<0.001) and were nearly three times more active than controls. During the two first hours of the recording which can be considered as an exploratory period, the behavior of the two groups was also different (Figure 2). Firstly,

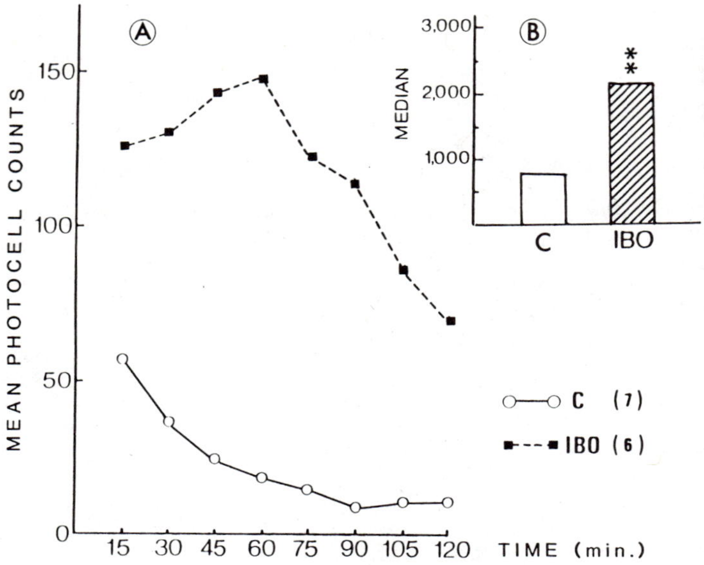

Fig.2 : Locomotor activity during the first two hours of the experiment. A) Time course. B) Medians of the total counts for the first two hours of record. C : Controls. IBO : Experimental group. (**P < 0.01 Mann & Whitney U test).

the animals with lesions were more active than controls over the first 15 minutes ($P<0.001$). Secondly, in control rats the locomotor activity decreased regularly as they habituated to the experimental situation. Conversely, the rats with lesions did not habituate and appeared both hyperactive and hyper-reactive in this test.

Exploratory behavior : A study of exploratory behavior seemed to be a logical extension to these initial results on locomotor activity. The rats were put for 30 minutes in a 8-hole box and the number and duration visits to each hole were recorded. There was no statistical difference between control and experimental groups for total number and duration of visits recorded over the 30 minute period. However, the time courses of the exploratory behavior (Figure 3) for these two groups were different (ANOVA 2, $F(1,11) = 5.33$, $P < 0.05$). In the control rats the number of visits decreased over time i.e. as the novelty diminished (ANOVA 1, time-factor effect $F(2,18) = 4.39$, $P<0.05$). On the other hand, the experimental group did not habituate in this test (Figure 3A). There was no time-factor effect for these rats ($F(2,15) = 0.68$, n.s.). These results supported the data on locomotor activity. Nevertheless the deficit in exploratory behavior cannot be reduced solely to a disturbance in locomotor activity. The time course of the mean duration for one visit (Figure 3 B) differed between control and experimental group. This parameter which is calculated as the mean ratio of the duration to the number of visits is independent of locomotor activity. For the control rats the mean duration for any hole visit increased with time ($F(2,18) = 8.39$, $P < 0.05$) while this parameter remained stable for the experimental animal ($F(2,15) = 0.02$, n.s.).

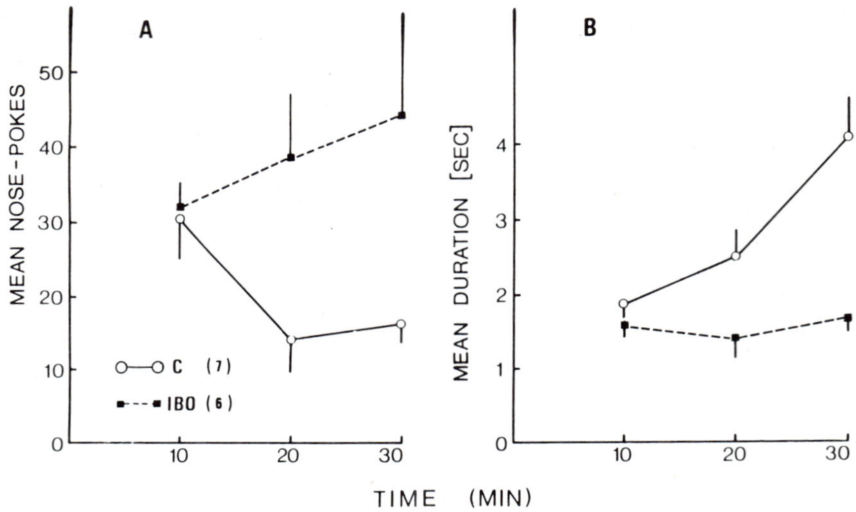

Fig.3. Time course of exploration (mean ± SEM) during 30 minutes in a 8-hole box. A) Number of visits. B) Mean duration for one visit.

Learned behaviors

Animals were submitted to a spatial discrimination task in a T-maze, and a spatial memory task in a radial 8-arm maze.

Spatial discrimation in a T-maze : This experiment lasted two days using food deprived rats. On day 1 the rats were trained in a

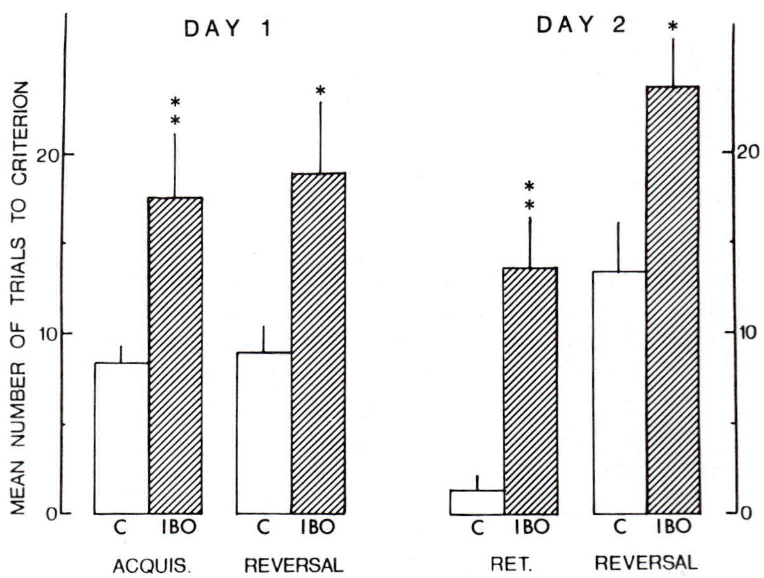

Fig.4. Spatial discrimination and reversal in a T-maze.
*P < 0.05 ; **P < 0.01 Student's t test.

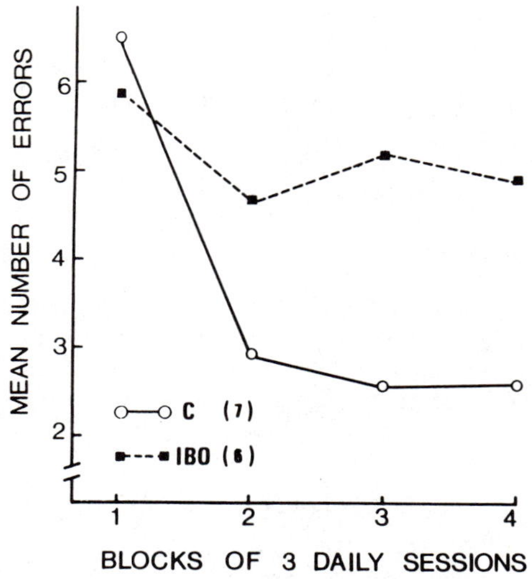

Fig.5. Time course of the number of errors during learning in the radial maze.

spatial discrimination task with food reinforcement placed in the same goal box for each trial. Immediately after reaching a learning criterion of 5 consecutive correct trials, the food was placed in the opposite arm for reversal learning until the same criterion was reached. On day 2 the same procedure was repeated, i.e. rats were submitted to a retention of the reversal habit and then to another reversal. In this task the rats with lesions needed more trials than the control group (Figure 4) to reach the criterion over all sessions : acquisition and reversal on the two days (day 1-acquisition : $t = 3.3$, $df = 11$, $P < 0.01$; reversal : $t = 2.3$, $df = 11$, $P < 0.05$. Day 2-acquisition : $t = 3.9$, $df = 11$, $P < 0.01$; reversal : $t = 2.9$, $df = 11$, $P < 0.05$). it would be worth investigating whether the learning and reversal deficits are just restricted to spatial aspects aspects of the task or whether they represent a more generalized deficit.As a reversal deficit is a classical indication of a perseverative behavior, the experimental rats may have had difficulty in suppressing a previously learned habit.

Spatial memory task in a radial maze : Food deprived rats were placed in the central arena and they were allowed to search for food placed at the extremity of each arm. The experiment was terminated when the rats obtained the 8 rewards. We found that rats with lesions made more retracing errors as defined by a re-entry in a previously visited arm (ANOVA 1, group-effect, $F(1,37) = 8,03$, $P < 0.001$) (Figure 5). The experimental animals did not seem to learn the task since the number of errors did not decrease over trials (ANOVA 1, $F(11,66) = 1.15$, n.s.). This learning deficit in the radial maze which is regarded as a spatial task, supported the preceding results for the T-maze task. This learning deficit could be interpreted in terms of a working memory disturbance along the lines of Olton's hypothesis about the cognitive nature of the radial maze test (Olton and Samuelson, 1976). It should be pointed out that the performance of the two groups of rats were not different in the first three sessions of the experiment.

BIOCHEMICAL EFFECTS OF NBM LESIONS

Regional CAT activity was assayed at the end of the experiments (Table III). There was a significant reduction in CAT in all cortical areas as well as in the amygdala. However, the reduction observed was no more than 50% and could be explained by the presence of intrinsic cholinergic neurons in the cortex and in the amygdala. The greatest depletion was in the prefrontal cortex, which confirmed our previous results. Our ibotenic acid lesions could be considered as selective since both dopaminergic and noradrenergic fibers passing through the injection area and projecting to the prefrontal cortex were not destroyed (DA : -1% ; NA : -14%). Moreover, diffusion of this excitotoxin was relatively limited since we did not observe any change of CAT activity in the striatum or in the cholinergic neuron projection of the septum-diagonal band, i.e. in the ventral tegmental area and hippocampus.

Table III : Regional brain CAT activity (mmoles/h/mg prot) after ibotenic acid lesions (IBO) of the NBM.

Structures	Controls (7)	IBO (6)
Prefrontal Cx	63.7 ± 1.3	43.2 ± 2.1***
Sensori-motor Cx	49.6 ± 2.1	40.7 ± 1.3**
Amygdala	121.3 ± 2.8	84.6 ± 5.6***
Dorsal Hippocampus	80.8 ± 3.8	81.7 ± 4.7
Ventral tegmentum	215.6 ± 25.4	198.7 ± 8.5
Striatum	234.3 ± 17.0	221.8 ± 18.8

**P < 0.01 ; ***P < 0.001 Student's t test

CONCLUSIONS

NBM lesions give rise to a complex behavioral syndrome which includes : locomotor hyperactivity, exploratory deficits, impairment in hoarding behavior, perseveration and memory disturbances. These results can be regarded as a first step in the investigation of the functional role of the NBM and it would be premature to draw definitive conclusions. However, it may be hypothesized that the learning and memory deficits could be the result of an exaggerated response to environmental stimuli. The locomotor hyperactivity, the reduced habituation of the exploratory response and the abnormal pattern of feeding behavior consisting of partial bouts of eating might all be indicative of an increased susceptibility to interfering stimuli.

Perseveration and memory disorders are always observed in the clinical syndrome of SDAT. Moreover, recent work (Davis and Mumford, 1984) has shown that memory disorders in senile dementia cannot be explained by a retrieval deficit, but are suggestive of a processing deficit at the acquisition stage. Although an abnormal sensitivity to interfering stimuli could exist in both senile dementia and in NBM lesioned rats, we would not suggest that these rats represent an animal model for senile dementia. It should be remembered that ibotenic acid destroys cells bodies other than cholinergic systems ; somatostatin has also been found to be depleted (Rossor et al., 1980). Nevertheless, a better knowledge of the role of the NBM would certainly be valuable in furthering our understanding of SDAT.

REFERENCES

Bartus, R.T., Dean, R.L., Beer, B., and Lippa, A.S., 1982, The cholinergic hypothesis of geriatric memory dysfunction, Science, 217 : 408.

Bowen, D.M., Smith, C.B., White, P., and Davidson, A.N., 1976, Neurotransmitter-related enzymes and indices of hypoxia in senile dementia and other abiotrophies, Brain, 99 : 459.

Coyle, J.T., Price, D.L., and Delong, M.R., 1983, Alzheimer's disease : a disorder of cortical cholinergic innervation, Science, 219 : 1184.

Davies, P., and Verth, A.H., 1978, Regional distribution of muscarinic acetylcholine receptor in normal and Alzheimer's type dementia brains, Brain Res., 138 : 395.

Davis, P.E., and Mumford, S.J., 1984, Cued recall and the nature of the memory disorder in dementia, Brit. J. Psychiat., 144 : 383.

Dubois, B., Mayo, W., Agid, Y., Le Moal, M., and Simon, H., 1985, Profound disturbances of spontaneous and learned behaviors following lesions of the nucleus basalis magnocellularis in the rat, Brain Res., in press.

Flicker, C., Dean, R.L., Watkins, D.L., Fisher, S.K., and Bartus, R.T., 1983, Behavioral and neurochemical effects following neurotoxic lesions of a major cholinergic input to the cerebral cortex in the rat, Pharmacol. Biochem. Behav., 18 : 973.

Friedman, E., Lerer, B., and Kuster, J., 1982, Loss of cholinergic neurons in the rat neocortex produces deficits in passive avoidance learning, Pharmacol. Biochem. Behav., 19 : 309.

Johnston, M.V., McKinney, M., and Coyle, J.T., 1979, Evidence for a cholinergic projection to neocortex from neurons in the basal forebrain, Proc. Nat. Acad. Sci., 76 : 5392.

Lo Conte, G., Bartolini, L., Casamenti, F. and Marconcini-Pepeu, G., 1982, Lesions of cholinergic forebrain nuclei : changes in avoidance behavior and scopolamine actions. Pharmacol. Biochem. Behav., 17 : 933.

Mayo, W., Dubois, B., Ploska, A., Javoy-Agid, F., Agid, Y., Le Moal, M., and Simon, H., 1984, Cortical cholinergic projections from the basal forebrain of the rat, with special reference to the prefrontal cortex innervation, Neurosci. Lett., 47 : 149.

Olton, D.S., and Samuelson, R.J., 1976, Remembrance of places passed : spatial memory in rats, J. Exp. Psychol., 2 : 97.

Perry, E.K., Perry, R.H., Blessed, G., and Tomlinson, B.E., 1977, Necropsy evidence of central cholinergic deficits in senile dementia, Lancet, 1 : 189.

Perry, E.K., Tomlinson, B.E., Blessed, G., Bergmann, K., Gibson, P.H., and Perry, R.H., 1978, Correlation of cholinergic abnormalities with senile plaques and mental test scores in senile dementia, Br. Med. J., 2 : 1457.

Pilleri, G., 1966, The Kluver-Bucy syndrome in man. A clinico-anatomical contribution to the function of the medial temporal lobe structures, Psychiat. Neurol., 152 : 65.

Rossor, M.N., Emson, P.C., Mountjoy, C.Q., Roth, M. and Iversen, L.L., 1980, Reduced amounts of immunoreactive somatostatin in the temporal cortex in senile dementia of the Alzheimer type, Neurosci. Lett., 20 : 373.

Rossor, M.N., Svendson, C., Hunt, C.P., Mountjoy, C.Q., Roth, M. and Iversen, L.L. 1982, The substantia innominata in Alzheimer's disease : an histochemical and biochemical study of cholinergic marker enzymes, Neurosci. Lett., 28 : 217.

Salamone, J.D., Beart, P.M., Alpert, J.E., and Iversen, S.D., 1984, Impairment in T-maze reinforced alternation performance following nucleus basalis magnocellularis lesions in rats, Behav. Brain Res., 13 : 63.

Sims, N.R., Bowen, D.M., Allen; S.J., Smith, C.C.T., Neary, D., Thomas, D.J., and Davison, A.N., 1983, Presynaptic cholinergic dysfunction in patients with dementia, J. Neurochem., 40 : 503.

Whitehouse, P.J., Price, D.L., Clark, A.W., Coyle, J.T., and Delong, M.R., 1981, Alzheimer's disease : evidence for selective loss of cholinergic neurons in the nucleus basalis, Ann. Neurol., 10 : 122.

EARLY UNDERNUTRITION AND RECOVERY FROM ACUTE BRAIN DAMAGE LATER IN LIFE

Stanley Finger, Abdelaziz Bouzrara, and Mario Silva

Department of Psychology
Washington University
St. Louis, Missouri

INTRODUCTION

One of the most perplexing problems in the neurological sciences is why lesions that look the same are not always associated with equivalent signs and symptoms of brain damage (Finger and Stein, 1982). In fact, it is not rare to find one person severely debilitated by the same lesion that may have only minimal or transient effects in another individual. In an attempt to understand the basis of this variability, scientists have been working to manipulate and isolate certain factors that they think can interact with the locus of the damage itself. Some potential intervening variables are now much better appreciated than they were even a generation ago, such as age at the time of trauma, speed of lesion growth and experiential history (Finger and Stein, 1982). Others, such as gender differences, are only beginning to be recognized, and without question still additional factors remain to be discovered and elucidated.

In this laboratory, we have been interested in the effects of brain damage and recovery from brain lesions for more than 15 years. The model which has guided our work is one which holds that the response to a brain lesion depends not only upon the structure damaged, but also on the condition of the remaining brain. It is in this context that we have begun to ask whether early dietary conditions can also affect performance after acute brain lesions sustained later in life. We hypothesized that this was quite likely since many investigators have shown that when a pregnant or lactating rat, cat, monkey or human is given inadequate amounts of food, or suboptimal quantities of protein, anatomical, physiological and behavioral abnormalities often can be detected in the offspring. For example, on a structural level, undernutrition during the period of rapid brain growth has been found to affect cellular growth and differentiation, synapse formation and myelination, while on a behavioral level differences in emotional responsivity, exploration and learning functions also tend to be correlated with poor early dietary conditions (Dodge, Prensky and Feigin, 1975; Winick, 1976).

In the series of experiments that we will now describe using the laboratory rat as a model, we will show that early undernutrition can affect the response to brain damage, in some cases by potentiating the symptoms, but in at least one very interesting preparation, by attenuating the expected deficits. It will also be shown that some lesion x nutrition interactive effects may even appear when poor early nutrition does not appreciably affect the performance scores of animals without focal brain lesions. Finally, as with other potential interacting factors, such as age and the speed of lesion growth, the point will be made that these interactive and additive effects do not generalize across all lesions and tasks.

POSTERIOR CORTEX

Our first study employed lesions of the posterior cortex of the rat (Mangold et al., 1981). In this experiment the animals in the underfed group came from dams who received only half of the normal amount of rat chow consumed by control dams from the time of birth through to weaning (Day 21). Both groups then received standard laboratory chow ad libitum. This frequently used model for undernourishing laboratory animals reduces the quantity of milk, but does not affect the fat or protein content of the milk (Crnic and Chase, 1978). Litter size was kept constant at 10 pups per dam and when the animals were about 2 months old they were tested for the ability to make a brightness discrimination using a modified Thompson-Bryant two-choice apparatus. The well-fed and previously underfed animals performed equivalently in this stage of the experiment (means of 28 and 26 trials to criterion).

Upon completion of training the rats received either sham operations or very large lesions of the posterior cortex and were retested on the brightness discrimination. The well-fed and underfed animals with sham operations continued to show perfect or near-perfect retention scores. As expected, the animals with lesions performed significantly worse, but while the well-fed rats with lesions took about as long to relearn the task as they needed originally to learn it, those who had been undernourished prior to posterior cortical lesions were more severely impaired. The relearning means of the well-fed and underfed groups with lesions were 29 and 48 trials to criterion, respectively (Figure 1). An analysis of variance and multiple comparisons between means showed that these differences were statistically significant at at least the .05 level.

That the animals in this experiment were in fact undernourished was confirmed by their approximately 50% body weight deficiencies at the time of weaning (which were largely overcome by the start of testing) and by brain weight differences of about 8% among the sham operated animals upon completion of testing. This initial study thus showed that early undernutrition can affect performance after a later, acute brain lesion even when the learning scores of the underfed animals without focal brain damage are within the normal range.

HIPPOCAMPUS

Because the results of the first experiment were so intriguing, we next set out to see whether these findings would generalize to another lesion and some very different tasks. A decision was made to look at the hippocampus because of the sizeable literature on the behavioral effects of hippocampal damage (Isaacson, 1974; O'Keefe and Nadel, 1978) and the knowledge that the hippocampus is one of the latest developing structures in the central nervous system (Fish and Winick, 1969). The fact that some cells are still migrating to the rat hippocampus 15 days after birth is a feature that would lead one to predict that the hippocampus could be more markedly affected by postnatal undernutrition than most other brain structures.

With this in mind we undernourished rats using the same procedure as had been used in the posterior cortex study--reducing food to the dam by 50% during the period of lactation. The animals then were fed standard laboratory chow ad libitum and received sham operations or large, bilateral electrolytic lesions of the dorsal hippocampus and the pathways between the hippocampus and the septum when they were 70 days of age. The male animals were the subjects for one experiment (Finger and Green, 1983), and the females were used in another experiment (Laughlin et al., 1983) which involved a very different set of tasks.

The males were tested on an increasing series of DRL (differential reinforcement of low rates of responding) schedules in an operant conditioning apparatus. Under a DRL schedule, a lever press response is reinforced only if it occurs after a minimum period of time has elapsed since the occurrence of the last response. The rats were first trained on a DRL 5-sec. schedule and maintained on it for 15 days, after which the DRL value was increased every 15 days in the following sequence: DRL 10, 15, 20, 25, 30, 35, 40, 45, 50, 60, and 75-sec. Analyses of variance were conducted on response rate, responses per reinforcement and efficiency.

As expected, the rats with lesions performed worse than those with sham operations, especially at the higher DRL values. However, for the most part the scores of the well-fed and underfed sham operated animals were comparable, just as had been found in the posterior cortex study. The surprising finding was that although both lesion groups were impaired, the scores of the underfed lesion group were significantly better than those of the well-fed lesion group in efficiency and responses per reinforcement, especially on the higher DRL values (Figure 2).

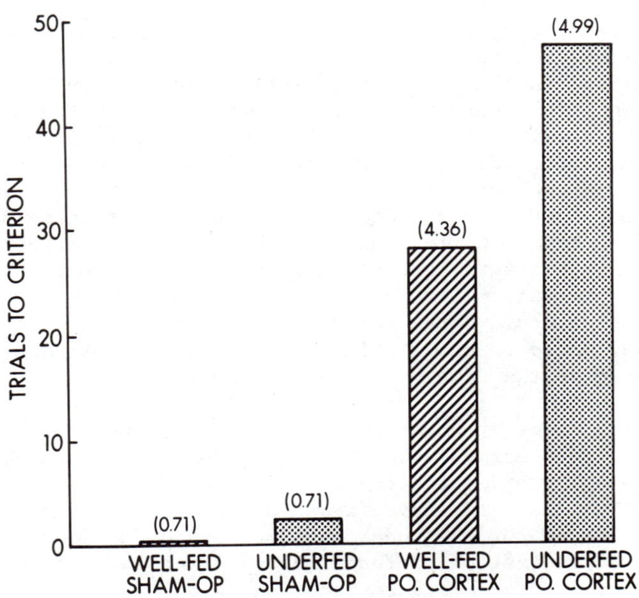

Fig. 1. Mean performance scores of well-fed and underfed rats with posterior cortical lesions or sham ablations on retention of a brightness discrimination. Standard deviations in brackets. Well-fed and underfed groups performed equivalently prior to surgery. (Based on data of Mangold et al., 1981, and presented with the permission of the publisher.)

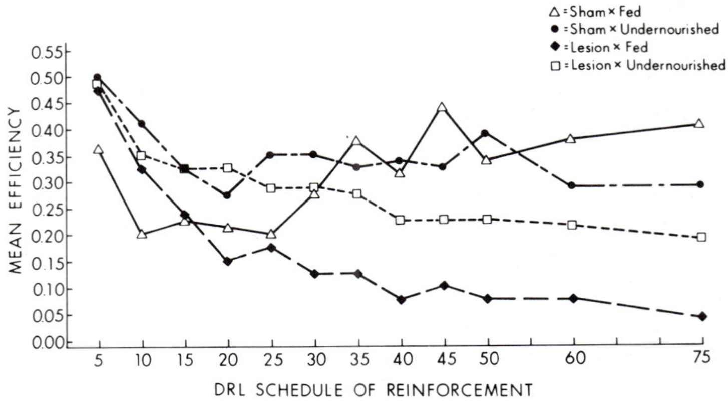

Fig. 2. Mean efficiency values (number of reinforcements obtained divided by number of reinforcements possible) at each DRL schedule value for well-fed and underfed animals with hippocampal lesions or sham operations. The means are based on the medians from the last 5 days on each condition for each rat. (From Finger and Green, 1983, and reproduced with the permission of the publisher.)

Although these results confirmed the finding that early nutritional history can affect the response to a later, acute brain lesion, the effect was in fact in the opposite direction--poor early nutrition attenuated the lesion effect here while it potentiated it in the posterior cortex study. One possible explanation for the hippocampal result might be that the early undernutrition so severely affected the functioning of the late-developing hippocampus that other less-affected, but related brain structures "took over" the critical hippocampal function as sometimes seems to occur after focal lesions early in life. The hippocampi of the underfed control rats were in fact small and pale although all cell layers appeared to be present. If this did happen, the effects of a later hippocampal lesion would not be expected to be as devastating to these animals as they would be to animals showing normal brain development and greater reliance at the time on the hippocampus itself.

In contrast, it might be argued that the opposite would be seen if the areas that would normally be used to compensate for focally-damaged regions were equally or more severely affected by undernutrition than the target area. In this case, the effects of the focal lesion would be expected to be worse for the undernourished subjects since there may not be any early reorganization of function and since they would now have to rely on other areas that are also functioning suboptimally. This may be what happened after the posterior cortical lesions in the cortical study where the underfed animals may have been forced to rely on functionally affected, remaining cortical areas to mediate the relearning process.

Although these ideas are obviously speculative, they should not detract from the important finding that in our first two experiments undernutrition did in fact affect the response to brain damage while having little effect on the animals sustaining control operations.

In our third study, which involved the female animals with hippocampal lesions or sham operations, a third type of finding emerged (Laughlin et al., 1983). In this case there was little evidence for any nutrition x lesion interaction effects. In this experiment the animals were tested for open field behavior, spontaneous alternation, spontaneous exploration, and spatial discrimination and reversal learning. Main effects of nutrition were found in the open field, and on the spatial learning and reversal tasks, and significant lesion effects were found in the open field and in reversal learning, as well as in spontaneous alternation. However, in contrast to the earlier studies, early diet had little effect on the response to later brain damage. That is, even where both diet and lesion effects appeared on the same task (e.g., open field), they tended to affect different performance measures.

Although the tasks used with the female rats did not have the same cognitive-loading as the DRL tasks used in the aforementioned study, these data are important because they show that the finding that early diet can affect the behavioral response to focal brain damage later in life may not generalize across all experimental conditions. In this sense, the role of early undernutrition may be very much like that of age at the time of trauma, experiential history or speed of lesion growth in that all of these variables seem to be important in some but not all situations (Finger and Stein, 1982).

FRONTAL CORTEX

In our latest experiments we have examined the effects of lesions of the frontal cortex anterior to the motor zones. These studies, however, differ from those described above in more ways than just the locus of the lesion. In recent experiments we have begun to look at diets limited only in protein. In these experiments female rats were placed on low-protein diets 5 weeks prior to mating and remained on the diets until their pups were removed 3-5 weeks after birth. The pups, who were raised in litters of 8, were allowed free access to a normal protein diet at the time of weaning or soon after. These procedures resulted in lower body weights at the time of weaning and surgery; indications that the rats were in fact undernourished.

The first of these studies (Silva et al., 1984) involved 8 and 25% isocaloric protein diets; 8% being considered restrictive, and 25% being within the normal range for laboratory rats. Rats raised on 6% isocaloric protein diets (severely undernourished) were used in a second study (Bouzrara et al., in press). These rats were operated upon and tested under conditions identical to those used by Silva et al. (1984). Hence, we will describe the results of both experiments together.

When the animals were 3 months old they were given bilateral lesions of the frontal cortex or control operations and were enucleated to prevent them from using visual cues while testing. After a 10-day recovery period they were tested in a T-maze for the ability to make a simple two-choice discrimination--in this case distinguishing between a rough and a smooth floor. After mastering the rough-smooth discrimination the valances of the surfaces were switched such that an animal previously rewarded on the rough surface now had to choose the smooth surface to obtain a food reward. Two more reversals of the rewarded cue followed this manipulation.

Figure 3 shows the results obtained by Silva et al. (1984) for original learning using animals in the 8% and 25% protein conditions. As can be seen, the underfed animals with frontal cortical lesions had the worst scores, while those without lesions and with a good nutritional history performed the best. An ANOVA on these data revealed a significant effect of the lesion and of nutrition, but no interaction between these two factors. The later testing of the 6% animals (Bouzrara et al., in press; not shown in Fig. 3) showed that they had the poorest scores (mean = 36 errors to criterion) of the 3 groups without frontal cortical lesions. Among the groups with acute lesions, however, the 6% rats performed only as poorly as the 8% rats (mean = 42 errors to criterion) in original learning.

The results on the series of three reversals also showed that there was a main effect of nutrition, and one of lesion, but no interactions between these factors. Here too the effects of nutrition and brain lesion were found to be additive, with the worst scores coming from those animals with a history of severe undernutrition (i.e., 6% undernutrition) plus acute brain lesions (see Bouzrara et al., in press).

CONCLUSIONS

In summary, we have now conducted a number of studies on the role that early nutrition might play in affecting the behavioral response to later, focal brain damage. The results of some of these experiments show very clearly that a poor early diet has the potential to modify the effects of acute brain lesions sustained at maturity. In two cases, both involving cerebral cortical areas, the net result of undernutrition was to further impair performance. In one experiment which involved the hippocampus, however, undernutrition during the period of rapid brain growth actually attenuated the deficit that appeared among brain damaged animals that were tested on DRL schedules with long delays. We believe that these findings are even more significant because in some of

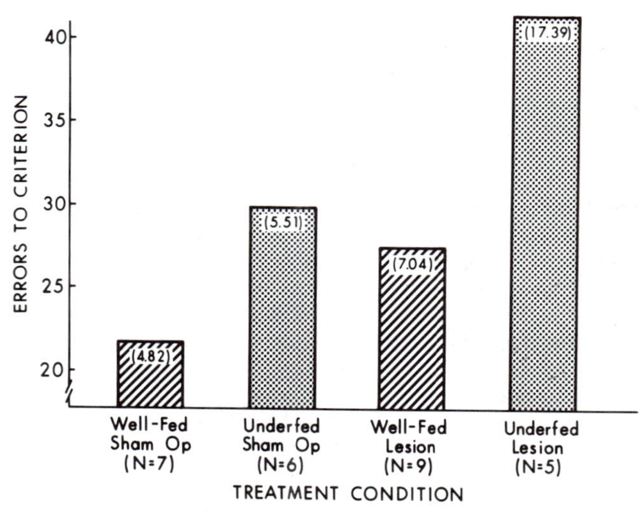

Fig. 3. Mean performance scores of well-fed and early underfed rats (8% protein diets) with frontal cortex lesions or sham ablations on acquisition of a simple tactile discrimination. Standard deviations appear in brackets. (From Silva et al., 1984, and reproduced with permission of the publisher.)

our studies early undernutrition had only minimal behavioral effects in the absence of focal brain damage. These results thus might help to explain why two individuals who appear to be quite healthy prior to acute head injuries or strokes may show very different signs and symptoms of brain damage afterward.

While the implications of these data are intriguing, it should still be recognized that early undernutrition is not the only factor that can affect the response to a brain injury. As we have noted, developmental status of the nervous system, whether the lesion is evolving rapidly or slowly, and other aspects of the experiential history of the subject can also affect performance after acute brain lesions. Controlled laboratory studies on these other phenomena have shown that none is capable of generalizing across all organismic, task and lesion conditions (Finger and Stein, 1982). For example, staged ("serial") lesions of the hypothalamus seem to be associated with excellent sparing of function when rats are required to swim in water close to body temperature; partial recovery of activity in an open field; and no better recovery than that seen among the animals with one-stage bilateral lesions in body temperature regulation (Finger et al., 1981). Our present findings with nutrition are comparable in terms of specificity as evidenced by the fact that nutrition x lesion interactive and additive effects were found on the DRL battery but not on certain other tasks following hippocampal damage, and in that early undernutrition attenuated the hippocampal lesion effect on the DRL tasks, but potentiated it on other measures following frontal cortical ablations. Nevertheless, since only these few studies have looked at the relationship between nutrition and the response to later brain damage, the generality and limitations of these effects still remain to be adequately determined.

Finally, we want to state that we are aware that our efforts will probably raise more questions than they answer about nutrition, and about brain damage and behavior. Hopefully, these first few experiments will stimulate other types of studies, such as research on the roles that specific diets might play in the response to brain damage in organisms past the period of rapid brain growth. For instance, to the best of our knowledge, no one has asked how a diet high or low in serotonin might affect the response to an incomplete lesion of a serotonergic structure. Nor has there been any work on "super-enriched" diets and recovery from brain damage. Future experimentation might also deal with different quantities of specific amino acids in the diets of pregnant or lactating organisms, or less-severe dietary deficiencies which may be thought of as more appropriate models for "subclinical malnutrition."

In summary, early dietary conditions should be thought of as but one newly-recognized factor that can affect performance after a brain lesion. In this context, one of the most important tasks still facing scientists is to try to find, analyze and attempt to understand still other factors that may also be capable of accounting for the variability that is so characteristic of clinical populations.

ACKNOWLEDGEMENTS

This research and the preparation of this presentation were supported by Grant BRSG SO7 RR07054-17 awarded by the Biomedical Research Support Grant Program, Division of Research Resources, National Institutes of Health, and by a Faculty Grant from Washington University, both to the first author. The second author was supported by the Scientific Mission of Tunisia, and the third author by National Institute of Aging Grant AG 00030-07. The authors would like to thank Peter Morgane and the Worcester Foundation for Experimental Biology (Grant NICHD HD-06364) for providing the animals and special diets for some of these studies, and C. Robert Almli for his assistance on those projects.

REFERENCES

Bouzrara, A., Silva, M., Waksman, D., Finger, S., and Almli, C. R., in press, Relation between the severity of early malnutrition and the effects of later frontal cortical lesions in rats. Physiol. Psychol.

Crnic, L. C., and Chase, H. P., 1978, Models of infantile undernutrition in rats: effects on milk, J. Nutr., 108:1755.

Dodge, P. R., Prensky, A. L., and Feigen, R. D., 1975, "Nutrition and the Developing Nervous System," Mosby, St. Louis.

Finger, S., and Green, L., 1983, Early undernutrition and later hippocampal damage: I. DRL performance in rats, Nutr. Beh., 1:195.

Finger, S., Gruenthal, M., and Bell, J., 1981, Some perspectives on the serial lesion effect, in: "Functional Recovery from Brain Damage," M. W. van Hof & G. Mohn, eds, Elsevier, Amsterdam.

Finger, S., and Stein, D. G., 1982, "Brain Damage and Recovery of Function: Research and Clinical Perspectives," Academic Press, New York.

Fish, I., and Winick, M., 1969, Cellular growth in various regions of the developing rat brain, Ped. Res., 3:407.

Isaacson, R. L., 1974, "The Limbic System," Plenum, New York.

Laughlin, N. K., Finger, S., and Bell, J., 1983, Early undernutrition and later hippocampal damage: Open field, spontaneous alternation and exploration and spatial reversal learning measures, Physiol. Psych., 11:269.

Mangold, R. F., Bell, J., Gruenthal, M., and Finger, S., 1981, Undernutrition and recovery from brain damage: a preliminary investigation, Brain Res., 230:406.

O'Keefe, J., and Nadel, L., 1978, "The Hippocampus as a Cognitive Map," Clarendon Press, Oxford.

Silva, M., Bouzrara, A., Finger, S., and Almli, C. R., 1984, Effects of early protein undernutrition and later frontal cortex damage on habit acquisition and reversal learning in the rat, Physiol. Psych., in press.

Winick, M., 1976, "Malnutrition and Brain Development," Oxford University Press, New York.

BEHAVIOURAL EFFECTS OF PREOPERATIVE AND POSTOPERATIVE

DIFFERENTIAL HOUSING IN RATS WITH BRAIN LESIONS: A REVIEW

John C. Dalrymple-Alford and Christian R. Kelche

Laboratoire de Neurobiologie des Comportements
7 rue de l'Université
67000 Strasbourg, France

INTRODUCTION

It is now widely recognized that the specific behavioural outcome of experimental brain lesions depends on much more than the size and locus of the damage (Finger and Stein, 1982; and see Finger et al., Jarrard, and Toniolo et al., this volume). This chapter reviews studies that examined the effects of the quality of the preoperative or postoperative housing environment (differential housing) on behavioural function in rats given experimental brain lesions. Some aspects of this research have been described previously (Greenough et al., 1976; Finger, 1978a; Will, 1981; Will and Eclancher, 1984). In the present article, we describe and evaluate the progress that has been made in this field, emphasising the research on learning and memory in rats with experimental lesions to septal and hippocampal pathways.

Differential housing is operationally defined as the comparison of "enriched," "grouped" and "individual" housing conditions. In our case, the enriched condition houses 12 rats in a large double cage and includes a daily change of about 10-15 "junk" objects (e.g., Kelche and Will, 1978). In the grouped condition, sometimes labelled as a "social" or "standard" condition, simply no objects are placed in the cage which may otherwise be the same as the enriched environment or a smaller cage that contains fewer (usually 3) rats, depending on the experimental comparisons being made. In the individually housed condition, often referred to as an "impoverished" condition or as "social isolation," a rat is housed alone in a small cage without any objects, although the effects of hippocampal lesions in isolated rats placed in a novel individual cage either with or without objects have also been studied (Will et al., in preparation; see also, Wells et al., 1972). The onset and duration of differential housing have varied in different studies, as have the age at operation, the age at testing, the locus of the lesions and the behavioural tasks examined. In the subsequent sections we address the importance of these variables, taking into account some of our recent findings.

There are three major reasons for investigating the influence of differential housing in brain-damaged subjects. As alluded to at the beginning of this article, the effects of differential housing provide us with information on one important factor or, perhaps more accurately, one group of factors that influence the effects of brain dysfunction induced by experimental brain lesions. Indeed the major impetus for studying such effects has been that these studies generate some basic information on the value and in a notional sense the clinical value of the environment as a therapy for brain dysfunction (Greenough et al., 1976). It is often the case that animals are placed in social isolation after surgery and throughout behavioural testing and thus, especially in studies in which the sparing or recovery of function is the major interest, suboptimal conditions for promoting improved behavioural function may have been employed. Rosenzweig (1980) has suggested that differential housing provides a useful animal model to study factors that may promote rehabilitation in humans with brain damage.

Secondly, the findings produced by studies on differential housing in subjects with brain lesions are pertinent to the more general brain lesion literature because there is evidently a wide variation in housing conditions routinely employed in different laboratories (Greenberg and Burdsall, 1982). Thus the findings of one laboratory might not always generalize to those of another and it may be the case that the use of different housing conditions (as with other variables that influence the impact of or the recovery from a brain lesion) explains in part some of the discrepancies prevalent in the literature.

Thirdly, work in this field may reveal information on the neural representations of housing effects in intact animals (Einon et al., 1980; Will et al., 1981; Dalrymple-Alford and Benton, 1984a) which have been of interest to many workers studying a wide range of behavioural measures (Dalrymple-Alford, 1983). Stated simply, if a group of animals given damage to a neural structure or pathway is found to be resistant to the effects of differential housing that are normally apparent in intact animals, then these findings would suggest that the damaged structures or pathways and/or those related structures altered by that damage are, if not sufficient, at least necessary neural elements for the elicitation of such effects. Conversely, when clear environmental effects are present in animals previously subjected to brain damage then these findings would constitute strong evidence that the damaged pathways are not necessary for the presence of these housing effects, although they may be sufficient in the intact animal. Similar arguments can be elaborated for the effects of preoperative environments, although many workers feel that the effects of preoperative and postoperative housing may be different (e.g., Greenough et al., 1976).

To reiterate a general assertion evident in much of the literature, it is considered likely that many behavioural deficits produced by brain damage can be reduced or (seemingly) even absent in animals housed in a socially and physically enriched environment compared with deficits in animals housed in a relatively impoverished one. Some of the evidence to support this belief was provided by housing studies in which a "general" brain dysfunction was present due to a genetic abnormality (e.g., Bouchon and Will, 1982, 1983) or in which a dysfunction was produced in rats by inducing an experimental hypothyroid (Davenport, 1976) or hyperthyroid condition (Sjödén and Söderberg, 1976; but see Sjödén, 1976, and Sjödén and Lindqvist, 1978 for evidence of converse effects), by prenatal or infantile malnutrition (e.g., Tanabe, 1972; Swanson et al., 1983; but see Tachibana, 1979, and Osborne et al., 1980), by lead poisoning (Petit and Alfano, 1979), or by X-irradiation (Shibagaki et al., 1981). Nonetheless, a major source of such support were studies on the effects of postoperative differential housing in rats with experimental brain lesions.

TABLE 1. Effects of postoperative environments in brain-damaged rats : Early studies on Hebb-Williams maze learning

Authors	Strain	Lesion locus	Age at lesion	Period of differential housing	Lesion effect E.C rats	Lesion effect I.C rats	Environment effect control rats	Environment effect lesioned rats
SMITH 1959	Long Evans	anterior temporal cortex	20-25 or 95-100	18-90	–	–	+	+
		posterior temporal cortex	20-25 or 95-100		–	O	+	+
SCHWARTZ 1964	Hooded rats	posterior cortex	1	6-50	–	–	+	+
WILL and ROSENZWEIG 1976	Fischer	occipital cortex	120	124-184	–	–	+	+
WILL et al. 1976	Berkeley S1	♂ large lesions occipital cortex	1	5-65	–	–	+	+
		♂ small lesions occipital cortex		22-65	–	–	+	+
		♀ small lesions occipital cortex		22-65	–	–	O	+
WILL et al. 1977a	Fisher	large lesions occipital cortex	31	35-95	–	–	+	+
	Berkeley S1	small lesions occipital cortex			–	–	O	+
WILL et al. 1977b	Fisher	occipital cortex (restricted lesions)	31	35-95 (2h/day)	O	-?	+?	+
	Fisher		31	35-65 (2h/day)	O	-?	+?	+
KELCHE and WILL 1978	August	dorsal hippocampus	120	125-185	–	–	O	+

E.C., enriched ; I.C., impoverished ; +, facilitation or increase ; –, impairment or reduction ; O, no difference ; ?, unclear effects.

EFFECTS OF POSTOPERATIVE HOUSING

Early Studies

The impressive evidence produced by early studies on the effects of postoperative differential housing in sham-operated rats and in rats given brain lesions is summarised in Table 1. All the studies listed in Table 1 examined Hebb-Williams maze learning and found that the lesion-induced impairment was reduced in rats housed postoperatively in an enriched environment compared with the deficits found in isolated (Will and Rosenzweig, 1976; Will et al., 1976, 1977a, 1977b) or in grouped counterparts (Smith, 1959; Schwartz, 1964). However, Bland and Cooper (1969, 1970) found that enriched as opposed to standard (group) housing was ineffective in improving the deficits in visual pattern and brightness discrimination learning in rats subjected to visual cortex lesions as infants or as adults.

The second feature in Table 1 is that with the exception of one study the early positive results were obtained from rats given lesions in the occipital cortex (see also Rosenzweig, 1980). After Smith (1959), who reported only the combined effects of pre- and postoperative housing, and following Schwartz's (1964) lead, Will and Rosenzweig employed occipital cortex lesions in a series of studies that have received wide recognition.

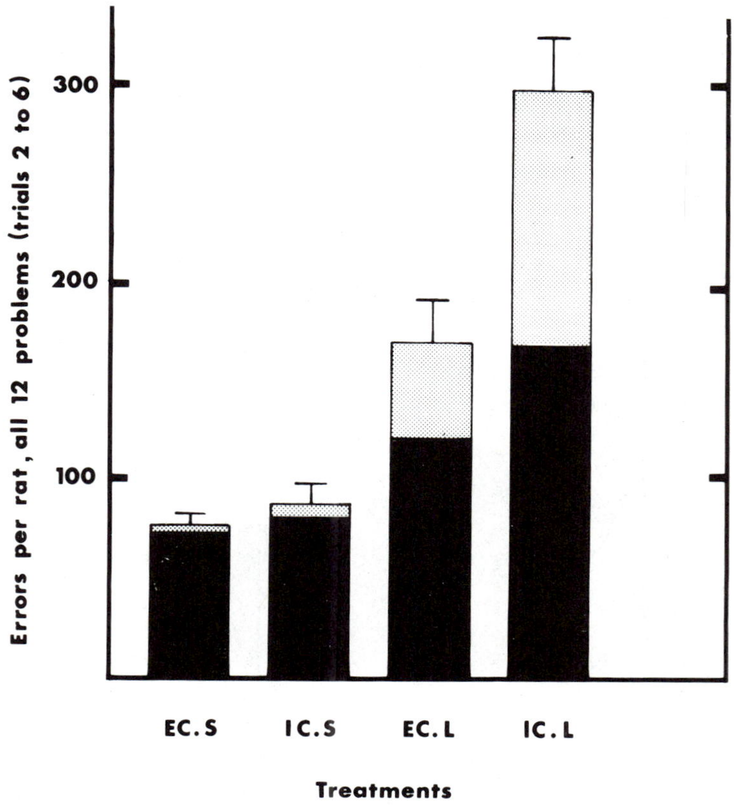

Fig. 1. Mean errors in the Hebb-Williams maze made by sham-operated rats (NL) and rats with dorsal hippocampus lesions (L) housed postoperatively in enriched (E) or isolated conditions (I). The lower hatched part and the upper shaded part of the histograms represent, respectively, the contribution of "initial" and "repetitive" errors to the total. Data taken from Kelche and Will (1978).

The occipital cortex lesions in most of Will and Rosenzweig's studies always included some damage to the underlying hippocampus (usually unilaterally) or were accompanied by a hernia of the hippocampal formation. In such studies, more reliable environmental effects were found with more restricted occipital cortex lesions (Will et al., 1976, 1977a). The study in which the hippocampus was found to be undamaged also revealed a greater impairment in isolates with occipital cortex lesions (Will et al., 1977b). While the overall effects of environment were significant in the latter study, the lesion effects were less pronounced than those previously found in studies in which some dysfunction of the hippocampus could be assumed to have occurred.

The obvious next step was taken by Kelche and Will (1978) to see whether a recovery or sparing of function might be modified by postoperative differential housing in rats given lesions restricted to the dorsal hippocampus. The findings in that study were encouragingly clear in that the severe deficits in Hebb-Williams maze learning produced by lesions were considerably smaller in enriched than in isolated rats (Fig. 1). It is impossible to evaluate with any certainty the influence of hippocampal damage on the effects reported by Will and collaborators in their studies on occipital cortex lesions. What does seem warranted is that the maze performance of rats with occipital cortex or hippocampal damage is influenced by postoperative differential housing (but see below). On the basis of the evidence summarised thus far, there is support for the generality of such housing effects in that the lesions were made at several ages, differential housing was imposed at different ages and for various periods either immediately or at some later time after surgery, and the effects were present in both male and female rats and in different strains.

Hippocampal System Lesions

The impact of Kelche and Will's study was to shift their emphasis of research to examine the effects found after damage to limbic system structures, the hippocampal formation and its anatomically related structures in particular. The incentive for examining these structures is that the hippocampal system has proven to be a rewarding model for an impressive array of plastic events including lesion-induced changes such as collateral and terminal sprouting and reactive synaptogenesis (Cotman and Nieto-Sampedro, 1982; Frotscher, 1983) and phenomena such as long-term potentiation (McNaughton, 1983; Buzsaki, and Laroche this volume) that may be related to memory processes. It should also be recalled that while the neuronal and glial changes induced by differential housing in intact animals occur throughout the brain, especially the cerebral cortex (e.g., Rosenzweig and Bennett, 1977), some admittedly mixed evidence is available that changes also occur in the hippocampal formation (Walsh et al., 1971; Rosenzweig and Bennett, 1978; Susser and Wallace, 1982; Jones and Smith, 1983; Katz and Davies, 1983). It is especially interesting that the dendritic complexity of dentate gyrus granule cells, which receive a major input via the perforant path from the entorhinal cortex, is highly sensitive to enriched environments in female rats, though apparently not in males (Juraska et al., 1983).

Recent published studies on the effects of hippocampal system lesions used rats operated on at 30-33 days of age (Table 2) primarily to avoid interference with the late development of the dentate gyrus (Altman and Bayer, 1975) and the proposed corresponding delay in behavioural maturation (Douglas, 1975). Inspection of Table 2 reveals that the attenuation or smaller magnitude of the deficits produced by dorsal hippocampus lesions in enriched as opposed to isolated rats is not restricted to the Hebb-Williams maze task. Thus enriched rats with hippocampal lesions showed a reduced impairment of reponse perseveration during extinction of a shuttle-box task (Will and Kelche, 1979) and a reduced impairment of radial-arm maze performance (Einon et al., 1980). Moreover, enriched rats with hippocampal

TABLE 2. Effects of postoperative environments in rats with hippocampal system lesions

Authors	Strain	Lesion locus	Age at lesion	Period of differential housing	Test used	Lesion effect E.C. rats	Lesion effect I.C. rats	Environment effect control rats	Environment effect lesioned rats
WILL and KELCHE 1979	August	dorsal hippocampus	31	34-65	shuttle box .acquisition	+	+	O	O
				34-95	.extinction	-	-	+	+
EINON et al. 1980	Long Evans	dorsal hippocampus	30	36-65	radial maze test	-	-	+	+
					response transfer	-	-	O	O
WILL et al. 1981	Hooded rats ♀	dorsal hippocampus	30	33-40	spontaneous alternation	-	-	O	+
				33-48		-	-	O	+
				33-56		-	-	O	+
WILL et al. 1981	Berkeley S1	entorhinal cortex	31	36-72 and 36-96	spontaneous alternation	-	-	O	O
						-	-	O	O
				36-96	Hebb-Williams maze test	-	-	+?	O

E.C., enriched ; I.C., impoverished ; +, facilitation or increase ; -, impairment or reduction ; O, no difference ; ?, unclear effects.

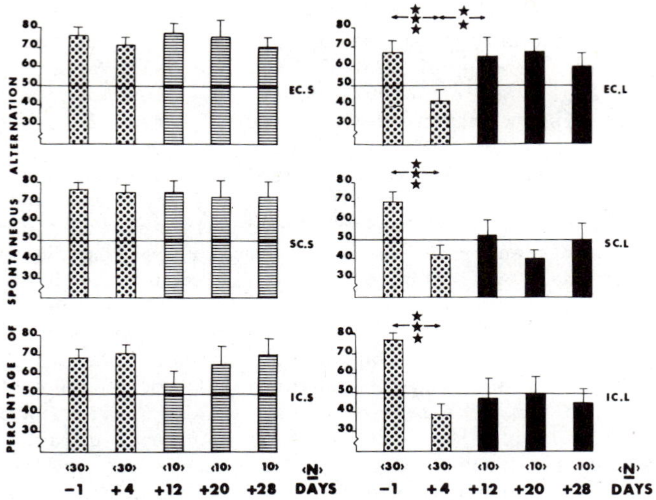

Fig. 2. Mean alternation before (-1) and after (+4, and 12, 20, or 28 days) surgery in control rats (S; left side) and in rats with dorsal hippocampus lesions (L; right side) housed postoperatively in enriched (EC; top row), social (SC; middle row) or isolated conditions (IC; bottom row). ** $P<0.01$; *** $P<0.001$. From Will et al., (1983), with permission.

lesions showed a complete recovery from a postoperative deficit in spontaneous alternation (Will et al., 1983), an instance that bears merit not only because the design permits, at least in operational terms, the use of the word "recovery" but also because the effects were produced even with brief periods of differential housing (Fig. 2). However, a lesion-induced impairment in response transfer was not affected by housing conditions (Einon et al.,1980). Also, enriched and isolated rats with dorsal hippocampal lesions did not markedly differ in their acquisition of a two-way active avoidance task, whereas grouped counterparts showed comparatively facilitated acquisition (Will and Kelche, 1979).

The aforementioned studies give more support for the generality of the phenomena across different lesion loci as well as different tasks. However, impairments produced by lesions of the entorhinal cortex were found to be the same in enriched and isolated rats even though two tasks were used (Hebb-Williams maze learning and spontaneous alternation) that are sensitive to the effects of housing in rats given dorsal hippocampus damage (Will et al., 1981; Table 2). Another negative result was that the effects of fimbria-fornix lesions were recently found to be also resistant to the influence of differential housing, using tests of spontaneous alternation and radial-arm maze performance (Kelche and Will, in preparation). It seems that the picture observed after damage to hippocampal connections is different from that found after damage to the hippocampus itself, although the latter lesion presumably involved less severe damage to extrinsic connections of the hippocampal formation and its anatomically related structures. To the extent that the fimbria-fornix, the hippocampal formation and the entorhinal cortex are regarded as an integrated system with an assumed common function (e.g., Numan, 1978; Gray, 1982; Olton et al., 1982; but see Jarrard, this volume, and Buzsaki et al., 1980, 1982), differences between the influence of differential housing on the apparently similar effects produced by lesions to these structures are surprising.

One explanation of these differences derives from a viewpoint of hippocampal system function like the one proposed by Olton et al. (1982). Olton contends that a dysfunction of the hippocampal system, whatever the "true" nature of its function, occurs only when a lesion or a set of lesions produces a neural disconnection of the elements integral to the preservation of its normal function. From this viewpoint, the positive effects obtained with differential housing may somehow depend on the existence of the two major extrinsic connections of the hippocampal formation. The assumption is that the dorsal hippocampus lesions produced only a partial disruption of the hippocampally-transmitted (-modified) information passing between the cortex and subcortex and that sufficient pathways remained throughout the system on which, directly or indirectly, the modifications by environmental experience depend. These assumptions lead to the expectation that partial rather than more complete disruption of the hippocampal formation should permit at least greater if not the only sparing or recovery of function on tasks such as maze learning and alternation. We recently tested this prediction in two studies, one of which employed different extents of hippocampal damage, and the other a disconnection of the hippocampus which left the structure itself almost undamaged (full descriptions of the following results are in preparation).

The effects of differential housing on radial-arm maze performance were examined in 90-day old rats given dorsal, ventral or a combination of dorsal and ventral electrolytic lesions of the hippocampal formation. However, disconfirming the hypothesis under investigation, only an overall effect of enriched housing (fewer errors) with no significant interaction between housing status and brain state was found. Moreover, planned comparisons revealed that the effects of housing were reliable only in the subgroups given the combination of dorsal and ventral lesions and were negligible in the remaining comparisons (a subsequent test with a modified Krech maze by Dr. C. Pacteau did reveal a greater deficit in isolated as opposed to enriched rats with dorsal hippocampus lesions; the two other lesion subgroups were not tested).

The neural disconnection of the hippocampal formation was achieved using the method of making a combination of a lesion of the left fimbria-fornix and the hippocampal commissures with a lesion of the right entorhinal cortex. One possible benefit of using such a combination of unilateral lesions is that the subsequent impairments are likely to be long-term in contrast to the transient effects of each unilateral lesion placed alone (Olton et al., 1982) and that such a lesion model may therefore be useful in the study of neuronal plasticity that may be correlated with environmental effects in animals with brain damage. This study also differed from earlier ones in that the rats were given hippocampal lesions at weaning (24 days of age) prior to placement in the environments 1-2 days later. It has been established that many though not all the deficits induced by social isolation occur only when this condition is imposed at this age (Rosenzweig, 1971; Einon and Morgan, 1977; Einon, 1980; Dalrymple-Alford and Benton, 1984b). Some behavioural, e.g., maze learning, and anatomical changes, e.g., occipital cortex, that can be attributed to the use of enriched conditions are not dependent on such early treatment (Rosenzweig, 1971) but are none the less more pronounced when enrichment is imposed early rather than later in life (Forgays and Read, 1962; Uphouse, 1980; but see Katz and Davies, 1983, and Walsh, 1981).

We compared the effects of the environment on the deficits in open field behaviour, spontaneous alternation and radial-arm maze performance produced by the hippocampal disconnection with those produced by bilateral lesions of the dorsal hippocampus. In terms of open field ambulation, the increase in activity produced to an equal extent by both lesions was greater in isolated than in enriched rats. Thus an effect of postoperative enrichment was evident despite a severe disruption of the hippocampally-mediated flow of neural information from cortex to subcortex that was produced with minimal damage to the hippocampus itself. Such a housing effect on the activity of rats with lesions may have arisen from multiple causes and have multiple neural representations including perhaps the exchange of information still possible in remaining septo-hippocampal pathways (in addition, some ventral aspects of the fimbria remained intact in many animals in this study). When the animals were tested for spontaneous alternation and for performance in the radial maze, the disconnection study revealed no housing-induced changes in the impaired performance of rats with either type of hippocampal dysfunction used, findings that conflict with those of previous studies (Table 2) in which dorsal hippocampus lesions and differential housing were imposed at 30 or more days of age. Even more recently, another study using ibotenic acid lesions of the dorsal hippocampus made in 25 day old animals also failed to find reduced deficits in the Olton maze in enriched rats (Einon, personal communication). However, when our rats were subsequently tested in the radial-arm maze with only 5 of the 8 arms baited at the beginning of each trial, isolates with lesions immediately committed a greater number of errors in both the baited and unbaited arms than did enriched counterparts. Our data indicated that these latter effects were probably a response to the changed reinforcement contingency as they were negligible by the end of 8 sessions. As with the open field test, the environment effects in the baited/unbaited task were not restricted to the subgroups with dorsal hippocampus damage.

Other Limbic Lesions

The fact that septal lesions too can be influenced by the postoperative housing condition received early support from the finding that the hyper-reactivity syndrome produced by large lesions dissipated more slowly in adult rats that were subsequently isolated as opposed to group-housed (Ahmad and Harvey, 1968). More recently, it has been reported that post-weaning isolation considerably enhanced social interaction ("gregariousness") not only in intact rats but also in rats given large septal lesions at 7 days of age (Johnson et al., 1984). We examined the effects of more restricted septal lesions made at weaning on the series of tasks used in the hippocampal disconnection study described in the previous section. Preliminary analysis of these data unexpectedly revealed that enriched rats with septal lesions tended to be more active in an open field and tended to make more errors in the standard radial maze task than did isolated counterparts. The only "benefit" that enrichment produced was to increase within-trial habituation of activity, while the deficits in spontaneous alternation were unchanged.

Eclancher and Karli (1979a, 1979b) reported a greater increase in ambulation and rearing in an open field in isolation-reared as opposed to group-reared rats that received large septal lesions at 7 days of age, and the inverse effect of environment on the increase in these parameters produced by amygdala lesions made at the same age. These workers also reported that the facilitated acquisition of shuttle-box avoidance produced by early septal lesions was greater in rats reared in postweaning isolation compared with group-reared rats (Eclancher and Karli, 1981b), whereas large amygdala lesions at the same age did not change acquisition of this task (except to reduce response latencies in isolates) and lesions of the centromedial amygdala produced impaired acquisition in group-reared rats but facilitated acquisition in isolation-reared rats (Eclancher and Karli, 1980). However, the increased incidence of muricide in rats with such lesions and in rats with ventromedial hypothalamic lesions was unaffected by differential rearing, whereas the increase in muricide produced by olfactory bulbectomy and dorsomedial thalamus lesions was augmented by isolation (Eclancher and Karli, 1981a). Hallonquist and Brandes' (1981) data give another example of the fact that housing may influence some but not all the changes induced by a brain lesion. These workers found that the persistent hyperactivity present in adult rats with ventromedial hypothalamic lesions was considerably greater in isolated as opposed to grouped rats, but housing had no effect on the lesion-induced obesity.

The research on the effects of preoperative differential housing differs from that on postoperative housing in that fewer studies have been made by a relatively greater number of research groups using a relatively wider variety of lesion loci and behavioural tasks in the former case. As a result, it is more difficult to integrate the data and, while several interesting reports are available, the overall evidence that the more enriched preoperative conditions improve behavioural function in rats with brain damage is not yet convincing.

There are two seemingly conflicting reports on the effects of preoperative housing in rats with hippocampal lesions. The first, an often cited study by Hughes (1965), produced data that were comparable to many findings in postoperative studies in that rats placed in enriched conditions from 33 to 66 days of age, as opposed to grouped or isolated, showed a reduction in the considerably impaired Hebb-Williams maze performance produced by dorsal hippocampus lesions. Furthermore, the less severe deficit produced by posteroventral hippocampus lesions was greater in isolated rats than in grouped rats housed in standard or enriched conditions. Hughes also found that lesions restricted to the occipital cortex above the hippocampus produced a deficit in the isolated rats only. In the second study, in which differential housing lasted 90 days post-weaning, lesions of the dorsal hippocampus produced behavioural changes in the socially-reared rats that were not evident in isolation-reared rats, including an increase in ambulation and object contact in an open field, poorer acquisition and retention of footshock-motivated passive avoidance in an alley and facilitated avoidance acquisition in a shuttle-box, effects that were primarily though not entirely due to the poor performance evident in sham-operated isolates (Dalrymple-Alford and Benton, 1984a). However, both intact isolates and isolates with dorsal hippocampus damage made relatively fewer "manipulatory" versus "non-manipulatory" contacts and reduced habituation more slowly in the open field than did group-reared counterparts; these two measures were unaffected by the lesions.

Nisenbaum et al. (1983) found that, in terms of total errors, enriched but not isolated rats with anterior cingulate cortex lesions were impaired in a radial maze task. These effects were apparently due to the superior performances of enriched controls as enriched rats in general made more correct arm choices before the first error. These data provide further support for the speculation (Dalrymple-Alford and Benton, 1984a) that frontal cortex/hippocampal systems are sensitive to modulation by the housing environment.

Three further studies on preoperative housing and brain lesions have also produced weak or inconsistent support for the "beneficial" value of preoperative enrichment. Somatosensory cortex lesions resulted in comparable impairments in tactile discrimination learning in enriched enucleated rats and group-housed counterparts (Finger, 1978b). Kemble and Davies (1981) found no clear effects of enrichment in rats with lesions of the amygdala on tests of intake of water and novel foods, open field activity and predation, although amygdala lesions facilitated shuttle-box avoidance more strongly in isolated than in enriched rats. In the only study to investigate the effects of housing on brain-stem lesions, Weldon and Smith (1979) found that enriched housing improved Hebb-Williams maze learning and increased initial activity in an open field in both intact rats and rats with superior colliculus lesions; these lesions impaired maze-learning but increased activity.

Contrary to the inconclusive impression created by the mixed findings listed thus far in this section, it has been claimed by Donovick and his collaborators that the impact of enriched presurgical experience in brain-damaged subjects is established (Donovick et al., 1979). This claim comes from their work on preoperative environments and septal lesions but is better supported by measures of reactivity rather than learning ability. In their first experiment, Donovick et al. (1973) found that preoperative postweaning enrichment as opposed to social isolation increased exploratory activity (reduced emergence latencies and increased preference for the more complex of two chambers) in rats with septal lesions and reduced their enhanced consumption of water and their enhanced reactivity to preferred and non-preferred solutions. However, the exploration differences were minimised on a second test suggesting that the housing variable influenced chiefly defensive reactions. Moreover, the impaired alternation learning in rats with septal lesions was unaffected by these housing conditions. Three other studies by this group examined the effects of preoperative differential housing in mice. The first such study showed no straightforward effects of the environment in tests of rotorod performance, water and novel fluid intake and shuttle-box avoidance (Donovick et al., 1975). The next study found that septal lesions substantially augmented the postoperative reactivity to handling in mice housed in isolation prior to and after surgery but increased to a much lesser extent such reactivity in enriched counterparts, whereas there was also a considerable increase on this measure in mice reared in enrichment but placed in isolation after surgery (Goodlett et al., 1982). Their third study revealed that an impairment in acquisition and reversal learning in a water maze was shown only by isolation-reared mice with septal lesions (Engellenner et al., 1982), but it is possible that these effects were due to increased reactivity to the aversive situation in these animals.

DISCUSSION

Overall, the evidence summarised in this review constitutes a strong body of data to support the conclusion that the effects of localised brain damage in rodents are often modified by environmental constraints defined by preoperative or postoperative housing conditions. Nonetheless, it is clear that any sweeping statement or generalized assumption on the expected therapeutic value of any single housing treatment is too simplistic a representation of the evidence. The interesting task is to establish when and if possible why such modifications by housing variables occur. The available evidence is still somewhat sketchy and recent findings have only added to rather than subtracted from the misleadingly superficial and inaccurate picture reflected by a simple count of "positive" and "negative" findings. It is more worthwhile to specify as far as possible the factors responsible for the different findings that have emerged.

The most obvious and expected difference between conflicting reports is that lesions at different loci do not produce the same results with respect to differential housing. However, the age when animals receive brain damage and perhaps also the age when differential housing is imposed, are more subtle determinants of the effects observed. In both preoperative and postoperative studies that employed dorsal hippocampal lesions, generally improved performance in brain-damaged subjects was found after group-housing or enrichment at 33 days whereas fewer and more mixed changes occurred in subjects differentially housed at 21-26 days. But it should be recalled that differential housing imposed at 23 days of age was found to improve behavioural function after occipital cortex lesions in neonatal rats (Will et al., 1976). Donovick et al. (1973) also used postweaning conditions in their study on preoperative differential housing in which they found housing effects on reactivity but not learning ability in rats with septal lesions.

A clear example of how the age factor can be crucial to determine the impact of housing conditions in brain-damaged subjects comes from studies on the effects of social isolation on sexual behaviour in male rats with medial preoptic area (MPOA) lesions. MPOA lesions produced impaired sexual behaviour in adult rats whether these were grouped or isolated (Twiggs et al., 1978). In contrast, a severe deficit was observed in isolates but there were no effects in grouped rats when these animals were given MPOA lesions prior to differential housing at 25 days (Meisel, 1982) or at 30-33 days (Twiggs et al., 1978; Leedy et al., 1980).

When the behaviour of a brain-damaged subject is influenced by housing experience, the effects may be additive rather than interactive. Whether housing induces compensatory processes, affects spared attributes or promotes functional recovery (in the narrow sense, restitution of specific function), this does not appreciably alter the fact that such experience is an influential factor for the behaviour of brain-damage subjects in many instances. Nevertheless, such questions have theoretical implications that may provide clues as to which processes are likely to change as a function of differential housing. The decision on whether experience has interacted with the sequelae of brain damage is usually based on the statistical one. The examples in Figs. 1 and 2 visibly show that enriched housing reduced the deficit produced by hippocampal lesions but did not change the behaviour of intact animals. However, even when interactions occur there are often floor or ceiling effects or problems of scales of measurement. Yet it is also possible that housing effects in intact and brain-damaged subjects are sometimes modulated by different processes.

Brain measures have thus far provided only weak evidence to complement the suggestion that housing interacts with the consequences of brain damage. Weldon and Smith (1979) noted that the size of extracollicular damage was correlated with ambulation and with maze errors in enriched but not isolated rats, whereas the size of collicular damage and the number of maze errors were correlated in isolated but not in enriched rats. Data reported by Petit and Alfano (1979) indicated that enrichment increased hippocampal weight in controls but reduced it in groups given experimental lead exposure (the authors stated that there were no significant effects). Will et al. (1977a) found that the reduction in brain DNA in cortical regions distant from the occipital damage was greater in isolated than in enriched rats but later studies suggest a more complicated picture with remote loss being less after large but greater after small occipital cortex lesions in enriched as opposed to isolated rats (Rosenzweig, 1980). Two studies examined exclusively anatomical effects. In Kelche and Will's study (1982), the number of basilar dendritic spines in occipital cortex pyramidal neurons was found to be reduced by hippocampal lesions but increased by enrichment, whereas a concentric ring analysis implicated that higher order branching of these dendrites was reduced to an equal extent by lesions in enriched and isolated rats but increased by enrichment in controls only. Katz and Davies (1983) reported that nutritionally-induced reductions in the weight and size of the cerebrum and in the area of the dorsal hippocampus were greater in isolated as opposed to enriched rats but there were no housing by nutrition interactions and other cerebral measures were each affected exclusively by one or other of these two treatments.

There are also interpretative problems with regard to the housing treatments themselves, e.g., possible social differences between the enriched and grouped conditions; problems of sensory deprivation in addition to social isolation . Furthermore, it is important to recognize that there is evidence in some studies of absent or comparatively reduced deficits in isolated animals (Sjöden, 1976; Eclancher and Karli, 1979b; Weldon and Smith, 1979; Nisebaum et al., 1983; Dalrymple-Alford and Benton, 1984a). In the majority of studies, however, improved behavioural scores have been found in enriched rats compared with the scores in isolated rats, and enriched rats have been either similar to (Hughes, 1965; Will and Kelche, 1979) or on more occasions

(Hughes, 1965; Einon et al., 1980; Rosenzweig, 1980; Will et al., 1983) better than grouped counterparts. Davenport (1976) has found it necessary to use a "super-enriched" environment to reliably reduce the effects of experimental hypothyroidism. Often, however, we cannot be sure whether enrichment, isolation or both factors contributed to the effects observed. The many reports of an increase in reactivity in isolated as opposed to socially housed rats with brain damage are examples where the effects are most likely due to the isolation variable, whereas reductions in maze-learning deficits seem to reflect the operation of both enrichment and isolation in some cases.

We have yet to explain adequately why we have some postoperative housing effects after dorsal hippocampus damage at all ages examined but the nature of the effects at different ages at surgery coupled with the age at differential housing varies as a function of the task, and why the presence of some effects depends on the lesion locus irrespective of the age at lesion and, it seems, of the task employed. It may be that measures taken in the Olton maze are less sensitive to environmental effects than are those taken from other maze tasks, and we plan to look at this question further in experiments on age factors and on more localised damage to the hippocampal formation. We are especially keen to see whether the effects of dentate gyrus damage and subicular damage are susceptible to housing manipulations. It is also necessary to address the problem of whether a postoperative environment is in fact more efficacious than a preoperative one to reduce the effects of brain damage, and to consider the persistence of housing effects particularly in the context of interactions between presurgical and postsurgical treatments. The need for answers to the latter questions is underlined by the fact that postoperative enrichment is often described as post-traumatic "therapy" and preoperative enrichment is labelled as "protective" (Stein et al., 1983).

The absence of environmental effects in the studies by Bland and Cooper (1969, 1970) and by Finger (1978b) and the proliferation of positive results in studies on maze learning justifiably led some workers to conclude that the housing environment modifies a deficit when the measures chosen reflect general rather than specific behavioural attributes and when the structure damaged is regarded as non-critical for the hypothesised function(s) under study (Greenough et al., 1976; Finger, 1978a; Will, 1981). Thus it is asserted that housing produces its effects via an influence on "general adaptive" or "general learning" capacities. Even though this reasoning seems little more than a redescription of the fact that enriched rats make fewer maze errors, its value is improved by the assumption that to solve problems such as the Hebb-Williams maze many information channels may be substituted to compensate for damaged ones (Stein et al., 1983). However, the weakness of this position is that there are too many discrepancies for it to accommodate particularly, for example, those related to the use of different lesion loci. It may be that housing influences recovery only at the behavioural level, i.e. few specific effects on neuronal plasticity per se, perhaps in conjuncton with or via changes in neuromodulatory factors such as those described in this volume. But whatever explanations one may suggest, we emphasise that they must be qualified by reference to the variables of lesion locus, task and age.

ACKNOWLEDGEMENTS

J. C. Dalrymple-Alford gratefully acknowledges The Royal Society of London for his European Science Exchange fellowships and the European Science Foundation for his ETP fellowship. C. R. Kelche is thankful to the C.N.R.S. for support (grant Nos. 6580085 and 4784).

REFERENCES

Ahmad, S. S., and Harvey, J. A., 1968, Long-term effects of septal lesions and social experience on shock-elicited fighting in rats, J. Comp. Physiol. Psychol. , 66:596.

Altman, J., and Bayer, S., 1975, Post-natal development of the hippocampal dentate gyrus under normal and experimental conditions, in : "The Hippocampus, Vol. 1," R. L. Isaacson and K. H. Pribram, eds., Plenum Press, New York.

Bouchon, R. and Will, B. E., 1982, Effects of early enriched and restricted environments on the exploratory and locomotor activity of Dwarf mice. Behav. Neural Biol. , 35:174.

Bouchon, R., and Will, B. E., 1983, Effects of post-weaning environment and apparatus dimension on spontaneous alternation as a function of phenotype in "Dwarf" mice, Physiol. Behav. , 30:213.

Bland, B. H., and Cooper, R. M., 1969, Posterior neodecortication in the rat: Age at operation and experience, J. Comp. Physiol. Psychol. , 69:345.

Bland, B. H., and Cooper, R. M., 1970, Experience and vision of the posterior neodecorticate rat, Physiol. Behav. , 5:211.

Buzsaki, G., Acsadi, G., and Jani, A., 1980, Differential contribution of fimbria and fornix fibers to behavior, Behav. Neural Biol. , 28:79-88.

Buzsaki, G., Bors, L., and Nagy, F., 1982, Spatial mapping, working memory, and the fimbria-fornix system, J. Comp. Physiol. Psychol. , 96:26.

Cotman, C. W., and Nieto-Sampedro, M., 1982, Brain function, synapse renewal, and plasticity, Ann. Rev. Psychol. , 33:371.

Dalrymple-Alford, J. C., 1983, Social isolation in rats and the role of the septo-hippocampal system, Unpublished PhD. Thesis, Swansea, U.K.

Dalrymple-Alford, J. C., and Benton, D., 1984a, Preoperative differential housing and dorsal hippocampal lesions in rats, Behav. Neurosci. 98:23.

Dalrymple-Alford, J. C., and Benton, D., 1984b, Behavioural inhibition and the age at social isolation in the rat, Quart. J. Exp. Psychol. , 36:27.

Davenport, J. W., 1976, Environmental therapy in hypothyroid and other disadvantaged animal populations, in :"Environments as a Therapy for Brain Dysfunction," R. N. Walsh and W. T. Greenough, eds., Plenum Press, New York.

Donovick, P. J., Burright, R. G., and Swidler, M. A., 1973, Presurgical rearing envionment alters exploration, fluid consumption, and learning of septal lesioned and contol rats, Physiol. Behav. , 11:543.

Donovick, P. J., Burright, R. G., Fuller, J. L., and Branson, P. R., 1975, Septal lesions and behavior: Effects of presurgical rearing and strain of mouse, J. Comp. Physiol. Psychol. , 89:859.

Donovick, P. J., Burright, R. G., and Bengelloun, W. A., 1979, The septal region and behavior: An example of the importance of genetic and experiential factors in determining effects of brain damage, Neurosci. Biobehav. Rev. , 3:83.

Douglas, R. J., 1975, The development of hippocampal function: Implications for theory and for therapy, in :"The Hippocampus, Vol. 2," R. L. Isaacson and K. H. Pribram, eds., Plenum Press, New York.

Eclancher, F., and Karli, P., 1979a, Septal damage in infant and adult rats: Effects on activity, emotionality, and muricide, Aggres. Behav. , 5:389.

Eclancher, F., and Karli, P., 1979b, Effects of early amygdaloid leions on the development of reactivity in the rat, Physiol. Behav. , 22:1123.

Eclancher, F., and Karli, P., 1980, Effects of infant and amygdaloid lesions upon acquisition of two-way active avoidance by the adult rat: Influence of rearing conditions, Physiol. Behav. , 24:887.

Eclancher, F., and Karli, P., 1981a, Environmental influences on behavioural effects of early and late limbic and diencephalic lesions in the rat, in :"Functional Recovery from Brain Damage," M. W. Van Hof and G. Mohn, eds., Elsevier, Amsterdam.

Eclancher, F., and Karli, P., 1981b, Influence of rearing conditions on the acquisition of two-way active avoidance responses by rats septalectomised at an early age, Behav. Brain Res. , 3:83.

Einon, D. F., and Morgan, M. J., 1977, A critical period for social isolation in the rat, Devel. Psychobiol. , 10:123.

Einon, D. F., 1980, Spatial memory and response strategies in rats: Age, sex and rearing differences in performance, Quart. J. Exp. Psychol. , 32:473.

Einon, D. F., Morgan, M. J., and Will, B. E., 1980. Effects of postoperative environment on recovery from dorsal hippocampal lesions in young rats: Tests of spatial memory and motor transfer, Quart. J. Exp. Psychol. , 32:137.

Engellenner, W. J., Goodlett, C. R., Burright, R. G., and Donovick, P. J., 1982, Environmental enrichment and restriction: Effects on reactivity, exploration and maze learning in mice with septal lesions, Physiol. Behav. , 29:885.

Finger, S., 1978a, Environmental attenuation of brain-lesion symptoms, in : "Recovery from Brain Damage," S. Finger, ed., Plenum Press, New York.

Finger, S., 1978b, Post-weaning environmental stimulation and somesthetic performance in rats sustaining cortical lesions at maturity, Devel. Psychobiol. , 11:5.

Finger, S., and Stein, D. G., 1982, "Brain Damage and Recovery," Academic Press, New York.

Forgays, D. G., and Read, J. M., 1962, Crucial periods for free-environmental experience in the rat, J. Comp. Physiol. Psyhol. , 55:816.

Frötscher, M., 1983, Dendritic plasticity in response to partial deafferentation, in : "Neurobiology of the Hippocampus," W. Seifert, ed., Academic Press, London.

Goodlett, C. R., Engellenner, W. J., Burright, R. G., and Donovick, P. J., 1982, The influence of environmental rearing history and postsurgical environmental change on the septal rage syndrome in mice, Physiol. Behav. , 28:1077.

Gray, J. A., 1982, "The Neuropsychology of Anxiety," Clarendon Press, Oxford.

Greenberg, G. A., and Burdsall, C., 1982, Animal colony practices in North American institutions: A survey, J. Gen. Psychol. , 106:165.

Greenough, W. T., Fass, B., and DeVoogd, T. J., 1976, The influence of experience on recovery following brain damage in rodents: Hypotheses based on developmental research, in: "Environments as a Therapy for Brain Dysfunction," R. N. Walsh and W. T. Greenough, eds., Plenum Press, New York.

Hallonquist, J. D., and Brandes, J. S., 1981, Housing affects hyperreacivity but not obesity induced by medial hypothalamic lesions, Physiol. Behav. , 26:1025.

Hughes, K. R., 1965, Dorsal and ventral hippocampus lesions and maze learning: Influence of preoperative environment, Can. J. Psychol. , 19:352.

Jones, D. G., and Smith, B. J., 1980, Morphological analysis of the hippocampus following differential rearing in environments of varying social and physical complexity, Behav. Neural Biol. , 30:135.

Johnson, D. A., Spiker, M., and Carlson, K., 1984, Open field social behavior in rats as a function of septal lesions in infancy and group versus isolated rearing conditions, Physiol. Psychol. , 12:14.

Juraska, J. M., Fitch, J., Henderson, C., and Rivers, N., 1983, Sex differences in the dendritic response to differential environments in the granule cells of the dentate gyrus, Soc. Neurosci. Abstracts , 9:947.

Katz, H. B., and Davies, C. A., 1983, The separate and combined effects of early undernutrition and environmental complexity at different ages on cerebral measures in rats, Devel. Psychobiol. , 16:47.
Kelche, C. R., and Will, B. E., 1978, Effets de l'environnement sur la restauration fonctionnelle après lésions hippocampiques chez des rats adultes, Physiol. Behav. , 21:935.
Kelche, C. R., and Will, B., 1982, Effects of postoperative environments following dorsal hippocampal lesions on dendritic branching and spines in rat occipital cortex, Brain Res. , 245:107.
Kemble, M. D., and Davies, V. A., 1981, Effects of prior environmental enrichment and amygdaloid lesions on consumatory behavior, activity, predation, and shuttle-box avoidance in male and female rats, Physiol. Psychol. , 9:340.
Leedy, M. G., Vela, E. A., Popolow, H. B., and Gerall, A. A., 1980, Effect of prepuberal medial preoptic area lesions on male rat sexual behaviour, Physiol. Behav. , 24:341.
McNaughton, B. L., 1983, Activity dependent modulation of hippocampal synaptic efficacy: Some implications for memory processes, in : "Neurobiology of the Hippocampus," W. Seifert, ed., Academic Press, London.
Meisel, R. L., 1982, Effects of postweaning rearing conditions on recovery of copulatory behavior from lesions of the preoptic area, Devel. Psychobiol. , 15:331.
Nisenbaum, E., Juraska, J., and Henderson, C., 1983, Differential environments and radial maze performance: The effects of anterior cingulate cortex lesions, Soc. Neurosci. Abstracts , 9:640.
Numan, R., 1978, Cortical-limbic mechanisms and response control: A theoretical review, Physiol. Psychol. , 6:445.
Olton, D. S., Walker, J. A., and Wolf, W. A., 1982, A dis- connection analysis of hippocampal function, Brain Res. , 233:241.
Osborne, G. L., Caul, W. F., and Fernandez, K., 1980, Behavioral effects of prenatal ethanol exposure and differential early experience, Pharmacol. Biochem. Behav. , 12:393.
Petit, T. L., and Alfano, D. P., 1979, Differential experience following developmental lead exposure: Effects on brain and behavior, Pharmacol. Biochem. Behav. , 11:165.
Rosenzweig, M. R., 1971, Effects of environment on brain and behavior, in : "The Biopsychology of Developments," E. Tobach, L. R. Aronson and E. Shaw, eds., Academic Press, New York.
Rosenzweig, M. R., 1980, Animal models for effects of brain lesions and for rehabilitation, in : "Recovery of Function: Theoretical Considerations for Brain Injury Rehabilitation," P. Bach-y-Rita, ed., Huber, Bern.
Rosenzweig, M. R., and Bennett, E. L., 1977, Effects of environmental enrichment or impoverishment on learning and brain values in rodents, in : "Genetics, Environment and Intelligence," A. Oliverio, ed., Elsevier, Amsterdam.
Schwartz, S., 1964, Effect of neonatal cortical lesions and early environmental factors on adult rat behavior, J. Comp. Physiol. Psychol. , 57:72.
Shibagaki, M., Seo, M., Asano, C., and Kiyono, S., 1981, Environmental enrichment to alleviate maze performance deficits in rats with microcephaly induced by X-irradiation, Physiol. Behav. , 27:797.
Sjödén, P. O., 1976, Effects of noenatal thyroid hormone stimulation and differential preweaning rearing on spatial discrimination learning in rats, Physiol. Psychol. , 4:515.
Sjödén, P. O., and Lindqvist, M., 1978, Behavioral effects of neonatal thyroid hormones and differential postweaning rearing in rats, Devel. Psychobiol. , 11:371.

Sjödén, P. O., and Söderberg, U., 1976, Effects of neonatal thyroid hormone stimulation and differential preweaning rearing on open-field behavior in adult rats, Devel. Psychobiol. , 9:413.
Smith, C. L., 1959, Mass action and early environment in the rat, J. Comp. Physiol. Psychol. , 52:154.
Stein, D. G., Finger, S., and Hart, T., 1983, Brain damage and recovery: Problems and perspectives, Behav. Neural Biol. , 37:185.
Susser, E. R., and Wallace, R. B., 1982, The effects of environmental complexity on the hippocampal formation of the adult rat, Acta Neurobiol. Exp. , 42:203.
Swanson, H. H., McConnell, P., Uylings, H. B. M., Van Oyen, H. G. and Van de Poll, N. E., 1983, Interaction between pre-weaning undernutrition and post-weaning environmental enrichment on somatic development and behaviour in male and female rats, Behav. Proc. , 8:1.
Tachibana T., 1979, Effects of early nutritional and environmental conditions on later runway test behavior in rats, Japan. Psychol. Res. , 2:99.
Tanabe, G., 1972, Remediating maze deficiencies by the use of environmental enrichment, Devel. Psychol. , 7:224.
Twiggs, D. G., Popolow, H. B., qnd Gerall, A. A., 1978, Medial preoptic lesions and male sexual behaviour: Age and environmental interactions, Science , 200:1414.
Uphouse, L., 1980, Reevaluation of mechanisms that mediate brain differences between enriched and impoverished animals, Psychol. Bull. , 88:215.
Walsh, R., 1981, Sensory environments, brain damage and drugs: A review of interactions and mediating mechanisms, Intern. J. Neuroscience , 14:129.
Walsh, R. N., Budtz-Olsen, D. E., Penny, J. E., and Cummins, R. A., 1971, The effects of environmental complexity on the histology of the rat hippocampus, J. Comp. Neurol. , 137:361.
Weldon, D. A., and Smith, C. J., 1979, Superior colliculus lesions and environmental experience: Non visual effects on problem solving and locomotor activity, Physiol. Behav. , 23:159.
Wells, A. M., Geist, C. R., and Zimmermann, R. R., 1972, Influence of environmental and nutritional factors on problem solving in the rat, Percept. Motor Skills , 35:235.
Will, B. E., 1981, The influence of environment on recovery after brain damage in rodents, in : "Functional Recovery from Brain Damage," M. W. Van Hof and G. Mohn, eds., Biomedical Press, Amsterdam.
Will, B. E., Deluzarche, F., and Kelche, C. R., 1983, Does post- operative environment attenuate or exacerbate symptoms which follow hippocampal lesions in rats? Behav. Brain Res. , 7:125.
Will, B., and Eclancher, F., 1984, Early brain damage and early environment, in : "Early Brain Damage, Vol 2," C. R. Almli and S. Finger, eds., Academic Press, New York.
Will, B. E., and Kelche, C. R., 1979, Effects of different postoperative environments on the avoidance behavior of rats with hippocampal lesions: Recovery or improvement of function? Behav. Neural Biol. , 27:96.
Will, B. E., Kelche, C. R., and Deluzarche, F., 1981, Effects of postoperative environment on functional recovery after entorhinal cortex lesions in the rat, Behav. Neural Biol. , 33:303.
Will, B. E., and Rosenzweig, M. R., 1976, Effets de l'environnement sur la récupération fonctionnelle après lésions cérébrales chez les rats adultes, Biol. Behav. , 1:5.
Will, B. E., Rosenzweig, M. R., and Bennett, E. L., 1976, Effects of differential environments on recovery from neonatal brain lesions, measured by problem-solving scores and brain dimensions, Physiol. Behav. , 16:603.

Will, B. E., Rosenzweig, M. R., Bennett, E. L., Herbert, M., and Morimoto, H., 1977a, Relatively brief environmental enrichment aids recovery of learning capacity and alters brain measures after postweaning brain lesions in rats, J. Comp. Physiol. Psychol. , 91:33.

Will, B. E., Sutter, A. R., and Offerlin, M. R., 1977b, Effets de la méthamphétamine et d'un environnement complexe sur la récupération comportementale après atteinte cérébrale, Psychopharmacol. , 51:273.

TEMPORALLY SPACED LESIONS AND RECOVERY OF FUNCTION

Guy Toniolo*, Bruno Will*, Anant Dravid** and Claude Destrade***

* Laboratoire de Neurobiologie des Comportements, Université Louis Pasteur, Strasbourg, France
** Sandoz Ltd, Preclinical Research, Basel, Switzerland
*** Laboratoire de Psychophysiologie, Université de Bordeaux I, Bordeaux, France

As documented by many clinical reports, it has long been recognized that slow growing lesions, i.e. lesions of "slow momentum" (Riese, 1948), cause less impairments and fewer symptoms than would be expected after similar brain damage that occurs more acutely (e.g., Finger, 1978; Jeannerod and Hecaen, 1979 for rewiews). In order to develop animal models of "lesion momentum" phenomena numerous studies have examined the effects of brain lesions produced in more than one stage, with interoperative intervals ranging from several days to several months (e.g., Finger and Stein, 1982 for a review). Substantial evidence from these experimental studies indicates that behavioral deficits following one-stage cerebral lesions are relatively attenuated or even abolished when virtually equivalent but temporally spaced lesions are produced. Such attenuations of behavioral symptoms due to multi-stage surgery have generally been termed "serial lesion phenomena" (Finger et al., 1973). The first part of this chapter outlines the various approaches utilized in this field and presents a brief summary of the major findings. Particular attention is given to some of the less frequently employed experimental methodologies and we report a study with temporally and structurally spaced lesions in which we included the use of colchicine to produce a "transient" lesion at the time of the first surgery. The second part of this chapter considers the processes, especially the biological processes, that may underlie the effects of temporally spaced lesions.

PARADIGMS AND FINDINGS

Temporally spaced lesions

The large majority of experimenters have chosen two-stage lesion procedures to explore temporally spaced lesion effects although more complex designs have also been utilized (Stein and Lewis, 1975 for a review). Generally, two basic paradigms have been tried. The classical, serial unilateral lesion paradigm (Fig. 1A), involves a complete unilateral destruction of a given brain structure at the time of the first operation, followed by an equivalent destruction of the contralateral structure in a second operation performed some time later. The second, progressive

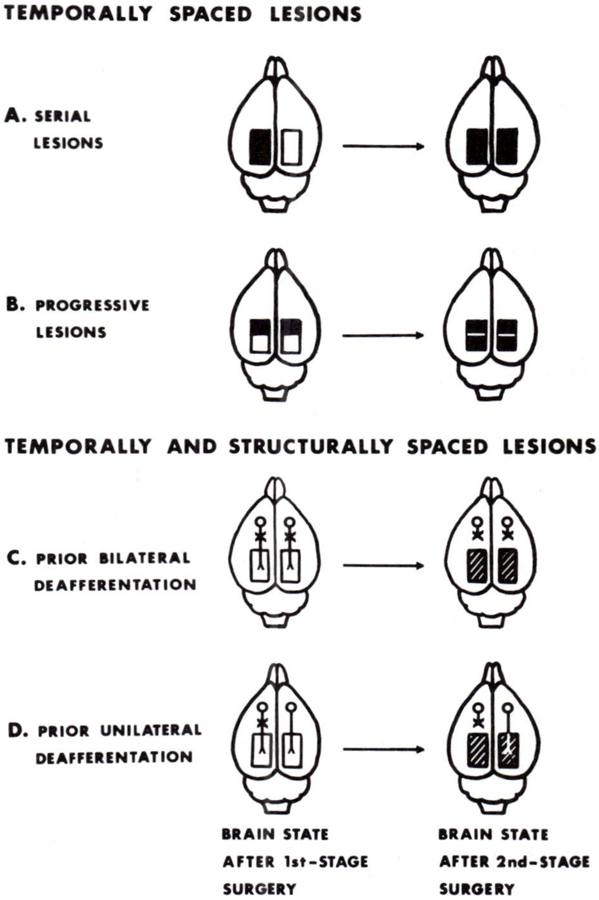

Figure 1. Paradigms in studies on temporally spaced lesions. In paradigms A and B, control groups sustain one-stage lesions at the time of either first-stage or second-stage surgery. In paradigms C and D, control groups sustained one-stage lesions at the time of second-stage surgery, either with concomitant bilateral deafferentation (Glick and Greenstein, 1972; paradigm C) or without concomitant unilateral deafferentation (Toniolo et al., 1985; paradigm D). Black areas = complete lesion; hatched areas = partial lesion.

bilateral* lesion paradigm (Fig. 1B) involves a bilateral but partial lesion of a given brain structure, followed by a destruction of the remaining part of that structure at a later time. The progressive bilateral lesion paradigm has been utilized only in a few studies with most experimenters chosing the serial unilateral lesion paradigm.

Serial unilateral lesions. Serial lesions were initially employed in behavioral research because they resulted in lower mortality than would be expected after large, one-stage lesions (Flourens, 1824; Lashley, 1929; Kleitman and Camille, 1932; Adametz, 1959). Today, the range and diversity of sparing and/or recovery effects observed after serial lesions provide strong evidence that more is at stake than just the survival from injury. Serial lesion effects have been observed in various behavioral tasks that tested brain functions ranging from sensory, motor and motivational, to cognitive ones (e.g., Finger and Stein, 1982). Although serial lesions effects have not been observed in some studies, beneficial effects of multi-stage surgery have been reported after sensory, motor and associational cortex lesions, as well as after lesions of limbic and other subcortical structures (e.g., Jeannerod and Hecaen, 1979, for a review). Such beneficial effects have been reported for primates, carnivores, rodents and birds.

Progressive bilateral lesions. As with serial unilateral lesions, progressive bilateral lesions were also generally followed by reduced impairments compared with the effects of one-stage lesions (Finger et al., 1971; Treichler, 1975; Barbas and Spear, 1976; Isseroff et al., 1976; Curtis and Nonneman, 1977; Fass, 1983). Three of these studies deserve special mention because they directly compared the effects of serial and progressive lesions. Each of these reports showed that progressive bilateral lesions were at least as efficient as serial unilateral lesions for enhancing recovery of behavioral function. Barbas and Spear (1976) found that rats with either progressive bilateral or serial unilateral visual cortex lesions exhibited considerable savings when relearning a brightness discrimination task in comparison to rats with one-stage lesions. Isseroff et al. (1976) found that rats that sustained progressive bilateral lesions of the hippocampus showed superior performances than did one-stage lesion controls in both passive avoidance and reversal learning tasks whereas animals with serial unilateral lesions differed from such controls on passive avoidance only; however, neither progressive bilateral nor serial unilateral lesions facilitated recovery of performance on spontaneous and learned alternation tasks. More recently, Fass (1983) reported that progressive bilateral but not serial unilateral lesions of the entorhinal cortex enhanced both the extent and rate of recovery of open-field ambulation.

In the studies by Isseroff et al. and Fass, the time intervals between successive surgeries in the progressive bilateral and the serial unilateral paradigms were the same. Thus, the enhanced behavioral recovery found only after the progressive lesions in these two studies was likely to be due to other variations inherent to the two paradigms employed. An obvious major difference was that the structures involved and their connections with other brain structures in each partial lesion at the time of the first and second surgery was not exactly the same in the two paradigms. These differences in the configuration of the lesions and in the subsequent denervation (including differences in interhemisheric pathways that remain after the first stage lesions) suggest that temporal characteristics, in addition to the time interval between lesions, need to be considered for understanding the phenomena associated with temporally spaced brain damage. For example, the bilateral lesion in the progressive paradigm may produce a deficit after

---

* Also labelled serial bilateral ; e.g., Barbas and Spear, 1976.

the first-stage surgery whereas the unilateral lesions in the serial paradigm may produce a deficit only after the second-stage surgery, as appeared to be the case in Fass' study. In such instances, the animals in the serial-lesion group would have had behavioral deficits for a shorter period when all the lesion subgroups are tested after the second-stage surgery.

Temporally and structurally spaced lesions

Prior bilateral deafferentation. The approach in this paradigm is exemplified by the study by Glick and Greenstein (1972). Although this investigation was not conceived in the conceptual frame of temporally spaced lesion studies, the experimental design is quite close to the progressive bilateral lesion paradigm (Fig. 1C). In this study, rats that sustained frontal cortex lesions 30 days prior to lateral hypothalamic lesions were found to be able to regulate their body weight, whereas rats that received frontal cortex and lateral hypothalamic lesions at the same time, or only 10 days apart, were aphagic and adipsic. This experiment thus demonstrates that prior deafferentation of a given cerebral structure is adequate to induce a serial lesion effect, if it is produced at a sufficiently long time prior to a one-stage bilateral lesion of the already denervated structure.

Prior unilateral deafferentation. Bilateral destruction of afferents to the CNS structure that will be lesioned at a later stage is not essential; we found that a unilateral first-stage deafferentation is sufficient to induce serial lesion effects as well (Fig. 1D; Toniolo and Will, 1983; Will and Toniolo, 1983). Adult rats with a unilateral fimbria-fornix lesion induced electrolytically or with 5ug colchicine 30 days before receiving bilateral dorsal hippocampus lesions, were less impaired in a maze learning task than were rats that received hippocampal lesions only (Toniolo, Dravid and Will, 1985; Table 1).

Colchicine at all doses (5, 10 and 15 ug/rat; groups C5, C10 and C15) decreased theta amplitude in the ipsilateral hippocampus but only the lowest dose of colchicine, the only dose that facilitated behavioral recovery, produced an alteration of theta amplitude that recovered to the preoperative level within a time interval comparable to the interoperative interval employed in the behavioral experiment (Table 1). However, it seems unlikely that the serial lesion effect we observed was due to the recovery of normal theta amplitude, since we also found that rats with unilateral electrolytic lesions of the fimbria-fornix (group E) did not show any recovery of theta amplitude, at least within the 23 days of postoperative recording.

Colchicine, as other neurotoxins, is known to induce long lasting changes in synaptic density through axonal sprouting in the transiently deafferented target area (Goldowitz and Cotman, 1980; Guth et al., 1980). Thus we made a complementary neurochemical study to assess the postsurgical reorganisation of cholinergic fibers, one of the major extrinsic afferent pathways to the hippocampal formation. We measured acetylcholinesterase (AChE) and choline acetyltransferase (ChAT) activities in the anterior and posterior halves of the hippocampus from rats that had sustained either a unilateral fimbria-fornix lesion alone eight days before sacrifice, or a unilateral fimbria-fornix lesion and a bilateral dorsal hippocampal lesion three and two months before sacrifice, respectively (Table 1). Eight days following unilateral fimbria-fornix electrolytic or colchicine lesions, ChAT and AChE activities in the ipsilateral side were significantly depressed by about 20 to 50 % relative to control levels. On the contralateral side, ChAT and AChE activities were also slightly decreased after colchicine injection by about 20 to 25 % relative to control levels but only in the anterior half (with the exception of AChE activity of group C5 which was also reduced in the posterior part). Three months after fimbria-fornix lesions a partial though significant recovery of ChAT activity was found on the ipsilateral side. This recovery was apparent in the anterior hippocampus in rats of the E and C5 groups, and in the posterior hippocampus in rats belonging to C5 and C10 groups. On the contralateral side, only AChE activity varied at three months, being increased in the anterior half of the hippocampus for the C5 and C10 groups and decreased in the posterior half for the C15 group. Recovery of AChE activity followed a trend closely similar to that of ChAT activity. In all hippocampal pieces analysed, there were highly significant positive correlations between ChAT and AChE activities.

Table 1. Functional improvement or impairment in the two-stage operated groups.

Group	E	C5	C10	C15
a) Behavior	+35	+40	- 8	-21
b) Electrophysiology	+ 3	+92	+ 3	+10
c) Biochemistry				
- Lesion Side				
. Anterior Part				
AChE	+14	+31	+ 9	- 7
ChAT	+19	+30	+ 6	- 8
. Posterior Part				
AChE	-28	+17	+20	-21
ChAT	-23	+18	+18	-14
- Contralateral Side				
. Anterior part				
AChE	+19	+18	+ 6	+ 3
ChAT	+ 1	+11	+ 7	- 7
. Posterior part				
AChE	- 1	- 1	- 6	- 2
ChAT	-12	0	-12	-13

Percentage of functional improvement (+) or impairment (-) in the two-stage operated groups (E = electrolytic lesions ; C5, C10 and C15 = colchicine injections with 5, 10 and 15 ug respectively).
a) errors in maze learning (100 % = difference between mean of one-stage operated control rats and mean of sham-operated rats).
b) variations in the theta amplitude in the ipsilateral hippocampus between 2 and 23 days after surgery (100 % = difference between preoperative level and the level reached two days after surgery).
c) variations in hippocampal ChAT and AChE activities between 8 and 90 days after first-stage surgery (100 % = difference between each group and sham operated controls for each postoperative time considered).

The partial recovery of ChAT activity observed in some parts of the hippocampus after some of the treatments used might be explained by the restitution of axonal flow initially blocked by the colchicine treatment. It might also be due to sprouting of spared fibers after electrolytic or chemical lesions. Cholinergic fibres coursing through the fimbria-fornix constitute the major (ca 85%) proportion of cholinergic input to the hippocampus. Since greater than 50 % of the hippocampal ChAT activity was still present eight days after fimbria-fornix lesions, it can be assumed that there was a partial preservation of this subcallosal pathway which might be the source of collateral sprouts reinnervating the hippocampus. This interpretation is supported by recent studies demonstrating a parallel recovery of hippocampal ChAT and AChE activities following partial lesions of the fimbrial bundle (Dravid and Van Deusen, 1983, 1984). The possible involvement of the ventral and supracallosal cholinergic pathways (Gage et al., 1984) can not, however, be ruled out. Since we found that the groups of rats which displayed a behavioral serial lesion effect also exhibited recovery of cholinergic enzyme activities in the anterior part of the hippocampus, we have suggested that the behavioral recovery might partly be explained by time-dependent modifications of the activity (perhaps through sprouting) of cholinergic neurons that project onto the anterior hippocampus (Toniolo et al., in preparation).

INTERPRETATIONS

From a behavioral point of view, functional improvement after CNS damage might occur through processes such as changes in cue selection or in strategy formation (Le Vere, 1980) which involve compensatory mechanisms rather than a genuine sparing and/or recovery of function. It is not clear whether two-stage lesions encourage compensation rather than sparing and/or recovery, although some evidence indicates that two-stage lesions in contrast to one-stage lesions may be followed more often by recovery and less often by compensatory mechanisms (Gentile et al., 1978). The workers made single or two-stage lesions of the motor-parietal cortex in rats to determine whether the animals could recover the ability to negotiate a narrow elevated platform. Both groups eventually appeared to recover this ability, but the rats suffering one-stage lesions had learned to navigate the runway with postoperative response patterns that were markedly different from those used preoperatively whereas many of the rats with serial lesions returned to a movement pattern similar to that shown preoperatively. Whatever behavioral description best fits experimental data showing temporally spaced lesion effects, it may be of value to look for a biological explanation of these effects.

As in most cases the effects of temporally spaced lesions are not attributable to differences in lesion size, site or extent of retrograde and transynaptic degeneration, and as regeneration after injury remains very limited under normal circumstances in the mammalian central nervous system (CNS ; e.g. Black, 1985), it is necessary to look for functional and/or structural changes in the undamaged CNS. Two possibilities will be briefly considered : 1) Non-plastic functional changes and 2) plastic changes which involve both structural and functional reorganization (see Will et al., this volume).

Non-plastic changes

Compensatory behavioral processes do not necessarily imply structural reorganization, but may merely be realized through functional, non-plastic changes. Additional, apparent recovery of behavioral functions might simply be explained by a reduction of functional anomalies such as seizure activity in spared neuronal tissue. Several processes have been considered as likely candidates including a take over of the lost functions by the spared neural tissue following a reduction of the vascular and/or neural shock (Von Monakow, 1910) and/or a reduction in "emergence trauma" (Fuller, 1967), if one conceives the control of brain functions as being redundantly (Teitelbaum, 1971) or hierarchically (Jackson, 1881) achieved at a structural level.

Plastic changes

Although it is possible that non-plastic processes may underly some aspects of behavioral compensation, sparing or recovery found after CNS injury it is more probable that, because of the involvement of temporal factors, plastic events such as axonal sprouting underlie the temporally spaced lesion effects. First, behavioral improvement has been found after two-stage lesions even when the one-stage animals benefited from an overall longer post-operative interval than did the two-stage animals (e.g., Finger, 1978). Thus, in these cases at least, the more rapid behavioral improvement observed in the two-stage animals could not be attributed to a longer post-operative period conducive to a higher probability for the disappearence of "shock" or "emergence trauma" (e.g., Gentile et al., 1978). Secondly, as temporally spaced lesion effects were observed even after very long post-operative delays (Stein and Lewis, 1975), it seems very likely that initial post-operative "shock" and "emergence" trauma should have disappeared long before the beginning of behavioural testing. Moreover, Reyes (1977) reported that better recovery was observed when a long post-operative period was combined with a long interoperative interval. Thirdly, temporally spaced lesion effects were generally observed only when the interoperative interval consisted of at least ten days (Patrissi and Stein, 1975). This time is considerably longer than that required for inducing denervation supersentivity which could be observed after delays as short as 2 days, and thus supersensitivity is most probably not the sole process involved in temporally spaced lesion effects. The interoperative interval, however, approximately corresponds to the time necessary for axonal sprouting to occur in the brain regions examined (Cotman and Nieto-Sampedro, 1982). In the Glick and Greenstein's (1972) study in which serial effects were observed with a 30-day interoperative interval, even a 10-day interoperative interval was not long enough to facilitate recovery. Although subsequent findings by Glick and coworkers (1972) provide some support to the supersensitivity interpretation, a complementary explanation involving sprouting and reactive synaptogenesis also deserves consideration.

If temporally spaced surgery induces better recovery than one-stage surgery, one can assume that the structural reorganization occuring after the first stage lesion alters the structure-function relationships in the CNS such that the second-stage lesion (in a two-stage lesion paradigm) and the corresponding aspect of a one-stage lesion do not impinge on functionally equivalent structures. One can also assume that the structural reorganization through sprouting and through other processes that follow the second-stage lesion is quantitatively and/or qualitatively different from that produced by one-stage lesions.

Reactions to first-stage surgery. Following partial deafferentation of a given brain structure, intact neurons sprout new axon collaterals and terminals to reinnervate vacated targets (Cotman and Nadler, 1978) and display a reactive synaptogenesis by an "axonization" of dendrites in CNS regions apparently devoid of axonal sprouting reactions (Hamori and Silakov, 1980). Axonal sprouting induced by partial deafferentation has been observed in both the peripheral system (PNS) and the CNS, in many different brain areas (Cotman and Nieto-Sampedro, 1982). In many species, sprouting processes were also observed in adult animals although their latency, speed and extent was lower than that found in younger animals (Gall and Lynch, 1978). In some instances it has been shown that the new projections might mediate the recovery of function (Azmitia et al., 1978; Goldberger and Murray, 1974; Murray and Goldberger, 1974; Dieringer and Precht, 1977; Scheff and Cotman, 1977; Steward et al, 1977; Crutcher et al, 1983; Gage et al., 1983); however in other cases they presumably lead to the development of abnormal behavior (Mc Couch et al., 1958; Schneider, 1973).

Electrophysiological data to support the functional significance of such neuronal plasticity was provided by Scheff and Wright (1976) who tested rats for the retention of a black-white discrimination after serial or one-stage visual cortex ablations. Recovery of behavioral function was observed in rats with serial lesions which received interoperative training. Scheff and Wright also observed new evoked responses to light in the cortex adjacent to the damaged zone and suggested that cortical reorganization followed serial but not one-stage lesions. However, it is impossible to know whether the electrophysiological changes observed by Scheff and Wright were due to training or to a serial lesion effect which may be due either to reactions to first- or to second-stage surgery.

<u>Reactions to second-stage surgery.</u> In the "conditioning lesion" paradigm, axonal regeneration following axotomy (the testing lesion) has been reported to be changed in neurons which have undergone a previous axonal injury (the conditioning lesion) (e.g., Forman et al., 1981). These "conditioning lesion effects" have been found in several types of neurons of the mammalian PNS and in neurons of the visual system of lower vertebrates. Although the details of the response to conditioning surgery are not the same in different types of neurons, the most consistent effects of conditioning lesions appear to be a reduction of the initial delay for sprout outgrowth and an increase in speed of that outgrowth after the second-stage surgery. The amount and nature of materials synthetized in the cell body and transported to axon terminals may be changed by conditioning lesions (Black, 1985). Conditioning lesions may also hasten the morphological maturation of sprouts and, in some systems, increase remyelinisation after the second surgery. It has also been shown that only the motor or sensory axons which received the conditioning lesion showed accelerated growth, while nearby nonconditioned axons were unaffected (Mc Quarrie, 1978; Forman et al., 1980) and that a conditioning axotomy lesion of peripheral axons originating in the dorsal root ganglion increased the rate of axonal regeneration of only the peripheral, but not the central axons (Oblinger and Lasek, 1984). Interestingly, such conditioning lesions have also been reported to accelerate recovery of motor functions following PNS damage in rats (Gutman, 1942; Hall-Craggs and Brand, 1977; Sebille and Bondoux-Jahan, 1980) and of visual functions after optic nerve crush in goldfishes (Edwards et al., 1981). These conditioning lesions experiments, however, differ markedly from that of serial and progressive lesion paradigms in that the two (or more) temporally spaced lesions concern the same neuronal fibers at each stage, and not different (ipsi- or contralateral) parts of a given CNS structure.

Scheff et al. (1977, 1978), using an experimental paradigm similar to the progressive lesion paradigm, observed that the neuronal reactions to the second surgery were dependent on the first surgery. That is, they found in the mammalian CNS a phenomenon paralleling the conditioning lesion effect found in the mammalian PNS and in the CNS of lower vertebrates. However, this phenomenon concerns axonal sprouting rather than (but perhaps also) axonal regeneration. Scheff et al. made partial lesions of the entorhinal cortex (EC) and found that the spared fibers of the perforant path which project onto the granule cells of the dentate gyrus showed sprouting and expansion of their projection field. This sprouting phenomenon appears only four or five days after a single-stage lesion and is rather sluggish and limited in its extent. However, when a small first-stage lesion (termed "priming lesion" by Scheff et al., 1977) of the most medial EC was produced before the subsequent removal of the remaining medial EC, the delay for the onset of the growth response following the second lesion was greatly shortened (within two days) and the ultimate extent of sprouting increased. Such "conditioning" lesion effects were also observed when the medial EC lesion was preceded by a lateral (instead of a limited medial) EC lesion, but they were not observed after a commissurotomy or a lesion of the crossed perforant path. These data suggest that sprouting of axonal collaterals from contralateral (homotypic ?) neurons might be a prerequisite for conditioning lesion effects in the mammalian CNS.

Conditioning lesions were shown to produce significant effects only when unexpectedly stringent temporal requirements were used. Conditioning lesion effects have not been observed when the interoperative interval was shorter than 2 days in the case of axonal regeneration (Forman et al., 1980; Edwards et al., 1981) and shorter than 4 days in that of axonal sprouting (Scheff et al, 1978). In either case, the conditioning lesion effects were optimal with a two-week interoperative interval and clearly diminished with a one month interoperative interval.

However, as shown by Scheff et al. (1977, 1978), there are not only temporal requirements but also spatial requirements for conditioning lesion effects to occur. The observations by Scheff and coworkers as well as those by Fass (1983) lead one to consider interhemispheric relationships and, particularly, first- and second-stage reactions of contralateral (homotypic ?) neurons as an important feature for the explanation of temporally spaced lesion effects. Such considerations led us to use only unilateral deafferentation for the first-stage surgery of our temporally spaced lesion experiments (in contrast to what was done by Glick and Greenstein).

DIRECTIONS FOR FUTURE RESEARCH

Studies with serial and progressive lesions have demonstrated the importance of temporal as well as spatial factors for the outcome of neurosurgery, and have provided strong arguments against too rigid a view of localization of brain functions (e.g., Will and Stein, 1981). However, these experimental paradigms are perhaps not the most well suited for specifying physiological and anatomical changes which may underly serial lesion effects, as these changes may be produced anywhere in the spared neural tissue. Furthermore, in the classical serial and progressive lesion paradigms, the quantity of neural tissue damaged in one-stage and in two-stage operated groups differs at any stage considered, so that these groups might differ also in terms of non-plastic alterations such as recovery from vascular and neural shock or from emergence trauma.

If one assumes that plastic changes such as sprouting do contribute to behavioral recovery, then not only enhanced (second-stage) sprouting but also, as suggested by Fass (1983), the manipulation of the spatial and temporal pattern of sprouting might alter the rate and extent of recovery. Such plastic changes, whether quantitative or qualitative, might be detected in one or several CNS structures physically connected to the serially or progressively damaged structure. However, another perhaps more convenient and efficient way to look for such plastic changes is to use the temporally and structurally spaced lesion paradigm illustrated in Figures 1C and 1D. In this way, we preserve a genuine possibility for enhanced homotypic collateral sprouting to occur where it may have functional significance, plus some indication of the cerebral locus where the structural and/or functional changes that underlie serial lesion effects might be found.

Moreover, it would be of both theoretical and practical value to pursue the use of first-stage unilateral deafferentation to investigate the effects of multi-stage lesions. First, as the neural damage inflicted on the target structure at stage two in the two-stage groups is the same as that used in the one-stage lesion group, and as the overall damage is greater, the behavioral recovery we have observed cannot be due to less severe shock or emergence trauma. Secondly, we have shown that although electrolytic and low dose colchicine treatments produced similar behavioral effects, the colchicine treatment produced better recovery of biochemical and electrophysiological parameters. Thus, as direct brain damage produced by DC current is not essential for the initiation of sprouting (Cotman et al, 1981b), it can be assumed that one critical process sufficient to induce serial lesion effects is the interruption or reduction of axonal transport by neurotoxins such as colchicine. The hypothesis (Diamond et al., 1976; Goldowitz and Cotman, 1980; Cotman et al., 1981b) that propensity of nerves for collateral growth which is ordinarily repressed (in adult animals) by factors that depend on axonal transport, could in part explain our results. Indeed, the interruption or reduction of the supply of these putative anti-growth factors transported by neurons could in effect enhance the efficacy of endogenous trophic factors. The gradual accumulation of these trophic factors in the damaged area reaches a maximum at a time (15 days) similar to the optimal interoperative interval for inducing behavioral serial lesion effects as well as conditioning lesion effects.

ACKNOWLEDGEMENTS

This research was supported in part by a grant to B. Will from the Ministère de la Recherche et de la Technologie (ATP.N84.C.1320). The authors express their appreciation to Dr. J. Dalrymple-Alford for his helpful suggestions and are grateful for the skillful assistance of M. de Dios, F. Deluzarche, Y. Jeantet and M. Klughertz.

REFERENCES

Adametz, J.H., 1959, Rate of recovery of functioning in cats with rostral reticular lesions. An experimental study, J. Neurosurg. , 16:85.

Avrith, D., Haas, H.L., and Mogenson, G.J., 1979, Behavioral and electrophysiological changes following microinjections of colchicine into the substantia nigra, Neuroscience , 4:227.

Azmitia, E.C., Buchan, A.M., and Williams, J.H., 1978, Structural and functional restoration by collateral sprouting of hippocampal 5-HT axons, Nature , 274:374.

Barbas, H., and Spear, P., 1976, Effects of serial unilateral and serial bilateral visual cortex lesions on brightness discrimination relearning in rats, J. Comp. Physiol. Psychol. , 90:279.

Black, M.M., 1985, Intrinsic determinants of regenerative vigor, in : "Central Nervous System Plasticity and Repair", A. Bignami, F.E. Bloom, C.L. Bolis and A. Adeloye, eds., Raven Press, New-York.

Cotman, C.W., Lewis, E.R., and Hand, D., 1981, The critical afferent theory: A mechanism to account for septohippocampal development and plasticity, in : "Lesion - Induced Neuronal Plasticity in Sensorimotor Systems", H. Flohr and W. Precht, eds, Springer-Verlag, Berlin.

Cotman, C.W., and Nadler, J.V., 1978, Reactive synaptogenesis in the hippocampus, in : "Neuronal plasticity", C.W. Cotman, ed., Raven Press, New-York.

Cotman, C.W., and Nieto-Sampedro, M., 1982, Brain function, synapse renewal, and plasticity, Ann. Rev. Psychol. , 33:371.

Cotman, C.W., Nieto-Sampedro, M., and Harris, E.W., 1981b, Synapse replacement in the nervous system of adult vertebrates, Physiol. Rev., 61:684.
Crutcher, K.A., Kesner, R.P, and Novak, J.M., 1983, Medial septal lesions, radial arm maze performance, and sympathetic sprouting: A study of recovery of function, Brain Res. , 262:91.
Curtis, S.D., and Nonneman, A.J., 1977, Effects of successive bilateral hippocampectomy on DRL performance in rats, Physiol.Behav. , 19-707.
Diamond, J., Cooper, E., Turner, C., and MacIntyre, L., 1976, Trophic regulation of nerve spouting, Science , 193:371.
Dieringer, N., and Precht, W., 1977, Modification of synaptic input following unilateral labyrinthectomy, Nature (London), 269:431.
Dravid, A.R., and Van Deusen, E.B., 1983, Recovery of choline acetyltransferase and acetylcholinesterase activities in the ipsilateral hippocampus following unilateral, partial transection of the fimbria in rats, Brain Res. , 277:169.
Dravid, A.R., and Van Deusen, E.B., 1984, Recovery of enzyme markers for cholinergic terminals in septo-temporal regions of the hippocampus following selective fimbrial lesions in adult rats, Brain Res., 324:119.
Edwards, D.L., Alpert, R.M., and Grafstein, B., 1981, Recovery of vision in regeneration of goldfish optic axons: Enhancement of axonal outgrowth by a conditioning lesion, Exp. Neurol. , 72:672.
Fass, B., 1983, Temporal changes in open-field activity following progressive lesions of entorhinal cortex: Evidence for enhanced recovery , Behav. Neural Biol. , 37:108.
Field, P.M., Coldham, D.E., and Raisman, G., 1980, Synapses formation after injury in the adult rat brain: Preferential reinnervation of denervated fimbrial sites by axons of the contralateral fimbria, Brain Res. , 18:103.
Finger, S., 1978, Lesion momentum and behavior, in : Recovery from brain damage: Research and theory", Finger, ed., Plenum, New-York.
Finger, S., Marshak, R.A., Cohen, M., Scheff, S., Trace, R., and Niemand, D., 1971, Effects of successive and simultaneous lesions of somatosensory cortex on tactile discrimination in the rat, J. Comp. Physiol. Psychol. , 77:221.
Finger, S., and Stein, D.G., Fast-versus slow-growing lesions and behavioral recovery, in :"Brain Damage and Recovery. Research and Clinical Perspectives", S. Finger and D.G. Stein, eds., Academic Press, New-York.
Finger, S., Walbran, B., and Stein, D.G., 1973, Brain damage and behavioral recovery: Serial lesion phenomena, Brain Res. , 63:1.
Flourens, P., 1824, Recherches expérimentales sur les propriétés et les fonctions du système nerveux chez les animaux vertébrés, Crevot, Paris.
Forman, D.S., McQuarrie, I.G., Grafstein, B., and Edwards, D.L., 1981, Effect of a conditioning lesion on axonal regeneration and recovery of function, in : "Lesion - Induced Neuronal Plasticity in Sensorimotor Systems", H. Flohr and W. Precht, eds., Springer-Verlag, Berlin.
Forman, D.S., McQuarrie, I.G., Labore, F.W., Wood, D.K., Stone, L.S., Braddock, C.H., and Fuchs, D.A., 1980, Time course of the conditioning lesion effect on axonal regeneration, Brain Res. , 182:180.
Frötscher, M., 1983, Dendritic plasticity in response to partial deafferentation, in : "Neurobiology of the Hippocampus", W. Seiffert, ed., Academic Press, London.
Fuller, J.L., 1967, Experiential deprivation and later behavior, Science , 158:1645.

Gage, F.H., BJörklund, A., and Stenevi, U., 1983, Reinnervation of the partially deafferented hippocampus by compensatory collateral sprouting from spared cholinergic and noradrenergic afferents, Brain Res. , 268:27.

Gage, F.H., Björklund, A., and Stenevi, U., 1984, Cells of origin of the ventral cholinergic septo-hippocampal pathway undergoing compensatory collateral sprouting following fimbria-fornix transection, Neurosci. Lett. , 44:211.

Gage, F.H., Björklund, A., Stenevi, U., and Dunnett, S.B., 1983, Functional correlates of compensatory collateral sprouting by aminergic and cholinergic afferents in the hippocampal formation, Brain Res. , 268:39.

Gall, C., and Lynch, G., 1981, The regulation of sprouting in the adult hippocampus: Some insights from developmental studies, in : "Lesion - Induced Neuronal Plasticity on Sensorimotor Systems", H. Flohr and W. Precht, eds., Springer-Verlag, Berlin.

Gentile, A.M., Green, S., Nieburgs, A., Schmelzer, W., and Stein, D.G., 1978, Disruption and recovery of locomotor and manipulatory behavior following cortical lesions in rats, Behav. Biol. , 22:417.

Glick, S.D., and Greenstein, S., 1972, Facilitation of recovery after hypothalamic damage by prior ablation of frontal cortex, Nature ,239:187.

Glick, S.D., Greenstein, S., and Zimmerberg, B., 1972, Facilitation of recovery by alpha-methyl-p-tyrosine after lateral hypothalamic damage, Science , 177:534.

Goldberger, M.E., and Murray, M., 1974, Restitution of function and collateral sprouting in the cat spinal cord: The deafferented animal, J. Comp. Neurol. , 158:37.

Goldowitz, D., and Cotman, C.W., 1980, Do neurotrophic interactions control synapse formation in the adult rat brain, Brain Res. , 181:325.

Goldschmidt, R.B., and Steward, O., 1982, Neurotoxic effects of colchicine: Differential susceptibility of CNS neuronal populations, Neuroscience, 7:695.

Grafstein, B., and McQuarrie, I.G., 1978, Role of the nerve cell body in axonal regeneration, in : "Neuronal Plasticity", C.W. Cotman, Ed., Raven Press, New-York.

Guth, L., Smith, S., Donati, E.J., and Albuquerque, E.X., 1980, Induction of intramuscular collateral nerve sprouting by neurally applied colchicine, Exp. Neurol , 67:513.

Gutman, E., 1942, Factors affecting recovery of motor function after nerve lesions, J. Neurol. Psyhiatr. , 5:81.

Hall-Graggs, E.C.B., and Brand, P., 1977, Effects of previous nerve injury on the regeneration of free autogenous muscle grafts, Exp. Neurol. , 57:275.

Hamori, J., and Silakov, V.L., 1980, Plasticity of relay neurons in dorsal lateral geniculate nucleus of the adult cat: Morphological evidence, Neurosci. , 5:2073.

Isseroff, A., Leveton, L., Freeman, G., Lewis, M.E., and Stein, D.G., 1976, Differences in the behavioral effects of single-stage and serial leions of the hippocampus, Exp. Neurol. , 53:339.

Jackson, J.H., 1881, Remarks on the dissolution of the nervous system as exemplified by certain post-epileptic conditions, Medical Press and Circular , 1:329.

Jeannerod, M., and Hecaen, H., 1979, Adaptation et restauration des fonctions nerveuses, Simep, Villeurbanne.

Kleitman, N., and Camille, N., 1932, Studies on the physiology of sleep. VI. The behavior of decorticated dogs, Am. J. Physiol. , 100:474.

Lashley, K.S., 1929, "Brain mechanisms and intelligence", University of Chicago Press, Chicago.

Loesche, J., and Steward, O., 1977, Behavioral correlates of denervation and reinnervation of the hippocampal formation of the rat: Recovery of alternation performance following unilateral entorhinal cortex lesions, Brain Res. Bull. , 2:31.
Le Vere, T.E., 1980, Recovery of function after brain damage: A theory of the behavioral deficit, Physiol. Psychol. , 8:297.
Mc Couch, G.P., Austin, G.M., Lin, C.N., and Lun, C.Y., 1958, Sprouting as a cause of spasticity, J. Neurophysiol. , 21:205.
Mc Quarrie, I.G., 1978, The effect of a conditioning lesion on the regeneration of motor axons, Brain Res. , 152:597.
Murray, M., and Goldberger, M.E., 1974, Restitution of function of collateral sprouting in the cat spinal cord: The partially hemisected animal, J. Comp. Neurol. , 158:19.
Oblinger, M.M., and Lasek, R.J., 1984, A conditioning lesion of the peripheral axons of dorsal root ganglion cells accelerates regeneration of only their peripheral axons, J. Neurosci. , 4:1736.
Panksepp, J., Meeker, R., Reilly, P., and Vilberg, T., 1976, Reversible CNS lesions and disruption of self-stimulation by inhibition of axoplasmic flow, in : "Brain-Stimulation Reward", A. Wauquier and E.T. Rolls, eds., Elsevier, New-York.
Patrissi, G., and Stein, D.G., Temporal factors in recovery of function after brain damage, Exp. Neurol. , 47:470.
Perisic, M., and Cuenod, M., 1972, Synaptic transmission depressed by colchicine blockade of axoplasmic flow, Science , 175:1140.
Pilar, G., and Landmesser, L., 1972, Axotomy mimicked by localized colchicine application, Science , 171:1116.
Riese, W., Aphasia in brain tumors, Confin.Neurol. , 9:64.
Reyes, R., 1977, Effects of interlesion interval and postoperative recovery period on recovery of function following serial lesions of somatic cortex in rats, Unpublished doctoral dissertation , Washington University.
Scheff, W., Bernardo, L., and Cotman, C.W., 1977, Progressive brain damage accelerates axon sprouting in the adult rat, Science , 197:795.
Scheff, S.W., Bernardo, L.S., and Cotman, C.W., 1978, Effect of serial lesions on sprouting in the dentate gyrus: Onset and decline of the catalytic effect, Brain Res. , 150:45.
Scheff, S.W., and Cotman, C.W., 1977, Recovery of spontaneous alternation following lesions of the entorhinal cortex in adult rats: Possible correlation to axon sprouting, Behav. Biol. , 21:286.
Scheff, S., and Wright, D.C., 1976, Behavioral and electro- physiological evidence for cortical reorganization of function in rats with serial lesions of the visual cortex, Physiol. Psychol. , 5:103.
Schneider, G.E., 1973, Early lesions of superior colliculus: Factors affecting the formation of abnormal retinal pojections, Brain Behav. Evol. , 8:73.
Sebille, A., and Bondoux-Jahan, M., 1980, Effects of electric stimulation and previous nerve injury on motor function recovery in rats, Brain Res. , 193:362.
Stein, D.G., 1974, Some variables influencing recovery of funcion in the rat, in : "CNS Plasticity: Recovery of function", D.G. Stein, J. Rosen and N. Butters, eds., Academic Press, New-York.
Stein, D.G., and Lewis M.E., 1975, Functional recovery after brain damage in adult organisms, in : "Aspects of Neural Plasticity", F. Vital-Durand and M. Jeannerod, eds., Colloques INSERM, Paris.
Steward, O., Loesche, J., and Horten, W.C., 1977, Behavioral correlation of denervation and reinnervation of the hippocampal formation of the rat: Open-field activity and cue utilization following bilateral entorhinal cortex lesions, Brain Res. Bull. , 2:41.
Teitelbaum, P., 1971, The encephalization of hunger, in : "Progress in Physiological Psychology", E. Stellar and J.M. Sprague, eds., Academic Press, New-York.

Toniolo, G., and Will, B., 1983, Effects of fimbria-fornix "conditioning" lesions on behavioral recovery from subsequent hippocampal lesions in rats, III European Winter Conf., Abstracts.
Toniolo, G., Dravid, A., Destrade, C., and Will, B., 1984, Conditioning lesions and recovery of cognitive functions, EBBS-SFECA Workshop, Abstracts.
Treichler, R.F., 1975, Two-stage frontal lesion influences upon severity of delayed-response deficit, Behav. Biol., 13:35.
Von Monakow, C., 1910, Neue Gesichtspunkte in der Frage nach der Lokalization im Grosshirn, J.F. Bergman, Wiesbaden.
Will, B., and Stein, D.G., 1981, La récupération après lésions cérébrales, Pour la Science, 43:66.
Will, B., and Toniolo, G., 1983, Both transitory and irreversible fimbria/fornix "conditioning" lesions facilitate recovery from bilateral hippocampal lesions, Soc. Neurosci. Abstracts, 9:946.
Zolovick, A.J., Avrith, D., and Jalowiec, J.E., 1977, Reversible colchicine-induced disruption of amygdaloid function in sodium appetite, Brain Res. Bull., 5:35.

NEUROPEPTIDES AND FUNCTIONAL RECOVERY AFTER BRAIN DAMAGE

H. Dick Veldhuis, Csaba Nyakas and David De Wied

Rudolf Magnus Institute for Pharmacology
Medical Faculty, University of Utrecht
Vondellaan 6, 3521 GD Utrecht, The Netherlands

INTRODUCTION

Recent studies have shown that systemic administration of ACTH-(1-39) can enhance functional and structural changes after crushing of peripheral nerves in the rat. Animals, having elevated ACTH-plasma lvels, due to adrenalectomy, or after peripheral ACTH-administration, recovered more quickly from disturbed foot movements (normal and heat-evoked) after nerve crush than control animals (Strand and Kung, 1980; Saint-Côme et al., 1982). Such beneficial effects of ACTH on functional recovery after peripheral nerve crush were also reported by Bijlsma and colleagues (1981). They found that sensorimotor function in rats after sciatic nerve crush was facilitated by ACTH-related peptides, especially those containing the melanotropic moiety (Bijlsma et al., 1983a). ACTH-(4-10) and the ACTH-(4-9) analogue, ORG 2766 were active in this respect, but α-MSH was shown to be the most potent peptide influencing recovery speed (Bijlsma et al., 1983a). Treatment with the N- and C-terminal parts of ORG 2766, ACTH-(4-7) and the tripeptide Phe^{7}-D-Lys^{8}-Phe^{9} (PDLP) respectively, and with ACTH-(11-24) remained ineffective. Thus, sofar, the effective fragments having beneficial effects on recovery include those having behavioral but not adrenocortical potency. Interestingly, although ORG 2766 exhibits a thousand times more potent effect than ACTH-(4-10) on avoidance behavior (De Wied et al., 1975; Greven and De Wied, 1977), it was only as active as the ACTH-(4-10) fragment in the test of functional recovery. Besides functional recovery, Bijlsma et al. (1983b) found that the number of outgrowing neurites in the regenerating sciatic nerve was facilitated by ACTH-related peptides. Using the compensation process following unilateral labyrinthectomy in Rana temporaria as a preparation to study the influence of ACTH-related peptides in lesion-induced plastic processes, it was found that chronic ACTH-(4-10) administration accelerated the acquisition of the compensation state, while D-Phe^{7}-ACTH-(4-10) inhibited vestibular compensation (Flohr and Lüneberg, 1982; Flohr, 1983). Repeated administration of ORG 2766 has been shown to ameliorate some of the usual behavioral consequences of septal area damage, such as hyperemotionality, by hastening a return to near normal emotional responsiveness (Isaacson and Poplawsky, 1983). Finally, long-term treatment with ORG 2766 of mid-aged rats reduced hippocampal morphological correlates of brain aging, such as neuronal loss and glial reactivity. Moreover, some behavioral deficits developing during the aging process could also be retarded by this treatment (Landfield et al., 1981).

The parafascicular area of the posteromedial thalamus plays an important role in the processes concerned with extinction of conditioned avoidance behavior. Extensive lesions in the posterior thalamus interfere with both acquisition and extinction of a conditioned avoidance response, while smaller lesions that destroy only the parafascicular area solely affect extinction of the response (Bohus and De Wied, 1967a; Van Wimersma Greidanus et al., 1974). Treatment with long-acting preparations of α-MSH or ACTH-(4-10) in rats given lesions of the parafascicular area failed to affect the rate of extinction of the avoidance response, indicating a role for this brain region as a site of action of ACTH-related peptides (Van Wimersma Greidanus et al., 1974; Bohus and De Wied, 1967b). Lesions of the parafascicular region have also been found to retard acquisition of spatial reversal tasks in the rat (Thompson, 1976; Thompson et al., 1981), and from these studies it was proposed that the parafascicular area contributed to the consolidation of spatial memory traces.

The present paper describes experiments in which lesions of the parafascicular area in rats are used as a model to assess the effect of chronic treatment with two ACTH-related peptides, i.e. ORG 2766 and α-MSH, on functional recovery from damage in the central nervous system, as measured by T-maze reversal learning ability. The results suggest that ORG 2766 and α-MSH have a stimulatory effect on recovery of cognitive functions as a result of their neurotropic influences, and not via a short-term effect on attention and motivation.

METHODS

Animals. Male Wistar rats of an inbred strain (Central Breeding Laboratory TNO, Zeist, The Netherlands) weighing 205 - 230 g at the time of operation were used. They were housed 5 per cage under controlled light-dark conditions (light on from 05.00 to 19.00 h) and had ad libitum access to food and tap water.

Surgery. Bilateral lesions of the parafascicular area were made stereotaxically under Hypnorm® anesthesia with a radio-frequency lesion maker (Grass LM4), using a current intensity of 10 mA for 10 sec. A stainless steel electrode of 150 μm diameter, with a 0.7 mm uninsulated tip was used. For sham-lesions the electrode was inserted to the parafascicular area, but no current was delivered. Reconstruction of the lesions was made on 100 μm thick sections, taken at 200 μm distance.

Behavior. Open-field behavior was tested in a circular arena as described by Weijnen and Slangen (1970). Ambulation, rearing, grooming episodes and defecation were recorded during three minutes. The animals were tested at post-operative days 7, 13 and 19 between 08.00 and 11.00 h.

The rod grasping response of rats was tested using a horizontally positioned rod. The time the animal hang suspended above the floor was measured to the nearest 0.1 sec. (Van Ree et al., 1980). Rod grasping was measured before (3rd postoperative day) and during the course of chronic peptide treatments. The test sessions were performed prior to injection.

Reversal learning in a T-maze was performed as described earlier by Thompson (1976). On the first day the rats were briefly pretrained by being allowed to explore the alleys and goal boxes and were then trained to run from the start box and into a goal box to avoid electric footshock. If the rat had not started to move in the correct direction within 15 sec after placement in the start box, a scrambled electric footshock (0.4 mA; 1 sec) was delivered repeatedly till the animal entered the correct goal box. The time allowed in the goal box varied between 20 - 25 sec. The training was continued till 5 consecutive positive responses had been made. An additional reinforcing trial was given one hour after the learning session. Reversal learning was performed on the second day, starting with a retention trial, which served to reinforce the response acquired on the learning day. For reversal learning the rat was forced to choose the opposite arm and goal box of the maze by closing the door of that goal box that was "correct" at the learning day. Footshocks (0.2 mA; 1 sec) were delivered if an animal did not leave the start box within 25 sec, or the choice-point of the maze and also when the rat chose the incorrect (previously correct) arm to a distance of 5 - 6 cm (or less) of the incorrect goal. Trials were repeated till 5 consecutive errorless responses had been made. An error was recorded when a rat approached the incorrect goal box to a distance of 5 - 6 cm.

Electric footshock sensitivity. One day after T-maze reversal learning the electric footshock sensitivity (EFS) was measured according to the procedure described elsewhere (Gispen et al., 1970). The lowest current intensity which elicited flinch and jerk served as an index of the response threshold.

Peptide treatments. Chronic peptide treatment consisted of 8 subcutaneous injections, administered on alternate days, starting at postoperative day 3. Either the ACTH-(4-9) analogue, ORG 2766 (H-Met(O_2)-Glu-His-Phe-D-Lys-Phe-OH) or α-MSH were used (peptides donated by H.M. Greven, Organon International B.V., Oss, The Netherlands). Each peptide was given in doses of 10 μg or 50 μg, dissolved in 0.5 ml physiological saline. In another experiment, the peptides were injected acutely one hour prior to the reversal learning session.

RESULTS

The localization and extent of the parafascicular area lesions are shown in Fig. 1. The maximal and minimal sizes of the lesion are indicated.

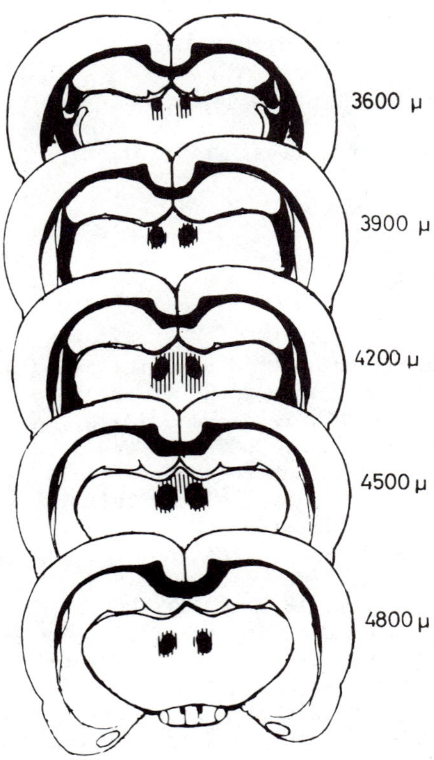

Figure 1:
Extent of parafascicular area lesion is shown on serial brain sections. Smallest (black area) and largest (hatched area) lesions are indicated, using modified pictures of the atlas of Palkovits (1980).

The majority of animals demonstrated an intermediate size of lesions. There was an acceptable overlap of the lesion sizes in the rats of the different experimental groups.

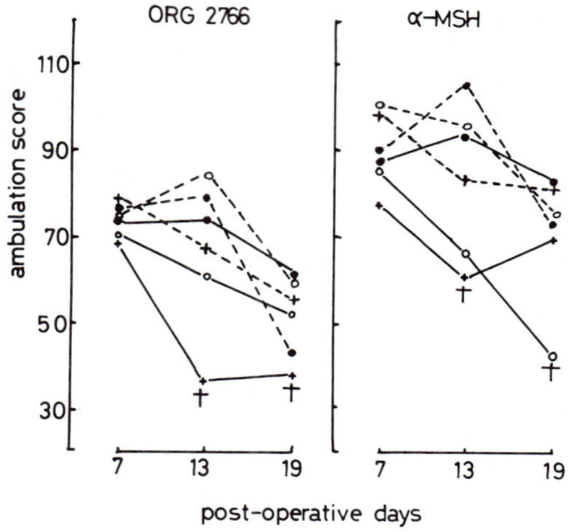

Figure 2:
Ambulatory activity at post-operative days 7, 13 and 19 of parafascicular and sham-lesioned rats, treated chronically with ORG 2766 (left) or α-MSH (right). Means (n = 8) are shown. S.E.M.'s (not shown) varied between 5% and 15% of the mean values. Sham rats are represented by solid lines; rats with lesions are represented by broken lines.
●, treated with saline; o, treated with 10 μg peptide;
+, treated with 50 μg peptide.
† $p < 0.05$ vs saline-treated controls (Dunnett test, two-tailed)

The reason for following open-field behavior and rod grasping responses in the course of treatment was to make an initial assessment of treatment effects on motor function in both lesioned and sham-lesioned animals. In the open-field test, treatment with the higher dose of ORG 2766 and both doses of α-MSH resulted, on days 13 and 19 (second and third open-field test), in a significantly decreased ambulation in sham-lesioned animals only (Fig. 2). Thus, the peptides apparently facilitated habituation in the sham-lesioned rats, but had no effects in rats with lesions. Rearing activity was significantly lower in lesioned rats (day 7: 12.7 ± 2.3) compared to that in sham-lesioned controls (day 7: 18.7 ± 2.1), but was not affected by the peptide treatments. Grooming activity as well as defecation were not affected by the lesion, nor by chronic peptide treatments (data not shown).

Fig. 3 shows the results obtained in the rod grasping test. The duration of rod grasping was markedly increased in lesioned rats in comparison to sham-lesioned controls. From post-operative day 3, the grasping time of lesioned rats gradually decreased till day 13. Peptide treatments affected neither the duration of rod grasping of lesioned rats, nor that of sham lesioned animals. Data analysis on acute effects of the peptides

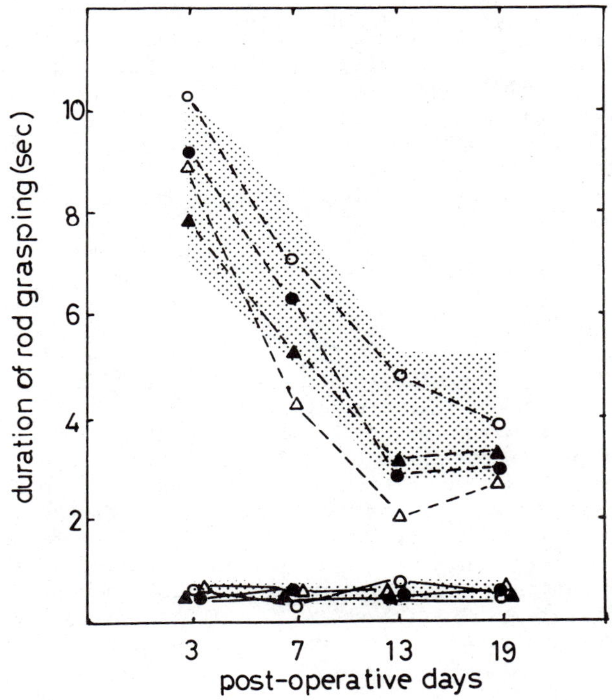

Figure 3:
Rod grasping of parafascicular- and sham-lesioned rats during and after chronic treatment with ORG 2766 and α-MSH. Means (n = 8) are shown. Dotted areas indicate mean ± S.E.M. of lesioned and sham-lesioned rats receiving saline.
Sham rats are represented by solid lines; rats with lesions are represented by broken lines.
o, treated with 10 μg ORG 2766; Δ treated with 50 μg ORG 2766; •, treated with 10 μg α-MSH; ▲ treated with 50 μg α-MSH.

on rod grasping also revealed no treatment effects in both experimental groups (data not shown).

The effects of chronic treatment with ORG 2766 and α-MSH on T-maze reversal learning (tested on post-operative day 21) are shown in Fig. 4. Sham-lesioned rats showed a relatively low number of errors and their performance was unaffected by peptide treatments. In lesioned animals, chronic treatment with ORG 2766 resulted in a dose-dependent reduction of the number of errors made during reversal learning. Treatment with the highest dose of ORG 2766 resulted in a reversal learning performance in

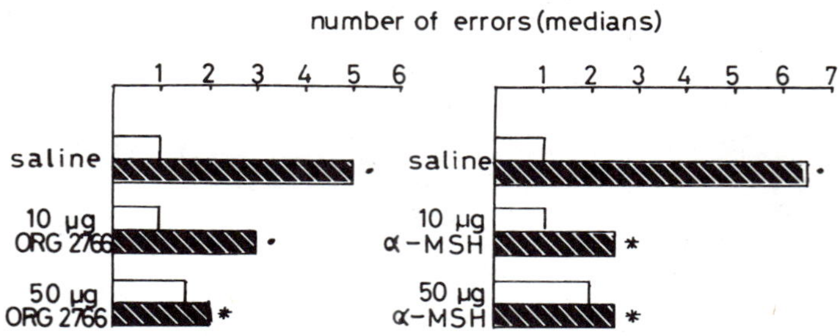

Figure 4:
Effect of chronic treatment with ORG 2766 (left) or α-MSH (right) on reversal learning performance of parafascicular area lesioned (▧) and sham-lesioned (□) rats. Median number of errors (n = 8) is given.
o: $p < 0.05$ vs appropriately treated sham-lesioned controls
*: $p < 0.05$ vs saline-treated lesioned rats
(Kruskal-Wallis analysis followed by Mann-Whitney U-test).

lesioned rats that was not significantly different from sham-lesioned controls treated with either saline or the ACTH-(4-9) analogue. Both doses of α-MSH reduced the number of errors in rats with lesions of the parafascicular area to the amount observed in similarly treated sham-lesioned rats. In contrast to chronic peptide treatment as shown in Fig. 4, acute treatment with either ORG 2766 or α-MSH did not affect the impaired reversal learning of lesioned rats (Fig. 5).

Finally, no differences were observed in electric footshock sensitivity between sham-lesioned and lesioned rats. Threshold current intensities eliciting flinch were 41.7 ± 1.5 and 42.2 ± 1.9 µA respectively, for sham-lesioned and lesioned rats; jerking was observed at threshold currents of

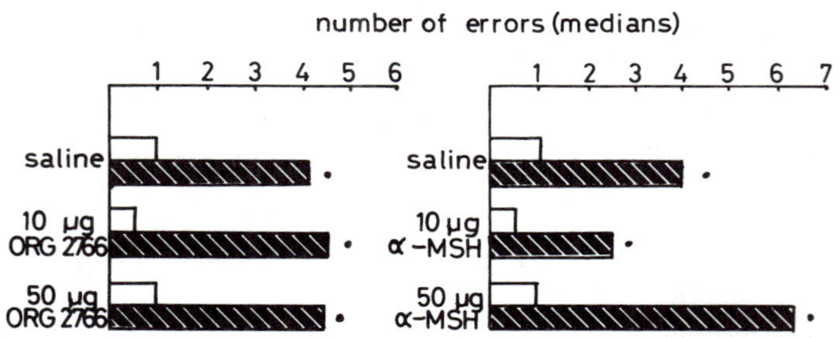

Figure 5:
Effect of acute peptide treatment (one hour prior to the reversal learning session) on reversal learning performance of parafascicular area lesioned (▧) and sham-lesioned (□) rats. Median number of errors (n = 8) is shown.
0: $p < 0.05$ vs appropriately treated sham-lesioned control
(Kruskal-Wallis analysis followed by Mann-Whitney U-test)

56.4 ± 3.7 and 67.4 ± 5.7 μA, respectively. Neither chronic ORG 2766 nor α-MSH treatment in sham-lesioned or lesioned rats affected these threshold current intensities.

DISCUSSION

The present study shows that chronic treatment with two ACTH-related peptides, i.e. the ACTH-(4-9) analogue ORG 2766 and α-MSH, results in a normalization of some impairments in cognitive function, but not of motor dysfunctions, in rats with bilateral parafascicular area lesions. This inference was made using reversal learning ability in a T-maze as a measure of cognitive functioning and activity in an open-field test and grasping responses as parameters of motor functions. The improved performance of reversal learning in lesioned rats could not be explained by a short-term effect of ACTH-related peptides on attention and motivation, as reported previously (Bohus and De Wied, 1980). Reversal learning ability was tested four days after the last peptide injection and, furthermore, acute treatment with ORG 2766 and α-MSH one hour prior to the reversal learning session did not affect the impaired performance of lesioned animals. These findings are in line with earlier studies, in which administration of ACTH-related peptides failed to affect impaired active avoidance learning in parafascicular area lesioned rats (Van Wimersma Greidanus et al., 1974; Bohus and De Wied, 1967b).

At present, the mechanism of action of ACTH-related peptides on recovery from a behavioral deficit induced by brain damage is still unclear. Histological examination of the lesioned area could give information about a possible facilitation of regeneration of spared neuronal elements by the peptide treatments. At the moment there is some evidence supporting this possibility: ACTH and related peptides have been shown to increase the number of regenerating axons (Bijlsma et al., 1983b), and also to facilitate axonal growth (Strand and Kung, 1980) after crushing of peripheral axons. In the brain, long-term treatment with ORG 2766 reduces morphological correlates of brain aging (Landfield et al., 1981). Alternatively, according to the hypothesis put forward by LeVere and LeVere (1982) (the compensation theory of behavioral deficits), recovery of function could also be enhanced by suppression of functionally antagonistic structures. In this respect it is noteworthy that the parafascicular area has close anatomical and functional relations with the hippocampus (Thompson et al., 1981).

In conclusion, the results of this study point to a beneficial effect of ORG 2766 and α-MSH on recovery of a cognitive function, likely caused by long-term neurotropic influences; the mechanism of action (influences on brain plasticity or nerve regeneration per se) needs to be elucidated.

REFERENCES

Bohus, B., and De Wied, D., 1967a, Avoidance and escape behavior following medial thalamic lesions in rats, J. Comp. Physiol. Psychol., 64: 26.

Bohus, B., and De Wied, D., 1967b, Failure of α-MSH to delay extinction of conditioned avoidance behavior in rats with lesions in the parafascicular nuclei of the thalamus, Physiol. Behav. 2: 221.

Bohus, B., and De Wied, D., 1980, Pituitary-adrenal system hormones and behavior, in: "General, Comparative and Clinical Endocrinology", I.C. Jones, and I.W. Henderson, eds., Academic Press, London.

Bijlsma, W.A., Jennekens, F.G.I., Schotman, P., and Gispen, W.H., 1981, Effect of corticotrophin (ACTH) on recovery of sensimotor function in the rat: structure-activity study, Eur. J. Pharmacol. 76: 73.

Bijlsma, W.A., Schotman, P., Jennekens, F.G.I., Gispen, W.H., and De Wied,

D., 1983a, The enhanced recovery of sensimotor function in rats is related to the melanotropic moiety of ACTH/MSH neuropeptides, Eur. J. Pharmacol., 92: 231.

Bijlsma, W.A., Jennekens, F.G.I., Schotman, P., and Gispen, W.H., 1983b, Stimulation by ACTH-(4-10) of nerve fiber regeneration following sciatic nerve crush, Muscle and Nerve 6: 102.

De Wied, D., Witter, A., and Greven, H.M., 1975, Behaviorally active ACTH analogues, Biochem. Pharmacol., 24: 1463.

Flohr, H., and Lüneberg, U., 1982, Effects of ACTH-(4-10) on vestibular compensation, Brain Res., 248: 169.

Flohr, H., 1983, Control of plastic processes, in: "Springer Series in Synergetics", E. Basar, H. Flohr, H. Haben, and A.J. Mandell, eds., Springer Verlag, Berlin.

Gispen, W.H., Van Wimersma Greidanus, Tj.B., and De Wied, D., 1970, Effects of hypophysectomy and ACTH-(4-10) on responsiveness to electric shock in rats, Physiol. Behav. 5: 143.

Greven, H.M., and De Wied, D., 1977, Influence of peptides structurally related to ACTH and MSH on active avoidance behavior in rats, in: "Frontiers of Hormone Research", Tj.B. Van Wimersma Greidanus, ed., Karger, Basel.

Isaacson, R.L., and Poplawsky, A., 1983, An ACTH-(4-9) analog (ORG 2766) speeds recovery from septal hyperemotionality in the rat, Behav. Neural Biol., 39: 52.

Landfield, P.W., Baskin, R.K., and Pitler, T.A., 1981, Brain aging correlates: Retardation by hormonal-pharmacological treatments, Science, 214: 581.

LeVere, N.D., and LeVere, T.E., 1982, Recovery of function after brain damage: Support for the compensation theory of the behavioral deficit, Physiol. Psychol., 10: 165.

Palkovits, M., 1980, Útmutató és Atlasz patkányagy sejtcsoportjainok izdált hivételére. Akadémiai Kiadó, Budapest.

Saint-Côme, C., Acker, G.R., and Strand, F.L., 1982, Peptide influences on the development and regeneration of motor performance, Peptides 3: 439.

Strand, F.L., and Kung, T.T., 1980, ACTH accelerates recovery of neuromuscular function following crushing of peripheral nerve, Peptides 1: 135.

Thompson, R., 1976, Stereotaxic mapping of brainstem areas critical for memory of visual discrimination habits in the rat, Physiol. Psychol., 4: 1.

Thompson, R., Kao, L., and Yang, S., 1981, Rapid forgetting of individual spatial reversal problems in rats with parafascicular lesions, Behav. Neural Biol., 33: 1.

Van Ree, J.M., Bohus, B., and De Wied, D., 1980, Similarities between behavioral effects of des-tyrosine-γ-endorphin and haloperidol and of α-endorphin and amphetamine, in: "Endogenous and Exogenous Opiate Agonists and Antagonists", E.L. Wang, ed., Pergamon Press, New York.

Van Wimersma Greidanus, Tj.B., Bohus, B., and De Wied, D., 1974, Differential localization of the influence of lysine vasopressin and of ACTH-(4-10) on avoidance behavior: A study in rats bearing lesions in the parafascicular nuclei, Neuroendocrinology 14: 280.

Weijnen, J.A.W.M., and Slangen, J.L., 1970, Effects of ACTH-analogues on extinction of conditioned behavior, Progr. Brain Res., 32: 221.

GANGLIOSIDES, NEUROPLASTICITY, AND BEHAVIORAL RECOVERY AFTER BRAIN DAMAGE

Bernhard A. Sabel,* Gary L. Dunbar, Barry Fass, and
Donald G. Stein

Department of Psychology and Brain Science, Massachusetts
Institute of Technology, Cambridge, MA 02139;* Department of
Psychology, Clark University, Worcester, MA 01610

INTRODUCTION

One of the central tenets of modern Neuroscience is that axons in the adult mammalian brain typically do not regenerate following injury. This tenet does not imply that axons in the brain are incapable of any growth at all. Rather, intact axons can grow new branches and/or terminals which replace those lost as a consequence of damage to other areas of the brain (Cotman, 1978; Cotman et al., 1981). In at least some situations, the newly established synapses have been shown to operate electrophysiologically and there is evidence that they might contribute to behavioral recovery (Steward, 1982).

If lesion-induced "sprouting" and "reactive synaptogenesis" are functionally and/or behaviorally meaningful phenomena, it may be beneficial to manipulate their rate and/or extent in order to optimize behavioral recovery. For instance, in those situations where axonal growth is accompanied by behavioral recovery, a manipulation which enhances sprouting would be desirable. Conversely, in those cases where the growth is accompanied by aberrant or maladaptive behavior, a manipulation which suppresses sprouting would be preferable. One pharmacological tool which might affect lesion-induced neuroplasticity and recovery of function is gangliosides.

Gangliosides are a family of glycosphingolipids which contain sialic acid, and they are normally found in abundant quantities in the brain (Wiegant, 1971; Ledeen, 1978), especially in the gray matter (Suzuki, 1965). They were first described by Klenk (1935), who named them on the basis of their accumulation in the ganglia of patients suffering from Tay Sachs disease. This and other ganglioside storage diseases involve the formation of meganeurites with dendritic-like growth cones (Purpura et al., 1978), which suggests that gangliosides might have growth-promoting characteristics. This discovery prompted further investigations into the possibility that exogenous gangliosides might promote neural growth, and therefore might ultimately be used to enhance behavioral recovery after brain damage.

EXOGENOUS GANGLIOSIDES AND NEURAL GROWTH IN VITRO

One of the first lines of evidence that exogenous gangliosides promote neural growth was provided by studies of cells in vitro. Although there is abundant evidence that exogenous gangliosides enhance neuritogenesis in vitro (see Ledeen, 1984), their precise mode of action in this process is still not understood. Three possible explanations for the enhancement of neural growth are that gangliosides: 1) are growth-promoting substances themselves; 2) act as receptors for growth-promoting substances; or 3) have an indirect synergistic interaction with growth-promoting substances.

Although none of these possibilities can be ruled out at present, the available data tend to favor the possibility that exogenous gangliosides indirectly interact with growth-promoting agents; for example, gangliosides have been shown to potentiate the effects of nerve growth factor (NGF) in PC12 cells (Ferrari et al., 1983; Katoh-Semba et al., 1984) and dorsal root ganglia from chick embryos (Leon et al., 1984; Roisen et al., 1981). Since gangliosides fail to elicit neuritic outgrowth in cultured PC12 cells in the absence of NGF (Ferrari et al., 1983; Katoh-Semba et al., 1984), it seems unlikely that in PC12 cells gangliosides themselves are growth-promoting substances. Furthermore, since the augmentation of NGF-induced neurite outgrowth occurs without altering the binding of NGF to these cells (Ferrari et al., 1983), it is also unlikely that gangliosides are receptors for growth factors.

A possible explanation of how exogenous gangliosides may be indirectly interacting with growth factors has been recently proposed by Katoh-Semba and colleagues (1984). They found evidence suggesting that exogenous gangliosides enhance neuritic outgrowth by overcoming existing factors which may limit such growth, rather than potentiating the inductive action of growth factors. If exogenous gangliosides do indeed interact with growth factors in this way, or in some other indirect manner, the elucidation of the precise mechanism of the interaction would provide valuable information for understanding how exogenous gangliosides might affect neuroplasticity in the living organism.

EXOGENOUS GANGLIOSIDES AND NEURAL GROWTH IN VIVO

Since exogenous gangliosides potentiate neuritogenesis in vitro, they conceivably might also serve a similar function in vivo. Evidence that ganglioside treatments do promote neural growth in the living organism comes from two lines of research.

The first line of research provided data which suggest that gangliosides play an important role during neural development. For example, it was shown that there are increases in brain gangliosides just prior to the onset of myelination (Suzuki, 1965; Rahmann, 1980), and increases in brain gangliosides also correspond to axodendritic proliferation and synapse formation during ontogeny (Bass, 1981). In addition, administration of antibodies to gangliosides impairs the normal course of neural development; they disrupt normal maturation of the cerebral cortex and impair performance of complex learning tasks (Kasarskis et al., 1981). Conversely, ganglioside treatments during early infancy accelerate CNS maturation in rats (Karpiak et al., 1984).

A second line of evidence indicating that exogenous gangliosides promote neural growth in vivo comes from studies of axonal regeneration in the peripheral nervous system. Ceccarelli and colleagues (1976) were the first to report that ganglioside treatments accelerate regenerative growth of the superior cervical ganglion in cats. Functional recovery (as measured by contractions of the nictitating membrane) was accelerated by two weeks, and was accompanied by an enhancement of regeneration of the severed nerves in ganglioside-treated animals. Similar results were obtained by Gorio and co-workers (1980), who found that exogenous gangliosides enhance sprouting of the peroneal nerve into the denervated extensor digitorium longus muscle of the rat. Subsequent research has generally supported the conclusion that exogenous gangliosides accelerate or promote sprouting of peripheral nerves (c.f., Ledeen, 1984).

EXOGENOUS GANGLIOSIDES AND LESION-INDUCED PLASTICITY IN THE BRAIN

Since exogenous gangliosides have been shown to promote neural growth in vivo and after peripheral nerve damage, considerable attention has been recently focused on the intriguing possibility that ganglioside treatments also promote growth in injured brains. This possibility is appealing because of its behavioral implications; if ganglioside treatments increase the rate and/or extent of lesion-induced sprouting, they also might improve recovery from behavioral aberrations. Before assessing the current literature pertaining to the potential utility of gangliosides as a treatment for brain damage, we will first consider whether lesion-induced sprouting is a behaviorally meaningful phenomenon, per se.

In order to assess the potential behavioral significance of lesion-induced sprouting in the CNS, Steward (1982) has proposed that two criteria be fulfilled. One is that the time course of sprouting should parallel the time course of behavioral recovery (i.e., a time-dependent return toward normal levels of performance; Laurence and Stein, 1978). The second is that a manipulation of the rate and/or extent of sprouting should result in a corresponding change in behavior. Fulfillment of these criteria would imply that the new circuitry is behaviorally meaningful, and that the growth of new connections is related to the observed change in behavior. One has to keep in mind that sprouting of homotypic fibers (e.g., those arising from a spared portion of the damaged system itself) may contribute to recovery of function in a different way than sprouting by heterotypic fibers (e.g., ones originating from an uninjured portion of the brain which coinnervate the deafferented structure) (Steward, 1982).

During the past twenty years, numerous investigations have provided evidence for a parallel between sprouting and behavioral recovery. However, in some situtations, postlesion axonal growth has been shown to result in maladaptive behavior. McCouch and co-workers (1958) observed, for example, that axonal growth in the damaged spinal cord is accompanied by the appearance of spasticity. Similarly, removal of the tectum during infancy results in the growth of aberrant connections into the opposite (wrong) tectum, and this growth is accompanied by incorrect orienting behaviors (Schneider and Jhaveri, 1974). Thus, lesion-induced sprouting is behaviorally meaningful, but detrimental, in certain cases.

There is, however, some evidence that ganglioside treatments may promote sprouting in the brain, and that this sprouting might facilitate behavioral recovery. Grafstein and colleagues (1983) found that both local application and intraocular injection of low doses of gangliosides increase the rate of optic tract regeneration and facilitate the recovery of startle reactions in goldfish. Evidence for similar effects has been reported within the past two years in the mammalian brain, also. The neuronal pathways which have been studied most extensively in this context are the nigrostriatal and septohippocampal systems.

Nigrostriatal system

Research on neuroplasticity after transections of the nigrostriatal pathway has examined whether ganglioside treatments promote sprouting and/or the survival of axotomized neurons. Studies on this system were initiated by Toffano and colleagues (1983), who showed that ganglioside treatments increase homovanillic acid content, tyrosine hydroxylase (TH) activity, and TH-related immunofluorescence in the denervated striatum after unilateral hemitransections of the nigrostriatal fibers. Since the treatments did not affect TH activity in the intact side of ganglioside-treated animals or in sham-operated controls, the increases were attributed to a ganglioside-induced enhancement of sprouting. Unfortunately, a monotonic relationship between the amount of TH activity and the number of axon terminals or synapses in the striatum has yet to be established. Thus, these findings also could reflect an increase in TH activity per se within the denervated striatum of ganglioside-treated animals, rather than an increase in the density of synapses or axon terminals.

We used the tracer horseradish peroxidase (HRP) in order to evaluate gangliosides' effects on sprouting and to determine the source of newly grown inputs to the striatum, without relying on a transmitter-related enzyme. In an initial study (Sabel et al., 1984a), we injected HRP into the denervated striatum of ganglioside or saline-treated rats which previously sustained unilateral hemitransections. At postlesion day 3, there was an equal reduction in the number of labelled neurons in the ipsilateral substantia nigra, pars compacta (iSNc) and in the ipsilateral ventral tegmentum (iVTA) for both groups of animals. By postoperative day 15, however, the number of labelled neurons within these two structures (and also the contralateral SNc) was greater in ganglioside-treated animals. In fact, the number of labelled neurons in the iVTA and cSNc were, respectively, 200 and 300 percent above those of the sham-operated controls ("hypertrophy"). Interestingly, by day 45, the number of labelled neurons in both structures returned to the level of both the saline and sham-operated controls. This effect may have been due to the termination of ganglioside treatments on day 15, or could have been a late-onset secondary compensatory event.

These results seemed to provide additional support for Toffano's suggestion that ganglioside treatments may enhance lesion-induced sprouting. However, the possibility still remained that gangliosides merely were affecting the rate of HRP transport and that this may have led to the increased neuronal labelling. In an attempt to rule out this possibility, we made secondary lesions in the iSNc and iVTA using 6-hydroxydopamine. Secondary lesions in these structures were expected to result in the degeneration of new terminals in the striatum, and we did find evidence for greater degeneration in the ganglioside-treated animals (Sabel et al., submitted). Thus, it appeared that ganglioside treatments indeed promoted reinnervation of the denervated striatum.

Although we favor the hypothesis that ganglioside treatments enhance lesion-induced sprouting by the nigrostriatal system, alternative interpretations of the data presented above cannot be excluded at present. For example, an equally interesting alternative possibility is that hemitransections spare some of the cells-of-origin of the nigrostriatal pathway and ganglioside treatments promote their survival. Recent evidence reported by Toffano's group (1984) lends support to this alternative. They found that ganglioside treatments are ineffective in elevating TH activity in the denervated striatum after total (rather than partial) lesions of the substantia nigra. Thus, exogenous gangliosides may function as a survival-promoting factor in the nigrostriatal system, rather than (or in addition to) functioning as a growth-promoting agent. This interpretation is further supported by Toffano's findings (Agnati et al., 1983; Toffano et al., 1984) that exogenous gangliosides preserve the number of TH immunoreactive neuronal somata and dendrites in the substantia nigra after hemitransections. If ganglioside treatments do promote the survival of spared nigrostriatal neurons, it is not clear why an increase in the number of retrogradely-labelled nigral neurons or an elevation of striatal TH activity would not appear until about two weeks postlesion. Further research will be needed to resolve these issues.

As mentioned previously in this chapter, lesion-induced sprouting has been shown to be behaviorally meaningful in some situations. Assuming that sprouting by the nigrostriatal pathway is behaviorally significant, one would expect to find that ganglioside-induced effects on sprouting are accompanied by a corresponding change in behavior. We tested this hypothesis by quantifying drug-induced rotational behavior after hemitransections. Unilateral cuts of the nigrostriatal pathway result in turning behavior toward the side of the lesion. This lesion-induced rotation is believed to be a manifestation of an asymmetric dopamine transmission, and it is more pronounced after a pharmacological manipulation which stimulates dopamine release (e.g., amphetamine) or receptors (e.g., apomorphine) (Ungerstedt, 1971). Rotational behavior typically recovers within a week, and the recovery is thought to be associated with the reorganization of nigrostriatal fibers (Pritzel et al., 1983).

We found that ganglioside treatments result in a significant reduction in the number of amphetamine-induced rotations beginning on postoperative day 2 (Sabel et al., 1984a). Whereas saline-treated animals showed some degree of recovery by day 12, ganglioside-treated counterparts showed no impairment at all. However, this normal-like behavior exhibited by ganglioside-treated animals was only temporary; by day 39, they again showed drug-induced rotation as pronounced as on day 2.

The temporary, total absence of amphetamine-induced rotational behavior at day 12 parallels the significant increase in HRP labelling ("hypertrophy") which we found in various mesencephalic structures of ganglioside-treated animals. Even at later postlesion intervals (day 39), the loss of these "hypertrophied" fibers is paralleled by the reappearance of behavioral deficits. Thus, the ganglioside-induced increase in HRP labelling corresponds in time with a reduction of behavioral aberrations, thereby fulfilling one of Steward's (1982) criteria. It is interesting to note that the significant reduction of amphetamine-induced rotational behavior on day 2 occurs <u>prior</u> to the time when increased retrograde labelling first appears. Such a dissociation between behavioral and anatomical phenomena at early but not at later stages after injury indicates that events <u>other</u> than sprouting which occur soon after injury may also be affected by ganglioside treatments. There is evidence that this indeed may be the case; exogenous gangliosides have been shown to reduce the extent of secondary degeneration (see above; Agnati et al., 1983; Toffano et al., 1984; Sabel et al., submitted) and cerebral edema (Karpiak and Mahadik, 1984) following brain damage.

Effects of ganglioside treatments on apomorphine-induced rotation after hemitransections also suggest that the treatments may promote neuronal survival. Apomorphine stimulates dopamine receptors in the denervated striatum (Ungerstedt, 1971), and Toffano's group (1983) hypothesized that if ganglioside treatments stimulate reinnervation of the denervated striatum, the sensitivity of striatal receptors to apomorphine should be reduced. In agreement with the findings of Toffano's group, we found a reduction of apomorphine-induced rotation. Whereas saline-treated animals grew into a deficit (i.e., vigorous rotation developed between 15 and 42 days postlesion), ganglioside-treated animals did not. Interestingly, animals in both groups rotated toward the side of the lesion rather than away from it (as is observed in the classic "denervation supersensitivity" paradigm; Ungerstedt, 1971). We interpreted this finding as indirect evidence that ganglioside treatments reduce the rotational behavior after apomorphine injections by preventing neuronal death and preserving dopamine receptors in the denervated striatum (see Sabel et al., 1984). Although we favor this interpretation, it is also possible that the reduction of behavioral deficits in ganglioside-treated animals may depend upon other neural events such as reduced degeneration (Sabel et al., submitted) or cerebral edema (Karpiak and Mahadik, 1984).

Since ganglioside treatments apparently are effective in reducing lesion-induced motor-function deficits, we became interested in whether they also reduce impairments in learning and memory. The model system we used was spatial reversal behavior after bilateral lesions of the caudate nucleus. These lesions render animals unable to solve a spatial reversal task in a two-choice discrimination-learning maze (e.g., Schultze and Stein, 1975). We found that ganglioside treatments are effective in reducing learning deficits on spatial reversal after such lesions (Sabel et al., 1984b). In fact, on most measures, ganglioside-treated animals were statistically indistinguishable from sham-operated controls. It is important to note that differences in behavioral performance were apparent as soon as testing began (postoperative day 9). Ganglioside treatments therefore reduced the severity of the deficit rather than enhancing recovery. Moreover, when retested at three months postlesion, the ganglioside-treated group was still able to solve the task without signs of deterioration. These findings indicate that ganglioside treatments provide longer-lasting behavioral benefits after caudate nucleus lesions than after nigral hemitransections.

In summary, the data presented above are consistent with the possibility that ganglioside treatments enhance lesion-induced sprouting. Since such data do not indicate the actual number of terminals or synapses in the denervated striatum, however, it will be important to employ other approaches (e.g., quantitative autoradiography, electron microscopy) to confirm whether ganglioside treatments do in fact enhance lesion-induced sprouting. The behavioral data, by contrast, are fairly compelling that ganglioside treatments reduce the severity of deficits after striatal lesions or hemitransections.

Septohippocampal system

Research on the effects of ganglioside treatments following brain damage has also been conducted using the hippocampus as the model structure. The hippocampus offers several important neuroanatomical advantages for investigating whether exogenous gangliosides enhance lesion-induced plasticity in the brain. For example, removal of one source of afferents to the hippocampus results in a well-documented reorganization and proliferation of surviving inputs; such reorganization can be revealed by various techniques, ranging from histochemical stains to quantitative electron microscopy (Cotman and Nadler, 1978).

Since there is behavioral evidence that exogenous gangliosides might accelerate and/or enhance lesion-induced sprouting in the dentate gyrus (DG) (Karpiak, 1983), we became interested in evaluating the effects of exogenous gangliosides on such sprouting. In an initial study (Fass and Ramirez, 1984), we focused on one of the afferent systems to the DG which participate in the sprouting response after lesions of the entorhinal cortex (EC); namely, the acetylcholinesterase (AChE)-containing septodentate pathway. Sprouting by this pathway can be revealed by a simple and quantifiable histochemical stain for AChE. Autoradiographic and electron microscopic studies have confirmed that the intensity of AChE staining in the denervated region of the DG is a reliable marker for terminal proliferation and synaptogenesis by the septodentate pathway (Cotman et al., 1973; Stanfield and Cowan, 1982).

We found that ganglioside treatments do <u>not</u> enhance the intensification of AChE staining in the denervated DG after EC lesions. Rather, there was a consistent reduction in staining intensity at 3,5,7, and 10 days postlesion (Fass and Ramirez, 1984). This finding is inconsistent with the hypothesis that ganglioside treatments accelerate or enhance lesion-induced sprouting in the brain. Our data instead raise the unanticipated possibility that ganglioside treatments might <u>suppress</u> the amount of sprouting by the septodentate pathway. We are currently investigating whether exogenous gangliosides similarly affect other afferents to the DG (i.e., the crossed temporodentate pathway) which sprout after EC lesions.

In more recent studies, we investigated the effects of fimbria-fornix lesions on cholinergic enzymes in the hippocampus of ganglioside-treated rats.

Such transections induce collateral sprouting by AChE-containing inputs to the hippocampus which originate in the nucleus of the diagonal band, and the sprouting begins within three months after the transections (Gage et al., 1983). In an attempt to ascertain whether ganglioside treatments might accelerate the onset of such sprouting, we assessed the intensity of AChE staining in the hippocampus at 2,4,14, and 18 days after unilateral transections of the fimbria-fornix (Fass et al., in preparation). There were no detectable differences between the treated and untreated groups at any of these survival intervals. Biochemical assays for AChE and choline acetyltransferase (ChAT) activity in the hippocampus also were performed at 19 days posttransection, and they showed that levels of both enzymes were lower in the treated animals. Taken together, our findings seem to imply that ganglioside treatments do not accelerate the onset of proliferation by surviving AChE-containing inputs to the hippocampus after fimbria-fornix transections. These findings do not indicate, however, whether the final extent of sprouting might be affected.

An alternative interpretation of our data is that exogenous gangliosides affect cholinergic enzymes per se, rather than affecting the number of enzyme-containing terminals or synapses. Several lines of evidence are consistent with this interpretation. For example, exogenous gangliosides enhance the kinetics of biochemically demonstrable AChE in the cerebral cortex of intact neonatal rats (Karpiak et al., 1984). Exogenous gangliosides also stimulate high affinity choline uptake in the cortex after lesions of nucleus basalis magnocellularis (Pedata et al., 1984). Thus, our data may signify that ganglioside treatments reduce cholinergic enzymes in the denervated hippocampus without altering the rate or extent of sprouting by enzyme-containing fibers of the septohippocampal pathway.

Contrary to our observations, Oderfeld-Nowak and co-workers (Wojcik et al., 1982; Oderfeld-Nowak et al., 1984) have reported evidence that ganglioside treatments enhance sprouting in the denervated hippocampus. These investigators found significantly higher biochemical levels of AChE and ChAT in the denervated hippocampus of ganglioside-treated rats than in untreated counterparts at 18 days postlesion, but not at 3 or 5. However, lesion size may account for the apparent discrepancy between the results. Whereas our fimbria-fornix transections were complete and effectively axotomized virtually all cells-of-origin of the septohippocampal pathway, Oderfeld-Nowak's lesions of the medioventral septum appeared to be partial (as illustrated by the lesion reconstruction in Wojcik et al., 1982). The effect of ganglioside treatments, then, could have been to promote the survival of spared cholinergic neurons which were metabolically impaired by the lesion (see Karpiak, 1983). The difference between our results and Oderfeld-Nowak's could alternatively be attributable to methodological differences. Whereas we used a range of substrate concentrations on formalin-fixed tissue (all of which yielded the same results as described in the preceding paragraph), Oderfeld-Nowak used one concentration on unperfused material. Karpiak (personal communication) has obtained evidence from in vitro studies that exogenous gangliosides stimulate AChE activity in unfixed tissue. Thus, it appears that the data can be reconciled in at least three ways: 1) lesion size and/or cell survival is a critical factor; 2) the analysis of AChE activity may depend upon how the tissue is processed, and consequently the effects of exogenous gangliosides on this measure may be difficult to interpret; and/or 3) ganglioside treatments affect enzyme kinetics rather than the sprouting response. At present, it is not possible to rule out any of these alternatives. Based upon the evidence currently available, however, we favor the interpretation that ganglioside treatments do not enhance lesion-induced sprouting by the septohippocampal pathway.

In summary, it is not entirely clear whether exogenous gangliosides do exert growth-promoting effects in the denervated hippocampus. Currently, we are conducting studies which do not rely upon enzyme histochemistry to evaluate the rate and extent of sprouting in the hippocampus of ganglioside-treated rats. An alternative, but perhaps equally important possibility is that exogenous gangliosides promote the survival of axotomized neurons. Data obtained from the nigrostriatal system (described above) are consistent with this possibility, and we are currently evaluating the survival of cells-of-origin of the septohippocampal pathway after fimbria-fornix transections in ganglioside-treated rats. Regardless of the outcome, the data reviewed above seem fairly convincing that exogenous gangliosides do affect transmitter-related enzymes in the damaged brain.

Although it is not completely clear whether ganglioside treatments enhance lesion-induced sprouting in the hippocampus, they evidently affect the behavioral recovery which accompanies such sprouting. For example, unilateral lesions of the EC result in impaired performance of spatial alternation which recovers during the same time when the crossed temporodentate pathway sprouts (Loesche and Steward, 1977; Steward and Loesche, 1977). Ganglioside treatments improve this recovery of alternation after unilateral EC lesions (Karpiak, 1983). Karpiak's data indicate that the treatments not only reduce the initial postlesion impairment, but also enhance the rate and extent of recovery. Since recovery (by definition) is a temporal phenomenon, and since learning paradigms confound acquisition with time (Steward, 1982), we subsequently decided to assess the effects of ganglioside treatments on an unlearned behavior which also normally recovers after EC lesions, namely, open-field activity. In an initial experiment (Fass and Ramirez, 1984), we found that ganglioside-treated rats with bilateral EC lesions showed a smaller lesion-induced increase in locomotor activity and were less active throughout the experiment than untreated counterparts. Thus, ganglioside treatments reduced the severity of the lesion-induced behavioral abnormality. We recently have replicated this experiment (Ramirez et al., in preparation), and confirmed that the treatments reduce the magnitude of the lesion-induced hyperactivity <u>without</u> affecting the rate or total amount of recovery.

We propose that the effects of ganglioside treatments on the behavioral sequellae of EC lesions most likely are unrelated to any effects on sprouting in the hippocampus. The reason is that behavioral effects of exogenous gangliosides have been observed <u>prior</u> to the time when sprouting begins (i.e., 2-4 days postlesion vs. 6-8 days). Indeed, the behavioral data imply that ganglioside treatments affect early events after EC lesions rather than later events. The possibility exists, for example, that the beneficial behavioral effects of ganglioside treatments are a manifestation of a reduction in cerebral edema (Karpiak and Mahadik, 1984). Additional time-course data are needed to determine the extent to which the various effects of exogenous gangliosides after EC lesions exhibit a temporal parallel.

DISCUSSION AND CONCLUSION

We believe the evidence presented in this chapter is fairly compelling that ganglioside treatments affect plasticity and behavioral recovery after some types of brain damage. It appears that, in some instances, exogenous gangliosides might promote lesion-induced sprouting and/or neuronal survival (Agnati et al., 1983; Sabel et al., 1984a; submitted; Toffano et al., 1983; 1984). Ganglioside treatments also appear to reduce the severity of behavioral abnormalities which occur soon after nigrostriatal hemitransections (Agnati et al., 1983; Sabel et al., 1984a; Toffano et al., 1983) and entorhinal lesions (Karpiak, 1983; Fass and Ramirez, 1984). Similar results have been obtained with lesions of the mediofrontal cortex (Sabel et al., 1984a) and early lesions of the septum (Massarelli et al., 1984). There are, however, instances of negative results: 1) ganglioside treatments evidently do not enhance sprouting by the septodentate pathway after EC lesions

(Fass and Ramirez, 1984); and 2) ganglioside-treated rats with bilateral lesions of the occipital cortex show a more severe impairment of brightness discrimination performance while deficits in pattern discrimination remain unaffected (Butler et al., in preparation). Thus, there may be some limits to the capabilities of exogenous gangliosides to promote axonal growth and ameliorate the behavioral deficits induced by certain types of brain lesions.

Based upon the current data, we propose that exogenous gangliosides exert effects on both early (i.e., prior to six days postlesion, after which time sprouting typically begins) and later (i.e., after six days) postlesion events.

Early events after injury

With regard to behavior, exogenous gangliosides have been shown to reduce the severity of the aberrations seen immediately after several types of brain lesions. For example, at 2-4 days postlesion, the deficit in learned alternation after unilateral EC damage (Karpiak, 1983) and the hyperactivity after bilateral EC damage (Fass and Ramirez, 1984) are reduced by ganglioside treatments. Similar effects have been observed in the case of amphetamine-induced rotation at 2 days after hemitransection (Sabel et al., 1984a).

These early behavioral effects of ganglioside treatments correspond in time with effects on neurochemical and pathological events. For example, ganglioside treatments have been shown to reduce cerebral edema on the day after cortical lesions (Karpiak and Mahadik, 1984). The treatments also appear to affect the kinetics of various transmitter enzymes, such as AChE and ChAT, in the hippocampus at early intervals after EC or supracallosal lesions (Fass and Ramirez, 1984; Oderfeld-Nowak et al., 1984). In addition, ganglioside treatments reduce the extent of secondary degeneration induced by lesions (Agnati et al., 1983; Sabel et al., submitted; Toffano et al., 1983). The possibility exists that the ganglioside-induced early reductions of behavioral abnormalities may be a manifestation of these early neurochemical or pathological effects.

Late events after injury

The only studies to date which demonstrated behavioral effects of ganglioside treatments at later postlesion intervals (i.e., after six days) are those of Karpiak (1983) and Sabel and co-workers (1984a). In the former study, ganglioside treatments improved both the rate and total amount of recovery of learned alternation after unilateral EC lesions. The study by Sabel and co-workers evaluated an unlearned behavior (drug-induced rotation), and showed that ganglioside treatments reduced rotational asymmetries at later stages after hemitransections. It will be important in future studies to assess whether the timing of the treatments (i.e., pre- vs. postoperative) is an important factor in whether early or late behavioral effects occur.

In addition, late but not early effects on transmitter-related enzymes have been observed. Ganglioside treatments evidently do not increase cholinergic enzymes in the hippocampus until 18 days after partial denervation (Wojcik et al., 1982; Oderfeld-Nowak et al., 1984). Similarly, such treatments do not stimulate TH activity until 14 days after nigrostriatal lesions (Agnati et al., 1983; Toffano et al., 1983). There also is some evidence that exogenous gangliosides might enhance lesion-induced sprouting after hemitransections, as shown by our retrograde-labelling studies (Sabel et al., 1984a). However, autoradiographic and/or EM data will be needed to verify that ganglioside treatments do affect the rate and/or total amount of reactive synaptogenesis.

Based upon the evidence presented in this chapter, we favor the interpretation that ganglioside treatments do promote lesion-induced sprouting in the brain. In the event that future studies do confirm this interpretation, the importance of lesion size will need to be examined. For example, the effects of exogenous gangliosides on sprouting might be apparent only after partial (but not after total) lesions. Whereas most studies of the nigrostriatal system have involved partial lesions, one study (Toffano et al., 1984) involved complete lesions and it showed no evidence for an enhancement of sprouting in ganglioside-treated rats. In studies of the septohippocampal system involving total lesions (Fass and Ramirez, 1984; Fass et al., in preparation), no evidence was found for an enhancement of sprouting. In the one instance where deafferenting lesions of the hippocampus appeared to be partial (Wojcik et al., 1982), the results were suggestive that ganglioside treatments promoted sprouting.

If future studies do not confirm our interpretation about sprouting, it will become important to determine whether ganglioside treatments affect the measures used as indices of sprouting, such as axoplasmic transport (Kalia and DiPalma, 1982) or enzyme kinetics (Karpiak et al., 1984), rather than the sprouting response itself. It may be that ganglioside treatments are more effective in promoting true regenerative growth in the peripheral nervous system than in promoting sprouting in the brain. Consequently, gangliosides may be a useful tool for obtaining clues about the fundamental cellular differences between regenerative growth and lesion-induced sprouting.

The mechanism of gangliosides' effects on neuroplasticity is unclear, at present. Since the measures of neuroplasticity are reported to be affected more dramatically in the damaged hemisphere (Oderfeld-Nowak et al., 1984; Pedata et al., 1984; Toffano et al., 1983), the possibility exists that the lesion disrupts the blood-brain barrier unilaterally and thereby permits higher concentrations of exogenous gangliosides to enter the brain and exert their effects on the lesion side. It therefore is becoming increasingly important to investigate the fate of peripherally administered gangliosides in intact and damaged brains. In addition, more attention needs to be directed to investigating the function of exogenous gangliosides in the brain. Recent evidence implies that gangliosides interact synergistically with growth-promoting substances (e.g., Leon et al., 1984), but the precise mechanism of this interaction remains to be elucidated.

In conclusion, exogenous gangliosides appear to be an effective pharmacological approach to treat various sorts of brain damage in rats. Although there may be limits to the capabilities of ganglioside treatments to promote neuroplasticity and recovery of function, the treatments evidently do affect both early and late sequellae of brain lesions. While our understanding of how exogenous gangliosides exert their behavioral effects is still unclear, they may nevertheless prove useful in the clinical management of brain injury.

ACKNOWLEDGEMENT

Funds for preparing this chapter were provided by Fidia Research Laboratories (Abano Terme, Italy) and Clark University.

REFERENCES

Agnati, L. F., Fuxe, K., Calza, L., Benfenati, F., Cavicchioli, L., Toffano, G., and Goldstein, M., 1983, Gangliosides increase the survival of lesioned nigral dopamine neurons and favour the recovery of dopaminergic synaptic function in striatum of rats by collateral sprouting, Acta Physiol. Scand., 119:347.

Bass, N. H., 1981, Ganglioside sialic acid as a quantitative neurochemical index of the integrity of synaptic function in cognitive disorders of development and aging, in: "Gangliosides in Neurological and Neuromuscular Function, Development, and Repair," M. M. Rapport and A. Gorio, eds., Raven Press, New York.

Ceccarelli, B., Aporti, F., and Finesso, M., 1976, Effects of brain gangliosides on functional recovery in experimental regeneration and reinnervation, in: "Ganglioside Function: Biochemical and Pharmacological Implications," G. Porcellati, B. Ceccarelli, and G. Tettamanti, eds., Plenum Press, New York.

Cotman, C. W., 1978, "Neuronal Plasticity," Raven Press, New York.

Cotman, C. W., Matthews, D. A., Taylor, D., and Lynch, G., 1973, Synaptic rearrangement in the dentate gyrus: histochemical evidence of readjustments after lesions in immature and adult rats, Proc. Natl. Acad. Sci., U. S. A., 70:3473.

Cotman, C. W. and Nadler, J. V., 1978, Reactive synaptogenesis in the hippocampus, in: "Neuronal Plasticity," C. W. Cotman, ed., Raven Press, New York.

Cotman, C. W., Nieto-Sampedro, M., and Harris, E. W., 1981, Synapse replacement in the nervous system of adult vertebrates, Physiol. Rev., 61:684.

Fass, B., and Ramirez, J. J., 1984, Effects of ganglioside treatments on lesion-induced behavioral impairments and sprouting in the CNS, J. Neurosci. Res., 12:445.

Ferrari, G., Fabris, M., and Gorio, A., 1983, Gangliosides enhance neurite outgrowth in PC12 cells, Devel. Brain Res., 8:215.

Gage, F. H., Bjorklund, A., and Stenevi, U., 1983, Reinnervation of the partially deafferented hippocampus by compensatory collateral sprouting from spared cholinergic and noradrenergic afferents, Brain Res., 268: 27.

Gorio, A., Carmignoto, G., Facci, L., and Finesso, M., 1980, Motor nerve sprouting induced by ganglioside treatment. Possible implications for gangliosides on neuronal growth, Brain Res., 197:236.

Grafstein, B., Yip, H. K., and Meiri, H., 1983, Techniques for improving axonal regeneration: assay in goldfish optic nerve, in: "Nervous System Regeneration," B. Haber, J. R. Perez-Polo, G. A. Hashim, and A. M. G. Stella, eds., A. R. Liss, New York.

Kalia, M., and DiPalma, J. R., 1982, Ganglioside-induced acceleration of axonal transport following nerve crush injury in the rat, Neurosci. Lett., 34:1.

Karpiak, S. E., 1983, Ganglioside treatment improves recovery of alternation behavior after unilateral entorhinal cortex lesion, Exp. Neurol., 81:330.

Karpiak, S. E., and Mahadik, S. P., 1984, Reduction of cerebral edema with GM1 ganglioside, J.Neurosci.Res., 12:485.

Karpiak, S. E., Vilim, F., and Mahadik, S. P., 1984, Gangliosides accelerate rat neonatal learning and levels of cortical acetylcholinesterase, Devel. Neurosci., 6:127.

Kasarskis, E. J., Karpiak, S. E., Rapport, M. M., Yu, R. K., and Bass, N. H., 1981, Abnormal maturation of cerebral cortex and behavioral deficit in adult rats after neonatal administration of antibodies to gangliosides, Devel. Brain Res., 1:25.
Katoh-Semba, R., Skaper, S. D., and Varon, S., 1984, Interaction of GM1 ganglioside with PC12 pheochromocytoma cells: serum- and NGF-dependent effects on neuritic growth (and proliferation), J. Neurosci. Res., 12:299.
Klenk, E., 1935, Uber die Natur der Phosphatide und anderer Lipoide des Gehirns und der Leber bei der Niemann-Pickschen Krankheit, Hoppe Seylers Z. Physiol. Chem., 235: 24.
Laurence, S., and Stein, D. G., 1978, Concepts of recovery, in: "Recovery From Brain Damage: Research and Theory," S. Finger ed., Plenum Press, New York.
Ledeen, R. W., 1978, Ganglioside structures and distribution: are they localized at the nerve ending?, J. Supramolec. Struct., 8:1.
Ledeen, R. W., 1984, Biology of gangliosides: neuritogenic and neuronotrophic properties, J. Neurosci. Res., 12:147.
Leon, A., Benvegnu, D., Dal Toso, R., Presti, D., Facci, L., Giorgi, O., and Toffano, G., 1984, Dorsal root ganglia and nerve growth factor: a model for understanding the mechanism of GM1 effects on neuronal repair, J.Neurosci. Res., 12:277.
Loesche, J., and Steward, O., 1977, Behavioral correlates of denervation and reinnervation of the hippocampal formation of the rat: recovery of alternation performance following unilateral entorhinal cortex lesions, Brain Res. Bull., 2:31.
Massarelli, A. C., Dreyfus, H., Eclancher, F., and Massarelli, R., 1984, Effect of gangliosides on behavioral recovery after early septal lesion in rats, Paper presented at the European Symposium on Cellular and Pathological Aspects of Glycoconjugate Metabolism, Strasbourg, France.
McCouch, G. P., Austin, G. M., Liu, C. N., and Liu, C. Y., 1958, Sprouting as a cause of spasticity, J.Neurophysiol., 21:205.
Oderfeld-Nowak, B., Skup, M., Ulas, J., Jezierska, M., Gradkowska, M., and Zaremba, M., 1984, Effect of GM1 ganglioside treatment on post-lesion responses of cholinergic enzymes in rat hippocampus after various partial deafferentations, J.Neurosci. Res., 12:409.
Pedata, F., Giovannelli, L., and Pepeu, G., 1984, GM1 ganglioside facilitates the recovery of high-affinity choline uptake in the cerebral cortex of rats with a lesion of the nucleus basalis magnocellularis, J. Neurosci. Res., 12:421.
Pritzel, M., Huston, J. P., and Sarter, M., 1983, Behavioral and neuronal reorganization after unilateral substantia nigra lesions: evidence for increased interhemispheric nigrostriatal projections, Neurosci., 9:879.
Purpura, D. P., Pappas, G. D., and Baker, H. J., 1978, Fine structure of meganeurites and secondary growth processes in feline GM1- gangliosidosis, Brain Res., 143: 1.
Rahmann, H., 1980, Gangliosides and thermal adaptation, in: "Structure and Function of Gangliosides," L. Svennerholm, H. Dreyfus, and P. F. Urban, eds., Plenum Press, New York.
Roisen, F. J., Bartfeld, H., Nagele, R., and Yorke, G., 1981, Ganglioside stimulation of axonal sprouting in vitro, Science, 214:577.
Sabel, B. A., Dunbar, G. L., and Stein, D. G., 1984a, Gangliosides minimize behavioral deficits and enhance structural repair after brain injury, J.Neurosci. Res., 12:429.
Sabel, B. A., Slavin, M. D., and Stein, D. G., 1984b, GM1 ganglioside treatment facilitates behavioral recovery following bilateral brain damage, Science, 225:340.

Schneider, G. E., and Jhaveri, S. R., 1974, Neuroanatomical correlates of spared or altered function after brain lesions in the newborn hamster, in: "Plasticity and Recovery of Function in the Central Nervous System," D. G. Stein, J. J. Rosen, and N. Butters, eds., Academic Press, New York.

Schultze, M. J., and Stein, D. G., 1975, Recovery of function in the albino rat following either simultaneous or seriatim lesions of the caudate nucleus, Exp. Neurol., 46:291.

Stanfield, B. B., and Cowan, W. M., 1982, The sprouting of septal afferents to the dentate gyrus after lesions of the entorhinal cortex in adult rats, Brain Res.. 232:162.

Steward, O., 1982, Assessing the functional significance of lesion-induced neuronal plasticity, in: "International Review of Neurobiology, vol. 23," J. R. Smythies and R. J. Bradley, eds., Academic Press, New York.

Steward, O., and Loesche, J., 1977, Quantitative autoradiographic analysis of the time course of proliferation of contralateral entorhinal efferents in the dentate gyrus denervated by ipsilateral entorhinal lesions, Brain Res., 125:11.

Suzuki, K., 1965, The pattern of mammalian brain gangliosides. III. Regional and developmental differences, J.Neurochem., 12:969.

Toffano, G., Savoini, G., Moroni, F., Lombardi, G., Calza, L., and Agnati, L. F., 1983, GM1 ganglioside stimulates the regeneration of dopaminergic neurons in the central nervous system, Brain Res., 261: 163.

Toffano, G., Savoini, G. E., Moroni, F., Lombardi, G., Calza, L., and Agnati, L. F., 1984, Chronic GM1 ganglioside treatment reduces dopamine cell body degeneration in the substantia nigra after unilateral hemitransection in rat, Brain Res., 296:233.

Ungerstedt, U., 1971, Striatal dopamine release after amphetamine or nerve regeneration revealed by rotational behavior, Acta Physiol. Scand., suppl 367:49.

Wiegant, H., 1971, Glycosphingolipids, Adv. Lipid. Res., 9:249.

Wojcik, M., Ulas, J., and Oderfeld-Nowak, B., 1982, The stimulating effect of ganglioside injections on the recovery of choline acetyltransferase and acetylcholinesterase activities in the hippocampus of the rat after septal lesions, Neurosci., 7:495.

EFFECTS OF NERVE GROWTH FACTOR ON CHOLINERGIC NEURONS OF THE RAT FOREBRAIN

F. Hefti[1], J. Hartikka[1], and B. Will[2]

[1]Sandoz Ltd., Preclinical Research, Basle Switzerland, and
[2]Laboratoire de Neurobiologie des Comportements Université Louis Pasteur, Strasbourg, France

INTRODUCTION

Nerve growth factor (NGF) is essential for development and function of peripheral sympathetic and sensory neurons (Thoenen and Barde, 1980). Administration of NGF to rodents produces hypertrophy of sympathetic neurons, whereas treatment with antibodies to NGF results in atrophy. Furthermore, NGF selectively stimulates the production of enzymes specific for the synthesis of catecholamine transmitters in sympathetic neurons. In contrast to these pronounced effects on peripheral sympathetic neurons, central catecholaminergic neurons are not affected by NGF or antibodies to NGF (Dreyfus et al., 1980; Konkol et al., 1978; Schwab et al., 1979). However, the following observations have suggested that NGF may play a role in the function of central cholinergic neurons located in septum and nucleus basalis and projecting to hippocampus and cortex. First, NGF injected into these cholinergic target areas is specifically taken up by cholinergic nerve terminals and is transported retrogradly to their cell bodies in septum and nucleus basalis (Schwab et al., 1979; Seiler and Schwab, 1984). Secondly, cells from the forebrain of fetal rats grown in aggregating cultures respond to NGF with an increase in choline acetyltransferase (ChAT) activity (Honegger and Lenoir, 1980). Third, destruction of the cholinergic input to the hippocampus results in an ingrowth of peripheral sympathetic fibers which matches the previous distribution of cholinergic terminals in the hippocampus (Crutcher et al., 1979; Loy and Moore, 1977; Stenevi and Björklund, 1978). As peripheral sympathetic neurons are known to react with fiber outgrowth to NGF, a similarity between the putative hippocampal signal and NGF has been postulated (Crutcher and Davis, 1981).

Effects of NGF on central cholinergic neurons are worth further investigation in light of the fact that these neurons seem to be specifically affected in Alzheimer's disease. Postmortem analysis of brains from such patients revealed a significant reduction in ChAT activity in cortex and hippocampus (Terry and Davis, 1980), and a selective loss of cell bodies in the nucleus basalis (Whitehouse et al., 1982). We therefore decided to characterize in studies with animals, in vivo and in vitro, the effects of NGF on forebrain cholinergic neurons. NGF's actions at the cellular level were investigated using cultured cholinergic neurons obtained from the septal area of fetal rat brains. Furthermore, we studied the effect of intraventricular application of NGF to rats with defined lesions in the cholinergic systems, i.e., in animals which can be considered models for Alzheimer's disease.

EFFECT OF NGF ON CHOLINERGIC NEURONS IN PRIMARY CULTURES OF DISSOCIATED SEPTAL NEURONS

The septal area was dissected from fetal rat brains of embryonic day 17. Cells were dissociated enzymatically and mechanically and then were plated on dishes coated with polyethyleneimine (a substitute for polyornithine; Ruegg and Hefti, 1984). Cholinergic neurons were identified by immunocytochemical visualization of ChAT and cytochemical demonstration of acetylcholinesterase (AChE). ChAT immunocytochemistry stained cell bodies and proximal processes, while AChE cytochemistry visualized the entire neuron. All of the ChAT-positive neurons were doublestained for AChE while only 6% of the AChE-positive cells were ChAT-negative in these cultures. These results indicate that AChE is a reliable marker for cholinergic neurons in cultures of dissociated septal cells. Being the more sensitive method, AChE staining was therefore used to identify cholinergic cells. For biochemical measurements cultures were homogenized in buffer and aliquots were taken for the determination of ChAT activity and protein content. The methods have been described in detail elsewhere (Hefti et al., 1984b).

Table 1. Effect of NGF on survival and fiber growth of cholinergic neurons in primary cultures of septal neurons. Cultures were grown for 10 days with or without the addition of NGF (100ng/ml). Growth of non-neuronal cells was suppressed by the addition of the cytostatic compound cytosine arabinoside (10^{-5}M) to the medium. Cholinergic neurons were identified by AChE cytochemistry. The number of processes was counted at a distance of 90um from the cell body.

	controls	NGF
total number of neurons per dish ($x10^{-3}$)	138 ± 22	153 ± 32
number of cholinergic neurons per dish	2150 ± 200	2450 ± 250
number of fibers per cholinergic cell	5.4 ± 0.2	5.6 ± 0.3

m ± S.E.M., 10-12 dishes were analyzed per group; the morphology was established on 30 cholinergic cells per group.

Table 2. NGF-mediated increase in ChAT activity in cultures of septal neurons. The cultures were grown for 10 days in presence of cytosine arabinoside. Thereafter, they were taken for the biochemical analysis of ChAT activity and proteins.

		ChAT activity (% of controls)
mouse-NGF	1ng/ml	103 ± 8
	10ng/ml	151 ± 12*
	100ng/ml	172 ± 18*
	1000ng/ml	175 ± 14*
mouse-NGF 100ng/ml + antibodies to NGF		109 ± 12
renin-free bovine NGF 100ng/ml		184 ± 7*
epidermal growth factor		99 ± 9
insulin		110 ± 5
cytochrome c		90 ± 4

ChAT activity in control cultures was 0.047 ± 0.003 pmol/min/μg protein = 100%. *Significantly different from controls, $p < 0.01$.

Addition of NGF to the medium failed to affect the number of cholinergic neurons surviving in culture. Furthermore, NGF did not influence the general morphological appearance and the number of processes of these neurons (Table 1). However, NGF elevated the biochemically measured activity of ChAT up to two-fold. The NGF-mediated increase in ChAT activity was dose-dependent with an ED_{50} of approximately 10ng/ml (4×10^{-10} M). The increase was highly specific for NGF. It was blocked by anti-NGF, and epidermal growth factor, insulin and other control proteins failed to exert a similar effect (Table 2). NGF had to be present for at least 3 days in the culture medium to increase ChAT activity, suggesting that the increase was due to an elevated ChAT synthesis rather than to an activation of the enzyme (data not shown).

The findings indicate that NGF stimulates the activity of the key enzyme of transmitter synthesis of cultured cholinergic neurons, but that it does not affect their survival or fiber growth. The effect of NGF on forebrain cholinergic neurons therefore appears to be very limited when compared to its actions on peripheral sympathetic neurons, where NGF is necessary for survival, fibre outgrowth and expression of the transmitter-specific enzyme, tyrosine hydroxylase (Thoenen and Barde, 1980).

EFFECT OF INTRAVENTRICULAR INJECTIONS OF NGF TO RATS WITH PARTIAL LESIONS OF THE SEPTO-HIPPOCAMPAL PATHWAY

Biochemical studies

To have animals with defined lesions of a cholinergic pathway, we lesioned the septo-hippocampal tract, because its anatomy is clearly described and because the cholinergic fibers coursing through the fimbria are easily accessible for sugery (Meibach and Siegel, 1977; Segal and Landis, 1979). Partial rather than total lesions were made in order to leave a population of cholinergic neurons able to react to exogenously administered NGF.

Lesions were made by making partial transections of the fimbria with a specially designed knife (8mm long, 0.7mm wide). It was inserted into the brain at the anteroposterior level A6100 and at a position of 1.0mm lateral to the midline; it was lowered

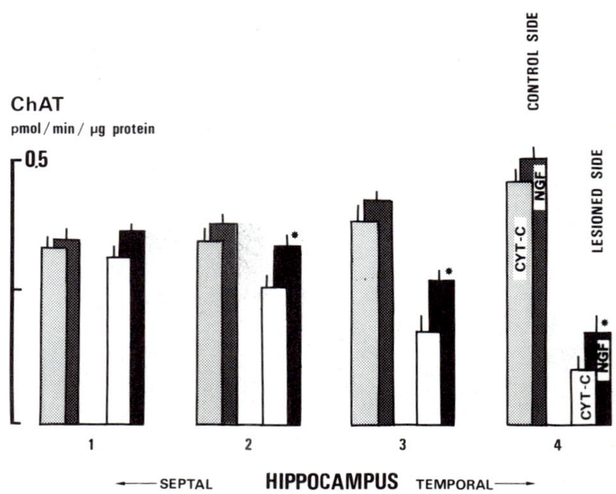

Fig. 1. Effect of intraventricular administration of NGF (10μg twice weekly during 4 weeks) on ChAT activity in hippocampus of rats with unilateral partial lesions of the cholinergic septo-hippocampal pathway. Lesions were made by cutting the lateral two-thirds of the fimbria. Hippocampus was dissected into 4 pieces along its septo-temporal axis. Lightly stippled and open bars represent unlesioned and lesioned sides of cytochrome c-treated controls, respectively. The lesions resulted in a gradual reduction of ChAT activity from the septal to the temporal hippocampal pole (significantly different in pieces 2,3,4, $p < 0.01$). Heavily stippled and black bars represent unlesioned and lesioned sides of NGF-treated animals, respectively. NGF treatment only slightly increased ChAT activity on the unlesioned side, whereas pronounced increases were seen on the lesioned side. Bars represent mean ± S.E.M. (n=14; *significantly different from corresponding control by $p < 0.01$).

Table 3. Effect of intraventricular injection of NGF (10ug twice weekly during 4 weeks) on ChAT activity in brains of rats with partial septo-hippocampal lesions.

	controls	NGF
septum	0.343 ± 0.028	0.550 ± 0.037*
cortex	0.101 ± 0.008	0.104 ± 0.005
rest of brain	0.464 ± 0.028	0.514 ± 0.042

ChAT activities are given in pmol/min/µg protein. Values represent means ± S.E.M. (n=10). *Significantly different from controls, $p<0.001$.

vertically so that its tip reached the dorso-ventral level 0.5mm (König and Klippel, 1963). The knife was then moved laterally to 5.0mm lateral to the midline and retrieved at this position. After having performed the lesion, a cannula was inserted into the lateral ventricle and permanently fixed with dental cement. For intraventricular injections, a hypodermic needle attached to a Hamilton syringe was inserted through the cannula into the ventricle. Half of the operated animals were injected twice weekly during 4 weeks with 10 micrograms of mouse submaxillary gland 2.5S NGF. Control animals received equal amounts of cytochrome c (a protein with similar molecular weight and isoelectric point as NGF). For biochemical determinations animals were lesioned unilaterally. Three days after the last intraventricular injection, they were decapitated and septum and hippocampus were removed. The hippocampus was further dissected into 4 pieces of equal size along its septo-temporal axis. Tissue pieces were homogenized and analyzed for ChAT activity and protein content. Some brains were taken for histochemical demonstration of AChE. The methods have been described in detail elsewhere (Hefti et al., 1984a).

Unilateral partial fimbrial transections reduced ChAT activities in the three most temporal regions of the hippocampus analyzed. The reduction in ChAT activity progressively increased from 27% to 77% in the most temporal hippocampal piece (Fig. 1). Repeated intraventricular injections of NGF resulted in increases of hippocampal ChAT activities on the lesioned side of 20-60% above the activities measured on the lesioned side of control aminals (Fig. 1). The relative increases were largest in the temporal parts of the hippocampus. On the unlesioned side, only small increases (approximately 10%) were observed which did not reach statistical significance. ChAT activity in the septum of NGF-treated rats was increased by 60% as compared to control animals. ChAT activity in cortex and the remaining parts of the brain was not affected by NGF-treatment (Table 3).

To assess whether the NGF-induced increase in hippocampal ChAT activity reflects a promotion of sprouting of cholinergic fibers, the effect of the NGF-treatment was analyzed with histochemical techniques. AChE, visualized histochemically, was used as a marker for hippocampal cholinergic fibers. In this brain area, AChE is predominantly localized in cholinergic fibers as indicated by the fact that fimbrial lesions reduce hippocampal AChE activity to a similar extent as ChAT activity (Björklund and Stenevi, 1977) and that the two enzyme activities develop in parallel during ontogenesis (Nadler et al., 1974). Histochemical analysis of lesioned animals confirmed the effectiveness of the fimbrial transection: AChE staining was strongly reduced in the temporal parts of the hippocampus, whereas only slight reductions were observed in the septal parts. In the temporal hippocampus, staining intensity was most strongly reduced in the CA3 field. We could not detect any difference in staining pattern or intensity between NGF-treated and control animals. This was even the case for rats chronically treated during 10 weeks (data not shown). This finding indicates that the NGF-mediated increase in hippocampal ChAT activity in animals with partial fimbria lesions is likely to be due to an activation or induction of the enzyme within cholinergic terminals rather than to a promotion of sprouting of cholinergic fibers.

Behavioral studies

Animals used for behavioral tests received bilateral partial transections of the fimbria. They otherwise were treated with NGF or cytochrome c during 4 weeks as the animals taken for biochemical studies. Three days after the last intraventricular injection, the rats were pretrained and then tested for 10 consecutive days in an 8-arm radial maze containing a food pellet in each arm. All rats with lesions, whether treated with NGF or cytochrome c (CYTO), were impaired in their ability to learn the radial maze. They visited less arms before committing their first error and made more errors (re-visits) than did unlesioned, or sham-operated control rats. However, the NGF- and cytochrome-treated animals differed in their capacity to increase performance over trials. The performance of only the NGF-treated group increased significantly over trials (comparison to the zero slope for performances of the NGF group: $p<0.02$ and $p<0.001$ for 'errors' and 'arms', respectively). During the two last sessions, only the cytochrome c-tested rats visited significantly fewer arms before the first error than control rats ($p<0.02$; Fig. 2).

CONCLUSIONS

The results presented indicate that NGF has the ability to trophically influence cholinergic neurons of the rat forebrain. The reaction of the cholinergic neurons apparently is limited to an elevation of the activity of the transmitter-specific enzyme, ChAT, and does not include a promotion of survival or fiber growth of these neurons. NGF elevated ChAT activity

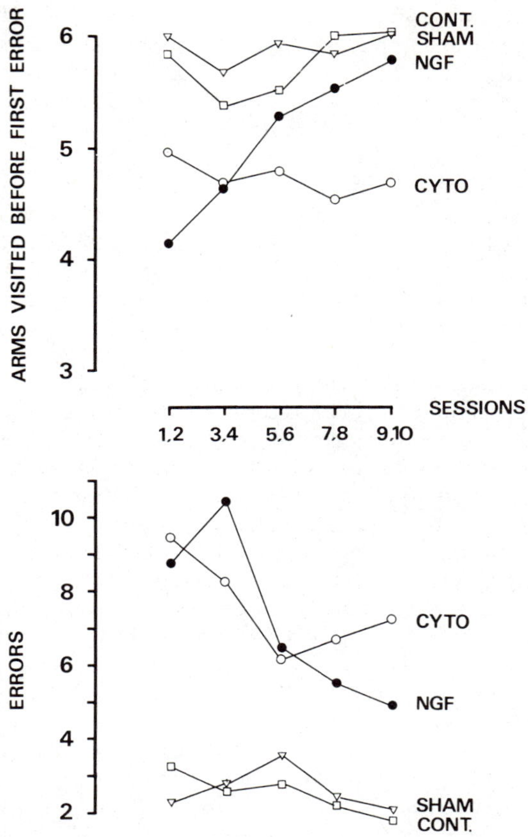

Fig.2. Effect of NGF-treatment in rats with bilateral partial fimbria lesions on their performance in an 8-arm radial maze task. Lesioned rats received either intraventricular injections of NGF (10µg twice weekly) during 4 weeks or equal amounts of cytochrome c (CYTO) control injections. CONT = Unlesioned control rats; SHAM = Sham operated rats.

in cholinergic neurons developing in culture. In an earlier study, it was shown that intraventricular injections of NGF to rats during their first postnatal week, i.e., during the time of development of the cholinergic septo-hippocampal pathway, increase ChAT activity in the forebrain of these animals (Gnahn et al., 1983). In intact adult rats, NGF failed to affect hippocampal ChAT activity; however, it was effective in animals with partial fimbria lesions. It therefore appears that NGF has the ability to elevate ChAT activity in cholinergic septohippocampal neurons during development and after partial lesions of these pathways. The mechanism by which a partial septohippocampal lesion induces the cholinergic neurons to become responsive to NGF is unknown. Such a lesion might induce the formation of functional receptors for NGF or might couple existing receptors to a functional process.

NGF-treatment of adult animals with partial fimbria lesions was found to accelerate the recovery from behavioral deficits produced by the lesion. This behavioral effect possibly reflects the trophic influence of NGF on cholinergic neurons. However, behavioral improvements in lesioned animals have also been reported to occur after single injections of NGF, which in our experience fail to stimulate ChAT activity, and in animals with lesions not affecting the septo-hippocampal or basalo-cortical cholinergic systems (Hart et al., 1978; Stein et al., 1984). It therefore cannot be excluded that the behavioral improvements mediated by NGF are due to an action other than the trophic influence on cholinergic neurons.

As stated in the preceding section cholinergic neurons of both the septo-hippocampal projection and the projection from the nucleus basalis to the cortex can respond to NGF. These projections apparently are homologous to those specifically affected in Alzheimer's disease (Terry and Davis, 1980; Whitehouse et al., 1982; Rossor et al., 1982). Given the trophic influence of NGF on cholinergic neurons as was observed in rats, NGF or compounds mimicking its action therefore might be of value in the treatment of this disease. However, it remains to be shown whether cholinergic neurons in the human brain are able to react to NGF.

It has been speculated that Alzheimer's disease might be caused by a lack of a specific neurotrophic factor for cholinergic neurons (Appel, 1981), and, furthermore, that such a factor might be identical to NGF (Hefti, 1983). Findings have been reported which provide indirect evidence for the existence of an endogenous neurotrophic factor for cholinergic neurons (Crutcher and Collins, 1982). However, until the unequivocal demonstration of an endogenous factor, it cannot be excluded that NGF has no physiological role in the function of central cholinergic neurons.

REFERENCES

Appel, S.H., 1981, A unifying hypothesis for the cause of amyotrophic lateral sclerosis, parkinsonism, and Alzheimer's disease, Ann. Neurol. 10:499.

Björklund, A. and Stenevi, U., 1977, Reformation of the severed septo-hippocampal cholinergic pathway in the adult rat by transplanted septal neurons, Cell. Tiss. Res., 185:289.

Crutcher, K.A., Brothers, L. and Davis, J.N., 1979, Sprouting of sympathetic nerves in the absence of afferent input, Exp. Neurol. 66:778.

Crutcher, K.A. and Collins, F. 1982, In vitro evidence for two distinct hippocampal growth factors: basis of neuronal plasticity? Science 217:67.

Crutcher, K.A. and Davis, J.N., 1981, Sympathetic noradrenergic sprouting in response to central cholinergic denervation, Trends Neurosci., 4:70.

Dreyfus, C.F., Peterson, E.R. and Crain, S.M., 1980, Failure of nerve growth factor to affect fetal mouse brain stem catecholaminergic neurons in culture, Brain Res., 194:540.
Gnahn, H., Hefti, F., Heumann, R., Schwab, M.E. and Thoenen, H., 1983, NGF-mediated increase of choline acetyltransferase (ChAT) in the neonatal rat forebrain; evidence for a physiological role of NGF in the brain? Develop. Brain Res., 9:45.
Hart, T., Chaimas, N., Moore, R.Y. and Stein, D., 1978, Effects of nerve growth factor on behavioral recovery following caudate nucleus lesions in rats, Brain Res. Bull., 3:245.
Hefti, F., 1983, Alzheimer's disease caused by a lack of nerve growth factor?, Ann. Neurol., 13:109.
Hefti, F., Dravid, A. and Hartikka J., 1984a, Chronic intraventricular injections of nerve growth factor elevate hippocampal choline acetyltransferase activity in adult rats with partial septo-hippocampal lesions, Brain Res., 293:305.
Hefti, F., Hartikka, J., Eckenstein, F., Gnahn, H., Heumann, R. and Schwab, M. 1984b, Nerve growth factor (NGF) increases choline acetyltransferase but not survival or fiber outgrowth of cultured septal cholinergic neurons, Neurosci., in press.
Honegger, P. and Lenoir, D., 1982, Nerve growth factor (NGF) stimulation of cholinergic telencephalic neurons in aggregating cell cultures, Develop. Brain Res., 3:229.
König, J.F.R. and Klippel, R.A., 1963, "The Rat Brain. A Stereotaxic Atlas of the Forebrain and Lower Parts of the Brain Stem", Williams and Wilkins, Baltimore, MD.
Konkol, R.J., Mailman, R.B., Bebdeich, E.G., Garrison, A.M., Müller, R.A. and Breese, G.R., 1978, Evaluation of the effects of nerve growth factor and anti-nerve growth factor on the development of central catecholamine-containing neurons, Brain Res., 144:277.
Leis, T., Pallage, V., Toniolo, G. and Will, B., 1984, Working memory theory of hippocampal function needs qualification, Behav. Neural. Biol., 42:140.
Loy, R. and Moore, R.Y., 1977, Anomalous innervation of the hippocampal formation by peripheral sympathetic axons following mechanical injury, Exp. Neurol., 57:645.
Meibach, R.C. and Siegel, A., 1977, Efferent connections of the septal area in the rat: an analysis utilizing retrograde and anterograde transport methods, Brain Res., 119:1.
Nadler, J.V., Mattews, D.A., Cotman, C.W. and Lynch, G.S., 1974, Development of cholinergic innervation in the hippocampal formation of the rat, Develop. Biol., 36: 142.
Rossor, M.N., Svendsen, C., Hunt, S.P., Mountjoy, C.Q., Roth, M. and Iversen, L.L., 1982, The substantia innominata in Alzheimer's disease: an histochemical and biochemical study of cholinergic marker enzymes, Neurosci. Lett., 28:217.

Ruegg, U.T. and Hefti, F., 1984, Growth of dissociated neurons in culture dishes coated with synthetic polymeric amines, Neurosci. Lett., 49:319.

Schwab, M.E., Otten, U., Agid, Y. and Thoenen, H., 1979, Nerve growth factor (NGF) in the rat CNS: absence of specific retrograde axonal transport on tyrosine hydroxylase induction in locus coeruleus and substantia nigra, Brain Res., 168:473.

Seiler, M. and Schwab, M.E., 1984, Specific retrograde transport of nerve growth factor (NGF) from neocortex to nucleus basalis in the rat. Brain Res., 300:33.

Segal, M. and Landis, S., 1979, Afferents of the hippocampus of the rat studied with the method of retrograde transport of horseradish peroxidase, Brain Res., 78:1.

Stein, D.G., Deluzarche, F.M. and Will, B., 1984, Nerve growth factor improves radial maze performance in adult rats with hippocampal lesions, in preparation.

Stenevi, U. and Björklund, A., 1978, Growth of vascular sympathetic axons into the hippocampus after lesions of the septo-hippocampal pathway: a pitfall in brain lesion studies, Neurosci. Lett., 7:219.

Terry, R.D. and Davies, P., 1980, Dementia of the Alzheimer type, Ann. Rev. Neurosci., 3:77.

Thoenen, H. and Barde, Y.A., 1980, Physiology of nerve growth factor, Physiol. Rev., 60:1284.

Whitehouse, P.J., Price, D.L., Stroble, R.G., Clark, A.W., Coyle, J.T. and DeLong, M.R., 1982, Alzheimer's disease and senile dementia: loss of neurons in the basal forebrain, Science, 215:1237.

REORGANIZATION AND RESTORATION OF CENTRAL NERVOUS CONNECTIONS AFTER INJURY: A LESION AND TRANSPLANT STUDY OF THE RAT HIPPOCAMPUS

Jens Zimmer, Niels Sunde and Torben Sørensen

Institute of Anatomy B (Neurobiology)
University of Aarhus
DK-8000 Aarhus C, Denmark

INTRODUCTION

The hippocampus and fascia dentata are cortical brain structures with a distinct laminar organization of both the neuronal cell layers and the terminations of extrinsic and intrinsic afferents in the neuropil. In addition, the dentate granule cells are among the latest forming neurons in the brain. It has thus been estimated that in the rat about 85% of the granule cells form after birth, mainly during the first postnatal days (Bayer, 1980). These structural and developmental features have prompted - and facilitated - a large number of studies of lesion-induced plasticity of the hippocampal and dentate nerve connections (see Cotman and Lynch, 1976; Cotman and Nadler,

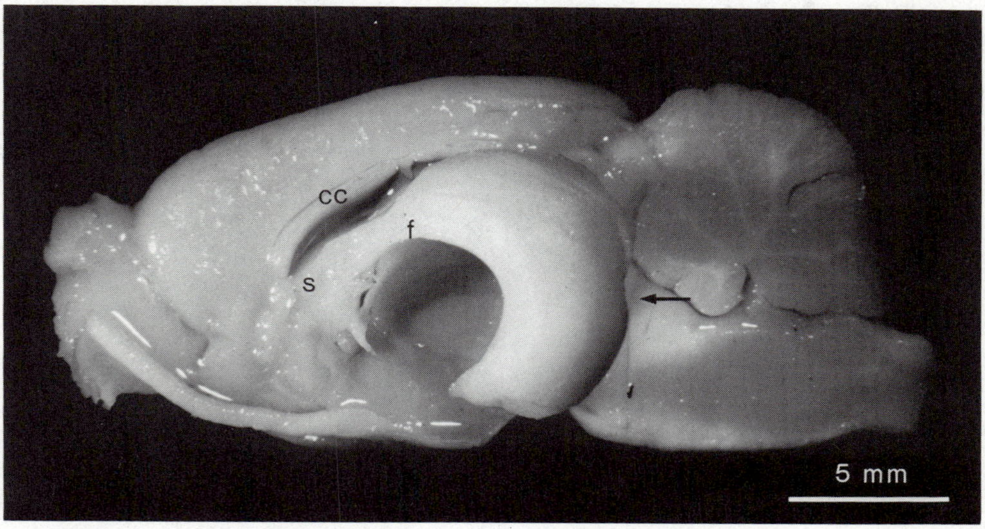

Fig. 1. The hippocampus and the fascia dentata of the rat brain in a lateral view freed from other overlying and adjacent cortical structures. Arrow indicates level shown in Fig. 2a,b. cc, corpus callosum; f, fimbria; s, septum.

1978; Zimmer, 1978; Laurberg and Zimmer, 1980, 1981; Gall and Lynch, 1980). More recently these studies have been extended to include studies of the cellular and connective organization of the hippocampus and fascia dentata after intracerebral and intraocular transplantations (Zimmer, 1978; Sunde and Zimmer, 1981, 1983; Goldowitz et al., 1982, 1984; Kromer et al., 1983; Zimmer and Sunde, 1984; Zimmer et al., 1985).

In this chapter we first briefly review some of the findings on the lesion-induced reorganization of hippocampal and dentate connections. These data will form the background for a description of the organization and connective patterns of hippocampal and dentate tissue transplanted to different regions of normal, immature and adult rat brains. In the final section we will demonstrate how transplantation of developing dentate granule cells can replace and help restore normal connections in rats with severe deficits of granule cells resulting from neonatal X-irradiation (Sunde et al., 1984, 1985).

Normal Anatomy and Lesion-Induced Plasticity of Hippocampal and Dentate Connections

The rat hippocampus and fascia dentata are intimately embraced structures which from the midline just posterior and dorsal to the septum curve in a dorsolateral and then ventral and ventromedial direction towards their temporal pole at the base of the brain (Fig. 1). Sections cut perpendicular

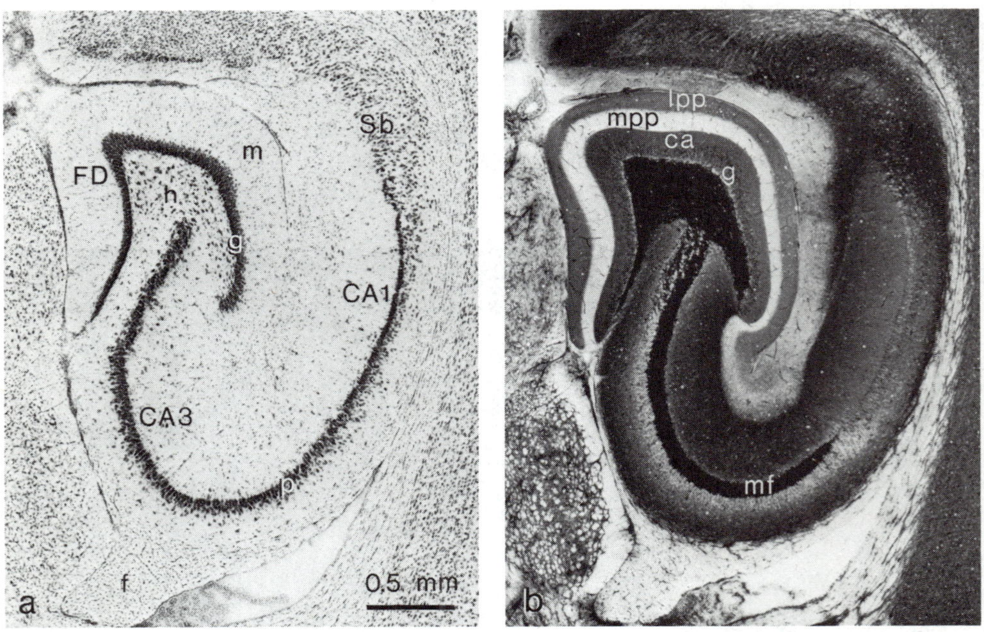

Fig. 2. Histological sections through the rat hippocampus and fascia dentata, showing the cell and neuropil layers in thionin staining (a), and the laminar terminal fields in Timm staining (b). ca, commissural-associational terminal zone; CA1, hippocampal field CA1; CA3, hippocampal field CA3; f, fimbria; FD, fascia dentata; g, dentate granule cell layer; h, hilus (CA4); lpp, lateral perforant path zone; m, dentate molecular layer; mf, mossy fiber zone; mpp, medial perforant path zone; p, hippocampal pyramidal cell layer; Sb, subiculum.

to this longitudinal septotemporal axis reveal the characteristic cellular organization of the hippocampus and fascia dentata with the main neuronal cell types arranged in two distinct layers, the dentate granule cell layer and the hippocampal pyramidal cell layer (Fig. 2a). The main extrinsic and intrinsic (associational) pathways terminate as separate or overlapping projections in distinct laminae covering proximo-distal segments along the dendritic trees in the adjacent neuropil. This is clearly illustrated in sections stained with the Timm sulphide silver method (Fig. 2b). This method, which is a histochemical stain for so-called heavy metals, including zinc (Timm, 1958a,b; Haug, 1973; Danscher and Zimmer, 1978; Danscher, 1984), is highly reliable for the monitoring of the distribution, development and lesion-induced reorganization of hippocampal pathways (see Zimmer, 1978). Recent studies using immunohistochemical techniques have shown a corresponding laminar distribution of different neuropeptides related to the afferent pathways. Enkephalin-reactivity has thus been found corresponding to the mossy fibers and the lateral perforant path (Stengaard-Pedersen et al., 1981, 1983; Gall et al., 1981; Fredens et al., 1984; Zimmer and Sunde, 1984), and cholecystokinin-reactivity to the medial perforant path and commissural-

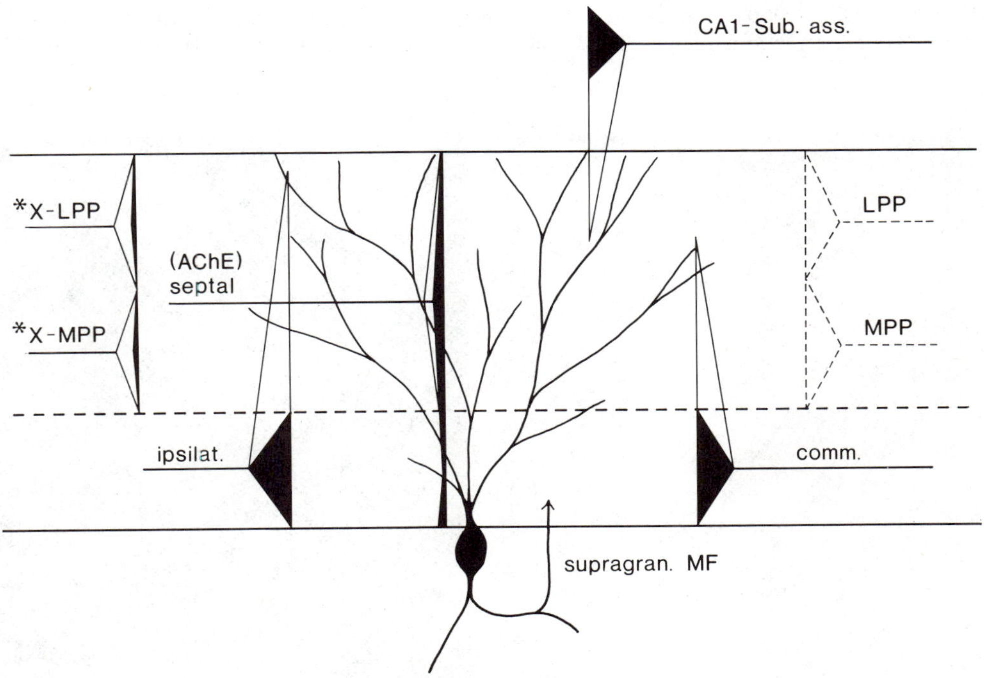

Fig. 3. Schematic illustration of the connective reorganizations which follow removal of the perforant path input to the dentate molecular layer. The molecular layer is between the solid lines; str. moleculare of CA1 and subiculum is above the upper line. Commissural (comm.) and ipsilateral associational fibers (ipsilat.) extend into the denervated perforant path zones where AChE-reacting septal afferents also proliferate. At the most septal levels the normal, but sparse, crossed perforant pathways (asterisks) also expand. More posteriorly and temporally CA1-subiculum associated afferents (CA1-Sub.ass.) expand across the hippocampal fissure. The formation of a supragranular mossy fiber projection (supragran. MF) by ingrowth of mossy fiber collaterals from the hilus is indicated by arrow. Reproduced from Zimmer (1978) with permission.

associational systems (Stengaard-Pedersen et al., 1983; Fredens et al., 1984; Zimmer and Sunde, 1984).

Both during the stage of normal ontogenic growth and differentiation early postnatally and in adulthood, various lesions, which remove extrinsic afferents to the hippocampus and fascia dentata have been found to induce extension and reorganization of the remaining afferent pathways. This aberrant axonal growth and collateral sprouting have most thoroughly been studied after lesions of the entorhinal cortex which remove the perforant path input to the distal parts of the dentate granule cell dendrites (Lynch et al., 1972, 1973; Nadler et al., 1973; Zimmer, 1973a,b, 1978; Zimmer and Hjorth-Simonsen, 1975; Gall and Lynch, 1980; Steward, 1980; Stanfield and Cowan, 1982; Caceres and Steward, 1983). The resulting connective reorganization is shown schematically in Fig. 3.

The lesion-induced changes were age-dependent in the sense that the growth and segregation of the individual, reorganizing pathways differed between rats lesioned during early development (0-9 days postnatally) and rats lesioned in adulthood. With regard to the denervation-induced aberrant growth and sprouting the differences appeared to involve only the speed of the reorganization and some mutual quantitative, but not qualitative differences (Lynch et al., 1973; Zimmer, 1974; Zimmer and Hjorth-Simonsen, 1975; Gall et al., 1980). In contrast, more pronounced age-dependent differences were found in the rerouting of the mossy fibers following lesions placed

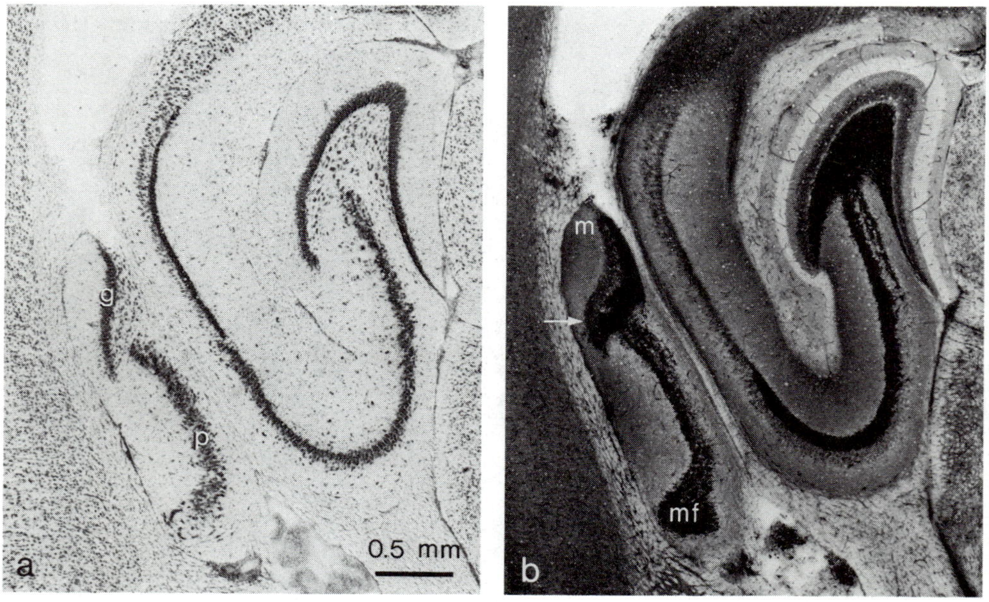

Fig. 4. Transplant consisting of fascia dentata and the hippocampal field CA3 located lateral to the host hippocampus. Donor, newborn; recipient, newborn; survival 2 months. The characteristic cell and neuropil layers are shown in thionin (a) and Timm staining (b). Distinct granule cell (g) and CA3 pyramidal cell layers (p) are present in the transplant. The mossy fibers (mf) from the dentate granule cells distribute as normally through the hilus along the CA3 pyramical cell layer. The Timm staining of the dentate molecular layer (m) illustrates the spread of intrinsic associational pathways throughout the dentate molecular layer in the absence of perforant path afferents. Arrow points to supragranular mossy fibers.

along the normal trajectory of these fibers (Laurberg and Zimmer, 1980). The resulting rerouting of the main axons in CA3 thus showed a clear decline with age during the first three postnatal weeks, while sprouting of mossy fiber collaterals in response to denervation of the fascia dentata persisted without decrease into adulthood (Laurberg and Zimmer, 1981; Frotscher and Zimmer, 1983).

Hippocampal and Dentate Transplants: Organization and Connections

Cells. Pieces of hippocampal tissue, including the fascia dentata, the hippocampus proper (CA3 and CA1) and sometimes adjacent parts of the subiculum survived well when taken from the brains of late embryonic or early postnatal rats and transplanted into the brain of other newborn or adult rats (Sunde and Zimmer, 1981, 1983; Kromer et al., 1983; Zimmer et al., 1983, 1985). Within the transplants the basic organotypic organization of the main cell and neuropil layers was usually preserved irrespective of the age of the recipient rats and the localization of the transplants (Fig. 4a). Also the presence and distribution of other neurons, including interneurons normally located both within and outside the main cell layers, appeared undisturbed when observed in transplants stained immunohistochemically for somatostatin- and CCK-reacting cells or stained with the DFP-AChE technique which demonstrates AChE-producing cells (see Zimmer et al., 1983 and Zimmer and Sunde, 1984 for detailed findings and discussion).

Connections. The distribution of transplant connections of intrinsic origin was of course primarily determined by the cellular content and size of the transplants. Within a given transplant the connections did, however, follow the patterns of distribution observed in lesion-studies where the same tissue was subjected to various types of denervation, but remained located in situ in the brain (Fig. 4b). The only exceptions were observed after transplantation to immature recipients in transplants which received a heavy input of afferents from the host brain (see below). In these transplants the intrinsic connections displayed limitations in their expansion, apparently because they had to compete for the synaptic sites with the ingrowing host afferents.

As just indicated above, the density and pattern of the host innervation of the transplanted dentate granule cells depended on the age of the recipient rats, the location of the transplants (determining the host afferent pathways available among other things), and the nerve type. Dense innervation by host afferents has thus so far only been observed after transplantation to recipients still in their normal developmental period. Depending on the location of the transplants these afferents encountered host perforant path fibers, including the enkephalin-reactive component of the lateral perforant path and the CCK-reactive component of the medial perforant path (Zimmer and Sunde, 1984), host hippocampodentate commissural fibers, host hippocampo-hippocampal commissural fibers and callosal afferents (Sunde and Zimmer, 1983). The first two types of afferents which normally innervate dentate granule cells terminated in and hence apparently recognized their normal terminal zones along the dentate granule cell dendrites in the transplants (cells 3 and 4 in Fig. 5). The latter two types of host afferents, which do not normally contact dentate granule cells, were confined to the distal portions of the dentate granule cell dendrites where they replaced the normal perforant path input (cells 5 and 6 in Fig. 5). The presence of a distinct laminar segregation of the extrinsic and intrinsic afferent termination in these transplants is of great interest, because it proves that the determinants of this segregation also act on "alien afferents" like for instance fibers from the corpus callosum, which do not normally project to dentate granule cells. This supports the idea of a mechanism of interaxonal interaction rather than axon-target specification (Changeux and Danchin, 1976; Diamond et al., 1976; Zimmer et al., 1983).

Referring to the role of the nerve type in innervation of transplants by host nerve fibers it is important to note that innervation by host cholinergic fibers appeared at all recipient ages. Transplants placed in a variety of different areas in both immature and adult recipient brains thus displayed a virtually normal density of AChE staining in the neuropil. Such staining was found irrespective of whether the transplants were located within the normal reach of the cholinergic septohippocampal system or within the normal projection field of the cholinergic projection from the basal nucleus of Meynert (Lehman et al., 1980; Wenk et al., 1980) as, for instance, the frontal or parietal neocortex (Zimmer et al., 1983; Sunde and Zimmer, 1983). The pattern of AChE staining in the transplants was, however, largely determined by the distribution of the other major, non-cholinergic extrinsic and intrinsic pathways (dotted panel to the left of the cells shown in Fig. 5), in agreement with observations from other lesion-studies (Zimmer and Laurberg, 1978; Zimmer et al., unpublished).

Ultrastructure of synaptic organization. Electron microscopy of the dentate molecular layer of the transplants revealed that irrespective of the

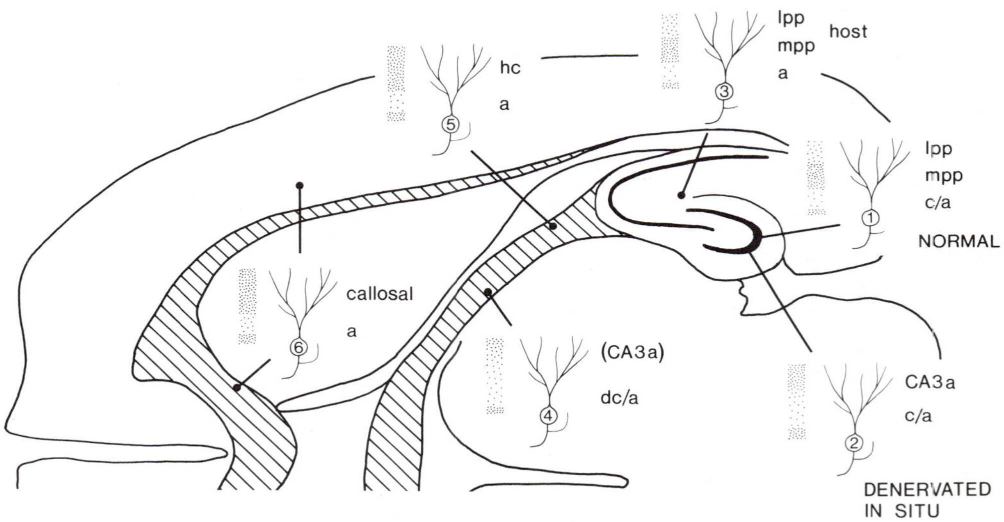

Fig. 5. Schematic illustration of the laminar innervation of dentate granule cells in normal rats (1), after in situ removal of the perforant pathways (2), and after transplantation to different sites in immature recipient brains, namely the hippocampus or fascia dentata, encroaching on host perforant path zones (3), the medial part of the fimbria, encroaching on the host commissural projection to the host dentate (4), the lateral part of the fimbria, encroaching on the commissural projection to the host hippocampus (5), and the neocortex or the corpus callosum (6). In the various situations the dentate granule cell afferents include normal (c) and host dentate commissural fibers (dc), associational fibers from the dentate hilus (a), aberrant associational fibers from the adjacent CA3 (CA3a), normal and host entorhinal lateral (lpp) and medial perforant path (mpp) fibers, host hippocampal commissural fibers (hc), and fibers from the host corpus callosum (callosal). The dotted panels to the left of the cells indicate the AChE patterns attained at the different locations. Reproduced with permission from Zimmer et al. (1985).

transplant site both symmetric (Gray's type II) and asymmetric (Gray's type I) synapses were present (Fig. 6a,b). The symmetric type associated with synaptic vesicles with a tendency for flattening is generally supposed to be inhibitory in function, while the asymmetric type associated with spherical synaptic vesicles is thought to be excitatory. Both types of synapses were found on dendritic shafts throughout the dentate molecular layer, but in greater numbers towards the granule cell layer where the larger shafts are found. The great majority of synapses were, however, of the asymmetrical type located on dendritic spines (Fig. 6a) and as in the normal fascia dentata it was possible to distinguish several types on the basis of the configuration of the zone of contact between the spines and the terminals (Blackstad, 1963, 1967; Laatsch and Cowan, 1966; Matthews et al., 1976). The majority were of the simple type with a straight synaptic contact between the terminal and the spine (Fig. 6a). The rest were of the so-called complex type where the postsynaptic spine partly surrounded the presynaptic terminal, and the synaptic contact was either U-shaped (U, Fig. 6a) or had the form of a W (W, Fig. 6a). The latter form was due to an indentation of the terminal in the middle of the U or a direct invagination of the terminal by a slender process arising from the spine, the so-called spinule (Tarrant and Routtenberg, 1977).

Mossy fiber terminals arising from the transplanted granule cells were found with all their normal characteristics corresponding to their usual distribution, that is below the dentate granule cells in the hilus (CA4) and along the CA3 pyramidal cell layer. In correspondence with the light micro-

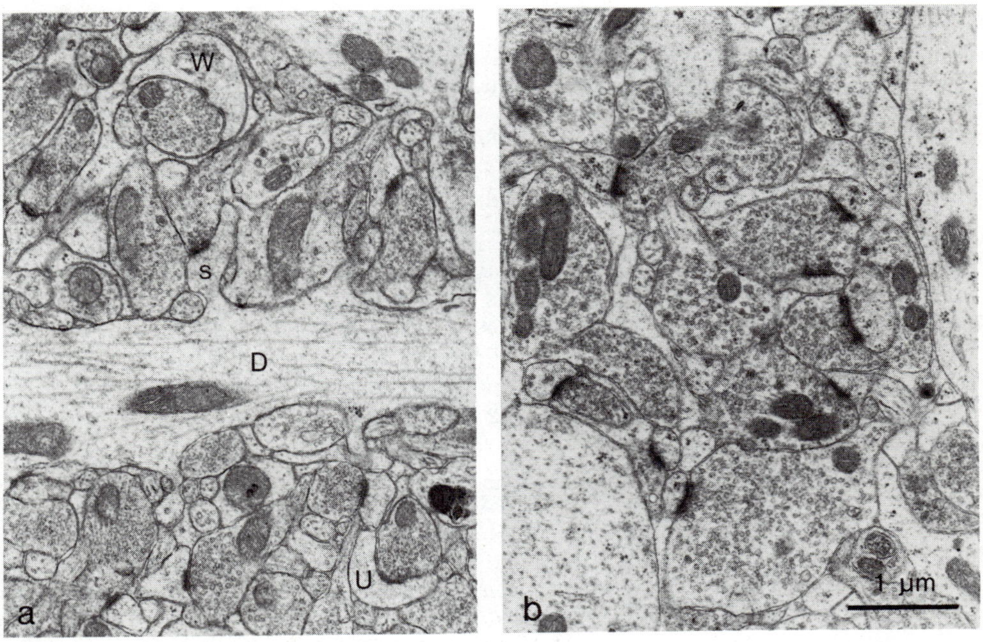

Fig. 6. Electron micrographs from the molecular layer of fascia dentata transplants. - (a) Different asymmetrical synapses on spines with configurations characteristic of the normal dentate molecular layer are found. U, U-shaped synapse; W, W-shaped synapse. One spine (s) is seen emerging from a dendritic shaft (D). Donor, embryonic day 21; recipient, 7 days old; survival 3 months. - (b) Aberrant, supragranular mossy fiber terminals arranged in cluster. Some of the terminal profiles form asymmetric synaptic contacts with spines.

scopic findings of Timm stained, aberrant supragranular mossy fibers in the transplants, terminals with the characteristics of mossy fiber terminals, but smaller, were also found innermost in the transplant dentate molecular layer. Their numbers varied from transplant to transplant apparently depending on the size, and cellular content and extrinsic connections of these (see above). They usually appeared in clusters as vesicle-filled presynaptic terminals making asymmetric synaptic contacts with spines (Fig. 6b). Some of these spines were quite large and complex with spinulae, resembling "en miniature" the large and very complex spines (excrescences) arising from the hippocampal pyramidal cell dendrites in the normal mossy fiber zones. The ultrastructure of these terminals in the transplants in other words clearly corresponded to that described for aberrant supragranular mossy fiber terminals in the in situ denervated fascia dentata (Laurberg and Zimmer, 1981; Frotscher and Zimmer, 1983).

One synaptic abnormality observed in the transplants relative to normal dentate tissue was the fairly frequent occurrence of spines arising directly from the granule cell bodies and forming asymmetrical synapses. Normally the cell bodies with the characteristic thin brim of cytoplasm around the nucleus are tightly packed leaving only limited space for some symmetric and asymmetric synaptic contacts directly on the cell body. In the transplants, however, the packing of the granule cells often was less dense, and this might have allowed the outgrowth of somatic spines in addition to the formation of the normal axosomatic symmetrical and asymmetrical synapses. Observations supporting this explanation have also been done in the Reeler mutant mouse where the granule cells are dispersed (Stirling and Bliss, 1978) and in displaced granule cells in slice cultures (Zimmer and Gähwiler, 1984).

Conclusion. Our observations on intracerebral hippocampal and dentate transplants in normal rats have shown that - 1) normal, developing dentate granule cells and hippocampal pyramidal cells can survive transplantation to various locations in normal immature and adult recipient brains, - 2) the transplanted tissue develops an organotypic organization of cell and neuropil layers which also include the ultrastructure of synaptic organization, - 3) the intrinsic transplant connections reorganize according to principles known from lesion-studies on the in situ located tissue with collateral sprouting and aberrant axonal growth and rerouting, - 4) host afferents can innervate the transplants but show a marked decrease in this capacity after the normal developmental period with the exception of host cholinergic afferents, - 5) homotypic host afferents like commissural hippocampodentate and perforant path afferents distribute to their normal terminal zones in the transplants, while heterotypic afferents like host callosal afferents always are confined to the "vacant" perforant path zones on the distal parts of the dendrites.

Restoration of Hippocampal Connections after Neonatal X-Irradiation

The experimental model. Due to the late, almost exclusively postnatal development of the dentate granule cells it is possible to prevent the formation of most of these cells by X-irradiating the hippocampal regions of newborn rats (Bayer and Altman, 1975; Laurberg and Hjorth-Simonsen, 1977). In such rats which grow up with only 10-20% of the normal amount of granule cells, the hippocampal neuronal circuitry is changed in two ways: - 1) the hippocampal mossy fiber system is attenuated corresponding to the reduction of granule cells, and - 2) the perforant pathways from the entorhinal cortex project aberrantly onto the basal dendrites of CA3 hippocampal pyramidal cells in the absence of sufficient target granule cells (Fig. 7a,b) (Laurberg and Hjorth-Simonsen, 1977).

Realizing that such irradiated animals might serve as a model for early brain damage we initiated a series of transplantation experiments with the

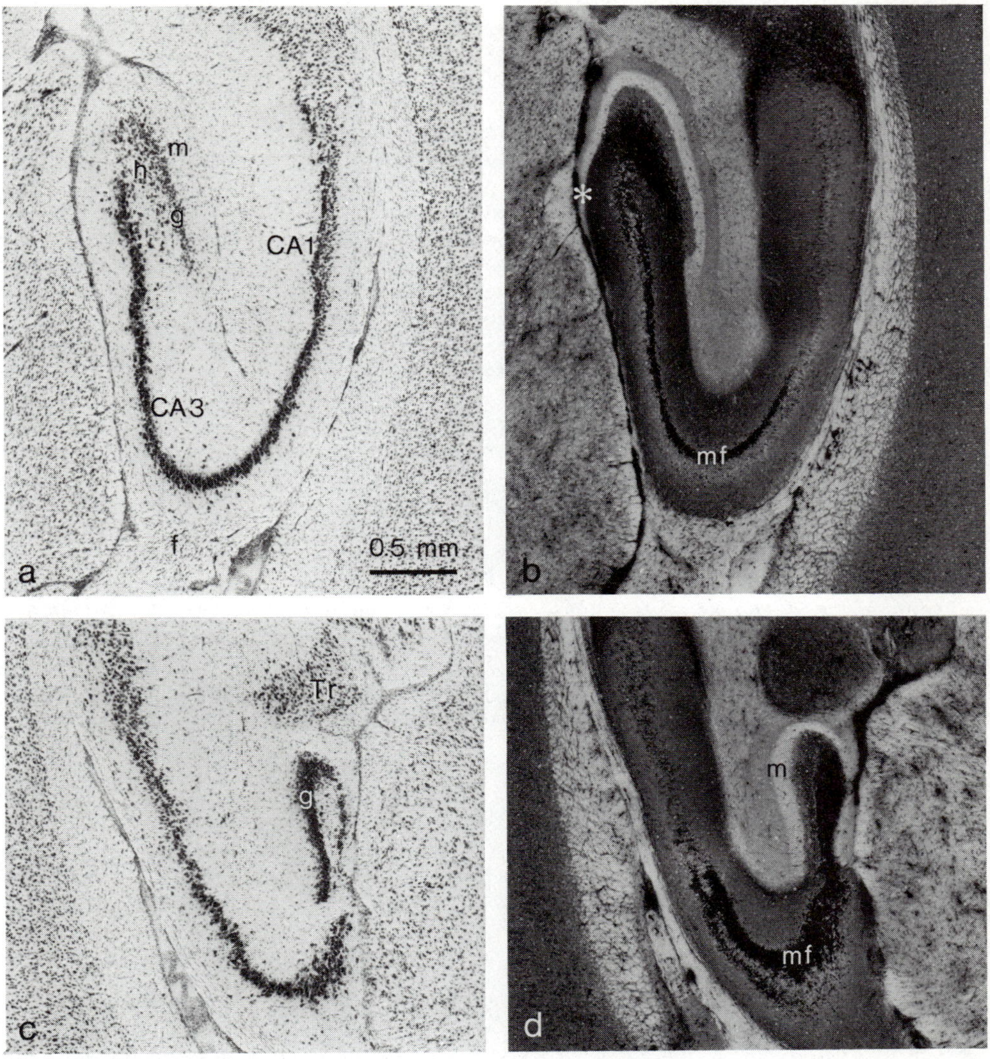

Fig. 7. Normalization of hippocampal connective patterns in rat X-irradiated bilaterally at the day of birth, transplanted on one side with normal developing granule cells the same day and allowed to survive for 4 months. - (a) Thionin stained section from the irradiated non-transplanted side showing the severe deficit of dentate granule cells. - (b) Timm stained section from the irradiated, non-transplanted side, showing marked attenuation of the mossy fiber projection (mf) and aberrant, but still laminar perforant path projections (asterisk) along the medial side of the reduced fascia dentata onto extended basal dendrites of the CA3 pyramidal cells. - (c) Thionin stained section from transplanted side showing transplant (Tr) with dentate granule cells in the anterior part arranged in a characteristic layer (g), replacing the irradiated host fascia dentata at these levels. - (d) Timm stained section of the dentate transplant showing a normalized dentate molecular layer with distinct perforant path and commissural-associational zones and normalization of the mossy fiber projection along the host CA3 pyramidal cell layer (mf) (cp. Fig. 6a for control of irradiation).

primary goal to restore completely or in part the normal neuronal circuitry, and then later extend the studies to include possible beneficial behavioral effects of the anatomical restoration. Results of the anatomical restoration are now available (Sunde et al., 1984, 1985) and are summarized below.

In our attempts to repair the neonatally induced damage of dentate granule cells we used the following experimental paradigm. Newborn rats were irradiated (600 rad) bilaterally over the hippocampal regions while the rest of the brain, head and body was shielded by lead. Blocks of dentate tissue from normal, newborn rats containing developing granule cells were then grafted unilaterally into the hippocampal region of the irradiated rats either just after the irradiation or after a delay of 3, 5 and 10 weeks. The histological analysis of the transplant-host interactions was performed after a survival time of at least 6 weeks. Then the recipients were sacrificed, their brains cut and the sections stained for cells (thionin), terminal fields (Timm, AChE) and tracing of axonal connections (silver impregnation of acute anterograde axonal degeneration) (Sunde et al., 1984, 1985).

Immediate transplantation. Transplantation to newborn recipients on the day of irradiation resulted in the formation of specific, serial and precisely laminar connections from the host into and out of the dentate transplants, provided that these were located in or encroached on the irradiated host fascia dentata (Fig. 7c). Host lateral and medial perforant path and commissural hippocampodentate fibers thus innervated their respective appropriate laminae along the granule cell dendrites (Fig. 7d,e). Transplant mossy fibers in turn invaded the adjacent host CA3 tissue and projected along the CA3 pyramidal cell layer, normalizing the number of Timm stained mossy fiber terminals here (cp. Fig. 7b and 7d). The same transplants also received a normal cholinergic input from the septohippocampal system, judged by the organotypic distribution of AChE staining (Fig. 7f). The position of the

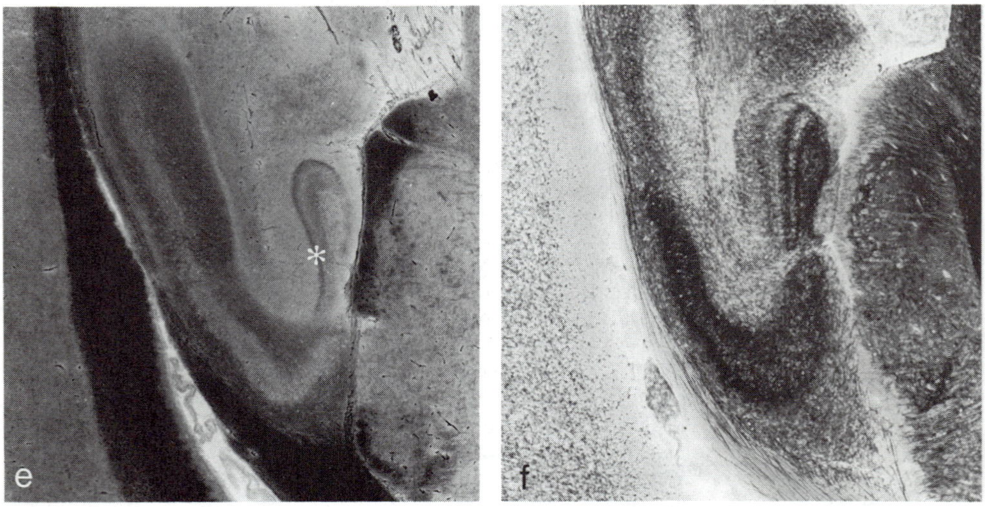

Fig. 7. (e) Host commissural projection (asterisk) to the commissural-associational zone in the dentate transplant revealed by Fink-Heimer staining of the anterograde terminal degeneration which followed transection of the contralateral fimbria 3 days before sacrifice. - (f) The transplant displays a characteristic organotypic laminar distribution of AChE corresponding to innervation by host cholinergic septohippocampal fibers.

dentate transplants is as mentioned essential for the establishment of the serial host-transplant-host connections. Several additional transplants participated, however, in either the afferent host-transplant or the efferent transplant-host part of the serial connection. Some more posteriorly located transplants thus received distinctly laminated host perforant path projections, but did not send mossy fibers to the host. Other more anteriorly placed transplants, outside the reach of the host perforant pathways, sent mossy fibers to the appropriate layer in the host hippocampus.

Delayed transplantation. When transplantation to the irradiated rats was delayed by 3 weeks, extensive and precise serial host-transplant-host connections still formed. Increasing the delay, and hence the age of the recipient rats, to 5 and 10 weeks did, however, disclose an age-dependent decrease in the growth of host fibers into the transplants. By using the Fink-Heimer technique in combination with acute anterograde degeneration, we were thus unable to demonstrate ingrowth of commissural hippocampodentate fibers into transplants grafted into either 5 or 10 week old irradiated rats (Sunde et al., 1985), and with these recipient ages the host perforant path afferents to the transplants occupied even narrower zones in the transplant dentate molecular layer as if they were losing the competition for terminal space to the intrinsic associational systems which widened their terminal zone correspondingly. In correspondence with the observations in normal recipient rats the host cholinergic fibers showed no apparent age-dependent decrease in the innervation of the transplants as judged by the AChE staining. Outgrowth of transplant mossy fibers into the host CA3 still occurred when the transplantation was delayed until 5 and 10 weeks of age, but the extent and density of this projection was significantly reduced relative to the extensive outgrowth observed after immediate transplantation to the newborn and even the 3 week old recipients.

Conclusion. From our transplant studies on rats subjected to neonatal irradiation-induced damage of dentate granule cells we conclude that - 1) developing dentate granule cells can normalize dentate afferent and efferent connections when transplanted to appropriate positions next to or in the irradiated host fascia dentata, - 2) the optimal period for such repair corresponds to the early postnatal period, when the hippocampal region and its connections are still developing, - 3) similar host-transplant-host serial connections can also form after delayed transplantation to 5 and 10 week old recipient rats irradiated at birth, but then to a much more limited extent.

REFERENCES

Bayer, S.A., 1980, Development of the hippocampal region in the rat. I. Neurogenesis examined with ^{3}H-thymidine autoradiography, J. Comp. Neurol., 190:87.

Bayer, S.A., and Altman, J., 1975, Radiation-induced interference with postnatal hippocampal cytogenesis in rats and its long-term effects on the acquisition of neurons and glia, J. Comp. Neurol., 163:1.

Blackstad, T.W., 1963, Ultrastructural studies on the hippocampal region, Progr. Brain Res., 3:122.

Blackstad, T.W., 1967, Cortical grey matter. A correlation of light and electron microscopic data, in: "The Neuron", H. Hydén, ed., Elsevier Publishing Company, Amsterdam.

Caceres, A., and Steward, O., 1983, Dendritic reorganization in the denervated dentate gyrus of the rat following entorhinal cortical lesions: A Golgi and electron microscopic analysis, J. Comp. Neurol., 214:387.

Changeux, J.-P., and Danchin, A., 1976, Selective stabilisation of developing synapses as a mechanism for the specification of neuronal networks, Nature (Lond.), 264:705.
Cotman, C.W., and Lynch, G.S., 1976, Reactive synaptogenesis in the adult nervous system. The effects of partial deafferentation on new synapse formation, in: "Neuronal Recognition", S.H. Barondes, ed., Chapman and Hall, London.
Cotman, C.W., and Nadler, J.V., 1978, Reactive synaptogenesis in the hippocampus, in: "Neuronal Plasticity", C.W. Cotman, ed., Raven Press, New York.
Danscher, G., 1984, Do the Timm sulphide silver method and the selenium method demonstrate zinc in the brain?, in: "The Neurobiology of Zinc. Part A: Physiochemistry, Anatomy, and Techniques", C.J. Frederickson, G.A. Howell and E.J. Kasarskis, eds., Alan R. Liss, Inc., New York.
Danscher, G., and Zimmer, J., 1978, An improved Timm sulphide silver method for light and electron microscopic localization of heavy metals in biological tissues, Histochemistry, 55:27.
Diamond, J., Cooper, E., and Macintyre, L., 1976, Trophic regulation of nerve sprouting, Science, 193:371.
Fredens, K., Stengaard-Pedersen, K., and Larsson, L.-I., 1984, Localization of enkephalin and cholecystokinin immunoreactivities in the perforant path terminal fields of the rat hippocampal formation, Brain Res., 304:255.
Frotscher, M., and Zimmer, J., 1983, Lesion-induced mossy fibers to the molecular layer of the rat fascia dentata: Identification of postsynaptic granule cells by the Golgi-EM technique, J. Comp. Neurol., 215:299.
Gall, C., Brecha, N., Karten, H.J., and Chang, K.-J., 1981, Localization of enkephalin-like immunoreactivity to identified axonal and neuronal populations of the rat hippocampus, J. Comp. Neurol., 198:335.
Gall, C., and Lynch, G., 1980, The regulation of fiber growth and synaptogenesis in the developing hippocampus, in "Current Topics in Developmental Biology", Vol. 15, A.A. Moscona and A. Monroy, eds., Academic Press, New York.
Gall, C., McWilliams, R., and Lynch, G., 1980, Accelerated rates of synaptogenesis by "sprouting" afferents in the immature hippocampal formation, J. Comp. Neurol., 193:1047.
Goldowitz, D., Seiger, Aa., and Olson, L., 1982, Anatomy of the isolated area dentata grown in the rat anterior eye chamber, J. Comp. Neurol., 208:382.
Goldowitz, D., Seiger, Aa., and Olson, L., 1984, Regulation of axonal ingrowth into area dentata as studied by sequential double intraocular brain tissue transplantation, J. Comp. Neurol., 227:50.
Haug, F.-M. S., 1973, Heavy metals in the brain. A light microscope study of the rat with Timm's sulphide silver method. Methodological considerations and cytological and regional staining patterns. Adv. Anat. Embryol. Cell Biol., 47, Fasc. 4:1.
Kromer, L.F., Björklund, A., and Stenevi, U., 1983, Intracephalic embryonic neural implants in the adult rat brain. I. Growth and mature organization of brainstem, cerebellar, and hippocampal implants, J. Comp. Neurol., 218:433.
Laatsch, R.H., and Cowan, W.M., 1966, Electron microscopic studies of the dentate gyrus of the rat. I. Normal structure with special reference to synaptic organization, J. Comp. Neurol., 128:359.
Laurberg, S., and Hjorth-Simonsen, A., 1977, Growing central axons deprived of normal target neurones by neonatal X-ray irradiation still terminate in a precisely laminated fashion, Nature, 269:158.
Laurberg, S., and Zimmer, J., 1980, Lesion-induced rerouting of hippocampal mossy fibers in developing but not in adult rats, J. Comp. Neurol., 190:627.
Laurberg, S., and Zimmer, J., 1981, Lesion-induced sprouting of hippocampal mossy fiber collaterals to the fascia dentata in developing and adult rats, J. Comp. Neurol., 200:433.

Lehman, J., Nagy, J.I., Atmajda, S., and Fibiger, H.C., 1980, The nucleus basalis magnocellularis: the origin of a cholinergic projection to the neocortex of the rat, Neuroscience, 5:1161.

Lynch, G.S., Matthews, D.A., Mosko, S., Parks, T., and Cotman, C.W., 1972, Induced acetylcholinesterase-rich layer in rat dentate gyrus following entorhinal lesions, Brain Res., 42:311.

Lynch, G., Stanfield, B., and Cotman, C.W., 1973, Developmental differences in post-lesion axonal growth in the hippocampus, Brain Res., 59:155.

Matthews, D.A., Cotman, C., and Lynch, G., 1976, An electron microscopic study of lesion-induced synaptogenesis in the dentate gyrus of the adult rat. I. Magnitude and time course of degeneration, Brain Res., 115:1.

Nadler, J.V., Cotman, C.W., and Lynch, G.S., 1973, Altered distribution of choline acetyltransferase and acetylcholinesterase activities in the developing rat dentate gyrus following entorhinal lesion, Brain Res., 63:215.

Stanfield, B.B., and Cowan, W.M., 1982, The sprouting of septal afferents to the dentate gyrus after lesions of the entorhinal cortex in adult rats, Brain Res., 232:162.

Stengaard-Pedersen, K., Fredens, K., and Larsson, L.-I., 1981, Enkephalin and zinc in the hippocampal mossy fiber system, Brain Res., 212:230.

Stengaard-Pedersen, K., Fredens, K., and Larsson, L.-I., 1983, Comparative localization of enkephalin, cholecystokinin and heavy metals in the hippocampus, Brain Res., 273:81.

Steward, O., 1980, Trajectory of contralateral entorhinal axons which reinnervate the fascia dentata of the rat following ipsilateral entorhinal lesions, Brain Res., 183:277.

Stirling, R.V., and Bliss, T.V.P., 1978, Observations on the commissural projection to the dentate gyrus in the Reeler mutant mouse, Brain Res., 150:447.

Sunde, N., Laurberg, S., and Zimmer, J., 1984, Brain grafts can restore irradiation-damaged neuronal connections in newborn rats, Nature, 310:51.

Sunde, N., and Zimmer, J., 1981, Transplantation of central nervous tissue. An introduction with results and implications, Acta Neurol. Scandinav., 63:323.

Sunde, N.Aa., and Zimmer, J., 1983, Cellular, histochemical and connective organization of the hippocampus and fascia dentata transplanted to different regions of immature and adult rat brains, Develop. Brain Res., 8:165.

Sunde, N., Zimmer, J., and Laurberg, S., 1985, Repair of neonatal irradiation-induced damage to the rat fascia dentata. Effects of delayed intracerebral transplantation, in: "Neural Grafting in the Mammalian CNS", A. Björklund and U. Stenevi, eds., Elsevier, Amsterdam (in press).

Tarrant, S.B., and Routtenberg, A., 1977, The synaptic spinule in the dendritic spine: electron microscopic study of the hippocampal dentate gyrus, Tissue & Cell, 9:461.

Timm, F., 1958a, Zur Histochemie der Schwermetalle. Das Sulfid-Silber-Verfahren, Dtsch. Z. Gesamte Gerichtl. Med., 46:706.

Timm, F., 1958b, Zur Histochemie des Ammonshorngebietes, Z. Zellforsch., 48:548.

Wenk, H., Bigl, V., and Meyer, U., 1980, Cholinergic projections from magnocellular nuclei of the basal forebrain to cortical areas in rats, Brain Res. Rev., 2:295.

Zimmer, J., 1973a, Extended commissural and ipsilateral projection in postnatally deentorhinated hippocampus and fascia dentata demonstrated in rats by silver impregnation, Brain Res., 64:293.

Zimmer, J., 1973b, Changes in the Timm sulfide silver staining pattern of the rat hippocampus and fascia dentata following early postnatal deafferentation, Brain Res., 64:313.

Zimmer, J., 1974, Long term synaptic reorganization in rat fascia dentata deafferented at adolescent and adult stages: observations with the Timm method, Brain Res., 76:336.

Zimmer, J., 1978, Development of the hippocampus and fascia dentata: Morphological and histochemical aspects, Progr. Brain Res., 48:171.

Zimmer, J., and Gähwiler, B.H., 1984, Cellular and connective organization of slice cultures of the rat hippocampus and fascia dentata, J. Comp. Neurol., (in press).

Zimmer, J., and Hjorth-Simonsen, A., 1975, Crossed pathways from the entorhinal area to the fascia dentata. II. Provokable in rats, J. Comp. Neurol., 161:71.

Zimmer, J., and Laurberg, S., 1978, Postlesional changes in the acetyl cholinesterase (AChE) staining pattern of the hippocampus and fascia dentata may depend on trophic interactions with major afferent pathways and not denervation alone, Neurosci. Letts., Suppl. 1: S49.

Zimmer, J., Laurberg, S., and Sunde, N., 1983, Neuroanatomical aspects of normal and transplanted hippocampal tissue, in: "Neurobiology of the Hippocampus", W. Seifert, ed., Academic Press, London.

Zimmer, J., and Sunde, N., 1984, Neuropeptides and astroglia in intracerebral hippocampal transplants: An immunohistochemical study in the rat, J. Comp. Neurol., 227:331.

Zimmer, J., Sunde, N., Sørensen, T., Møller, A.G., and Gähwiler, B.H., 1985, The hippocampus and fascia dentata. An anatomical study of intracerebral transplants and intraocular and in vitro cultures, in: "Neural Grafting in the Mammalian CNS", A. Björklund and U. Stenevi, eds., Elsevier, Amsterdam (in press).

COMPENSATION OF LESION-INDUCED CHANGES IN CEREBRAL METABOLISM AND BEHAVIOUR BY STRIATAL NEURAL IMPLANTS IN A RAT MODEL OF HUNTINGTON'S DISEASE

O. Isacson, P. Brundin, F.H. Gage and A. Björklund

Department of Histology, University of Lund
Biskopsgatan 5
S-223 62 Lund, Sweeden

INTRODUCTION

There now exists ample evidence that the capacity of the adult CNS for functional recovery after large long-term lesions can be promoted by implants of foetal brain tissue (see, e.g., Björklund and Stenevi, 1979; Björklund et al., 1980; Dunnett et al., 1981a,b, 1982; Perlow et al., 1979; Gash and Sladek, 1979; Freed et al., 1980; Krieger et al., 1980; Gage et al., 1983, 1984; Deckel et al., 1983; Labbe et al., 1983; Isacson et al., 1984; Fine et al., this volume). The use of neural grafting as an experimental technique complements lesion and stimulation experiments in neurobiology. In the assessment of lesion-induced changes in the rat it has been investigated to what extent the grafting of neural tissue to the young lesioned or aged impaired animal can create a sufficient condition for functional recovery (see Björklund et al., 1983; Gage et al., 1983, 1984). Functional recovery after lesion-induced changes has been correlated with histological, neurochemical, physiological and metabolic parameters sometimes giving new insights into the mode of operation of certain neural circuitries or transmitter systems.

The grafting technique has also provided a new tool for the analysis of structural plasticity at the cellular level, such as, for instance, trophic interactions following lesions of the septo-hippocampal system in the rat (Björklund and Stenevi, 1981; Gage et al., 1984) and conditions for axonal regeneration (Aguayo et al., 1984; Kromer et al., 1981; David and Aguayo, 1981). Neural grafting studies are also complementary to in vivo test systems for neuroembryological questions normally addressed by cell-culture techniques (see, e.g. McLoon et al., 1981) of the CNS.

The neural grafting technique employed in most of the aforementioned studies involves the insertion of embryonic neural donor tissue into adult lesioned or aged host rat brains. The donor tissue can be implanted as solid pieces into pre-made cavities with enriched vascular supply or directly into the brain or the ventricular system. Another way of grafting nervous tissue, employed in the present studies, is in the form of a dissociated cell suspension, injected directly into the host-brain parenchyma. The neural grafting by the injection of cell suspensions allows small volumes of cells to be placed stereotaxically at almost any desired site within the host CNS (see Björklund et al., 1983). Experimentally, the cell suspension technique makes possible the control of such parameters as cell viability and cell density and, moreover, the separation, mixing, culturing or other manipulations of the neural tissue prior to grafting.

The ibotenic acid lesion model of Huntington's disease

We have investigated the effects of intracerebral neural grafts in animals with lesions that reproduce neuropathological and neurochemical defects seen in certain neurodegenerative diseases (Björklund and Stenevi, 1979; Dunnett et al., 1981; Gage et al., 1983, 1984; Isacson et al., 1984). One such model is represented by rats subjected to ibotenic acid (IA) lesions of the caudate-putamen (CP), causing severe striatal neuronal cell loss and shrinkage (see Coyle and Schwarcz, 1983). IA is a short-acting glutamate analogue, which is believed to derive its neurotoxic properties from an overexcitation of receptive cells (Olney, 1979) with only minor damage to the axons or other structures within the injected brain area. In comparison with a similar so-called excitotoxin, kainic acid (KA), IA causes much less cell damage distal to the injection site, and does not share the potent epileptogenic effect of KA (see Coyle and Schwarcz, 1983). Following the IA or KA injections in the CP there are reductions in GABA-ergic, cholinergic and peptidergic neurochemical markers in both the CP and its primary projection areas, the globus pallidus and the substantia nigra, similar to the neuropathology seen in Huntington's disease (Coyle and Schwarcz, 1976, 1983; McGeer and McGeer, 1976). The neurodegenerative actions of excitotoxic amino acids may have some bearing also on the pathogenetic mechanisms underlying neural cell losses seen in clinical conditions. Thus it has been suggested that abnormal formation of endogenous amino acids with excitotoxic potency (such as quinolinic acid) may cause progressive cell damage in certain diseases. Alternatively hyperexcitation of glutamatergic receptors, e.g., due to presynaptic hyperactivity (Aldinio et al., 1983) or enzyme deficiencies (Plaitakis et al., 1982; Duvoisin et al., 1983), could conceivably have a similar effect.

Animals with striatal lesions by IA or KA show behavioural disturbances such as locomotor hyperactivity, learning impairments, sensori-motor integration deficits and weight loss (Sanberg and Fibiger, 1979; Mason and Fibiger, 1979; Dunnett and Iversen, 1981; Pisa et al., 1981; Isacson et al., 1984). This behavioural syndrome is accompanied by a substantial increase in local cerebral metabolic activity in several primary or secondary targets of the CP within the extrapyramidal system (Kimura et al., 1980; Kelly et al., 1982; Isacson et al., 1984). In agreement with findings in the Huntingtonian patients (Kuhl et al., 1981, 1984), the metabolism of the lesioned CP itself is markedly reduced, whereas the primary and secondary striatal target areas have, as yet, not been studied in detail clinically.

The deficits in learning and memory that have been demonstrated after excitotoxic lesion of the CP (Dunnett and Iversen, 1981; Pisa et al., 1981) and the cognitive deterioration occurring in Huntingtonian patients could be explained as a direct effect of damage and altered activity of the CP itself. Alternatively, the symptoms in HD and the animal model could be due to long lasting changes in the functional characteristics of remaining connections directly or indirectly related to the CP. In fact, in IA lesioned rats there is metabolic hyperactivity within limbic and cortical systems, e.g. in the ventral tegmental area, nucleus accumbens and the sensorimotor cortex (Kelly et al., 1982; Isacson et al., 1984) after the damage to the CP. Changes can thus occur in some structures several synapses away from the lesion, in which little or no morphological damage has been detected , at the light-microscopic level. One should also be aware of the possibility of chronic changes mediated through trophic and metabolic neural interactions following the acute neuronal damage of the CP. Krammer (1980) suggested that long-term neural degeneration (transsynaptic) took place in the substantia nigra pars reticulata and the globus pallidus over a 3- to 7-week period following KA lesion of the CP.

This article summarizes some of our experimental findings in this rat model of Huntington's disease. Our experimental strategy is to use grafts of fetal striatal neurons to replace lost neurons in the previously damaged striatum in an attempt to compensate for the lesion-induced functional changes.

Neural grafting in IA lesioned rats

After a series of pilot experiments with intrastriatal injections of either KA or IA we chose a lesion made by stereotaxic injections of small volumes of IA into 4 sites in the rat neostriatum (Isacson et al., 1984). Within one week after the IA lesion, during which reactive gliosis takes place in the CP, we implanted a dissociated cell suspension prepared from foetal rat striatum (13-15 days gestational age) into the previously lesioned sites in the CP. Briefly, embryos are sequentially removed from the uterine horns of an anaesthetized pregnant rat. The embryonic brains are dissected out under a dissection microscope using microsurgery forceps and scissors. With the brain floating in a dish filled with a 0.6% glucose-saline medium, the cortex overlying the striatal primordia is incised and parted to the sides to expose the two anlages of the striatum bulging into the relatively large lateral ventricle (see Fig. 1). The embryonic striatal primordia are carefully cut out and stored for a short time in the medium prior to a brief incubation in trypsin at $37^{o}C$, followed by washing and gentle dissociation of the tissue with a fire-polished Pasteur pipette. The resulting cell suspension is made to yield concentrations of about 75×10^3 - 125×10^3 cells per ul, where 90-95% of the cells are viable as assessed by the vital stain acridine orange/ethidium bromide (Brundin et al., 1985) using a fluorescence microscope. Between 4 and 6 ul of cell-suspension are grafted by stereotaxic injection into the lesioned CP (Fig. 1) (Isacson et al., 1984, 1985; Dawbarn et al., 1985).

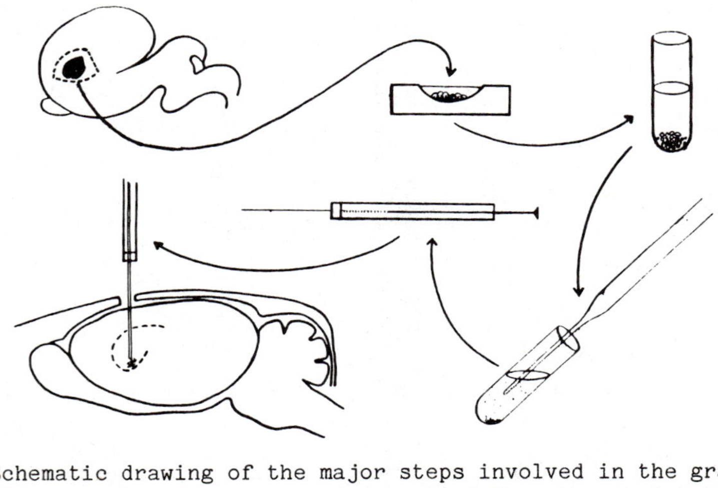

Fig. 1. Schematic drawing of the major steps involved in the grafting procedure of a neural cell suspension prepared from the striatal primordia of the rat (see text).

CHARACTERISTICS OF RATS WITH NEOSTRIATAL IA LESION

Morphology and biochemistry

Unilateral IA injections induced considerable neural cell loss and gliosis in the CP, leaving about 5-10% of the normal neural CP volume intact in the head of the CP, while the tail of the CP showed greater sparing. By 10 weeks after the lesion, the head of the CP appeared as a gliotic scar penetrated by densely packed fibre bundles of the internal capsule. In addition, the glutamic acid decarboxylase (GAD) and choline acetyltransferase (ChAT) enzyme activity levels were reduced to 20-30% and 30-50%, respectively of the contralateral control side (see Fig. 2). Immunohistochemistry indicated a marked reduction in cells positive for the neuropeptides substance P, met-enkephalin, somatostatin, CCK and neuropeptide Y in the CP. The primary striatal target areas, globus pallidus and substantia nigra pars reticulata, also showed a markedly reduced staining for met-enkephalin and substance P, respectively (Dawbarn et al., 1985). Both of these structures were markedly atrophic by 10 weeks after the IA lesion (c.f. Fig. 4b). Since there is no clear-cut atrophy at 1 week after IA lesion it seems likely that these atrophic changes are not due to the direct, primary toxicity of the drug, but that they occur as a secondary consequence of the IA lesion resulting in shrinkage due to loss of striatal efferents and atrophy or possibly transsynaptic degeneration of the substantia nigra reticulata neurons (Krammer, 1980; Isacson et al., 1985). The progressive atrophy of the CP which followed after the lesion resulted in a 65% reduction of the CP volume, due to the neural cell loss, and a compensatory 7-fold increase in the volume of the lateral ventricle at the same cross-sectional levels (Isacson et al., 1984). The magnitude of these degenerative and atrophic changes in the CP is similar to that reported in humans in advanced cases of Huntington's disease (see, e.g. Bird, 1980; Bruyn, 1982).

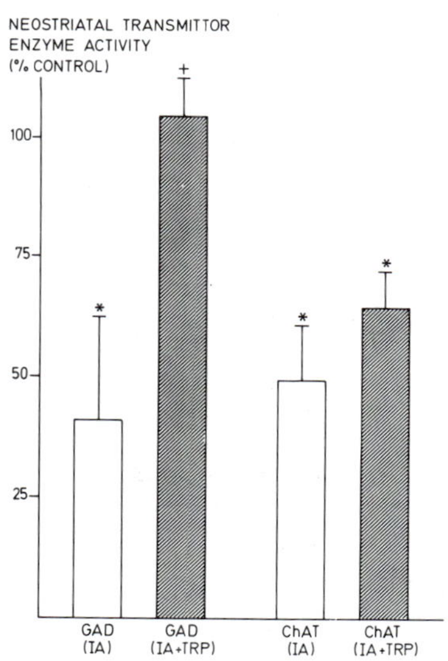

Fig. 2. ChAT and GAD activities were recorded from striatal homogenates in IA lesioned (n=6) and grafted rats (n=5) and contralateral sides (n=11). Enzyme activities were calculated by measurement of synthesized ^{14}C acetylcholine from ^{14}C-acetyl-CoA and choline, and the amount of $^{14}CO_2$ liberated in the synthesis of GABA from ^{14}C-L-glutamate according to the micromethods of Fonnum (1975) and Fonnum et al. (1970), respectively. * = $p<0.05$ against control, + = p<0 05 against IA lesion alone. (From Isacson et al., 1985).

Regional cerebral glucose metabolism

Local glucose utilization as determined by the autoradiographic (^{14}C)-2-deoxyglucose (2DG) method of Sokoloff et al. (1977) has proven to be a valuable tool in neurobiological research. It has been shown that regionally specific changes in local 2DG utilization occur in discrete brain regions during performance of, for example, motor tasks (Schwartzman et al., 1981; Matsunami et al., 1981) or during acquisition of different learning tasks (Martinez et al., 1982; Matsunami and Kubota, 1983). Manipulation of neurotransmission, for instance by axotomy or by transmitter-specific receptor activation or blockade, which result in behavioral syndromes, is also reflected in changes in local glucose use (see McCulloch, 1983; for review).

In experiments measuring the regional cerebral glucose metabolism in restrained, awake rats with unilateral striatal IA lesions using the fully quantitative autoradiographic 2DG method, an average decrease in glucose utilization of about 60% was found throughout the lesioned CP with lesser reductions in the medial and caudal aspects of the CP. The contralateral intact CP exhibited an average 20-30% increase in glucose use. Primary striatal projection areas (such as globus pallidus, pars reticulata of substantia nigra and subthalamic nucleus), and some secondary projection areas (including the deep superior colliculus and the ventrolateral and ventromedial thalamus) also demonstrated an increased glucose utilization. This metabolic hyperactivity in structures distal to the CP was in most cases manifested bilaterally, and it was greatest in the globus pallidus and substantia nigra pars reticulata. Interestingly, some areas closely associated with the classical extrapyramidal system, e.g. the ventral tegmental area and the nc. accumbens, also showed bilateral increases in glucose utilization.

Behavioural analysis

Following unilateral injection of 20 ug IA into the head of the CP under short acting barbiturate anaesthesia (methohexital), the rats wake up and display a motor excitation state with stereotypic limb movements and rotation or bent posture in a direction contralateral to the injected side. If anaesthetics with a longer duration are used (such as ketamine in combination with xylazine) the rats do not wake up directly after the IA injection. Under such conditions the rats can stay asleep for periods up to 24 hours, possibly as a consequence of the IA decarboxylation into muscimol (Schwarcz and Coyle, 1983), thus interacting with, and prolonging the sedative effects of the anaesthesia.

Rats with unilateral or serial bilateral (1-2 weeks interval) neostriatal IA lesions show several behavioural changes, following the neuronal cell loss in the striatum (Isacson et al., 1984, and unpublished data). During a 3-4 day post operative recovery-time, the rats show spontaneous ipsilateral turning and transient hypophagia. At later times when observed in their home cages, the lesioned rats show a fairly normal range of behaviours. However, under controlled experimental conditions the lesioned rats show long term deficits in motor coordination, sensori-motor integration, as well as learning and memory as assessed by the performance in the Morris water maze. As a rule, the unilaterally lesioned rats are less deficient than the bilaterally lesioned ones. The unilateral IA lesioned rats are hyperactive in terms of nocturnal locomotor activity, measured as horizontal movements in automated activity boxes, and about 10 weeks after the IA lesion they show a 2-fold increase in locomotor activity compared to control or sham operated rats. When injected with the dopamine receptor blocking agent, haloperidol, at doses which normally cause profound catalepsy, the IA lesioned rats show almost no signs of catalepsy on the side contralateral to the lesion, thus giving evidence for the critical role of striatal output neurons for eliciting this effect. As illustrated in Fig. 3, this effect is demonstrated by the rat's ability to correct a head-down position on an inclined grid. Under haloperidol, an intact rat will remain in the head-down position (i.e. long latency to correct), while the unilaterally IA lesioned rats will correct their position rapidly and always in the direction away from the lesioned (i.e., non-cataleptic) side.

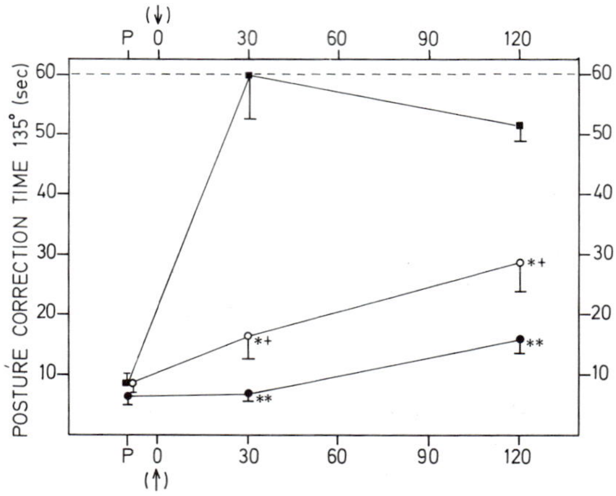

Fig. 3. Rats (controls n = 6, IA n = 9, IA + transplant n = 10) were tested on an inclined (50°) elevated grid. The direction and time to correct their posture from the initial head down position to near full up turn of the body (135°) was recorded. Mean values and SEM from 4 trials at pre- and post-injection times (1 mg/kg haloperidol, i.p.) were calculated. One way analysis of variance F (2,22) = 58.7; post hoc Newman-Keul's test, ** = $p<0.01$, * = $p<0.05$ against controls, + = $p<0.05$ against IA lesion alone.

CHARACTERISTICS OF IA LESIONED RATS WITH STRIATAL NEURAL GRAFTS

Morphology and biochemistry

The volume of the surviving grafts ranged between 4 and 8 mm^3, as assessed by 10-15 weeks after implantation. Under a surgical microscope the sliced CP grafts could be seen in unstained sections of the brain as tissue masses surrounded by dense fibre bundles and gliotic-like tissue (Fig. 4). Histological analysis revealed that the graft contained numerous neuronal populations. In fact, morphometric analysis has demonstrated that, although the average size of the total cell population (neurons plus glia) was slightly smaller than in the normal CP, the number of medium-sized to large neurons (defined as cells with a diameter larger than 10 um) was twice as high in the graft per unit area as in the normal CP. Of the peptidergic neurons lost by the IA lesion, several of the cell types normally found in intact CP were also identified within the graft (Dawbarn et al., 1985). Hence, immunoreactive met-enkephalin, substance P, CCK, somatostatin and neuropeptide Y containing neurons were demonstrable in the striatal grafts together with a number of acetylcholinesterase-positive cells (Fig. 5). This suggests that the cells of the foetal striatal eminence, taken at an early proliferative phase of striatal neurogenesis, will develop into a fairly normal range of neuronal cell types in their new environment in the lesioned adult CP (Dawbarn et al., 1985; Isacson et al., 1985b). However, the striosomal pattern, described by Graybiel et al. (see Graybiel and Ragsdale, 1983, for review) for the normal CP, seemed to retain a rather immature patchiness, of coinciding high acetylcholine-esterase activity and opiate immunoreactivity, in the grafted tissue.

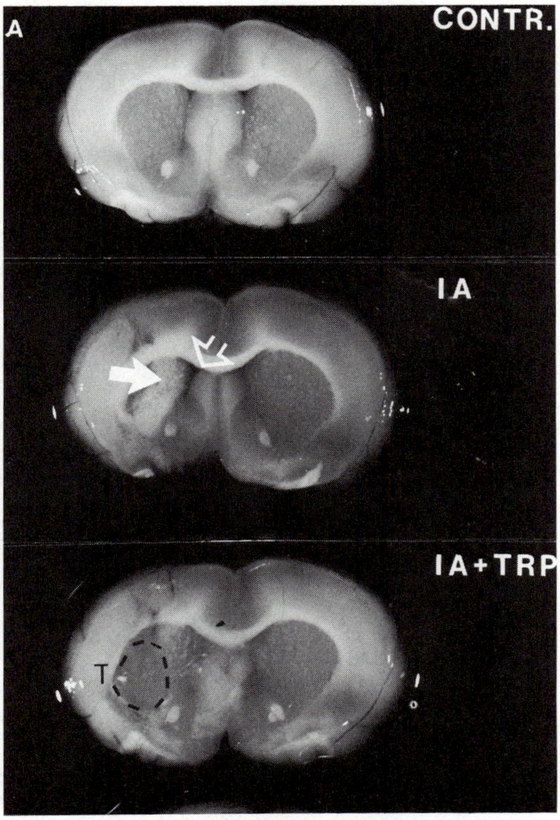

Fig. 4 (a) Photos of fresh unstained sections sliced through the head of the caudate-putamen (2 mm thickness), 20 weeks after IA lesion and grafting prior to routine dissection for biochemical analysis. Note the packed fibre bundles of the IA lesioned rat (IA) in a scar-like appearance (white arrow) and the increased size of the lateral ventricle (open arrow). There is also some local damage along the needle tract in the overlying cortex. In the lesioned and transplanted rat (IA + TRP) the transplant appears as a structure in the lateral part of the neostriatum (T), without the characteristic fibre-bundle arrangement of the normal caudate-putamen, though restoring the tissue volume lost after the IA lesion. (From Isacson et al., 1985).

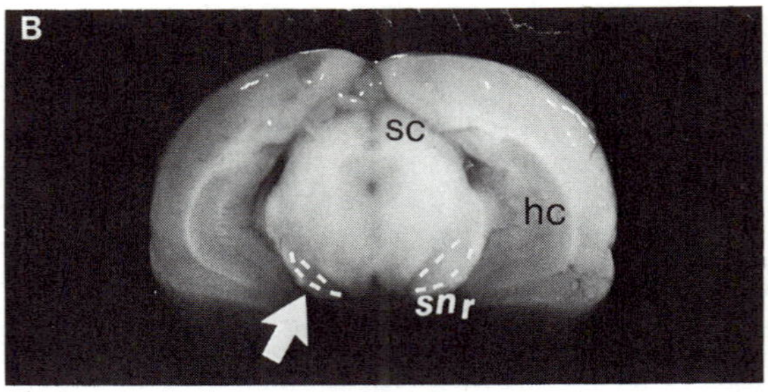

Fig. 4 (b) Photo of a fresh unstained brain, sectioned at the level of the substantia nigra, of a rat with a unilateral IA lesion of the CP (20 weeks prior to sacrifice). Note the marked atrophy that has evolved in the substantia nigra; pars reticulata (arrow). Snr = substantia nigra pars reticulata, sc = superior colliculus, hc = hippocampus.

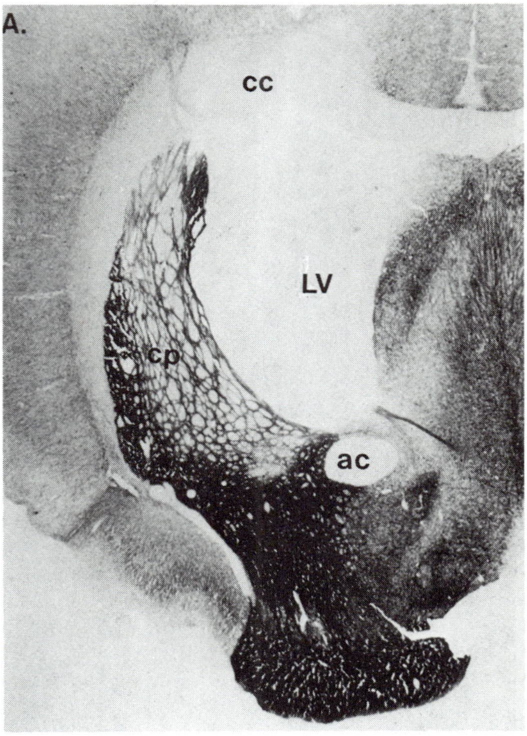

Fig. 5 (A) Acetylcholine-esterase stained section through the head of the caudate-putamen of a rat that had received 20 ug ibotenic acid, by multiple stereotaxic injections into the caudate-putamen. Abbreviations: see Fig. 5 (B).

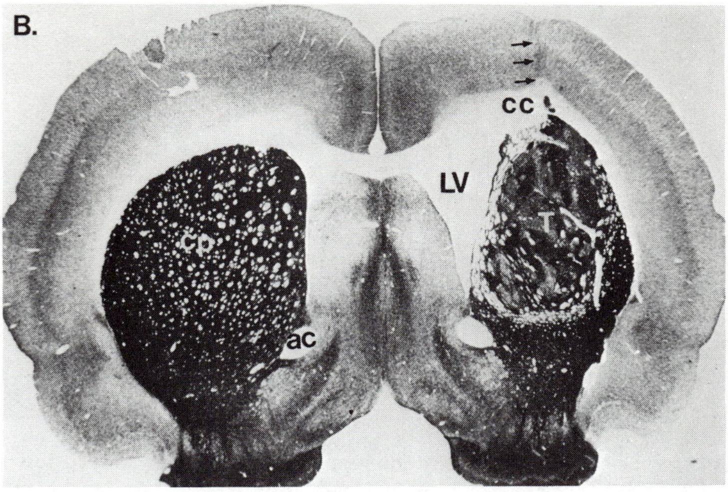

Fig. 5 (B) Acetylcholine-esterase stained section of a rat, treated as above, which also received a cellular suspension of dissociated embryonic striatal cells into the previously lesioned sites in the caudate-putamen 5 days after IA lesion and 20 weeks before sacrifice. (LV) lateral ventricle, (ac) anterior commissure, (cc) corpus callosum, (cp) caudate-putamen, (T) transplant, arrows indicate needle tract in the cortex.

The IA lesion caused an average of 58% reduction in the specific activity of the GABA synthesizing enzyme, GAD (expressed per ug protein), in the head of the CP. Considering the volume reduction of the lesioned CP, one can estimate the total GAD activity content to be approximately 15% of normal levels. ChAT enzyme activity (expressed per ug protein) was reduced in the IA lesioned CP between 30 and 50%. In the striatal graft, GAD activity levels were not different from controls (Fig. 2) (Isacson et al. 1984). The ChAT activity levels (nmoles/h/ug protein) in the grafted animals were increased compared to the IA-lesioned CP although the total levels remained significantly below control levels (Isacson et al., 1985).

Regional cerebral glucose metabolism

In the ^{14}C-2-DG autoradiography, the striatal graft area by itself showed a glucose utilization averaging about 70% of the intact control CP. The graft did not significantly influence glucose use in the surrounding gliotic or spared areas of the CP, while the contralateral increases in CP glucose utilization seen in animals with IA lesion alone were no longer present, suggesting a decreased asymmetry in brain activity in the grafted animals. The hypermetabolic responses in the primary and secondary striatal efferent areas seen in the IA-lesioned rats were normalized by the implant ipsilaterally in the globus pallidus and subthalamic nucleus, and bilaterally in the deep superior colliculi. A significant transplant-induced decrease in metabolic activity was also seen in the ventral tegmental area and nc. accumbens (Fig. 6) (Isacson et al., 1984).

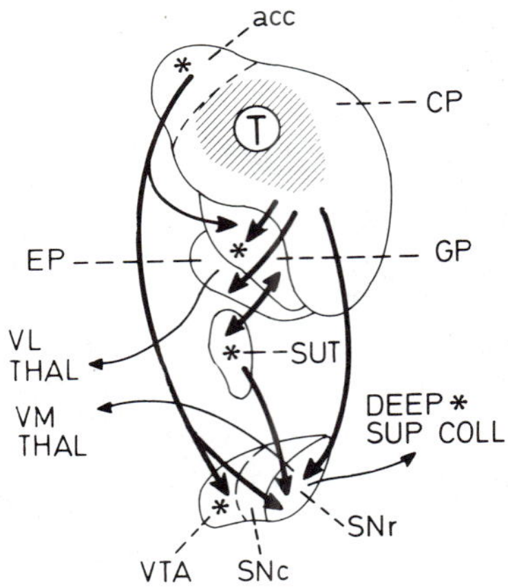

Fig. 6. Schematic drawing in the horizontal plane of the major output pathways from the neostriatum and nucleus accumbens. The primary and secondary target areas which exhibited graft-induced reductions in ^{14}C-2-DG utilization are marked by asterisks (*). acc, nucleus accumbens; CP, caudate-putamen; Deep Sup Coll, deep superior colliculus; entopeduncular nucleus; GP, globus pallidus; SN, substantia nigra (c, compacta, r, reticulata); SUT, subthalamic nucleus; VL and VM Thal, ventrolateral and ventromedial thalamic nuclei; VTA, ventral tegmental area.

Behavioural analysis

Spontaneous locomotor activity was measured in two different tests. Consistent with previous reports of excitotoxic CP lesions (Mason and Fibiger, 1979; Dunnett and Iversen, 1981), the IA lesioned rats were hyperactive during overnight/dark cycle testing, when rats normally express about 90% of their diurnal locomotor activity (Glick and Cox, 1978), or during food deprivation. The overnight locomotor hyperactivity was completely compensated for in the animals with IA lesions, by the striatal transplants (Isacson et al., 1984). These effects were confirmed also when locomotor activity was measured in rotometer bowls and the individual scores recorded in the two tests were highly correlated. These results are consistent with the previous finding of Deckel et al. (1983), reporting an effect of striatal grafts on behavioural habituation in an open field in bilaterally KA lesioned rats.

In addition, the decrease in neuroleptic-induced catalepsy, seen in IA lesioned animals, was significantly changed in the grafted animals compared to lesion-alone (unpublished data) (Fig. 3).

DISCUSSION

The present results show that the striatal suspension grafts not only will survive and differentiate within the host neostriatum previously depleted of its intrinsic neurons by IA, but also that they are capable of compensating for lesion induced regional cerebral metabolic hyperactivity, behavioural locomotor hyperactivity and changes in neuroleptic induced catalepsy.

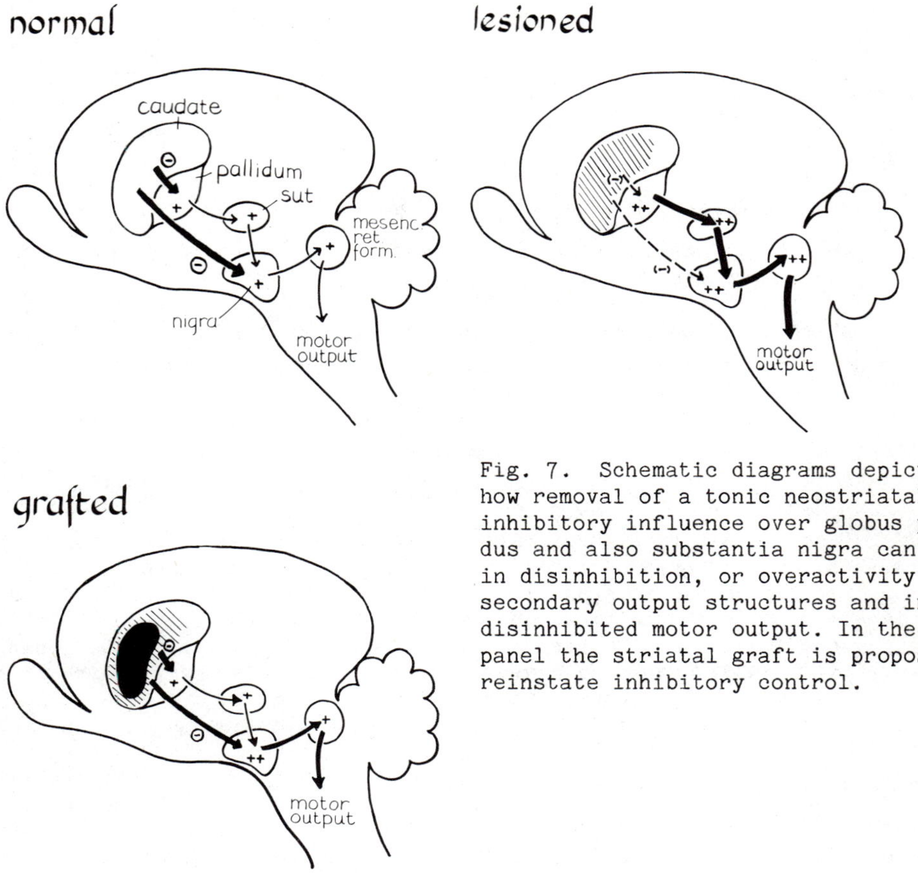

Fig. 7. Schematic diagrams depicting how removal of a tonic neostriatal inhibitory influence over globus pallidus and also substantia nigra can result in disinhibition, or overactivity in secondary output structures and in disinhibited motor output. In the bottom panel the striatal graft is proposed to reinstate inhibitory control.

There are little data at this point to tell how the striatal grafts exert their functional effects in the IA lesioned rats. In fact, the exact cellular mechanisms underlying the behavioural and hypermetabolic effects of the striatal IA lesion are not yet fully understood. In Fig. 7 we have proposed that the neostriatum normally exerts a tonic inhibitory control over the principal primary output stations, the globus pallidus and also pars reticulata of the substantia nigra. This is supported not only by the observation that glucose metabolism in these target structures is markedly increased after a major destruction of the striatal output neurons (see above), but also by the fact that many of the striatal output neurons are GABA-ergic (see Graybiel and Ragsdale, 1983), and that the principal electrophysiological effects in the pallidum and substantia nigra of neostriatal stimulation is inhibitory (see Kitai, 1981). Although it is known that the functions of the neostriatum are highly complex, we will for the purpose of the present discussion view the effects of the neostriatal IA lesion on spontaneous locomotion and baseline glucose metabolism as a result of disinhibition of the neostriatal output structures through removal of a tonic, probably GABA-ergic, inhibition. If this is correct, it may be sufficient for the implanted striatal neurons to reinstate a relatively nonspecific tonic inhibitory action over the primary striatal target structures in order to normalize the rats' hyperlocomotor and hypermetabolic responses, as illustrated in the lower panel in Fig. 7.

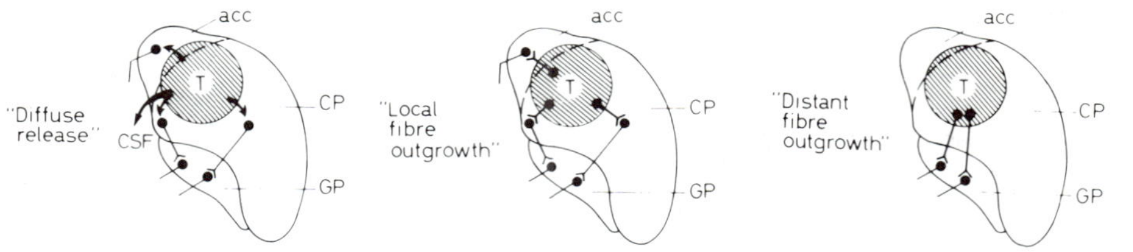

Fig. 8. Proposed alternative modes of action of the striatal grafts in compensating for altered cerebral metabolism and behavioural changes in the IA-lesioned animals (see discussion). Schematic drawing in the horizontal plane of the caudate-putamen (CP), globus pallidus (GP), nucleus accumbens (acc) and a striatal transplant (T, hatched area).

This tonic inhibition model raises several questions. First, the mechanism(s) by which the striatal implants could reinstate inhibitory control over globus pallidus and/or substantia nigra is unclear. Previous experiments with implants of catecholamine-producing cell systems (Björklund et al., 1980, 1981, 1983; Dunnett et al., 1983; Freed et al., 1980, 1981; Freund et al., 1985) have provided evidence that intracerebral grafts can exert their functional effects in the host brain both via direct axonal connections with previously denervated elements in the host target and via a more diffuse release of the active compound into the host tissue or the cerebrospinal fluid. As illustrated diagramatically in Fig. 8, both these mechanisms are also conceivable in the action of intrastriatal implants. "Diffuse Release" (left panel) implies that the grafted neurons can act in a neurohumoral manner through the secretion of active compounds either locally in the tissue or into the cerebrospinal fluid. "Local" and "Distant Fibre Growth" (middle and right panels in Fig. 8) require that the grafted neurons have established fibre connections with neurons in the host brain, and that these connections are sufficient to mediate the graft effect. The existence of such connections has not yet been demonstrated, but it seems possible that they could occur either with spared neurons in the surrounding parts of the caudate-putamen or nc accumbens, or (over somewhat greater distances) with neurons in, for example, the globus pallidus or the entopeduncular nucleus.

Secondly, even if the transplanted neurons can reinstate the release of transmmitters in the initially disinhibited and denervated neostriatal target structures, it remains to be clarified how this may influence basic motor behaviour, such as spontaneous locomotion. In Huntington patients with destruction of the neostriatum, it is known that the dyskinetic symptoms can be ameliorated by dopamine receptor blocking agents (neuroleptics) (Whittier et al., 1968; Chase, 1976). It has been speculated therefore, that the loss of the striato-nigral inhibitory pathway innervating the nigral dopamine neurons would result in a dopaminergic hyperactivity, which in turn would be responsible for the hypermotoric behaviour. Although Melamed et al. (1982) failed to detect any such dopaminergic hyperactivity in the caudate-putamen in Huntington patients or in rats with neostriatal KA lesions, it still seems possible that it could occur in the nc accumbens, which is the striatal structure most intimately associated with dopamine-induced locomotion in the rat (Kelly et al., 1975; Kelly, 1977). In Fig. 9A this effect is suggested to be mediated via a lesioned-induced loss of strio-nigral neurons normally regulating the activity in dopaminergic neurons innervating the accumbens area.

An alternative explanation stems from the fact that the outputs from the dorsal "motoric" striatum (nc. caudatus-putamen) and the ventral "limbic" striatum (including the nc. accumbens and the ventral pallidum) converge onto the same downstream target areas and are likely to interact there. Thus, nc. accumbens, which is likely to play an important role in the inhibition of locomotion, projects both into the globus pallidus (Mogenson and Nielsen, 1983) and (via the so-called ventral pallidum) into the substantia nigra and the mesencephalic reticular formation (Swanson et al., 1984; cf. Fig. 9B). This provides a possibility for the primary or secondary output neurons from the caudate-putamen to regulate (reduce or increase) the magnitude of the motor output initiated via the nc. accumbens output routes. Fig. 9B summarizes in a simplified form how the strio-nigral and strio-pallidal output pathways can regulate the "gain" in the accumbens-related locomotor initiation system, and where the points of possible interaction are indicated by asterisks.

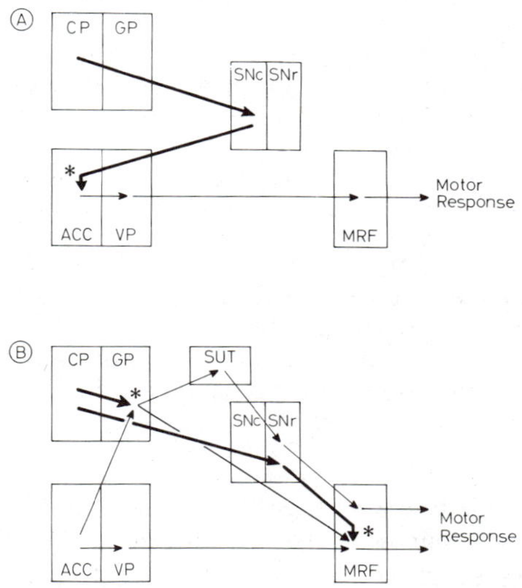

Fig. 9. Points of interaction between the output pathways from the dorsal ("motoric") and ventral ("limbic") parts of the striatum which may explain the hyperlocomotor responses in rats with IA lesions of the caudate-putamen. In A the effect is mediated via the dopamine neurons in the medial substantia nigra; in B via points of interaction (asterisks) in the globus pallidus or the mesencephalic reticular formation (for further explanation, see text).

Knowledge of the extent of efferent connections from the grafted striatal neurons will be essential in order to evaluate whether such possible interactions may occur also in the grafted animals. If they indeed are important for the observed behavioural effects it may be possible to obtain the same effects with striatal neurons implanted directly into the sites of presumed interactions, i.e., in the globus pallidus or the substantia nigra. In fact, grafts of embryonic mesencephalic dopaminergic neurons have been found to be functionally efficient only if they are grafted into or immediately adjacent to their normal target areas, i.e., the caudate-putamen or the nc. accumbens. Placements of such grafts in their normal location in the substantia nigra region or along the course of the nigrostriatal pathway were thus inefficient, both in terms of dopamine fibre connections and functional effects (see Dunnett et al., 1983; Björklund et al., 1983). It will be interesting to see to what extent the same principles may also hold also for striatal implants in the IA lesion model.

Acknowledgement

The research summarized here was supported by grants from the Swedish MRC and the NIH. We thank Siv Carlson for excellent assistance in typing the manuscript.

REFERENCES

Aguayo, A.J., Björklund, A., Stenevi, U., and Carlstedt, T.., 1984, Fetal mesencephalic neurons survive and extend long axons across PNS grafts inserted into the adult rat striatum. Neurosci. Lett., 45:53.

Aldinio, C., French, E.D. and Schwarcz, R., 1983, The effects of intra hippocampal ibotenic acid and their blockade by (-)2-amino-7 phosphonohepatonic acid: morphological and electroencephalographical analysis. Exp. Brain Res., 51:36.

Bird, E.D., 1980, Chemical pathology of Huntington's disease, Ann. Rev. Pharmacol. Toxicol., 20:533.

Björklund, A., Schmidt, R.H., and Stenevi, U., 1980, Functional reinnervation of the neostriatum in the adult rat by use of intraparenchymal grafting of dissociated cell suspensions from the substantia nigra, Cell Tiss. Res., 212:39.

Björklund, A., and Stenevi, U., 1979, Reconstruction of the nigro striatal dopamine pathways by intracerebral nigral transplants, Brain Res., 177:555.

Björklund, A., and Stenevi, U., 1981, In vivo evidence for a hippocampal adrenergic neurotrophic factor specifically released on septal deafferentation. Brain Res., 229:403.

Björklund, A., Stenevi, U., Schmidt, R.H., Dunnett, S.B. and Gage, F.H., 1983, Intracerebral grafting of neuronal cell suspensions. Acta Physiol. Scand., Suppl. 522.

Brundin, P., Isacson, O., and Björklund, A., 1985, Monitoring of cell viability in suspensions of embryonic CNS tissue and its use as a criterion for intracerebral graft survival. Brain Res. (in press).

Bruyn, G.W., 1982, Neurotransmitters in Huntington's Chorea - a clinician's view. Prog. Brain Res., 55:445.

Chase, T.N., 1979, Rational approaches to the pharmacotherapy of chorea. in: "The basal ganglia", M.D. Yahr, ed., Raven Press, New York.

Coyle, J.T., and Schwarzc, R., 1976, Lesions of striatal neurones with kainic acid provides a model for Huntington's chorea. Nature, 263:244.

Coyle, J.T., and Schwarzc, R., 1983, The use of excitatory amino acids as selective neurotoxins. in: "Handbook of Chemical Neuroanatomy", Vol. 1, A. Björklund and T. Hökfelt, eds., Elsevier Science Publishers, Amsterdam.

David, S., and Aguayo, A.J., 1981, Axonal elongation into PNS "bridges" after CNS injury in adult rats. Science, 214:931.

Dawbarn, D., Brundin, P., Isacson, O., Gage, F.H., Emson, P.C., and Björklund, A., 1985, Striatal transplants in ibotenic acid lesioned rats: survival of neurones and development of peptide immunoreactivity (submitted to J. Neurosci.).

Deckel, A.W., Robinson, R.G., Coyle, J.R., and Sanberg, P.R., 1983, Reversal of long-term locomotor abnormalities in the kainic acid model of Huntington's disease by day 18 fetal striatal implants. Eur. J. Pharm., 93:287.

Dunnett, S.B., Björklund, A., Stenevi, U., and Iversen, S.D., 1981a, Behavioural recovery following transplantation of substantia nigra in rats subjected to 6-OHDA lesions of the nigrostriatal pathway. I. Unilateral lesions. Brain Res., 215:147.

Dunnett, S.B., Björklund, A., Stenevi, U., and Iversen, S.D., 1981b, Grafts of embryonic substantia nigra reinnervating the ventrolateral striatum ameliorate sensorimotor impairments and akinesia in rats with 6-OHDA lesions of the nigrostriatal pathway. Brain Res., 229:209.

Dunnett, S.B., and Iversen, S.D., 1981, Learning impairments following selective kainic acid-induced lesions within the neostriatum of rats. Behav. Brain Res., 2:189.

Dunnett, S.B., Low, W.C., Iversen, S.D., Stenevi, U., and Björklund, A., 1982, Septal transplants restore maze learning in rats with fornix-fimbria lesions. Brain Res., 251:335.

Duvoisin, R.C., Chokroverty, S., Lepore, F., and Nicklas, W., 1983, Glutamate dehydrogenase deficiency in patients with olivopontocerebellar atrophy. Neurology, 33:1322.

Fonnum, F., 1975, A rapid radiochemical method for the determination of choline acetyltransferase. J. Neurochem., 24:407.

Fonnum, F., Storm-Mathisen, J., and Walberg, F., 1970, Glutamate decarboxylase in inhibitory neurons. A study of the enzyme in Purkinje cell axons and boutons in the cat. Brain Res., 20:259.

Freed, W.J., Perlow, M.J., Karoum, F., Seiger, A., Olson, L., Hoffer, B.J., and Wyatt, R.J., 1980, Restoration of dopaminergic function by grafting of fetal rat substantia nigra to the caudate nucleus: long-term behavioural, biochemical amd histochemical studies. Ann. Neurol., 8:510.

Gage, F.H., Björklund, A., and Stenevi, U., 1984, Denervation releases a neuronal survival factor in adult rat hippocampus. Nature, 308:637.

Gage, F.H., Dunnett, S.B., Björklund, A., and Stenevi, U., 1983, Aged rats: Recovery of motor coordination impairments by intrastriatal nigral grafts. Science, 221:966.

Gash, D., Sladek, J.R., and Sladek, C.D., 1980, Functional development of grafted vasopressin neurons. Science, 220:1367.

Glick, S.D., and Cox, R.D., 1978, Nocturnal rotation in normal rats: correlation with an amphetamine-induced rotation and effects of nigrostriatal lesions. Brain Res., 150:149.

Graybiel, A.M., and Ragsdale, C.W., 1983, Biochemical anatomy of the striatum. in: "Chemical neuroanatomy", P.C. Emson, ed., Raven Press, New York.

Isacson, O., Brundin, P., Dawbarn, D., Kelly, P.A.T., Gage, F.H., Emson, P.C., and Björklund, A., 1985b, Striatal grafts in the ibotenic acid lesioned striatum. in: "Neural grafting in the mammalian CNS", A. Björklund and U. Stenevi, eds., Elsevier Biomedical Press, Amsterdam (in press).

Isacson, O., Brundin, P., Kelly, P.A.T., Gage, F.H., and Björklund, A., 1984, Functional neuronal replacement by grafted striatal neurones in the ibotenic acid-lesioned rat striatum. Nature, 311:458.

Isacson, O., Brundin, P., Gage, F.H., and Björklund, A., 1985, Neural grafting in a rat model of Huntington's disease: Progressive neurochemical changes after neostriatal ibotenate lesion and striatal tissue grafting (submitted to Neuroscience).

Kelly, P.A.T., Graham, D.I., and McCulloch, J., 1982, Specific alterations in local cerebral glucose utilization following striatal lesions. Brain Res., 233:157.

Kelly, P.H., 1977, Drug-induced motor behaviour. in: "Handbook of Psychopharmacology", Vol. 8, L.L. Iversen and S.D. Iversen, eds., Plenum Press, N.Y.

Kelly, P.H., Seviour, S.D., and Iversen, S.D., 1975, Amphetamine and apomorphine responses in the rat following 6-OHDA lesions of the nucleus accumbens septi and corpus striatum. Brain Res., 94:507.

Kimura, H., McGeer, E.G., and McGeer, P.L., 1981, Metabolic alterations in an animal model of Huntington's disease using the ^{14}C-deoxyglucose methods. J. Neurol. Transmiss. Suppl., 16:103.

Kitai, S.T., 1981, Electrophysiology of the corpus striatum and brain stem integrating systems. in: "Handbook of Physiology. Nervous System II, Motor Control", V. Brooks, ed. Williams and Wilkins, Baltimore.
Krammer, E.B., 1980, Anterograde and transsynaptic degeneration 'en cascade' in basal ganglia introduced by intrastriatal injection of kainic acid: an animal analogue of Huntington's disease. Brain Res., 196:209.
Krieger, D.T., Perlow, M.J., Gibson, M.J., Dames, T.F., Zimmerman, E.A., Ferin, M., and Charlton, H.M., 1982, Brain grafts reverse hypogonadism of gonadotropin releasing hormone deficency. Nature, 298:468.
Kromer, L.F., Björklund, A., and Stenevi, U., 1981, Regeneration of the septohippocampal pathway in adult rats is promoted by utilizing embryonic hippocampal implants as bridges. Brain Res., 210:173.
Kuhl, D.E., Metter, E.J., Riege, W.H., and Markham, Ch., 1984, Patterns of cerebral glucose utilization in Parkinson's disease and Huntington's disease. Ann. Neurol., 15:419.
Kuhl, D.E., Phelps, M., Markham, C., Winter, J., Metter, J., and Riege, W., 1981, Local cerebral glucose metabolism in Huntington's disease determined by emission computed tomography of ^{18}F flurorodeoxyglucose. Cerebral Blood Flow Metab., 1: Suppl. 1, 459.
Labbe, R., Firl, A., Mufson, E.J., and Stein, D.G., 1983, Fetal brain transplants: reduction of cognitive deficits in rats with frontal cortex lesions. Science, 221:470.
Martinez, J.L., Petty, C., and Messing R.B., 1982, Regional brain uptake of 2-deoxy-D-glucose following training in a discriminated Y-maze avoidance task. J. Comp. Neurol. Physiol. Psychol. 96:721.
Mason, S.T., Fibiger, H.C., 1979, Kainic lesions of the striatum in rats mimic the spontaneous motor abnormalities of Huntington's disease. Neuropharmacol., 18:403.
Matsunami, K., Kageyama, T., and Kubota, K., 1981, Radioactive 2-deoxy-D-glucose incorporation into the prefrontal and premotor cortex of the monkey performing a forelimb movement. Neurosci. Lett., 26:37.
Matsunami, K., and Kubota, K., 1983, Radioactive deoxyglucose uptake into the prefrontal cortex during a delayed response task of the monkey. Neurosci. Lett., 36:329.
McLoon, L.K., McLoon, S.C., and Lund, R.S., 1981, Cultured embryonic retinae transplanted to rat brain: Differentiation and formation of projections to host superior colliculus. Brain Res., 226:15.
Melamed, E., Hefti, F., and Bird, E.D., 1982, Huntington's chorea is not associated with hyperactivity of nigrostriatal dopaminergic neurons: studies in post-mortem tissues and in rats with kainic acid lesions. Neurology, 32:640.
McCulloch, J., 1982, Mapping functional alterations in the CNS with (^{14}C)-deoxyglucose. in: "Handbook of Psychopharmacology", Vol. 15, L.L. Iversen, S.D. Iversen and S.H. Snyder, eds., Plenum Publishing Corp., New York.
McGeer, E.G., and McGeer, P.L., 1976, Duplication of biochemical changes of Huntington's chorea by intrastriatal injection of glutamic and kainic acids. Nature, 263:517.
Mogenson, G.J., and Nielsen, M.A., 1983, Evidence that an accumbens to subpallidal GABA-ergic projection contributes to locomotor activity. Brain Res. Bull., 11:309.
Olney, J.W., 1979, Excitotoxic amino acids and Huntington's disease. in: "Advances in Neurology", Vol. 23, Chase, Wexler, Barbeau, eds., Raven Press, N.Y.
Perlow, M.J., Freed, W.J., Hoffer, B.J., Seiger, A., Olson, L., and Wyatt, R.J., 1979, Brain grafts reduce motor abnormalities produced by destruction of nigrostriatal dopamine system. Science, 204:643.

Pisa, M., Sanberg, P.R., and Fibiger, H.C., 1981, Striatal injections of kainic acid selectively impair serial memory performance in the rat. Exp. Neurol., 74:633.
Plaitakis, A., Berl, S., and Yahr, M.D., 1982, Abnormal glutamate metabolism in an adult onset degenerative neurological disorder. Science, 216:193.
Sanberg, P.R., and Fibiger, H.C., 1979, Body weight, feeding and drinking behaviours in rats with kainic acid-induced lesions of striatal neurons -with a note on body weight symptomatology in Huntington's disease. Exp. Neurol., 66:444.
Schwartzman, R.J., Greenberg, J., Revich, M., Klose, K.J., and Alexander, G.M., 1981, Functional metabolic mapping of a conditional motor task in primates utilizing 2-(^{14}C) deoxyglucose. Exp. Neurol., 72:153.
Sokoloff, L., 1977, Relation between physiological function and energy metabolism in the central nervous system, J. Neurochem., 29:13.
Swanson, L.W., Mogenson, G.J., Gerfen, C.R., and Robinson, P., 1984, Evidence for a projection from the lateral preoptic area and substantia innominata to the 'mesencephalic locomotor region' in the rat. Brain Res., 295:161.
Whittier, J.R., and Korenyi, C., 1968, Effect of oral fluphenazine on Huntington's chorea. Int. J. Neuropsychiatr., 4:1.

FUNCTIONS OF NUCLEUS BASALIS MAGNOCELLULARIS INPUT TO THE NEOCORTEX: A RECONSTITUTION ANALYSIS USING INTRACEREBRAL TRANSPLANTATION

Alan Fine[1], Stephen B. Dunnett[2] and Guy Toniolo[3]

(1) MRC Neurochemical Pharmacology Unit, Medical Research Council Centre, Medical School, Hills Road, Cambridge, CB2 2QH, U.K. (2) Department of Experimental Psychology, Downing Street, Cambridge CB2 3EB, U.K. and (3) Laboratoire de Neurobiologie des Comportements, Universite Louis Pasteur, 67000-Strasbourg, France

INTRODUCTION

Pharmacological studies have, over the past two decades, established the involvement of acetylcholine in learning and memory processes (Deutsch, 1971; Bartus et al., 1982). This link has been strengthened by the discovery that senile dementia of the Alzheimer's type (SDAT), which is characterized by progressive debilitating memory loss, is associated with reduction of cholinergic markers throughout much of the neocortex and hippocampus (Bowen et al., 1976; Perry et al., 1978).

The medial septal nucleus and vertical limb of the diagonal band of Broca supply the major cholinegic input to the hippocampus (Lewis and Shute, 1967; Fibiger, 1982; Mesulam et al., 1983b); the major source of cholinergic innervation to the neocortex is the nucleus basalis magnocellularis (NBM: Divac, 1975; Lewis and Shute, 1967; Mesulam and van Hoesen, 1976; Mesulam et al., 1983a). Indeed, several studies have recently noted a marked loss or atrophy of cholinergic NBM cells in patients with SDAT (Whitehouse et al., 1982; Candy et al., 1983; Pearson et al., 1983).

The functions of this cholinergic input to the neocortex are not known. Lesions of the NBM have been shown to impair performance on simple learning tasks such as passive or active avoidance (LoConte et al., 1982; Flicker et al., 1983; Friedman et al., 1983). Attribution of these impairments to destruction of the cortical cholinergic projection is confounded, however, by the inevitable damage to non-cholinergic elements of the globus pallidus, substantia innominata and adjacent diencephalic nuclei. In the present experiments, we have attempted to circumvent this difficulty by reconstituting the lesioned inputs to the neocortex by intracortical transplantation of embryonic neurones from the cholinergic ventral forebrain. We have then examined the ability of such grafts to ameliorate the lesion-induced impairments.

LESION AND TRANSPLANTATION SURGERY

The subjects of these studies are young adult female Sprague-Dawley rats. Experimental animals receive large unilateral lesions of the NBM by stereotaxic injection of ibotenic acid at two sites. Such lesions reduce ipsilateral cortical choline acetyltransferase (ChAT) levels by approximately 60% as compared with the oposite, unlesioned side. While recovery of cortical ChAT has been reported after small NBM lesions (Wenk and Olton, 1984), these large lesions are stable in the absence of cholinergic grafts.

One week after lesioning, the animals are allocated to three groups for transplantation surgery: one group receives cortically-placed grafts of cholinergic-rich (rostral-ventral forebrain) tissue, a second receives control grafts of non-cholinergic (hippocampal) tissue; and a third "lesioned-only" group receives no graft. Graft tissue is dissected from embryos after 16-17 days of gestation. The collected tissue is then incubated in trypsin, washed and mechanically dissociated into a single-cell suspension. In the present studies, all transplanted rats receive stereotaxic injections of either cholinergic ventral forebrain or control cholinergic-poor hippocampal tissue cell suspension into two sites in dorsolateral frontal and parietal cortex, ipsilateral to the lesion. Full methodological details are presented elsewhere (Fine et al., in press, a).

CHOLINERGIC RECONSTITUTION BY TRANSPLANTS

Acetylcholinesterase (AChE) histochemistry and staining for Nissl substance have confirmed the placement of lesions, the profound reduction in number of large AChE-positive cells in the area of the lesioned NBM, and the massive loss of laminar AChE-positive terminal staining throughout ipsilateral neocortex, sparing cingulate and entorhinal regions (see Fig. 1). Histochemical analysis of grafts one and six months after transplantation has shown that in all cases transplanted cells survive and proliferate. Only the cholinergic grafts, however, are densely comprised of AChE-positive cells, from which AChE-positive fibres extend progressively farther over time. By six months after transplantation, AChE-positive staining is frequently seen in host cortex beyond 3 mm from the border of the graft. In many cases, the pattern of graft-derived AChE-positive fibres resembles the normal laminar distribution of cortical staining.

The cholinergic nature of the AChE-positive transplanted cells has been confirmed by ChAT immunohistochemistry and by enzyme assay (Fine et al., in press, a). In the biochemical ChAT assays, the lesion-induced deficit in ChAT activity was significantly ameliorated (from nearly 60% to less than 40% reduction) in the cortex adjacent to cholinergic grafts, six months after transplantation.

BEHAVIOURAL STUDIES

Bilateral lesions of the NBM sufficient to produce profound and stable cortical deafferentation are frequently lethal. We have therefore concentrated on characterizing the behavioural consequences of full unilateral lesions, which have the additional advantage of providing internal controls for subsequent histological and biochemical analysis. The performance of the three experimental groups described above has been compared with that of unlesioned control animals. At no point in these tests has the performance of animals with control grafts of hippocampal tissue differed from that of the "lesioned-only" group; for clarity of presentation these groups have therefore been combined into a single "lesion" group.

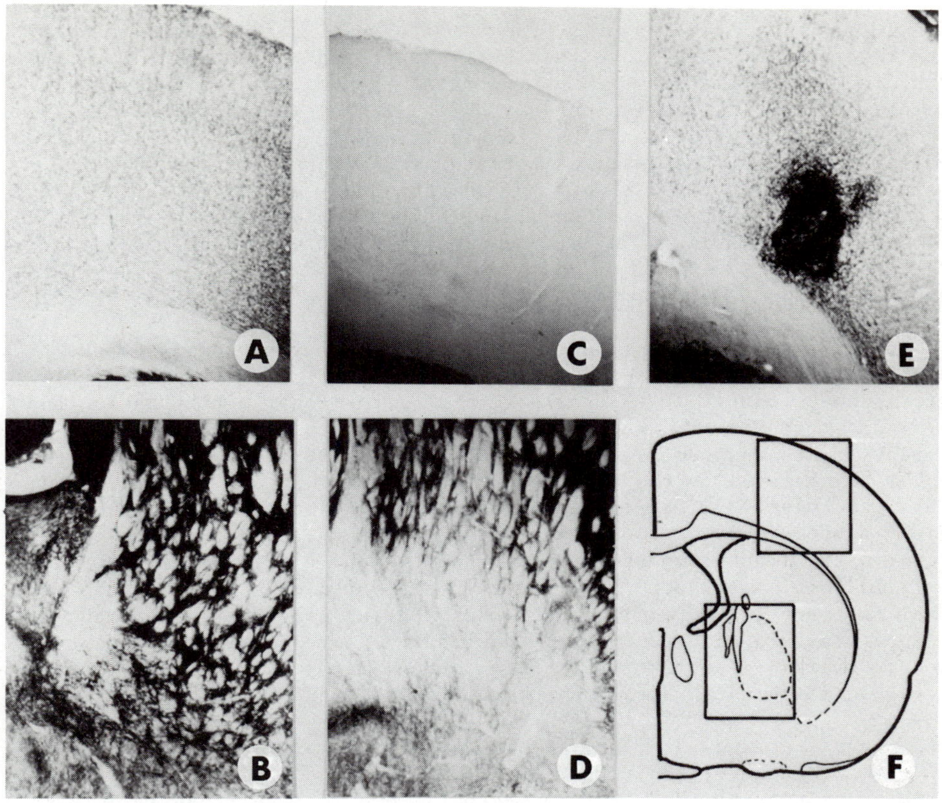

Fig.1. AChE histochemistry of NBM lesions and grafts of cholinergic-rich tissue to the neocortex. A: Normal terminal distribution of AChE-positive fibres in the cortex of an intact rat. B: AChE-positive cells in the NBM of a normal rat. Following ibotenic acid injection into NBM, AChE-positive cells are lost from the lesioned NBM (D), which is accompanied by profound loss of cortical fibres and terminals (C). A ventral forebrain graft in the lesioned neocortex (E) is densely AChE-positive, and gives rise to AChE-positive fibres reinnervating the deafferented host neocortex. F: Schematic diagram of the field of view of photomicrographs of the neocortex (A,C,E) and NBM/globus pallidus (B,D).

Sensorimotor tests

In one experiment (Dunnett et al., in press), all animals have been given a battery of sensorimotor tests both prior to, and 1, 2 and 3 months following transplantation surgery. All the tests have revealed sensory and motor asymmetries induced by the lesion. Thus, for example, the lesioned rats showed a reduced orientation to lateralized somaesthetic stimuli applied to the contralateral body surface, and they showed significant turning biases to the lesion side, either spontaneously, or when induced by pinching the tail or placing the animal head-down on an inclined grid.

In many of these measures, the animals with grafts of cholinergic ventral forebrain cells showed a significant reinstatement of symmetrical responding to the two sides. Thus, whereas the orientation asymmetry remained constant in the lesioned rats, it was reduced progressively in each successive test on the rats with cholinergic grafts. Similarly, on the inclined plane, the lesioned rats turned on approximately 80% of occasions to the lesioned side. By contrast, the ventral forebrain graft group, which had initially shown a similar bias, reestablished turning to the contralateral side on 50% of occasions, following transplantation.

Learning tasks

We have exmained behaviour in several learning paradigms. Taste aversion learning is unaffected by these unilateral NBM lesions or grafts. Performance on maze learning tasks (T-maze forced-choice alternation and radial arm maze) is impaired by the lesions, but the sensory-motor asymmetry makes assessment of any learning deficit per se difficult. This problem is obviated by the use of an inhibitory avoidance paradigm (Fine et al., in press, b). Experimental and unlesioned control animals are readily habituated to a two-chamber test apparatus so that upon being placed in the light compartment they promptly enter the dark compartment. If a mild (0.5 mA x 2 sec) foot-shock is delivered upon entering the dark, normal rats are inhibited from such entry during subsequent placement in the apparatus; latency to re-enter the dark in immediately subsequent tests provides a measure of task acquisition, and after various delays provides a measure of retention, of this inhibitory avoidance.

A predetermined re-entry latency (120 sec) was established as the criterion for acquisition. Unlesioned rats learn the avoidance after one or two shock trials. NBM-lesioned animals require significantly more trials for acquisition (range 2-4), regardless of the presence or absence of cholinergic grafts. Retention was examined 24 and 72 h after training to criterion. The lesion group manifested a significantly shorter latency (i.e., poorer retention) than did either the ventral forebrain graft group or unlesioned controls, which did not differ significantly from one another. Thus NBM lesions inhibit both acquisition and retention of inhibitory avoidance learning. The retention deficit, however, is specifically ameliorated by transplantation of embryonic cholinergic-rich grafts to the neocortex, while control, hippocampal, grafts are without effect.

DISCUSSION

In the absence of specific cholinergic neurotoxins, efforts to reveal functions of the cholinergic NBM-cortical projection by studying the consequence of lesions of the NBM are hampered by the inevitable non-specific destruction of adjacent structures. However, restitution of impaired function by transplantation of embryonic ventral forebrain (including presumptive NBM) cells to the denervated cortical target of the lesioned NBM offers stronger evidence that restored functions are indeed mediated by NBM in the normal animal. We believe that such reconstitution analysis is potentially powerful and useful. The present results suggest that the NBM is important not only for retention (consolidation and/or retrieval) of certain forms of memory, but also for sensory-motor function. The ability of transplants, presumably deprived of appropriate normal presynaptic input to NBM, to restore such behaviours, suggests that the NBM cholinergic input to the cortex may serve to control the behaviours by regulating intrinsic cortical mechanisms. Such an "enabling" function may not require a discretely time-patterned release of acetylcholine. The NBM may in this way allow limbic system activity to modulate patterns of cortical function.

Further refinements of this approach, currently underway, include systematically varied graft placement to probe the cortical topography of these NBM-mediated functions, and the further purification of the dissociated embryonic cells, to allow more certain specification of the chemical mediation and physiology of these behaviours.

ACKNOWLEDGEMENTS

We thank Miss L. Evans and Mrs M. Wynn for typing the manuscript. This research was supported by the Medical Research Council (A.F. and S.B.D.) and the European Training Programme in Brain and Behaviour research (G.T.)

REFERENCES

Bartus, R.T., Dean, R.L., Beer, B. and Lippa, A.S., 1982, The cholinergic hypothesis of geriatric memory dysfunction, Science, 217:408.

Bowen, D.M., Smith, C.B., White, P. and Davison, A.N., 1976, Neurotransmitter-related enzymes and indices of hypoxia in senile dementia and other abiotrophies, Brain, 99:459.

Candy, J.M., Perry, R.H., Perry, E.K., Irving, D., Blessed, G., Fairbairn, A.F. and Tomlinson, B.E., 1983, Pathological changes in the nucleus of Meynert in Alzheimer's and Parkinson's diseases, J. Neurol. Sci., 54:277.

Deutsch, J.A., 1971, The cholinergic synapse and the site of memory, Science, 174:788.

Divac, I., 1975, Magnocellular nuclei of the basal forebrain project to neocortex, brainstem and olfactory bulb. Review of some functional correlates, Brain Res., 93:385.

Dunnett, S.B., Toniolo, G., Fine, A., Björklund, A. and Iversen, S.D., ransplantation of embryonic ventral forebrain neurons to the neocortex of rats with lesions of nucleus basalis magnocellularis. II. Sensorimotor and learning impairments. Neuroscience, in press (b).

Fibiger, H.C., 1982, The organization and some projections of cholinergic neurons of the mammalian forebrain, Brain Res. Rev., 4:327.

Fine, A., Dunnett, S.B., Björklund, A. and Iversen, S.D. Transplantation of embryonic ventral forebrain neurons to the neocortex of rats with lesions of nucleus basalis magnocellularis. I. Anatomical and biochemical studies, Neuroscience, in press (a).

Fine, A., Dunnett, S.B., Björklund, A. and Iversen, S.D. Cholinergic grafts reinnervating the neocortex improve memory in a rat model of Alzheimer's disease, Brain Res., in press (b).

Flicker, C., Dean, R.L., Watkins, D.L., Fisher, S.K. and Bartus, R.T., 1983, Behavioral and neurochemical effects following neurotoxic lesions of a major cholinergic input to the cerebral cortex in the rat, Pharm. Biochem. Behav., 18:973.

Friedman, E., Lerer, B. and Kuster, J., 1983, Loss of cholinergic neurons in the rat neocortex produces deficits in passive avoidance learning, Pharm. Biochem. Behav., 19:309.

Lewis, P.R. and Shute, C.C.D., 1967, The cholinergic limbic system: projections to hippocampal formation, medial cortex, nuclei of the ascending cholinergic reticular system, and the subfornical organ and supra-optic crest, Brain, 90;521.

LoConte, G., Bartolini, L., Casamenti, F., Marconcini-Pepeu, I. and Pepeu, G., 1982, Lesions of cholinergic forebrain nuclei: changes in avoidance behavior and scopolamine actions, Pharm. Biochem. Behav., 17:933.

Mesulam, M.M. and Van Hoesen, G.W., 1976, Acetylcholinesterase-rich projections from the basal forebrain of the rhesus monkey to neocortex, Brain Res., 109:152.

Mesulam, M.M., Mufson, E.J., Levey, E.J. and Wainer, B.H., 1983, Cholinergic innervation of cortex by the basal forebrain: cytochemistry and cortical connections of the septal area, diagonal band nuclei, nucleus basalis (substantia innominata) and hypothalamus in the rhesus monkey, J. comp. Neurol., 214:170.

Mesulam, M.M., Mufson, E.J., Wainer, B.H. and Levey, A.I., 1983, Central cholinergic pathways in the rat: an overview based on an alternative nomenclature (Ch 1 - Ch 6). Neurosci., 10:1185.

Pearson, R.C.A., Sofroniew, M.V., Cuello, A.C., Powell, T.P.S., Eckenstein, F., Esiri, M.M. and Wilcock, G.K., 1983, Persistence of cholinergic neurons in the basal nucleus in a brain with senile dementia of the Alzheimer's type demonstrated by immunohistochemical staining for choline acetyltransferase, Brain Res., 289:375.

Perry, E.K., Tomlinson, B.E., Blessed, G., Bergmann, K., Gibson, P.H. and Perry, R.H., 1978, Correlation of cholinergic abnormalities with senile plaques and mental test scores in senile dementia, Brit. Med. J., 2:1475.

Wenk, G.L. and Olton, D.S., 1984, Recovery of neocortical choline acetyltransferase activity following ibotenic acid injection into the nucleus basalis of Meynert in rats, Brain Res., 293:184.

MATERNAL CORTICOIDS DURING PERINATAL LIFE INFLUENCE HIPPOCAMPUS-DEPENDENT BEHAVIORAL AND ENDOCRINE RESPONSES IN THE ADULT RAT

L. Angelucci, F.R. Patacchioli, S. Scaccianoce and A. Di Sciullo

Farmacologia 2a, Facoltà di Medicina, Università di Roma "La Sapienza" P.le A. Moro 2, 00185 Roma

Most pharmacological and environmental stimuli able to produce subtle long-term neurochemical and behavioral effects in the offspring also produce hormonal responses in the mother organism, which can propagate to the fetus or newborn through the placenta or milk. In particular, the rat is an excellent model, due to the fact that brain ontogenesis is accomplished during perinatal life, and the stringent dependence upon the mother for maturational processes via hormones and other regulators in the milk. Inducing variations of maternal plasma corticosterone during pregnancy and/or lactation produced neurochemical, endocrine and behavioral effects ascertainable as late as three-four month age. These long-term consequences can be attributed to actions on the ontogenesis and maturation of the hippocampus, hypothalamus, pituitary and adrenals. The hippocampal glucocorticoid receptor showed permanent variations of the binding capacity, detectable as early as 40 days or as late as 100 days of age. Absence of maternal corticoids during fetal or neonatal life reduced in vitro adrenocortical response to ACTH, and increased in vitro pituitary response to hypothalamic extract. Perinatal lack of maternal corticoid markedly reduced in vitro adrenal activity with no change in in vitro pituitary activity. Partial or total deprivation of maternal corticoids increased basal plasma levels of corticosterone but reduced adrenocortical response to environmental stress. Because of the integration between hippocampus and pituitary-adrenocortical activity in behavioral responses, such permanent variations might have kindled subtle anomalies in the expression of innate or learned behavior and of endocrine responses after appropriate environmental stimuli. The functional manifestations of these anomalies may be overcome by vicariance; however, they may not be detrimental in themselves, their adaptive value depending on circumstances. Moderate impacts on the hormonal status of the mother-conceptus unit during ontogenesis and maturation of brain and endocrine pace-makers may contribute to the origin of individual traits.

Acknowledgement: Grant CNR 82.02039.04 and M.P.I. 1982.

VERTEBRATE MEMORY MODELS: ALTERATIONS IN NEURONAL PHOSPHOPROTEINS

B. Bank, D.L. Chute and J. Gurd

University of Toronto, Scarborough College
Division of Life Sciences
West Hill, Ontario, Canada

The intent of this research is to demonstrate the role of synaptic phosphoproteins in neuronal plasticity. We believe this research represents the first demonstration of a long-term alteration in phosphorylational state of a post-synaptic membrane protein as a consequence of neuronal activation. The model a long-term plasticity employed in this study is the kindling model (Goddard et al., 1969). Specifically, we have used the "rapid kindling" method whereby full motor convulsions can be elicited by stimulation of the hippocampus in 3-6 hrs. as opposed to 30-60 days in normal kindling. Rapid kindling displays all the classical characteristics of normal kindling (Zorumski and Hatlelid, 1982). Essentially, rats are stimulated in biphasic square wave at a frequency of 10 Hz. for 10 seconds. Five days after the kindling program, rats were decapited, their brains removed and the hippocampus-entorhinal cortex and frontal cortex dissected out. Synaptic membranes were isolated and then incubated with $[\gamma-^{32}P]$ATP. The degree of phosphate incorporation into various protein bands was determined by one and two dimensional SDS polyacrylamide gel electrophoresis. Synaptic glycoproteins were isolated by concanavilin A affinity chromatography and then separated by SDS polyacrylamide gel electrophoresis. We found that the hippocampus-entorhinal cortex of kindled animals exhibits a significant decrease in *in vitro* phosphorylation of a 180 K protein as compared to controls. While this difference is much less pronounced in frontal cortex. We subsequently identified this protein as GP 180. The phosphorylation of this protein is stimulated by cAMP (Gurd et al., 1983) or Ca^{++}-Calmodulin (Gurd, 1984). This glycoprotein is localized post-synaptically in the post-synaptic density. The decreased *in vitro* phosphorylation suggests the opposite has occured in vivo perhaps by an increase in kinase, decrease in phosphatase or an increase in the protein substrate. Hence it can be shown that post-synaptic phosphoproteins can be altered as a consequence of neuronal activation possibly by neurotransmission (via cAMP) and subsequent calcium influx (via calmodulin) to elicit long term changes in neuronal functioning which may be the basis of learning and memory.

EVENT-RELATED POTENTIALS, LEARNING AND MEMORY

J.P. Banquet and R. El Massioui

LENA-CNRS La Salpêtrière Paris

Previous work (Banquet et al., 1984) showed an opposite long term evolution of the N2 and P3 components of the event-related potentials (ERPs) in a passive learning situation, with N2 decreasing and P3 increasing in amplitude during the learning process. It was assumed in this experiment that the comparison of short- and long-range effects of practice could yield cures on the participation of short- and long term memory in the ERPs generation.

Eight subjects received high and low pitch tone series requiring a go/no go discrimination. The order of the stimuli at constant ISI was random in 3 types of Bernouilli series:. 8/2, .5, .2/8. The short and long range passive learning situations were respectively realised by delivering 5 consecutive blocks (60 stimuli each) of the different probability conditions in 2 identical sessions a week apart. EEG was recorded from 6 midline electrodes starting from frontal Fz, spaced 1/10th the inion-nasion distance and referred to linked ears.

Behavioral results confirmed the law of the stimulus probability effect i.e. an inverse relation between stimulus frequency and reaction time (RT). Also, they demonstrated a learning process: The law became effective only by the last blocks of the 1st session.

Short-term effect of practice concerned only N2 amplitude in a similar way during the 2 sessions. This within-session evolution was reflected by a between-blocks difference: The N2 component evolved from a state of no amplitude difference in relation to prior probability to a state of maximal differentiation, e.e. minimal amplitude for frequent and maximal for rare stimuli.

Long-term between-sessions effect was reflected by an evolution of P3 amplitude opposite to that of N2. Amplitude did not significantly change from block to block by increased in the second session. Conversely, N2 reduced its amplitude in the 2nd session, mostly for frequent stimuli responses, suggesting an habituation.

In conclusion: 1. Short-term practice in the absence of any long-term memorisation, affected the mechanisms responsible for N2 but not for P3 generation. These changes did not durvive the long-term interval.

2. Long-term practice affected the processes reflected by changes in P3, suggesting long-term memorisation and a cumulative process involved in its generation. Since P3 amplitude has been shown to increase as a task becomes easier, it can be assumed that the subjects completed the learning of the cognitive task only by the second session.

3. Conversely to N2, the absence of P3 amplitude decrement during practice suggests that the degree of learning or the nature of the corresponding cognitive task did not permit habituation to take over.

4. The identification of the probability law did not take any more than a few minutes of practice as demonstrated by its indexing by both P3 and N2.

5. The evolution of RT paralleled that of N2- but not on P3, supporting the view of several authors that RT depends on N2- and not on P3- related processing

REFERENCES

Banquet, J.P., Baribeau, J., Lesèvre, N., 1984, in: "Brain and Information: Event-Related potentials, " R. Karrer, J. Cohen, and P. Tueting, eds. Ann., N.Y. Acad, Sci., New York.

MAMMILLARY BODY LESIONS IN MICE INDUCE MEMORY DEFICITS WHICH RESEMBLE THOSE OF THE AMNESIC KORSAKOFF SYNDROME

Daniel Berocochea and Robert Jaffard

Laboratoire de Psychophysiologie,U.A. CNRS n°339, Univ. Bordeaux I, avenue des Facultés, 33405 Talence, France

Two experiments were designed to investigate the effects of either kaïnic acid or electrolytic lesions of the median mammillary nucleus (MM) on spatial memory in mice. In the 1st experiment, animals were tested for spontaneous alternation (SA) in a T-maze in a sequential test procedure using 6 to 9 successive free trials with an intertrial interval (I.T.I.) of either 5 or 30 sec. No differences were observed between MM lesioned and control subjects at the 5 sec I.T.I.. For an I.T.I. of 30 sec, experimental subjects exhibited normal SA rates on the second trial but were progressively impaired on subsequent trials (54.6% and 49.1% on the 5th and 6th trials) as compared to controls (80.0% and 79.7%; $p < 0.001$). This deficit was reversed by adding an intra-maze cue (white carboard) on the 5th trial (from 55.4 to 81.9%; $p < 0.001$),but additional trials resulted again in a progressive decrease of SA until the cardboard was removed (9th trial : from 52.7% to 74.7%; $p < 0.05$). The 2nd experiment was aimed at studying long-term retention of the same spatial information. Thus, animals were given two forced trials (acquisition) followed by a free trial (test) at varying acquisition-test intervals. Results showed that MM lesioned animals had normal SA rates at short (5 min) but were considerably impaired at the longest interval (6 hr)(55.2% vs 76.3%; $p < 0.001$). Our results resemble those reported for Korsakoff amnesic patients i) required to discriminate which of two highly familiar items was seen more recently and which ii) exhibit an accelerated rate of forgetting for lists of frequent words (Huppert and Piercy, 1976). The fact that a context shift (cardboard added and removed) readily improved performance suggests that, as in amnesics (see Winocur et al.,1981), our mice suffered from an increased vulnerability to proactive interference.

This research was supported by the CNRS (U.A. n°339).

ACKNOWLEDGEMENT

This research was supported by the CNRS (U.A. n°339).

REFERENCES

Huppert, F.A., and Piercy, M., 1976, Recognition memory in amnesic patients: effect of temporal context and familiarity of materials. Cortex, 12:3.

Winocur, G., Kinsbourne, M., and Moscovitch, M., 1981, The effect of cuing on release from proactive interference in amnesic patients. J. Exp. Psychol., 7:56.

NON ASSOCIATIVE AND ASSOCIATIVE LEARNING IN THE NAPLES HIGH (NHE) AND LOW EXCITABLE (NLE) RAT STRAINS

L. Bivolarska, A. Cerbone, A. Grimaldi, C. Lamberti
G. Manzi, M. Sacco and A.G. Sadile

Institute of Human Physiology & Medical Physics
University of Naples, 1st Medical School
I-80138 Naples, Italy

The purpose of the present research was to investigate the contribution of non associative processes (excitability, emotionality, habituation) to behavioural plasticity. To this end a series of non associative and associative behavioural tasks was carried out on rats of the Naples High (NHE) and Low Excitable (NLE) strains, which have been selected genetically for divergent behavioural arousal to novel environments in a modified Làt's maze (2).

NON ASSOCIATIVE TASKS: Behavioural habituation of locomotor reactivity to novelty was studied with two 10min-exposures to an Animex box, an hole-board and a square enclosure, the Làt's maze (1). Preference for novel, sweet taste was tested in two 15min-tests in a two-bottle water-saccharin choice task. Each series of tasks was given on successive days.

ASSOCIATIVE TASKS: A) Acquisition to criterion of two-way avoidance in a shuttle-box with a retention test of 30 trials on 4 consecutive days; B) Conditioned taste aversion (CTA) (induced by LiCl) tested in a two-bottle water-saccharin (0.1%) choice system, followed by forced extinction in a saccharin-saccharin no-choice system, on five consecutive days.

In the non associative behavioural tasks, NHE and NLE rats differed in activity in a novel situation by 20% in the Animex, by 40% in the holeboard and by a factor of two in the square enclosure, with NHE showing a lower habituation rate in the latter test only. No NHE/NLE difference was found in the defecation score nor in the taste preference. In associative behavioural tasks, NHE and NLE differed in that, in the shuttle-box, almost all rats from both strains reached criterion but 50% of NHE showed retention of avoidance whereas NLE rats only freezing. In the CTA paradigm, both lines established aversion, but only NHE rats showed resistance to extinction.

The results suggest a) these strains to differ in reactivity to spatial but not to gustatory novelty; b) a shift in behavioural coping strategies, task (in shuttle-box and not in CTA) and genotype-dependent (in NLE and in 50% of NHE). It remains to be ascertained whether and how the differences in behaviour are related to the differences in neuronal circuitry (5), in α-adrenergic system (3) and in responsiveness to vasopressin and opioids (4), in the hippocampal regio inferior of these strains.

REFERENCES

1. A.G. Sadile, B. De Luca, L.A. Cioffi, 1978, Abh.Akad.Wiss.DDR, 5: 203.
2. A.G. Sadile, A. Cerbone, G. Manzi, A. Grimaldi, 1983, Soc.Neurosci.Abstr., 9: 643.
3. A.G. Sadile, A. Cerbone, C. Lamberti, L.A. Cioffi, 1984, Behav.Brain Res., 12: 228.
4. A.G. Sadile and A. Cerbone, 1984, in: Frontiers in Neuroscience Series, E.E. Müller and A.R. Genazzani, eds., Raven Press, New York.
5. A.G. Sadile, A. Cerbone, H-P. Lipp, B. Heimrich and H. Schwegler, 1984, Trabajos del Instituto Cajal, 75: 53.

THE SPECTRUM OF THE MONKEY'S SACCADIC REACTION TIMES

R. Boch & B. Fischer

Dept. Neurophysiol., Univ. Freiburg
Hansastr. 9, 7800 Freiburg F.R.G.

During the preparation of voluntary goal directed saccades different processes take place: a) break of actual fixation; b) attention-shift to the new target; c) decision to move the eye; and d) computation of the coordinates for the saccade. We manipulated the visual and behavioural aspects of tasks in which **monkeys** always had to saccades to the same target. The idea was that the duration of these above processes should be reflected by the saccadic reaction times (SRTs). Monkeys were trained to fixate a central fixation point for randomly varying periods of time. If the fixation point was turned off the monkeys had to saccade to a peripheral target which could occur up to 1 sec before (task A) or 200 msec after (task B) the fixation point disappeared. In task C, the monkeys had to saccade to the new target, while the fixation point remained visible. The spectrum of SRTs contained at least four different peaks: express-saccades with reaction times of 75 $\pm$ 5 msec in task B and 90 - 100 msec in task C; there were also fast regular saccades with SRTs around 130 msec in task B, slow regular saccades around 180 msec in task C and a very slow peak around 220 msec in task A. At the beginning of training the SRTs were longer: about 20 msec for the express-saccades, 50 msec for the fast regular and 250 msec for the slow and very slow regular saccades. Assuming afferent and efferent delays of 35 and 25 msec respectively the time for the computation of coordinates for express-saccades was of the order of 10 - 20 msec; decision making adds about 50 msec to the SRT of the fast regular saccades; break of previous fixation delays the SRTs by about 80 msec for slow regular saccades. Using a task where the animal had to saccade to remembered target locations revealed an extra delay of about 50 msec for memory recall. The results show, that the oculomotor reaction can be used to identify different brai processes and to measure the time of their completion.

TESTOSTERONE AND IMPRINTING

J.J. Bolhuis, B.J. McCabe and G. Horn

University of Cambridge, Department of Zoology
Downing Street, Cambridge EB2 3EJ, U.K.

Testosterone has previously been implicated in memory and attentional processes in the domestic chick (Andrew et al., 1981). Imprinting is a type of learning by which exposure of a chick to an object results in the restriction of the chick's preference to that object. Is imprinting influenced by testosterone?

Ross I chicks were hatched and reared in darkness. Between 3 and 8 h after hatching, each was injected with 0.1 ml oil (controls), or 0.1 ml oil containing 0.5, 1, 3 or 5 mg testosterone enanthate. Seventeen hours after injection, the chicks were trained by exposure for 2 h either to a rotating stuffed jungle fowl or to a rotating, internally illuminated red box. Two hours after the end of training, each chick's preference was measured by exposing the chick sequentially to its own training stimulus and to the other, novel stimulus. A measure of preference for the training object (preference score) was then derived. Thirty minutes after the test, the chicks were killed and blood was taken for testosterone assay.

Plasma testosterone concentration in control female chicks was not significantly different from that in control males. The mean concentration of plasma testosterone in chicks trained on the jungle fowl was not significantly different from that in chicks trained on the red box.

In control chicks trained on the jungle fowl, there was a significant correlation between preference for the training object and plasma testosterone concentration ($P < 0.01$). There was no such correlation in control chicks trained on the red box. Injection of testosterone significantly increased preference for the jungle fowl in chicks trained on this object ($P < 0.05$) but was without significant effect on chicks trained on the box.

The results suggest that plasma testosterone levels may impose a constraint on the formation of a preference for the jungle fowl.

REFERENCES

Andrew, R.J., Clifton, P.G. and Gibbs, M.E. (1981). J. Comp. Physiol. Psychol. 95, 406.

THE MATURATION OF EYE MOVEMENT IN THE CAT

Leopoldo Bon and Cristina Lucchetti

Istituto di Fisiologia
Università di Trieste
via Fleming, 22 - 34127 Trieste, Italy

The aim of this report was to verify whether maturation can influence eye movement in the alert kitten. The experiments were carried out on 5 adult cats and on 5 kittens aged 35-50 days. The maturation of visuomotor behaviour (Van Hof 1976) and of the vestibular-ocular reflex (Flandrin et al. 1979) is incomplete at this age. Animals were operated under general anaesthesia one week before the experiment to allow painless fixation of the head. The experiments were executed in a darkened cage. To define the behavioural state EEG and EMG were recorded. Horizontal and vertical components of eye movement were recorded by a modified Robinson's search coil using the magnetic field technique. Briefly, the coil was inserted in a suction contact lens that was placed on the ocular globe on the day of the experiment. We have found that in awake kittens the spatio-temporal pattern develops in a prevailing vertical plane; on the contrary, in the cat it is developed in a horizontal plane. The vertical orientation reminded us of what we had found in REM's episodes in kittens as well as in cats (Bon et al. 1983). Generally, a rapid eye movement is followed by a fixation. In the present experiment we have computed the percentage of movements followed by fixation. In the kitten, the percentage was low (26.7%), in the cat, it was higher (68.7%). Without any doubt, the spatio-temporal organization changes during development and a defined differentiation in the oculomotor activity in young age does not exist during wakefulness and paradoxical sleep. As far as the low number of fixations in the kitten is concerned, we suggest that the maturation can modify the neural structures that subserve fixation.

REFERENCES

Bon. L., Benassi. C., and Morigi, M., 1983, Rapid eye movements in the kitten, Electroenceph. clin. Neurophysiol., 55:601.

Flandrin, J.M., Courjon, J.K., and Jeannerod, M., 1979, Development of vestibulo-ocular response in the kitten, Neurosci. Letters, 12:295.

Van Hof-Van Duin, J., 1976, Development of visuomotor behaviour in normal and dark-reared cats, Brain Res., 104:233.

^{3}H-MUSCIMOL BINDING TO MEMBRANES OF CHICK FOREBRAIN: CHANGES DURING DEVELOPMENT AND AFTER PASSIVE AVOIDANCE TRAINING

R.C. Bourne and M.G. Stewart

Brain Research Group, Department of Biology
The Open University, Milton Keynes, MK7 6AA, U.K.

Little is known of the role of inhibitory neurotransmitters in the chick brain, or how they might be affected by passive avoidance training. Because of this we decided to use ^{3}H-muscimol, a potent GABA-agonist, to investigate the development of GABA receptors and the effects of passive avoidance training on receptor density in chick forebrain. In view of the considerable evidence of hemispheric asymmetry of avian brain function our experiments were also designed to examine whether any changes in ^{3}H-muscimol binding might be lateralized. For the developmental study, forebrains from chicks aged from 9 days prehatch (12 days *in ovo*) to 21 days post hatch, were dissected into left and right hemispheres and stored at -20°C under buffer. For passive avoidance training, chicks at 24-36 hours after hatching were trained to avoid pecking at a bead coated with methyl anthranilate (MeA) using the method of Rose et al. (1980). At intervals after training (10, 30 and 180 minutes), the brains were divided into anterior and posterior forebrain roof and forebrain base and then subdivided into left and right hemispheres. Washed membrane preparations were assayed in 50mM Tris-HCl buffer pH 7.3 containing 55 nmol l^{-1} ^{3}H-muscimol (and 10 μm GABA to estimate non-specific binding) and expressed as pmol muscimol bound per mg membrane protein. The number of binding sites was estimated in MeA-trained birds and compared with birds trained to peck at a water-coated bead (W-control).

Receptor density increases rapidly in pre-hatch chicks, and then declines slowly following hatching. Data from the passive avoidance studies show a significant but transient increase ($p < 0.05$) in muscimol bingind in the left hemisphere of the posterior forebrain roof 30 min after training, implying an alteration in the number of GABA receptors in this region. These results demonstrate that changes occur not only in excitatory neurotransmitter systems (Rose et al., 1980; Aleksidze et al., 1981), but also involve an inhibitory neurotransmitter, GABA.

REFERENCES

Aleksidze, N., Potempska, A., Murphy, S., and Rose, S.P.R., 1981, Passive avoidance training in the young chick affects forebrain α-bungarotoxin and serotonin binding, Neurosci. Lett. Suppl. 7:244.

Rose, S.P.R., Givvs N.E., and Hambley, J., 1980, Transient increase in forebrain muscarinic cholinergic receptor binding following passive avoidance learning in the young chick. Neurosci. 5:169.

HABITUATION OF RESPONSES RECORDED IN THE RETICULAR FORMATION: THE POSSIBLE INVOLVEMENT OF OPIATES

Margaret I. Boxer*

Neurobiology Laboratory
Department of Anatomy
The University of Liverpool L69 3BX (U.K.)

The amplitude of a response can decrease or even disappear upon repeated stimulation. This change is known as habituation and is a form of learning. Cells in the reticular formation of an intact frog tend to habituate to repeated peripheral stimuli. In decerebrate frogs this is not the case. Electrical stimuli of 0.2 msec. duration were applied every 5 sec. to the skin surface at intensity 1.25 x threshold. Potentials evoked were recorded using monopolar glass capillary microelectrodes of low resistance and filled with 4M NaCl. The recordings were made in the reticular formation of the medulla oblongata, 0.5mm lateral to the midline, 1.5mm rostral to the obex.

Opiates appear to be involved in this habituation mechanism, as naloxone, administered to an intact frog was found to produce complete dishabituation, and D-ala^2-met^5 enkephalinamide administered to a decerebrate frog led to a reduction in response amplitude which increased again upon a subsequent administration of naloxone.

INTRASTRIATAL GRAFTING OF DOPAMINE-CONTAINING CELL SUSPENSIONS: EFFECTS OF MIXING WITH TARGET AND NON-TARGET CELLS

P. Brundin, O. Isacson, F.H. Gage and A. Björklund

Department of Histology, Biskopsgatan 5
S-223 62 Lund, Sweden

Grafting of dissociated cell suspensions prepared from embryonic ventral mesencephalon (VM), rich in dopamine (DA) containing cells, into 6-hydroxydopamine (6-OHDA) denervated rat striatum has previously been shown to survive and restore some functional deficits induced by the lesion (Björklund et al., 1983). Dissociated VM cells allowed to re-aggregate and develop in rotatory cultures *in vitro* have a greater number of surviving DA-cells when co-aggregated with tissue from a DA target area, i.e. the striatal primordium (Hoffman et al., 1983, Shalaby et al., 1984) as compared to cultures of DA cells alone or of DA cells re-aggregated with non-target cells. In addition, *in vitro* culturing of DA neurons mixed with striatal target cells has been shown to induce an increase DA-synthesis and uptake (Di Porzio et al., 1980) as well as induce changes in the morphology of the DA neurons (Denis-Donini et al., 1983). The present experiment was designed to investigate to what extent the addition of embryonic striatal target cells to VM cell suspensions might influence the survival, fibre outgrowth and functional effects of VM cell suspension grafts in the denervated striatum of adult host rats. Rats with unilateral 6-OHDA lesions of the ascending DA mesostriatal pathway were tested for amphetamine-induced rotation, prior to receiving embryonic suspension grafts of one of three different types; all containing an approximately equal number of VM cells. Rats receiving VM grafts mixed with striatal cells showed a significant potentiation of graft function, expressed in terms of asymmetry-reduction, as compared to pure VM grafts of VM grafts mixed with spinal cord cells. The three types of grafts were on subsequent histochemical analysis found to contain approximately equal mean numbers of cells, however there was a significantly greater outgrowth of fluorescent fibers in the VM + striatum group. In addition, the grafts containing the striatal target cells exhibited a novel outgrowth pattern consisting of several isolated round islands of intense catecholamine fluorescence with catecholamine fluorescent perikarya located mainly at the border of the islands. The significance of this rather striking morphological finding is not known, but it may bear some relationship to previous findings of altered outgrowth patterns of DA-neurons co-cultured with striatal neurons *in vitro* (Denis-Donini et al., 1983).

REFERENCES

Björklund et al., 1983, Acta Physiol, Scand., Suppl. 522.
Denis-Donini et al., 1983, J. Neurosci.,3:2292.
Di Porzio et al., 1980, Nature, 288:370.
Hoffman et al., 1983, Brain Res., 174:275.
Shalaby et al., 1984, Brain Res., 307:347.

THE ROLE OF THE CORPUS CALLOSUM IN VISUOMOTOR RUNNING PATTERNS

Deborah Christie and I. Steele Russell

MRC Unit on Neural Mechanisms of Behaviour
3 Malet Place
London, WC1E 7JG

Rats adopt a highly stereotyped running pattern during acquisition and testing in a two-choice visual discrimination task. The running pattern does not alter during testing for angular difference threshold. This is true when the degree of tilt is subthreshold, or when visual input is reduced by patching one eye or blocked by binocular patching. This suggests that the running pattern is independent from the visual comparisons required in the discrimination task. The running pattern interferes with the visual discrimination task by producing a negligence for the field contralateral to the preferred approach side (Christie and Steele Russell, 1984). This is similar to the attentional neglect reported in primates following a unilateral Frontal Eye Field lesion (Crowne et al, 1981). Unilateral lesions of rat anteromedial cortex produce a visual neglect for stimuli in the contralateral field to the lesion but do not produce any alteration of the running pattern (Christie and Steele Russell, 1984).

It has been reported by Mohn (1981,1984) that section of the Corpus Callosum produces a disorganisation of the running pattern during testing for interocular transfer. Despite this change in the running pattern, transfer of the discrimination performance was good.

The present study showed that callosal section produces a clear disruption of the running pattern during choice behaviour. Animals trained under monocular occlusion developed a marked ocular asymmetry where the animal using the eye ipsilateral to the side preference performs at chance. With the contralateral eye, the animal retains its normal angular difference threshold.

As the running pattern is not altered by visual manipulation alone this suggests the running pattern is a pure motor habit. The loss of discrimination of the stimuli in the field contralateral to the side preference must therefore be a result of the disruption of the motor habit.

REFERENCES

Christie, D., and Steele Russell, I., 1984, Frontal Eye Field Lesions, Visual Neglect and Oculomotor Strategies. Behav. Brain Res., 12. 141.

Crowne, D., Yeo, C., and Steele Russell, I., 1981, The effects of Unilateral Frontal Eye Field Lesions in the Monkey. Visual Motor Guidance and Avoidance Behaviour. Behav. Brain Res., 2. 165-187.

Mohn, G., 1981, Visuomotor Behaviour of Callosum Sectioned Rats. Behav. Brain Res., 2. 266.

Mohn, G., 1984, Behav. Brain Res., 11. 95.

EVOKED POTENTIAL CORRELATES OF CLASSICAL AND INSTRUMENTAL CONDITIONING

Valéria Csépe, G. Karmos, M. Molnár and I. Winkler

Institute for Psychology of the Hungarian
Academy of Sciences
Budapest, Hungary, H-1394

In our earlier study (Karmos et al., 1982) we analysed the changes of auditory cortical evoked potentials elicited by clicks which were kept indifferent by applying them as background stimuli during classical conditioning. It was shown that the modification of the late components could most probably be related to the fluctuation in the general motivational state of the animal.

In the present experiments the auditory cortical evoked potentials (AEPs) elicited by clicks used as warning stimuli were studied both in classical and instrumental aversive conditioning paradigms in freely moving cats. The recording electrodes were chronically implanted epidurally above the auditory cortex. The AEP changes were analysed by horizontal and vertical averaging within and throughout the trials and by sorting of AEPs according to the amplitudes of different components.

It was found that the changes of AEPs manifested themselves in the middle latency waves while the first components showed only slight fluctuation. Significant and dynamic changes of the 40-50 ms peak latency negative component appeared both in the classical and instrumental conditioning paradigms. The observed changes of this latter component suggest that they are closely related to the early attentional effect. However the temporal pattern of the amplitude changes of this negative wave was different in the two paradigms. At the end of classical conditioning all AEPs of the trials appeared with high amplitude negative component. In the instrumental paradigm in the last phase of conditioning the AEPs with highest N50 wave appeared only before the execution of the avoidance response. The changes of the late positive wave were also different in the two paradigms. At the end of the classical conditioning the positivity increased during the whole time period of the signal simultaneously with the arousal level of the animal. At the end of the instrumental conditioning the increased positivity appeared only at the beginning of trials and gradually decreased during the acquisition of the avoidance response.

REFERENCES

Karmos, G., Csépe, V. and Molnár, M., 1982, Auditory cortical evoked potential changes during classical conditioning in cats, in: "Psychophysiology 1980," R. Sinz and M.R. Rosenzweig, eds., VEB Gustav Fischer Verlag, Jena.

CEREBRAL SYSTEMS INVOLVED IN A CLASSICAL CONDITIONING IN RATS

Jean Delacour

Laboratoire de Psychophysiologie
Université Paris 7
75251 Paris Cedex 05 France

Unit activity of the hippocampus, the centrum-medianum-parafacicularis (CM) and medialis dorsalis (MD) nuclei of the thalamus, was recorded in chronic rats during a classical conditioning; neocortical EEG and somatic responses were also recorded.

Conditioned unit responses (CUR) of the different groups of neurons were compared according to the precocity of their appearance, their stability, their latency, ans their CS^{+}/CS^{-} differentiation.

CUR of type I hippocampal neurons (probably pyramids) and of CM neurons did not differentiate between the CSs; they developped rapidly, then declined and disappeared. They were contemporary with an initial conditioning stage which was characterized by undifferentiated arousal responses (EEG desynchrony) to both CSs.

CUR of type II hippocampal neurons (probably granule cells or inter-neurons of the fascia dentata) and MD neurons developped slowly and differentiated between the CSs during a late conditioning stage which was characterized by the regression of arousal responses and the differentiation of somatic responses.

These data strongly suggest that two neuronal systems, each having a different role, are involved in a classical conditioning in the rat.

THE INFLUENCE OF ELECTROCONVULSIVE SHOCK ON LEARNING AND RETENTION OF MEMORY IN RAT

Nenad Denić, Slavko Janković, Rosica Veskov,
Ljubisav Rakić and Julius Ivanuš

Institute for Biological Research
ul. 29. novembra br.142
11000 Belgrade, Yugoslavia

The influence of maximal electroconvulsive shock (MECS) on the ability to learn and on memory retention was investigated in male Wistar rats, 3o-8o days of age. Animals were kept under conditions of unrestrained food and water intake and a 12 hours light-day-darkness cycle. Conditioning of animals with thirty trials per day over ten days was performed in a shuttle-box using light as the conditioned stimulus and foot shock (1 mA, 4 s) as the unconditioned stimulus. Electroconvulsive activity was elicited with a current (up to 85v, loo Hz) applied through ear clips. A correct response was taken to be an escape reaction to the other part of the shuttle box within three minutes following light onset which preceded the painful electrical stimulus. The rats that by the seventh day of training did not produce a single positive avoidance response were eliminated from the study. Rats were separated into three groups: the control group did not receive electric ear shock; the second group received ECS thirty minutes before conditioning and the third group received ECS immediately after conditioning, respectively.

Results obtained show that control animals, which did not receive ECS, had a linear increase in the number of positive active avoidance responses with the time course of training. In the second group receiving ECS before the conditioning the process of learning was not substantially influenced compared to the learning of control group. Only on the seventh day the learning ability dropped in a degree that reached statistical significance ($p < o,o5$). Subsequent day showed an improvement of learning which closely approached the control values. The speed of learning was slowed down in respect to control but the diference never reached statistical significance. The third group, with ECS treatment performed after the conditioning, demonstrated decrease in their ability to reproduce learned reaction to the conditioned stimulus. After four day the rate of correct answers was significantly low ($p < o,o5$), somewhat improving afterwards but still to be low ($p < o,ol$) on the tenth day, as compared to the control group. The speed of learning of these animals was significantly slowed down.

AN INTERDISCIPLINARY STUDY OF THE EFFECTS OF 6-OHDA LESIONS OF THE SEPTUM ON CHOLINERGIC SEPTO-HIPPOCAMPAL ACTIVITY IN INBRED MOUSE STRAINS

Thomas Durkin, Daniel Galey, Yannick Jeantet,
Georgeos Sifakis and Robert Jaffard

Laboratoire de Psychophysiologie,U.A. CNRS n° 339, Univ
Bordeaux I, avenue des Facultés, 33405 Talence, France

The functional activity of the cholinergic neurones of the septo-hippocampal pathway has been postulated to play an important role in learning and memory processes (Jaffard and Destrade, 1982) and is trans-synaptically regulated by numerous neurotransmitter systems, including dopamine, in the septum (Costa et al.,1983). A10 dopamine terminals mediate a tonic inhibitory control on these cholinergic neurones and Durkin et al. (1983) reported that significant genotypic variation exists for this dopaminergic control in the C57BL/6 and BALB/c mouse strains and that this variation may underly strain differences in learning and memory performance. We therefore decided to undertake an interdisciplinary study of the effects of 6-OHDA lesions of the lateral septum (3μg/1 μl, bilateral) in C57BL/6 and BALB/c mice. Between 6 and 12 days post-operation their alternation behaviour, acquisition, retention and reversal of a spatial learning task was tested in a T-maze and compared to vehicle-injected and intact controls. 6-OHDA injected mice exhibited higher rates of spontaneous alternation and a facilitation of both acquisition and reversal of the spatial discrimination, these facilitative effects being significantly greater in C57BL/6 mice as compared to BALB/c. Electrophysiological investigations into the spectral characteristics of hippocampal theta rhythm showed that 6-OHDA lesions produced, within 48 hours, the disappearance of the synchronisation normally produced by haloperidol (1 mg/kg i.p.). Neurochemical investigations using measures of high affinity sodium dependent choline uptake revealed quantitative strain differences for the 6-OHDA induced activation (disinhibition) of hippocampal cholinergic activity which paralleled the interstrain differences in behavioral measures. Thus septal dopamine terminals via their inhibitory control of septo-hippocampal cholinergic activity appear to play an important functional role in the modulation of learning behaviors observed in these mouse strains.

This research was supported by the CNRS (UA n° 339) and by grant from MRI (n° 84.C.1306).

REFERENCES

Costa, E., Panula, P., Thompson, H., Cheney, D., 1983, The transynaptic regulation of the septal-hippocampal cholinergic neurones. Life Sci., 32:165.

Durkin, T., Hashem-Zadeh, H., Mandel, P., Kempf, J., Ebel, A., 1983, Genotypic variation in the dopaminergic inhibitory control of striatal and hippocampal cholinergic activity in mice. Pharmacol. Biochem. Behav., 19:63.

Jaffard, R., Destrade, C., 1982, Learning and memory processes as related to genotypic or experimental variations of hippocampal cholinergic activity in inbred strains of mice. in: "The Genetics of the Brain," E. Lieblich, Ed., Elsevier, Amsterdam.

ΓHE ENTORHINAL CORTEX IN MEMORY PROCESSES :

N OF LATERAL AND MEDIAL PARTS

Monique Gauthier and Claude Destrade

Laboratoire de Psychophysiologie, U.A.CNRS n° 339, Univ. Bordeaux I, avenue des Facultés 33405 Talence, Cedex, France

It was previously shown, using lesions (Gauthier and Soumireu-Mourat, 1981) or stimulation techniques (Gauthier et al., 1982; Gauthier and Destrade, 1984), that the lateral entorhinal cortex is involved in memory processes. However, the entorhinal cortex consists of lateral and medial parts which each give rise to separate projections of the perforant path with specific electrophysiological characteristics. Do these two parts play the same role in memory processes ?

The effects of bilateral lesions or subconvulsive stimulations of the medial entorhinal cortex (MEC) or the lateral entorhinal cortex (LEC) in mice were examined during acquisition, retention and extinction (only for the lesioned animals) of an operant CRF conditioning in a Skinner-box.

Compared with the control animals, lesions in the MEC had no behavioral effects whereas lesions in the LEC enhanced retention and increased resistance to extinction. However, the acquisition of this task was not affected by either lesion.

Post-training electrical stimulation was applied to MEC or LEC at different time intervals ranging from 30 sec to 3 h after the acquisition session in the Skinner-box. Surprisingly, LEC greatly improved retention when applied 30 min after the acquisition session, whereas MEC stimulation improved retention when applied between 30 sec and 15 min after training.

These results show :

- a dissociation of MEC and LEC in terms of behavior which is in agreement with evidence from anatomical and electrophysiological studies;
- that the MEC and LEC have different temporal gradients of intervention in memory processes.

This research was supported by the CNRS (UA n°339) and by a grant from the INSERM (n° 81.6.014).

RERERENCES

Gauthier, M., Somireu-Mourat, B., 1981, Behavioral effects of bilateral entorhinal cortex lesions in the BALB/c mouse, Behav. Neural Biol., 33:419.

Gauthier, M., Destrade, C., Soumireu-Mourat, B., 1982, Late post-learning participation of entorhinal cortex in memory processes, Brain Res., 233:255.

Gauthier, M., Destrade, C., 1984, Late post-learning effect of entorhinal cortex stimulation persits despite destruction of perforant path, Brain Res., 310

MEMORY REORGANIZATION OVER TIME AS REVEALED BY INTERACTION BETWEEN TYPE OF PRE-TEST CUEING AND LENGTH OF RETENTION INTERVAL

Pascale Gisquet-Verrier, Anne Dekeyne, and Tatiana Alexinsky

Département de Psychophysiologie - PN2 - C.N.R.S.
91190-Gif-sur-Yvette, France

According to a multidimensional conception, memory must be considered as a collection of attributes, internal representations of the different events constituting an episode to be remembered.

In a Y maze, rats were trained in a visual discrimination to escape or to avoid a footshock by running into the lighted arm. The performance during retention test was significantly affected by the delay interposed between acquisition and testing phases. Deficits were observed after 1, 2 or 4 hours (Kamin effect) and 21 days (spontaneous forgetting). An enhancement occured after 3 days (reminiscence) whereas performance was at the same level as at the end of acquisition after 1 and 5 days.

We hypothetized that the spontaneous fluctuations of memory accessibility previously described might reflect changes in the relative role of each individual attribute in memory organization over time. In order to test this hypothesis, a pre-test cueing procedure was used. Such a procedure allows estimation of the effectiveness of different cues in triggering reactivation leading to restitution of the whole memory, as a function of the retention interval. The 9 selected cueing treatments were related either to contextual stimuli or to the reinforcement or to a compound of these categories.

Our results indicated that prior cueing enhances performance only when performance is impaired (1 hr or 21 days). Furthermore, there is no systematic additive effect when the treatment consisted of the presentation of several cues (compound cues). Finally, the results suggest that there is no fixed hierarchy of efficacy among cues.

Three categories of prior cueing treatments, based on their efficacy on retrieval processes, are suggested by our results: 1) Cues which alleviated impaired performance independently of the length of retention interval (those related to the reinforcement) 2) Cues whose efficacy decreased over time (discriminative stimulus) 3) Cues whose efficacy increased over time (more complex cues such as the context or compound stimuli).

In all, these results clearly demonstrated changes in the role of different memory attributes and strongly suggest reorganization of memory over time. Such a reorganization would be characterized by two phases: a differentiation phase, shortly after training, where attributes corresponding to simple cues have a determinant role in memory organization and a long term generalization phase, where relations between attributes seem to be more important than the specific content of the representation itself.

PAVLOVIAN LEVER-DIRECTED ACTIVITY IN RATS WITH MICROCEPHALY DUE TO GAMMA IRRADIATION

Laura H. Goldstein and David A. Oakley

Department of Psychology and Centre for Neuroscience
University College London
Gower Street, London WC1E 6BT, G.B.

Rats with microcephaly induced by prenatal exposure to ionizing irradiation have been reported to be capable of good performance on tests of associative learning despite being impaired on tests of cognitive learning ability. As a measure of associative learning, six adult rats, prenatally exposed to a single dose of 150rad of gamma irradiation on the sixteenth day of gestation, were compared with six sham-irradiated rats in a Pavlovian autoshaping sitation that involved testing the animals in conditioning chambers in which there were two perspex levers which could be illuminated from behind. The illumination of one lever (CS+) was predictive of food reinforcement, while the illumination of the other lever (CSo) was not. The delivery of food was not contingent upon a response being made by the animal. Twenty illuminations of each lever occurred per session. Fifteen autoshaping sessions were given followed by ten in which the predictive significance of the levers was reversed for the animals. Despite the overall reduction in brain size (mass, volume and encephalization quotient) and varying degrees of internal disruption seen in the prenatally irradiated rats' brains (for example, rudimentary corpus callosum, ectopic tissue masses, thin cortical mantle, dilation of ventricles) there were no differences during the initial fifteen autoshaping sessions in the total number of contacts or presses made by the two groups to the CS+ when it was illuminated. Both groups directed more activity to the CS+ than to the CSo and differentiated between the levers equally well. During the reversal phase, the two groups directed the same level of activity to the new CS+ and decreased the activity directed towards the new CSo, thereby demonstrating good response inhibition. The data are consistent with the view that associative learning is spared (or in some instances may be enhanced) despite abnormal brain development (Goldstein & Oakley, 1984).

REFERENCES

Goldstein, L.H., and Oakley, D.A., 1984, Expected and actual behavioural capacity after diffuse reduction in cerebral cortex: A review and suggestions for rehabilitative techniques with the mentally handicapped and head injured, Br. J. clin. Psychol., 23:

CNS PLASTICITY AFTER EARLY CEREBELLAR HEMISPHERECTOMY IN THE RAT

Albert Gramsbergen

Department of Developmental Neurology
Oostersingel 59, Groningen, The Netherlands

Experiments in rats demonstrated that cerebellar hemispherectomy performed on the 5th or 10th day after birth has more severe effects on locomotor behaviour at adult age than similar lesions on the 30th day (Gramsbergen, A., 1982). Extensive neuronal remodelling appears to occur in rats hemicerebellectomized before the 11th day (for summary: Gramsbergen, A., and IJkema-Paassen, J., 1984). One of the aberrant projections is a projection from the remaining cerebellar lateral and interposed nuclei in the remaining hemisphere into the ipsilateral red nucleus. The density of this aberrant projection appears to decrease with the age at lesioning: between 20 and 30% of the cerebello-rubral neurones are rerouted to the ipsilateral side after operation on the 2nd day and between 6 and 8% after operation on the 10th day. In control rats less than 2% ipsilaterally projecting cerebello-rubral neurones occur. Another finding was that the aberrantly projecting fibres are not collaterals from normal contralaterally projecting fibres but stem from separate parent cells (Gramsbergen, A., and IJkema-Paassen, J., 1982). This latter finding raised the possibility that the aberrantly projecting cells stem from supernumerary cells which would have died in normal undisturbed development. We studied this by investigating cell death during ontogeny in the lateral and interposed nuclei of control rats and rats hemicerebellectomized at the 2nd postnatal day. The live to dying cell ratio in the cerebellar nuclei of control rats is around 20 promille in the first 12 days and decreases to less than 4 promille thereafter. In rats lesioned at the 2nd day cell death comes to a halt allready between the 5th and 8th day (Gramsbergen, A., and IJkema-Paassen, J., 1985). Early cerebellar hemispherectomy in rats disturbs neuronal integration and secondary cell death. Taken in conjunction with the more impaired locomotor behaviour this leads us to conclude that the dogma attributing a greater compensational capacity to the young brain is not generally valid.

REFERENCES

Gramsbergen, A., 1982, Behav. Brain Res., 6:85.
Gramsbergen, A., and IJkema-Paassen, J., 1984, in: "Cerebellar Functions," Bloedel, Dichgans, Precht, eds., Springer, Berlin.
Gramsbergen, A., and IJkema-Paassen, J., 1982, Neurosci. Lett., 33:129.
Gramsbergen, A., and IJkema-Paassen, J., 1985, Ergebn. exp. Med., Berlin, in press.

RESTORATIVE PLASTICITY IN THE BEHAVIOUR OF THE MUTANT MOUSE STAGGERER

Jean-Marie Guastavino* and Knut Larsson**

*Laboratoire d'Ethologie, U.A. CNRS N°667, Université Paris-Nord, av. JB Clément, 93430, Villetaneuse, France
**Dept. of Psychology, Unit of Psychobiology, University of Göteborg Fack S-400 20, Goteborg 14, Sweden

The staggerer mutation occurs in the C 57/BL strain of mice. It is due to a single recessive gene on chromosome 9, which prevents normal cerebellar development. Behaviorally, the mutant is recognized from day 9-10 postnatally by its abnormal gait and impaired body balance, hence the denotation "staggerer". The mutant does not survive beyond weaning under standard laboratory conditions and was until recently only maintained by intercrossing heterozygous sg/+ parents. We report here that by adaptating the physical and social environment to the staggerer's peculiarities, this mutant can be maintained until adulthood and even reproduce. The following experimental results are reported:

1) The mutant does not rise on his hindlegs, cannot eat solid food and does not build nests. To compensate for these behaviour deficits, special arrangements were made in the cage. The mutant has a faulty body temperature regulation and was therefore provided shelters made out of a thermoinsulating material. The shelters permitted the entrance of only one individual at a time, thus giving the animal an opportunity to escape aggressive encounters with normal mice, thus improving the social life of the staggerer. Under these environmental conditions the mutant can live a fairly normal life until old age.

2) Sexually inexperienced staggerers do not mate with other staggerers, but may do so with normal mice. However, once a mutant has acquired sexual experience, by copulating with normal mice it may mate with other staggerers and sire litters.

3) The normal maternal behaviour of the mutant is greatly impaired : the mother does not eat the placenta, nor lick her pups or retrieve them. She even actively rejects her pups. A procedure was developped, involving constraint of the mother with pups and crossfostering with normal pups, which partly compensated for the behavioural deficits.

4) By stimulating the motor-vestibular activity of the staggerer in a tilted rotational device at an early age, its locomotor activity was temporarily improved.

It is concluded that changes in the social and physical environment which are of negligible importance for a normal mouse of this strain drastically improve the survival rate of the staggerer and restore basic behaviours such as sexual and maternal behaviour, which otherwise would not occur. The mutant staggerer thus represents a tool to investigate behavioural plasticity. The results obtained call for further research on the physiological mechanisms mediating these behavioural effects.

ACTIVITY OF RAT HIPPOCAMPAL NEURONES RELATED TO PERFORMANCE OF A RECOGNITION MEMORY TASK

M.H. Harries and M.W. Brown

Department of Anatomy, The Medical School
University of Bristol
Bristol BS8 1TD, U.K.

Lesions disrupting the hippocampal region produce impairments of recognition memory. Single unit activity has been recorded from the hippocampal region of the rat during learning and performance of a visual recognition memory task (delayed matching to sample).

A rat was placed in a start-box and exposed to one of six stimulus patterns (e.g. dots). After a variable interval (<1s to 80s) the start-box doors were opened and the rat moved to push either a right or a left panel. Pushing the panel bearing the previously presented (target) stimulus led to a food pellet reward. Pushing the other panel bearing one of the other five patterns (distractor) was not rewarded. Stimuli were pseudo-randomly interchanged from one trial to the next.

Recordings during both learning and performance of the tasks were obtained for 16 hippocampal units in 12 rats by using 25 μm diameter platinum wire microelectrodes. All but one (6%) of the 16 units were found to have statistically significant changes in activity related to one or more events during the trials. During learning of the task using a <1s delay interval none of these changes was specific to a particular stimulus pattern. During later trials using a >1s delay interval 3 (19%) units were found to have activity which varied significantly depending on the particular stimulus used as target and on the rat's behaviour. In each case the specific activity occurred after the start-box doors had opened, when the rat had to choose between the target and distractor stimuli, i.e. when the rat had to retrieve and use the memory of the target stimulus. No stimulus-specific changes in activity were found at the time the rat had to acquire the information in the start-box. For each unit the specific activity was not determined by either the sensory stimulus or the rat's motor response alone; the activity differed dependent on whether the stimulus was the target or distractor and on whether it was on the right or left panel. Thus the activity encoded the sensory stimulus in the context of the task. Context-dependent responses similar to these have been described for hippocampal units recorded in the monkey (Brown, 1982).

REFERENCES

Brown, M.W., 1982, Effect of context on the response of single units recorded from the hippocampal region of behaviourally trained monkeys, in: "Neuronal Plasticity and Memory Formation," C. Ajmone Marsan and H. Matthies, eds., Raven Press, New York.

INFLUENCE ON MEMORY BY A NEW NOOTROPIC COMPOUND

Franz J. Hock*,** and James L. McGaugh**

*HOECHST AG, P.O. Box 800320
D-6230 Frankfurt/M. 80, Germany

**Center for the Neurobiology of
Learning and Memory, University of
California, Irvine, CA 92717, USA

The new compound Hoe 175 [3,8-dimethyl-4-phenyl-5,6,7,8-tetrahydro-pyrazolo(3,4-b)(1,5)diazepine-lH,4H-5,7-dione] shows marked nootropic activity. Though it is a pyrazolodiazepine, the compound produced only slight effect in classical anxiolytic tests. We investigated in mice the effect of Hoe 175 on (1) aquisition and retention in a step-through inhibitory (passive) avoidance task and (2) retention of an active avoidance task. The apparatus used for the experiments consisted of a light (L) and a dark (D) compartment, connected by a guillotine door.

(1) Step-through inhibitory (passive) avoidance. During the acquisition trial an unavoidable footshock (FS, 600 μA, 2 sec) was delivered immediately after entering D. The intraperitoneal (i.p.) injection of either the test compound or vehicle was given immediately after entering D or was delayed 2 or 6 hrs. Retention (the latency for entering D) was tested 24 hrs later. At doses from 3 - 12 mg/kg i.p. Hoe 175-treated mice had retention latencies significantly higher than those of controls if administered immediately after training. The step-through latencies during the retention trial in animals with delayed injections were significantly longer with doses from 6 - 12 mg/kg i.p. if administered 2 hrs after the acquisition trial. A 6 hrs delayed injection was not effective.

(2) Active avoidance. Animals were trained (8 trials) on the first day to avoid a footshock in D by moving to L. Each trial consisted of 10 sec to avoid the footshock followed by 20 sec footshock (350 μA) after which there was a 20 sec intertrial interval in L. The retention test trial was run 24 hrs later under the same parameters. During training and testing the time for entering L and the avoidance numbers were measured. Hoe 175 was administered i.p. immediately, 2 or 6 hrs after the last trial on training day. In comparison with controls, Hoe 175 in the dose range of 1 - 25 mg/kg i.p. significantly increased avoidances on the retention test if administered immediately after the last trial. If the injection was delayed, the effects were only significant with a 2 hrs delayed injection.

BRAIN LESIONS, IMPRINTING, AND OPERANT LEARNING

M.H. Johnson and G. Horn

University of Cambridge, Department of Zoology
Downing Street, Cambridge CB2 3EJ, England

Bilateral lesions placed in a restricted part of the chick forebrain (IMHV) impair the acquisition (McCabe et al., 1981) and the retention (McCabe et al., 1982) of imprinting preferences. It is not known however, whether IMHV is specifically involved in imprinting or if it is also involved in more general learning processes. Consequently the present study was designed to investigate the effects of lesions to IMHV on performance in an operant conditioning task.

Twenty eight domestic chicks received bilateral lesions to IMHV 12h after hatching and 32 chicks served as sham operated controls. On the following day all birds underwent two sessions in an operant conditioning apparatus in which the reward for pressing the correct one of two pedals was the illumination and rotation of a conspicuous stimulus. For half of the chicks this stimulus was a stuffed jungle fowl and for half it was a red box. Each session was terminated after 45 min or earlier if a pre-set criterion for learning had been reached. Two hours after the termination of the operant task all the birds were subjected to a preference test involving the simultaneous presentation of the red box and the jungle fowl.

Lesioned birds were not impaired on the operant task or on measures of general activity. The sham operated birds showed a significant preference for the stimulus to which they had been exposed during the operant task i.e. they had become imprinted on it. In contrast, the lesioned birds showed no preference for the previously exposed stimulus i.e. they had not imprinted on it. This finding, taken together with the observation that lesions to IMHV do not effect performance in a simultaneous visual discrimination task in which chicks worked for a reward of warm air (McCabe et al., 1982), suggest that IMHV is not required for the performance of an associative learning task but that it is involved in object recognition memory.

REFERENCES

McCabe, B.J., Horn G., and Bateson, P.P.G., 1981, Brain Res., 205: 29.

McCabe, B.J., Cipolla-Neto, J., Horn, G., and Bateson, P., 1982, Exp. Brain Res., 48: 13.

EFFECT OF POST-NATAL SENSORY STIMULATION ON SLEEP PATTERNS AND COGNITIVE PROCESSES IN PRETERM INFANTS

Claire Lambert and Pierre Leconte

Labacolil Université de Lille 3
B.P. 149 59653 Villeneuve d'Ascq Cedex

It has been previously shown that REM sleep is involved in information processing. In animals, acquisition sessions have been found to be followed by an increase in the frequency of REM sleep periods (Hennevin & Leconte, 1977). In adult man, learning has been found to produce a significant increase of the high frequency (<1hz) EM number during REM sleep (Spreux & al., 1982).

Two sets of data provide evidence that preterm newborns show deficits (1) in sleep pattern maturation and (2) in cognitive processes. Other experiments suggest that sensory stimulation could attenuate these deficits. In the present study we tested (1) the efficiency of sensory stimulation on sleep pattern maturation and cognitive processes and (2) the possibility of a correlation between sleep pattern quality and cognitive processes in 16 preterm newborns, weighing below 1600g., studied from birth (30 to 34 weeks of gestational age (G.A.) to six months of conceptional age (C.A.). The preterm newborns were selected on the basis of a neurological and physiological investigation that revealed no obvious pathology. A control group of 8 such infants received the current hospital care whereas an experimental group of 8 infants received 3 daily sessions of a visual and auditory stimulation program from the day following the first sleep recording to the day of leaving hospital. A sleep cycle was analyzed a few days after admission to the hospital (32 to 36 weeks (G.A.) then just before leaving the hospital (36-38 weeks G.A.), and lastly at normal term age (41 weeks G.A.). After the last sleep recording day, a test was performed for the 16 infants to investigate the visual stimuli recognition. At 6 months of C.A., neurological, physiological and psychological investigations were made. Results showed a faster decrease of indeterminate sleep, an increase of quiet sleep, the appearance of slow waves in quiet sleep, the appearance of low voltage waves in activated sleep in the experimental group : these changes were observed just after stimulation period and were maintained at term age. The stimulation period involved an increase of EM frequency of <1" intervals, effects that were not maintained at term age. These changes were correlated with the results of recordings made at 6 months. Thus, sensory stimulation performed in preterm newborns seemed to compensate for deficits in sleep pattern maturation and in cognitive processes.

REFERENCES

Hennevin, E. and Leconte, P., 1977, Etude des relations entre le sommeil paradoxal et les processus d'acquisition, Physiol Behav., 18 : 307-319.

Spreux, F., Lambert, C., Chevalier, B., Mériaux, H., Freixa i Baque, E., Grubar, J.C., Lancry, A., and Leconte, P., 1982, Modification des caractéristiques du sommeil paradoxal consécutif à un apprentissage chez l'homme, Cahiers de Psychologie Cognitive, 2 : 327-334.

MEMORY FOR WORDS AND FACES - A CLINICAL STUDY

J. Lanares and G. Assal

Centre de Neuropsychologie et de Pathologie du Langage
Chuv - 1011 Lausanne - Switzerland

Studies dealing with memory loss in patients having undergone surgery of the temporal lobes, usually demonstrate an interaction between the nature of the task and the side of the lesion. In this study we investigated the specificity of the deficit for verbal and non-verbal material in a population of patients with unilateral cerebral lesions (mostly vascular). 39 of them had a left hemispheric lesion (LHL), 30 a right hemispheric one (RHL). Two tests were presented to the patients. Each test was composed of 100 cards presenting either faces (test 1) or words (test 2). Word length, frequency and imagery were controlled. Full-face photographs with identical centering were drawn from press-books. The cards were presented successively for 2 seconds each, each item (of 30) appearing twice, according to the following pattern : 10 after 5 other items (I), 10 after 15 items (II), 10 after 30 items (III) and the 10 which come after 5, appear a third time after 30 other cards (IV). Patients had to point to the card seen for the second time. In order to take errors into account in the same analysis, an index (good answers-false recognitions + 20) was used for global results. An ANOVA performed on these data showed an effect of the task (WORDS>FACES, $p < .05$) and an interaction between the side of the lesion and the nature of the material ($F = 18{,}7$ $p < .01$). RHL patients are much more impaired for faces (31,00) than for words (45,8). Multiple comparisons revealed that the interaction was mainly due to effects in the RHL group, and there was no difference between words (38,5) and faces (34,7) in the LHL group. There were no significant differences between the four conditions in the LHL group for the words. But, as for faces, those appearing a third time after the longest delay are better recognized (IV>III). Delay had an effect upon face recognition (III<I). In the RHL group, delay between targets had an effect upon word but not upon face recognition. Repetition had a positive effect for both. We did not find any contextual effect. Classification of the patients according to their results for a test of facial recognition did not modify the main effects. Thus there was a clear dissociation between the two tasks in the RHL group but such a difference was not observed in LHL patients. RHL difficulties in remenbering faces were apparently due neither to perceptual disorders nor to a short-term memory deficit and imply higher processes such as perceptual categorization. The poor scores of the LHL group in the face memory test suggested that general abilities in processing to-be-remembered materials were impaired. It appears then, that RHL patients have specific difficulties related to the kind of material, whereas, in LHL patients, memorization mechanisms seem globally impaired.

AGEING AND COGNITIVE PERFORMANCE IN THE NATURAL ENVIRONMENT

Maryanne Martin

Department of Experimental Psychology
University of Oxford
Oxford Oxl 3UD, England

General changes in the central nervous system with increasing age (e.g., in neuronal metabolism) appear to be reflected in experimental demonstrations of concomitant decrements in human learning, memory, and other cognitive abilities. An alternative possibility, however, is that functional reorganization allows the effects of these changes at the substrate level to be combatted, and that observed decrements in cognitive abilities have other origins. In particular, it may be suggested that the aged population does less well in standard laboratory tasks because it is in general relatively lacking in practice at performing under the examination-like conditions of these tests. According to this hypothesis, less pronounced differences in overall abilities should be evident if performance within the natural rather than the laboratory environment is assessed. Recent work has shown that the difficulties inherent in studying people's cognitive performance in the natural environment may to a considerable extent be overcome by means of either diary-based or questionnaire-based studies. Accordingly, an experiment of the latter type was carried out which employed sixty subjects, half of whom were young (18-30 years) and half of whom were old (60-78 years) adults. Three existing instruments were used to provide measures of everyday cognitive failure (CFQ), everyday attention (EAQ), and everyday memory (EMQ) performance. The results failed to provide evidence that the widely reported deterioration in the level of cognitive performance with ageing necessarily occurs when performance within the natural environment is observed. Similarly, in this environment there was also no evidence for an increase with ageing in the degree of variation of performance level over individuals.

The implication of these results is that adverse consequences of relatively general changes with ageing in the hardware of the central nervous system can be counteracted by specific improvements that occur presumably at a more abstract level of organization. Such preservation of function may be expected to occur via the development of adaptive cognitive strategies over appreciable lengths of time. Durations of exposure to laboratory tasks are, however, likely to be insufficient to allow such development to occur. Further, such tasks are often designed expressly to constrain the subject's potential range of strategies. In consequence, it is in the natural environment that - as observed here - functional preservation is the more likely to be found.

ADAPTATION OF OCULOMOTOR BEHAVIOUR IN READING:

A STUDY OF SOME PATIENTS WITH VISUAL FIELD DEFECTS AFTER BRAIN DAMAGE

M.N. Metz-Lutz, A.M. Eber, N. Cremel, D. Strubel, and P. North

Service de Neuropsychologie et de Rééducation du Langage
Clinique Neurologique, C.H.U., 67091 Strasbourg Cédex, France

For the normal reader, the recording of eye movements in reading shows a well-known common oculomotor pattern with saccadic movements and fixations, whose length and duration are controlled by perceptual and linguistic factors. It has been demonstrated that while reading from the left to the right side, the perceptual span is asymetric and extends further to the right of fixation than to the left. The processing of this right non-central information would "prime" the identification of words (McConkie and Rayner, 1976, 1983). How does a reduced visual information influence the processes required in reading as the oculomotor computational processes and the language processing ?

A study of the oculomotor scanning behaviour and the reading performances of eight left-brain damaged patients with right homonymous hemianopia, as compared with 10 normal readers, revealed two groups of patients. The first group (6 Ss) they kept the same "staircase pattern" in the successive saccadic movements and the reading eye movements as was observed in the control group. The only changes consisted of an increased number of shorter saccades. Their mean reading time was increased but these patients were able to read without paralexic errors. In the second group (2 Ss), the successive saccadic movements to the right side were dysmetric, but the "staircase pattern" was still recognizable in the saccadic movements to the left. Their oculomotor behaviour in reading was completely disorganized and their reading abilities were impaired.

A possible adaptation of reading oculomotor behaviour to a right visual defect is demonstrated. It depends on a preserved saccadic movements to the right side. The disturbance of the right saccadic movements itself prevents the adaptation and impairs the different reading processes.

REFERENCES

Rayner K. and McConkie G.W. What guides a reader's eye movements ? Vision Research, 16, 829-837.

Eye Movements in reading. Perceptual and language processes, 1983. Ed. K. Rayner, New-York, Academic Press.

CHANGES IN HIPPOCAMPAL MULTIUNIT ACTIVITY DURING CONDITIONING: A NEURAL CORRELATE OF THE PREDICTIVE VALUE ACQUIRED BY THE CS?

Nicole Neuenschwander-El Massioui, Gérard Dutrieux
Jean-Marc Edeline and Serge Laroche

Département de Psychophysiologie, LPN2, C.N.R.S.
91190 Gif-sur-Yvette, France

Electrophysiological recording of unit or multiunit activity in the hippocampal formation during classical conditioning has been widely used as a model system to study the neural mechanisms involved in learning and memory. Indeed, it is now well documented that hippocampal neurons exhibit a marked increase in activity in response to a sensory stimulus after its pairing with an unconditioned stimulus.

Although considerable effort has been done to assess the associative nature of hippocampal cellular responses, the question still remains as to whether these learning-induced changes in hippocampal activity mainly reflect the stimulus-reinforcement or stimulus-response association. This question was addressed in an experiment in which we examined hippocampal (CA3) multiunit activity to a conditioned stimulus (CS) presented to animals while they were engaged in two previously acquired bar-pressing tasks.

Rats were first trained in a non-signalled bar-pressing dependent shock omission task (Sidman avoidance schedule) and in a food-motivated lever-pressing task (60-sec VI) until they maintained a stable rate of responding in both situations. Then, 5 sessions with presentations of a previously habituated tone (CS) paired with footshock as a US were given. Test sessions consisted in superimposing the CS alone on the baselines of instrumental responding, i.e. with the shock-avoidance schedule in effect or with the variable-interval schedule for food in effect, on alternative days. As expected, the presentation of the CS induced a marked increase in shock-avoidance responding as well as a marked suppression of lever-pressing for food. The analysis of multiunit activity to the CS revealed that hippocampal responses established during classical conditioning were maintained while two different responses (conditioned acceleration and conditioned suppression) were exhibited. The finding that the conditioned acceleration obtained in the avoidance task does not extinguish (after 7 days of presentations of the CS) while the cellular hippocampal conditioned response disappeared over days, constitutes further evidence that hippocampal cellular and behavioural responses are dissociated.

These results are taken to support the hypothesis that hippocampal cellular responses may best be viewed as an index of the predictive value acquired by the CS rather than a correlate of the specific conditioned response established during classical conditioning.

PARADOXICAL SLEEP AUGMENTATION FOLLOWING

SPATIAL STRATEGY REVERSAL IN BALB/c MICE

Chantal Pacteau

Laboratoire de Psychophysiologie
LPN-CNRS
91190 Gif/Yvette, FRANCE

The phenomenon of paradoxical sleep (PS) augmentation has been observed during the acquisition of complex learning in several species (mouse, rat and man). Experiments performed to study the relationship between this phenomenon and the time-course of learning have shown that: 1) PS increase is not a consequence of non-cognitive factors associated with the experimental conditions or the quantity of information received by the subjects; 2) it appears only if adaptive responses are elaborated and 3) it is concomitant with the stage of learning that preceeds effective coping. It has been hypothesized that PS augmentation could be involved in a crucial period of learning during which elaboration and memorization of information occurs and would reflect an increase in CNS activity (Bloch, V., Hennevin, E., Leconte, P., 1979 ; Pearlman, C.A., 1979). In order to assess this hypothesis, we chose an original task which consists of the reversal of a previously acquired rule. We predicted that in such a situation PS increase would occur during the critical period of reorganization of the previously learned information, identified in a preliminary experiment as immediately following rule-inversion.

In a 3-path maze, 9 BALB/c mice were run on an elimination task (they had to visit the 3 goal-arms without repeating any previous choice during each learning session); when criterion was reached, the rule was reversed so that they had to acquire a position habit (they had to choose the same goal-arm in all trials). Sleep was monitored for the 8 hours following training sessions. Results showed that: 1) as in our previous experiment, the maximum improvement in performance occured at the start of position learning (the number of errors diminished by about 50% in 2 sessions) and 2) PS increase - due to an augmentation of the number of phases - occured after the first 2 sessions of rule reversal, and only after these 2 sessions. Slow wave sleep was not modified.

In previous experiments with rodents, post-training modifications have been found to occur soon after learning and to be short-lived in the rat. However, in the mouse, such post-training modifications occur later and are longer-lasting. Here, we observed an intermediate picture: PS modifications appeared more precociously than is generally found in the mouse but began later and were more protracted than in the rat. As hypothesized, the phenomenon of PS augmentation occured consecutive to the reversal of the elimination rule. This augmentation was punctual and reflected the critical period of elaboration of the new strategy. Therefore, our results confirm that PS modifications appear at a stage of intense information processing.

REFERENCES

Bloch, V., Hennevin, E., and Leconte, P., 1979, Relationship between paradoxical sleep and memory processes, in "Brain Mechanisms and Learning from the single Neuron to Man," M.A. Brazier, ed., Raven Press.

Pearlman, C.A., 1979, REM sleep and information processing: evidence from animal studies, Neurosci. Biobehav. Rev., 3:57.

SOCIAL CONSTRAINTS ON LEARNING IN A GROUP OF BABOONS REARED IN AN ENCLOSURE

Berthille Pallaud & Hugues Lepoivre

C.N.R.S.-I.N.P.9

31, chemin Joseph-Aiguier
13402 Marseille cedex 9, France

A study of reversal-learning involving discrimination between two zones was conducted in a group of Baboons (*Papio papio*) living in an enclosure. The task was presented to the group as a whole and not to isolated individuals. The animals had to learn to forage near small poles to find corn which had been previously hidden at the bottom of a hole close to each pole. The results showed that only three dominant animals acquired the task. The rest of the group seemed to learn nothing but nevertheless could have access to the food as a result of animals' foraging behavior. The latter did not eat all the unearthed grains but moved from one pole to the other in order to explore other holes ; thus corn was left for the other individuals. This study does not corroborate results obtained on the influence of dominance factors upon learning capacities but reveals the importance of such factors as a social constraint restricting the possibility of acquisition (in as much as revealed by performance) to some individuals among the group. On the other hand the study shows that one animal (the dominant female) acquired this reversal learning faster than both males and that this result was due to a vicarious learning mechanism. As early as the first day after the reversal of the reinforcement contingency, the success rate of this female was better than random, which was not the case for the other two dominant individuals. Other behavioural parameters corroborate the hypothesis according to which this female learned through observation.

COMPARISON OF DORSAL AND VENTRAL HIPPOCAMPUS IN BALB/C MICE : BEHAVIORAL AND ANATOMICAL STUDIES

François Roman, Béatrice Alescio-Lautier,
Colette Devigne and Bernard Soumireu-Mourat

Lab. Neurobiologie des Comportements, CNRS U.A. 372,
Université de Provence, Marseilles, France

Behavioral studies in rodents have focused on the dorsal part of the hippocampus and there are only a few reports concerning the ventral part of this structure.

In the present experiments, we compared in BALB/c mice the effects of bilateral stimulation of the dorsal and ventral hippocampus on memory retention. The bilaterally implanted subjects were stimulated at subseizure current intensities during 80 seconds, just after a learning session. Several behavioral tasks were used. Firstly, the subjects were trained to perform a Go-NoGo discrimination task in parallel runways using food reinforcement; each mouse underwent 6 S+ and 6 S- randomized trials per daily session for 3 days and was bilaterally stimulated just after the first learning session. Compared to controls, implanted (and not stimulated) animals were impaired across sessions. Stimulated subjects were improved on the days following stimulation compared to implanted and even to non-implanted controls. Dorsal and ventral stimulation improved the discrimination ratio to the same degree, but through a different way (by increasing S- running times in the case of dorsal stimulation and by decreasing S+ running times in the case of ventral stimulation). For CRF operant appetitive conditioning and for passive avoidance conditioning, we found a clear improvement in retention after post-trial subseizure stimulation of the dorsal hippocampus, but only a slight effect of ventral hippocampal stimulation. In contrast, we observed a clear facilitation after the ventral stimulation, but no effect of dorsal stimulation in a one-way active avoidance task.

The present investigation shows that stimulation of the ventral hippocampus is not as efficient as that of the dorsal hippocampus. Taken together, these results suggest that the dorsal part of the hippocampus is mainly involved in memory processes through modulation of inhibition. On the other hand, the ventral hippocampus could be rather involved in behavioral adaptation by modulating the significance of stimuli in relation to attentional mechanisms.

In order to study relations between the dorsal and ventral parts of the hippocampus, we undertook an anatomical study of the hippocampal system, using retrograde transport of horseradish peroxydase after injection either into the dorsal or the ventral part of the hippocampus. This preliminary study showed differences in connections between the dorsal and the ventral hippocampus, and also between different structures and these two parts of the hippocampus. They could support the differences shown by the behavioral analysis.

A NEW BRAINSTEM TRANSCOMMISSURAL PATHWAY : THE BILATERAL PONTOCEREBELLAR SYSTEM

Annamaria Rosina and Luciano Provini

Istituto di Fisiologia dei Centri Nervosi, CNR
Milano, Italy

The existence of a commissural system connecting the two cerebellar hemispheres was proposed several years ago on the basis of degeneration experiments. Recent neuroanatomical studies have shown that there are pontocerebellar neurons which connect the two cerebellar hemispheres by intracerebellar axonal branching(Rosina et al.,1980).Apart from their functional implication,the interest in these cerebellar transcommissural neurons arises from the possibility that they sustain the compensatory sprouting in the neonatal cerebellar hemisphere found after unilateral lesions of the brachium pontis(Angaut et al.,1982).Any cerebellar transcommissural pathway surviving the unilateral pedunculotomy could be the basis for the observed rearrangement. The subject has now been reexamined to evaluate the origin, termination and incidence of any precerebellar neuronal set projecting bilaterally by transcommissural pathways. Experiments were performed on cats using the double fluorescent tracing method(Rosina et al., 1980).

The results indicated that pontine neurons may project bilaterally by intracerebellar branching not only to Crus I and II of the hemisphere (Rosina and Provini, 1984) but also to the forelimb areas of the anterior lobe-pars intermedia. The pathway in mainly composed of neurons with recrossing axons, and account for 5-10% of the pontine neurons projecting to the lateral parts of the cerebellum. Only a few neurons projecting bilaterally by intracerebellar branching were observed in other brainstem nuclei, i.e. in the nucleus reticularis tegmenti pontis, the perihypoglossal complex and the paramedian medullary reticular formation.

The pontocerebellar transcommissural pathway described here most likely represents one of the brainstem pathways by which unilateral cortical commands can still control the activities of both hands, in the absence of callosal or other higher commissures (Myers et al., 1982). In addition this cerebellar transcommissural system could sustain the compensatory sprouting produced by unilateral lesion of the cerebellar afferent pathways in the newborn rat (Angaut et.al., 1982).

REFERENCES

Angaut, P., Alvarado-Mallart, R.M., and Sotelo, C., 1982, J. comp. Neurol., 205: 101.

Myers, R.E., Sperry, R.W., and McCurdy, N.M., 1962, Amer. Med. Ass. Arch. Neurol., 7: 195.

Rosina, A., Provini, L., Bentivoglio, M., and Kuypers, H.G.J.M., 1980, Brain Res., 195: 461.

Rosina, A., and Provini, L., 1984, Brain Res., 296: 365.

INFLUENCE OF TRAINING STRENGTH ON LONG-TERM RETENTION OF A CONDITIONED SACCHARIN AVERSION IN RATS

G. Schmaltz and P. Marcant

Laboratoire de Psychophysiologie
Université de Lille I
59655 Villeneuve d'Ascq cedex, France

The retention of a conditioned taste aversion was tested under two training strengths, over a 14-day period. On training day, water was replaced by a 0.1 % saccharin solution (novel taste). Thirty minutes later, the rats received an i.p. injection of saline or were poisoned with LiCl (110 mg/kg). Fourteen independent groups (7 saline and 7 LiCl groups) were tested for retention either 1, 2, 3, 4, 7, 10 or 14 days following training. On retention day, each rat was submitted to a two-bottle choice procedure (saccharin vs water). Preference for saccharin was evaluated through the following ratio : 100 x (saccharin intake/total intake). When 30 minutes elapsed between saccharin (CS) and LiCl (US), the animals developed a significant aversion thereafter (p at least < 0.05) : the grand mean saccharin preference index was 5.3 % within the LiCl groups and 64.4 % within the saline groups. No significant fluctuations in the magnitude of aversion appeared between different retention intervals. A previous work (Biederman et al., 1974) showed that memory of a conditioned saccharin aversion followed a U-shaped function. However, the authors used a one-bottle procedure which does not seem to be the most appropriate to investigate the memory of aversion. In a second experiment, our general procedure was employed except that a 3-hour CS-US delay was used in order to decrease the strength of training. Three retention intervals were examined, namely 2, 7 and 14 days. Saline groups displayed a constant overall preference score of 68.6 %. By contrast, the mean saccharin preference index in the LiCl groups tested at either 2, 7 or 14 days was 52.3 %, 33.1 % and 9.8 % respectively, the strongest aversion being observed at the longest retention interval. Thus, when the time interval between CS and US was increased, a monotonic retention function appeared reflecting an improvement in memory over time.

In conclusion, memory of a conditioned saccharin aversion was dependent on the training strength but not in an intuitively obvious fashion. Unexpectedly, when the degree of training was decreased, an "incubation effect" was observed, i.e. an increase in strength of the aversion over time. This last result suggests that the retention of a conditioned taste aversion may share common features with the retention of more classical aversive conditioning (Logue, 1979).

REFERENCES

Biederman, G.B., Milgram, N.W., Heighington, G.A., Stockman, S.M., and O'Neill, W., 1974, Quart. J. Exp. Psychol., 26 : 610.

Logue, A.W., 1979, Psychol. Bull., 86 : 276.

EFFECTS OF EARLY HYPERTHYROIDISM ON SHUTTLE BOX BEHAVIOR AND HIPPOCAMPAL MOSSY FIBER DISTRIBUTION

Herbert Schwegler[1], Bernd Heimrich[1], and Hanspeter Lipp[2]

[1]Institut für Anthropologie und Humangenetik, D-6900 Heidelberg

[2]Anatomisches Institut der Universität CH-8057 Zürich

By means of 3 genetic approaches we have shown that rat and mouse strains that perform poorly in a two-way avoidance task have comparatively more mossy fiber synapses upon the basal dendrites of the hippocampal pyramidal neurons than strains that show higher levels of performance in that task (Schwegler and Lipp, 1983). There is a strong negative correlation between these two parameters. Early hyperthyroidism is known to increase the number of these synapses in a dose-dependent way (Lauder and Mugnaini, 1977). Therefore, we injected thyroxine during the first two postnatal weeks in pups of rat and mouse strains (Roman High-avoidance rats and DBA/2 mice). Thyroxine was applied in three different groups: daily in group 1; every second day in group 2; every third day in group 3; a control group was treated with saline. At the age of three months the animals were tested for two-way avoidance acquisition (rats) and performance (mice). After behavioral testing, the animals were killed and their brains were cut in horizontal sections, stained by means of the Timm method, and analyzed on a graphics tablet connected to a desk computer. Thyroxine treated animals showed both a decreased acquisition and performance of the shuttle-box task, and these effects were dose-dependent in rats. Early hyperthyroidism reduces inherited high performance in two-way avoidance and increases the number of mossy fiber synapses on the basal dendrites of the pyramidal cells. Since about 70% of the behavioral variability can be explained by the anatomical variability we believe that there is at least a functional if not a causal relationship between these two parameters.

REFERENCES

Lauder, J.M. and Mugnaini, E., 1977, Nature 268: 335

Schwegler, H. and Lipp, H.P., 1983, Behav. Brain Res. 7: 1

CORRELATION OF PRE- AND POSTSYNAPTIC ACTIVITIES: ITS ROLE IN VISUAL CORTICAL PLASTICITY

D.Shulz, Y.Frégnac, S.Thorpe and E.Bienenstock

Laboratoire de Neurobiologie du Développement
Bât.440, Univ. Paris XI, 91405 Orsay Cedex, France

Functional modifications of visual cortical neurons take place during a critical period of postnatal development. A plausible mechanism accounting for this cortical plasticity is that the efficiency of transmission of a given synapse will increase if the temporal correlation between pre- and post-synaptic activity is high (Hebb, 1949).

The aim of the present study was to modify ocular dominance and/or orientation preference of single visual cortical neurons in anaesthetized and paralyzed cats. Since extracellular recordings do not provide direct information on pre-synaptic activity, the strategy is to control the temporal correlation between the presentation of a visual stimulus in the receptive field and the post-synaptic activity of the recorded cell. By means of ionophoretic techniques, neuronal firing was clamped at a "high" level while presenting a stimulus to one eye, and at a "low" level while presenting the same stimulus to the other eye (or a different oriented stimulus to the same eye). Neuronal firing was controled by applying a retention/ejection current through the potassium acetate (or K+ chloride) recording electrode. In the optimal situation, the current control resulted in enhancement (ejection) or suppresion (retention) of the visual response.

Control recordings using constant levels of retention or ejection currents during exploration of the receptive field showed that the absolute level of current had no effect on the prefered orientation of the neuron which remained stable over time. For ocular dominance a higher variability was found. After the orientation pairing procedure, significant changes in the relative orientation preference, the temporal profile of the response and/or the orientation tuning curve were observed in normally reared cats as old as 5 months, although the clearest effects were seen in deprived kittens recorded at 6 weeks of age. Concerning ocular dominance, some neurons showed a modification of ocularity or changes in the temporal pattern of the response, even if the extent of the effect was less marked than for orientation preference. In one case the modification of ocularity was shown to be potassium independent.

All the functional modifications observed are consistent with Hebb's hypothesis, i.e. the stimulus parameter (orientation or ocularity) associated with the higher correlation of pre- and post-synaptic activity becomes more effective. It seems to be easier to induce clear changes in orientation preference than ocular dominance using such a procedure.

REFERENCES

Hebb, K.O., 1949, The Organization of Behaviour, N.Y.: John Wiley and Sons.

DEFEAT-INDUCED DEFENSE REACTIONS AND ANALGESIA IN MICE: EFFECT OF GENOTYPE

B. Siegfried, H.R. Frischknecht, P. Külling & P.G. Waser

Institute of Pharmacology, University of Zürich,
CH-8006 Zürich, Switzerland.

Mice react upon defeat with defense reactions which can be considered as naturally occurring avoidance responses. In an attempt to characterize the effect of genotype on adaptive capabilities of interacting male mice, a learning model was used, which measures behavioral reactions in two inbred strains of mice (C57BL/6 and DBA/2) during defeat and 24 h later during a confrontation with a nonaggressive opponent.

A first experiment evaluated the effects of a single defeat of various intensities (10, 30, 50 bites) on the behavior of C57 and DBA mice. Upon being bitten, mice of the DBA strain reacted with significantly more escape reactions, while C57 mice showed more immobility, crouch, and defensive sideways and upright postures. Clear genotype-dependent patterns were also evident from the conditioned responses recorded 24 h after defeat. DBA mice showed more escape and defensive sideways and upright postures upon contact with a nonaggressive partner; in contrast, C57 mice reacted with more immobility and crouch. With increasing number of bites the sum of learned responses increased in C57 mice, while it decreased in mice of the DBA strain. This decrement was paralleled by an increase in the defeat-induced analgesic response measured on the hot plate. No defeat-induced analgesia was obtained in C57 mice. To further investigate an eventual relationship between analgesic mechanisms and adaptive behavior, a second experiment examined the effects of repetitive daily defeats (4x30 bites; 3x50 bites; one defeat per day) on the behavior of the two mouse strains. From the first to the last defeat session, DBA mice, which reacted with analgesia upon defeat, maintained their behavioral response pattern during defeat, while C57 mice, which showed no defeat-induced analgesia, significantly changed their behavior (decrease of escape responses, increase of crouch and defensive upright postures). During a nonaggressive confrontation 24 h after the last defeat, repeatedly defeated C57 mice showed higher levels of learned responses when compared to animals that experienced a single defeat. In contrast, no difference was found in DBA mice.

From the two studies it is concluded that the genotype determines the pattern of defense reactions as well as the change in nociception induced by the stress of defeat. A link between processing of aversive experience and analgesia is suggested: In the analgesic DBA strain increased defeat resulted in an impaired retention of conditioned responses and prevented the animals from a change of the defense strategy, while the non-analgesic C57 strain reacted with increased conditioned responses and a modified defense pattern.

FETAL BRAIN TISSUE TRANSPLANTS PROMOTE BEHAVIORAL RECOVERY FROM BILATERAL LESIONS OF THE CEREBRAL CORTEX IN ADULT RATS

Donald G. Stein, Randy Labbe, Arthur Firl
Holly Rakowski and Michael Attella

Department of Psychology, Clark University
Worcester, Massachusetts 01610

We recently demonstrated that frontal cortex taken from embryonic rats and transplanted directly into the damaged frontal cortex of adult animals, could dramatically improve their performance on a spatial alternation task. Implants of fetal cerebellum had no such effect. Although recovery of function was more rapid in the rats with cortical transplants, they were not able to perform the learning task as well as intact controls. Since the recovery was not complete, we decided to examine some of the factors that could enhance the effectivness of the transplants.

In the second series of experiments, we examined the question of whether the age of the embryonic material was a critical factor in promoting functional recovery after frontal lesions in adult rats. Accordingly,we used fetal brain tissue taken from embryos at 17, 18, 19, or 20 days gestation. In brief, the older the fetal tissue at the time of implantation, the more effective the functional recovery we observed. However, once again, while the rats with transplants were always better than the lesion controls, they were still unable to perform as well as intact counterparts.

Next, we asked whether nerve growth factor or GM-1 ganglioside administration would enhance the effectiveness of the transplants, reasoning that these neurotrophic substances would facilitate the growth of connections between the host brain and the transplants. In brief, as we had seen previously, the implants of fetal cells improved the performance of the brain-damaged adults but the administration of neurotrophic substances conferred no additional benefits over the transplants alone.

Another question we addressed was whether fetal brain tissue taken from a different species, could be used to promote functional recovery. We took pieces of embryonic frontal cortex from hamsters, placed them into the damaged frontal cortices of adult rats, and tested the animals on a spatial alternation task. The interspecies transplants were as effective in promoting recovery as fetal tissue taken from conspecifics.

Finally, we wanted to learn whether transplants could be used to promote recovery from lesions placed in sensory, rather than 'association' cortex. We thus gave groups of adult rats lesions of occipital cortex and then implanted either fetal, frontal cortex or occipital cortex. The rats were then tested on learned brightness or pattern discrimination tasks. To our surprise, the transplants of occipital cortex had no effects on recovery at all. In contrast, embryonic frontal cortex transplanted into the damaged occipital cortex resulted in improved brightness discrimination but had no effects on a more complex pattern discrimination task. Taken together, our findings show that fetal brain tissue transplants can be an effective, but not a complete, means to promote functional recovery from severe brain injuries. Many of the factors that could influence the success of the transplants still remain to be investigated, but the technique itself may hold considerable promise for helping the victims of C.N.S. trauma.

MORPHOLOGICAL STUDIES OF PLASTICITY IN THE CHICK FOREBRAIN FOLLOWING PASSIVE AVOIDANCE TRAINING

Michael G. Stewart

Brain Research Group
The Open University
Walton Hall, Milton Keynes, MK7 6AA

The paradigm of one-trial passive avoidance training (PAL), in which day-old chicks are trained to avoid pecking at a bright bead coated with an aversive tasting substance such as methyl anthranilate (MeA), has been used extensively in our laboratory as a model for the cellular correlates of memory formation. Autoradiographic studies with 2-deoxyglucose have indicated an elevation in incorporation of radioactivity in 3 regions of the forebrain in MeA-trained birds compared to those of water trained controls (W-control); (1) the medial hyperstriatum ventrale (MHV), (2) the palaeostriatum augmentatum (PA) and (3) the lobus parolfactorius (LPO).

A quantitative morphological analysis was made of synapses in the left and right hemispheres of each of these three regions in MeA-trained and W-control chicks. The parameters examined were ($N_{v.syn}$), the number of synapses per unit volume of neuropil and ($\bar{D}$) the mean length of the post-synaptic thickening. There are no significant differences in $N_{v.syn}$ or $\bar{D}$ in either hemisphere of the PA in W-control or MeA-trained chicks, nor are there any differences in $N_{v.syn}$ in either hemisphere of the MHV and LPO in W-control or MeA-trained birds. However, there is a hemispheric asymmetry of $\bar{D}$ in both the MHV and LPO of W-control chicks (in both regions $\bar{D}$ in the right hemisphere is greater than in left by approximately 10%) and these differences disappear on MeA-training. An attempt has been made to classify synapses in the LPO of W-control and MeA-trained chicks according to the nature of the synaptic junction, (i.e.) asymmetrical (Gray type I with spherical vesicles) and symmetrical (Gray type II with pleomorphic vesicles). Preliminary data indicate that in the LPO hemispheric asymmetry in W-control chicks exists mainly in symmetrical synapses. In the MHV measurements of additional synaptic parameters were made in each hemisphere of W-control and MeA-trained chicks; ($V_{v.syn}$) the volume density of the pre-synaptic bouton, and ($\overline{ves}.syn$), the number of synaptic vesicles per synapse . No significant differences exist in $V_{v.syn}$ or $\overline{ves}.syn$ between the left and right hemispheres of W-control chicks; however, following MeA-training both $V_{v.syn}$ and $\overline{ves}.syn$ are greater in the left than in the right hemisphere (by 23% and 61% respectively).

Taken together these data (a) confirm the considerable evidence of lateralization of brain function in birds and (b) demonstrate that the morphological effects of PAL are not restricted to a single region of the chick brain.

PASSIVE AVOIDANCE LEARNING AND MEMORY-STORAGE IN DECEREBRATE RATS

C. Tomaz and J.P. Huston

Institute of Psychology III
University of Düsseldorf
4000 Düsseldorf 1, West Germany

Two experiments were undertaken to investigate learning and memory of an inhibitory avoidance response in rats devoid of the whole forebrain, except for the hypothalamus.

Experiment I (learning): 24 hours after removal of all telencephalic structures plus the thalamus, rats were given a learning trial in the up-hill avoidance task (in this task rats are placed on a slanted platform with the head placed down-hill, a 90° turn up-hill is punished with a tail-shock). When tested 2, 8 and 24 hours later the decerebrate rats showed significant increases in step-up latencies in comparison with control groups that had received non-contingent tail-shock or no shock.

Experiment II (memory): As in experiment I rats received either tail-shock contingent on the up-hill response, no tail-shock or non-contingent shock. Eight hours after the training they were decerebrate or sham-operated. When tested 24 hours later the group that had received up-hill contingent tail-shock showed significant increases in latencies to ascend the platform.

Experiment I indicated that the telencephalon plus thalamus are not essential structures for inhibitory avoidance learning. Experiment II demonstrated a transfer of training from the intact to the post-lesion state. This indicates that the site of learning and storage for this avoidance task takes place subtelencephalically in rats with intact brains.

LISTER AND WISTAR RATS DIFFER IN THE EXTENT TO WHICH EARLY LIFE UNDERNUTRITION AFFECTS LATER SPATIAL DISCRIMINATION LEARNING

J. Tonkiss and J.L. Smart

Department of Child Health, University of Manchester,
The Medical School, Oxford Road, Manchester M13 9PT, U.K.

The field of research on the behavioural effects of early life undernutrition is strewn with equivocal and contradictory findings. For instance, early life undernutrition has been found to impair, enhance or to have no effect upon the later learning performance of rats (see Smart 1977; Katz 1980). It is suggested that some of the confusion may have arisen by the use of rats of different genetic stocks in different laboratories; i.e. the nature and extent of the developmental response to undernutrition may depend on the genetic constitution of the animals concerned. This possibility was investigated in the following experiments using rats of the Lister (pigmented) and Wistar (albino) stocks.

Developing rats of the two outbred stocks were either well-nourished or undernourished by underfeeding their mothers throughout pregnancy and lactation. All rats were fed ad libitum from 25 days. After more than 18 weeks of nutritional rehabilitation, adult male rats were tested for learning in a water T-maze. Two experiments were performed, each using different animals. In both experiments, rats were required to learn to swim to an escape platform located in the opposite arm to that for which they had shown a preference during a pre-test session. After 5 days in Experiment 1, and 6 days in Experiment 2, the rats were required to reverse their learned response and swim to the opposite arm. Reversal learning continued for 3 days in both experiments. In Experiment 1, rats were allowed to correct incorrect responses without being removed from the water, and could therefore make more than one error per trial. In Experiment 2, when a rat made an incorrect choice it was retained within the wrong arm for 15 sec. before being replaced at the start arm for the next trial. Five trials and 10 trials per day were given in Experiments 1 and 2 respectively.

The number of errors made by previously undernourished (PU) rats was compared with that of well-nourished control (C) rats for each stock separately, using 2-way analyses of variance (nutrition X day) with one repeated measure (day). For the Lister stock it was found that PU rats made significantly more errors than controls during the initial learning phase of both experiments. In contrast, the error scores of PU and C rats of the Wistar stock were not significantly different.

During reversal learning in Experiment 1, there were no differences between PU and C rats of either stock in the number of errors committed. In Experiment 2, however, PU rats of the Lister stock showed a more rapid reduction in errors over days (nutrition X day interaction) than controls by virtue of having committed more errors on the first day of reversal learning. PU and C rats of the Wistar stock did not differ in their reversal learning performance in Experiment 2.

It is concluded that early life undernutriiton impaired the spatial learning performance of rats of the Lister stock but had a much smaller, or even negligible, effect in Wistar rats. Hence there is evidence that the extent of the effect of undernutrition on behavioural development depends on the genotype of the animals concerned.

REFERENCES

Katz, H. B. 1980, The influence of undernutrition on learning performance in rodents, Nutr. Abs. Rev. A., 50: 767.

Smart, J. L., 1977, Early life malnutrition and later learning ability: a critical analysis, in: "Genetics, Environment and Intelligence", A. Oliverio, ed., Elsevier/North-Holland, Amsterdam.

VISUAL NEGLECT IN PRIMATES DUE TO FRONTAL EYE FIELD DAMAGE: EYE AND HEAD MOVEMENT STRATEGIES DURING RECOVERY

J. van der Steen,* I. Steele Russell and G.O. James

MRC Unit on Neural Mechanisms of Behaviour
University College London
London, U.K.

* Dept. Physiology 1
Erasmus University
Rotterdam, Netherlands

The visual effects of unilateral damage to the frontal eye fields (FEF) in monkey include neglect of stimuli in the contralateral hemifield and conjugate deviation of the eyes towards the side ipsilateral to the lesion. These effects are temporary and gradual recovery typically occurs within a few weeks. The processes and structures involved in this recovery are unknown.

This experiment was designed to monitor the pattern of eye and head movements during the recovery process. Three monkeys (Macaca fascicularis) were trained to respond to briefly presented illuminated push buttons located in a horizontal perimeter. The task required the animal to fixate the centre button of the perimeter and press it. This response initiated a trial, which consisted of a brief illumination of another button somewhere else on the perimeter. The location of this target button was randomly chosen. To obtain a reward the animal had to press the target button within one second of the start of the trial. During each trial eye, and head movements were recorded with scleral induction coils using the phase difference technique in a homogeneous magnetic field.

Our results show that, although recovery on the behavioural task was complete within 14 days, permanent changes in eye-head coordination occurred. In all animals, head movements to the contralateral side became faster after the lesion, whereas saccadic gaze changes of the eyes were unaffected. Therefore the process of recovery from the lesion was seen to involve compensatory mechanisms rather than functional restitution.

THE EFFECTS OF EARLY VS LATE CEREBRAL LESIONS ON LEARNING AND MEMORY IN CHILDREN

Faraneh Vargha-Khadem and Elizabeth Isaacs

The Wolfson Centre, Institute of Child Health
and Department of Neurology, Hospital for Sick Children
Mecklenburgh Square, London W.C.1, England

The goal of this study was to determine whether memory and learning skills would be influenced by the variables of hemispheric side of lesion and age at injury during childhood. Hemiplegic children were divided into groups on the basis of congenital vs postnatally acquired lesions. Sixteen children, age and IQ matched to one of the patient groups, served as normal controls.

All patients received complete neurological evaluation. As part of the diagnostic procedures, all patients had computed tomography scans which were rated for severity of structural abnormality. Additional measures of the severity of cerebral damage was provided by quantitative testing of somatosensory and motor function.

All subjects received neuropsychological testing aimed at assessing intelligence quotient, memory quotient, verbal learning, immediate and delayed verbal recall, and immediate and delayed visuo-perceptual recall.

Analyses of variance carried out on the rated scores of CT scans, somatosensory and motor measures, indicated that there were no significant differences between patient groups with respect to severity of cerebral lesions. Furthermore, the subject groups were not significantly different on any of the intelligence measures. When the scores on all measures of memory and learning were subjected to analyses of variance, however, children with left cerebral lesions demonstrated deficits on all measures of verbal memory and learning. The deficits were relative to the normal control group and to the patient group with congenital right hemisphere injuries. No significant differences were found between groups on measures of non-verbal memory.

These results do not support the notion of "early plasticity" (Kennard's principle) and suggest that despite the early nature of left hemisphere injuries, verbal learning and verbal memory functions are not spared.

POSSIBLE MEMORY ENHANCING EFFECTS OF β-CARBOLINES

Patrice Venault, Lia Prado de Carvalho, Jean Rossier, and Georges Chapouthier

Depts I et II - Laboratoire de Physiologie Nerveuse - C.N.R.S. - 91190 Gif-sur-Yvette, France

Methyl-β-carboline 3 carboxylate (β-CCM) is a benzodiazepine receptor ligand that has been shown to have pharmacological properties that are opposite to those of benzodiazepines in all situations tested. Thus this molecule is a convulsant whereas benzodiazepines are known anticonvulsants; β-CCM is anxiogenic whereas benzodiazepines are known anxiolytics. β-CCM can therefore be considered as an inverse agonist of benzodiazepines. Since benzodiazepines have been shown to be amnesic drugs, we have hypothetized that β-CCM could have memory enhancing effects. This hypothesis was tested in two different behavioral situations. 1 - Imprinting in chicks. According to the technique used in our laboratory, chicks placed in front of a moving decoy during an acquisition session learn to follow it. This behavior of following the decoy is tested in the same situation 24 h later (test session). When chicks were injected with β-CCM (2.5 mg/kg i.p.) before the acquisition session, following of the decoy was enhanced in the test session (without drug). On the other hand, the classical benzodiazepine diazepam (DZ) (0.25 mg/kg i.p.), administered before the acquisition session, reduced following of the decoy during the test session. 2 - Latent learning in mice. According to the technique described by Soubrié et al. (1976), rodents previously deprived of food and placed for the first time in a new environment with food *ad libitum* eat only small quantities. When placed in the same environment for a test session a few days later, the same food deprived animals eat more than during the first session. The inhibition observed in the first session is assumed to be due to the novelty of the environment. β-CCM (0.1 to 0.3 mg/kg s.c.) when administered before the first session, induced an enhancement of food consumed during the test session (without drug); on the other hand, DZ (1 mg/kg s.c.) reduced food consumption. Our results suggest that, in both situations, β-CCM enhances memory whereas DZ has amnesic effects. More generally, we assume that a moderate level of anxiety is necessary for memory consolidation to occur. Anxiogenic drugs at low doses would therefore facilitate memory when anxiolytics would impair it. (Lia Prado de Carvalho was supported by a Fyssen Fellowship).

REFERENCES

Soubrié, P., Simon, P., and Boissier, J.R., 1976, Amnesie induite chez le rat par les benzodiazépines lors d'une situation ne faisant pas intervenir de composante nociciptive, C. R. Acad. Sc. Paris (D), 283:203.

INDEX